GUIDE TO TEXAS GRASSES

Texas A&M System

AgriLife Research and Extension Service Series

Craig Nessler and Edward G. Smith,
General Editors

Publication of this book was made possible by:

TEXAS A&M | Institute of Renewable Natural Resources

GLCI

Grazing Lands Conservation Initiative in Texas

TWAF

Texas Wildlife Association Foundation

TEXAS A&M UNIVERSITY

Department of Ecosystem Science and Management, Texas A&M University

NRCS

Natural Resources Conservation Service, US Department of Agriculture

TEXAS A&M UNIVERSITY PRESS   *College Station*

Guide
to Texas
Grasses

Robert B. Shaw

Photographs by Paul Montgomery

and Robert B. Shaw

A⫟M nature guides

This paper meets the requirements of
ANSI/NISO Z39.48–1992
(Permanence of Paper).
Binding materials have been chosen
for durability.

LIBRARY OF CONGRESS
CATALOGING-IN-PUBLICATION DATA

   Shaw, Robert Blaine, 1949–
Guide to Texas grasses / Robert B. Shaw ; pho-
tographs by Paul Montgomery and
Robert B. Shaw.—1st ed.
         p. cm.—(AgriLife Research and Exten-
sion Service series)
   "Institute of Renewable Natural Resources."
Includes bibliographical references and index.
   ISBN-13: 978-1-60344-186-5
(pb-flexibound : alk. paper)
   ISBN-10: 1-60344-186-7
(pb-flexibound : alk. paper)
   ISBN-13: 978-1-60344-674-7 (e-book)
   ISBN-10: 978-1-60344-674-5 (e-book)
   1. Grasses—Texas
Identification. I. Texas A & M University. Insti-
tute of Renewable Natural Resources. II. Title.
III. Series: AgriLife Research and Extension
Service series.
 QK495.G74S544   2011
 584.'909764—dc22
 2011005238

General editors for this series are Craig Nessler,
director of Texas AgriLife Research, and
Edward G. Smith, director of the Texas AgriLife
Extension Service.

"Why Grasses Are Important" is updated and
modified from Shaw (2008) to reflect information specific to Texas. Used with permission of
the University of Colorado Press.

"The Grass Plant" closely follows that presented
in Shaw (2008), Hignight, Wipff, and Hatch
(1988), and Stubbendieck, Hatch, and Landholt
(2003). Used with permission of the University
of Colorado Press and Texas AgriLife Research
and University of Nebraska Press, respectively.

The section on collecting grass specimens and
the glossary are adapted from Gould and Shaw
(1983).

Grass illustrations are from Barkworth et al.
(2003, 2007). Used with permission of Utah
State University.

The physiographic and major river maps follow
those of Stephens and Holmes (1989) and are
used with permission of the University of Oklahoma Press.

Figures on the grass plant, spikelet and floret
structure, inflorescence types, and ligule
forms are from *North American Wildland
Plants* (Stubbendieck, Hatch, and Landholt
2003). Used with permission of the University
of Nebraska Press.

Dedicated to
my teacher and mentor
Frank Walton Gould
(1913–1981)

Dr. Gould was the preeminent US agrostologist from the 1950s until his death in 1981. He was educated at Northern Illinois Teachers College (BE), University of Wisconsin (MS), and University of California at Berkeley (PhD) where he worked under Lincoln Constance. He taught at Dixie Junior College, Compton Junior College, University of Arizona, and Texas A&M University, where he concluded his academic career as a distinguished professor, earning emeritus status. His major books include *Grasses of the Southwestern US* (1951), *Texas Plants—a Checklist and Ecological Summary* (1962), *Grasses of the Texas Coastal Bend* (with Box, 1965), *Grass Systematics* (1968), *Grasses of Texas* (1975), *Common Grasses of Texas* (1978), and *Grasses of Baja California, Mexico* (with Moran, 1981). Some of the grass genera he researched and revised were *Elymus, Andropogon, Bouteloua, Panicum,* and *Dichanthelium.* His doctoral students (including Arshad Ali, Kelly Allred, Lynn Clark, Steve Hatch, Zarir Kapadia, Robert Lonard, Floyd Waller, Robert Webster, and Jose Valls) have made significant contributions to systematics, and their students continue the tradition.

The photograph is courtesy of Lucile Gould Bridges, his wife of 40 years, and the Gould family. Funds to purchase the illustrations for this book were made available through the Frank W. Gould Award for Graduate Research Support in Plant Systematics administered by the Texas A&M Development Foundation.

. . . all flesh is grass . . .

Isaiah 40:6

CONTENTS

Preface   ix

1.   Why Grasses Are Important   1
2.   The Grass Plant   10
3.   Ecoregions of Texas   26
4.   Classification of the Grasses of Texas   66
5.   Generic Keys and Species Accounts   157

Appendix: Collection, Preparation, Handling,
and Storage of Grass Specimens   1035
Glossary   1045
Literature Cited   1053

Index   1059

PREFACE

This compilation is an attempt to bring together all the taxonomic information about grasses in Texas. I tried to include every grass reported for, or thought to occur in, the state. My desire was to assure users that if they collected a grass within the borders of Texas, it would be found somewhere in this work. This is an impossible task, of course, because the grass flora in Texas is still very under collected and poorly documented (Shaw, Rector, & Dube 2011). I am convinced there are numerous new species and new introductions in the state. I fancy that numerous species only collected once or a few times within Texas, and suggested as perhaps not a permanent member of the flora, can still be found.

All works of this type are based on the dedicated efforts of those who came before us. Credit goes to all collectors who braved the elements and other obstacles to fulfill their passion for collecting grasses in Texas. Often overlooked and frequently forgotten is the first compilation of grasses for the entire state completed by W. A. Silveus, who self-published *Texas Grasses* in 1933. Silveus said he traveled over 60,000 miles during a 3-year period collecting and photographing as many species as possible. Certainly the roads in the late 1920s and early 1930s were not nearly as conducive to rapid and comfortable automobile travel as today. He accomplished his tasks without air conditioning or the luxury of a digital camera. It would still be a strenuous effort and take a very dedicated individual to accomplish what he did. Amazingly he was not a trained agrostologist, he was lawyer!

Albert S. Hitchcock's *Manual of the Grasses of the United States* (1935), and revised by Mary Agnes Chase in 1951, was the authority for

Texas grass identification for nearly 40 years. That changed with the publication of Gould's *Grasses of Texas* (1975a). This monumental work presented a new taxonomic classification of the grasses and provided keys, descriptions, distributions, and numerous illustrations. It covered 523 grass species reported for the state but excluded most ornamentals and a few cultivars. The recent volumes on the Poaceae in the Flora of North America series by Barkworth et al. (2003, 2007) have replaced Hitchcock's manual as the new authority on grasses north of Mexico. Obviously, Gould (1975a) and Barkworth et al. (2003, 2007) were used extensively in this compilation. Also consulted were state floras (Lundell 1961, Correll and Johnston 1970) and regional treatments: *Grasses of the Texas Coastal Bend* (Gould and Box 1965); *Grasses (Poaceae) of the Texas Cross Timbers and Prairies* (Hignight, Wipff, and Hatch 1988); *Guide to Grasses of the Lower Rio Grande Valley, Texas* (Lonard 1993); *Grasses of the Trans-Pecos and Adjacent Areas* (Powell 1994); *Grasses of the Texas Gulf Prairies and Marshes* (Hatch, Schuster, and Drawe 1999); *Flora of North Central Texas* (Diggs, Lipscomb, and O'Kennon (1999); and *Grasses of the Texas Hill Country: A Field Guide* (Loflin and Loflin 2006). While this book was in production, a *Grasses of South Texas: A guide to identification and value* (Everitt et al. 2011) was published, but it was impossible to incorporate their information into this work.

Checklists, a valuable tool in themselves, were used extensively in compilation of this book. Cory and Parks (1937) made an early attempt at a complete catalog of the Texas flora. Gould's (1962, 1969, 1975b) *Texas Plants—A*

Checklist and Ecological Summary was a standard for over 25 years until updated by Hatch, Gandhi, and Brown (1990). An excellent list that included many ornamental grasses is *Vascular Plants of Texas* (Jones, Wipff, and Montgomery 1997). The blending of old and new sources of information has increased the total number of grass taxa reported for the state to 9 subfamilies, 19 tribes, 723 species, 79 subspecies, and 110 varieties. The county distribution information came from over 17,500 records found mostly in Barkworth et al. (2003, 2007) and Turner (2003). These records were integrated with the revised 12 ecoregions defined by the Environmental Protection Agency to yield the expanded ecological information in this book's checklist. During the production of this volume, a *Distribution of Texas Grasses* (Shaw, Rector, and Dube 2011) was published. It includes lists of species by county, counties by species and summaries by ecoregions. The astute reader will notice slight variations in the total number of species by ecoregion, county records, etc. The distributional data in Shaw, Rector, and Dube (2011) were based on several thousand additional county distribution records that were impossible to incorporate into this book.

The uses of grasses, grass structures, generic characteristics, diagnostic features, and terminology comes from *Grass Systematics* (Gould and Shaw 1983), *Grasses (Poaceae) of the Texas Cross Timbers and Prairies* (1988), volume 23 and 24 of the *Flora of North America* (Barkworth et al. 2003, 2007), and *Grasses of Colorado* (Shaw 2008). Species descriptions and extensive literature sources were omitted to keep this volume to a manageable size, but that information can be accessed through the Utah State University Intermountain Herbarium website www.herbarium.usu.edu/webmanual.

A special thanks to the Land Information Systems lab at the Texas A&M Institute of Renewable Natural Resources (IRNR) for assistance with the Texas maps, county distribution maps, and readying the illustrations for publication. The efforts of Amy Grones Snelgrove (Geospatial/Technology Manager), Amanda Dube (Research Assistant), Ross Anderson (Web Developer), Brent Stevener (Software Applications Developer), and Will Brademan (GIS Technician) are greatly appreciated. Also, Brian Hays (Extension Program Specialist) and Amy Hays (Emerging Technologies Specialists) at the IRNR Gatesville office produced the initial drafts of the ecoregion descriptions. Student workers Lindsey Franklin and Kelsey Thurman were instrumental in building the county and ecoregion distribution databases. Again, thanks to all at IRNR who supported this work.

Grasses are difficult to identify, but pictures certainly help in verification. Permission to use the excellent illustrations was granted by Utah State University. Abbreviations for the illustrators are listed below. Thanks to Mary E. Barkworth, Director of the Intermountain Herbarium and editor of the Grass Manual Project, for her support and encouragement. Also, the excellent photographs by Paul M. Montgomery augment the usefulness of this book and point out the often overlooked beauty of grasses.

Thank you to Shannon M. Davies, Louise Lindsey Merrick Editor for the Natural Environment at Texas A&M University Press, for securing financial resources to publish this book and for support and encouragement throughout this effort. Thanks also to Patricia A. Clabaugh, Associate Editor, Texas A&M Press for her support and patience. I am afraid I have exposed them to more information about grasses then they ever wanted to know. They were excellent to work with over the last several years.

Illustrators

BFG   Bee F. Gunn

JRJ   J. R. Janish

Bellamy Parks Jansen   Bellamy Parks Jansen

K   Karen Klitz

SL   Sandy Long

M   Annaliese Miller

B   Linda Bea Miller

HP   Hana Pazdírková

CR   Christine Roberts

CR   Cindy Roché

S   Andy Sudkamp

V   Linda Ann Vorobik

Tω   Tracy Wager

New Combinations:

Chondrosum barbatum (Lag.) Clayton var. *rothrockii* (Vasey) R. B. Shaw
Chonsrosum hirsutum (Lag.) Sweet var. *pectinatum* (Featherly) R. B. Shaw
Chondrosum ramosa (Scribn. *ex* Vasey) R. B. Shaw
Chondrosum trifidum (Thurb.) Clayton var. *burkii* (Scribn. *ex* Watson) R. B. Shaw

GUIDE TO
TEXAS GRASSES

The Poaceae (Gramineae) consists of approximately 785 genera and 11,000 species (Watson and Dallwitz 1992; Chen et al. 2006). Based on genera, the grasses are the third-largest family of flowering plants after the Asteraceae (sunflowers) and Orchidaceae (orchids). Grasses are fifth in the number of species behind the Asteraceae, Fabaceae (legumes), Orchidaceae, and Rubiaceae (madders) (Good 1953). Two-thirds of the earth's land surface is used for grazing, and one-third is composed of grasslands (Schantz 1954). Grasses occur on every continent and within nearly every terrestrial ecosystem. Hartley (1954) estimates that there are more individual grass plants than all other vascular plants combined. Based on completeness of representation in all regions of the world and percentage of the world's vegetation, grasses far surpass all other plant families (Gould and Shaw 1983). The economic, ecological, and geographic importance of grasses cannot be overestimated or overemphasized (Clark and Kellogg 2007).

Food for Human Consumption

The cereal grains (barley, corn, millet, oat, rice, rye, sorghum, wheat) supply the bulk of food that humans consume (fig. 1.1a). Rice feeds more people than any other food product. Wheat cultivation covers more area than any other crop. No crop covers a wider geographic range than corn. A major portion of the world's sugar comes from sugarcane (*Saccharum officinarum*). Even the "woody" grasses are a source of nutrition (bamboo shoots and caryopses during mass flowering).

The major dietary substance found in grass caryopses is carbohydrates. These carbohydrates are stored in the endosperm, nutritive tissue used during seed germination and seedling establishment, which along with the embryo forms the major part of the grass caryopsis. The seed containing the endosperm is of most value to humans. From the seed comes flour, corn meal, rice, oats, and intoxicants (beer, rice and barley wine, corn and rye liquor, etc.). For this reason all cereal species are almost exclusively annual plants. Annuals put most of their energy into reproduction (more seeds) rather than roots and/or vegetative structures, which die at the end of the growing season. Humans harvest and put to multiple uses the seeds that ensure propagation of the annual species. Carbohydrates in the endosperm are the substance upon which most civilizations have developed and been maintained. The ability of farmers to feed many individuals, not just themselves and their families, allows others to pursue such activities as the arts, manufacturing, trade, education, bureaucracies, and, alas, war—all the fabric of human civilizations. The classic example of endosperm is the large, white, "exploded" portion of a piece of popcorn. The soft, sweet portion of a partially popped kernel, sometimes referred to as "old maids" or "duds," is the embryo. The golden covering of the endosperm and embryo, which more often than not gets stuck between one's teeth, is the seed coat.

1.1a

1.1b

1.1c

1.1d

Figure 1.1a. Grasses are the major source of grain, including corn, used for human consumption. b. Texas produces more hay than any other state. c. Texas produces more cattle and livestock products than any other state. d. Turfgrasses for lawns, athletic fields, and so on, as well as goods and services to maintain them are a major industry. e. Bamboo (Phyllostachys aurea) culm showing diverse branching pattern. Bamboo has long been grown as an ornamental. f. Pampasgrass (Cortaderia selloana) is one of the most common and easily recognized ornamental grasses. (Photographs by Robert B. Shaw)

1.1e

1.1f

Grasses and grass by-products dominate the Texas agricultural economy. Over $4.5 billion in grass crops (hay [dry and silage], grain [corn, rice wheat, sorghum], sugar) is produced annually (fig. 1.1b). More land is used for hay, grass silage, and greenchop in Texas than in any other state (5 million acres). Texas ranks second in sorghum for grain and total value of agricultural products sold (USDA-NASS 2007).

Forage for Wild and Domestic Animals

The majority of large herbivores characteristic of the expansive grasslands of the world are dependent upon grasses as a major portion of their diet. Humans, in turn, depend upon wild and domestic herbivores as a major source of protein and nutrients in the form of meat, blood, and milk (fig. 1.1c). These animals are also significant as the source of leather (hides) and animal fiber (wool, mohair, etc.).

In Western societies large infrastructures have been developed to supply these animal products to an ever-enlarging and demanding population. Corn and sorghum silage is harvested for livestock in feedlots. Also, much of the corn, millet, and sorghum grain is used as feed for cattle, swine, and poultry.

Over 14 million head of cattle and calves and 1 million sheep are maintained on Texas rangelands and feedlots. Texas ranks first among the states in value of livestock, poultry, and their products; cattle and calves; and sheep, goats, and their products. Cash receipts of over $11.3 billion were from livestock or livestock-related products in Texas in 2006 (USDA-NASS 2006). All these animals survive primarily on grasses and grass products.

While it is easy to envision the role and importance of grasses to the large herds of herbivores that once occupied the grasslands of

Africa, Eurasia, and North America, one often forgets about the importance of grasses to other wildlife. There are many graminivorous upland birds and small mammals. Also, grasses compose a significant portion of waterfowl diets. Mannagrasses (*Glyceria*), cutgrasses (*Leersia*), and wildrices (*Zizania, Zizaniopsis*) are of special importance to birds that utilize swamps, lakes, and marshes. Some migratory birds not only use wetlands dominated by grasses but also overwinter on grain fields and pastures (Gould and Shaw 1983).

Soil Conservation and Land Improvements

A majority of the world's most productive soils now used for grain production were developed under perennial grassland cover (Gould and Shaw 1983). Removal of this native, perennial grass cover by plowing or poor grazing management has led to both wind and water erosion, and a large amount of topsoil has been lost over the centuries. Reestablishment of a perennial grass cover is a common practice in soil conservation and range improvements. A "good" cover of grasses not only stabilizes the site, reducing erosion, but helps in replacing depleted nutrients from overutilized soils. Numerous state and federal agencies developed during the twentieth century to assist landowners in reducing erosion and returning lands to a more productive and economically viable state. Specific industries have developed that focus on improving the condition of rangelands and restoring other disturbed sites (land rehabilitation, mine reclamation).

Some grasses adapted to restricted ecological niches have been used for specific erosion control purposes. For example, vetiver (*Vetiveria*) is used extensively in southern Asia and Oceania for erosion control and terrace production (National Research Council 1993). Beachgrass (*Ammophila*) with its extensive rhizomes has been used to stabilize sand dunes. The same is true for the native blowout grass (*Redfieldia flexuosa*), which is used to stabilize shifting sands of the Great Plains region.

Turf and Ornamentals

Often overlooked, but not insignificant, is the use of grasses for turf and ornamentals. Grasses with rhizomes, stolons, or both (sod formers) have been used for centuries to produce a dense, thick cover that will resist use (primarily foot traffic) and be aesthetically pleasing (generally of uniform color and density) (fig. 1.1d). Most common uses of turf grasses are for lawns, parks, highway rights-of-way, golf courses, and athletic fields of all sorts (Gould and Shaw 1983). Maintenance of turf supports a multi-billion-dollar industry supplying seed, fertilizer, herbicides, insecticides, specialized machinery, sprinklers, hoses, and so on, as well as lawn care services. In Texas major turfgrasses recognized by most people are St. Augustine (*Stenotaphrum secundatum*), bahiagrass (*Paspalum notatum*), Bermudagrass (*Cynodon dactylon*), zoysia (*Zoysia japonica*), buffalograss (*Buchloë dactyloides*), centipedegrass (*Eremochloa ophiuroides*), ryegrass (*Lolium multiflorum*), and Kentucky bluegrass (*Poa pratensis*). Included in this book are the many turfgrasses that have escaped cultivation and occur as permanent members of Texas grass flora.

Recently, grasses have become much more popular as ornamentals. Bamboos (e.g., *Bambusa, Phyllostachys*) have traditionally been cultivated, but now many herbaceous grass species are being incorporated into human landscapes (fig. 1.2a, b). This popularity is based on the relative cost-effective propaga-

1.2a

Figure 1.2a. Inflorescences of Miscanthus sinensis, *a showy and often-used ornamental grass. b. Decorative use of fountaingrass (*Pennisetum *spp.). c. Comparison of different cultivars of* Sorghum bicolor; *grain sorghum on the left; "giant" sorghum being developed for bioenergy on the right. d.* Zizania texensis, *a federally listed endangered species, being artificially grown in specially designed tanks at the U.S. Fish & Wildlife Fish Hatchery at Uvalde, Texas. e. Giant cane (*Arundo donax*) has become an undesirable grass species around culverts, railroad and road rights-of-way, and woodland openings. f. Johnsongrass (*Sorghum halepense*) introduced as a forage species is now one of the most common perennial "weedy" grasses along roadsides, fallow fields, and pastureland. (Photographs by Robert Shaw)*

1.2b

1.2c

1.2d

1.2e

1.2f

tion of herbaceous grasses, their ease of maintenance, adaptability to a wide range of soil textures and nutritive levels, and versatility. Their uses can vary from ground cover; single-specimen plants such as *Erianthus* (*Saccharum*) and *Miscanthus;* mass planting (*Festuca, Pennisetum*); and erosion control. Historically, mostly exotic, introduced grasses were used as ornamentals, but native species (e.g., *Panicum, Andropogon, Sorghastrum, Chondrosum, Bouteloua*) are becoming more and more popular.

Biofuels

As the world's petroleum supply diminishes and/or is controlled by unstable and/or hostile governments, many countries are developing alternative energy sources. The four major alternatives are solar, wind, geothermal, and biological energy. Use of alternative energy sources also reduces greenhouse gas emissions, which in turn assists in mitigation of global climate change. Ethanol, sometimes referred to as grain alcohol, is a primary bioenergy source. It has long been known as a product of fermentation and distillation. Ethanol is now being used as an additive to gasoline, and the newer flex vehicles are capable of using up to 85% ethanol. Corn and sugarcane are the most commonly used plants for ethanol production. Of the world's projected 12 billon bushels of corn produced in 2007, some 3.2 billion bushels, or 26%, will be used for ethanol production. However, production from sugar-based ethanol yields 8 times more alcohol per acre than corn does. There are certain to be social consequences and conflicts between developed and underdeveloped countries when people are starving and crops are being used for fuel rather than food. If every kernel of corn produced in the United States were used to produce ethanol, only about 15%–20% of our current petroleum consumption could be met. Obviously, production of bioenergy from grains is only a temporary solution to the energy problem.

Cellulosic biomass has the potential to diminish the demand for petroleum and increase bioenergy while reducing the conflict between food and fuel. Bioenergy from cellulosic biomass uses the carbon within the cellulose and hemicellulose of plant cell walls to produce biofuels, again primarily ethanol. Cellulosic biomass can come from municipal waste, forest waste products (slash), crop residue (corn stover), native herbaceous vegetation, and "designer" crops such as cultivars of switchgrass (*Panicum virgatum*) and giant sorghum (*Sorghum bicolor*) (fig. 1.2c). Although this is still an emerging industry, it has great potential to help the environment while meeting a significant portion of the world's energy needs.

Other Uses of Grasses

It has been said that more structures are composed of bamboo than stone, brick, and wood combined! Although this may be an exaggeration, it certainly points out the importance of bamboo to billions of people living in Central and South America, Africa, Asia, and Oceania. Not only is bamboo used in construction but it can be a source of nutrition (bamboo shoots) as well as an occasional source of grain in time of famine (Gould and Shaw 1983). The non-food uses of bamboo are endless, but here are a few examples: poles and posts for building construction, fence posts, bridges, boat masts, ladders, cages, flooring; shafts, spears, bows and arrows, fishing poles; handles for tools, whips, knives; furniture; window shades; woven articles such as mats, roofs, baskets; rope and cordage; water conduits and drainpipes; musical instruments; toys; cooking utensils; and a number of miscellaneous items such as chopsticks, pipe stems, sieves, writing paper, facial tissue, and cigarette papers (Gould and Shaw 1983; Chapman 1996). Bamboos also are used in religious ceremonies; as artwork; and as a source of medicine for asthma, coughs, fevers, and kidney problems. Fabric made from bamboo is becoming more and more popular and environmentally "acceptable."

Lemongrass (*Cymbopogon*) is used as a flavoring in cooking and, along with vetiver, is harvested for essential oils (National Research Council 1993). Other aromatic grasses (*Hierochloa, Anthoxanthum*), which contain coumarin, are used for perfumes, hair tonics, and flavoring vodka.

As mentioned previously, grasses are one of the largest families and, from a biodiversity standpoint, of great importance. Species composition, especially of grasses, in plant communities can be an indicator of past use, grazing capacity, wildlife habitat, and "ecological health." Some grasses have even been included on the endangered species list (*Zizania texensis*) (fig. 1.2d).

Harmful Grasses

Grasses are fairly benign, doing little harm to humans or animals compared to the benefits that they afford. That said, some fungi that use grasses as a host have caused serious problems throughout history; the alkaloids produced by the fungus *Claviceps purpura* L. has caused many deaths when ingested in significant quantities. "Ergot" is the term most commonly used for these endophytic fungal diseases. Ergot outbreaks caused hundred of thousands of deaths in the Middle Ages after contaminated grain (rye for the most part) was ingested (Lorenz 1979). "St. Anthony's fires" is the term often used for spontaneous hysteria caused by ergot poisoning. Some have even attributed the hallucinations and hysteria during the Salem witch episode in 1692 to ergot poisoning. Poisoning by ergot is very rare in modern times because of high agricultural grain inspection requirements. Surprisingly, the alkaloids produced by these fungi have shown potential for medicinal use (Chapman 1996).

Infected kodo millet (*Paspalum scrobiculatum* L.) and pearl millet (*Pennisetum glaucum* (L.) R. Br.) have caused intoxication and poisoning. The fungus *Aspergillus tamaril* Kita growing on these plants produces cyclopiazonic acid, which is toxic to humans (Krishnamachari and Bhat 1976). Maize (corn) contaminated with aflatoxins also has been reported to cause deaths in India (Bhat, Nagarajan, and Tulpule 1978). Consumption of any moldy grain should be considered a serious health hazard.

Allred (2005) lists the following harmful compounds produced by grasses that can cause poisoning: coumarin, cyanide, nitrates, and oxalates. He also lists other harmful effects of grasses as dermatitis, hay fever, asthma, and photosensitivity. Often forgotten is the harmful impact caused by mechanical injury. Generally the awns or stiff bristles that can penetrate soft tissues of the eyes, nose, throat, and underbelly cause problems among grazing animals (Allred 2005). One needs only a single experience of wandering into a "sticker" patch barefooted to marvel at the mechanical adaptations of *Cenchrus* for dispersal. Some think that the accumulation of secondary compounds, mutualism with fungi, and mechanical structures capable of injury are adaptations to combat herbivory (Vicari and Bazely 1993).

One should probably include negative economic impacts of grasses under the heading "Harmful Grasses." By definition, a weed is any plant growing where it is not wanted, so any plant could be considered a weed depending on the situation. Consider Bermudagrass for example: it is an important turf species for lawns, athletic fields, and road rights-of-way and a highly desirable forage species used for hay and domestic livestock grazing, but it also is a nuisance and difficult to control and eradicate from flower beds and St. Augustine lawns.

Many annual grasses and a few perennials would fit into the "weedy" category (fig. 1.2e, f). Cost of control of annual grasses such as crabgrass (*Digitaria* spp.), barnyardgrass (*Echinochloa* spp.), and goosegrass (*Eleusine indica*) is extremely high, and the chemicals used for control have negative impacts on the environment. The same applies to control of annual grasses in cultivated crops. Either through the cost of chemicals and application for herbicides or the fuel and labor costs in cultivation, "weedy" grasses negatively impact human food and fiber production and costs.

Grasses also play a role in the realm of "unintended consequences." In an effort to increase national security by reducing dependence on foreign petroleum, as well as reduce carbon emissions by using cleaner fuels, the addition of ethanol to gasoline has been on the increase over the last decade. Most if not all of us would agree this is a "good" thing. In the United States, ethanol has been primarily made from an annual cultivated grass—corn. To increase production of ethanol, farmers are planting more acres into corn, thus increasing the amount of nitrogen and phosphate fertilizer going into the environment. These chemicals accumulate in watersheds and eventually make their way to the Gulf of Mexico, where they create "dead zones" in the marine environment. Fish and shrimp harvests have been diminished—increasing prices for us all—and harvesters have to search farther from shore for longer periods of time to find their catch, increasing fuel consumption and carbon emissions, and further increasing commodity prices. Most of us would agree these are "bad" things.

Grasses impact our lives every day, in obvious and sometimes not so obvious ways. Grasses have been inseparable from human social structure since its beginnings. As go the grasses, so goes human society.

The individual flowering grass tiller consists of extremely simple structures. A phenomenal amount of diversity, however, occurs within these structures, making up the characteristics that differentiate over 700 genera and 11,000 species of grasses.

Roots

At the time of germination, the grass seed (caryopsis), extends along 2 axes. The radicle, protected by the coleorhiza, produces the primary root system that begins to stabilize the seedling and absorbs water and nutrients from the soil. At the other end of the embryo the plumule, protected by the coleoptile, elongates upward to become the primary shoot system (fig. 2.1). This aboveground portion is responsible for photosynthesis, supplying the plant with energy to grow and reproduce. The primary root system functions for only a few short weeks or months and soon withers and dies. There is no taproot in grasses as is typically found in herbaceous forbs or woody plants. For the remainder of the life of the grass, belowground function is performed by a series of adventitious roots that compose the secondary root system. These adventitious roots are fibrous and arise from the lowermost nodes of the grass culm (stem). Occasionally, some grasses (e.g., *Zea, Sorghum, Rottboellia*) develop stout "prop" roots from the lower aboveground culm nodes (fig. 2.2). Prop roots are necessary for mechanical support of the large aboveground portion of the plant.

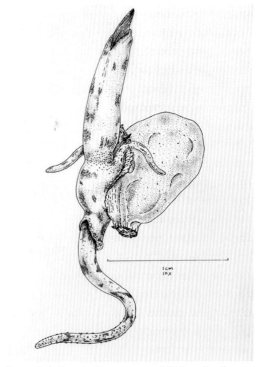

Figure 2.1. Germinating corn seed (University of Colorado Press).

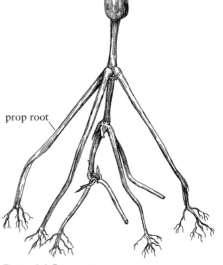

prop root

Figure 2.2 Prop roots.

Culms

Grass culms are made up of swollen nodes and smooth internodes. Internodes are typically round or elliptical in cross section. Nodes are always solid, but the internodes may be solid, semisolid, or hollow. Grass leaves, branches, and adventitious roots arise from meristematic tissue near the nodes.

It is convenient to think of the aboveground portion of a grass plant as composed of a series of "building blocks," which are called phytomeres. A phytomere, or basic unit of the grass shoot, is composed of a node, internode, meristematic tissue, and leaf (fig. 2.3). Usually, the internodes are very short during the major portion of the life cycle, particularly in grasses with a cespitose growth form. This low growth form also maintains the apical meristem (growing point) of the plant near the soil surface, protecting it from grazing animals and wildfires. In this stage the grass plant appears to be nothing more than a clump of blades. However, when flowering occurs, the internodes elongate and push the reproductive portion of the plant upward. Meristematic tissue (vegetative bud) located just above the basal node of the phytomere in the axil of the leaf can develop into several structures. In most cases it produces a tiller or lateral branch; less frequently, a horizontal culm develops. If the horizontal culm is underground, it is called a rhizome (fig. 2.4a–c), but if the horizontal culm is located aboveground, it is referred to as a stolon (fig. 2.5a–d). Some grasses may have both stolons and rhizomes (*Cynodon*).

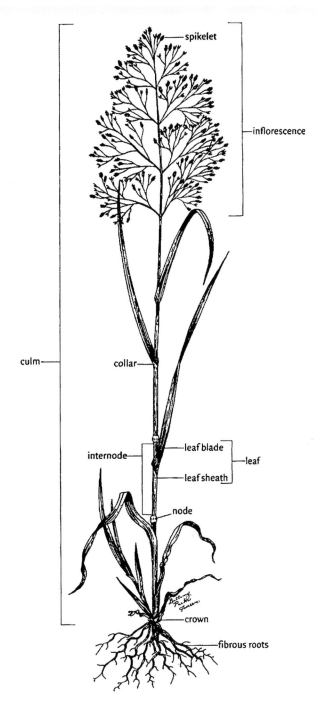

Figure 2.3. Characteristics of the grass plant (used with permission of Stubbendieck et al. 2003).

Leaves

Grass leaves are 2-ranked and alternate. Thus, at each node a leaf is oriented 180° from the previous one. The grass leaf is differentiated

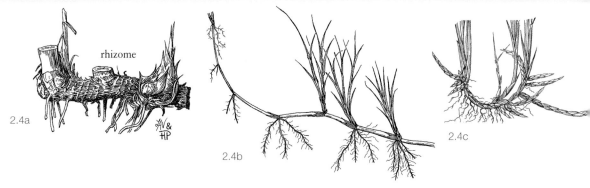

2.4a

rhizome

2.4b

2.4c

Figure 2.4a. Stout, woody rhizome. b. Long, slender rhizome. c. Scaly rhizome.

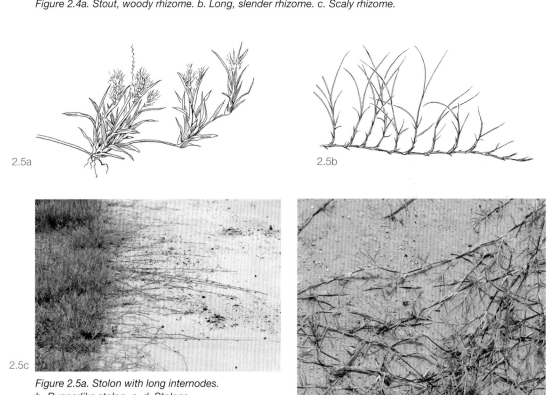

2.5a

2.5b

2.5c

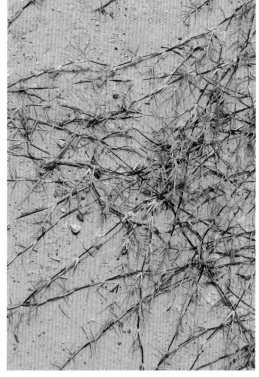

2.5d

Figure 2.5a. Stolon with long internodes.
b. Runnerlike stolon. c, d. Stolons.
(Photographs by Robert B. Shaw)

into a basal portion (sheath), which encloses the culm, and an upper portion (blade or lamina). The sheath typically has free margins, but in some genera the sheath margins are connate, forming a tubular-like structure (*Bromus, Glyceria, and Melica*) (fig. 2.6a). The blade is characteristically flat and elongated. In arid environments the blade may be terete and very narrow or have variously rolled margins, whereas in forested areas the blades may become broad and ovate. Generally there is a ligule at the inside (adaxial) junction of the leaf sheath and the leaf blade. The ligule may be a ring of hairs (fig. 2.6b), membranous (fig. 2.6c), or a ciliated membrane (fig. 2.6d). Rarely is it absent (in most species of *Echinochloa*). The ligule is the most often used and reliable vegetative character for grass identification. Occasionally, there are auricles or fingerlike appendages present at the top of the sheath or at the base of the blade (fig. 2.6e).

Inflorescence

The inflorescence, or flowering portion of the plant, is an excellent diagnostic characteristic in grass identification. The stalk upon which the inflorescence sits is called the peduncle, and the central axis of the grass inflorescence is the rachis (fig. 2.3). There are three basic types of grass inflorescences: the spike, the raceme, and the panicle. The spike, the simplest of the inflorescence types (fig. 2.7a, b), consists of a rachis and sessile flowering bodies. The next level of complexity occurs in the raceme, which has a rachis, and flowering bodies are borne upon a pedicel (fig. 2.7c, d). A spicate raceme is a combination of the first two in which both sessile and pediceled flowers occur on the same rachis (fig. 2.7e, f). The most common inflorescence type in grasses is the panicle, an inflorescence that is at least branched and rebranched, or any inflorescence

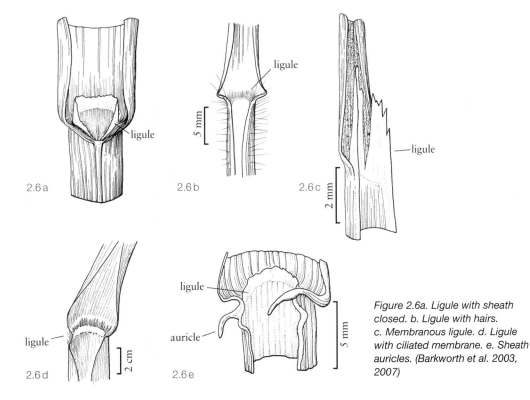

Figure 2.6a. Ligule with sheath closed. b. Ligule with hairs. c. Membranous ligule. d. Ligule with ciliated membrane. e. Sheath auricles. (Barkworth et al. 2003, 2007)

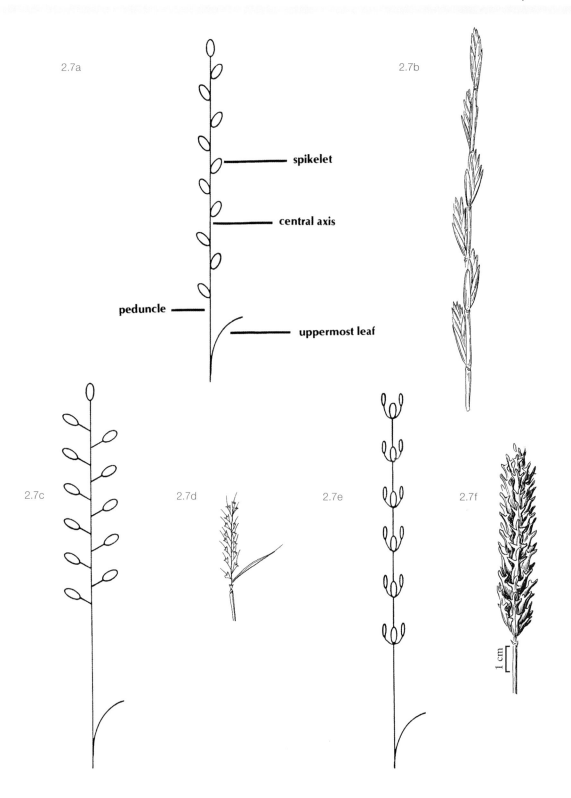

Figure 2.7a, b. Spike inflorescence. c, d. Raceme inflorescence. e, f. Spicate raceme inflorescence. (Hignight, Wipff, and Hatch 1988)

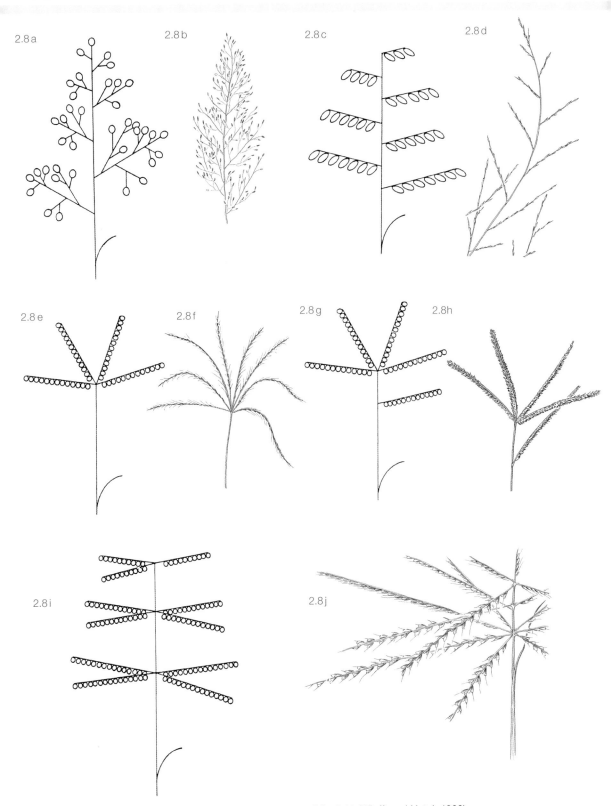

Figure 2.8. Variations in the panicle-type grass inflorescence (Hignight, Wipff, and Hatch 1988).

where the flowering bodies are not sessile or individually pediceled on the rachis. Figure 2.8 illustrates the enormous variation that occurs within this inflorescence type.

There is also considerable variation in inflorescence form within the tribe Andropogoneae (fig. 2.9a–f). The basic dispersal unit in the Andropogoneae consists of a spikelet pair (one pediceled and one sessile) and an internode. These dispersal units are stacked together to form rames, which are variously grouped to form inflorescences. Other unusual forms found in the Andropogoneae are tightly cylindrical panicles (*Imperata*) (fig. 2.10a), large plumose panicles (*Saccharum*) (fig. 2.10b), monoecious spikelets either on the same (*Tripsacum*) (fig. 2.10c) or separate (*Coix*) rame (fig. 2.10d), and spikelets embedded in a thickened branch (*Coelorachis*) (fig. 2.10e).

Caution should be used when determining inflorescence type—especially with extremely contracted panicles. Always examine the lower most flowering bodies to determine if they are indeed single and sessile on the rachis. Many misidentifications occur when genera with a reduced or contracted panicle (e.g., *Alopecurus*, *Phleum*, *Cenchrus*, *Pennisetum*) are keyed as if they had a spike inflorescence.

Figure 2.9. Variation in inflorescences within the Andropogoneae: a. Diagram of inflorescence structures and composition of various types; b. Inflorescence a spikate raceme; c. Compound inflorescence with 2 or more rames; d. Open panicle inflorescence ; e. Inflorescence of subdigitate rames; f. Inflorescence a false panicle. (Barkworth et al. 2003)

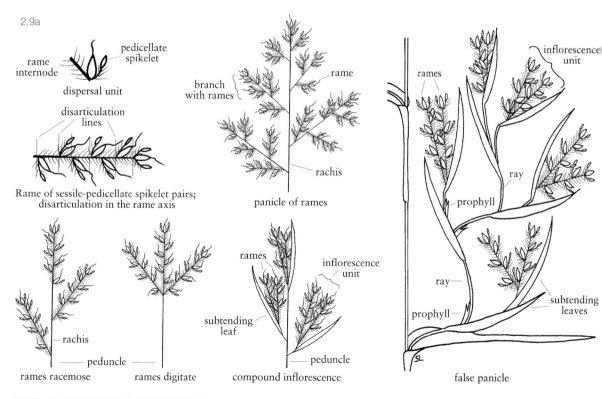

2.9a

rame internode — pedicellate spikelet

dispersal unit

disarticulation lines

Rame of sessile-pedicellate spikelet pairs; disarticulation in the rame axis

branch with rames — rame

rachis

panicle of rames

rames — inflorescence unit

rames racemose   rames digitate

rachis — peduncle

subtending leaf — peduncle

compound inflorescence

inflorescence unit

ray — prophyll

ray

prophyll — SL

subtending leaves

false panicle

INFLORESCENCE STRUCTURES

2.9b

2.9c

2.9d

2.9e

2.9f

Spikelets

Spikelets, the flowering bodies in grasses, are the basic unit of the grass inflorescence (fig. 2.11a, b). Typically the grass spikelet consists of a short axis (rachilla) upon which 2 series of floral bracts (modified leaves) subtend the grass flower. The first (lower) and second (upper) glumes compose the initial series of bracts. Both glumes are usually present, but occasionally the first may be reduced in size (e.g., *Digitaria*, *Paspalum*) (fig. 2.11c) or completely absent (*Eriochloa*). The lower and up-

per glumes may be partially fused (*Alopecurus* and some species of *Polypogon*). The glumes become setaceous in *Critesion* and some *Elymus*. Very rarely both glumes are lacking (*Leersia*) (fig. 2.11d).

The second series of bracts consists of the lemma and palea (fig. 2.12a). The lemma is probably the most reliable and frequently used character in grass identification. It is always present and has a high degree of stability within a genus. Lemma texture, shape, nervation, awn development, and surface features are used extensively in identification

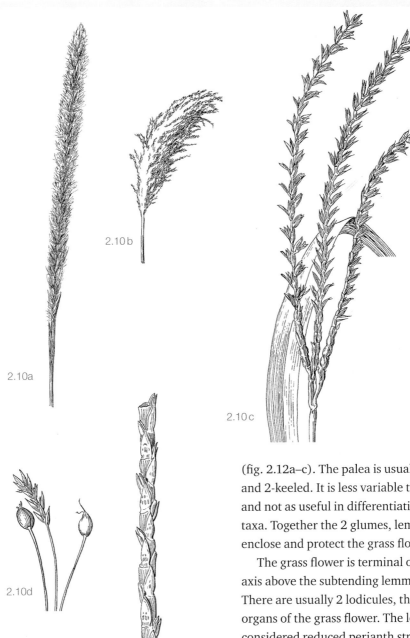

2.10b

2.10a

2.10c

2.10d

2.10e

Figure 2.10. Other unusual inflorescence forms found in the Andropogoneae. a. Tightly cylindric panicles. b. Large, plumose panicles. c, d. Monoecious spikelets, together and separate. e. Embedded spikelets. (Barkworth et al. 2003)

(fig. 2.12a–c). The palea is usually 2-nerved and 2-keeled. It is less variable than the lemma and not as useful in differentiating between taxa. Together the 2 glumes, lemma, and palea enclose and protect the grass flower.

The grass flower is terminal on a short axis above the subtending lemma and palea. There are usually 2 lodicules, the lowermost organs of the grass flower. The lodicules are considered reduced perianth structures, and they function to help open the grass spikelet at anthesis. Above the lodicules is the anthecium (male floral parts) composed of the stamens. There are generally 3 stamens consisting of a long, slender filament and a pollen-producing anther. Occasionally stamen number is reduced to 1 or 2; conversely, the number can reach as many as 120 or more in some bamboos. The gynecium (female portion

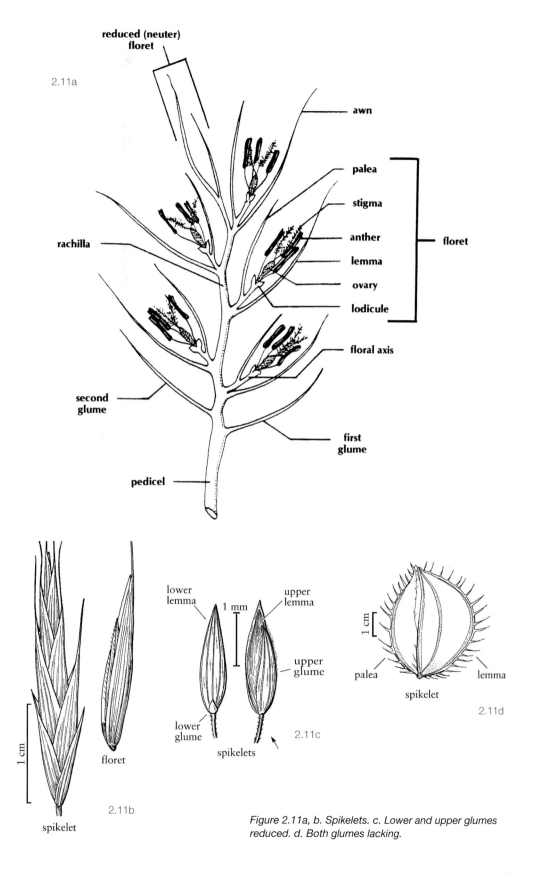

2.11a

reduced (neuter) floret

awn

palea

stigma

anther

lemma

ovary

lodicule

floret

floral axis

rachilla

second glume

first glume

pedicel

lower lemma

upper lemma

1 mm

upper glume

lower glume

spikelets

2.11c

1 cm

palea

lemma

spikelet

2.11d

1 cm

floret

spikelet

2.11b

Figure 2.11a, b. Spikelets. c. Lower and upper glumes reduced. d. Both glumes lacking.

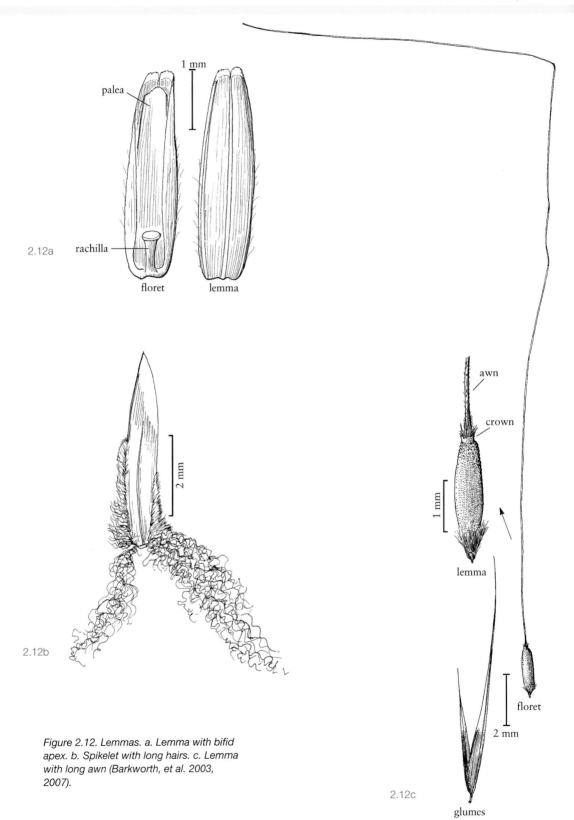

Figure 2.12. Lemmas. a. Lemma with bifid apex. b. Spikelet with long hairs. c. Lemma with long awn (Barkworth, et al. 2003, 2007).

of the flower) is a 1-locular ovary with a single ovule and usually 2 styles and stigmas. The grass fruit, a caryopsis, is a dry, indehiscent, 1-seeded structure in which the ovary wall and pericarp are fused. Some genera (*Sporobolus*, *Eleusine*) have an achene-type of fruit where the pericarp does not fuse with the ovary wall.

Spikelet sexuality is variable in the Poaceae. For the most part, grass spikelets are hermaphroditic or perfect (both male and female parts present and functional); however, occasionally they are unisexual or imperfect (lacking either the male or female reproductive parts). If the staminate and pistillate spikelets are present on the same plant, as in *Zea*, the plant is monoecious. If the staminate and pistillate spikelets are on separate plants, the plant is dioecious. *Buchloë dactyloides*, *Scleropogon brevifolius*, and *Distichlis spicata* are examples of dioecious grasses. Complete sterility of some spikelets in an inflorescence also occurs. In some species of *Critesion* setaceous glumes are all that remain of lateral spikelets. Within the tribe Andropogoneae, members have paired spikelets, one sessile, and one pedicellate. Often the pedicellate spikelet is staminate, sterile, or in some cases completely lacking, leaving only a naked pedicel.

Disarticulation, or how the spikelets separate from the plant at maturity, has long been a diagnostic characteristic used in grass taxonomy. There are 2 general types of disarticulation: above the glumes or below the glumes. Disarticulation above the glumes means that the glumes remain in the inflorescence and the lemma, palea, and flower separate from the plant. Conversely, disarticulation below the glumes means that the glumes fall with the lemmas, paleas, and flowers, leaving nothing in the inflorescence except a naked pedicel or rachis. Rarely, the entire inflorescence will break away from the plant at maturity (*Schedonnardus*, and some species of

Eragrostis). Also in genera such as *Cenchrus* and *Pennisetum* several spikelets are clustered together, subtended by an involucre of bristles or spines, and disarticulation is below the involucre. Members of the Triticeae tribe have spike or spicate raceme inflorescences, and disarticulation in some species occurs at the rachis nodes. Additionally, several members of the Cynodonteae tribe have spikelets that fall in clusters on short inflorescence branches (*Buchloë*, *Bouteloua*, and *Chondrosum*). Theoretically, these extreme forms of disarticulation are considered below the glumes because no glumes remain on the plant.

Florets

The basic unit of the grass spikelet is the floret. A floret consists of a lemma, palea, and flower. Thus, the simplest grass spikelet that has all the parts would consist of a pair of glumes, lemma, palea, and perfect flower and is called a spikelet with 1 perfect floret (fig. 2.13a). Because grass flowers may be "perfect," "staminate," "pistillate," or "sterile," these terms are also applied to spikelets and florets. Often the term "fertile flower," "fertile spikelet," or "fertile floret" will be used to refer to a flower that has a functioning pistil and is capable of producing a fruit, regardless of the presence or absence of stamens. The next level of complexity after a spikelet with 1 perfect floret is a spikelet with 2 perfect florets, which has a pair of glumes; one floret consisting of a lemma, palea, and flower; plus a second floret with a lemma, palea, and flower (fig. 2.13b). Complexity can increase to spikelets with dozens of perfect florets. As in spikelets, reduction in the reproductive capacity of florets also occurs (fig. 2.14a–d). Reduced florets are either staminate or sterile. A staminate floret has a lemma, palea, and functioning stamen; a sterile floret lacks any functioning reproductive organs. Often reduced florets consist of "empty" lemmas and paleas, lacking flowering parts. Occasionally, reduced florets lack the lemma and palea and are recognized by a naked rachilla extending past the uppermost fertile floret. In the Paniceae and Andropogoneae tribes the spikelet consists of 2 florets. The upper floret is fertile, whereas the lower is reduced and either staminate or sterile. In most other cases the reduced florets are typically above the fertile florets. *Chasmathium,* however, has reduced florets above and below.

The typical spikelet of a member of the Paniceae consists of 2 florets; the lower is rarely perfect, often staminate, or most frequently represented only by a lemma (fig. 2.15). Thus, the panicoid spikelet has 2 florets, the lower reduced and the upper perfect. Dorsal compression of the spikelet and disarticulation below the glumes are also characteristic of this spikelet type. Figure 2.15a shows a schematic of the typical panicoid spikelet. Figure 2.15c–d illustrate examples of a staminate lower floret, a lower floret represented by a lemma and reduced palea, and represented only by a lemma.

The spikelets and inflorescences in the Andropogoneae are very complex and can be confusing. The typical Andropogoneae spikelet is similar to the panicoid spikelet in that they both have 2 florets, the lower reduced and the upper perfect. However, in the Andropogoneae the lower and upper glumes are well developed and tightly enclose the florets. The lemma and palea of the florets are greatly reduced and membranous rather than large and indurate as in the Paniceae. Figure 2.16a shows a schematic of the typical spikelet form and various ways that the spikelets can be paired. Most frequently the sessile and pedicellate spikelet pairs are heteromorphic (the sessile one perfect and the pedicellate one reduced). Other

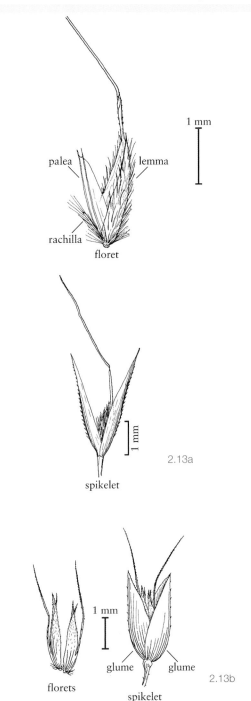

Figure 2.13a. Spikelet with 1 floret. b. Spikelet with 2 florets (Barkworth, et al. 2003, 2007).

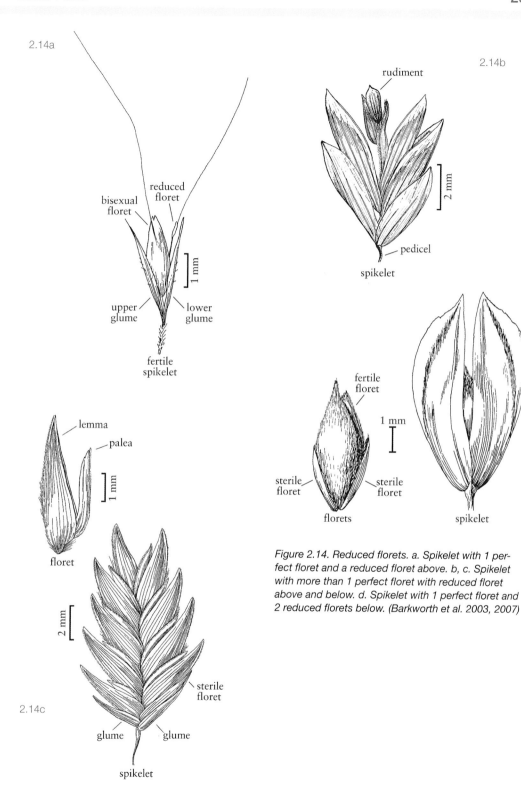

2.14a

reduced floret

bisexual floret

upper glume

lower glume

fertile spikelet

1 mm

2.14b

rudiment

pedicel

spikelet

2 mm

lemma

palea

1 mm

floret

2.14c

2 mm

sterile floret

glume

glume

spikelet

fertile floret

sterile floret

sterile floret

florets

1 mm

spikelet

2.14d

Figure 2.14. Reduced florets. a. Spikelet with 1 perfect floret and a reduced floret above. b, c. Spikelet with more than 1 perfect floret with reduced floret above and below. d. Spikelet with 1 perfect floret and 2 reduced florets below. (Barkworth et al. 2003, 2007)

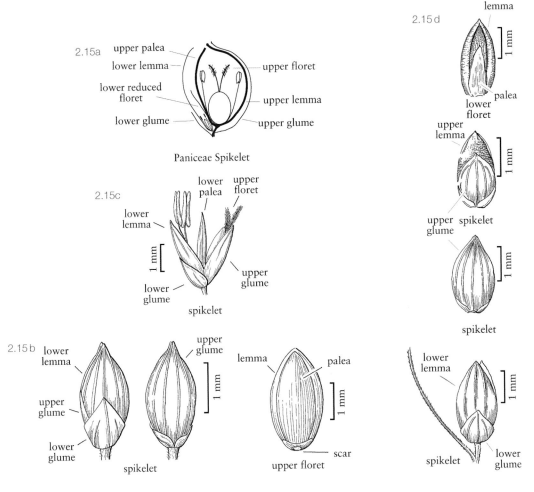

Figure 2.15. Typical panicoid spikelets. a. Schematic of panicoid spikelet. b. Typical panicoid spikelet. c. Panicoid spikelet with staminate lower floret. d. Panicoid spikelet subtended by a single bristle. (Barkworth et al. 2003)

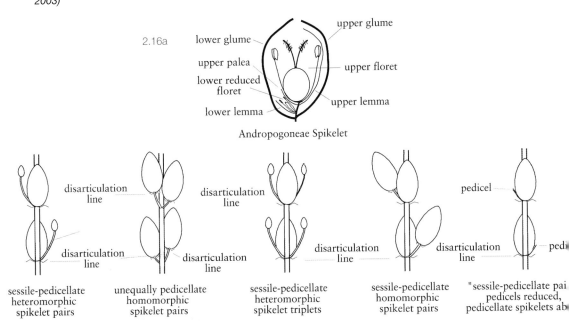

SPIKELETS and SPIKELET UNITS

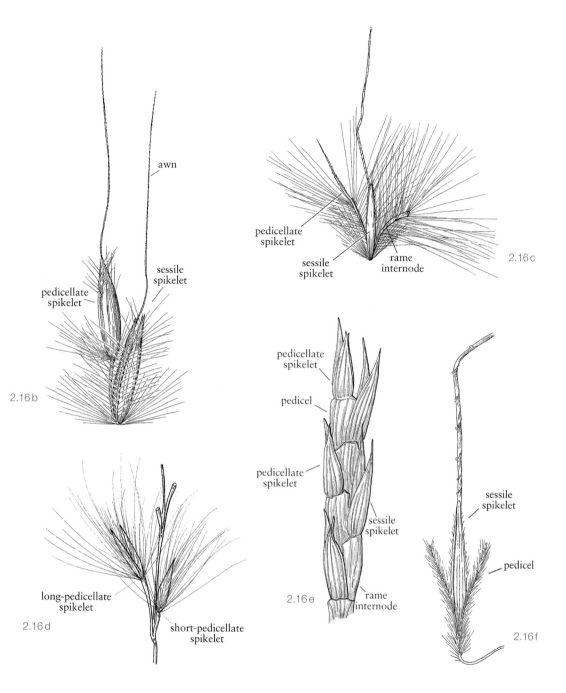

awn

pedicellate
spikelet

sessile
spikelet

2.16 b

long-pedicellate
spikelet

short-pedicellate
spikelet

2.16 d

pedicellate
spikelet

sessile
spikelet

rame
internode

2.16 c

pedicellate
spikelet

pedicel

pedicellate
spikelet

sessile
spikelet

rame
internode

2.16 e

sessile
spikelet

pedicel

2.16 f

Figure 2.16. Spikelets in the Andropogoneae. a. Diagram of the Adropogoneae spikelet and various forms of spikelet units. b. Spikelets paired and both perfect. c. Spikelets paired, sessile one perfect, pediceled one reduced. d. Spikelets both perfect and pediceled. e. Spikelets paired and embedded in a thickened rachis. f. Terminal spikelet unit: 1 is sessile and perfect; the 2 lateral ones are represented by a pedicel. (Barkworth et al. 2003).

variations in spikelet pairing are illustrated (fig. 2.16b–f).

Number of florets per spikelet, presence or absence of reduced florets, and location of reduced florets (above or below the fertile ones) are important diagnostic characteristics. Unfortunately, these are some of the most frequently misinterpreted and overlooked features in grass identification.

Anyone who has been in Texas understands that the state is huge! It is 801 mi (1,282 km) from the northernmost point in Dallam County to the southern tip of Cameron County, and 773 mi (1,237 km) from the Sabine River in Newton County to the Rio Grande in El Paso County. The state covers over 268,000 mi^2 (694,120 km^2) in surface area, which converts to roughly 172 million ac (69 million ha) (Poole et al. 2007). Texas has over 3,800 mi (6,080 km) of border, with over 2,011 mi (3,218 km) along the Rio Grande separating Texas from Mexico (Dallas Morning News 2008).

The size of the state affords the opportunity for an amazing number of combinations of natural features referred to in various ways, such as vegetation types, natural regions, ecoregions, and major land resource areas. The term "ecoregions" has been selected for use in this book.

Ecoregions are typically large land (for our purposes) or water areas that have a homogenous set of natural features. Geologic, physiognomic, topographic, climatic, edaphic, and biotic (vegetative composition and structure in this case) features should be relatively similar across an ecoregion. Each of the natural features that contribute to ecoregions will be briefly discussed and illustrated at the state level to help visualize their relationship to the 12 ecoregions identified for Texas. Each ecoregion will then be discussed individually.

Geology

A curved line from Marathon around the big bend in the Rio Grande to Del Rio along the Balcones Escarpment past San Antonio, Austin, Waco to Dallas, and extending to the Red River delineates the Gulf of Mexico coastline during the Cretaceous Period (60–100 million years ago [mya]) (fig. 3.1; Spearing 1991). Surface rocks east and south of that line are younger than those of the Cretaceous and become progressively more recent toward the Gulf. During the Cenozoic Era (66 mya) erosive forces have continued to deposit large amounts of sediments, extending the state farther and farther into the Gulf. Cretaceous-aged limestone dominates the central part of the state. Triassic rocks (245–208 mya) are exposed along the Caprock Escarpment. Permian deposits (286–245 mya) cover the area between Amarillo to Abilene and the Guadalupe, Delaware, and Apache mountains. Pennsylvanian formations (320–286 mya) occur farther east toward Fort Worth. Most of the rocks in the western Panhandle on the Caprock represent materials eroded from the Rocky Mountains during the Pliocene, Miocene, and Oligocene epochs (35–2 mya) of the Cenozoic Era. Volcanism has been fairly restricted in Texas. Remnants of a small volcanic episode during the Cretaceous are visible around Austin and Uvalde. A much larger and more dramatic episode that occurred in the early Tertiary Period (around 65 mya) in West Texas formed lava fields and mountains (i.e., Davis and Chisos mountains). The most ancient of rocks (Precambrian, up to 1 billion years old) are found in the Llano area and several uplifts in West Texas

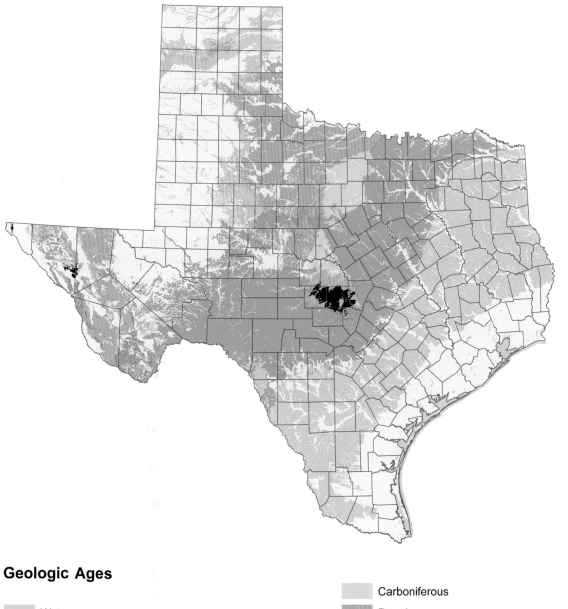

Geologic Ages

Water	Carboniferous
Lower Paleozoic: Cambrian, Ordovician, Silurian, Devonian	Permian
Cretaceous	Tertiary
Jurassic	Quaternary
Triassic	Tertiary & Cretaceous volcanic rocks
	Precambrian

Figure 3.1. Geologic ages in Texas (adapted from Geologic Map of Texas. Bureau of Economic Geology, University of Texas at Austin).

(i.e.,Franklin Mountains). Holocene, or the most recent, deposits are represented by the sand dunes in the Coastal Bend region and in the upper Pecos River watershed.

Physiography

Figure 3.2 illustrates the complexity of Texas physiography, or the physical expression of landforms. As expected, most features are closely aligned with geology. For example, the slightly sloping Gulf Coastal Plains are composed of "recent" Cenozoic deposits. The Edwards, Stockton, and Comanche plateaus are hard Cretaceous limestones that have resisted erosion for millions of years. Pennsylvanian rocks occur in the Palo Pinto Basin. The Osage Plains or Red Bed Plains are all Permian in age and often have a distinctive reddish color. The Llano Estacado (High Plains) consists entirely of Tertiary-aged materials eroded from the Rocky Mountains and deposited to form a relatively flat and featureless plain. The Llano Basin (for the most part) is really an uplift of Precambrian-aged rocks. Recent Holocene deposits are the sandy lands in the Coastal Sand Plains and Toyah Basin. The Big Bend region of the state is most complex with pockets of nearly all rock ages and types.

Topography

Texas is also complex topographically. The Gulf Coastal Plains are relatively flat and gently tilted toward the Gulf of Mexico. Two discernible bands, the Carrizo Sand Ridge and the Bordas-Oakville Escarpment, run from southwest to northeast across the plain (Spearing 1991). Two other distinctive escarpments are better known in Texas. The Balcones Escarpment delineates the southern and part of the eastern boundary of the Edwards Plateau. The canyonlands and hills along this boundary are known as the "Texas Hill Coun-

try." The Caprock Escarpment delineates the eastern boundary of the High Plains and is well known for Palo Duro Canyon and Caprock Canyon state parks. The plateaus are characterized by mesas and grasslands. Most of the plains are gently rolling or undulating regions, with the exception of the Llano Estacado, which is mostly a flat and somewhat featureless expanse. The topography of West Texas is mountainous with basins and ranges.

Elevation is, obviously, lowest at sea level along the 624 mi (998 km) Texas Gulf Coast tidewater line stretching from the mouth of the Sabine River to the mouth of the Rio Grande (Dallas Morning News 2008). Guadalupe Peak is the highest point in Texas at 8,749 ft (2,651 m) above sea level. This rise in elevation has a significant impact on biotic diversity. Many plant species found in the Guadalupe and Davis mountains occur nowhere else in the state and usually have a much greater affinity to the Rocky Mountains flora.

Texas has 6,783 mi^2 (17,636 km^2) of water surface. The major Texas watersheds drain from the northwest toward the southeast and empty into the Gulf of Mexico (fig. 3.3) (Dallas Morning News 2008). Large reservoirs are found on most of the major rivers. From east to west, the major river basins of the state are the Sabine, Neches, Trinity, Brazos, Colorado, Guadalupe, Nueces, and Rio Grande. The Pecos River, an important river that drains much of west-central Texas, joins the Rio Grande in Val Verde County. The Canadian, Red, and Sulfur rivers flow from west to east and are part of the Mississippi River system (Schuster and Hatch 1990).

Climate

As one would expect, precipitation varies considerably across Texas. Rainfall in East Texas is highest in Orange County (56 in, or 1,422 mm, per year) and lowest in West Texas in El Paso

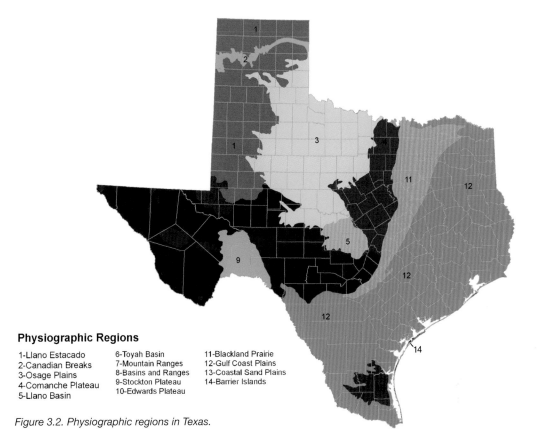

Physiographic Regions

1-Llano Estacado
2-Canadian Breaks
3-Osage Plains
4-Comanche Plateau
5-Llano Basin

6-Toyah Basin
7-Mountain Ranges
8-Basins and Ranges
9-Stockton Plateau
10-Edwards Plateau

11-Blackland Prairie
12-Gulf Coast Plains
13-Coastal Sand Plains
14-Barrier Islands

Figure 3.2. Physiographic regions in Texas.

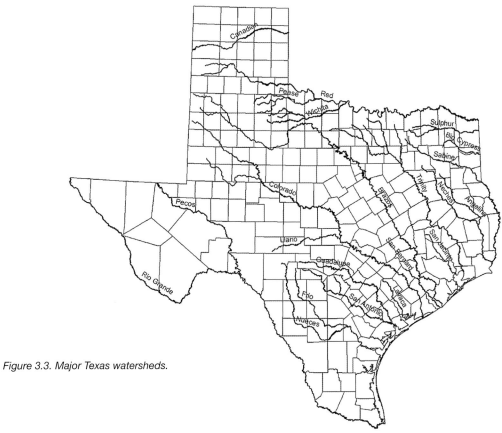

Figure 3.3. Major Texas watersheds.

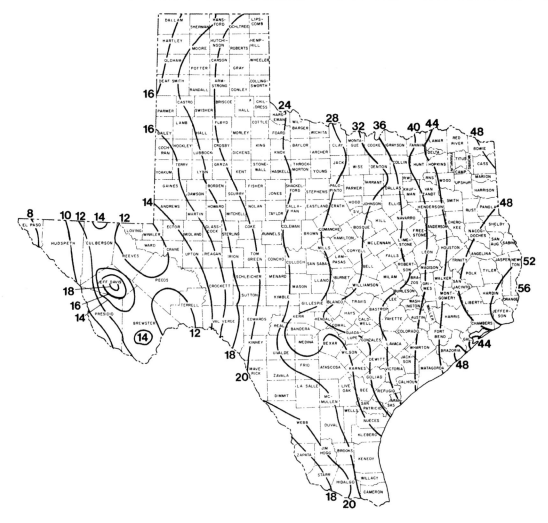

Figure 3.4. Mean annual total precipitation (inches) in Texas (adapted from Griffiths and Orton 1968, in Schuster and Hatch 1990).

County (8 in, or 203 mm).[1] This is a 48 in (1,219 mm) decrease in annual precipitation across the 773 mi (1,237 km) width of the state. Roughly, a decrease of 1 in (25.4 mm) of precipitation occurs for every 16 mi (25 km) traveled from east to west (fig. 3.4). Average annual precipitation for the state is approximately 32 in (812 mm). The 32-inch isoline on Figure 3.4 is skewed toward the eastern part of the state. The 24-inch isoline is more aligned with the geographic center of the state, which runs through McCulloch County.

When precipitation falls during the year is equally as important as the amount. Most of Texas, except the far eastern side and the Trans-Pecos, receives the majority of the rain bimodally, with a peak in May and a peak in September or October (Schuster and Hatch 1990). It can be assumed that the early fall peak is related to the occurrence of hurricanes during this period. The forests of East Texas typically receive precipitation more or less

1. The Texas Almanac puts the extremes at a high of 60.57 in (1,538 mm) in Jasper County to a low of 9.43 in (240 mm) in El Paso (Dallas Morning News 2008).

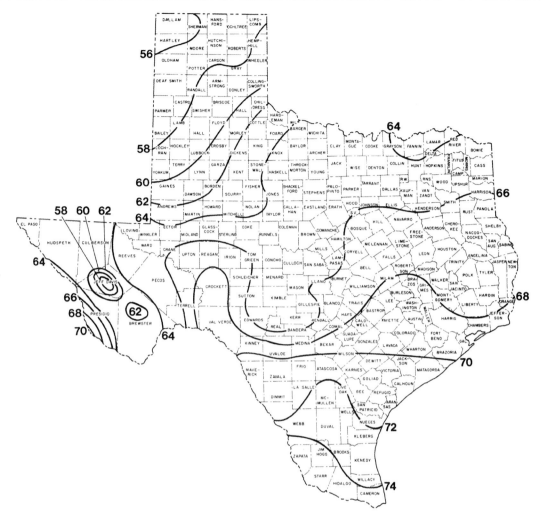

Figure 3.5. Mean annual temperature (°F) (adapted from Griffiths and Orton 1968, in Schuster and Hatch 1990).

evenly throughout the year. Conversely, the desert areas of the Trans-Pecos receive most of their precipitation in the form of monsoonal rains during the months of July, August, and September.

Temperature also varies across the state along an irregular north-to-south gradient (fig. 3.5). Much of the irregularities in the gradient result from elevational change. Mean annual temperatures vary from 56°F (13°C) in Dallam County in the Panhandle to 74°F (23°C) in Cameron County in the Rio Grande Valley—an 18°F (10°C) difference over 800 mi (1,280 km), or a 1°F increase for every 45 mi one

travels south (1°C change for every 130 km). Average temperature for the state is about 65°F (18°C). The geographic center of Texas lies almost exactly between the two average temperature extremes.

The number of frost-free days is closely aligned with average temperature. The lowest number of frost-free days is 185 in the northwestern portion of the Panhandle, and the greatest is 340 in South Texas (Poole et al. 2007), a 155-day (5-month) difference in growing season. Thus, there is an average decrease of 1 frost-free day for every 5 mi (8 km) traveled north from the Lower Rio Grande Valley, or each 160 mi (257 km) traveled equates to about 30 fewer frost-free days (fig. 3.6).

Numerous climatic classification schemes integrating precipitation, timing of precipitation, and temperature have been produced. The 2 widely recognized schemes are the Köppen system, as adapted and modified by Trewartha (1968), which constructs regional climatic groups and types; and the Bailey system (1995), which adds vegetation to cli-

mate and recognizes domains, divisions, and provinces. Bailey's system basically divides Texas into 4 major longitudinal provinces: from east to west, Southeastern Mixed Forest Province, Prairie Parkland Province, Southwest Plateau and Plains Dry Steppe and Shrub Province, and Chihuahuan Semi-Desert Province (fig. 3.7). Four other provinces constitute a minor fraction of the Texas land area. Table 3.1 presents a comparison of these two systems.

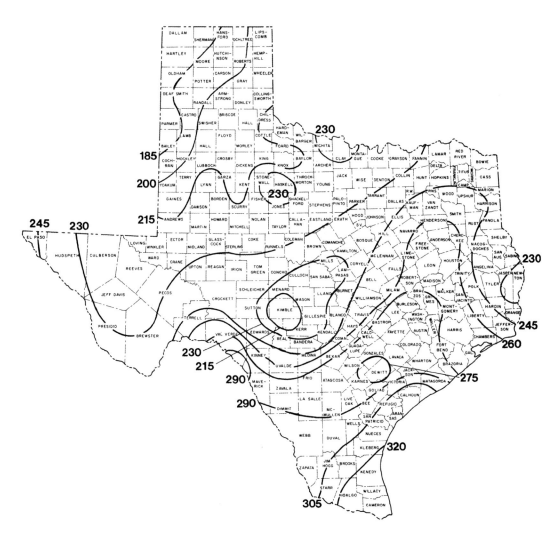

Figure 3.6. Mean length (days) of frost-free period (adapted from Griffiths and Orton 1968, in Schuster and Hatch 1990).

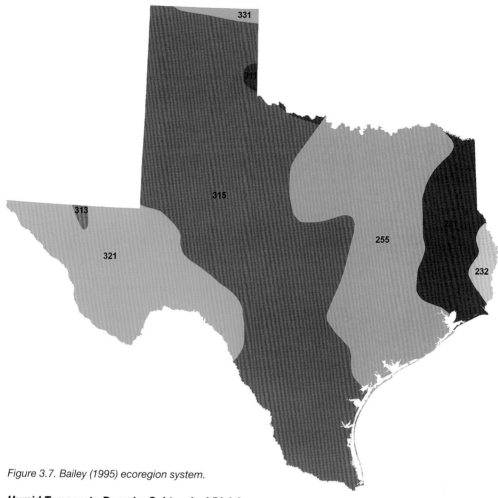

Figure 3.7. Bailey (1995) ecoregion system.

Humid Temperate Domain, Subtropical Division
 231, Southeastern Mixed Forest Province
 232, Outer Coastal Plain Mixed Forest Province
Prairie Division
 255, Prairie Parkland (Subtropical) Province
Dry Domain, Tropical/Subtropical Steppe Division
 311, Great Plains Steppe and Shrub Province
 315, Southwest Plateau and Plains Dry Steppe and Shrub Province
Tropical/Subtropical Steppe Regime Mountains
 M313, Arizona–New Mexico Mountains Semi-Desert–Open Woodland–Coniferous Forest–Alpine
 Meadow Province
Tropical/Subtropical Desert Division
 321, Chihuahuan Semi-Desert Province
Temperate Steppe Division
 331, Great Plains–Palouse Dry Steppe Province (Bailey 1995)

Table 3.1. Comparison between Köppen Climatic Classification System and the Bailey Ecoregion System.

Köppen	Bailey
SUBTROPICAL CLIMATES	HUMID TEMPERATE DOMAIN (200)
Humid subtropical (8 months 50°F [10°C], coldest month <64°F [18°C])	230 Subtropical Division
	231 Southeastern Mixed Forest Province
	232 Outer Coastal Plain Mixed Forest Province
	250 Prairie Division*
	255 Prairie Parkland (Subtropical) Province
TEMPERATE CLIMATES	
Temperate continental, warm summer (4–7 months >50°F [10°C)], coldest month <32°F [0°C], warmest month >72°F [22°C])	250 Prairie Division
	255 Prairie Parkland Province
Dry Climates	Dry Domain (300)
	310 Tropical/Subtropical Steppe Division
Tropical/subtropical semiarid Potential evaporation exceeds precipitation, and all months >32°F (0°C)	311 Great Plains Steppe and Shrub Province
	315 Southwest Plateau and Plains Dry Steppe and Shrub Province
	320 Tropical/Subtropical Desert Division
	321 Chihuahuan Semi-Desert Province
	330 Temperate Steppe Division
	331 Great Plains–Palouse Dry Steppe Province
	M310 Tropical/Subtropical Steppe Regime Mountains
	M313 Arizona-New Mexico/Mountains Semi-Desert–Open Woodland–Coniferous Forest–Alpine Meadow Province

Sources: The Köppen system has been adapted and modified by Trewartha (1983); Bailey (1995).
* Köppen included Prairies in both the Subtropical and Temperate Climates.

Soils

Soils are in part the result of long-term physical and chemical weathering of geologic deposits and the erosion and deposition of materials by wind (eolian) and water (alluvial). Soils are diverse across the state and represent 7 of the 12 major soil orders (table 3.2). Some general distributional patterns become apparent when the orders are mapped across the state (fig. 3.8). Alfisols and Ultisols are leached soils of forested areas and are most common in the eastern forests. These 2 forest soil orders account for almost 55 million ac (22 million ha), or about 32% of the soils within the state (table 3.2).

Aridisols and Entisols are soils of arid, semi-desert, and desert regions and most commonly occur in the Trans-Pecos region of West Texas. These 2 soil orders of arid areas account for only about 16% (27 million ac, or 11 million ha) of Texas soils. Vertisols and Mollisols, mostly found in grassland or prairie areas, are characteristic of the central portion of the state and account for 43%, or about 72 million ac (29 million ha). These observations are a simplification of a very complex soil distribution pattern, and the diversity of soils within the state is enormous. Over 1,200 different soil-mapping units have been identified and described for Texas (Dallas Morning News 2008; Poole et al. 2007).

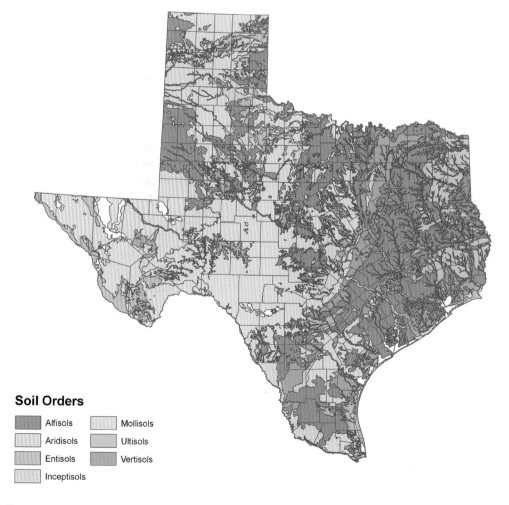

Soil Orders

▓	Alfisols	▒	Mollisols
▒	Aridisols	▒	Ultisols
▒	Entisols	▓	Vertisols
▒	Inceptisols		

Figure 3.8. Soil orders across Texas. (adapted from General Soils map of Texas. Bureau of Economic Geology, University of Texas at Austin).

Table 3.2. Soil orders in Texas.

Soil order	Area	Total Area (%)	Description
Alfisols	48,200,000 ac 19,521,000 ha	28	Moderately leached forest soils, high native fertility; clay pan in subsurface horizon
Aridisols	16,840,000 ac 6,820,200 ha	10	Calcium carbonate soils of arid regions, some horizon development, dry most of the year
Entisols	9,918,000 ac 4,016,790 ha	6	Soils of recent origin, only A horizon (top layer) developed
Inceptisols	12,590,000 ac 5,099,000 ha	7	Soils with minimal horizon development, often of steep slopes
Mollisols	55,940,000 ac 22,655,700 ha	33	Soils of grassland ecosystems with thick, dark horizons
Ultisols	6,708,000 ac 2,716,740 ha	4	Strongly leached, acid forest soils with relatively low native fertility
Vertisols	16,537,000 ac 6,698,000 ha	10	Clay-rich soils, shrink and swell with changes in soil moisture, horizons well developed
Unclassified	4,500,000 ac 1,822,500 ha	2	

Source: Adapted from University of Idaho, College of Agriculture & Life Sciences online (http://soils.ag.uidaho.edu/soilorders).

Vegetation Areas of Texas

Probably the most frequently cited and recognized map of vegetation areas in Texas is the one originally presented by Gould, Hoffman, and Rechenthin (1960). It has been the foundation of the extremely useful and popular series of checklists published by the Texas Agricultural Experiment Station (now AgriLife Research) titled *Texas Plants—A Checklist and Ecological Summary* (Gould 1962, 1969, 1975b) and *Checklist of the Vascular Plants of Texas* (Hatch, Gandhi, and Brown 1990).

Comparison of Gould's vegetation map (fig. 3.9) with the geology, physiography, average annual precipitation, and soils maps illustrates the strong relationship between these features. Forests and woodlands (Piney Woods, Post Oak Savannah) typically occur on recent geologic areas and have Alfisols and Ultisols, with annual precipitation generally more than 32 in (813 mm). Semidesert and desert Aridisols occur in the Trans-Pecos region with annual precipitation less than 16 in (406 mm). Shrublands, grasslands, and prairies (Blackland Prairies, South Texas Plains, Edwards Plateau, Rolling Plains, High Plains) are generally on Mollisols and Vertisols and lie within the middle of the precipitation extremes. Entisols and Inceptisols, or young and underdeveloped soils, are often found in areas of recent parent-material deposition, such as along rivers, in floodplains and deltas, and along shorelines (Gulf Prairies and Marshes). Vegetation on these younger soils tends to be adapted to establishing on and

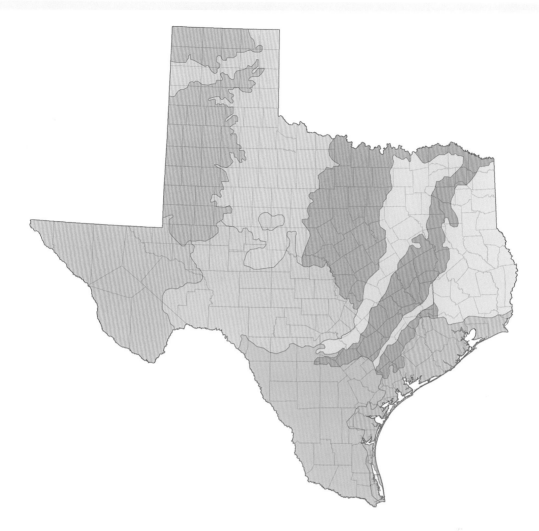

Vegetation Regions of Texas

- Piney Woods
- Gulf Prairies and Marshes
- Post Oak Savannah
- Blackland Prairies
- Cross Timbers and Prairies
- South Texas Plains
- Edwards Plateau
- Rolling Plains
- High Plains
- Trans-Pecos, Mountains and Basins

Figure 3.9. Vegetation regions of Texas (from Gould 1962).

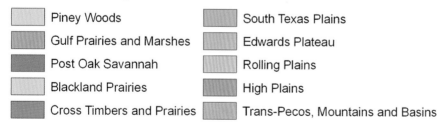

stabilizing these barren sites. Moisture is not generally a limiting factor in plant establishment and growth on these younger soils because deposition is primarily by water. Thus, Entisols and Inceptisols occur across the full range of climates found in the state. Texas Parks and Wildlife has produced an excellent map delineating 45 vegetation types plus urban areas and lakes (Frye, Brown, and McMahan 1984).

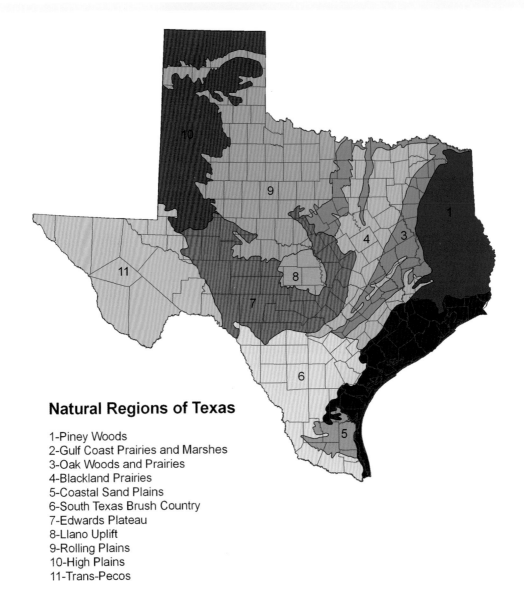

Natural Regions of Texas

1-Piney Woods
2-Gulf Coast Prairies and Marshes
3-Oak Woods and Prairies
4-Blackland Prairies
5-Coastal Sand Plains
6-South Texas Brush Country
7-Edwards Plateau
8-Llano Uplift
9-Rolling Plains
10-High Plains
11-Trans-Pecos

Figure 3.10. Natural regions of Texas (from Poole et al. 2007).

Natural Regions of Texas

Poole et al. (2007) follow the classification of natural regions first developed by the Johnson School of Public Affairs (1978) (fig. 3.10). It differs from Gould, Hoffman, and Rechenthin's (1960) map by presenting 11 rather than 10 vegetational areas. Major differences are that the map in Poole et al. (1) delineates the Llano Uplift from the Edwards Plateau, (2) separates the Coast Sand Plains from the South Texas Brush Country, and (3) combines the Cross Timbers and Post Oak Savannah into a single group called the Oak Woods and Prairies.

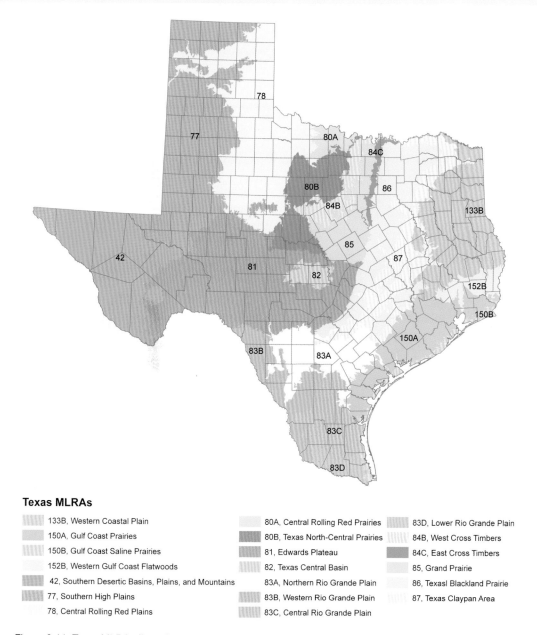

Texas MLRAs

133B, Western Coastal Plain	80A, Central Rolling Red Prairies
150A, Gulf Coast Prairies	80B, Texas North-Central Prairies
150B, Gulf Coast Saline Prairies	81, Edwards Plateau
152B, Western Gulf Coast Flatwoods	82, Texas Central Basin
42, Southern Desertic Basins, Plains, and Mountains	83A, Northern Rio Grande Plain
77, Southern High Plains	83B, Western Rio Grande Plain
78, Central Rolling Red Plains	83C, Central Rio Grande Plain

83D, Lower Rio Grande Plain
84B, West Cross Timbers
84C, East Cross Timbers
85, Grand Prairie
86, TexasI Blackland Prairie
87, Texas Claypan Area

Figure 3.11. Texas MLRAs (from Scott and McKimmey 2000).

Major Land Resource Areas

The Natural Resource Conservation Service (NRCS) uses Major Land Resource Areas (MLRAs) to delineate land resource units to be used in local, state, regional, and national agricultural planning and natural resource assessments (fig. 3.11) (Scott and McKimmey 2000). Table 3.3 presents this scheme in comparison to those previously discussed.

Table 3.3. Classification schemes describing Texas ecoregions based on natural features.

Ecoregions (12)	Vegetation regions (10) (Gould 1962; Hatch, Gandhi, and Brown 1990)	Natural regions (11) (Poole et al. 2007)	MLRAs (19) (Scott and McKimmey 2000)	EPA ecoregions (12) (Griffith et al. 2004)
Trans-Pecos	Trans-Pecos, Mountains and Basins	Trans-Pecos	Southern Desert Basins, Plains, and Mountains	Chihuahuan Desert
High Plains Rolling Plains	High Plains Rolling Plains	High Plains Rolling Plains	Southern High Plains Central Rolling Red Plains (in part)	High Plains Southwestern Tablelands
			Texas North-Central Prairies (in part)	
Central Great Plains	Rolling Plains	Rolling Plains	Central Rolling Red Prairies Texas North-Central Prairies (in part)	Central Great Plains
	Cross Timbers and Prairies (in part)			
Cross Timbers and Prairies Blackland Prairies	Cross Timbers and Prairies Blackland Prairies	Oak Woods and Prairies Blackland Prairies	Central Rolling Red Prairies Grand Prairie, East Cross Timbers, West Cross Timbers Texas Blackland Prairie	Cross Timbers and Prairies Texas Blackland Prairies
Post Oak Savannah (in part) Post Oak Savannah	Post Oak Savannah	Oak Woods and Prairies	Texas Claypan Area	East Central Texas Plains (in part) East Central Texas Plains (in part)
Piney Woods	Piney Woods	Piney Woods	Western Coastal Plain	South Central Plains
Gulf Prairies and Marshes	Gulf Prairies and Marshes	Gulf Coast Prairies and Marshes Coastal Sand Plains	Western Gulf Coast Flatwoods Gulf Coast Prairies Central Rio Grande Plain (in part)	Western Gulf Coastal Plain
South Texas Plains	South Texas Plains	South Texas Brush Country	Lower Rio Grande Plain Northern, Central, Western Rio Grande Plain	South Texas Plains
Edwards Plateau	Edwards Plateau	Edwards Plateau	Lower Rio Grande Valley Edwards Plateau	Edwards Plateau
Guadalupe Mountains	Trans-Pecos, Mountains and Basins	Llano Uplift Trans-Pecos	Texas Central Basin Southern Desertic Basins, Plains, and Mountains	Arizona/New Mexico Mountains

Note: The ecoregions listed in the first column are the ones used in this book.

Ecoregions

Ecoregions denote areas of similarity in type, quality, and quantity of environmental resources (Griffith et al. 2004). Biotic and abiotic features are used to identify these ecoregions. Level III ecoregions will be used in this book. The conterminous United States has 84 Level III ecoregions, 12 of which are represented in Texas (Griffith et al. 2004).

The major difference among the vegetational, natural areas and ecoregions maps is that Griffith et al. (2004) separate the Rolling Plains into the Central Great Plains and Southwestern Tablelands. Another difference is that the Llano Uplift and Coastal Sand Plains are not delineated on the Level III ecoregions map as they were on the natural regions map. Finally, a sliver of the Arizona/New Mexico Mountain ecoregion is recognized in Culberson County. Based on area, this is a rather insignificant addition; however, it is important because this is the only location in the state for numerous plants, including grasses, that have affinities with those of the Southern Rocky Mountains. It is extremely important from a biodiversity standpoint. Each ecoregion will be discussed individually. This book will use the more common nomenclature for vegetational areas and natural regions, which may differ from the EPA name. EPA nomenclature will be shown in parentheses. Information for each ecoregion comes from Griffith et al. (2004), Poole et al. (2007), Schuster and Hatch (1990), Godfrey, McKee, and Oakes (1973), and Gould (1975b).

Piney Woods (South Central Plains)

SIZE: 15,626,627 ac (6,323,871 ha)
TEXAS LAND AREA: 9%
RANK IN LAND AREA: 5th
ANNUAL PRECIPITATION: 40–56 in (1,000–1,425 mm)
FROST-FREE DAYS: 235–270
CLIMATE: Humid subtropical with hot summers and mild winters
ELEVATION: 25–700 ft (8–215 m)
SOILS: Ultisols (41%), Alfisols (40%), Vertisols (6%)

The Piney Woods ecoregion occupies the extreme eastern portion of the state (fig. 3.12). It is bordered on the west by the Post Oak Savannah and a small section of the Blackland Prairies and to the south by the Gulf Prairies and Marshes. The Piney Woods extends northward into Oklahoma and Arkansas and eastward into Louisiana. The Piney Woods is the major ecoregion in 30 counties: Anderson (161), Angelina (118), Bowie (76), Camp (37), Cass (55), Cherokee (59), Gregg (73), Hardin (97), Harrison (80), Houston (106), Jasper (110), Liberty (138), Marion (44), Montgomery (117), Morris (39), Nacogdoches (161), Newton (78), Panola (48), Polk (119), Rusk (71), Sabine (49), San Augustine (92), San Jacinto (108), Shelby (50), Smith (85), Trinity (83), Tyler (111), Upshur (66), Walker (153), and Wood (74).[2]

About 75% of the ecoregion is used for agricultural production. Timber production is the leading agricultural land use and accounts for nearly half the acreage (7.2 million ac, or 2.9 million ha). Native rangeland and introduced pastures amount to over 2.1 million ac (0.8 million ha) each. Bermudagrass (*Cynodon*

---

2. The number in parentheses is the number of documented grass species reported for the county in each region.

dactylon), bahiagrass (*Paspalum nodatum*), and dallisgrass (*P. dilatatum*) are the major introduced grass species. Only about 200,000 ac (78,740 ha) of the ecoregion are used as cropland. Feed grains, fruits, and vegetables are the major crops. This ecoregion is important for oil and gas production.

The Piney Woods is dominated by pine forests and woodlands. Loblolly pine (*Pinus taeda*) is the most common species and occurs throughout the area. Shortleaf pine (*P. echinata*) is found on drier sites, while longleaf pine (*P. palustris*) is restricted to the southeastern portion of the ecoregion. Slash pine (*P. elliottii*) has been introduced from the southeastern United States and is used extensively in pine plantations. Hardwoods are frequent associates in the pine forests. Major hardwoods are various oaks (*Quercus* spp.), hickories (*Carya* spp.), red maple (*Acer rubrum*), and sweetgum (*Liquidambar styraciflua*). Bottomland forests are predominantly oaks: swamp chestnut oak (*Q. michauxii*), overcup oak (*Q. lyrata*), and water oak (*Q. nigra*) being the most common.

Clayey soils in the southern portion of the ecoregion support American beech (*Fagus grandifolia*) and magnolia (*Magnolia grandiflora*). Bald cypress (*Taxodium distichum*) and tupelo (*Nyssa aquatica*) occur in swamp forests throughout the ecoregion.

Numerous grasses are found in the forest, woodlands, pastures, and clearings throughout the Piney Woods. Dominant species include Canada wildrye (*Elymus canadensis*), Virginia wildrye (*E. virginicus*), purpletop (*Tridens flavus*), woodoats (*Chasmanthium* spp.), little bluestem (*Schizachyrium scoparium*), giant cane (*Arundinaria gigantea*), carpetgrasses (*Axonopus* spp.), panicgrasses (*Panicum* spp.), paspalums (*Paspalum* spp.), and rosettegrasses (*Dichanthelium* spp.).

The grass flora in the Piney Woods ecoregion is very diverse. There are 410 grass species in 126 genera documented from the area. Panicoideae, Chloridoideae, and Poöideae are the major subfamilies with 184, 112, and 83 species, respectively (table 3.4).

Table 3.4. Taxa by grass subfamilies documented for the Piney Woods ecoregion.

Subfamilies	Tribes	Genera	Species	Subspecific taxa
Aristidoideae	1	1	11	10
Arundinoideae	1	2	2	-
Bambusoideae	1	2	4	1
Centothecoideae	1	1	3	-
Chloridoideae	2	30	112	35
Danthonioideae	1	2	2	-
Ehrhartoideae	1	4	9	-
Panicoideae	2	41	184	76
Poöideae	6	43	83	17
Total	16	126	410	139

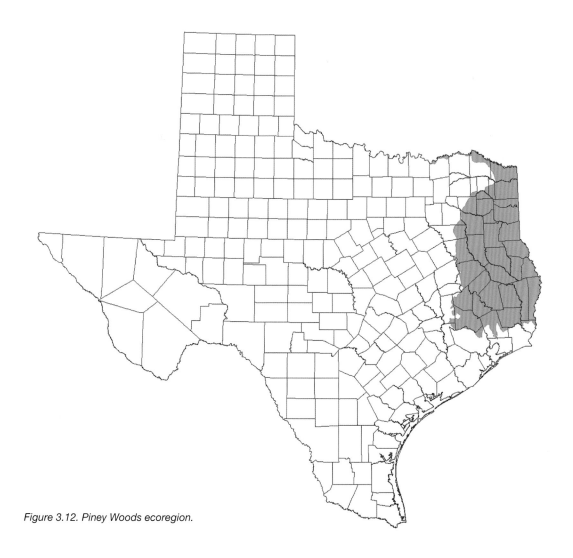

Figure 3.12. Piney Woods ecoregion.

Gulf Prairies and Marshes (Western Gulf Coastal Plain)

SIZE: 17,047,291 ac (6,898,793 ha)
TEXAS LAND AREA: 10%
RANK IN LAND AREA: 4th
ANNUAL PRECIPITATION: 26–56 in
(660–1,420 mm)
FROST-FREE DAYS: 245–340
CLIMATE: Humid subtropical with warm to hot
summers and mild winters
ELEVATION: 0–250 ft (0–75 m)
SOILS: Alfisols (46%), Vertisols (21%),
Mollisols (19%)

The Gulf Prairies and Marshes ecoregion is bordered to the north by the Piney Woods and Post Oak Savannah and to the west by the South Texas Plains (fig. 3.13). It extends eastward into Louisiana and southward into Mexico. The majority of the following 27 counties occur within the ecoregion: Aransas (185), Austin (86), Brazoria (164), Brooks (80), Calhoun (86), Cameron (191), Chambers (140), Colorado (83), Fort Bend (85), Galveston (188), Harris (237), Hidalgo (164), Jackson (65), Jefferson (123), Jim Hogg (42), Jim Wells (91), Kenedy (109), Kleberg (172), Matagorda (80), Nueces (132), Orange (52), Refugio (143), San Patricio (172), Victoria (76), Waller (68), Wharton (57), and Willacy (152).

About 70% of the Gulf Prairies and Marshes ecoregion is used for agricultural production. Native rangeland accounts for the largest acreage (6.5 million ac, or 2.6 million ha), and cattle grazing is the predominant use. Introduced pasture, principally Bermudagrass, amounts to about 1 million ac (0.4 million ha). Nearly 3 million ac (1.2 million ha) of dry and irrigated cropland in the ecoregion, much of it in the Lower Rio Grande Valley, produce mainly fruits, vegetables, rice, sorghum, and corn. The Gulf Prairies and Marshes is a major oil- and gas-producing area and an important petrochemical-refining region.

The sand dunes on the outer-barrier islands that parallel most of the southern coastal area are stabilized by sea oats (*Uniola paniculata*) and bitter panicum (*Panicum amarum*). Usually behind the dunes is a strip of upland prairie dominated by little bluestem and gulfdune paspalum (*Paspalum monostachyum*). Coastal prairies on the mainland side are composed of marshhay cordgrass (*Spartina patens*) and gulf cordgrass (*S. spartinae*). Tidal marshes are often dominated by smooth cordgrass (*S. alternifolia*). Farther inland, coastal prairies are composed of big bluestem (*Andropogon gerardii*), little bluestem, yellow indiangrass (*Sorghastrum nutans*), brownseed paspalum (*P. plicatulum*), Pan American balsamscale (*Elionurus tripsacoides*), crinkleawn (*Trachypogon spicatus*), gulf muhly (*Muhlenbergia capillaries*), and tanglehead (*Heteropogon contortus*). Marshmillet (*Zizaniopsis miliacea*) and maidencane (*Panicum hemitomon*) are the most important grass species of freshwater marshes of the upper coast. Live oak and post oak woodlands are common.

Poole et al. (2007) separate the Coastal Sand Plains from the Gulf Prairies and Marshes. The former is an area of eolian sands deposited during the Holocene by the prevailing western Gulf winds. The sands cover most of Kenedy and Brooks counties and extend west to Jim Hogg and Zapata counties. Major plant communities on sandy soils include oak (*Quercus* spp.) mottes with tallgrass species, while woodlands dominated by mesquite (*Prosopis glandulosa*) occur on clayey soils with midgrasses.

Grass diversity is high in the Gulf Prairies and Marshes, and 118 genera and 402 grass species have been documented from the ecoregion (table 3.5). Panicoideae, Chloridoideae, and Poöideae are the major subfamilies with 194, 119, and 64 species, respectively.

Table 3.5. Taxa by grass subfamilies documented for the Gulf Prairies and Marshes ecoregion.

Subfamilies	Tribes	Genera	Species	Subspecific taxa
Aristidoideae	1	1	8	10
Arundinoideae	1	2	2	-
Bambusoideae	1	2	3	1
Centothecoideae	1	1	3	-
Chloridoideae	2	30	119	37
Danthonioideae	1	1	1	-
Ehrhartoideae	1	4	8	-
Panicoideae	2	43	194	78
Poöideae	5	34	64	12
Total	15	118	402	138

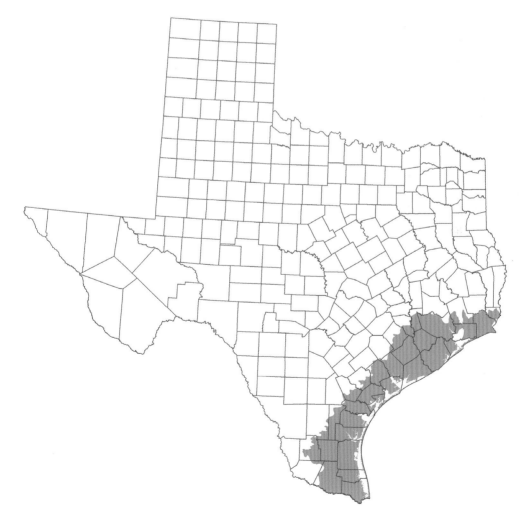

Figure 3.13. Gulf Prairies and Marshes ecoregion.

Post Oak Savannah (East Central Texas Plains)

SIZE: 10,849,486 ac (4,390,631 ha)
TEXAS LAND AREA: 6%
RANK IN LAND AREA: 11th
ANNUAL PRECIPITATION: 30–45 in (762–1,143 mm)
FROST-FREE DAYS: 235–270
CLIMATE: Humid subtropical with hot summers and mild winters
ELEVATION: 25–700 ft (8–215 m)
SOILS: Ultisols (41%), Alfisols (40%), Vertisols (6%)

The Post Oak Savannah ecoregion stretches from the Red River in the north to a line between the San Antonio and Frio rivers in the south (fig. 3.14). It is bordered on the east by the Piney Woods and Coastal Prairies and Marshes and on the west by the Blackland Prairies. The Sulfur, Sabine, Trinity, Navasota, Brazos, Colorado, Guadalupe, and San Antonio rivers dissect the Post Oak Savannah and produce a characteristic topography of terraces and floodplains. The remainder of the ecoregion is gently rolling hills. The narrow San Antonio Prairie, sometimes considered part of the Blackland Prairies, is included here. The Camino Real (Kings Highway, Old Spanish Road or Trail) runs through the middle of this prairie. Much of the ecoregion has an underlying clay hardpan that impacts movement of moisture through the soil and water availability to plants. The Post Oak Savannah is the major ecoregion in 22 counties: Bastrop (135), Bee (59), Brazos (276), Burleson (111), Franklin (25), Freestone (92), Goliad (60), Gonzales (107), Grimes (104), Henderson (84), Hopkins (20), Karnes (107), Lee (44), Leon (143), Madison (102), Milam (85), Rains (9), Red River (46), Robertson (178), Titus (34), Van Zandt (61), and Wilson (59). The ecoregion is occasionally grouped with the Cross Timbers and Prairies because of the dominance of post oak (*Quercus stellata*) and blackjack oak (*Q. marilandica*) in each. They are separated here because the former is developed on Eocene substrate, whereas the latter is primarily on Cretaceous limestones. The Post Oak Savannah is restricted to Texas.

Roughly 85% of the ecoregion is in some type of agricultural use. Native rangeland accounts for the majority of agricultural acreage (5.5 million ac, or 2.2 million ha). Introduced pasture, often seeded after fields are abandoned, accounts for an additional 2.4 million ac (0.9 million ha). Cropland (0.5 million ac, or 0.2 million ha) is generally restricted to floodplains and terraces. Major row crops include corn, sorghum, cotton, and soybeans. There is some, but not extensive, oil and gas production in this ecoregion.

Vegetation is characteristically post oak and blackjack oak mixed with tallgrass prairie herbaceous species. Hackberries (*Celtis* spp.), elms (*Elmus* spp.), and hickories (*Carya* spp.) are associated with the oaks. Dominant understory species include yaupon (*Ilex vomitoria*), greenbriar (*Smilax* spp.), American beautyberry (*Callicarpa americana*), and grapes (*Vitis* spp.). Major grasses are little bluestem, yellow indiangrass, switchgrass (*Panicum virgatum*), bluestems (*Andropogon* spp., *Bothriochloa* spp.), Texas wintergrass (*Nassella leucothrica*), rosettegrasses, and threeawns (*Aristida* spp.). Major woodlands along rivers and streams have been mostly removed to allow for cultivation; water oak (*Quercus nigra*), pecan (*Carya illinoinensis*), walnuts (*Juglans* spp.), mulberries (*Morus* spp.), and others once dominated.

This ecoregion has the greatest grass diversity reported for the state. A total of 434 grass species in 130 genera have been documented from the ecoregion (table 3.6). The Panicoideae, Chloridoideae, and Poöideae are the dominant subfamilies with 185, 129, and 90 species, respectively. Brazos County, home of Texas A&M University, has 276 species documented. Collecting for botany, agronomy, and agrostology classes has certainly added to the knowledge of grasses in the county. Also, numerous species have been introduced into the university for testing, and some have escaped and been collected outside of cultivation within the county.

Table 3.6. Taxa by grass subfamilies documented for the Post Oak Savannah ecoregion.

Subfamilies	Tribes	Genera	Species	Subspecific taxa
Aristidoideae	1	1	11	10
Arundinoideae	1	2	2	-
Bambusoideae	1	2	3	1
Centothecoideae	1	1	3	-
Chloridoideae	2	32	129	37
Danthonioideae	1	2	2	-
Ehrhartoideae	2	5	9	-
Panicoideae	2	42	185	79
Poöideae	5	43	90	16
Total	16	130	434	143

Figure 3.14. Post Oak Savannah ecoregion.

Blackland Prairies (Texas Blackland Prairies)

SIZE: 13,916,808 ac (5,631,932 ha)
TEXAS LAND AREA: 8%
RANK IN LAND AREA: 8th
ANNUAL PRECIPITATION: 30–45 in (760–1,145 mm)
FROST-FREE DAYS: 227–277
CLIMATE: Subtropical (north and west) to humid subtropical (south and east) with warm to hot summers and mild winters
ELEVATION: 250–700 ft (75–215 m)
SOILS: Vertisols (58%), Alfisols (26%), Mollisols (13%)

The Blackland Prairies ecoregion is like a wedge separating the Cross Timbers and Prairies from the Post Oak Savannah (fig. 3.15). The tip extends a bit southwest of San Antonio, and the broad end approaches the Red River on a line from Sherman to Paris. Interstate 35 from San Antonio to Dallas lies along the western edge of this ecoregion. Also included in the Blackland Prairies is the Fayette Prairie, which extends from Karnes and DeWitt counties on a northeasterly line through Gonzales and Fayette counties to Grimes and Montgomery counties. The San Antonio Prairie and Grand Prairie are not included here but are found in the Post Oak Savannah and Cross Timbers and Prairies ecoregions, respectively. The following 25 counties have a majority of their area within the Blackland Prairies: Bell (126), Bexar (216), Caldwell (65), Collin (42), Dallas (160), Delta (11), DeWitt (64), Ellis (36), Falls (28), Fannin (50), Fayette (92), Grayson (113), Guadalupe (43), Hill (39), Hunt (61), Kaufman (42), Lamar (87), Lavaca (46), Limestone (47), McLennan (100), Navarro (50), Rockwall (29), Travis (168), Washington (85), and Williamson (90).

Over 75% of the Blackland Prairies ecoregion is used for agricultural production; approximately half is cropland, and half is introduced pasture or rangeland. Major crops include cotton, sorghum, corn, oats, wheat, and soybeans. This is a relatively minor oil- and gas-producing ecoregion.

Original prairie grass vegetation consisted of big bluestem, yellow indiangrass, switchgrass, eastern gamagrass (*Tripsacum dactyloides*), Texas wintergrass, and tall dropseed (*Sporobolus compositus* var. *macer*). Poor range sites or overutilized areas now support gramagrasses (*Bouteloua* spp., *Chondrosum* spp.), threeawns, and buffalograss (*Buchloë dactyloides*). Mesquite, huisache (*Acacia farnesiana*), oaks, and elms invade abandoned fields. Characteristic post oak vegetation is found along drainage areas and along fence lines. Pecans, oaks, elms, and cottonwood (*Populus deltoides*) are found along riparian areas.

A large number of grass species occur in this ecoregion (table 3.7), and 119 genera and 376 species have been documented. Panicoideae, Chloridoideae, and Poöideae are the major subfamilies with 154, 114, and 79 species, respectively.

Table 3.7. Taxa by grass subfamilies documented for the Blackland Prairies ecoregion.

Subfamilies	Tribes	Genera	Species	Subspecific taxa
Aristidoideae	1	1	12	12
Arundinoideae	1	2	2	-
Bambusoideae	1	2	2	1
Centothecoideae	1	1	3	-
Chloridoideae	2	30	114	35
Danthonioideae	1	2	2	-
Ehrhartoideae	2	4	8	-
Panicoideae	2	38	154	78
Poöideae	6	39	79	16
Total	17	119	376	142

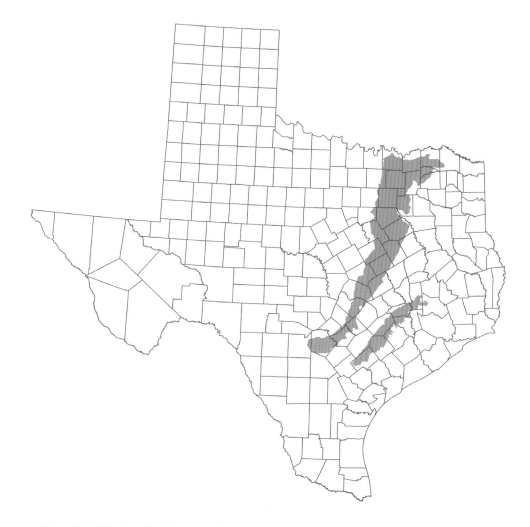

Figure 3.15. Blackland Prairies ecoregion.

Cross Timbers and Prairies

Size: 12,102,267 ac (4,897,613 ha)
Texas land area: 7%
Rank in land area: 9th
Annual precipitation: 25–35 in (635–890 mm)
Frost-free days: 230–280
Climate: Subtropical with hot summers and generally mild winters
Elevation: 500–1,500 ft (150–460 m)
Soils: Alfisols (45%), Mollisols (36%), Vertisols (10%)

The southern border of the Cross Timbers and Prairies ecoregion is slightly south of a line from Brownwood to Temple and extends northward, including Fort Worth and Denton, to the Red River (fig. 3.16). It is bordered on the east by the Blackland Prairies, on the west by the Central Great Plains, and on the south by the Edwards Plateau. The Limestone Cut Plain (sometimes referred to as the Lampasas Cut Plain) is included here, but some references place it as the northernmost part of the Edwards Plateau ecoregion. The Cross Timbers and Prairies ecoregion also occurs in Oklahoma. This is the major ecoregion in 22 counties: Bosque (49), Brown (90), Comanche (45), Cooke (46), Coryell (66), Denton (82), Eastland (23), Erath (51), Hamilton (37), Hood (38), Jack (53), Johnson (37), Lampasas (75), Mills (37), Montague (40), Palo Pinto (46), Parker (82), Somervell (25), Stephens (19), Tarrant (153), Wise (47), and Young (46). The Post Oak Savannah is occasionally grouped with the Cross Timbers and Prairies because of the dominance of post oak and blackjack oak in each. They are separated here because the former is developed on Eocene substrate, whereas the latter is primarily on Cretaceous limestones.

Almost 85% of the Cross Timbers and Prairies is used for agricultural purposes. Over 8 million ac (3 million ha) are native rangeland or introduced pasture and used for hay or livestock production. Cropland acreage, both irrigated and dry, amounts to 1.3 million ac (0.5 million ha). Wheat, oats, sorghum, and cotton are major row crops. This is not a major oil- and gas-producing ecoregion.

The wooded regions of the Cross Timbers and Prairies occur on deeper, sandy soils (Alfisols). Post oak and blackjack oak are the predominant woody plant species. Bottomlands are characterized by hackberry, elm, and pecan. Prairies occur on Mollisols and Vertisols and are dominated by big bluestem, little bluestem, switchgrass, indiangrass, Canada wildrye, Texas wintergrass, and buffalograss. Poor grazing management leads to invasion of annual threeawns, red lovegrass (*Eragrostis secundiflora*), little barleys (*Critesion* spp.) and paspalums. Shrub live oak (*Q. filiformis*), mesquite, and redberry juniper (*Juniperus pinchoti*) invade abandoned fields and mismanaged upland sites.

The Cross Timbers and Prairies ecoregion is represented by 102 genera and 291 species of documented grass collections (table 3.8). Panicoideae, Chloridoideae, and Poöideae are the dominant subfamilies with 112, 88, and 69 species, respectively.

Table 3.8. Taxa by grass subfamilies documented for the Cross Timbers and Prairies ecoregion.

Subfamilies	Tribes	Genera	Species	Subspecific taxa
Aristidoideae	1	1	11	12
Arundinoideae	1	2	2	-
Bambusoideae	1	2	2	1
Centothecoideae	1	1	2	-
Chloridoideae	2	23	88	37
Danthonioideae	1	2	2	-
Ehrhartoideae	1	2	3	-
Panicoideae	2	31	112	69
Poöideae	6	36	69	14
Total	16	102	291	133

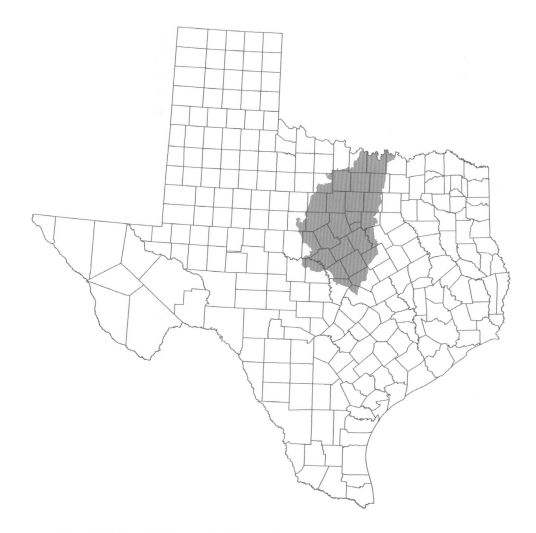

Figure 3.16. Cross Timbers and Prairies ecoregion.

South Texas Plains

SIZE: 13,926,638 ac (5,635,910 ha)

Texas land area: 8%

RANK IN LAND AREA: 7th

ANNUAL PRECIPITATION: 17–30 in (430–760 mm)

FROST-FREE DAYS: 263–340

CLIMATE: Subtropical with mild winters and hot, humid summers

ELEVATION: 0–1,000 ft (0–305 m)

SOILS: Mollisols (35%), Alfisols (29%), Inceptisols (20%)

The South Texas Plains ecoregion is clearly defined to the north by the Balcones Escarpment of the Edwards Plateau and to the west by the Rio Grande (fig. 3.17). The eastern boundary is less well defined and gradually blends with the Blackland Prairies, Gulf Prairies and Marshes, and Post Oak Savannah. This area extends across the border into Mexico, where it has been referred to as the Tamaulipan Thorn Scrub. The Nueces and Frio rivers are the major watersheds in the area. The South Texas Plains includes 15 counties: Atascosa (62), Dimmit (65), Duval (61), Frio (47), Kinney (45), LaSalle (51), Live Oak (55), Maverick (45), McMullen (38), Medina (69), Starr (125), Uvalde (84), Webb (87), Zapata (41), and Zavala (42).

About 94% of the South Texas Plains is used for agricultural production. Over 10 million ac (4.1 million ha) are native rangeland, and 1 million ac (0.4 million ha) are introduced pasture. Kleingrass (*Panicum coloratum*),

Bermudagrass, buffelgrass (*Cenchrus ciliaris*), and rhodesgrass (*Chloris gayana*) are major introduced pasture grasses. About 0.8 million ac (0.3 million ha) are dry or irrigated cropland. Cotton and sorghum are major dryland crops, and citrus and vegetables are grown on irrigated lands. This is a significant oil and gas producing area.

Original vegetation in South Texas was thought to be a more open grassland or savannah type with thick concentrations of woody plants restricted to ridges and along streams. Altered normal fire regimes and poor grazing management have led to the increase in woody plants. The area is now dominated by numerous species such as mesquite, acacias (*Acacia* spp.), whitebrush (*Aloysia gratissima*), lotebush (*Condalia obtusifolia*), and pricklypear (*Opuntia* spp.). Sandy lands support little bluestem, bristlegrasses (*Setaria* spp.), tanglehead, and big sandbur (*Cenchrus myosuroides*). Heavier soils are characterized by Arizona cottontop (*Digitaria californica*), pappusgrass (*Pappophorum* spp.), bristlegrasses, buffalograss, curly mesquite (*Hilaria belangeri*), and sideoats grama (*Bouteloua curtipendula*).

Diversity of grasses is relatively high in this ecoregion with over 102 genera and 320 species documented (table 3.9). Panicoideae, Chloridoideae, and Poöideae are the dominant subfamilies with 128, 118, and 54 species, respectively. Bambusoideae has yet to be documented, but members of this subfamily undoubtedly occur as ornamentals within the region.

Table 3.9. Taxa by grass subfamilies documented for the South Texas Plains ecoregion.

Subfamilies	Tribes	Genera	Species	Subspecific taxa
Aristidoideae	1	1	8	9
Arundinoideae	1	2	2	-
Bambusoideae	-	-	-	-
Centothecoideae	1	1	2	-
Chloridoideae	2	35	118	33
Danthonioideae	1	2	2	-
Ehrhartoideae	1	3	6	-
Panicoideae	2	32	128	52
Poöideae	5	26	54	15
Total	14	102	320	109

Figure 3.17. South Texas Plains ecoregion.

Central Great Plains

SIZE: 11,811,487 ac (4,779,939 ha)

TEXAS LAND AREA: 7%

RANK IN LAND AREA: 10th

ANNUAL PRECIPITATION: 18–28 in (455–711 mm)

FROST-FREE DAYS: 185–235

CLIMATE: Warm, temperate steppe with dry, moderate winters and hot, humid summers

ELEVATION: 1,000–3,000 ft (330–990 m)

SOILS: Mollisols (54%), Alfisols (29%), Inceptisols (10%)

The southern boundary of the Central Great Plains ecoregion is south of a line from San Angelo to Brownwood and extends northward, including Abilene and Wichita Falls, to the Red River (fig. 3.18). The Rolling Plains are to the west, and the Cross Timbers and Prairies are to the east. The Central Great Plains is the major ecoregion in 18 counties: Archer (33), Baylor (16), Callahan (39), Clay (28), Coleman (29), Concho (24), Fisher (9), Hardeman (46), Haskell (9), Jones (25), McCulloch (32), Runnels (27), Shackelford (21), Taylor (96), Throckmorton (37), Tom Green (64), Wichita (54), and Wilbarger (29). Two more or less disjunct areas are found in most of Wheeler County (47) and through central Collingsworth (39) and Donley (45) counties. This ecoregion also extends into Oklahoma, Kansas, and Nebraska.

Nearly 97% of this ecoregion is used for agricultural production. Over 7.5 million ac (3 million ha) are native pasture with another 0.5 million ac (200,000 ha) reported as introduced pasture. About 3 million ac (1.2 million ha) are used as dry cropland; cotton and grain sorghum are the major row crops. There is significant oil and gas production in the ecoregion.

The Central Great Plains is a transitional area between the tallgrass prairie to the east and the shortgrass steppe farther west. It is often referred to as the midgrass prairie and is dominated by sideoats grama, Texas wintergrass, and little bluestem. Big bluestem, switchgrass, yellow indiangrass, Canada wildrye, and western wheatgrass (*Pascopyrum smithii*) also are abundant on deeper soils and where moisture is more plentiful. Shorter grasses, curly mesquite, tobosagrass (*Pleuraphis mutica*), buffalograss, blue grama (*Chondrosum gracile*), sand dropseed (*Sporobolus cryptandrus*), and hooded windmillgrass (*Chloris cucullata*) are common on xeric or overgrazed sites.

Settlement, overgrazing, and the altered fire regimes has allowed for the increase in woody vegetation in the Central Great Plains. Mesquite, lotebush, pricklypear, and tasajillo (*O. leptocaulis*) invade almost all soils. Shinnery oak (*Q. havardii*) and sand sagebrush (*Artemisia filifolia*) are indicative of sandy lands. Junipers, particularly redberry juniper, are common on breaks and steeper slopes. Redberry juniper has spread onto overgrazed rangeland and abandoned cropland.

A total of 82 genera and 210 species of grasses have been documented from the Central Great Plains ecoregion in Texas (table 3.10). The Chloridoideae, Panicoideae, and Poöideae, have the most species with 76, 70, and 51, respectively. The Danthonioideae is the only subfamily not represented in this ecoregion.

Table 3.10. Taxa by grass subfamilies documented for the Central Great Plains ecoregion.

Subfamilies	Tribes	Genera	Species	Subspecific taxa
Aristidoideae	1	1	7	7
Arundinoideae	1	2	2	-
Bambusoideae	1	1	1	-
Centothecoideae	1	1	1	-
Chloridoideae	2	26	76	30
Danthonioideae	-	-	-	-
Ehrhartoideae	1	2	2	-
Panicoideae	2	20	70	39
Poöideae	5	29	51	13
Total	14	82	210	89

Figure 3.18. Central Great Plains ecoregion.

Rolling Plains (Southwestern Tablelands)

SIZE: 15,324,112 ac (6,201,448 ha)
TEXAS LAND AREA: 9%
RANK IN LAND AREA: 6th
ANNUAL PRECIPITATION: 18–30 in (455–760 mm)
FROST-FREE DAYS: 185–235
CLIMATE: Warm, temperate steppe with mild, dry winters and hot, humid summers
ELEVATION: 1,000–3,000 ft (305–915 m)
SOILS: Inceptisols (29%), Mollisols (26%), Alfisols (19%)

The Rolling Plains ecoregion lies almost entirely between the 100th and 102nd meridians on the eastern side of the Texas Panhandle (fig. 3.19). The southern boundary is along a line from Big Spring to the western edge of Runnels County. The eastern border grades into the Central Great Plains, and the western border is delineated by the Caprock Escarpment of the High Plains. The northern boundary is the Canadian River breaks stretching from Oldham to Upton and Hemphill counties. The Rolling Plains ecoregion extends northward into Oklahoma and Kansas and westward into New Mexico and Colorado. The Rolling Plains is the major ecoregion in 27 counties: Armstrong (49), Borden (10), Briscoe (44), Childress (42), Coke (7), Collingsworth (39), Cottle (36), Dickens (51), Donley (45), Foard (14), Garza (57), Gray (44), Hall (38), Hemphill (82), Hutchinson (48), Kent (16), King (25), Knox (34), Lipscomb (40), Mitchell (43), Motley (47), Oldham (74), Potter (69), Roberts (57), Scurry (37), Stonewall (24), and Wheeler (47).

Over 90% of the Rolling Plains is in agricultural production. Native rangeland accounts for roughly 10.5 million ac (4.25 million ha). Dry cropland amounts to nearly 3 million ac (1.2 million ha). Major crops are sorghum, cotton, and forages. This ecoregion is a major oil- and gas-producing area.

Original vegetation was a mixture of tall-, mid-, and shortgrasses, depending on moisture conditions. Tallgrass species of the wettest sites are big bluestem, yellow indiangrass, and switchgrass. Major midgrasses are sideoats grama, Canada wildrye, and western wheatgrass. Grasses of the more xeric sites are blue grama, buffalograss, curly mesquite, tobosagrass, sand dropseed, hooded windmillgrass, and threeawns.

Mesquite, lotebush, pricklypear, agarito (*Mahonia trifoliolata*), and tasajillo are the most common invaders on abandoned fields and overutilized rangelands. Sand sagebrush, yucca (*Yucca* spp.), and shinnery oak are characteristic of sandy soils. Red juniper, once only abundant on slopes in canyons and breaks, has greatly expanded its range.

The Rolling Plains, along with the High Plains and Central Great Plains, although predominantly grasslands, has some of the lowest diversity of grasses. Only 108 genera and 199 species have been documented from the area (table 3.11). The Chloridoideae, Panicoideae, and Poöideae are the most represented subfamilies, with 84, 58, and 52 species, respectively. The Bambusoideae, Centothecoideae, and Danthonioideae are yet to be documented from the ecoregion.

Table 3.11. Taxa by grass subfamilies documented for the Rolling Plains ecoregion.

Subfamilies	Tribes	Genera	Species	Subspecific taxa
Aristidoideae	1	1	6	7
Arundinoideae	1	2	2	-
Bambusoideae	-	-	-	-
Centothecoideae	-	-	-	-
Chloridoideae	2	28	84	33
Danthonioideae	-	-	-	-
Ehrhartoideae	1	1	1	-
Panicoideae	2	47	58	39
Poöideae	5	29	52	15
Total	12	108	199	94

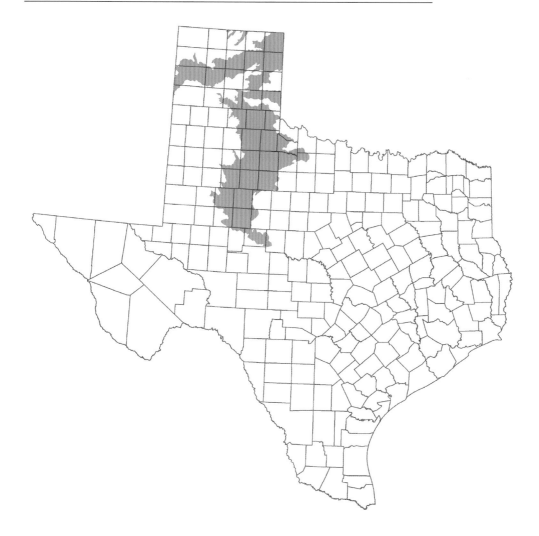

Figure 3.19. Rolling Plains ecoregion.

High Plains

SIZE: 21,679,699 ac (8,773,462 ha)
TEXAS LAND AREA: 13%
RANK IN LAND AREA: 1st
ANNUAL PRECIPITATION: 14–21 in (355–533 mm)
FROST-FREE DAYS: 180–220
CLIMATE: Temperate steppe with hot summers and moderate winters
ELEVATION: 2,400–4,000 ft (730–1,220 m)
SOILS: Mollisols (49%), Alfisols (28%), Aridisols (18%)

The High Plains ecoregion dominates the western half of the Texas Panhandle (fig. 3.20). The Caprock Escarpment clearly delineates the border on the eastern edge. The southern border is less well defined, and it intergrades with the Edwards Plateau and the Toyah Basin of the Trans-Pecos. The High Plains extends westward into New Mexico and stretches northward into Oklahoma, Kansas, Colorado, Nebraska, and Wyoming. This ecoregion is also referred to as the Staked Plains or Llano Estacado. The High Plains is the major ecoregion in 34 counties: Andrews (47), Bailey (62), Carson (56), Castro (29), Cochran (14), Crane (62), Crosby (17), Dallam (54), Dawson (32), Deaf Smith (85), Ector (37), Floyd (26), Gaines (25), Glasscock (15), Hale (27), Hansford (29), Hartley (58), Hockley (25), Howard (48), Lamb (42), Lubbock (78), Lynn (18), Martin (7), Midland (25), Moore (32), Ochiltree (49), Parmer (29), Randall (77), Sherman (40), Swisher (26), Terry (14), Upton (12), Winkler (56), and Yoakum (17).

A porous, caliche-capped sandstone formed during the Miocene and Pliocene (the Ogallala Formation) underlies the High Plains. This porous substrate enabled the development of a drainage system that forms isolated playa lakes rather than a typical stream network. The playas are an important resource for migratory birds. The only major watershed that dissects the High Plains in Texas is the Canadian River. The topography, geology, soils, and vegetation are so different from those of the surrounding High Plains that the Canadian River drainage or breaks is included in the Rolling Plains ecoregion.

Over 95% of the High Plains is used for agricultural production. More than 11 million ac (4.5 million ha) are cropland, one-third of which is irrigated. Prominent row crops are cotton, corn, sorghum, wheat, and sugar beets. Roughly 9 million ac (3.6 million ha) are native rangeland utilized primarily for cattle grazing. This is an important area for production of oil, natural gas, and helium.

Composition of grass species varies depending upon soil texture. Clayey soils support shortgrasses, such as buffalograss, blue grama, and galleta (*Pleuraphis jamesii*). Loamy sites tend to support more midgrasses, such as sideoats grama, little bluestem, western wheatgrass, and sand dropseed. Mesquite, yucca, pricklypear, and sand sagebrush invade sandier sites.

A zone of very sandy soils occurs along the southwestern boundary of the High Plains in Andrews, Bailey, Cochran, Cranes, Gaines, Lamb, Winkler, and Yoakum counties. It is characterized by sand shinnery oak, sand sagebrush, sand bluestem (*Andropogon hallii*), big sandreed (*Calamovilfa gigantea*), and giant dropseed (*Sporobolus giganteus*).

Grass diversity is not as great on the High Plains as it is on some more mesic areas or ecoregions. Only 78 genera and 203 species have been documented (table 3.12). The Chloridoideae, Panicoideae, and Poöideae are the dominant subfamilies with 84, 60, and 49 species, respectively. The Bambusoideae and Centothecoideae have not been documented from the High Plains, although some ornamental members of these subfamilies may be in the ecoregion.

Table 3.12. Taxa by subfamilies documented for the High Plains ecoregion.

Subfamilies	Tribes	Genera	Species	Subspecific taxa
Aristidoideae	1	1	6	7
Arundinoideae	1	2	2	-
Bambusoideae	-	-	-	-
Centothecoideae	-	-	-	-
Chloridoideae	2	28	84	33
Danthonioideae	1	1	1	-
Ehrhartoideae	1	1	1	-
Panicoideae	2	18	60	39
Poöideae	5	27	49	15
Total	13	78	203	94

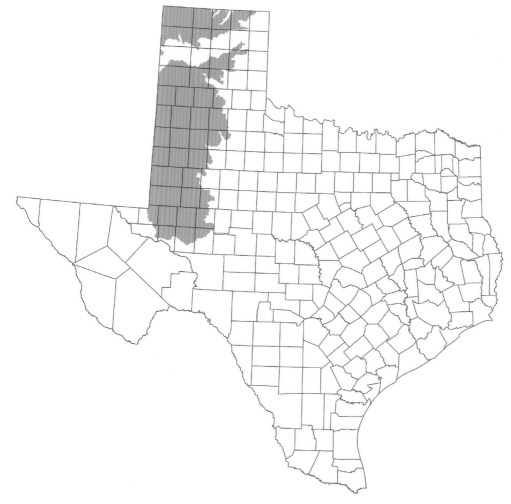

Figure 3.20. High Plains ecoregion.

Edwards Plateau

SIZE: 17,587,598 ac (7,117,448 ha)

TEXAS LAND AREA: 10%

RANK IN LAND AREA: 3rd

ANNUAL PRECIPITATION: 12–32 in (300–800 mm)

FROST-FREE DAYS: 220–300

CLIMATE: Subtropical steppe with hot summers and mild winters

ELEVATION: 500–3,000 ft (150–990 m)

SOILS: Mollisols (76%), Alfisols (8%), Inceptisols (8%)

The Edwards Plateau ecoregion dominates the west-central portion of the state (fig. 3.21). The southern and eastern boundaries are well defined by the Balcones Escarpment fault line. The northern boundary is less well defined, and the plateau grades into the Lampasas Cut Plain of the Cross Timbers and Prairies, Central Great Plains, Rolling Plains, and High Plains. There is a small disjunct portion of the Edwards Plateau in Mitchell, Nolan, and Taylor counties on the Callahan Divide. On the west, the Edwards Plateau integrates with the Stockton Plateau of the Trans-Pecos ecoregion. The Llano Uplift (Granitic Basin, Central Mineral Basin) is included here. The majority of the Edwards Plateau has a Cretaceous limestone substrate, with the exception of the Llano Uplift, which is primarily Precambrian granites, gneisses, and schists. The Edwards Plateau is the major ecoregion in 23 counties: Bandera (62), Blanco (90), Burnet (100), Comal (72), Crockett (79), Edwards (74), Gillespie (120), Hays (94), Irion (18), Kendall (52), Kerr (104), Kimble (78), Llano (113), Mason (81), Menard (50), Nolan (23), Reagan (29), Real (41), San Saba (77), Schleicher (29), Sterling (27), Sutton (72), and Val Verde (128).

Nearly 90% of the ecoregion is used for agricultural production. Native rangeland ac-counts for about 14 million ac (5.7 million ha) of agricultural land in the Edwards Plateau. Cattle, sheep, and goats are raised in this major U.S. wool- and mohair-producing area. Cropland and introduced pasture each account for about 0.5 million ac (0.2 million ha), and peach and plum orchards are common. Hunting of white-tailed deer and exotic animals is a major source of income in the ecoregion. Recreation and tourism are also large uses of the natural resources in the Edwards Plateau.

The Edwards Plateau is a complex mixture of oak-juniper savannah and woodlands and prairies. Topography is composed of deeply dissected canyonlands along the escarpment. Many hardwoods occur, such as oaks, walnuts, Texas ash (*Fraxinus texensis*), and hackberries. Bigtooth maple (*Acer grandidentatum*), chinkapin oak (*Quercus muhlenbergia*), and basswood (*Tilia americana*) occur on the more mesic sites. Tallgrass prairie grass species (big bluestem, little bluestem, yellow indiangrass, switchgrass) are common in this area. Riparian zones in this ecoregion are characterized by bald cypress. Upland woodlands are dominated by live oak and ashe juniper (*Juniperus ashei*). Shrubs include mountain laurel (*Sophora secundiflora*), persimmon (*Diospyros texana*), and agarito. Midgrasses such as sideoats grama and hairy grama (*Chondrosum hirsutum*), as well as tridens (*Tridens* spp.), threeawns, Texas cupgrass (*Eriochloa sericea*), and Texas wintergrass, replace the tallgrass species. Farther north and west, redberry juniper replaces ashe juniper. Shortgrasses (buffalograss, curly mesquite) predominate in drier habitats.

There is a high diversity of grasses represented in the Edwards Plateau (table 3.13). A total of 101 genera and 324 species have been documented from the area. Chloridoideae, Panicoideae, and Poöideae are the major subfamilies with 129, 91, and 86 species, respectively.

Table 3.13. Taxa by grass subfamilies documented for the Edwards Plateau ecoregion.

Subfamilies	Tribes	Genera	Species	Subspecific taxa
Aristidoideae	1	1	11	11
Arundinoideae	1	2	2	-
Bambusoideae	-	-	-	-
Centothecoideae	1	1	1	-
Chloridoideae	2	35	129	34
Danthonioideae	1	2	3	-
Ehrhartoideae	1	1	1	-
Panicoideae	2	26	91	39
Poöideae	5	33	86	20
Total	14	101	324	104

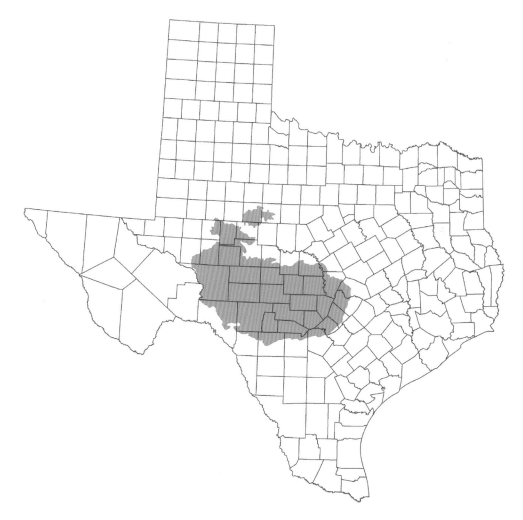

Figure 3.21. Edwards Plateau ecoregion.

Trans-Pecos (Chihuahuan Desert)

SIZE: 21,143,751 ac (8,556,572 ha)
TEXAS LAND AREA: 12%
RANK IN LAND AREA: 2nd
ANNUAL PRECIPITATION: 8–18 in (200–455 mm)
FROST-FREE DAYS: 220–245
CLIMATE: Arid subtropical (lowlands) to humid cool-temperate (mountains)
ELEVATION: 1,100–8,300 ft (330–3,360 m)
SOILS: Aridisols (51%), Mollisols (28%), Entisols (13%)

The Trans-Pecos ecoregion is delineated by the Rio Grande along its southern boundary with Mexico, the 32nd parallel along the northern border with New Mexico, and a diagonal line from the junction of the 32nd parallel with the 103rd meridian to Del Rio (fig. 3.22). This diagonal roughly follows the Pecos River watershed. The Trans-Pecos ecoregion includes most or all of 11 counties: Brewster (229), Culberson (162), El Paso (159), Hudspeth (116), Jeff Davis (179), Loving (42), Pecos (113), Presidio (164), Reeves (75), Terrell (104), and Ward (82). This ecoregion extends into New Mexico and slightly into Arizona and covers a large portion of northern Mexico.

Almost the entire region is used for agricultural production except areas used for state and national parks, urban development, and recreation. The vast majority of the ecoregion (15.2 million ac, or 6.2 million ha) is used for livestock production. Only about 200,000 ac (81,000 ha) are used as irrigated cropland along the Rio Grande to grow cotton, alfalfa, and sorghum as well as various vegetable and fruit crops.

A complex physiography of isolated mountain ranges with interceding basins (except the Stockton Plateau) characterizes the ecoregion. This basin and range topography is not as pronounced in the Trans-Pecos as in the Great Basin ecoregion (Utah and Nevada). Trans-Pecos mountain ranges are much more irregularly scattered than

in the Great Basin and are a mixture of limestone reefs with dolomite, shale, and gypsum outcrops (Guadalupe Mountains), Precambrian exposures (Franklin Mountains), exposed formations from the Pennsylvanian Period (Marathon Uplift), and volcanoes with associated basalt, rhyolite, tuff, and igneous intrusions (Davis Mountains). The Chisos Mountains contain a mix of igneous, sedimentary, and metamorphic parent material with limestones and other types exposed at lower elevations. The complex topography and geology found in the Trans-Pecos contribute to an interesting and sometimes confusing array of plant communities.

The Stockton Plateau is located in northern Terrell and southern Pecos counties. It is part of the Great Plains physiographic region and not a portion of the basin and range topography characteristic of the rest of the Trans-Pecos. The predominant landform is open hills with smaller areas of tablelands. It was formed by deposition of erosive materials from adjacent mountains. This area has elevation ranges of 2,600–4,500 ft (800–1,300 m). Some references place the Stockton Plateau within the Edwards Plateau ecoregion rather than the Trans-Pecos.

The general vegetation pattern in the ecoregion consists of basins with creosotebush (*Larrea tridentate*), tarbush (*Flourensia cernua*), catclaw acacia (*Acacia greggii*), Mescat acacia (*A. constricta*), yucca, and tobosagrass flats. Plateaus and canyon areas are dominated by ocotillo (*Fouquieria splendens*), candelilla (*Euphorbia antisyphilitica*), lechuguilla (*Agave lechuguilla*), and sotols (*Dasylirion* spp.). Poorly drained saline areas support saltbush (*Atriplex* spp.) and alkali sacaton (*Sporobolus airoides*). Lower slopes are typically characterized by various junipers that intergrade at midslope with Mexican pinyon (*Pinus cembroides*) and pinyon pine (*P. edulis*). Ponderosa pine (*P. ponderosa*) occurs at higher elevations.

Desert grasslands are composed of warm-season grasses such as the muhlys, lovegrasses, dropseeds, gramas, and threeawns. Tobosagrass is also very common. Improper stocking has led to an increase in annual threeawns, burrograss (*Scleropogon brevifolius*), and fluffgrass (*Dasyochloa pulchella*) at these lower elevations. Grasses found at midelevations include bluestems, sideoats grama, blue grama, and needlegrasses (*Heterostipa* spp., *Nassella* spp., *Achnantherum* spp.). Fescues (*Festuca* spp.) and bromes (*Bromus* spp., *Bromopsis* spp., *Anisantha* spp.) are common at the higher elevations.

Rare grass species reported from the region are *Allolepis texana*, *Chondrosum kayi* (= *Bouteloua kayi*), *Chloris texensis* (unverified report, but highly unlikely that it would be so disjunct from its typical habitat along the Texas Gulf Coast), *Festuca ligulata*, and *Poa strictiramea* (Poole et al. 2007).

A total of 353 species from 112 grass genera have been documented for the Trans-Pecos ecoregion (table 3.14). Chloridoideae, Panicoideae, and Poöideae are the three largest grass subfamilies with 127, 126, and 79 species, respectively.

Table 3.14. Taxa by grass subfamilies documented for the Trans-Pecos ecoregion.

Subfamilies	Tribes	Genera	Species	Subspecific taxa
Aristidoideae	1	1	11	10
Arundinoideae	1	2	2	-
Bambusoideae	1	1	1	-
Centothecoideae	1	1	2	-
Chloridoideae	2	34	127	33
Danthonioideae	1	2	2	-
Ehrhartoideae	1	2	4	-
Panicoideae	2	33	126	62
Poöideae	5	33	79	16
Total	15	112	353	121

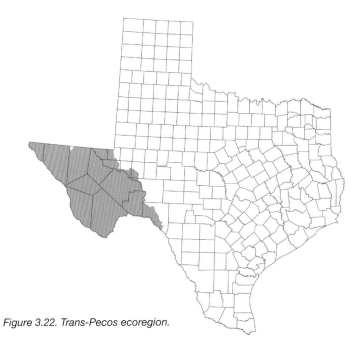

Figure 3.22. Trans-Pecos ecoregion.

Guadalupe Mountains (Arizona/New Mexico Mountains)

SIZE: 54,129 ac (21,905 ha)
TEXAS LAND AREA: Less than 0.001%
RANK IN LAND AREA: 12th
ANNUAL PRECIPITATION: 8–18 in (200–455 mm)
CLIMATE: Arid subtropical (lowlands) to humid cool-temperate (mountains)
ELEVATION: 2,000–8,500 ft (610–2,590 m)
SOILS: Mollisols (50%), Aridisols (5%)

The Guadalupe Mountains ecoregion is the southern extension of the Southern Rocky Mountains into Texas. The Apache, Delaware, Glass, and Guadalupe mountains, which are reefs and deep-water deposits of sandstone, limestone, and shale, were laid down during the Permian (245–285 mya) along the margins of the Delaware Basin (Spearing 1991). All the other prominent West Texas mountain ranges are of a different age and origin. This ecoregion is restricted to northwestern Culberson County in Texas, but it extends across much of New Mexico and Arizona (fig. 3.23). A total of 183 grass species have been collected in Guadalupe Mountains National Park, which encompasses most of this ecoregion in Texas.

Lower-elevation vegetation, composed of creosotebush and tarbush, is similar to that in the Trans-Pecos. Midslopes support pinyon-juniper-oak woodlands, and higher elevations can support ponderosa pine and Douglasfir (*Pseudotsuga menziesii*). Grasses at these higher elevations are dominated by bluegrasses (*Poa* spp.), bromes, and the occasional fescue.

Taxonomically this is a very diverse area because of the rapid increase in elevation and the occurrence of species more commonly found in the Southern Rocky Mountains. There are 77 genera and 184 species of grasses documented for the ecoregion (table 3.15). Major subfamilies are Chloridoideae, Poöideae, and Panicoideae with 75, 54, and 44 species, respectively.

Table 3.15. Taxa by grass subfamilies documented for the Guadalupe Mountains ecoregion.

Subfamilies	Tribes	Genera	Species	Subspecific taxa
Aristidoideae	1	1	7	7
Arundinoideae	1	2	2	-
Bambusoideae	-	-	-	-
Centothecoideae	-	-	-	-
Chloridoideae	2	24	75	20
Danthonioideae	1	2	2	-
Ehrhartoideae	-	-	-	-
Panicoideae	2	20	44	28
Poöideae	5	28	54	11
Total	12	77	184	66

Figure 3.23. Guadalupe Mountains ecoregion.

The classification of Texas grasses reflects the subfamilies and tribes presented by the Grass Phylogeny Working Group (2001) and adopted by Barkworth et al. (2003, 2007). The current trend in grass classification has been to increase (e.g., "split") the number of subfamilies. Historically, 2 subfamilies were recognized, "Panicoid" (Panicoideae) and "Festicoid" (Poöideae). A steady increase in the number of subfamilies since the 1930s has occurred, and currently 16 subfamilies are recognized. The last subfamily added was the Micrairoideae (Barkworth 2007).

Within the borders of Texas, 9 of the recognized 16 subfamilies are represented (table 4.1). As throughout North America, 3 subfamilies dominate. Based on the number of species, the Panicoideae (242) is the largest subfamily, followed by the Chloridoideae (213), and the Poöideae (176). These "big three" subfamilies represent 87% of the grass species within the state. Members of the Panicoideae are typically of more mesic habitats and most abundant in Central, South, and East Texas. The Chloridoideae are warm-season grasses characteristic of the more xeric environments and dominate the grass flora in west-central Texas and the Trans-Pecos. Throughout the state, winter and spring annuals are characteristically members of the cool-season Poöideae; cultivated species of this subfamily—*Triticum*, wheat; *Avena*, oats; *Hordeum*, barley; *Secale*, rye—are most common in the early spring as well.

In contrast to the splitting of subfamilies, the number of tribes has been diminishing in number ("lumped") from the time of Hitchcock's manuals (1935, 1950). For example, the well-known Eragrosteae and Chlorideae tribes have been submerged into the Cynodoneae; and the Aveneae is included within the Poeae. Most subfamilies in the state are represented by only 1 or 2 tribes, except the Poöideae, which has 8 (table 4.1). The "big three" subfamilies contain 63% of all the tribes within the state.

Gould's monumental work *The Grasses of Texas* (1975a) listed 122 genera and 523 species. Hatch, Gandhi, and Brown (1990) reported 131 genera, 545 species, and 60 infraspecific taxa. Jones, Wipff, and Montgomery (1997) increased these numbers significantly (160 genera, 655 species, and 104 infraspecific taxa), primarily by adding ornamentals to the list. This current work lists 181 genera, 723 species, and 189 infraspecific taxa (79 subspecies and 110 varieties). All grass species reported as growing in the state as ornamentals, permanent members of the flora (native or naturalized), or as chance introductions (waifs) have been included. All reported introduced species are included in the checklist. Whether an introduction has become established is uncertain, and all are included here in the off chance that someone might collect one.

As previously mentioned, 723 grass species have been reported as occurring in the state and are included in the checklist. There are 668 species keyed with illustrations, short descriptions, and in some cases photographs. There are 6 species included at the end of the species accounts that are late entries. Thus, there are

Table 4.1. Composition of Texas grass subfamilies.

Subfamilies	Tribes	Genera	Species	Subspecies	Varieties
Aristidoideae	1	1	18	0	13
Arundinoideae	1	3	3	0	0
Bambusoideae	1	10	51	2	6
Centothecoideae	1	1	3	0	0
Chloridoideae	2	47	213	8	35
Danthonioideae	1	3	6	0	0
Ehrhartoideae	2	6	11	0	0
Panicoideae	2	50	242	52	37
Poöideae	8	60	176	17	19
Total	19	181	723	79	110

49 species (all ornamentals and primarily bamboos) that are found only in the checklist.

There are several reasons for the increase in the number of genera reported for the state. The primary reason is new generic concepts in the Triticeae, Stipeae, and Paniceae. Following are the most recent modifications accepted in this work:

1. *Agropyron* to *Agropyron*, *Elymus*, *Leymus*, *Pascopyrum*, *Psathrostachys*, *Pseudoroegneria*, *Thinopyrum*
2. *Stipa* to *Achnatherum*, *Amelichloa*, *Hesperostipa*, *Nassella*
3. *Panicum* to *Dichanthelium*, *Hopia*, *Phanopyrum*, *Megathyrsus*, *Moorochloa*, *Steinchisma*, *Urochloa*, *Zuloagaea*

The second reason for an increase in the number of genera is that the compiler of this work has a very narrow generic concept, drawing the lines between genera often on 1 or 2 very stable characteristics. Some of the generic "splittings" currently accepted are the following:

1. *Agrostis* to *Agrostis*, *Lachnagrostis*
2. *Bouteloua* to *Bouteloua*, *Chondrosum*
3. *Bromus* to *Anisantha*, *Bromopsis*, *Bromus*, *Ceratochloa*

4. *Festuca* to *Festuca*, *Schedonorus*
5. *Hordeum* to *Critesion*, *Hordeum*
6. *Phalaris* to *Phalaris*, *Phalaroides*

The addition of a number of species to the list for the state is significant and is primarily the result of increased distributional information contained in the 2 recent volumes (24, 25) of the Flora of North America series. Also, new species concepts based on increased study of certain large and diverse groups (i.e., *Dichanthelium*, *Panicum*) have resulted in the recognition of more species and over 60 additional infraspecific taxa.

Based on the checklist, 65% of the state's grass flora are natives, and introduced species account for the remaining 35%. A significant number of the 723 species reported for Texas are perennial (77%), whereas annuals make up 23% of the total. Texas is predominantly an arid, semiarid state with a subtropical/tropical climate, so it is no surprise that 470 species are warm season. The 253 cool-season species are primarily winter/spring annuals or perennials of higher elevations in West Texas. There are 87 ornamental grasses listed for the state, but this number would certainly be much larger if all the native species used for aesthetics were considered. Cultivated grasses used for human consumption, livestock production, or land reclamation total 71 species; this number would also increase significantly if all the native grasses found in various seed mixtures were considered.

The genera with the largest number of species are *Muhlenbergia* (50), *Eragrostis* (32), *Paspalum* (32), *Dichanthelium* (26), *Panicum* (24), and *Sporobolus* (23). These 6 large genera represent over 25% of the grass species in the state. Anyone studying Texas grasses will soon become very familiar with the characteristics of these genera and learn to readily recognize them. But for some odd reason, if it is easy to determine the genus, it is generally difficult to determine the species—or perhaps it just seems so.

Checklist

On page 130 are the classification and checklist of the grasses of Texas.Capital Roman numerals designate subfamilies; Lowercase Roman numerals designate tribes, followed by genus, species, subspecies, or variety (see Table 5.1 on page 130). The checklist is alphabetized by subfamily, tribe within subfamily, genus within tribe, species within genus, and subspecies or variety within species. Subfamilies, tribes, genera, and species are numbered consecutively throughout the checklist, but subspecies and varieties are not.

Similar to checklists in Gould (1962, 1969, 1975b) and Hatch, Gandhi, and Brown (1990), this one includes, for each species, information concerning the origin (N = native, I = introduced), longevity (A = annual, P = perennial, B = both, either annual or perennial), and season of growth (C = cool season, W = warm season). Also designated is whether the species is a cultivar (C) or ornamental (O), and if the species is considered a threatened, endangered species or species of special concern (E), or a federally listed noxious weed (W). Those species not keyed are designated by an X. Occurrence within ecoregion was determined by integrating species county distribution and ecoregions by county (fig 4.1) (table 4.2). Numbers 1–12 correspond to the 12 ecoregions described in the previous chapter. A number in the column for an ecoregion that follows a species indicates that a documented reference verifies the occurrence of the species within a county in that particular ecoregion. Undoubtedly, most or all the grasses in this book have a wider distribution than indicated either in the ecological checklist or distribution maps. Distributional information for ornamentals and cultivars is sparse, and sometimes none exists for an ecological classification.

Table 4.2. Summary of grass species documented by Texas ecoregions. Ecoregion delineations equate to Environmental Protection Agency Level III Ecoregions (USEPA 2000). Names have been modified to reflect more commonly used natural vegetation nomenclature. If more than one ecoregion occurs within a county, the grass species for that county were included in each ecoregion.

ECOREGIONS	Number of Species
1 - Piney Woods	410
2 - Gulf Coast Prairies & Marshes	402
3 - Post Oak Savannah	435
4 - Blackland Prairies	375
5 - Cross Timbers	290
6 - Southern Texas Plains	319
7 - Central Great Plains	209
8 - Southwestern Tablelands	198
9 - High Plains	201
10 - Edwards Plateau	322
11 - Trans-Pecos	351
12 - Guadalupe Mountains	183

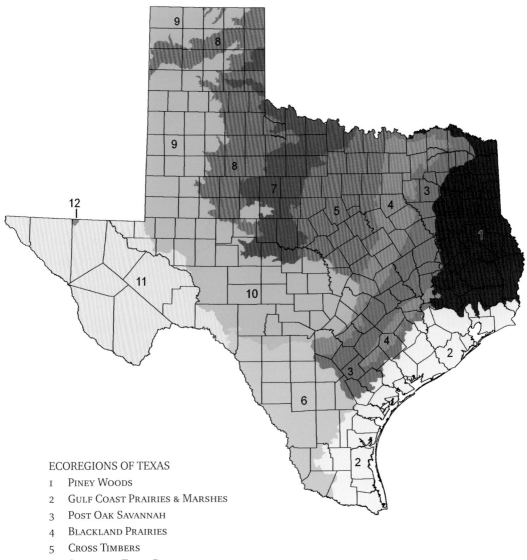

ECOREGIONS OF TEXAS

1 PINEY WOODS
2 GULF COAST PRAIRIES & MARSHES
3 POST OAK SAVANNAH
4 BLACKLAND PRAIRIES
5 CROSS TIMBERS
6 SOUTHERN TEXAS PLAINS
7 CENTRAL GREAT PLAINS
8 SOUTHWESTERN TABLELANDS
9 HIGH PLAINS
10 EDWARDS PLATEAU
11 TRANS-PECOS
12 GUADALUPE MOUNTAINS

Figue 4.1. Ecoregions of Texas based on Griffith et al. (2004) but with the more widely recognized names used in this book.

Sometimes county distributional data are not available for a species; this is particularly true of ornamentals and cultivars, and the map will be blank. Illustrations (permission granted from Utah State University) for the most part are from Barkworth et al. (2003, 2007). A few are from Hitchcock's manual (1935, 1950) and are in the public domain.

Descriptions and information about each species are from Gould (1975b), Barkworth et al. (2003, 2007), Allred (2005) and personal observation. Much of the information about ornamentals and cultivars is from Jones, Wipff, and Montgomery (1997).

GENERIC KEYS AND SPECIES ACCOUNTS

Keys, Distribution, Illustrations, and Short Descriptions

Keys to the grass genera follow (for the most part) Gould (1975b), Gould and Shaw (1983), and Shaw (2008). All keys are artificial, and unrelated taxa are often grouped together. Figure 5.1 presents a diagrammatic guide to major key groups.

The keys are solely for identification purposes and are not intended to express phylogenetic relationships. However, in some cases closely related genera are grouped together (i.e., Andropogoneae in Group 3, and the Paniceae in Group 4). Subfamilial, tribal, and generic relationships are best viewed in the checklist. Dichotomous keys are numbered and indented. Under "Key to the Groups," genera with similar morphological characteristics are placed in numbered keys (Groups 1–10). Groups 1–10 keys then lead to specific genera. Genera may key out in more than one numbered key and/or more than once in a single numbered key. The page number on which the generic description and the species key (if the genus has more than one species) is provided.

A brief generic description giving distinguishing morphological characteristics is provided. Also included are basic chromosome number (x) and photosynthetic pathway (C_3, C_4). Genera are alphabetized and numbered consecutively throughout the book. Subfamily and tribe are included for each genus.

Species and infraspecies keys follow those presented for each genus in volumes 24 and 25 in the Flora of North America series (Barkworth 2003, 2007) and in Gould (1975a). Author names are cited immediately after the species. Species are alphabetical and numbered consecutively within a genus. Included are the scientific name (genus, species, author citation) and vernacular names (sometimes referred to as the common name, which is not common at all). A brief description pointing out specific features of the plant or information concerning the species follows the names. A list of synonyms can be found in Jones, Wipff, and Montgomery (1997).

Included here are 668 grass species, including their infraspecies taxa (if warranted); 49 ornamental species have been omitted due to lack of distributional data, descriptive information, and distinguishing vegetative or floristic characteristics. The majority of the omitted ornamentals are "bamboos" that rarely flower, making them more difficult to accurately identify. Hopefully, a more inclusive section on ornamentals can be added in later editions or by other authors, as the number of forms, varieties, and species being introduced and grown as ornamentals continues to increase.

Species county distributional maps are based on 17,795 county records of occurrence provided by the Intermountain Herbarium, Utah State University and Turner et al. (2003).

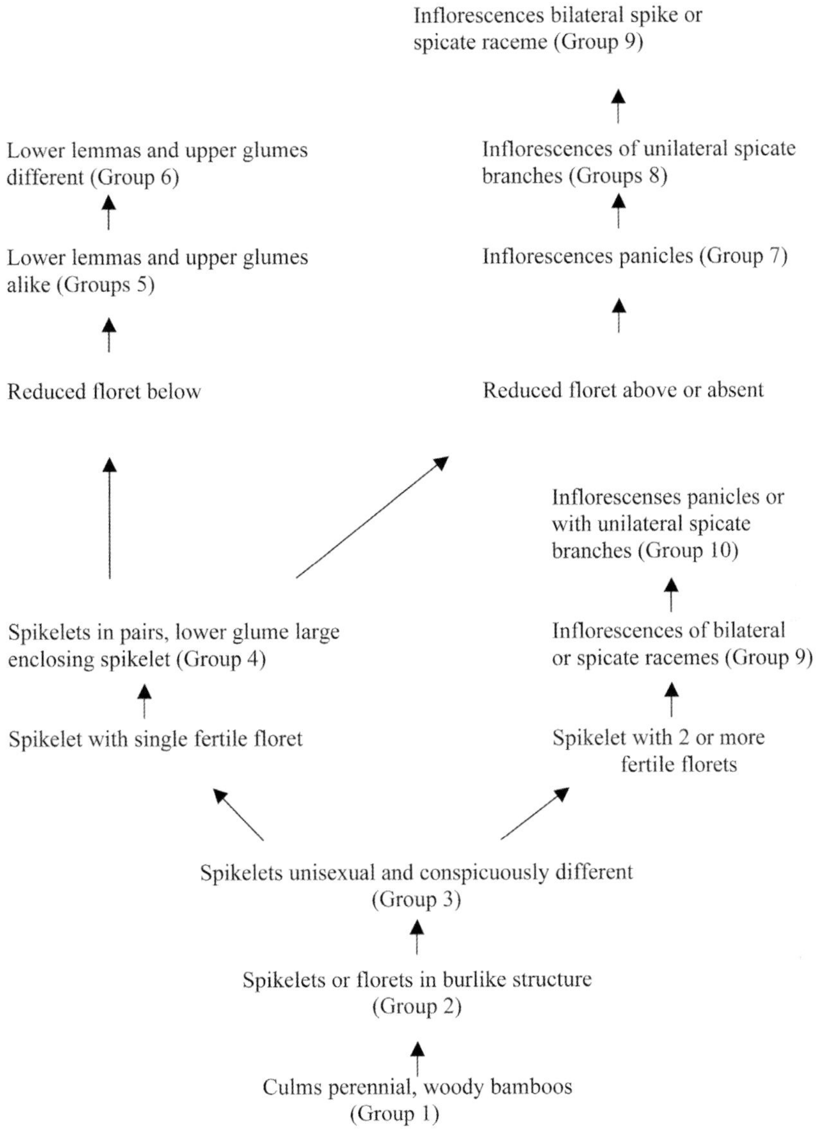

Inflorescences bilateral spike or
spicate raceme (Group 9)

↑

Lower lemmas and upper glumes   Inflorescences of unilateral spicate
different (Group 6)   branches (Groups 8)

↑   ↑

Lower lemmas and upper glumes   Inflorescences panicles (Group 7)
alike (Groups 5)

↑   ↑

Reduced floret below   Reduced floret above or absent

↑   ↗   Inflorescenses panicles or
                                        with unilateral spicate
                                        branches (Group 10)

                                        ↑

Spikelets in pairs, lower glume large   Inflorescences of bilateral
enclosing spikelet (Group 4)   or spicate racemes (Group 9)

↑   ↑

Spikelet with single fertile floret   Spikelet with 2 or more
                                        fertile florets

        ↖   ↗

        Spikelets unisexual and conspicuously different
                        (Group 3)

                        ↑

        Spikelets or florets in burlike structure
                        (Group 2)

                        ↑

        Culms perennial, woody bamboos
                        (Group 1)

Figure 5.1. Diagrammatic guide to major key groups of grasses in Texas. If the grass matches the
characteristics, go to that group; if the grass does not fit, follow the arrow.

Key to the Groups

1. Culms perennial, usually woody; often developing complex branching systems; upper leaves usually pseudopetiolate (Bambuseae) .... Group 1
1. Culms annual, rarely woody; sometimes branching above but not developing complex branching systems; leaves not pseudopetiolate.
  2. Spikelets or florets contained within burlike structure, disarticulation below that structure. .... Group 2
  2. Spikelets or florets not contained within burlike structure.
    3. Spikelets unisexual, staminate, and pistillate spikelets usually conspicuously different ... . Group 3
    3. Spikelets perfect, or if unisexual, then staminate and pistillate spikelets not readily distinguishable.
      4. Spikelets with a single fertile floret, with or without reduced florets.
        5. Spikelets in pairs of 1 sessile and 1 pediceled (sometimes 2 pediceled spikelets at branch tips); sessile spikelets fertile; pedicellate spikelets staminate, rudimentary, or absent with pedicel still obvious; lower glume large, firm, and tightly clasping, or enclosing the entire spikelet (Andropogoneae) .... Group 4
        5. Spikelets not as above.
          6. Inflorescences with 2 morphologically distinct spikelet forms............... Group 6
          6. Inflorescences with all spikelets morphologically similar.
            7. Reduced floret or florets present below fertile floret.
              8. Reduced floret 1; lemma of reduced floret similar to upper glume in shape, size, and texture; disarticulation below the glumes (Paniceae) .... Group 5
              8. Reduced floret 1 or 2; lemma of reduced floret(s) not similar to upper glumes in shape, size, or texture; disarticulation above the glumes.......... .... Group 6
            7. Reduced floret absent or above the fertile florets.
              9. Inflorescence paniculate, never spicate .... Group 7
              9. Inflorescence a spike, spicate, or with 2 to several spicate primary branches.
                10. Inflorescence of 1 to several unilateral spicate primary branches ... . Group 8
                10. Inflorescence a terminal, bilateral spike or spicate raceme Group 9
      4. Spikelets with 2 or more fertile florets.
            11. Inflorescence a spike or spicate raceme, or with 2 to several spicate primary branches
              12. Inflorescence with 2 (rarely 1) to several unilateral spicate primary branches.. Group 10
              12. Inflorescence a terminal, bilateral spike or spicate raceme.... .... Group 9
             11. Inflorescence an open or contracted panicle or raceme with spikelets on well-developed pedicels........................... Group 10

Group 1

(Culms perennial, usually woody; often developing complex branching systems; upper leaves usually pseudopetiolate) (Bambuseae)

1. Rhizomes pachymorphic, short, thicker than the culms..................................*Bambusa* (p. 248)
1. Rhizomes leptomorphic, long, thinner than the culms.
    2. Culm internodes grooved their entire length; branches usually without compressed inter ....
       nodes; spikelets sessile ....*Phyllostachys* (p. 801)
    2. Culm internodes usually terete, not grooved their entire length; branches usually with 1–5 .
       compressed internodes; spikelets pediceled..*Arundinaria* (p. 236)

(Stapleton 1999)

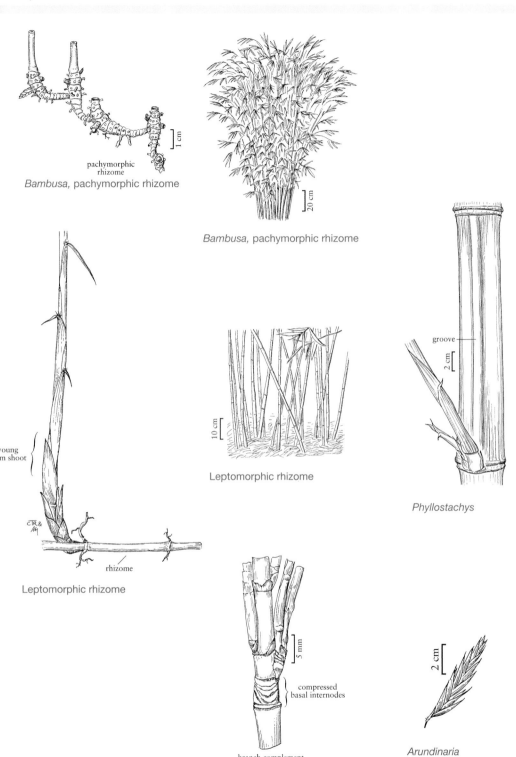

Bambusa, pachymorphic rhizome

pachymorphic
rhizome

1 cm

Bambusa, pachymorphic rhizome

20 cm

groove

2 cm

Phyllostachys

young
culm shoot

rhizome

Leptomorphic rhizome

10 cm

Leptomorphic rhizome

5 mm

compressed
basal internodes

branch complement
at midculm node

Arundinaria

2 cm

Arundinaria

Group 2

(Spikelets or florets contained within burlike structure, disarticulation below that structure)

1. Glumes with hooked spines...*Tragus* (p. 957)
1. Glumes without hooked spines.
    2. Plants dioecious; burlike structure formed by a cluster of spikelets whose lower portion of upper glumes are indurate, white, terminating in several awnlike teeth ....*Buchloë* (p. 299)
    2. Plants not as above.
        3. Burlike structure a reduced inflorescence branch.
            4. Axes of the branches extending beyond the base of the distal spikelet.
                5. Panicle branches deciduous .... *Bouteloua* (p. 268)
                5. Panicle branches persistent.. *Chondrosum* (p. 334)
            4. Axes of the branches terminating at the base of the distal spikelet.
                6. Spikelets in pairs; panicle branches often fused to the rachises.......................... .... *Lycurus* (p. 612)
                6. Spikelets in triplet; panicle branches sometimes appressed but not fused to rachises.
                    7. Branches straight at the base.
                        8. Glumes thickened, hardened, and conspicuously fused at base.................. ....*Hilaria* (p. 561)
                        8. Glumes papery or membranous, not fused at base.................................... ....*Pleuraphis* (p. 810)
                    7. Branches sharply curved at the base............................. *Cathestecum* (p. 303)
        3. Burlike structure (fascicle) composed of bristles or flattened spines.
            9. Fascicles composed of flattened spines fused for more than ½ their length; spines sharp enough to puncture flesh.................................... ....*Cenchrus* (p. 305)
            9. Fascicles composed of bristles; bristles free for more than ½ their length; spines not sharp enough to puncture flesh .... *Pennisetum* (p. 771)

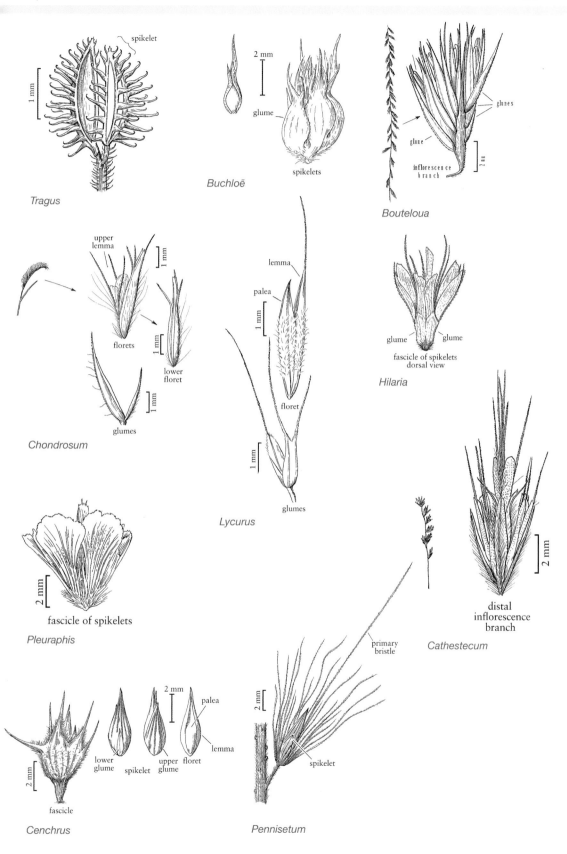

Tragus

spikelet

Buchloë

glume

spikelets

Bouteloua

glumes

glume

inflorescence branch

upper lemma

florets

lower floret

glumes

Chondrosum

lemma

palea

floret

glumes

Lycurus

glume

glume

fascicle of spikelets
dorsal view

Hilaria

Pleuraphis

fascicle of spikelets

distal
inflorescence
branch

Cathestecum

lower
glume

spikelet

upper
glume

palea

floret

lemma

fascicle

Cenchrus

primary
bristle

spikelet

Pennisetum

Group 3

(Spikelets unisexual, staminate; pistillate spikelets usually conspicuously different)

1. Plants monoecious.
   2. Pistillate spikelets concealed with an indurate, globose, beadlike structure ....*Coix* (p. 353)
   2. Pistillate spikelets not as above.
      3. Plants low, stoloniferous ....*Scleropogon* (p. 876)
      3. Plants usually 1 m or more in length; cespitose, with a single culm or submerged with floating culms.
         4. Plants submerged with floating culms .... *Luziola* (p. 609)
         4. Plants usually 1 m or more in length; cespitose, with a single culm.
            5. Pistillate and staminate spikelets in the same inflorescence; staminate spikelets distal to the pistillate spikelets .... *Tripsacum* (p. 978)
            5. Pistillate and staminate spikelets in different inflorescences; staminate inflorescences terminal on culm and branches; pistillate inflorescences terminal on axillary peduncles, sometimes clustered in false panicles .... *Zea* (p. 1014)
1. Other couplet on page 84.

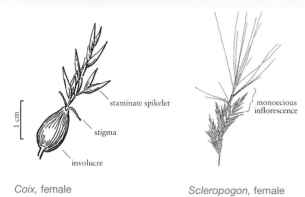

Coix, female

staminate spikelet

stigma

involucre

Scleropogon, female

monoecious
inflorescence

Luziola

1 mm

staminate
floret

1 mm

immature   mature

pistillate
florets

achene

1 mm

staminate
spikelets

pistillate
spikelets

Tripsacum

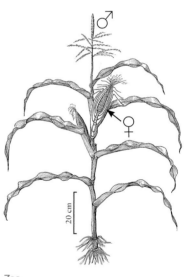

♂

♀

20 cm

Zea

1. Plants, at least a majority, dioecious.
    4. Staminate and pistillate spikelets markedly different in structure.
      5. Pistillate spikelets in burlike clusters hidden in the leafy portion of the plant, staminate spikelets sessile on 1–4 short spicate branches of a well-exserted inflorescence ....*Buchloë* (p. 299)
      5. Pistillate spikelets long-awned from the nerves of each lemma; staminate spikelets awnless; both spikelet types in contracted, usually spikelike racemes .... *Scleropogon* (p. 876)
    4. Staminate and pistillate spikelets similar in structure but may vary in size and sometimes number of florets per spikelet.
      6. Plants stoloniferous; leaf blades up to 1.5 cm long.. *Monanthochloë* (p. 631)
      6. Plants rhizomatous; leaves >1.5 cm long ....*Distichlis* (p. 440)

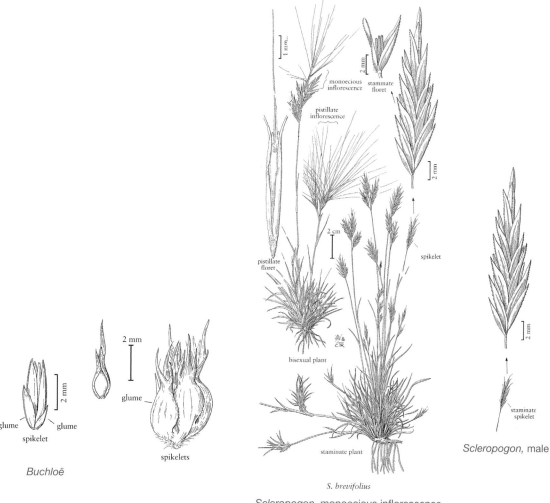

monoecious inflorescence

staminate floret

pistillate inflorescence

1 mm

2 mm

2 mm

2 cm

spikelet

pistillate floret

2 mm

bisexual plant

staminate spikelet

staminate plant

S. brevifolius

Scleropogon, monoecious inflorescence

Scleropogon, male

glume

spikelet

glume glume spikelet spikelets

2 mm

2 mm

Buchloë

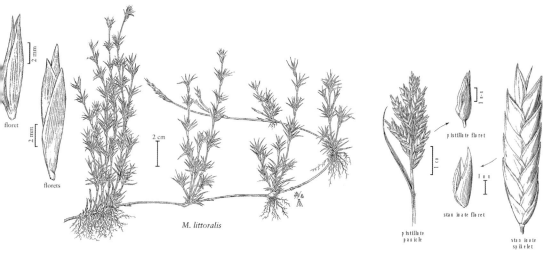

floret

2 mm

florets

2 mm

2 cm

M. littoralis

Monanthochloë

pistillate floret

1 cm

1 mm

staminate floret

1 mm

pistillate panicle

staminate spikelet

Distichlis

Group 4

(Spikelets in pairs of 1 sessile and 1 pediceled; lower glumes large and enclosing spikelet) (tribe Andropogoneae)

1. Spikelets all alike and fertile (this includes those genera where the pediceled spikelet is completely reduced and only a naked pedicel remains and those with both sessile and pediceled spikelets fertile and alike) .... A
1. Spikelets not all alike; the pediceled one (or less frequently the sessile one) is staminate or sterile.. .... AA

A (Spikelets all alike and fertile)

1. Pediceled spikelet completely reduced, only sessile remaining.
   2. Plants sprawling annuals; blades ovate; pedicels absent or <3 mm long ....*Arthraxon* (p. 234)
   2. Plants erect perennials; blades linear to lanceolate; pedicels always present, >3 mm long.
      3. Peduncles and branches not subtended by a modified leaf; inflorescences large terminal panicles .... *Sorghastrum* (p. 898)
      3. Peduncles subtended, and often partially enclosed, by a modified leaf; inflorescences terminal and axillary, composed of digitate clusters of rames on a common peduncle.......... ....*Andropogon* (p. 192)
1. Pediceled spikelets present and like sessile spikelets.
      4. Plants annual; blades 3–10 cm long; inflorescence branches 2–6 ....*Microstegium* (p.627)
      4. Plants perennial; blades elongate, at least some mature ones >10 cm long; inflorescence branches numerous.
         5. Spikelets falling separately from a persistent rachis.
            6. Spikelets awned; inflorescence branches usually 7–35 cm long ....*Miscanthus* (p. 629)
            6. Spikelets unawned; inflorescence branches 1–7 cm long .... *Imperata* (p. 573)
         5. Spikelet falling in pairs with sections of a disarticulating rachis (rames) .... *Saccharum* (p. 847)

AA (Spikelets not all alike; sessile or pediceled ones staminate or sterile)

1. Spikelet awnless.
   2. Flowering culms terminating in panicles with numerous branches ....*Sorghum* (p. 901)
   2. Flowering culms terminating in spicate racemes.
      3. Pedicels, at least partially, fused with rachises.
         4. Plants annual.

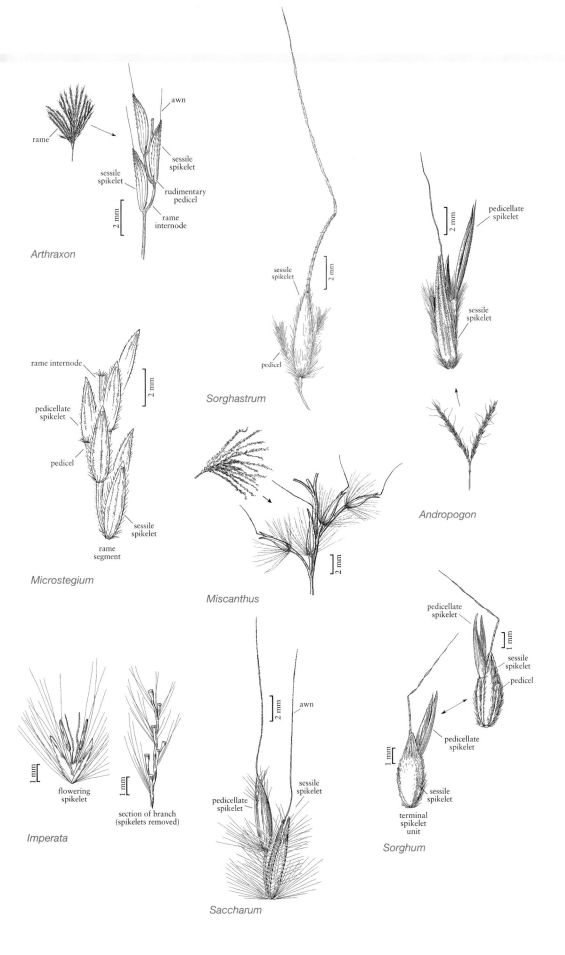

rame

awn

sessile
spikelet

sessile
spikelet

rudimentary
pedicel

rame
internode

2 mm

Arthraxon

rame internode

pedicellate
spikelet

pedicel

sessile
spikelet

rame
segment

2 mm

Microstegium

sessile
spikelet

pedicel

2 mm

Sorghastrum

2 mm

Miscanthus

2 mm

pedicellate
spikelet

sessile
spikelet

Andropogon

flowering
spikelet

1 mm

section of branch
(spikelets removed)

1 mm

Imperata

2 mm

awn

pedicellate
spikelet

sessile
spikelet

Saccharum

pedicellate
spikelet

1 mm

sessile
spikelet

pedicel

pedicellate
spikelet

1 mm

sessile
spikelet

terminal
spikelet
unit

Sorghum

    5. Sessile spikelets globose, conspicuously alveolate ....*Hackelochloa* (p. 549)

    5. Sessile spikelets not globose or alveolate.

        6. Sheaths with stiff, papillose-based hairs 1–3 mm long, causing skin irritation ....
        ....*Rottboellia* (p. 845)

        6. Sheaths without irritating hairs, often inflated...................*Hainardia*[1] (p. 551)

    4. Plants perennial ....*Hemarthria* (p. 553)

3. Pedicles appressed but not fused to the rachises.

            7. Keels of the lower glumes with spinelike projections ....
            ....*Eremochloa* (p. 514)

        7. Keels of the lower glumes without spinelike projections.

            8. Pedicellate spikelets 1–3 mm long ....*Coelorachis* (p. 350)

            8. Pedicellate spikelets 4–8 mm long ....*Elionurus* (p. 455)

1. Spikelets, at least some, awned.

                9. Terminal inflorescences with elongated rachises.

                10. Rame internodes and pedicels with a translucent median line, central groove, or membranous area....................*Bothriochloa* (p. 255)

                10. Rame internodes and pedicels with a translucent median line, central groove, or membranous area.

                    11. Sessile spikelets terete or laterally compressed ....
                    .... *Chrysopogon* (p. 345)

                    11. Sessile spikelets dorsally compressed ....
                    .... *Sorghum* (p. 901)

            9. Other couplet on page 90.

---

1. *Hainardia* is in the tribe Poeae, not Andropogoneae. It is often mistaken for the latter, however, because its embedded spikelets and fleshy rachis are similar to those found within the Andropogoneae. If the spikelet structure of *Stenotaphrum* of the Paniceae is misinterpreted, it is sometimes mistakenly keyed to this vicinity as well

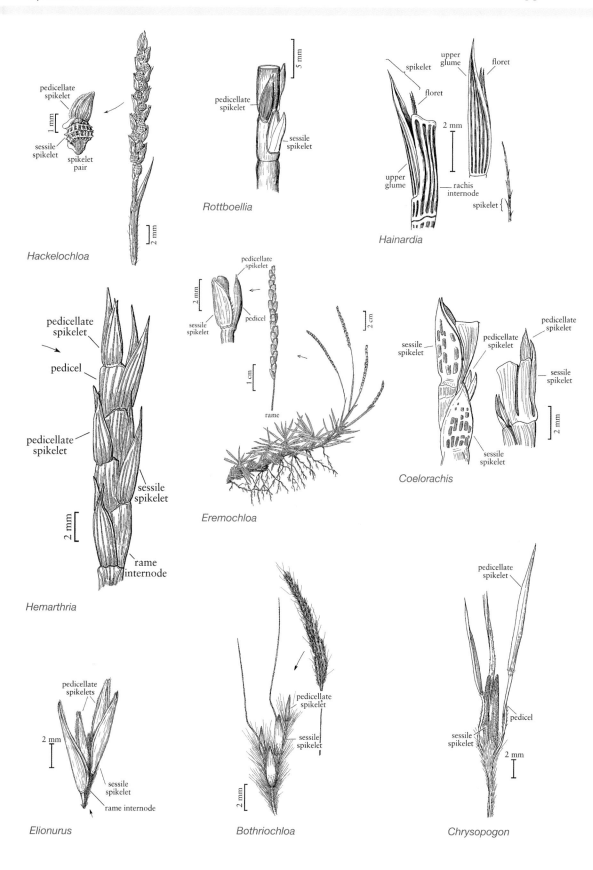

Hackelochloa

Rottboellia

Hainardia

Hemarthria

Eremochloa

Coelorachis

Elionurus

Bothriochloa

Chrysopogon

9. Terminal inflorescences without elongated rachises.

    12. Rame internodes and pedicels with a translucent median line, central groove, or membranous area.............................. ....*Bothriochloa* (p. 255)

    12. Rame internodes and pedicels without a translucent median line, central groove, or membranous area.

        13. Sessile and pedicellate spikelets differ in sexuality (heterogamous).

            14. Pediceled spikelets perfect, awned; sessile spikelets sterile or staminate, unawned ....*Trachypogon* (p. 955)

            14. Pediceled spikelets sterile or staminate, unawned; sessile spikelets perfect, awned.

                15. Rames 1(–2) on peduncles; lower glumes of sessile spikelets veined between keels................................ ....*Schizachyrium* (p. 867)

                15. Rames 2–15 on peduncles; lower glumes of sessile spikelets without veins between keels ....*Andropogon* (p. 192)

        13. Basal spikelet pairs on each rame with same sexuality (sterile or staminate) (homogamous).

            16. Rames with 3–10 homogamous spikelet units; awns 5–15 cm long, becoming twisted together at maturity............................*Heteropogon* (p. 558)

            16. Rames with 1–2 homogamous spikelet units; awns 1–5 cm (rarely up to 20 cm) long.

                17. Inflorescences terminal on culms, axillary inflorescences absent or rare........................... .... *Dichanthium* (p. 390)

                17. Inflorescences terminal and axillary, axillary numerous.

                    18. Homogamous spikelets distinct and forming an involucre around the rame base .... *Themeda* (p. 948)

                    18. Homogamous spikelets not distinct and not forming an involucre around the rame base ....*Hyparrhenia* (p. 570)

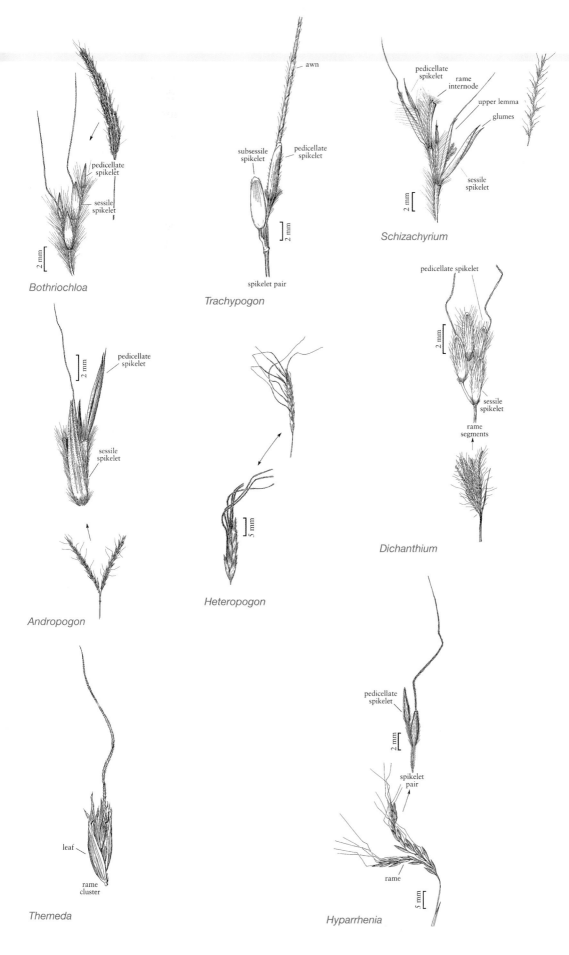

pedicellate spikelet

sessile spikelet

2 mm

Bothriochloa

awn

subsessile spikelet

pedicellate spikelet

2 mm

spikelet pair

Trachypogon

pedicellate spikelet

rame internode

upper lemma

glumes

sessile spikelet

2 mm

Schizachyrium

pedicellate spikelet

pedicellate spikelet

2 mm

sessile spikelet

2 mm

sessile spikelet

rame segments

Dichanthium

5 mm

Heteropogon

Andropogon

leaf

rame cluster

Themeda

pedicellate spikelet

2 mm

spikelet pair

rame

5 mm

Hyparrhenia

Group 5

(Reduced floret 1, below the perfect floret; lemma of reduced floret similar to upper glume in shape, size, and texture; disarticulation below the glumes) (tribe Paniceae)

1. Inflorescences spikelike with the branches embedded in a thick, flattened rachis ....
....*Stenotaphrum* (p. 946)
1. Inflorescences panicles, or if spikelike, branches not embedded or rachis not flattened.
   2. Spikelets in involucres of bristles or flattened spines, these disarticulating with spikelets.
      3. Bristles and/or spines (at least some) fused together for ½ their length ....
....*Cenchrus* (p. 305)
      3. Bristles and/or spines not fused together or fused less than ½ their length......................
....*Pennisetum* (p. 771)
   2. Spikelets not in involucres; bristles persistent; not disarticulating when present.
      4. Upper glumes awned; first glumes awned or unawnless.
         5. Spikelets and upper florets laterally compressed; lower glumes minute...................
....*Melinis* (p. 623)
         5. Spikelets and upper florets dorsally compressed; lower glumes well developed.
            6. Lower glumes much shorter than upper; blades linear to linear-lanceolate, midrib prominent, ligules usually absent .... *Echinochloa* (p. 442)
            6. Lower glumes about equal to upper; blades triangular to lanceolate, midrib not prominent; ligules present on all leaves .... *Oplismenus* (p. 695)
      4. Upper glumes unawned; lower glumes awnless.
         7. Upper lemmas and paleas cartilaginous and flexible at maturity; lemma margins flat, membranous, not involute, and clasping palea; lower glume absent or less than ¼ the length of spikelet.
            8. Inflorescence simple panicles with erect to ascending branches on elongate rachises; branches not spikelike ....*Anthenantia* (p. 204)
            8. Inflorescences usually panicles of digitate or subdigitate clusters of spikelike branches, or simple panicles with strongly divergent branches ....
....*Digitaria* (p. 423)
         7. Upper lemmas and paleas chartaceous to indurate and rigid at maturity; lemma margins involute and clasping paleas; lower glumes present or absent.
            9. Lower glumes absent on some or all spikelets.
               10. Spikelets subtended by a cuplike callus...............*Eriochloa* (p. 516)
               10. Spikelets not subtended by a cuplike callus.
                  11. Lemma of upper (fertile) florets with rounded back turned away from rachises; spikelets borne singly and widely spaced in 2 rows..*Axonopus* (p. 224)
                  11. Lemma of upper (fertile) florets with rounded back turned toward the rachises; spikelets closely spaced and often paired in 2 or 4 rows...*Paspalum* (p. 735)
            9. Other couplet on page 94.

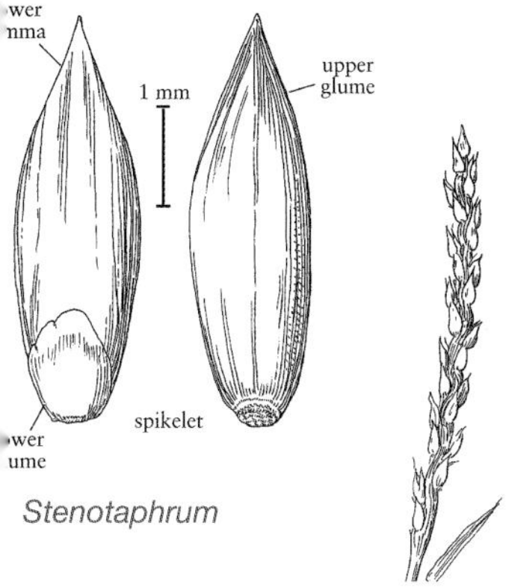

Stenotaphrum

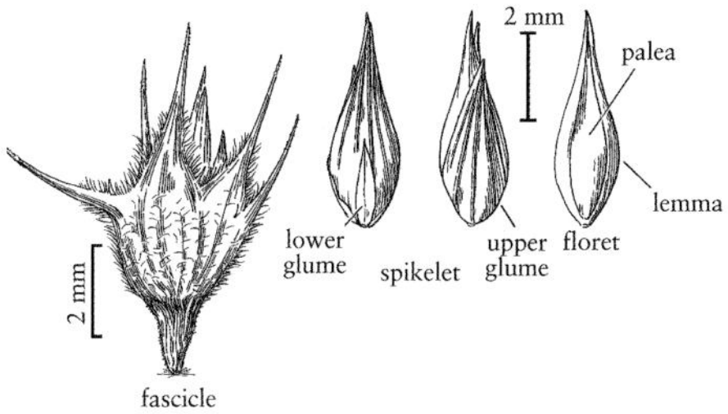

Cenchrus

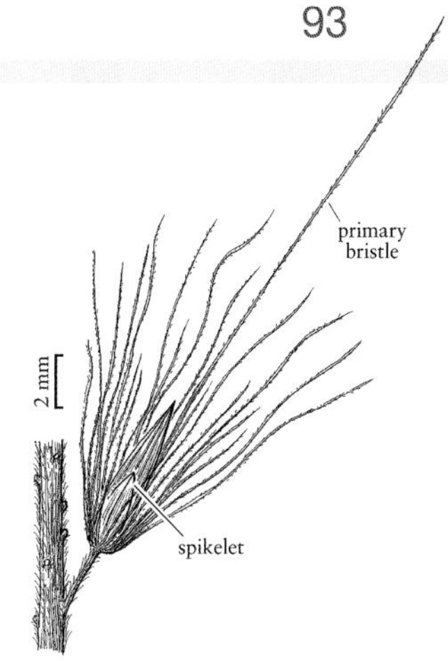

Pennisetum

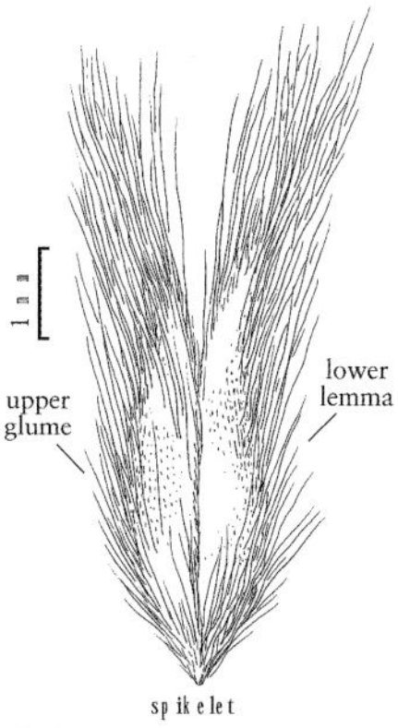

Melinis

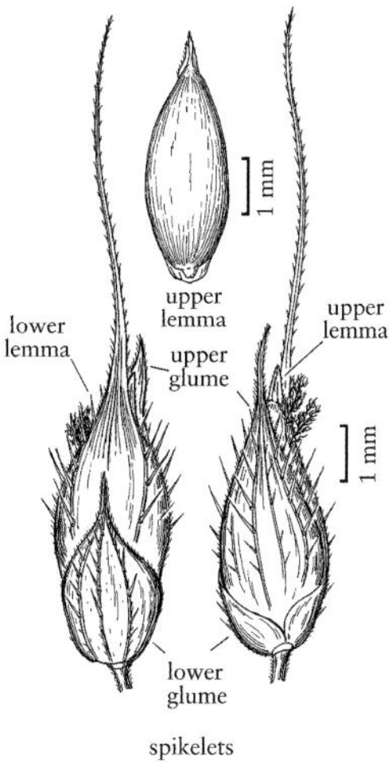

Echinochloa

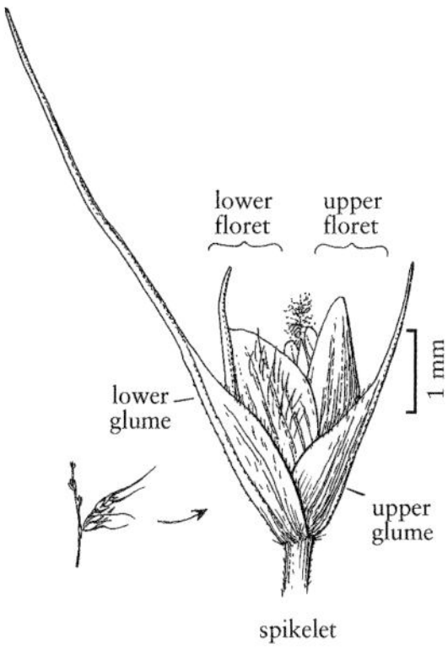

Oplismenus

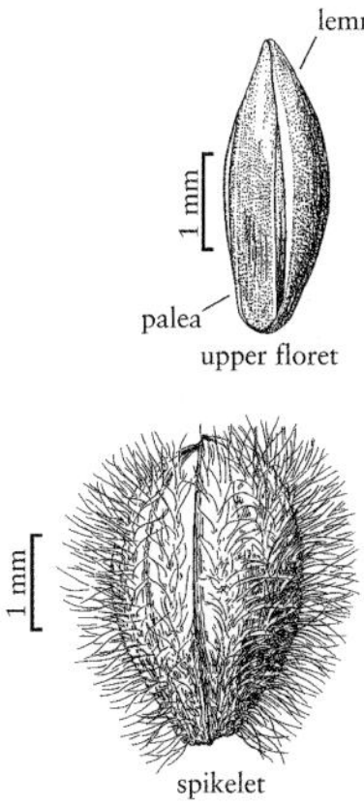

Anthenantia

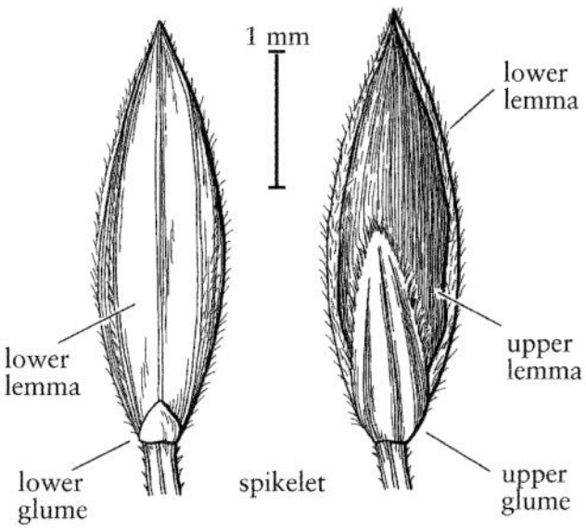

Digitaria

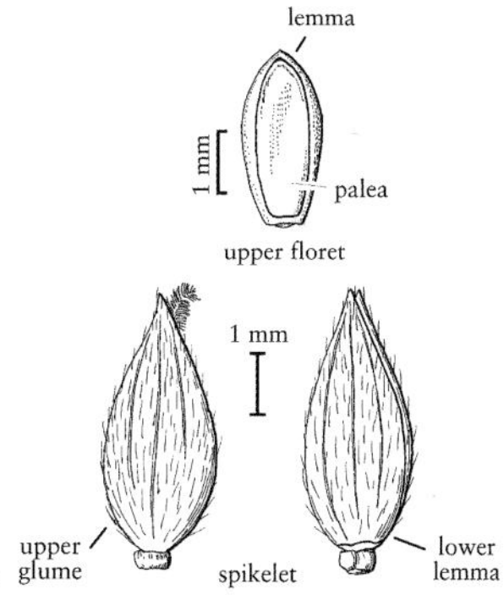

Eriochloa

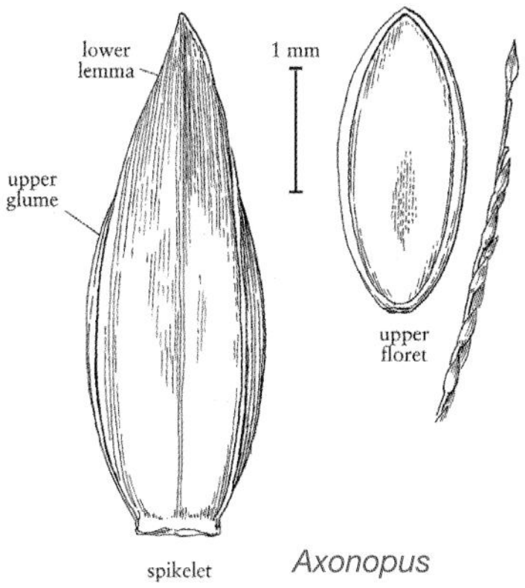

Axonopus

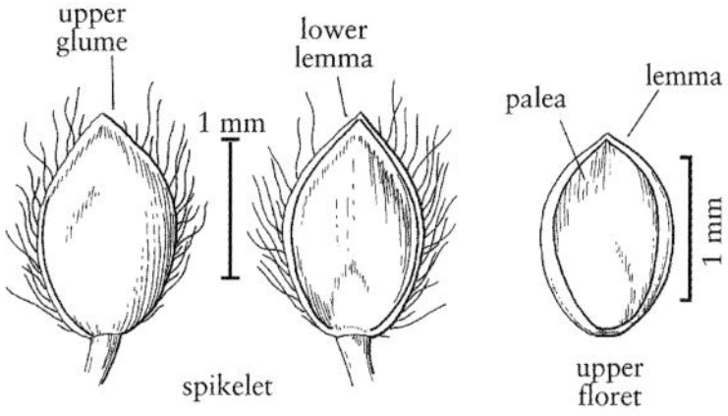

Paspalum

9. Lower glumes present on all or some spikelets.

    12. Ligules absent on most or all leaves ....
....*Echinochloa* (p. 442)

    12. Ligules present on most or all leaves.

        13. Palea of lower florets inflated at maturity; lower and upper florets separated at maturity ....
.... *Steinchisma* (p. 944)

        13. Palea of lower florets not inflated; lower and upper florets closely appressed at maturity.

            14. All or most spikelets subtended by bristles, or branches extending beyond distal spikelet as distinct bristle.......A

            14. All or most spikelets not subtended by bristles .... AA

A (spikelets, at least distal ones, subtended by 1 or more stiff bristles or extensions of panicle branch)

1. Spikelets, all or some (at least, those terminating branchlets), subtended by 1 to several stiff bristles.

    2. Spikelets subtended by 1 to many bristles; paleas of lower florets usually hyaline to membranous at maturity, rarely absent or reduced; palea veins not keeled ....
.... *Setaria* (p. 880)

    2. Spikelets subtended by a single bristle; paleas of lower florets coriaceous to indurate at maturity, keels thickened..*Ixophorus* (p. 576)

1. Spikelets not subtended by stiff bristles, but branch axes extending past terminal spikelets as a bristle 2–4 mm long ....*Paspalidium* (p. 733)

AA (spikelets not subtended by stiff bristles)

1. Upper lemmas rugose and verrucose.

    2. Inflorescences of 1-sided, spikelike primary branches...............................*Urochloa* (p. 990)

    2. Inflorescences a panicle with well-developed primary, secondary, and tertiary branches ......
....*Megathyrsus* (p. 615)

1. Upper lemmas usually smooth; if rugose, the panicle branches neither verticillate nor 1-sided and spikelike.

    3. Upper florets stipitate.

        4. Upper glumes gibbous... *Sacciolepis* (p. 856)

        4. Upper glumes not gibbous...*Moorochloa* (p. 633)

    3. Other couplet on page 96.

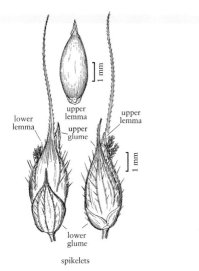

lower lemma
upper glume
upper lemma

upper lemma

lower glume

spikelets

Echinochloa

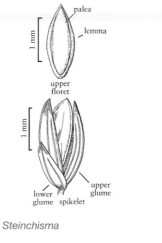

palea
lemma
1 mm
upper floret

1 mm

lower glume   spikelet   upper glume

Steinchisma

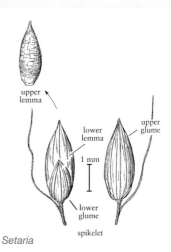

upper lemma

lower lemma
upper glume
1 mm
lower glume
spikelet

Setaria

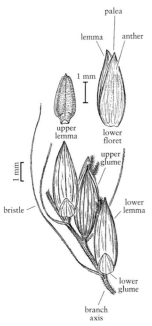

palea
lemma   anther
1 mm
upper lemma   lower floret
upper glume
1 mm
lower lemma
bristle
lower glume
branch axis

Ixophorus

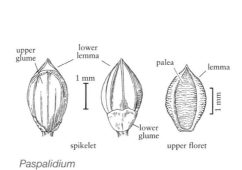

upper glume
lower lemma
1 mm
spikelet   lower glume
palea   lemma
1 mm
upper floret

Paspalidium

upper glume
1 mm
spikelet

lower lemma
1 mm
lower glume
spikelet

Urochloa

1 mm

upper floret

Megathyrsus

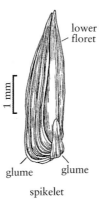

lower floret
1 mm
glume   glume
spikelet

Sacciolepis

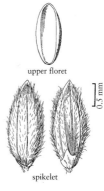

upper floret
0.5 mm
spikelet

Moorochloa

3. Upper florets not stipitate.

    5. Panicle branches 1-sided.

        6. Spikelets 5–7 mm long; upper floret less than ⅓ as long as spikelets ....
        ....*Phanopyrum* (p. 795)

        6. Spikelets not as above.

            7. Lower glumes about ¾ as long as spikelets; spikelets terete or only slightly laterally compressed; stoloniferous with villous nodes ....
            .... *Hopia* (p. 566)

            7. Lower glumes not more than ½ as long as spikelets; spikelets obviously laterally compressed; stoloniferous or not.

                8. Culm bases cormlike ....*Zuloagaea* (p. 1025)

                8. Culm bases varied, but not cormlike.............................. *Panicum* (p. 699)

    5. Panicles simple or well developed but not 1-sided.

            9. Basal leaves developing a distinctive winter rosette..............................
            ....*Dichanthelium* (p. 390)

            9. Basal leaves not developing a distinctive winter rosette.

                10. Primary inflorescences appearing midspring; axillary inflorescences appearing in summer and rebranching by fall; upper florets not disarticulating, plump ....*Dichanthelium* (p. 390)

                10. Primary inflorescences appearing midsummer; axillary inflorescences rare, but if present, usually not rebranching; upper florets either disarticulating or not plump.

                    11. Culm bases cormlike ....*Zuloagaea* (p. 1025)

                    11. Culm bases various, but not cormlike.......................................
                  .... *Panicum* (p. 699)

(Barkworth et al. 2007; Gould and Shaw 1983)

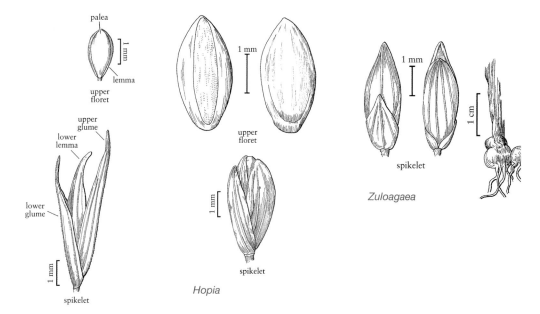

palea

1 mm

lemma

upper
floret

upper
glume

lower
lemma

lower
glume

1 mm

spikelet

Phanopyrum

1 mm

upper
floret

1 mm

spikelet

Hopia

1 mm

spikelet

Zuloagaea

1 cm

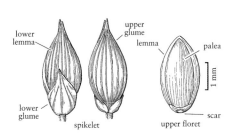

lower
lemma

upper
glume

lemma

palea

1 mm

lower
glume

spikelet

scar

upper floret

Panicum

1 mm

lower
lemma

lower
glume

spikelet

upper
glume

2 cm

Dichanthelium

Group 6

(Spikelets of 2 morphologically different types or reduced florets 1 or 2, below perfect florets; lemma of reduced florets not similar to upper glumes in shape, size, or texture; disarticulation various, but generally below the glumes)[2]

1. Spikelets of 2 morphologically different types.
    2. Spikelets in pairs, pedicels not fused at base; disarticulation above the glumes ....
    ....*Cynosurus* (p. 374)
    2. Spikelets in fascicles, pedicels fused at base; disarticulation at the base of fused pedicels .....
    ....*Lamarckia* (p. 582)
1. Spikelets of a single morphological type.
    3. Lemma of perfect floret with 9 subequal, plumose awns .... *Enneapogon* (p. 474)
    3. Lemma of perfect floret awnless or with a single awn.
        4. Spikelets with 3 well-developed, obvious florets.
            5. Lower glume much shorter than the second; foliage sweet smelling ....
            .... *Anthoxanthum* (p. 207)
            5. Lower glume equaling the second in length; foliage not sweet smelling ....
            .... *Ehrharta* (p. 450)
        4. Spikelets with 1 or 2 well-developed florets, reduced florets inconspicuous and attached below perfect floret.
            6. Well-developed florets 1, lemmas awnless.
                7. Plants annual .... *Phalaris* (p. 786)
                7. Plants perennial .... *Phalaroides* (p. 793)
            6. Well-developed florets 2, first or second lemma awned .... *Holcus* (p. 564)

(Gould and Shaw 1983; Barkworth et al. 2007)

---

2. This is a notoriously confusing and tricky group. The 2 spikelet types in *Cynosurus* and *Lamarckia,* one sterile and one bisexual, are confusing and difficult to distinguish. One of the sterile florets in *Ehrharta* is actually larger than the bisexual one, and the species is often misinterpreted as having 2 perfect florets rather than just 1. The 2 reduced florets below the upper floret in *Phalaris* and *Phalaroides* appear to be pubescent veins, and they usually are not correctly interpreted unless they are "teased" away from the fertile lemmas with a dissecting needle.

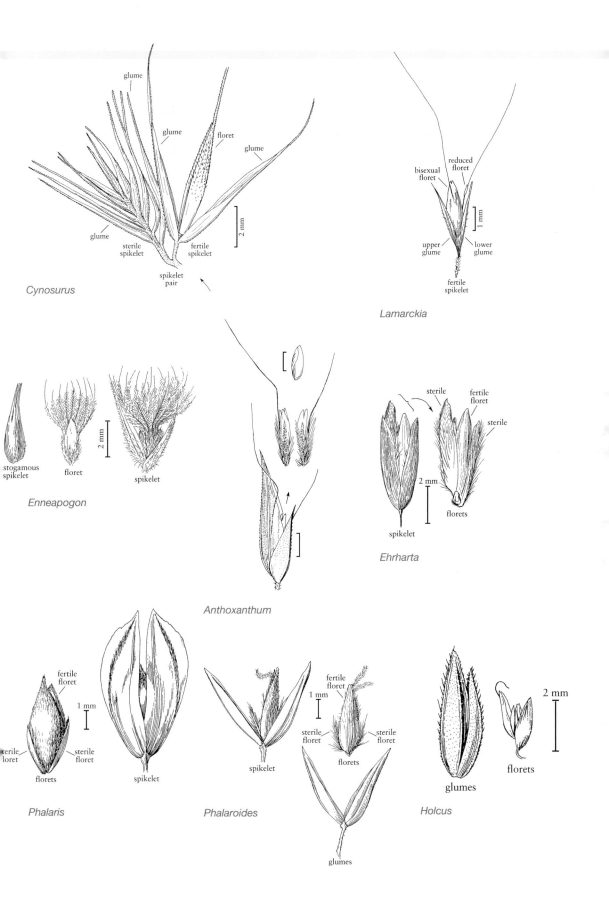

Cynosurus

glume
glume
floret
glume
glume
sterile spikelet
fertile spikelet
spikelet pair
2 mm

Lamarckia

bisexual floret
reduced floret
upper glume
lower glume
fertile spikelet
1 mm

Enneapogon

stogamous spikelet
floret
spikelet
2 mm

Anthoxanthum

Ehrharta

sterile
fertile floret
sterile
spikelet
florets
2 mm

Phalaris

fertile floret
sterile floret
florets
sterile floret
spikelet
1 mm

Phalaroides

fertile floret
sterile floret
sterile floret
florets
spikelet
glumes
1 mm

Holcus

glumes
florets
2 mm

Group 7

(One perfect floret; reduced florets if present above the perfect one; panicle inflorescence, never spicate)

1. Glumes absent or rudimentary.
   2. Lemma with an awn 1–2 cm long .... *Brachyelytrum* (p. 276)
   2. Lemma awnless or with awn <1 cm long.
      3. Spikelets perfect.
         4. Spikelets 7–10 mm long; plants annual, cultivated .... *Oryza* (p. 697)
         4. Spikelets <6 mm long; plants perennial, native .... *Leersia* (p. 584)
      3. Spikelets unisexual.
         5. Staminate and pistillate spikelets in different inflorescences, these inconspicuous; leaves 1–4 cm long .... *Luziola* (p. 609)
         5. Staminate and pistillate spikelets in the same inflorescences, these large and conspicuous; leaves >4 cm long.
            6. Staminate spikelets pendulous on spreading lower branches, pistillate spikelets appressed on the stiffly erect upper branches .... *Zizania* (p. 1017)
            6. Staminate and pistillate spikelets on same branches; pistillate at the tip, staminate below .... *Zizaniopsis* (p. 1019)
1. Glumes, at least the second, well developed.
      7. Glumes and lemmas awnless.
         8. Spikelets 5–8 mm long; lemmas with tuft of hair at base ....
         .... *Calamovilfa* (p. 301)
         8. Spikelets <5 mm long, or if >5 mm long, then lemmas without a tuft of hairs.
            9. Glumes both as long as, or longer than, lemmas ....
            .... *Agrostis* (p. 175)
            9. Glumes, at least the lower, longer than lemmas.
               10. Lemmas 3-veined.
                  11. Veins of lemmas densely pubescent ....
                  .... *Blepharoneuron* (p. 251)
                  11. Veins of lemmas glabrous or scabrous ....
                  .... *Muhlenbergia* (p. 635)
               10. Lemmas 1-veined .... *Sporobolus* (p. 918)
      7. Other couplet on page 102.

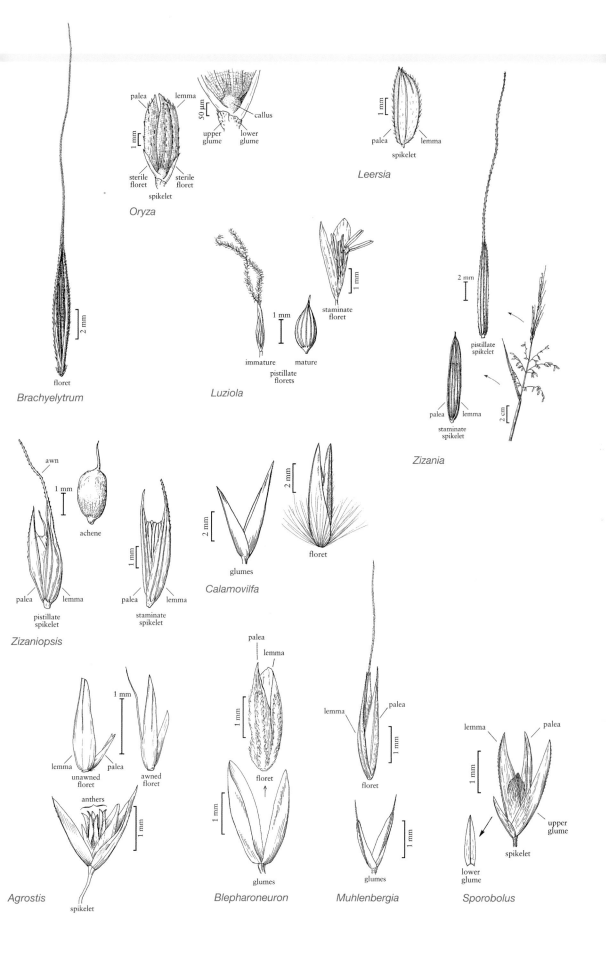

palea
lemma
50 µm
callus
upper glume
lower glume
1 mm
sterile floret
sterile floret
spikelet

Oryza

1 mm
palea
lemma
spikelet

Leersia

2 mm

floret

Brachyelytrum

1 mm
immature
mature
pistillate florets
staminate floret

Luziola

2 mm
pistillate spikelet
palea
lemma
staminate spikelet
2 cm

Zizania

awn
1 mm
achene
palea
lemma
pistillate spikelet
palea
lemma
staminate spikelet

Zizaniopsis

2 mm
2 mm
glumes
floret

Calamovilfa

1 mm
lemma
palea
unawned floret
awned floret
anthers
1 mm
spikelet

Agrostis

palea
lemma
1 mm
floret
1 mm
glumes

Blepharoneuron

lemma
palea
1 mm
floret
1 mm
glumes

Muhlenbergia

lemma
palea
1 mm
upper glume
spikelet
lower glume

Sporobolus

7. Glumes or lemmas awned.
    12. Lower glumes usually 2- or 3-awned, upper glumes usually 1-awned; spikelets in pairs, the lower of the pair sterile, the pair falling together...................................... *Lycurus* (p. 612)
    12. Lower and upper glumes not as above, or spikelets not falling in pairs.
        13. Disarticulation below glumes.
            14. Spikelets borne on stipes that fall with the spikelets .... *Polypogon* (p. 831)
            14. Spikelets borne on pedicels, disarticulation immediately below the glumes.
                15. Lemmas awned from the middle; glume bases often fused .... *Alopecurus* (p. 187)
                15. Lemmas awnless or awned from or near the apex; glume bases usually not fused.
                    16. Glumes usually awned; lemmas usually awnless . ... *Phleum* (p. 797)
                    16. Glumes usually awnless; lemmas awned.
                        17. Lemma awns up to 2.5 mm long; strong perennial.. *Cinna* (p. 348)
                        17. Lemma awns 5–15 mm long or longer; short-lived annual........................ *Limnodea* (p. 602)
        13. Disarticulation above glumes.
            18. Lemmas indurate, awned, with well-developed callus, usually enclosing paleas and caryopses ....A
            18. Lemmas not indurate or enclosing paleas and caryopses .... AA

A (Lemmas indurate, awned, with well-developed callus, usually enclosing paleas and caryopses) (Barkworth et al. 2007)

1. Awn of lemmas 3-branched, lateral branches rarely short or rudimentary............................... ....*Aristida* (p. 212)
1. Awn of lemmas not branched.
    2. Paleas sulcate, longer than lemmas; lemma margins involute, fitting into the paleal groove; lemma apices not lobed ....*Piptochaetium* (p. 806)
    2. Paleas flat, shorter or longer than lemmas; lemma margins convolute or not overlapping; lemma apices often lobed or bifid.
        3. Apices of leaf blades sharp and stiff; caryopses often with 3 ribs at maturity; cleistogenes usually present...*Amelichloa* (p. 190)
        3. Other couplet on page 104

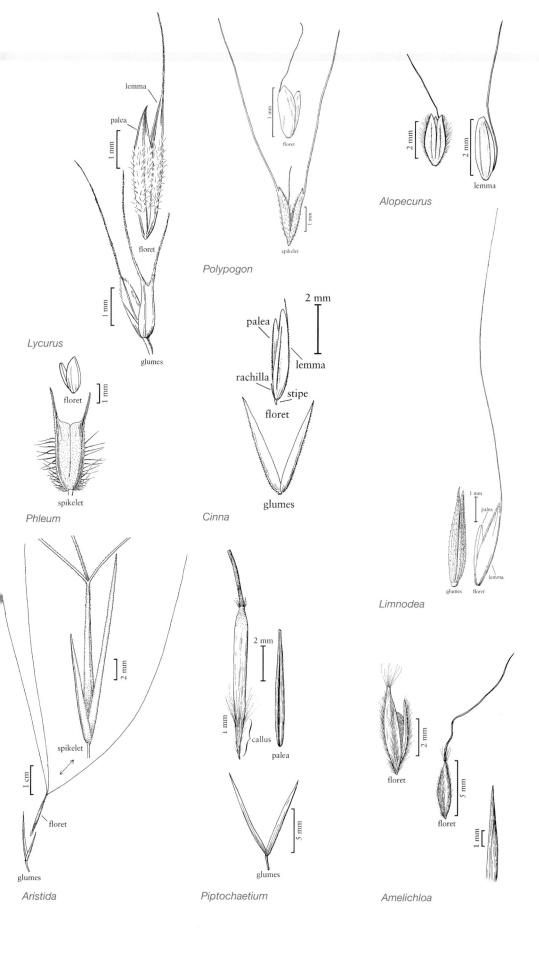

lemma

palea

1 mm

floret

Lycurus

1 mm

floret

spikelet

Phleum

glumes

1 mm

floret

spikelet

Polypogon

palea

2 mm

lemma

rachilla

stipe

floret

glumes

Cinna

2 mm

2 mm

lemma

Alopecurus

1 mm

palea

glumes

floret

lemma

Limnodea

2 mm

1 mm

spikelet

1 cm

floret

glumes

Aristida

2 mm

1 mm

callus

palea

5 mm

glumes

Piptochaetium

2 mm

floret

5 mm

floret

1 mm

Amelichloa

3. Apices of leaf blades acute or acuminate, never both sharp and stiff; caryopses without ribs; cleistogenes occasionally present.

    4. Lemma margins strongly overlapping their entire length at maturity, bodies usually rough, apices not lobed; paleas ¼ to ½ length of lemmas, veinless, glabrous .... *Nassella* (p. 691)

    4. Lemma margins usually not or only slightly overlapping, sometimes strongly overlapping in species with smooth lemmas, bodies usually smooth on lower portion; paleas from ⅓ to equaling or exceeding length of lemmas, 2-veined at least on lower portion, usually hairs on lemmas.

        5. Calluses 1.5–6.0 mm long, sharply pointed; lemma awns 65–500 mm long............. ....*Hesperostipa* (p. 555)

        5. Calluses <2 mm long, blunt or sharp-pointed; lemma awns shorter.

            6. Florets usually dorsally compressed or terete; paleas as long as or longer than lemmas and similar in texture and pubescence; lemma margins separate their entire length at maturity .... *Piptatherum* (p. 806)

            6. Florets usually laterally compressed; paleas usually shorter than lemmas; lemma margins often overlapping for part or all of their length at maturity .... *Achnatherum* (p. 161)

AA *(Lemmas not indurate or enclosing paleas and caryopses)*

1. Glumes equal, broad, abruptly short-awned from an obtuse apex; lemmas much shorter than glumes, awnless..*Phleum* (p. 797)

1. Glumes not equal, or if so, not abruptly awned.

    2. Upper glumes 4–5 times as long as lemmas; annual with densely contracted, spikelike panicles.. *Gastridium* (p. 539)

    2. Upper glumes shorter or only slightly longer than lemmas.

        3. Lemmas awned from an entire or minutely clefted apex; glumes, at least the lower, usually shorter than lemmas...*Muhlenbergia* (p. 635)

        3. Lemmas awned from the back, base, or distinctly cleft apex; glumes equaling or exceeding length of lemmas.

            4. Lemmas firm, awned from near the apex, awn straight or flexuous, 3–4 times as long as the lemma; annual ....*Apera* (p. 210)

            4. Lemmas thin, awned from the back or base, awn usually geniculate when long; annual and perennial.

                5. Rachillas not prolonged beyond base of floret; paleas absent, minute or subequal to lemmas ....*Agrostis* (p. 175)

                5. Rachillas prolonged beyond base of floret; paleas at least ½ as long as lemmas ....*Lachnagrostis* (p. 580)

Nassella

awn
crown
1 cm
floret
1 cm
glumes
2 mm
lemma

Hesperostipa

palea
1 cm
floret
5 mm
lemma
callus

Piptatherum

awn
floret
2 mm
2 mm
glumes

Achnatherum

2 mm
floret
1 cm
glumes
5 mm
floret

Phleum

floret
1 mm
spikelet

Gastridium

1 mm
lemma
upper glume
lower glume
spikelet

Muhlenbergia

lemma
palea
1 mm
floret
1 mm
glumes

Apera

1 mm
floret
glumes

Agrostis

1 mm
lemma
palea
unawned floret
awned floret
anthers
1 mm
spikelet

Lachnagrostis

1 mm
palea
lemma
rachilla
floret
1 mm
spikelet

Group 8

(Inflorescence of 1 to several unilateral spicate primary branches)

1. Glumes with hooked spines..*Tragus* (p. 957)
1. Glumes without hooked spines.
   2. Spikelets with 2 or more perfect florets.
      3. Inflorescence branches paired, verticillate, or clustered at the culm apex.
         4. Glumes and lemmas awnless..*Eleusine* (p. 452)
         4. Glumes or lemmas awned.
            5. Upper glumes short-awned or mucronate; rachis projecting stiffly beyond terminal spikelet.. *Dactyloctenium* (p. 378)
            5. Upper glumes not awned or mucronate; rachis not extending beyond terminal spikelet.
               6. Lower lemma with 3 awns.. *Trichloris* (p. 959)
               6. Lower lemma with a single awn ....*Chloris* (p. 321)
      3. Inflorescence branches distributed along culm axis, seldom more than 1 at each rachis node.
            7. Glumes >1 cm long, much longer than lower florets; lemmas long, ciliate on margins .... *Trichoneura* (p. 962)
            7. Glumes <1 cm long, 1 or both much shorter than lowermost florets; lemmas glabrous or puberulent on margins
               8. Lemmas 3-veined.
                  9. Lemmas glabrous, acute, and awnless at apex; spikelets widely spaced and not overlapping, on stiffly spreading branches..............................
                          ....*Eragrostis* (p.478)
                  9. Lemmas glabrous or puberulent on veins or at base, usually awned or mucronate; spikelets closely spaced and overlapping when lemmas glabrous or unawned ....*Leptochloa*[3] (p. 590)
               8. Other couplet on page 108.
   2. Other couplet on page 108.

---

3. See also *Tridens ambiguus* or *T. buckleyanus*

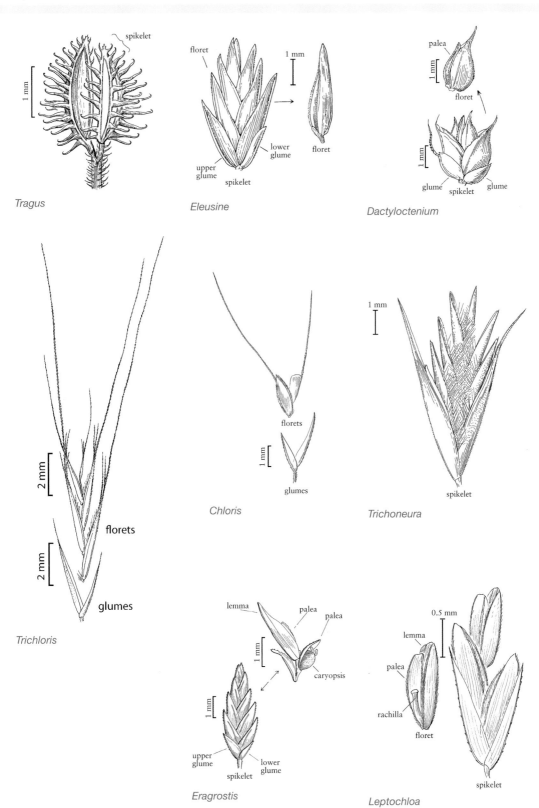

Tragus

Eleusine

Dactyloctenium

Trichloris

Chloris

Trichoneura

Eragrostis

Leptochloa

8. Lemmas 5- or more-veined.

    10. Plants perennial, with tall culms and slender, flexuous inflorescence branches..*Glyceria* (p. 541)

    10. Plants annual, with short, tufted culms and short, stiff inflorescence branches.

        11. Spikelets 3-flowered, disarticulation below glumes .... *Sclerochloa* (p. 874)

        11. Spikelets >3-flowered; disarticulation above glumes ....*Desmazeria* (p. 385)

2. Spikelets with 1 perfect floret.

      12. Spikelets on main axis as well as on branches.

        13. Glumes absent..*Leersia* (p. 584)

        13. Glumes, at least the upper, present.

          14. Lemmas awned .... *Gymnopogon* (p. 546)

          14. Lemmas awnless.

            15. Glumes stiff, strongly 1-veined ....*Schedonnardus* (p. 859)

            15. Glumes soft, veinless .... *Willkommia* (p. 1012)

      12. Spikelets all on branches and none on the inflorescence axis, the latter sometimes terminating in a single branch.

          16. Inflorescence branches 2 or more, digitate, clustered or in 2 or 3 verticels at culm apex.

            17. Reduced floret absent or a minute scale; spikelets awnless ....*Cynodon* (p. 368)

            17. Reduced floret or florets present above the perfect one; spikelets usually awned.

              18. Lowest lemma 3-awned ....*Trichloris* (p. 959)

            18. Other couplet on page 110.

          16. Other couplet on page 110.

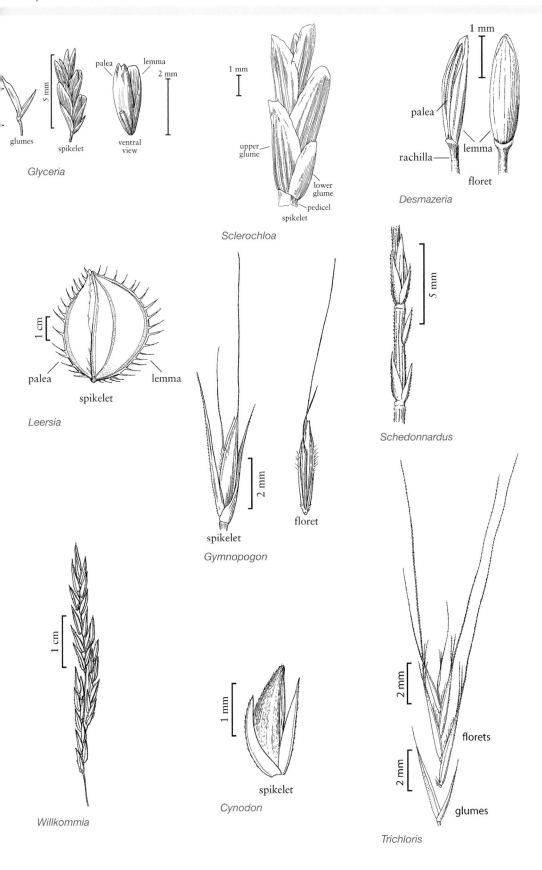

Glyceria

glumes
spikelet
palea — lemma
2 mm
5 mm
ventral view

Sclerochloa

upper glume
lower glume
pedicel
spikelet
1 mm

Desmazeria

1 mm
palea
rachilla —
lemma
floret

Leersia

1 cm
palea
spikelet
lemma

Schedonnardus

5 mm

Gymnopogon

spikelet
floret
2 mm

Willkommia

1 cm

Cynodon

1 mm
spikelet

Trichloris

2 mm
florets
2 mm
glumes

18. Lowest lemma with a single awn.
  19. Spikelets dorsally compressed ....
  ....*Enteropogon* (p. 476)
  19. Spikelets laterally compressed or terete.
    20. Upper glumes truncate or bilobed ....
    ....*Eustachys* (p. 527)
    20. Upper glumes acute or acuminate ....
    .... *Chloris* (p. 320)
16. Inflorescence 1 to several branches, not digitate, clustered or in verticels.
    21. Inflorescence with a single, stout, curved, unilateral branch; upper glume with a short, stout, dorsal awn ....
    .... *Ctenium* (p. 366)
    21. Inflorescence with 1 to several branches; upper glume awnless.
    22. Spikelets without reduced florets ....
    ....*Spartina* (p. 904)
    22. Spikelets with 1 or more staminate or reduced florets above the perfect ones.
      23. Spikelets in deciduous clusters of 3, middle spikelet perfect, lower 2 staminate.................. *Cathestecum* (p. 303)
      23. Spikelet not in clusters of 3, or if so, the lower 2 not staminate or sterile.
        24. Panicle branches deciduous, disarticulation occurring at their bases; spikelets usually 1–15 per branch, usually appressed rather than pectinate.............................
        ....*Bouteloua* (p. 268)
        24. Panicle branches persistent; disarticulation above the glumes, spikelets 6–130 per branch, pectinate ...
        ....*Chondrosum* (p. 334)

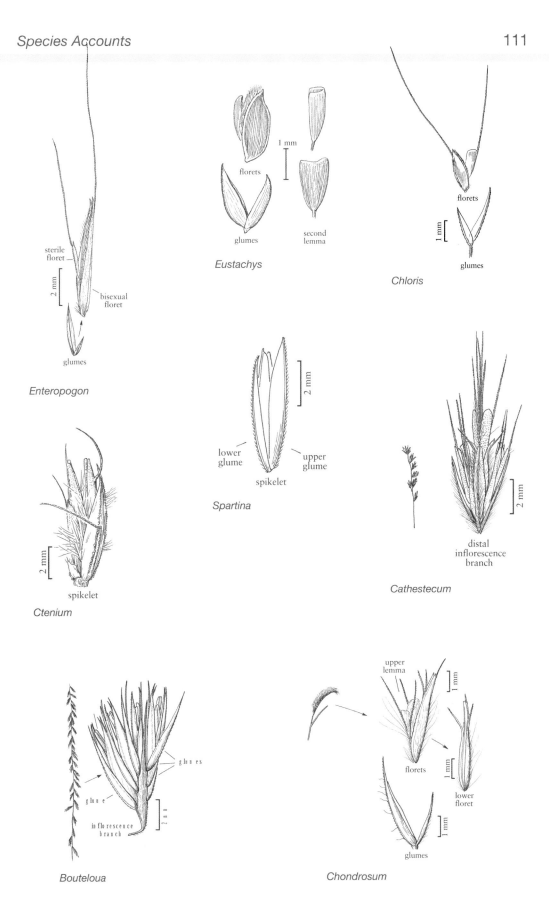

florets

glumes

second
lemma

1 mm

Eustachys

florets

1 mm

glumes

Chloris

sterile
floret

bisexual
floret

2 mm

glumes

Enteropogon

lower
glume

upper
glume

spikelet

2 mm

Spartina

distal
inflorescence
branch

2 mm

Cathestecum

spikelet

2 mm

Ctenium

upper
lemma

1 mm

florets

lower
floret

1 mm

glumes

1 mm

Chondrosum

glumes

glume

inflorescence
branch

2 mm

Bouteloua

Group 9

(Inflorescence a terminal, bilateral spike or spicate raceme)[4]

1. Spikelets in capitate clusters, these subsessile in leafy portion of plant; lemmas 3-veined; lower tufted or sod-forming grasses.
   2. Disarticulation below glumes, spikelets falling in burlike clusters; plants strongly stoloniferous...*Buchloë* (p. 299)
   2. Disarticulation above glumes, spikelets not falling in clusters.
      3. Lemmas with 3 stout, ciliate awns...*Blepharidachne* (p. 251)
      3. Lemmas with a single awn.
         4. Glumes much longer than lemmas; lemmas deeply bifid at apex..*Dasyochloa* (p. 383)
         4. Glumes shorter than lemmas; lemmas acuminate at apex, not bifid ....*Munroa* (p. 689)
1. Spikelets not in capitate clusters, or if so, these elevated above basal clump of leaves.
      5. Spikelets with 1 floret.
         6. Spikelets single at each node.
            7. Plants annual.
               8. Glumes both present on all spikelets............................ *Parapholis* (p. 729)
               8. Glumes both present only on terminal spikelet, lower glume lacking on all other spikelets...*Hainardia* (p. 551)
            7. Plants perennial.
               9. Margins of lemmas ciliate; inflorescence a slender, curved, unilateral spike..*Microchloa* (p. 625)
               9. Margins of lemmas glabrous.
                  10. Spikelets partially embedded in a thick, flattened rachis; plants with stout stolons ....*Stenotaphrum*[5] (p. 946)
                  10. Spikelets sessile or short-pediceled on a slender rachis ....*Zoysia* (p. 1021)
         6. Spikelets 3 at each node.
            11. Lateral spikelets pediceled and sterile ....*Critesion* (p. 360)
            11. Lateral spikelets sessile and perfect................ *Hordeum* (p. 568)
      5. Other couplet on page 114.

---

4. Misinterpretation of panicles with disarticulating branches (e.g., *Chondrosum, Hilaria, Pleuraphis*) or greatly reduced panicles (e.g., *Phleum, Alopecurus*) could erroneously lead one to this group.

5. *Stenotaphrum* is a member of the Paniceae, and the spikelet characters would key it out in Group 5. It is included here because the inflorescence is easily interpreted as a spike, and one might immediately come to Group 9 without close examination of the spikelet. If the specimen does not match any of the genera within this section, perhaps look at genera within the Andropogoneae (Group 4) that have spikelets embedded in a thickened rachis (e.g., *Hackelochloa, Rottboellia, Eremochloa, Coelorachis*).

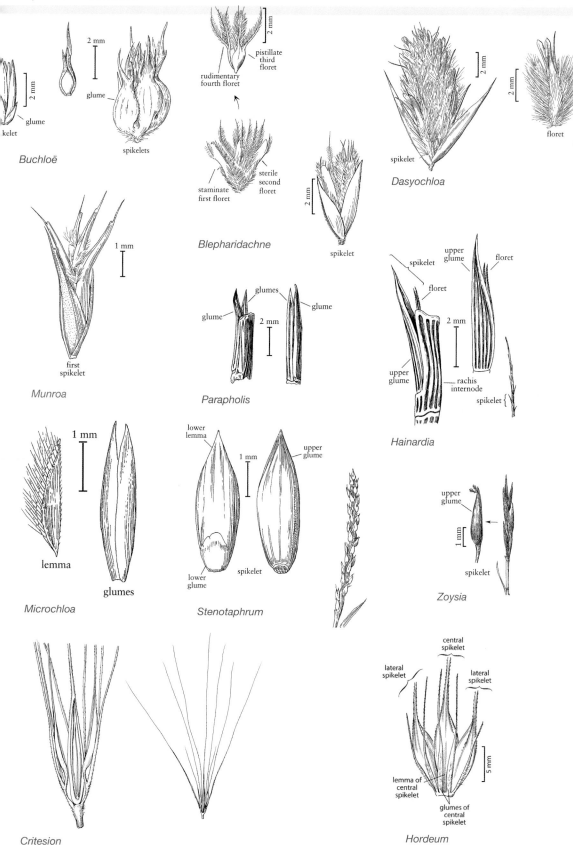

Buchloë

glume

kelet

glume

spikelets

rudimentary
fourth floret

pistillate
third
floret

staminate
first floret

sterile
second
floret

Blepharidachne

spikelet

Dasyochloa

spikelet

floret

Munroa

first
spikelet

glumes

glume

glume

glume

Parapholis

upper
glume

floret

spikelet

floret

upper
glume

rachis
internode

spikelet

Hainardia

Microchloa

lemma

glumes

Stenotaphrum

lower
lemma

upper
glume

lower
glume

spikelet

upper
glume

spikelet

Zoysia

Critesion

central
spikelet

lateral
spikelet

lateral
spikelet

lemma of
central
spikelet

glumes of
central
spikelet

Hordeum

5. Spikelets with 2 or more florets.

    12. Spikelets oriented edgewise to rachis, lower glumes absent except on terminal spikelets ....*Lolium* (p. 604)

    12. Spikelets not oriented edgewise to rachis; lower glumes present on all spikelets.

        13. Upper glume bearing a stout, divergent awn on the back .... *Ctenium* (p. 366)

        13. Upper glume not bearing a dorsal awn.

            14. Annuals or biennials; introduced, cultivated cereal or weed.

                15. Glumes ovate with 3 or more veins at midlength.

                    16. Spikelet sunken or embedded in rachis; spike narrow, <5 mm wide; rachis disarticulating at maturity....................................*Aegilops* (p. 171)

                    16. Spikelets not sunken into rachis; spike thick, >5 mm wide; rachis not disarticulating at maturity .... *Triticum* (p. 985)

                15. Glumes subulate to lanceolate with only 1 vein at midlength..*Secale* (p. 878)

            14. Perennial; native or introduced.

                17. Spikelets solitary at all, or almost all, rachis nodes A

                17. Spikelets 2–7 at all, or most, rachis nodes............ AA

A (Spikelets solitary at all, or almost all, rachis nodes)

1. Lemmas thin, awnless, distinctly 3-veined.

    2. Spikelets 1.5–3.0 cm long, staminate; stoloniferous perennial .... *Scleropogon* (p. 876)

    2. Spikelets <1 cm long.

        3. Lemmas awnless; spikelets unisexual, stoloniferous annual ....*Eragrostis reptans* (p. 504)

        3. Lemmas with delicate awn; spikelets perfect; cespitose perennial ....*Tripogon* (p. 976)

1. Lemmas thick or less frequently thin, 5- to several-veined; veins faint or distinct.

    4. Spikelets short-pediceled; thus, inflorescence a spicate raceme.

        5. Lemmas awnless.

        5. Other couplet on page 116.

    4. Other couplet on page 116.

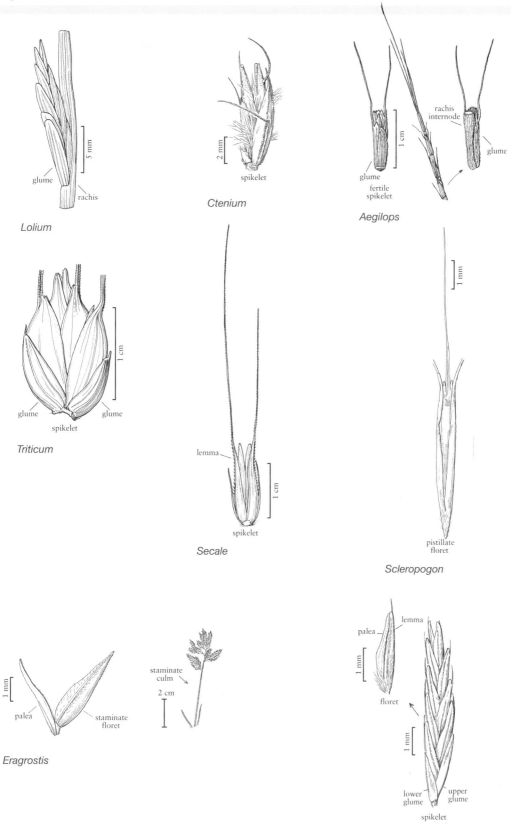

Lolium

Ctenium

Aegilops

Triticum

Secale

Scleropogon

Eragrostis

Tripogon

6. Disarticulation below the glumes; upper leaf sheaths enlarged and partially enclosing inflorescence..*Sclerochloa* (p. 874)

6. Disarticulation above the glumes; upper leaf sheaths not enlarged ....
....*Desmazeria* (p. 385)

5. Lemmas awned.

7. Glumes much longer than lemmas; lemmas with geniculate awn from a bifid apex .... *Danthonia* (p. 380)

7. Glumes not longer than lemmas; lemmas with a straight awn form an entire apex .... *Trachynia* (p. 953)

4. Spikelets sessile; thus, inflorescence a spike.

8. Rachis internodes <5 mm long; spikelets closely imbricate, more than 3 times the length of rachis internodes, diverging..
....*Agropyron* (p. 173)

8. Rachis internodes >5 mm long; spikelets appressed or ascending, 1–3 times the length of rachis internodes.

9. Glumes subulate to narrowly lanceolate, stiff, 1- or rarely 3-veined at midlength.

10. Glumes lanceolate, tapering to an acuminate tip, slightly curved to one side at apex... *Pascopyrum* (p. 731)

10. Glumes subulate from near the base, straight ....
....*Leymus* (p. 599)

9. Glumes usually ovate, rectangular, or lanceolate, narrowing above midrib; (1–)3–5(–7)-veined at midlength.

11. Glumes stiff, truncate, obtuse, or acute, unawned; glume keels smooth below, scabrous above ....*Thinopyrum* (p. 950)

11. Glumes flexible, acute to acuminate, sometimes awn-tipped; keels usually uniformly smooth or scabrous.

12. Spikelets distant, scarcely reaching base of spikelet above on same side of rachis ....*Pseudoroegneria* (p. 839)

12. Spikelets closely spaced, usually reaching at least the midlength of spikelet above on same side of rachis ....
....*Elymus* (p. 458)

AA (Spikelets 2–7 at all, or most, rachis nodes)

1. Glumes 4–18 mm long, subulate to narrowly lanceolate, 0- to 1-veined at midlength; blades usually with closely spaced, equally prominent veins on adaxial surface.

2. Disarticulation in the spikelet, beneath each floret; sometimes cespitose, often rhizomatous; ligules 0.3–8.0 mm long..*Leymus* (p. 599)

2. Disarticulation tardy, in rachises; plants cespitose, not rhizomatous; ligules 0.2–0.3 mm long.. *Psathyrostachys* (p. 837)

1. Glumes flat and with 3 veins at midlength or, if subulate and 1-veined, <4 mm long or >18 mm long; blades widely spaced, unequally prominent veins on adaxial surface ....
....*Elymus* (p. 458)

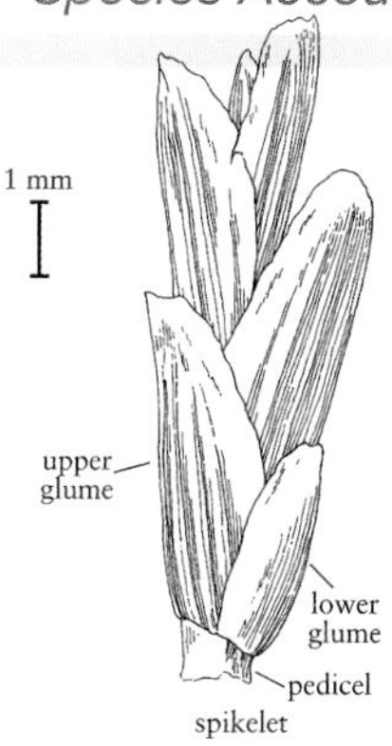

Sclerochloa

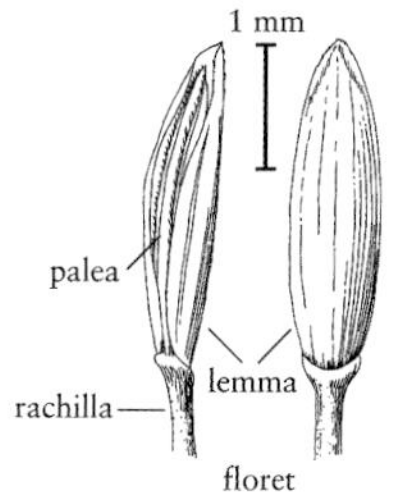

Desmazeria

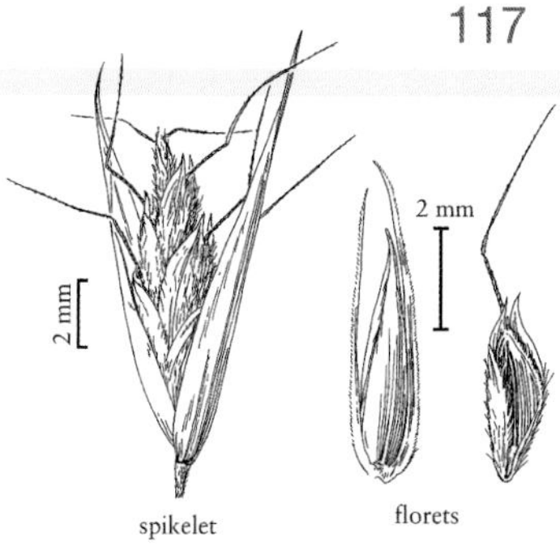

Danthonia

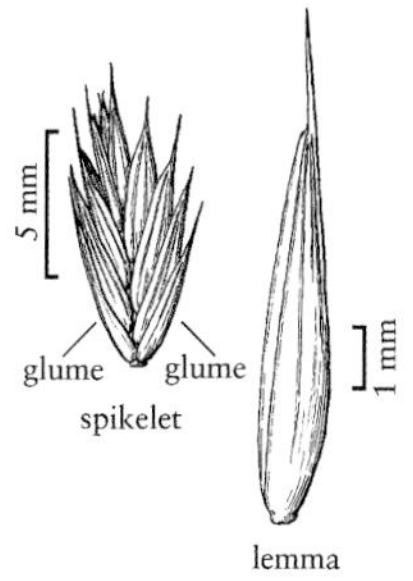

Trachynia

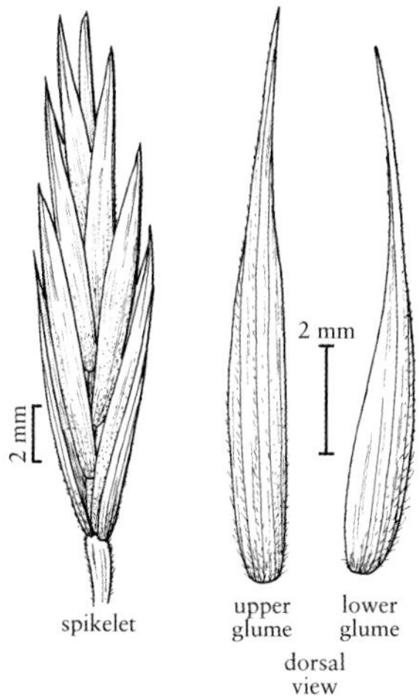

Agropyron

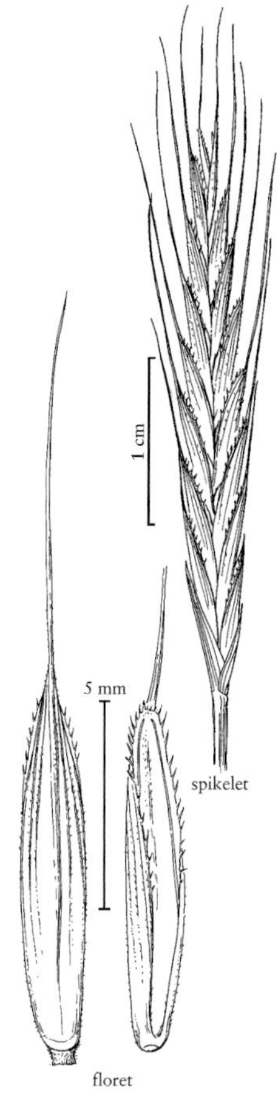

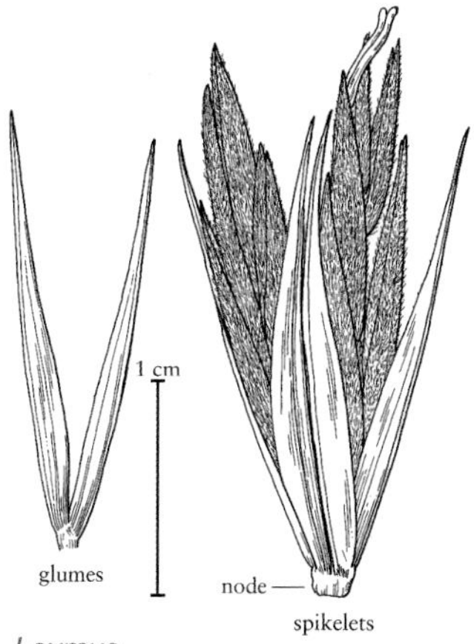

Leymus

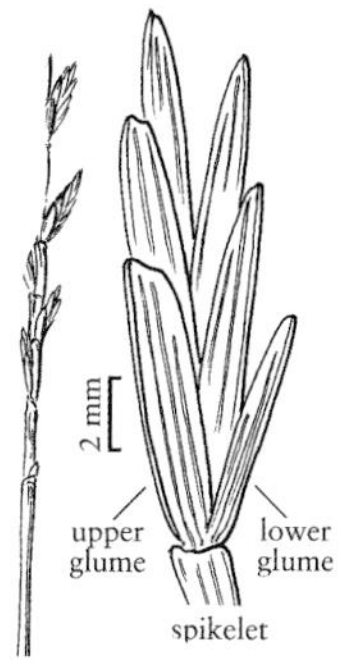

Thinopyrum

Pascopyrum

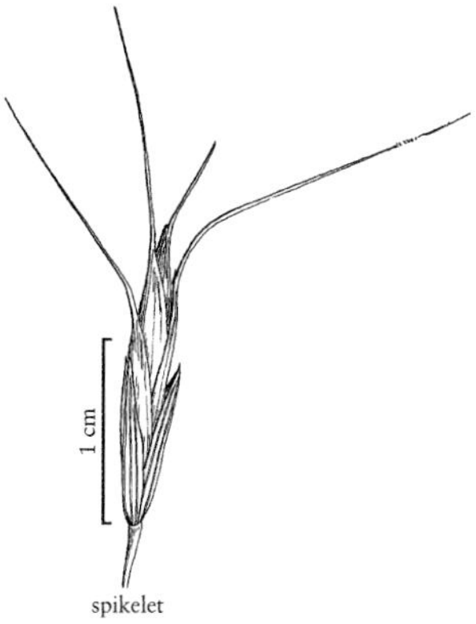

Pseudoroegneria

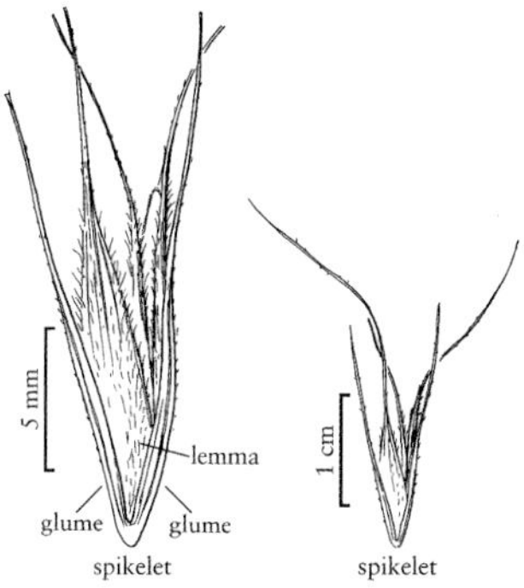

Elymus

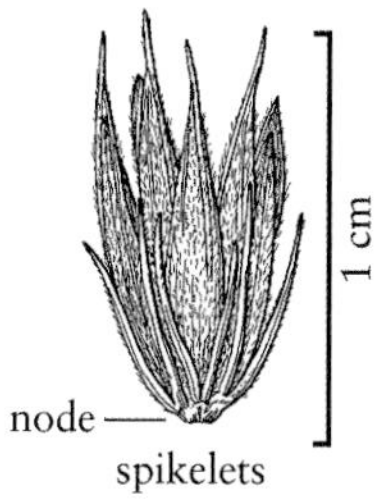

Psathyrostachys

Group 10

(2 or more fertile florets per spikelet; inflorescence an open or contracted panicle or raceme with spikelets on well-developed pedicels)

1. Plants 2–6 m tall.
    2. Spikelets mostly 3–7 cm long with 7–13 florets....................................*Arundinaria* (p. 236)
    2. Spikelets <2 cm long and with fewer than 7 florets.
        3. Leaves mostly basal, forming large, distinct clumps; without creeping rhizomes; blades 0.5–1.5 cm broad; ornamental, rarely escaping.................................*Cortaderia* (p. 355)
        3. Leaves evenly distributed; culms with creeping rhizomes forming large colonies; blades 2–6 cm broad.
            4. Lemmas villous; rachilla glabrous ....*Arundo* (p. 239)
            4. Lemmas glabrous, rachilla villous..*Phragmites* (p. 799)
1. Plants <2 m tall.
        5. Lemmas with 3, usually conspicuous, veins..A
        5. Lemmas with 5–15, conspicuous or obscure, veins... AA

A (Lemmas with 3, usually conspicuous, veins)

1. Lemma veins pubescent or puberulent, or base of lemma long-hairy.
    2. Plants rhizomatous; lemma veins glabrous, with tuft of hair at base ....
    ....*Redfieldia* (p. 841)
    2. Plants not rhizomatous; lemmas pubescent or puberulent on veins, at least below.
        3. Paleas densely long-ciliate on upper half...*Triplasis* (p. 980)
        3. Paleas not densely long-ciliate.
            4. Lemma veins each extending into an awn..*Triraphis* (p. 980)
            4. Lemma veins not extending into awns.
                5. Lemma veins conspicuously long-hairy, at least below ....
                .... *Erioneuron* (p. 523)
                5. Lemma veins inconspicuously puberulent...................................... *Tridens* (p. 964)
1. Lemma veins not pubescent, base not long-hairy.
        6. Lemmas 3-awned.
            7. Lemma awns 4–10 cm long .... *Scleropogon* (p. 876)
            7. Lemma awns <1 cm long... *Blepharidachne* (p. 251)
        6. Lemma awnless.

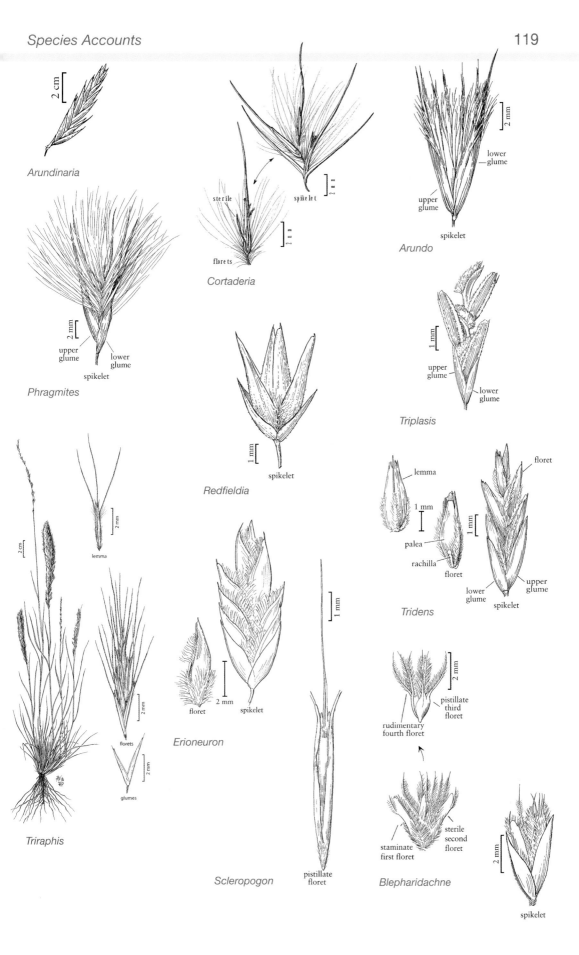

Arundinaria

Cortaderia

sterile

florets

spikelet

Arundo

lower glume

upper glume

spikelet

Phragmites

2 mm

upper glume

lower glume

spikelet

Triplasis

upper glume

lower glume

Redfieldia

spikelet

Tridens

lemma

palea

rachilla

floret

floret

lower glume

upper glume

spikelet

Triraphis

lemma

florets

glumes

Erioneuron

floret

spikelet

Scleropogon

pistillate floret

Blepharidachne

pistillate third floret

rudimentary fourth floret

staminate first floret

sterile second floret

spikelet

8. Caryopses large, turgid, and beaked..............................*Diarrhena* (p. 387)
8. Caryopses not large, turgid, and beaked.......................*Eragrostis* (p. 478)

AA (Lemmas with 5–15, conspicuous or obscure veins)

1. Lemmas awned.
   2. Lemmas with more than 5 awns.
      3. Glumes 1-veined, florets falling together.....................*Pappophorum* (p. 726)
      3. Glumes 5- to many-veined, florets falling separately.................*Cottea* (p. 358)
   2. Lemmas with a single awn.
      4. Culms woody, perennial; spikelets 3–7 cm long ....*Arundinaria* (p. 236)
      4. Culms not woody or perennial; spikelets rarely as long as 3 cm.
         5. Glumes up to 2 cm long; lemmas 1.5 cm long or longer; introduced annual.............
         ....*Avena* (p. 241)
        5. Glumes <2 cm long, or if longer, lemmas <1.5 cm long.
         6. Lemmas awned from about halfway down back.............................*Aira* (p. 183)
         6. Lemmas awned from a bifid or entire apex.
            7. Lower glumes longer than lowermost floret; lemmas awned from a bifid apex.
               8. Awns 5–15 mm long or longer; usually flattened at base...........................
               .... *Danthonia* (p. 380)
               8. Awns up to 2 mm long; not flattened at base ....
               .... *Schismus* (p. 864)
            7. Lower glumes about as long as or shorter than lowermost florets.
               9. Spikelets of 2 kinds, fertile and sterile; fertile spikelets sessile and nearly covered by sterile spikelets in a spicate or subcapitate panicle....
               .... *Cynosurus* (p. 374)
            9. Spikelets not as above.
               10. Paleas adherent to caryopses; lemmas usually awned from a bifid or bilobed apex; veins converging distally; ovary apices hairy........B
               10. Paleas not adherent to caryopses; lemmas not both bifid or bilobed and with convergent veins; ovary apices usually glabrous............BB
1. Other couplet on page 124.

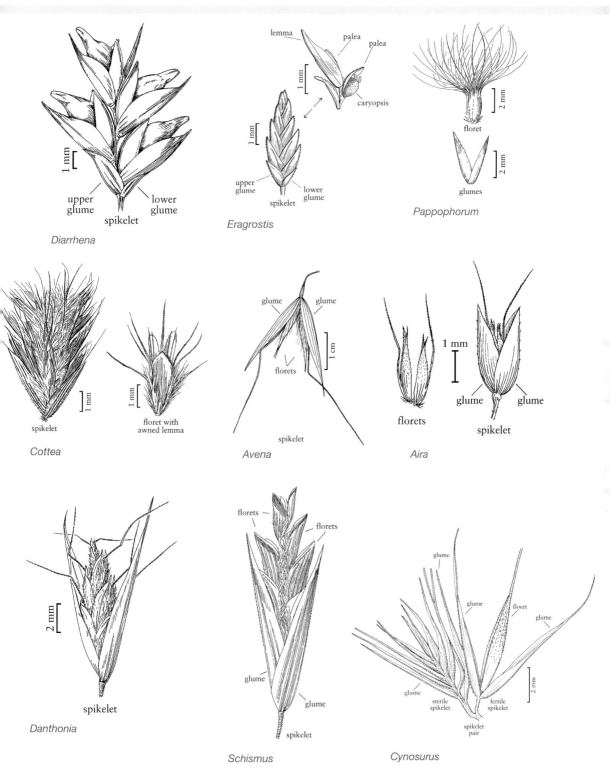

Diarrhena

upper glume
lower glume
spikelet

Eragrostis

lemma
palea
palea
caryopsis
1 mm
1 mm
upper glume
lower glume
spikelet

Pappophorum

2 mm
floret
2 mm
glumes

Cottea

spikelet
floret with awned lemma
1 mm
1 mm

Avena

glume
glume
florets
1 cm
spikelet

Aira

1 mm
florets
glume
glume
spikelet

Danthonia

2 mm
spikelet

Schismus

florets
florets
glume
glume
spikelet

Cynosurus

glume
glume
floret
glume
glume
sterile spikelet
fertile spikelet
spikelet pair
2 mm

B (Paleas adherent to caryopses)

1. Lemmas strongly keeled; spikelets strongly laterally compressed..
....*Ceratochloa* (p. 311)
1. Lemmas rounded over midrib; spikelets terete to elliptical.
    2. Awn arising <1.5 mm below lemma apices.
        3. Lower glumes 1–3-veined; upper glumes 3–5-veined; perennial, or if annual, lower
            glumes 1-veined and upper glumes 3-veined ....*Bromopsis* (p. 282)
        3. Lower glumes 3–5-veined; upper glumes 5–9-veined; annual and biennial ....
            .... *Bromus* (p. 291)
    2. Awn arising >1.5 mm or more below the lemma apices.
        4. Lemma apices bifid, teeth 1–5 mm long.. *Anisantha* (p. 199)
        4. Lemma apices entire or bifid, teeth <1 mm long............................... *Bromus* (p. 291)

BB (Palea not adherent to caryopsis)

1. Lower glumes with 3–5 distinct veins; glumes and lemmas rounded on back.
    2. Lower glumes 1-veined; disarticulation always above the glumes; lemmas never with hairs
       >1 mm long; culms never with bulbous bases; upper florets sometimes reduced but not
       forming a morphologically distinct rudiment; plants of wet habitats ....
       ....*Glyceria* (p. 451)
    2. Lower glumes 1–9-veined; disarticulation above and below glumes; sometimes with hairs
       >1 mm long; upper florets often reduced to form a morphologically distinct rudiment;
       culms sometimes with bulbous base; plants of drier, well-drained habitats .. *Melica* (p. 617)
1. Lower glumes with 1–3 distinct or indistinct veins; glumes and lemmas keeled or rounded on
  back.
    3. Paleas colorless.
        4. Upper glumes obovate, broadest above midlength; disarticulation below glumes ....
        ....*Sphenopholis* (p. 912)
        4. Upper glumes broadest below middle; disarticulation above glumes.
            5. Plants perennial; palea veins not extending into short awns...................................
            .... *Koeleria* (p. 578)
            5. Plants annual; palea veins extending into short awns..
            ....*Rostraria* (p. 843)
    3. Paleas green or brown, at least on veins.

1 cm
lemma

1 cm
spikelet

Ceratochloa

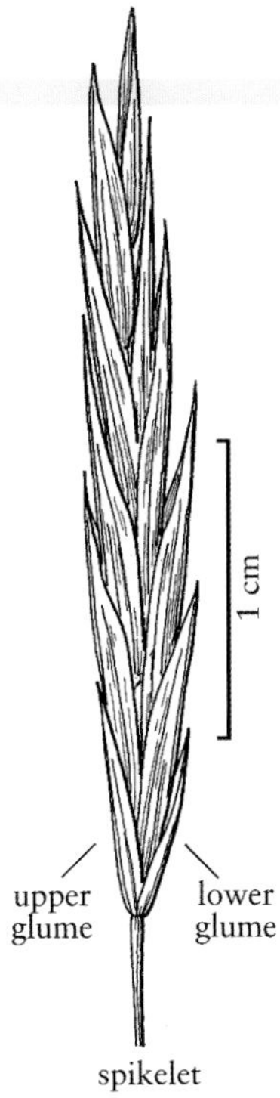

1 cm
upper glume
lower glume
spikelet

Bromopsis

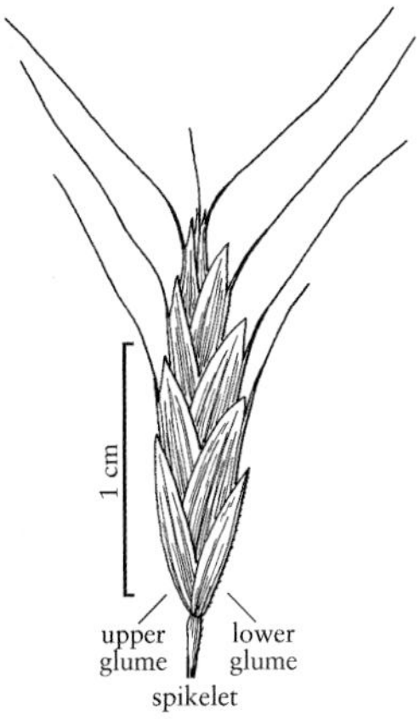

1 cm
upper glume
lower glume
spikelet

Bromus

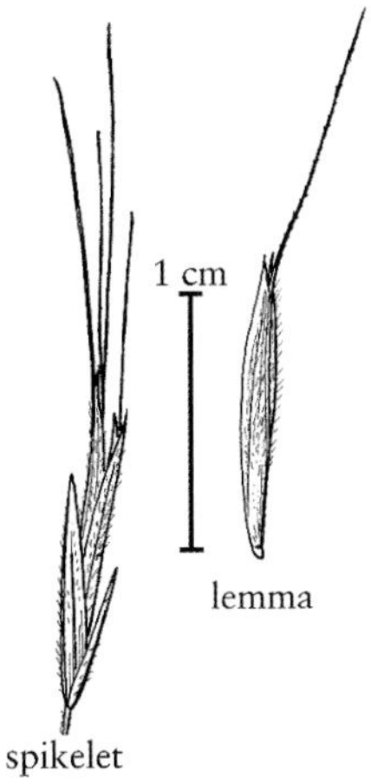

1 cm
spikelet
lemma

Anisantha

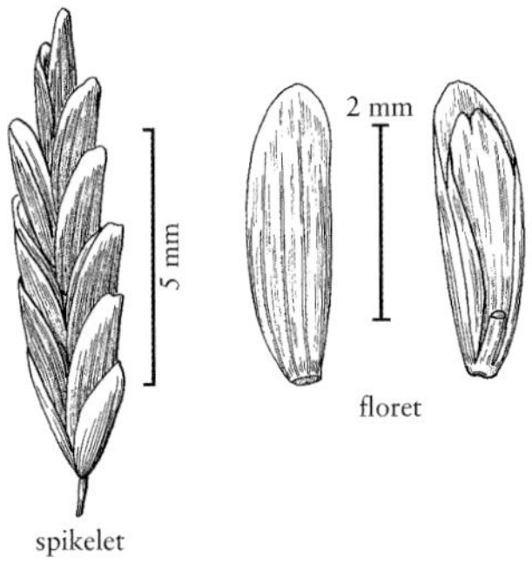

5 mm
spikelet
2 mm
floret

Glyceria

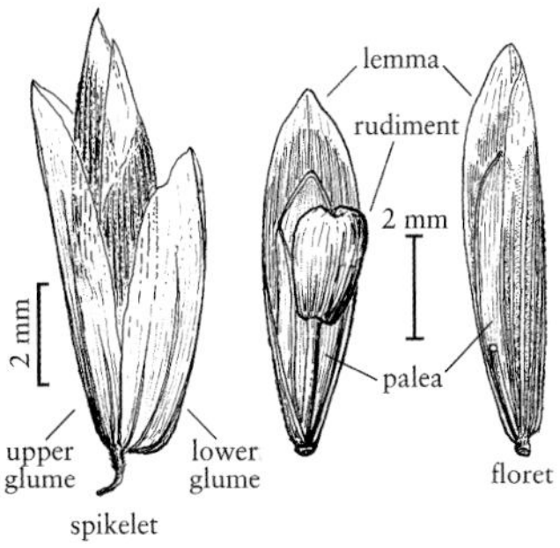

lemma
rudiment
2 mm
palea
2 mm
upper glume
lower glume
spikelet
floret

Melica

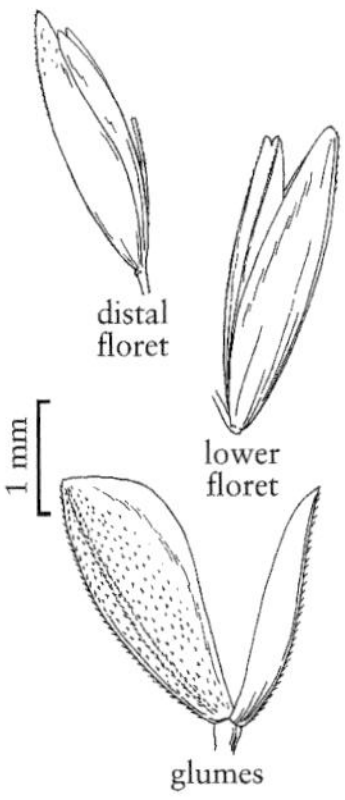

distal floret
lower floret
1 mm
glumes

Sphenopholis

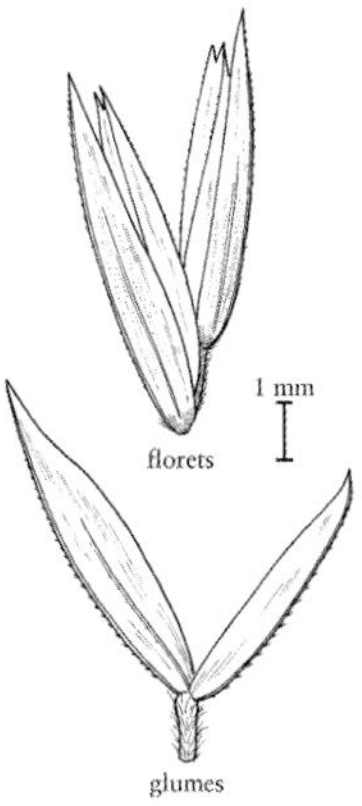

1 mm
florets
glumes

Koeleria

2 mm
florets
glumes

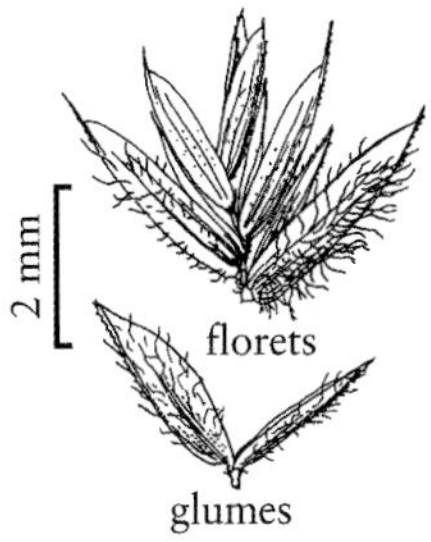

Rostraria

6. Lemmas awned from a distinctly bifid apex, awn straight or geniculate; upper glumes equaling or exceeding length of lowest floret.

    7. Plants perennial; disarticulation above glumes.................. *Trisetum* (p. 982)

    7. Plants annual; disarticulation below glumes ....
....*Sphenopholis* (p. 912)

6. Lemmas awned from an entire or minutely notched apex, awn straight; upper glumes usually shorter than lowest floret.

    8. Spikelets laterally compressed, asymmetrical, subsessile in dense clusters at apices of stiff, erect, or spreading branches; glumes and lemmas acute or irregularly short-awned; perennial............................*Dactylis* (p. 376)

    8. Spikelets not laterally compressed or asymmetrical, not in dense clusters at branch apices.

        9. Plants perennial...*Festuca* (p. 532)

        9. Plants annual.. *Vulpia* (p. 1007)

1. Lemmas awnless.

    11. Lemma veins strongly and uniformly developed and equally spaced; sheath margins connate, at least below...........................
....*Glyceria* (p. 451)

    11. Lemma veins not strongly and uniformly developed, or if so, not equally spaced.

        12. Glumes and lemmas spreading at nearly right angles to rachilla, inflated and papery; spikelets on slender pedicels .....
....*Briza* (p. 278)

        12. Glumes and lemmas not as above.

            13. Lower glumes distinctly longer than lowest florets ....
....*Schismus* (p. 864)

            13. Lower glumes equaling or shorter than lowest florets.

                14. Lowermost 1–3 florets reduced, sterile, about ½ as long as those above.

                    15. Disarticulation below glumes, spikelets falling entire; plants of coastal dunes.................................
....*Uniola* (p. 988)

                    15. Other couplet on page 126.

                14. Other couplet on page 126.

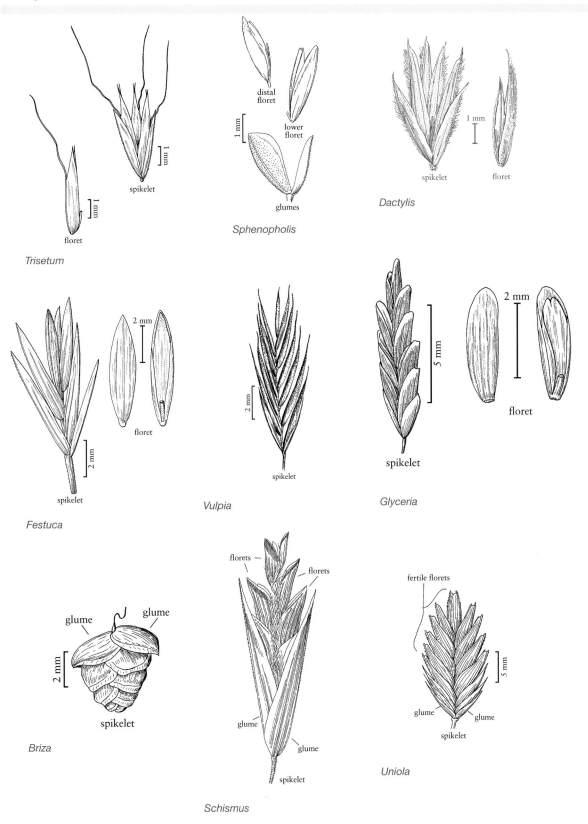

Trisetum

floret

1 mm

spikelet

1 mm

Sphenopholis

distal floret

lower floret

1 mm

glumes

Dactylis

spikelet

floret

1 mm

Festuca

spikelet

2 mm

floret

2 mm

Vulpia

spikelet

2 mm

Glyceria

spikelet

5 mm

floret

2 mm

Briza

glume

glume

2 mm

spikelet

Schismus

florets

florets

glume

glume

glume

spikelet

Uniola

fertile florets

glume

glume

spikelet

5 mm

15. Disarticulation above glumes and between florets; plants of woodland sites ....*Chasmanthium* (p. 316)

    14. Lowest florets not reduced, as large as or larger than those above.

16. Paleas colorless; lateral lemma veins indistinct.

17. Upper glumes obovate, broadest above mid-length; disarticulation below glumes ....*Sphenopholis* (p. 912)

17. Upper glumes broadest below middle; disarticulation above glumes ....*Koeleria* (p. 578)

    16. Paleas green or brown, at least on the veins.

18. Lemmas 7–13-veined C

18. Lemmas 3–5-veined CC

C (Lemmas 7–13-veined)

1. Spikelets unisexual, staminate and pistillate in separate inflorescences and usually on separate plants (dioecious); glumes and lemmas thick, firm, indistinctly veined.
   2. Plants with stolons, lacking rhizomes ....*Allolepis* (p. 185)
   2. Plants with rhizomes, lacking stolons ....*Distichlis* (p. 440)
1. Spikelets perfect; glumes and lemmas thin, lemma mostly with membranous margins.
   3. Sheath margins united at or near base; caryopsis without 2 hornlike styles.
      4. Paleas adherent to caryopses.
         5. Lemmas strongly keeled; spikelets strongly laterally compressed ....*Ceratochloa* (p. 311)
         5. Lemmas rounded over midrib; spikelets terete to slightly laterally compressed.
            6. Awn arising <1.5 mm below lemma apices.
               7. Lower glumes 1–3-veined; upper glumes 3–5-veined; perennial, or if annual, lower glumes 1-veined and upper glumes 3-veined ....*Bromopsis* (p. 282)
               7. Lower glumes 3–5-veined; upper glumes 5–9-veined; annual and perennial...*Bromus* (p. 291)
            6. Awn arising 1.5 mm or more below the lemma apices.
               8. Lemma apices bifid, teeth 1–5 mm long......................*Anisantha* (p. 199)
               8. Lemma apices entire or bifid, teeth <1 mm long ....*Bromus* (p. 291)
      4. Other couplet on page 128.
   3. Other couplet on page 128.

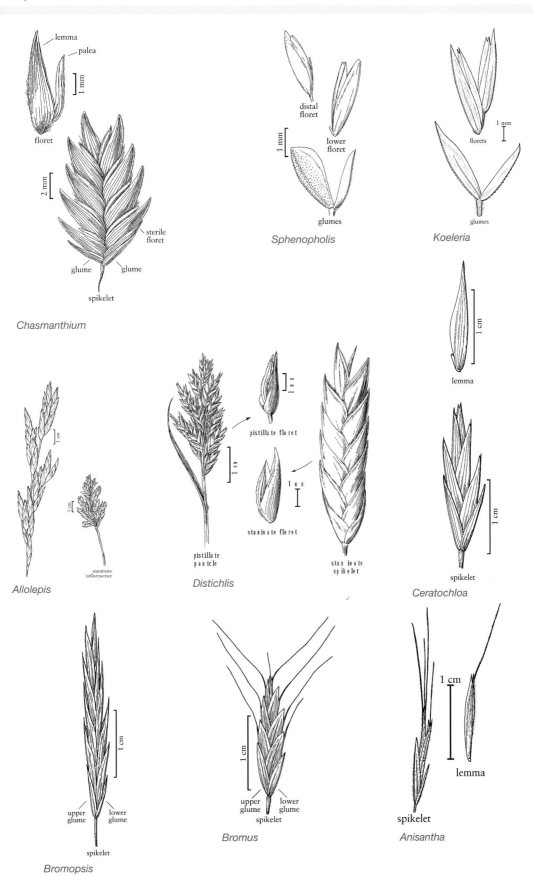

lemma
palea
1 mm
floret
2 mm
sterile floret
glume glume
spikelet

Chasmanthium

distal floret
1 mm
lower floret
glumes

Sphenopholis

1 mm
florets
glumes

Koeleria

1 cm
lemma

1 mm
pistillate floret
1 cm
1 mm
staminate floret
pistillate panicle
staminate spikelet

Distichlis

1 cm
1 cm
staminate inflorescence

Allolepis

1 cm
spikelet

Ceratochloa

1 cm
upper glume lower glume
spikelet

Bromopsis

1 cm
upper glume lower glume
spikelet

Bromus

1 cm
lemma
spikelet

Anisantha

　　4. Paleas not adherent to caryopses*Melica* (p. 617)

　3. Sheath margins free to base; caryopses suborbicular, with persistent hornlike styles ....
　　　...*Vaseyochloa* (p. 1005)

CC *(Lemmas 5-veined, rarely 3-veined)*

1. Lemmas thick, veins converging at apex to form a stout beak .... *Diarrhena* (p. 387)
1. Lemmas thin and/or firm, but not thick, veins not converging to form a beak.
　2. Lemma apices narrowly acute or attenuate, margins not scarious, usually short-awned.
　　3. Leaves without auricles ....*Festuca* (p. 532)
　　3. Leaves with auricles..*Schedonorus* (p. 861)
　2. Lemma apices broadly acute to obtuse, margins usually not scarious ....
　　.. *Poa* (p. 813)

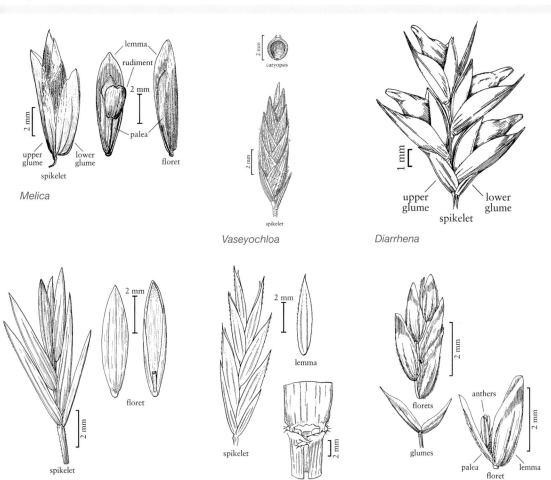

Melica

Vaseyochloa

Diarrhena

Festuca

Schedonorus

Poa

Table 5.1. Ecological checklist for Texas grass species.

Subfamily/Tribe/Genus/Specific Epithet/ Subspecific Taxa	1	2	3	4	5	Ecoregion Occurence										
I. Aristidoideae						1	2	3	4	5	6	7	8	9	10	1
i. Aristideae																
1. Aristida																
1. A. adscensionis	N	A	W			1	2	3	4	5	6	7	8	9	10	1
2. A. arizonica	N	P	W												10	
3. A. basiramea	N	A	W			1		3	4	5						
4. A. desmantha	N	A	W			1	2	3	4	5	6	7			10	11
5. A. dichotoma	N	A	W			1	2	3	4	5						11
var. dichotoma																
6. A. divaricata	N	P	W					3	4	5	6		8	9	10	11
7. A. gypsophila	N	P	W												10	
8. A. harvardii	N	P	W									7	8	9	10	
9. A. lanosa	N	P	W			1	2	3	4	5		7				
10. A. longespica	N	A	W			1	2	3	4	5	6	7				11
var. geniculata																
var. longespica																
11. A. oligantha	N	A	W			1	2	3	4	5	6	7				11
12. A. palustris	N	P	W			1										
13. A. pansa	N	P	W											9	10	11
14. A. purpurascens	N	P	W			1	2	3	4	5	6				10	11
var. purpurascens																
var. virgata																
15. A. purpurea	N	P	W			1	2	3	4	5	6	7	8	9	10	11
var. fendleriana																
var. longiseta																
var. nealleyi																
var. purpurea																
var. wrightii																
16. A. ramosissima	N	A	W			1		3	4		6					11
17. A. schiedeana	N	P	W												10	
var. orcuttiana																
18. A. ternipes	N	P	W						4	5			8	9	10	
var. gentilis																
var. ternipes																
II. Arundinoideae																
ii. Arundineae																
2. Arundo																
19. A. donax	I	P	W	O		1	2	3	4	5	6	7	8	9	10	11

1 – origin (N = native, I = introduced); 2 – longevity (A = annual, B = biennial, P = perennial); 3 – season (C = cool, W = warm); 4 – special designation (O = ornamental, E = endangered, C = cultivar, W = new additions); 5 – distributional data (X = no distributional data)

...mily/Tribe/Genus/Specific Epithet/ ...pecific Taxa	1	2	3	4	5	Ecoregion Occurence											
3. Hakonechloa																	
20. H. macra	I	P	C	O													
4. Phragmites																	
21. P. australis	N	P	W			1	2	3	4	5	6	7	8	9	10	11	12

...ambusoideae

iii. Bambuseae

	1	2	3	4	5	Ecoregion Occurence				
5. Arundinaria										
22. A. amabilis	I	P	C	O	X					
23. A. argenteostriata	I	P	C	O	X					
24. A. auricoma	I	P	C	O	X					
25. A. chino	I	P	C	O	X					
26. A. funghomii	I	P	C	O	X					
27. A. gigantea	N	P	C			1	2	3	4	5
subsp. gigantea										
28. A. pygmaea	I	P	C	O	X					
var. disticha										
var. pygmaea										
29. A. simonii	I	P	C	O	X					
30. A. tecta	N	P	C			1				
31. A. variegate	I	P	C	O	X					
6. Bambusa										
32. B. bambos	I	P	C	O						
33. B. beecheyana	I	P	C	O	X					
34. B. multiplex	I	P	C	O						
35. B. oldhamii	I	P	C	O	X					
36. B. textilis	I	P	C	O	X					
37. B. tuldoides	I	P	C	O	X					
38. B. ventricosa	I	P	C	O	X					
7. Chimonobambusa										
39. C. marmorea	I	P	C	O	X					
40. C. quadrangularis	I	P	C	O	X					
8. Indocalmus										
41. I. tessellates	I	P	C	O	X					
9. Otatea										
42. O. acuminate	I	P	C	O	X					
subsp. aztecorum										

1 – origin (N = native, I = introduced); 2 – longevity (A = annual, B = biennial, P = perennial); 3 – season (C = cool, W = warm); 4 – special designation (O = ornamental, E = endangered, C = cultivar, W = new additions); 5 – distributional data (X = no distributional data)

Subfamily/Tribe/Genus/Specific Epithet/ Subspecific Taxa	1	2	3	4	5	Ecoregion Occurence						
10. Phyllostachys												
43. P. angusta	I	P	C	O	X							
44. P. arcana	I	P	C	O	X							
45. P. aurea	I	P	C	O		1	2	3	4	5	7	11
46. P. aureosulcata	I	P	C	O		1	2	3				
47. P. bambusoides	I	P	C	O	X							
48. P. bissetii	I	P	C	O	X							
49. P. congesta	I	P	C	O	X							
50. P. decora	I	P	C	O	X							
51. P. dulcis	I	P	C	O	X							
52. P. edulis	I	P	C	O	X							
53. P. flexuosa	I	P	C	O	X							
54. P. glauca	I	P	C	O	X							
55. P. heteroclada	I	P	C	O	X							
56. P. makinoi	I	P	C	O	X							
57. P. meyeri	I	P	C	O	X							
58. P. nidularia	I	P	C	O	X							
59. P. nigra	I	P	C	O	X							
var. henonis												
var. nigra												
60. P. purpurata	I	P	C	O	X							
61. P. rubromarginata	I	P	C	O	X							
62. P. sulphurea	I	P	C	O	X							
var. sulphurea												
var. viridis												
63. P. viridi-glaucescens	I	P	C	O	X							
64. P. viridis	I	P	C	O	X							
65. P. vivax	I	P	C	O	X							
11. Pseudosasa												
66. P. japonica	I	P	C	O	X							
12. Sasa												
67. S. masumuneana	I	P	C	O	X							
68. S. palmate	I	P	C	O	X							
69. S. ramose	I	P	C	O	X							
70. S. veitchii	I	P	C	O	X							
13. Semiarundinaria												
71. S. fastuosa	I	P	C	O	X							
14. Shibataea												
72. S. kumasaca	I	P	C	O	X							

1 – origin (N = native, I = introduced); 2 – longevity (A = annual, B = biennial, P = perennial); 3 – season (C = cool, W = warm); 4 – special designation (O = ornamental, E = endangered, C = cultivar, W = new additions); 5 – distributional data (X = no distributional data)

family/Tribe/Genus/Specific Epithet/ specific Taxa	1	2	3	4	5	Ecoregion Occurence											
Centothecoideae																	
iv. Centotheceae																	
15. Chasmanthium																	
73. C. latifolium	N	P	W	O		1	2	3	4	5	6	7			10	11	
74. C. laxum	N	P	W			1	2	3	4								
75. C. sessiliflorum	N	P	W			1	2	3	4	5	6					11	
Chloridoideae																	
v. Cynodonteae																	
16. Allolepis																	
76. A. texana	N	P	W												10		
17. Blepharidachne																	
77. B. bigelovii	N	P	W												10	11	12
18. Blepharoneuron																	
78. B. tricholepis	N	P	W			1		3							10		12
19. Bouteloua																	
79. B. aristidoides var. aristidoides	N	A	W				2	3	4	5	6			9	10	11	12
80. B. chondrosoides	N	P	W					3	4	5	6			9	10	11	
81. B. curtipendula var. caespitosa var. curtipendula	N	P	W	O		1	2	3	4	5	6	7	8	9	10	11	12
82. B. repens	N	P	W			1	2	3	4	5	6	7	8	9		11	
83. B. rigidiseta var. rigidiseta	N	P	W			1	2	3	4	5	6	7	8	9	10	11	
84. B. uniflora var. uniflora	N	P	W			1	2	3	4	5	6	7	8		10	11	12
85. B. warnockii	N	P	W												10		12
20. Buchloe																	
86. B. dactyloides	N	P	W	C		1	2	3	4	5	6	7	8	9	10	11	12
21. Calamovilfa																	
87. C. gigantea	N	P	W									7	8	9	10		
22. Cathestecum																	
88. C. erectum	N	P	W												10		

1 – origin (N = native, I = introduced); 2 – longevity (A = annual, B = biennial, P = perennial); 3 – season (C = cool, W = warm); 4 – special designation (O = ornamental, E = endangered, C = cultivar, W = new additions); 5 – distributional data (X = no distributional data)

Subfamily/Tribe/Genus/Specific Epithet/ Subspecific Taxa	1	2	3	4	5	Ecoregion Occurence										
23. Chloris																
89. C. andropogonoides	N	P	W			1	2	3	4	5	6	7				11
90. C. barbata	N	A	W			1	2	3			6					11
91. C. canterae	I	P	W			1	2	3	4	5	6					
var. canterae																
var. grandiflora																
92. C. ciliata	N	P	W			1	2	3	4		6					11
93. C. cucullata	N	P	W			1	2	3	4	5	6	7	8	9	10	11
94. C. divaricata	I	P	W				2	3	4		6					11
95. C. gayana	I	P	W	C		1	2	3	4	5	6		8	9		11
96. C. pilosa	I	P	W	C												
97. C. submutica	N	P	W												10	
98. C. texensis	N	P	W			1	2	3							10	
99. C. verticillata	N	P	W			1	2	3	4	5	6	7	8	9	10	11
100. C. virgata	N	P	W			1	2	3	4	5	6	7	8	9	10	11
24. Chondrosum																
101. C. barbatum	N	A	W			1	2	3	4	5	6	7	8	9	10	11
var. barbatum																
var. rothrockii																
102. C. brevisetum	N	P	W								6			9	10	11
103. C. eriopodum	N	P	W			1		3	4	5	6	7	8	9	10	11
104. C. gracile	N	P	W	O		1	2	3	4	5	6	7	8	9	10	11
105. C. hirsutum	N	P	W			1	2	3	4	5	6	7	8	9	10	11
subsp. hirsutum																
subsp. pectinatum																
106. C. kayi	N	P	W												10	
107. C. ramosa	N	P	W								6				10	11
108. C. simplex	N	A	W										8	9	10	
109. C. trifidum	N	P	W			1	2	3	4	5	6	7	8	9	10	11
var. burkii																
var. trifidum																
25. Ctenium																
110. C. aromaticum	N	P	W	O		1										
26. Cynodon																
111. C. aethiopicus	I	P	W				2				6					
112. C. dactylon	I	P	W	C		1	2	3	4	5	6	7	8	9	10	11
var. dactylon																
113. C. nlemfuensis	I	P	W	C			2									
var. nlemfuensis																
114. C. plectostachyus	I	P	W	C		1		3	4		6					11
115. C. transvaalensis	I	P	W	C												

1 – origin (N = native, I = introduced); 2 – longevity (A = annual, B = biennial, P = perennial); 3 – season (C = cool, W = warm); 4 – special designation (O = ornamental, E = endangered, C = cultivar, W = new additions); 5 – distributional data (X = no distributional data)

Subfamily/Tribe/Genus/Specific Epithet/ Subspecific Taxa	1	2	3	4	5	Ecoregion Occurrence												
						1	2	3	4	5	6	7	8	9	10	11	12
27. Dactyloctenium																	
116. D. aegyptium	I	A	W			1	2	3	4	5	6					11	
28. Dasyochloa																	
117. D. pulchella	N	P	W								6		8	9	10	11	12
29. Distichlis																	
118. D. spicata	N	P	W			1	2	3	4	5	6	7	8	9	10	11	12
30. Eleusine																	
119. E. indica	I	A	W			1	2	3	4	5	6	7	8	9	10	11	12
120. E. tristachya	I	A	W			1		3									
31. Enteropogon																	
121. E. chlorideus	N	P	W			1	2	3			6						
32. Eragrostis																	
122. E. airoides	I	P	W			1		3									
123. E. amabilis	I	A	W														
124. E. barrelieri	I	A	W			1	2	3	4	5	6	7	8	9	10	11	12
125. E. capillaris	N	A	W			1	2	3	4	5						11	
126. E. cilianensis	I	A	W			1	2	3	4	5	6	7	8	9	10	11	12
127. E. ciliaris var. ciliaris	I	A	W			1	2	3									
128. E. curtipendicellata	N	P	W			1	2	3	4	5	6	7	8	9	10	11	
129. E. curvula	I	P	W	C		1	2	3	4	5	6	7	8	9	10	11	
130. E. elliottii	N	P	W			1	2	3	4	5		7	8				
131. E. erosa	N	P	W											9	10		12
132. E. hirsuta	N	P	W			1	2	3	4	5	6	7				11	
133. E. hypnoides	N	A	W			1	2	3	4	5	6					11	
134. E. intermedia	N	P	W			1	2	3	4	5	6	7	8	9	10	11	12
135. E. japonica	I	A	W			1	2										
136. E. lehmanniana	I	P	W	C		1	2	3	4		6	7	8	9	10	11	12
137. E. lugens	N	P	W			1	2	3	4	5	6				10	11	
138. E. mexicana subsp. mexicana	N	A	W												10		12
139. E. minor	I	A	W				2									11	12
140. E. palmeri	N	P	W				2				6				10	11	
141. E. pectinacea var. miserrima var. pectinacea	I	A	W			1	2	3	4	5	6	7	8	9	10	11	12
142. E. pilosa var. perplexa var. pilosa	I	A	W			1	2	3	4	5			8	9			

1 – origin (N = native, I = introduced); 2 – longevity (A = annual, B = biennial, P = perennial); 3 – season (C = cool, W = warm); 4 – special designation (O = ornamental, E = endangered, C = cultivar, W = new additions); 5 – distributional data (X = no distributional data)

Subfamily/Tribe/Genus/Specific Epithet/ Subspecific Taxa	1	2	3	4	5	Ecoregion Occurence										
143. E. refracta	N	P	W			1	2	3	4	5						
144. E. reptans	N	A	W			1	2	3	4	5	6	7			10	11
145. E. secundiflora subsp. oxylepis	N	P	W			1	2	3	4	5	6	7	8	9	10	11
146. E. sessilispica	N	P	W			1	2	3	4	5	6	7	8	9	10	11
147. E. silveana	N	P	W			1	2	3	4		6					11
148. E. spectabilis	N	P	W	O		1	2	3	4	5	6	7	8	9		11
149. E. spicata	N	P	W			1	2	3	4		6					11
150. E. superba	I	P	W	C		1	2	3	4	5		7			10	11
151. E. swallenii	N	P	W				2	3			6					
152. E. tef	I	A	W	C					4	5						
153. E. trichoides	N	P	W			1	2	3	4	5	6	7	8	9	10	11
33. Erioneuron																
154. E. avenaceum var. avenaceum	N	P	W												10	11
155. E. nealleyi	N	P	W												10	
156. E. pilosum	N	P	W			1	2	3	4	5	6	7	8	9	10	11
34. Eustachys																
157. E. caribaea	I	P	W			1	2	3	4		6					11
158. E. neglecta	N	p	W			1		3								
159. E. petraea	N	P	W			1	2	3			6					
160. E. retusa	I	P	W			1	2	3	4	5	6					
35. Gymnopogon																
161. G. ambiguus	N	P	W			1	2	3	4	5	6	7				11
162. G. brevifolius	N	P	W			1	2									
36. Hilaria																
163. H. belangeri var. belangeri var. longifolia	N	P	W				2	3	4	5	6	7	8	9	10	11
164. H. swallenii	N	P	W												10	
37. Leptochloa																
165. L. chloridiformis	I	P	W				2									
166. L. dubia	N	P	W			1	2	3	4	5	6	7	8	9	10	11
167. L. fusca subsp. fascicularis subsp. uninerva	N	A	W			1	2	3	4	5	6	7	8	9	10	11
168. L. nealleyi	N	A	W			1	2	3	4		6					
169. L. panicea subsp. brachiata subsp. mucronata	N	A	W			1	2	3	4	5	6	7	8	9	10	11

1 – origin (N = native, I = introduced); 2 – longevity (A = annual, B = biennial, P = perennial); 3 – season (C = cool, W = warm); 4 – special designation (O = ornamental, E = endangered, C = cultivar, W = new additions); 5 – distributional data (X = no distributional data)

family/Tribe/Genus/Specific Epithet/ specific Taxa	1	2	3	4	5	Ecoregion Occurence											
170. L. panicoides	I	A	W			1	2		4	5							
171. L. virgata	N	P	W			1	2	3	4		6					11	
172. L. viscida	N	A	W												10		
38. Lycurus																	
173. L. phleoides	N	P	W										8	9	10	11	12
174. L. setosus	N	P	W											9	10	11	12
39. Microchloa																	
175. M. kunthii	N	P	W												10		
40. Monanthochloe																	
176. M. littoralis	N	P	W			1	2	3			6						
41. Muhlenbergia																	
177. M. andina	N	P	W												10		12
178. M. arenacea	N	P	W					3	4		6		8	9	10	11	12
179. M. arenicola	N	P	W					3	4	5	6	7	8	9	10	11	12
180. M. asperifolia	N	P	W									7	8	9	10	11	12
181. M. brevis	N	A	W												10		
182. M. bushii	N	P	W			1		3	4	5		7				11	
183. M. capillaris	N	P	W			1	2	3	4	5	6	7				11	
184. M. crispiseta	N	A	W												10		
185. M. cuspidata	N	P	W											9			
186. M. depauperata	N	A	W												10		12
187. M. diversiglumis	I	A	W				2										
188. M. diversiglumis	N	P	W					3	4		6		8	9	10	11	12
189. M. dumosa	N	P	W	O	X												
190. M. eludens	N	A	W												10		
191. M. emersleyi	N	P	W			1									10	11	12
192. M. expansa	N	P	W			1											
193. M. filiformis	N	A	W			1		3									
194. M. fragilis	N	A	W												10		12
195. M. frondosa	N	P	W					3	4	5							
196. M. glabrifloris	N	P	W						4	5							
197. M. glauca	N	P	W												10		12
198. M. lindheimeri	N	P	W					3	4	5	6				10	11	
199. M. mexicana	N	P	W							5							
var. filiformis																	
var. mexicana																	
200. M. minutissima	N	A	W												10		
201. M. montana	N	P	W												10		12
202. M. pauciflora	N	P	W												10	11	12
203. M. polycaulis	N	P	W												10		12
204. M. porteri	N	P	W				2	3	4		6		8	9	10	11	12

1 – origin (N = native, I = introduced); 2 – longevity (A = annual, B = biennial, P = perennial); 3 – season (C = cool, W = warm); 4 – special designation (O = ornamental, E = endangered, C = cultivar, W = new additions); 5 – distributional data (X = no distributional data)

Subfamily/Tribe/Genus/Specific Epithet/ Subspecific Taxa	1	2	3	4	5	Ecoregion Occurrence 1	2	3	4	5	6	7	8	9	10	11
205. M. pungens	N	P	W										8	9		11
206. M. racemosa	N	P	W										8	9	10	
207. M. repens	N	P	W					3	4		6		8	9	10	11
208. M. reverchonii	N	P	W			1		3	4	5	6					11
209. M. rigens	N	P	W					3	4	5					10	11
210. M. rigida	N	P	W								6				10	11
211. M. schreberi	N	P	W			1	2	3	4	5	6	7			10	11
212. M. sericea	N	P	W	O			2									
213. M. setifolia	N	P	W								6				10	11
214. M. sobolifera	N	P	W					3	4	5		7	8	9		
215. M. spiciformis	N	P	W								6		8	9	10	11
216. M. straminea	N	P	W												10	
217. M. sylvatica	N	P	W													
218. M. tenuiflora	N	P	W							5						
219. M. tenuifolia	N	A	W			1	2	3							10	11
220. M. texana	N	A	W												10	
221. M. thurberi	N	P	W												10	
222. M. torreyi	N	P	W										8	9	10	
223. M. uniflora	N	P	W												10	
224. M. utilis	N	P	W			1		3	4	5	6				10	11
225. M. villiflora var. villosa	N	P	W					3					8	9	10	11
226. M. wrightii	N	P	W												10	
42. Munroa																
227. M. squarrosa	N	A	W													
43. Pleuraphis																
228. P. jamesii	N	P	W					3	4		6	7	8	9	10	11
229. P. mutica	N	P	W					3	4		6	7	8	9	10	11
44. Redfieldia																
230. R. flexuosa	N	P	W									7	8	9		
45. Schedonnardus																
231. S. paniculatus	N	P	W			1	2	3	4	5	6	7	8	9	10	11
46. Scleropogon																
232. S. brevifolius	N	P	W						4		6	7	8	9	10	11
47. Spartina																
233. S. alterniflora	N	P	W			1	2	3			6					
234. S. bakeri	N	P	W			1	2									
235. S. cynosuroides	N	P	W			1	2	3								
236. S. densiflora	I	P	W			1	2	3	4		6	7				11

1 – origin (N = native, I = introduced); 2 – longevity (A = annual, B = biennial, P = perennial); 3 – season (C = cool, W = warm); 4 – special designation (O = ornamental, E = endangered, C = cultivar, W = new additions); 5 – distributional data (X = no distributional data)

Family/Tribe/Genus/Specific Epithet/ specific Taxa	1	2	3	4	5	\<br\> Ecoregion Occurence											
237. S. patens	N	P	W			1	2	3									
238. S. pectinata	N	P	W			1	2	3	4	5		7	8	9	10	11	
239. S. spartinae	N	P	W			1	2	3		5	6					11	
48. Sporobolus																	
240. S. airoides	N	P	W			1	2	3	4		6	7	8	9	10	11	12
241. S. buckleyi	N	P	W				2				6						
242. S. clandestinus	N	P	W			1	2	3	4	5	6	7	8			11	
243. S. coahuilensis	N	P	W												10		12
244. S. compositus	N	P	W			1	2	3	4	5	6	7	8	9	10	11	
var. compositus																	
var. drummondii																	
var. macer																	
245. S. contractus	N	P	W				2						8	9	10		12
246. S. cryptandrus	N	P	W			1	2	3	4	5	6	7	8	9	10	11	12
247. S. diandrus	I	P	W												10	11	
248. S. domingensis	N	P	W				2										
249. S. flexuosus	N	P	W				2	3	4	5	6	7	8	9	10	11	12
250. S. giganteus	N	P	W			1		3	4	5		7	8	9	10	11	12
251. S. heterolepis	N	P	W			1	2	3									
252. S. indicus	N	P	W			1	2	3	4	5	6			9	10	11	
253. S. junceus	N	P	W			1	2	3	4		6					11	
254. S. nealleyi	N	P	W									7	8	9	10	11	12
255. S. neglectus	N	A	W					3	4	5	6		8	9		11	
256. S. purpurascens	N	P	W				2	3	4		6	.				11	
257. S. pyramidatus	N	A	W			1	2	3	4	5	6	7	8	9	10	11	
258. S. silveanus	N	P	W			1	2	3	4								
259. S. texanus	N	P	W								6	7	8	9	10	11	
260. S. vaginiflorus	N	A	W			1	2	3	4	5	6	7	8	9	10	11	
var. ozarkanus																	
var. vaginiflorus																	
261. S. virginicus	N	P	W			1	2	3	4		6					11	
262. S. wrightii	N	P	W			1	2	3			6	7	8	9	10	11	12
49. Tragus																	
263. T. berteronianus	N	P	W			1	2	3	4		6				10	11	12
50. Trichloris																	
264. T. crinita	N	P	W				2		4	5	6			9	10	11	12
265. T. pluriflora	N	P	W				2	3	4		6				10	11	
51. Trichoneura																	
266. T. elegans	N	A	W			1	2	3	4		6				10	11	
52. Tridens																	
267. T. albescens	N	P	W			1	2	3	4	5	6	7	8	9	10	11	12

1 – origin (N = native, I = introduced); 2 – longevity (A = annual, B = biennial, P = perennial); 3 – season (C = cool, W = warm); 4 – special designation (O = ornamental, E = endangered, C = cultivar, W = new additions); 5 – distributional data (X = no distributional data)

Subfamily/Tribe/Genus/Specific Epithet/ Subspecific Taxa	1	2	3	4	5	Ecoregion Occurence										
268. T. ambiguus	N	P	W			1	2									
269. T. buckleyanus	N	P	W	E				3	4	5	6					11
270. T. congestus	N	P	W			1	2	3	4	5	6	7	8			11
271. T. eragrostoides	N	P	W				2	3	4	5	6	7			10	11
272. T. flavus	N	P	W			1	2	3	4	5	6	7	8	9	10	11
var. chapmanii																
var. flavus																
273. T. muticus	N	P	W			1	2	3	4	5	6	7	8	9	10	11
var. elongatus																
var. muticus																
274. T. strictus	N	P	W			1	2	3	4	5		7			10	11
275. T. texanus	N	P	W			1	2	3	4	5	6	7			10	11
53. Triplasis																
276. T. purpurea	N	A	W			1	2	3	4	5	6	7	8	9	10	11
var. purpurea																
54. Tripogon																
277. T. spicatus	N	P	W					3	4	5		7	8			11
55. Triraphis																
278. T. mollis	I	P	W								6					
56. Uniola																
279. U. paniculata	N	P	W	O			2				6					
57. Vaseyochloa																
280. V. multinervosa	N	P	W				2	3	4		6					11
58. Willkommia																
281. W. texana	N	P	W			1	2	3	4		6					
var. texana																
59. Zoysia																
282. Z. japonica	I	P	W	C		1	2	3	4		6		8	9		
283. Z. matrella	I	P	W	C											10	
284. Z. pacifica	I	P	W	C												
vi. Pappophoreae																
60. Cottea																
285. C. pappophoroides	N	P	W												10	11
61. Enneapogon																
286. E. desvauxii	N	P	W								6	7	8	9	10	11

1 – origin (N = native, I = introduced); 2 – longevity (A = annual, B = biennial, P = perennial); 3 – season (C = cool, W = warm); 4 – special designation (O = ornamental, E = endangered, C = cultivar, W = new additions); 5 – distributional data (X = no distributional data)

Family/Tribe/Genus/Specific Epithet/Specific Taxa	1	2	3	4	5	\-	\-	\-	\-	\-	\-	\-	\-	\-	\-	\-	\-
						\-	\-	\-	\-	\-	Ecoregion Occurence	\-	\-	\-	\-	\-

Family/Tribe/Genus/Specific Epithet/Specific Taxa	1	2	3	4	5	1	2	3	4	5	6	7	8	9	10	11	12
62. Pappophorum																	
287. P. bicolor	N	P	W			1	2	3	4	5	6	7	8	9	10	11	
288. P. vaginatum	N	P	W			1	2	3	4		6	7	8	9	10	11	
Danthonioideae																	
vii. Danthonieae																	
63. Cortaderia																	
289. C. jubata	I	P	W	O													
290. C. selloana	I	P	W	O		1	2	3	4	5	6			9	10	11	12
64. Danthonia																	
291. D. sericea	N	P	C			1		3									
292. D. spicata	N	P	C			1		3	4	5	6					11	
65. Schismus																	
293. S. arabicus	I	A	C												10		
294. S. barbatus	I	A	C												10		12
Ehrhartoideae																	
viii. Ehrharteae																	
66. Ehrharta																	
295. E. calycina	I	P	C					3	4								
ix. Oryzeae																	
67. Leersia																	
296. L. hexandra	N	P	W			1	2	3	4		6						
297. L. lenticularis	N	P	W			1	2	3	4								
298. L. monandra	N	P	W			1	2	3	4		6					11	
299. L. oryzoides	N	P	W			1	2	3	4	5	6	7	8	9	10	11	
300. L. virginica	N	P	W			1	2	3	4	5	6					11	
68. Luziola																	
301. L. fluitans	N	P	C			1		3									
302. L. peruviana	I	P	C			1	2										
69. Oryza																	
303. O. sativa	I	A	C	C		1	2	3	4		6						
70. Zizania																	
304. Z. texana	N	P	C	E													

1 – origin (N = native, I = introduced); 2 – longevity (A = annual, B = biennial, P = perennial); 3 – season (C = cool, W = warm); 4 – special designation (O = ornamental, E = endangered, C = cultivar, W = new additions); 5 – distributional data (X = no distributional data)

Subfamily/Tribe/Genus/Specific Epithet/ Subspecific Taxa	1	2	3	4	5	Ecoregion Occurence										
71. Zizaniopsis																
305. Z. miliacea	N	P	C			1	2	3	4	5	6	7				11

VIII. Panicoideae

x. Andropogoneae

	1	2	3	4	5	Ecoregion Occurence										
72. Andropogon																
306. A. gerardii	N	P	W	O		1	2	3	4	5		7	8	9	10	11
307. A. glomeratus	N	P	W	O		1	2	3	4	5	6	7	8	9	10	11
308. A. gyrans	N	P	W			1	2	3	4	5		7				11
var. gyrans																
var. stenophyllus																
309. A. hallii	N	P	W							5		7	8	9	10	
310. A. ternarius	N	P	W			1	2	3	4	5		7				
var. ternarius																
311. A. virginicus	N	P	W			1	2	3	4	5						11
var. glaucus																
var. viginicus																
73. Arthraxon																
312. A. hispidus	I	A	W													
74. Bothriochloa																
313. B. alta	N	P	W												10	
314. B. barbinoidis	N	P	W			1	2	3	4	5	6	7	8	9	10	11
315. B. bladhii	I	P	W			1	2	3	4		6		8	9		11
316. B. edwardsiana	N	P	W								6					11
317. B. exaristata	N	P	W			1	2	3								
318. B. hybrida	N	P	W			1	2	3	4	5	6				10	11
319. B. ischaemum	I	P	W			1	2	3	4	5	6	7	8	9	10	11
var. ischaemum																
var. songarica																
320. B. laguroides	N	P	W			1	2	3	4	5	6	7	8	9	10	11
subsp. torreyana																
321. B. longipaniculata	N	P	W			1	2	3	4	5	6					11
322. B. pertusa	I	P	W				2									
323. B. springfieldii	N	P	W										8	9	10	11
324. B. wrightii	N	P	W									7			10	11
75. Chrysopogon																
325. C. pauciflorus	N	A	W			1	2									
326. C. zizanioides	I	P	W	C		1		3								

1 – origin (N = native, I = introduced); 2 – longevity (A = annual, B = biennial, P = perennial); 3 – season (C = cool, W = warm); 4 – special designation (O = ornamental, E = endangered, C = cultivar, W = new additions); 5 – distributional data (X = no distributional data)

Family/Tribe/Genus/Specific Epithet/Specific Taxa	1	2	3	4	5	Ecoregion Occurence											
76. Coelorachis																	
327. C. cylindrica	N	P	W			1	2	3	4	5	6	7				11	
328. C. rugosa	N	P	W			1	2	3									
77. Coix																	
329. C. lacryma-jobi	I	A	W	O		1	2	3									
78. Dichanthium																	
330. D. annulatum	I	P	W	C		1	2	3	4	5	6				10	11	
331. D. aristatum	I	P	W	C		1	2	3	4		6					11	
332. D. sericeum subsp. Sericeum	I	P	W			1	2	3	4		6					11	
79. Elionurus																	
333. E. barbiculmis	N	P	W												10		
334. E. tripsacoides	N	P	W			1	2	3	4		6				10	11	
80. Eremochloa																	
335. E. ophiuroides	I	P	W	C		1	2	3	4							11	
81. Hackelochloa																	
336. H. granularis	I	A	W														
82. Hemarthria																	
337. H. altissima	I	P	W				2	3			6				10	11	
83. Heteropogon																	
338. H. contortus	N	P	W			1	2	3	4	5	6				10	11	12
339. H. melanocarpus	I	A	W			1	2	3									
84. Hyparrhenia																	
340. A. hirta	I	P	W			1	2	3	4		6				10	11	
341. A. rufa	I	P	W			1		3									
85. Imperata																	
342. I. brevifolia	N	P	W	O			2	3							10		12
343. I. cylindrica	I	P	W	O		1		3									
86. Microstegium																	
344. M. vimineum	I	A	W			1		3									
87. Miscanthus																	
345. M. sinensis	I	P	W	O													
346. M. transmorrisonensis	I	P	W	O	X												

1 – origin (N = native, I = introduced); 2 – longevity (A = annual, B = biennial, P = perennial); 3 – season (C = cool, W = warm); 4 – special designation (O = ornamental, E = endangered, C = cultivar, W = new additions); 5 – distributional data (X = no distributional data)

Subfamily/Tribe/Genus/Specific Epithet/ Subspecific Taxa	1	2	3	4	5					Ecoregion Occurence						
88. Rottboellia																
347. R. cochinchinensis	I	A	W	W		1	2		4	5						
89. Saccharum																
348. S. alopecuroides	N	P	W			1		3								
349. S. baldwinii	N	P	W			1	2	3	4							
350. S. bengalense	I	P	W	O		1		3								
351. S. brevibarbe	N	P	W			1	2	3	4							
var. brevibarbe																
var. contortum																
352. S. coarctatum	N	P	W			1	2									
353. S. giganteum	N	P	W			1	2	3	4							11
354. S. officinarum	I	P	W	C			2				6					
355. S. ravennae	I	P	W	O		1		3		5						
90. Schizachyrium																
356. S. cirratum	N	P	W												10	
357. S. littorale	N	P	W				2				6					
358. S. sanguineum	N	P	W				2				6				10	
var. hirtiflorum																
359. S. scoparium	N	P	W			1	2	3	4	5	6	7	8	9	10	11
var. divergens																
var. scoparium																
360. S. spadiceum	N	P	W												10	
361. S. tenerum	N	P	W			1	2	3								
91. Sorghastrum																
362. S. elliottii	N	P	W			1	2	3	4	5	6			9	10	11
363. S. nutans	N	p	W	O		1	2	3	4	5	6	7	8	9	10	11
92. Sorghum																
364. S. bicolor	I	A	W	C		1	2	3	4	5	6	7	8	9	10	11
subsp. arundinaceum																
subsp. bicolor																
365. S. halepense	I	P	W	C		1	2	3	4	5	6	7	8	9	10	11
93. Themeda																
366. T. triandra	I	P	W	O												
94. Trachypogon																
367. T. secundus	N	P	W			1	2	3	4	5	6					
95. Tripsacum																
368. T. dactyloides	N	P	W			1	2	3	4	5	6	7	8	9	10	11
var. dactyloides																

1 – origin (N = native, I = introduced); 2 – longevity (A = annual, B = biennial, P = perennial); 3 – season (C = cool, W = warm); 4 – special designation (O = ornamental, E = endangered, C = cultivar, W = new additions); 5 – distributional data (X = no distributional data)

Family/Tribe/Genus/Specific Epithet/ specific Taxa	1	2	3	4	5	Ecoregion Occurence
96. Vetiveria						
369. zizanioides	I	P	W	C		
97. Zea						
370. Z. mays	N	A	W	C		1 2 3 4 5
subsp. mays						
371. Z. perennis	I	P	W			2
xi. Paniceae						
98. Anthenantia						
372. A. rufa	N	P	W			1 2 3 4
373. A. villosa	N	P	W			1 2 3 4
99. Axonopus						
374. A. compressus	N	P	W	C		1 2 3 4
375. A. fissifolius	N	P	W	C		1 2 3 4 11
376. A. furcatus	N	P	W			1 2 3 4
100. Cenchrus						
377. C. brownii	N	A	W			6 10 11
378. C. echinatus	N	A	W			1 2 3 4 5 6 10 11
379. C. longispinus	N	A	W			1 2 3 4 5 6 7 8 9 10 11 12
380. C. myosuroides	N	P	W			1 2 3 4 6 10 11
381. C. spinifex	N	B	W			1 2 3 4 5 6 7 8 9 10 11 12
101. Dichanthelium						
382. D. aciculare	N	P	C			1 2 3 4 5 7 11
subsp. aciculare						
subsp. angustifolium						
383. D. acuminatum	N	P	C			1 2 3 4 5 6 7 8 9 10 11 12
subsp. acuminatum						
subsp. lindheimeri						
subsp. longiligulatum						
subsp. spretum						
384. D. boscii	N	P	C			1 2 3 4
385. D. clandestinum	N	P	C			1 3 4 5
386. D. commutatum	N	P	C			1 2 3 4 5
subsp. ashei						
subsp. commutatum						
subsp. equilaterale						
subsp. joorii						
387. D. consanguineum	N	P	C			1 2 3
388. D. depauperatum	N	P	C			1 2 3 4 5
389. D. dichotomum	N	P	C			1 2 3 4 5 6 11

origin (N = native, I = introduced); 2 – longevity (A = annual, B = biennial, P = perennial); 3 – season (C = cool, W = warm); 4 – special ignation (O = ornamental, E = endangered, C = cultivar, W = new additions); 5 – distributional data (X = no distributional data)

Subfamily/Tribe/Genus/Specific Epithet/ Subspecific Taxa	1	2	3	4	5	Ecoregion Occurence										
subsp. dichotomum																
subsp. lucidum																
subsp. microcarpon																
subsp. nitidum																
subsp. roanokense																
390. D. ensifolium	N	P	C			1		3								
subsp. curtifolium																
subsp. ensifolium																
391. D. latifolium	N	P	C			1										
392. D. laxiflorum	N	P	C			1	2	3	4	5						
393. D. linearifolium	N	P	C			1	2	3	4	5		7				11
394. D. malacophyllum	N	P	C			1		3	4	5		7				
395. D. nodatum	N	P	C			1	2	3	4		6					11
396. D. oligosanthes	N	P	C			1	2	3	4	5	6	7	8	9	10	11
subsp. oligosanthes																
subsp. scribnerianum																
397. D. ovale	N	P	C			1	2	3	4	5						
subsp. ovale																
subsp. pseudopubescens																
subsp. villosissimum																
398. D. pedicellatum	N	P	C					3	4	5	6				10	11
399. D. polyanthes	N	P	C			1	2	3								
400. D. portoricense	N	P	C				2	3	4							
401. D. ravenelii	N	P	C			1	2	3	4	5						
402. D. scabriusculum	N	P	C			1	2	3	4							
403. D. scoparium	N	P	C			1	2	3	4							
404. D. sphaerocarpon	N	P	C			1	2	3	4	5	6	7		9	10	11
405. D. strigosum	N	P	C					3	4							
subsp. glabrescens																
subsp. leucoblepharis																
subsp. strigosum																
406. D. tenue	N	P	C			1										
407. D. wrightianum	N	P	C			1										
102. Digitaria																
408. D. arenicola	N	P	W			1	2									
409. D. bicornis	N	A	W			1	2	3	4		6					11
410. D. californica	N	P	W			1	2	3	4	5	6	7	8	9	10	11
411. D. ciliaris	N	A	W			1	2	3	4	5	6	7	8	9	10	11
var. ciliaris																
412. D. cognata	N	P	W			1	2	3	4	5	6	7	8	9		11
413. D. filiformis	N	A	W			1	2	3	4	5						11
var. filiformis																
var. villosa																
414. D. hitchcockii	N	P	W							5	6				10	11
415. D. insularis	N	P	W				2	3	4	5	6			9	10	11

1 – origin (N = native, I = introduced); 2 – longevity (A = annual, B = biennial, P = perennial); 3 – season (C = cool, W = warm); 4 – special designation (O = ornamental, E = endangered, C = cultivar, W = new additions); 5 – distributional data (X = no distributional data)

Family/Tribe/Genus/Specific Epithet/ Specific Taxa	*1*	*2*	*3*	*4*	*5*				*Ecoregion Occurence*							
416. D. ischaemum	I	A	W			1	2	3								
417. D. milanjiana	I	P	W													
418. D. patens	N	P	W				2	3	4		6				10	11
419. D. pubiflora	N	P	W			1	2	3	4	5	6	7	8	9	10	11	12
420. D. sanguinalis	I	A	W			1	2	3	4	5	6	7	8	9	10	11
421. D. texana	N	P	W			1	2	3								
422. D. violascens	N	A	W			1	2	3	4	5						
103. Echinochloa																
423. E. colona	I	A	W			1	2	3	4	5	6	7	8	9	10	11	12
424. E. crusgalli	I	A	W			1	2	3	4	5	6	7	8	9	10	11	12
425. E. crus-pavonis	N	A	W			1	2	3	4	5	6	7	8	9	10	11	12
var. crus-pavonis																
var. macra																
426. E. esculenta	I	A	W	C				3	4							
427. E. muricata	N	A	W			1	2	3	4	5	6	7	8	9	10	11	12
var. microstachya																
var. muricata																
428. E. polystachya	N	P	W			1	2				6					
var. polystachya																
var. spectabilis																
429. E. walteri	N	A	W			1	2	3	4	5	6				10	11
104. Eriochloa																
430. E. acuminata	N	A	W			1	2		4	5		7	8	9	10	11	12
var. acuminata																
var. minor																
431. E. contracta	N	A	W			1	2	3	4	5	6	7	8	9	10	11
432. E. polystachya	I	P	W	C			2									
433. E. pseudoacrotricha	I	P	W				2				6					
434. E. punctata	N	P	W			1	2	3			6					
435. E. sericea	N	P	W			1	2	3	4	5	6	7	8	9	10	11
105. Hopia																
436. H. obtusa	N	P	W			1	2	3	4	5	6	7	8	9	10	11	12
106. Ixophorus																
437. I. unisetus	I	P	W	C			2									
107. Megathyrsus																
438. M. maximus	I	P	W	C		1	2	3	4	5	6				10	11
108. Melinis																
439. M. repens	I	P	W			1	2	3	4	5	6					11

origin (N = native, I = introduced); 2 – longevity (A = annual, B = biennial, P = perennial); 3 – season (C = cool, W = warm); 4 – special signation (O = ornamental, E = endangered, C = cultivar, W = new additions); 5 – distributional data (X = no distributional data)

Subfamily/Tribe/Genus/Specific Epithet/ Subspecific Taxa	1	2	3	4	5	Ecoregion Occurence										
109. Moorochloa																
440. M. eruciformis	I	A	W							5		7				11
110. Oplismenus																
441. O. hirtellus	N	P	W			1	2	3	4	5	6					11
subsp. setarius																
111. Panicum																
442. P. amarum	N	P	W			1	2	3								
subsp. amarulum																
subsp. amarum																
443. P. anceps	N	P	W			1	2	3	4	5	6					
subsp. anceps																
subsp. rhizomatum																
444. P. antidotale	I	P	W	C		1	2	3	4	5	6	7	8	9	10	11
445. P. bergii	I	P	W			1	2	3	4		6					
446. P. brachyanthum	N	A	W			1	2	3	4							11
447. P. capillare	N	A	W			1	2	3	4	5	6	7	8	9	10	11
subsp. capillare																
subsp. hillmanii																
448. P. capillarioides	N	P	W			1	2	3	4		6					11
449. P. coloratum	I	P	W	C		1	2	3	4	5	6	7	8			11
450. P. dichotomiflorum	N	A	W			1	2	3	4	5	6	7	8	9		11
subsp. dichotomiflorum																
451. P. diffusum	N	P	W			1	2	3	4	5	6	7	8		10	11
452. P. flexile	N	A	W			1										
453. P. ghiesbreghtii	N	P	W				2				6					
454. P. hallii	N	P	W			1	2	3	4	5	6	7	8	9	10	11
subsp. filipes																
subsp. hallii																
455. P. hemitomon	N	P	W			1	2	3	4							
456. P. hirsutum	N	P	W				2				6					
457. P. hirticaule	N	A	W						4	5	6		8	9	10	11
subsp. hirticaule																
458. P. miliaceum	I	A	W	C		1	2	3	4	5	6	7			10	11
subsp. miliaceum																
459. P. philadelphicum	N	A	W			1		3	4	5			8	9		11
subsp. philadelphicum																
460. P. repens	I	P	W			1	2	3								
461. P. rigidulum	N	P	W			1	2	3	4	5	6					11
subsp. combsii																
subsp. elongatum																
subsp. pubescens																
subsp. rigidulum																
462. P. tenerum	N	P	W			1	2	3								
463. P. trichoides	I	A	W				2	3	4							11

1 – origin (N = native, I = introduced); 2 – longevity (A = annual, B = biennial, P = perennial); 3 – season (C = cool, W = warm); 4 – special designation (O = ornamental, E = endangered, C = cultivar, W = new additions); 5 – distributional data (X = no distributional data)

Family/Tribe/Genus/Specific Epithet/Specific Taxa	1	2	3	4	5	Ecoregion Occurence											
464. P. verrucosum	N	A	W			1	2	3	4								
465. P. virgatum	N	P	W			1	2	3	4	5	6	7	8	9	10	11	
112. Paspalidium																	
466. P. germinatum	N	P	W			1	2	3	4	5	6	7				11	
113. Paspalum																	
467. P. acuminatum	N	P	W			1	2	3	4								
468. P. almum	N	P	W			1	2	3	4		6					11	
469. P. bifidum	N	P	W			1	2	3	4	5							
470. P. boscianum	N	A	W			1	2	3									
471. P. conjugatum	N	P	W			1	2										
472. P. conspersum	I	P	W			1	2										
473. P. convexum	I	A	W			1											
474. P. dilatatum	I	P	W			1	2	3	4	5	6	7	8	9	10	11	
475. P. dissectum	N	P	W			1	2	3	4								
476. P. distichum	N	P	W			1	2	3	4	5	6	7	8	9	10	11	12
var. distichum																	
var. indutum																	
477. P. floridanum	N	P	W			1	2	3	4	5		7				11	
478. P. hartwegianum	N	P	W			1	2	3	4		6					11	
479. P. intermedium	I	P	W				2										
480. P. laeve	N	P	W			1	2	3	4								
var. circulare																	
var. laeve																	
var. pilosum																	
481. P. langei	N	P	W			1	2	3	4	5	6	7	8	9		11	
482. P. lividum	N	P	W			1	2	3	4	5	6	7				11	
483. P. malacophyllum	I	P	W	C		1		3	4		6					11	
484. P. minus	N	P	W			1	2	3	4		6					11	
485. P. modestum	I	P	W				2										
486. P. monostachyum	N	P	W			1	2	3			6						
487. P. notatum	I	P	W	C		1	2	3	4	5	6				10	11	
488. P. plicatulum	N	P	W			1	2	3	4	5	6				10	11	
489. P. praecox	N	P	W			1	2	3	4								
490. P. pubiflorum	N	P	W			1	2	3	4	5	6	7	8	9	10	11	
491. P. repens	N	A	W			1	2	3	4		6					11	
492. P. scrobiculatum	I	A	W	C			2										
493. P. setaceum	N	P	W			1	2	3	4	5	6	7	8	9	10	11	
var. ciliatifolium																	
var. muhlenbergii																	
var. setaceum																	
var. stramineum																	
var. supinum																	
494. P. unispicatum	N	P	W				2	3	4		6					11	

Origin (N = native, I = introduced); 2 – longevity (A = annual, B = biennial, P = perennial); 3 – season (C = cool, W = warm); 4 – special designation (O = ornamental, E = endangered, C = cultivar, W = new additions); 5 – distributional data (X = no distributional data)

Subfamily/Tribe/Genus/Specific Epithet/ Subspecific Taxa	1	2	3	4	5	Ecoregion Occurence										
495. P. urvillei	I	P	W			1	2	3	4	5	6				10	11
496. P. vaginatum	N	P	W			1	2	3		5	6				10	11
497. P. virgatum	I	P	W				2				6					
498. P. wrightii	I	P	W				2									
114. Pennisetum																
499. P. advena	I	P	W	C												
500. P. alopecuroides	I	P	W	O						5						
501. P. ciliare	I	P	W	C		1	2	3	4	5	6				10	11
502. P. flaccidum	I	P	W	O		1		3								
503. P. glaucum	I	A	W	C		1	2	3			6					
504. P. macrostachys	I	P	W	O												
505. P. nervosum	I	P	W			1	2	3	4		6					11
506. P. orientale	I	P	W	O		1	2	3								
507. P. polystachion subsp. Setosum	I	P	W			1	2	3								
508. P. purpureum	I	P	W	O		1	2	3	4		6					
509. P. setaceum	I	P	W	O		1	2	3				7			10	11
510. P. setigerum	I	P	W	C		1	2	3	4		6					11
511. P. villosum	I	P	W	O		1		3	4	5	6	7			10	11
115. Phanopyrum																
512. P. gymnocarpon	N	P	W			1	2	3	4							
116. Sacciolepis																
513. S. indica	I	A	W			1	2									
514. S. striata	N	P	W			1	2	3	4							
117. Setaria																
515. S. adhaerans	I	A	W			1	2	3	4	5	6	7	8		10	11
516. S. corrugata	N	A	W			1	2	3	4							11
517. S. grisebachii	N	A	W				2	3	4	5	6				10	11
518. S. italica	I	A	W	C		1	2	3	4	5	6	7	8	9	10	11
519. S. leucopila	N	P	W			1	2	3	4	5	6	7	8	9	10	11
520. S. macrostachya	N	P	W				2	3	4	5	6	7	8	9	10	11
521. S. magna	N	A	W			1	2		4	5	6				10	11
522. S. megaphylla	I	P	W	C												
523. S. palmifolia	I	P	W	O		1	2	3								
524. S. parviflora	N	P	W			1	2	3	4	5	6	7	8	9	10	11
525. S. pumila subsp. pumila	I	A	W			1	2	3	4	5	6	7	8	9	10	11
526. S. reverchonii subsp. firmula subsp. ramiseta subsp. reverchonii	N	P	W				2	3	4	5	6	7	8	9	10	11
527. S. scheelei	N	P	W			1	2	3	4	5	6	7		9	10	11

1 – origin (N = native, I = introduced); 2 – longevity (A = annual, B = biennial, P = perennial); 3 – season (C = cool, W = warm); 4 – special designation (O = ornamental, E = endangered, C = cultivar, W = new additions); 5 – distributional data (X = no distributional data)

Family/Tribe/Genus/Specific Epithet/Specific Taxa	1	2	3	4	5	Ecoregion Occurence											
528. S. texana	N	P	W				2	3	4		6	7		9	10	11	
529. S. verticillata	I	A	W			1	2	3	4	5	6	7	8		10	11	
530. S. villosissima	N	P	W			1		3		5					10	11	
531. S. viridis	I	A	W			1	2	3	4	5	6	7	8	9	10	11	12
var. major																	
var. viridis																	
118. Steinchisma																	
532. S. hians	N	P	C			1	2	3	4	5	6					11	
119. Stenotaphrum																	
533. S. secundatum	N	P	W	C		1	2	3	4	5	6			9	10	11	12
120. Urochloa																	
534. U. arizonica	N	A	W			1	2	3			6				10		12
535. U. brizantha	I	P	W					3	4		6	7	8			11	
536. U. ciliatissima	N	P	W			1	2	3	4	5	6	7	8	9	10	11	12
537. U. fusca	N	A	W			1	2	3	4	5	6	7	8	9	10	11	
538. U. mosambicensis	I	P	W	C			2				6						
539. U. mutica	I	P	W	C		1	2	3			6						
540. U. panicoides	I	A	W				2				6						
var. panicoides																	
541. U. plantaginea	I	A	W														
542. U. platyphylla	N	A	W			1	2	3	4	5	6						
543. U. ramosa	I	A	W	C		1	2	3	4	5	6						
544. U. reptans	I	A	W			1	2	3	4		6					11	
545. U. subquadripara	I	A	W				2				6						
546. U. texana	N	A	W			1	2	3	4	5	6	7	8	9	10	11	
121. Zuloagaea																	
547. Z. bulbosa	N	P	W				2	3	4	5	6			8	9	10	11	12
Pooideae																	
xii. Brachyelytreae																	
122. Brachyelytrum																	
548. B. erectum	N	P	C			1											
var. erectum																	
xiii. Brachypodieae																	
123. Trachynia																	

1 – origin (N = native, I = introduced); 2 – longevity (A = annual, B = biennial, P = perennial); 3 – season (C = cool, W = warm); 4 – special designation (O = ornamental, E = endangered, C = cultivar, W = new additions); 5 – distributional data (X = no distributional data)

Subfamily/Tribe/Genus/Specific Epithet/ Subspecific Taxa	1	2	3	4	5						Ecoregion Occurence					
						1	2	3	4	5	6	7	8	9	10	11
549. T. distachya	I	A	C		x											
xiv. Bromeae																
124. Anisantha																
550. A. diandrus	I	A	C			1	2	3	4	5					10	11
551. A. rubens	I	A	C								6				10	11
552. A. sterilis	I	A	C													
553. A. tectorum	I	A	C			1	2	3	4	5	6	7	8	9	10	11
125. Bromopsis																
554. B. anomalus	N	P	C					3	4		6				10	11
555. B. ciliatus	N	P	C												10	
556. B. inermis	I	P	C	C		1		3	4				8	9		
557. B. lanatipes	N	P	C										8	9	10	
558. B. porteri	N	P	C													
559. B. pubescens	N	P	C			1	2	3	4	5	6					11
560. B. richardsonii	N	P	C												10	
561. B. texensis	N	A	C				2	3	4		6				10	11
126. Bromus																
562. B. arenarius	I	A	C			1										
563. B. commutatus	I	A	C			1	2	3	4	5						
564. B. hordeaceus	I	A	C			1		3	4	5	6				10	11
subsp. hordeaceus																
subsp. molliformis																
565. B. japonicus	I	A	C			1	2	3	4	5	6	7	8	9	10	11
566. B. lanceolatus	I	A	C	C		1		3								
567. B. racemosus	I	A	C			1	2									11
568. B. secalinus	I	A	C			1	2	3	4	5	6	7	8	9	10	11
127. Ceratochloa																
569. C. arizonica	N	A	C													
570. C. carinata	N	A	C												10	
var. carinata																
571. C. cathartica	I	A	C	C		1	2	3	4	5	6	7	8	9	10	11
var. cathartica																
572. C. polyantha	N	P	C												10	
xv. Diarrheneae																
128. Diarrhena																
573. D. americana	N	P	C													
574. D. obovata	N	P	C						4	5						
xvi. Meliceae																

1 – origin (N = native, I = introduced); 2 – longevity (A = annual, B = biennial, P = perennial); 3 – season (C = cool, W = warm); 4 – special designation (O = ornamental, E = endangered, C = cultivar, W = new additions); 5 – distributional data (X = no distributional data)

family/Tribe/Genus/Specific Epithet/ specific Taxa	1	2	3	4	5	Ecoregion Occurence											
129. Glyceria																	
575. G. grandis	N	P	C			1											
576. G. notata	I	P	C														
577. G. septentrionalis	N	P	C			1	2	3	4								
var. arkansana																	
var. septentrionalis																	
578. G. striata	N	P	C			1		3	4	5					10	11	12
130. Melica																	
579. M. bulbosa	N	P	W												10	11	
580. M. montezumae	N	P	W								6				10	11	12
581. M. mutica	N	P	W			1	2	3	4	5	6	7	8	9		11	
582. M. nitens	N	P	W			1	2	3	4	5	6	7	8	9	10	11	12
583. M. porteri	N	P	W									7	8		10		12
var. laxa																	
var. porteri																	
xvii. Poeae																	
131. Agrostis																	
584. A. elliottiana	N	A	C			1	2	3	4	5						11	
585. A. exarata	N	P	C												10		12
586. A. gigantea	I	P	C														
587. A. hyemalis	N	P	C			1	2	3	4	5	6		8	9	10	11	
588. A. perennans	N	P	C			1	2	3	4								
589. A. scabra	N	P	C			1	2	3	4						10		
590. A. stolonifera	I	P	C			1	2			5		7	8	9	10		
132. Aira																	
591. A. caryophyllea	I	A	C			1	2	3	4	5							
var. capillaris																	
133. Alopecurus																	
592. A. carolinianus	N	A	C			1	2	3	4	5		7	8			11	
593. A. myosuroides	I	A	C														
134. Anthoxanthum																	
594. A. aristatum	I	A	C					3									
595. A. odoratum	I	P	C			1		3									
135. Apera																	
596. A. spica-venti	I	A	C					3	4								
136. Avena																	
597. A. fatua	I	A	C			1	2	3	4	5	6	7	8		10	11	12

Subfamily/Tribe/Genus/Specific Epithet/ Subspecific Taxa	1	2	3	4	5	Ecoregion Occurence										
						1	2	3	4	5	6	7	8	9	10	11
var. fatua																
598. A. sativa	I	A	C	C		1	2	3	4	5	6	7		9	10	11
var. sativa																
137. Briza																
599. B. maxima	I	A	C													
600. B media	I	P	C	O												
601. B. minor	I	A	C	O		1	2	3	4	5	6					11
138. Calamagrostis																
602. C. arundinacea var. brachytricha	I	P	C	O	X											
603. C. xacutiflora Karl Foerster'	I	P	C	O	X											
139. Cinna																
604. C. arundinacea	N	P	C			1		3				7				
140. Cynosurus																
605. C. echinatus	I	A	C			1		3								
141. Dactylis																
606. D. glomerata	I	P	C	C		1	2	3	4	5			8	9		11
142. Desmazeria																
607. D. rigida subsp. rigida	I	A	C			1	2	3	4	5						11
143. Festuca																
608. F. arizonica	N	P	C												10	
609. F. ligulata	N	P	C												10	
610. F. paradoxa	N	P	C			1		3	4							
611. F. rubra	N	P	C	C											10	
612. F. subverticillata	N	P	C			1		3	4	5						
613. F. versuta	N	P	C					3	4	5	6					11
144. Gastridium																
614. G. phleoides	I	A	C													
145. Hainardia																
615. H. cylindrica	I	A	C			1	2	3								
146. Holcus																
616. H. lanatus	I	P	C			1		3	4						10	11
147. Koeleria																

1 – origin (N = native, I = introduced); 2 – longevity (A = annual, B = biennial, P = perennial); 3 – season (C = cool, W = warm); 4 – special designation (O = ornamental, E = endangered, C = cultivar, W = new additions); 5 – distributional data (X = no distributional data)

Family/Tribe/Genus/Specific Epithet/Specific Taxa	1	2	3	4	5	Ecoregion Occurrence 1	2	3	4	5	6	7	8	9	10	11	12
617. K. macrantha	N	P	C					3	4	5		7			10		12
148. Lachnagrostis																	
618. L. filiformis	I	P	C												10		12
149. Lamarckia																	
619. L. aurea	I	A	C														
150. Limnodea																	
620. L. arkansana	N	A	C			1	2	3	4	5	6	7	8	9	10	11	
151. Lolium																	
621. L. multiflorum	I	A	C	C		1		3	4	5				9	10		
622. L. perenne	I	P	C	C		1	2	3	4	5	6	7	8	9	10	11	12
623. L. rigidum	I	A	C			1		3									
624. L. temulentum subsp. temulentum	I	A	C			1	2	3	4	5	6	7	8	9		11	
152. Parapholis																	
625. P. incurva	I	A	C			1	2	3			6						
153. Phalaris																	
626. P. angusta	N	A	C			1	2	3							10	11	
627. P. aquatica	I	P	C	C													
628. P. brachystachys	I	A	C														
629. P. canariensis	I	A	C	C		1	2	3	4	5	6	7		9		11	
630. P. caroliniana	N	A	C			1	2	3	4	5	6	7	8	9	10	11	12
631. P. minor	N	A	C					3	4								
154. Phalaroides																	
632. P. arundinacea	N	P	C	O													
155. Phleum																	
633. P. pratense susp. Pratense	I	P	C	C		1		3	4	5	6		8	9		11	
156. Poa																	
634. P. annua	I	A	C			1	2	3	4	5	6	7	8	9	10	11	12
635. P. arachnifera	N	P	C			1	2	3	4	5	6	7	8	9	10	11	12
636. P. arida	N	P	C			1	2	3				7	8	9	10		
637. P. autumnalis	N	P	C			1	2	3	4								
638. P. bigelovii	N	A	C					3	4	5	6	7	8		10	11	12
639. P. bulbosa	I	P	C	C					4	5							
640. P. chapmaniana	N	A	C			1		3	4	5							
641. P. compressa	I	P	C	C			2		4	5	6	7	8	9	10		
642. P. fendleriana	N	P	C												10		12

- origin (N = native, I = introduced); 2 – longevity (A = annual, B = biennial, P = perennial); 3 – season (C = cool, W = warm); 4 – special
signation (O = ornamental, E = endangered, C = cultivar, W = new additions); 5 – distributional data (X = no distributional data)

Subfamily/Tribe/Genus/Specific Epithet/ Subspecific Taxa	1	2	3	4	5						Ecoregion Occurence					
subsp. fendleriana																
643. P. interior	N	P	C													
644. P. leptocoma	N	P	C													
645. P. occidentalis	N	P	C												10	
646. P. pratensis	I	P	C	C		1		3	4	5		7	8	9	10	11
647. P. reflexa	N	P	C													
648. P. secunda	N	P	C					3							10	
649. P. strictiramea	N	P	C												10	
650. P. sylvestris	N	P	C			1	2	3	4	5	6					11
651. P. trivialis	I	P	C	C				3	4							11
157. Polypogon																
652. P. elongatus	N	P	C												10	
653. P. interruptus	N	P	C								6				10	11
654. P. maritimus	I	A	C					3	4		6					11
655. P. monspeliensis	I	A	C			1	2	3	4	5	6	7	8	9	10	11
656. P. viridis	I	P	C			1	2	3	4	5	6	7		9	10	11
158. Rostraria																
657. R. cristata	I	A	C			1	2	3								
159. Schedonorus																
658. S. arundinaceus	I	P	C	C		1	2	3	4	5		7	8	9	10	11
659. S. pratensis	I	P	C	C									8	9		
160. Sclerochloa																
660. S. dura	I	A	C			1		3	4	5	6					11
161. Sphenopholis																
661. S. filiformis	N	P	C			1	2	3								
662. S. intermedia	N	P	C			1	2	3	4				8	9	10	
663. S. longiflora	N	P	C			1	2	3	4							
664. S. nitida	N	P	W			1	2	3								
665. S. obtusata	N	P	C			1	2	3	4	5	6	7	8	9	10	11
162. Trisetum																
666. T. interruptum	N	A	C			1	2	3	4	5	6	7	8	9	10	11
667. T. spicatum	N	P	C												10	
163. Vulpia																
668. V. bromoides	I	A	C			1		3	4	5						
669. V. myuros	I	A	C			1	2	3	4	5						11
670. V. octoflora	N	A	C			1	2	3	4	5	6	7	8	9	10	11
var. glauca																
var. hirtella																
var. octoflora																

1 – origin (N = native, I = introduced); 2 – longevity (A = annual, B = biennial, P = perennial); 3 – season (C = cool, W = warm); 4 – special designation (O = ornamental, E = endangered, C = cultivar, W = new additions); 5 – distributional data (X = no distributional data)

family/Tribe/Genus/Specific Epithet/ specific Taxa	1	2	3	4	5	Ecoregion Occurence 1	2	3	4	5	6	7	8	9	10	11	12
671. V. sciurea	N	A	C			1	2	3	4	5	6					11	
xviii. Stipeae																	
164. Achnatherum																	
672. A. aridum	N	P	W												10	11	
673. A. curvifolium	N	P	C												10		12
674. A. eminens	N	P	W											9	10	11	12
675. A. hymenoides	N	P	W									7	8	9	10	11	12
676. A. lobatum	N	P	W												10	11	12
677. A. nelsonii	N	P	W												10	11	12
678. A. perplexum	N	P	C												10		12
679. A. robustum	N	P	W														
680. A. scribneri	N	P	C												10		12
165. Amelichloa																	
681. A. clandestina	N	P	C							5		7	8			11	
166. Hesperostipa																	
682. H. comata	N	P	C								6	7	8	9	10	11	12
subsp. comata																	
subsp. Intermedia																	
683. H. neomexicana	N	P	C									7	8	9	10	11	12
167. Nassella																	
684. N. leucotricha	N	P	C			1	2	3	4	5	6	7	8	9	10	11	
685. N. tenuissima	N	P	C	O							6	7	8		10	11	12
686. N. viridula	N	P	C					3		5						11	
168. Piptatherum																	
687. M. micranthum	N	P	C										8	9	10		12
169. Piptochaetium																	
688. P. avenaceum	N	P	C			1	2	3	4	5	6	7				11	
689. P. fimbriatum	N	P	C												10		12
690. P. pringlei	N	P	C												10		12
xix. Triticeae																	
170. Aegilops																	
691. A. cylindrica	I	A	C			1	2	3	4	5		7	8	9		11	
171. Agropyron																	
692. A. cristatum	I	P	C	C													
172. Critesion																	

1 – origin (N = native, I = introduced); 2 – longevity (A = annual, B = biennial, P = perennial); 3 – season (C = cool, W = warm); 4 – special designation (O = ornamental, E = endangered, C = cultivar, W = new additions); 5 – distributional data (X = no distributional data)

Subfamily/Tribe/Genus/Specific Epithet/ Subspecific Taxa	1	2	3	4	5	Ecoregion Occurence										
693. C. brachyantherum subsp. brachyantherum	N	P	C													
694. C. jubatum subsp. jubatum	N	P	C									7	8	9	10	11
695. C. marinum subsp. gussoneanum	I	A	C													
696. C. murinum subsp. glaucum subsp. leporinum	I	A	C			1		3	4	5	6	7	8	9	10	11
697. C. pusillum	N	A	C			1	2	3	4	5	6	7	8	9	10	11
173. Elymus																
698. E. arizonicus	N	P	C												10	
699. E. canadensis var. brachystachys var. canadensis	N	P	C			1	2	3	4	5	6	7	8	9	10	11
700. E. curvatus	N	P	C									7	8			
701. E. elymoides subsp. brevifolius	N	P	C							5	6	7	8	9	10	11
702. E. glabriflorus	N	P	C			1	2	3	4	5	6					11
703. E. interruptus	N	P	C			1	2	3	4	5	6	7	8	9	10	11
704. E. macgregorii	N	P	C			1	2	3	4	5	6	7			10	11
705. E. pringlei	N	P	C													
706. E. repens	I	P	C													11
707. E. riparius	N	P	C								6					11
708. E. texensis	N	P	C							5	6					11
709. E. trachycaulus var. trachycaulus	N	P	C									7	8	9	10	
710. E. villosus	N	P	C					3	4	5	6					11
711. E. virginicus var. jejunus var. virginicus	N	P	C			1	2	3	4	5	6	7	8	9	10	11
174. Hordeum																
712. H. vulgare subsp. Vulgare	I	A	C	C		1	2	3	4	5		7	8		10	11
175. Leymus																
713. L. arenarius	I	P	C	O												
714. L. triticoides	N	P	C													
176. Pascopyrum																
715. P. smithii	N	P	C			1	2	3	4	5		7	8	9	10	11
177. Psathrostachys																

1 – origin (N = native, I = introduced); 2 – longevity (A = annual, B = biennial, P = perennial); 3 – season (C = cool, W = warm); 4 – special designation (O = ornamental, E = endangered, C = cultivar, W = new additions); 5 – distributional data (X = no distributional data)

Family/Tribe/Genus/Specific Epithet/ specific Taxa	1	2	3	4	5	Ecoregion Occurence												
716. P. juncea	I	P	C	C														
178. Pseudoroegneria																		
717. P. spicata	N	P	C												10		12	
179. Secale																		
718. S. cereale	I	A	C	C		1	2	3	4	5								
719 S. strictum	I	P	C															
180. Thinopyrum																		
720. T. intermedium	I	P	C	C											10			
subsp. barbulatum																		
subsp. intermedium																		
721. T. ponticum	I	P	C	C											10			
181. Triticum																		
722. T. aestivum	I	A	C	C		1	2	3	4	5	6	7	8	9	10	11		
723. T. spelta	I	A	C	C														

Achnatherum P. Beauv.
(Poöideae: Stipeae)

Cespitose perennials, sometimes with short rhizomes. Inflorescence a terminal, usually contracted, panicle of 1-flowered, awned spikelets. Disarticulation above the glumes. Glumes longer than the florets, acute to acuminate and usually tapering to a hairlike tip. Lemma membranous to coriaceous, usually pubescent and terminating in a prominent awn. Palea $1/3$ to as long as the lemma, usually pubescent. Basic chromosome numbers, $x = 10$ and 11. Photosynthetic pathway, C_3. Represented in Texas by 9 species. At some time or another all the following have been included in *Stipa* and/or *Oryzopsis* (Gould 1975b; Hatch, Gandhi, and Brown 1990; Jones, Wipff, and Montgomery 1997).

1. Awn basal segments pilose, hairs >0.5 mm; blades forming circular arcs with age .... 2. *A. curvifolium*
1. Awn basal segments scabrous or hairs <0.5 mm; blades straight, not forming circular arcs.
  2. Lemmas evenly hairy, 1.2–6.0 mm; hairs on the body not any longer than those at apices .... 4. *A. hymenoides*
  2. Lemmas glabrous or with hairs 0.2–2.0 mm long at middle, glabrous or hairs distally, hairs at the middle markedly shorter than those at apices.
    3. Lemma apical hairs 2–7 mm long .... 9. *A. scribneri*
    3. Lemma apical hairs absent or <2 mm long.
      4. Terminal awn segment flexuous.
        5. Panicle contracted; ligules on flag leaves to 1.5 mm .... 1. *A. aridum*
        5. Panicle open; ligules on flag leaves to 4.5 mm .... 3. *A. eminens*
      4. Terminal awn segment straight or slightly arcuate.
        6. Flag leaves with a densely pubescent collar, hairs >0.5 mm .... 8. *A. robustum*
        6. Flag leaves glabrous or sparsely pubescent at the collar, hairs <0.5 mm.
          7. Glumes about equal, lower exceeding the upper by <1 mm ..... 6. *A. nelsonii*
          7. Glumes unequal, lower exceeding the upper by 1–4 mm.
            8. Lemma apical hairs erect, lobes 0.5–1.2 mm .... 5. *A. lobatum*
            8. Lemma apical hairs divergent to ascending, lobes 0.2–0.5 mm ....
            .... 7. *A. perplexum*

(Barkworth 2007a)

1 cm

5 mm

floret

2 mm

2 mm

C̄R̄

floret

1. *Achnatherum aridum* (M. E. Jones) (Mormon needlegrass). Cespitose perennial without rhizomes. Panicle contracted with appressed branches, appearing "*Elymus*-like" at first glance. Ligule minute. Reported from Texas (Jones, Wipff, and Montgomery 1997, as *S. arida* M. E. Jones), but no specimens have been located to verify these reports (Barkworth 2007a). If it does occur in Texas, it would be expected from the pinyon-juniper woodlands of the Trans-Pecos area.

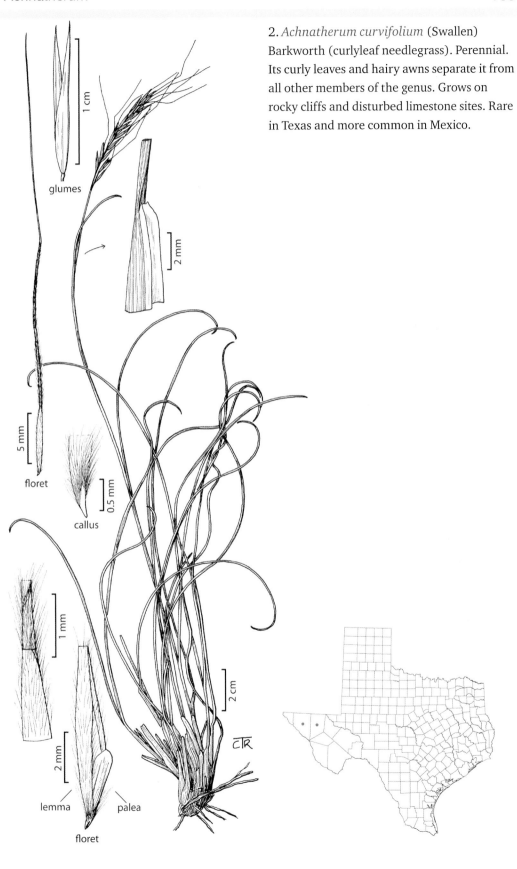

glumes

floret

callus

lemma   palea

floret

2. *Achnatherum curvifolium* (Swallen) Barkworth (curlyleaf needlegrass). Perennial. Its curly leaves and hairy awns separate it from all other members of the genus. Grows on rocky cliffs and disturbed limestone sites. Rare in Texas and more common in Mexico.

3. *Achnatherum eminens* (Cav.) Barkworth (southwestern needlegrass). Perennial, sometimes with rhizomatous, knotty bases. Distinguished by its open panicle and flexuous branches and awns. Generally found on dry, rocky slopes and in mountain valleys in the Trans-Pecos region. Considered by Powell (1994, as *Stipa eminens* Cav.) to be the most widespread "needlegrass" in the Trans-Pecos, but it is of limited forage value.

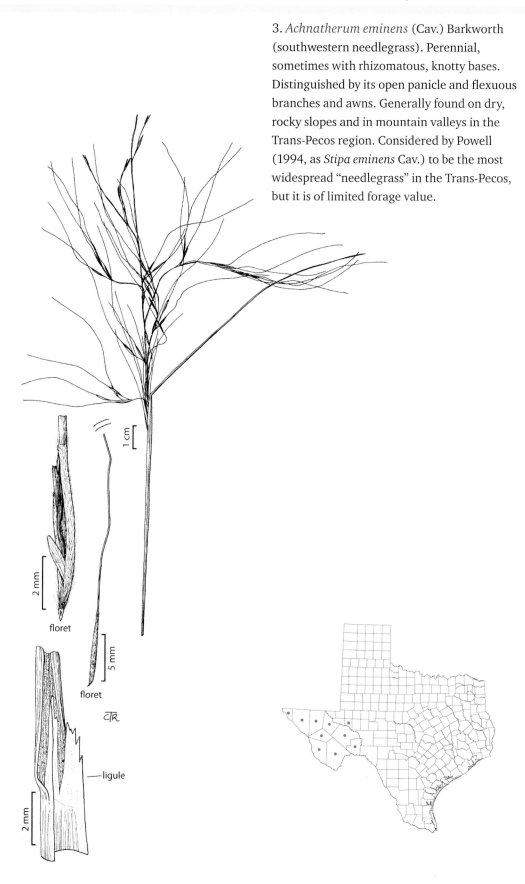

floret

floret

ligule

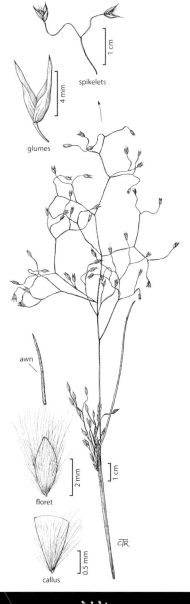

spikelets

4 mm

glumes

1 cm

awn

2 mm

floret

0.5 mm

callus

1 cm

CTR

4. *Achnatherum hymenoides* (Roem. and Schult.) Barkworth (Indian ricegrass). Perennial typically found on sandier soils. Characteristic inflorescence branching pattern that resembles "chicken wire" (Allred 2005). Seeds used as food source by indigenous peoples of southwestern North America. Highly palatable forage species that has been used in restoring roadsides and other disturbed sites.

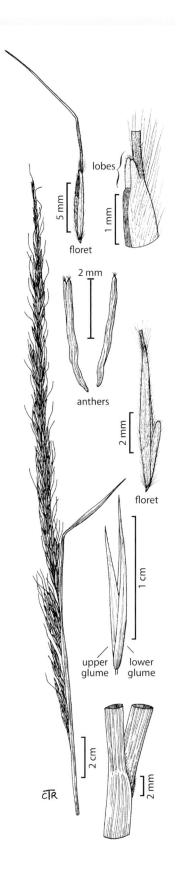

lobes

5 mm

1 mm

floret

2 mm

anthers

2 mm

floret

1 cm

upper
glume

lower
glume

2 cm

2 mm

CTR

5. *Achnatherum lobatum* (Swallen) Barkworth (lobed needlegrass). Cespitose perennial that grows up to 1 m tall. Scattered on rocky slopes in pinyon woodlands up to fir communities at elevations of 6,900–8,900 ft (2,100–2,700 m). It differs from *A. scribneri* by its shorter apical lemma hairs and blunt calluses, and from *A. perplexum* in having longer lemma lobes and erect apical hairs (Barkworth 2007a). Locally abundant but usually of limited forage value (Powell 1994).

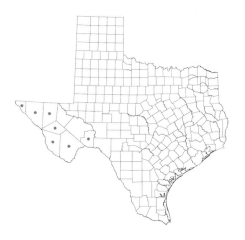

6. *Achnatherum nelsonii* (Scribn.)
Barkworth (Nelson's needlegrass). Cespitose
perennial without rhizomes. It can grow to
1.75 m tall. Typically flowers in late spring to
early summer. The Texas species belongs to
subsp. *nelsonii.* Found in sagebrush steppes,
pinyon-juniper woodlands, and up to
subalpine forests.

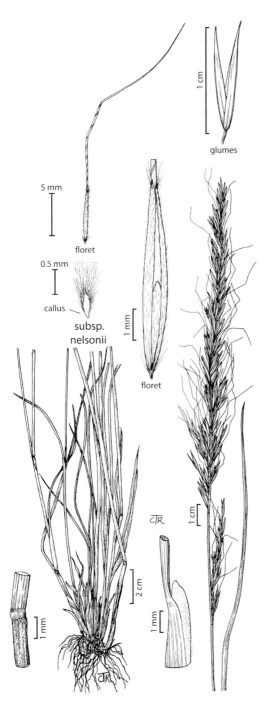

1 cm

glumes

5 mm

floret

0.5 mm

callus

subsp.
nelsonii

1 mm

floret

2 cm

1 mm

1 cm

1 mm

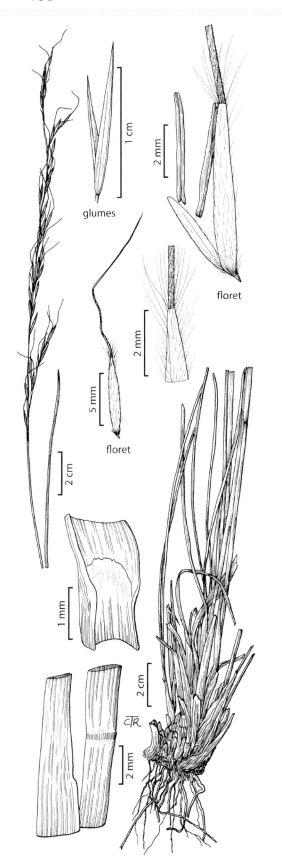

glumes

1 cm

2 mm

floret

2 mm

5 mm

floret

2 cm

1 mm

2 cm

2 mm

CTR

7. *Achnatherum perplexum* Hoge and Barkworth (perplexing needlegrass). Perennial growing up to 1 m tall. Typically flowers in late summer and early fall. Generally found in pinyon-juniper woodlands. Within the state, it has been collected only in El Paso and Culberson counties.

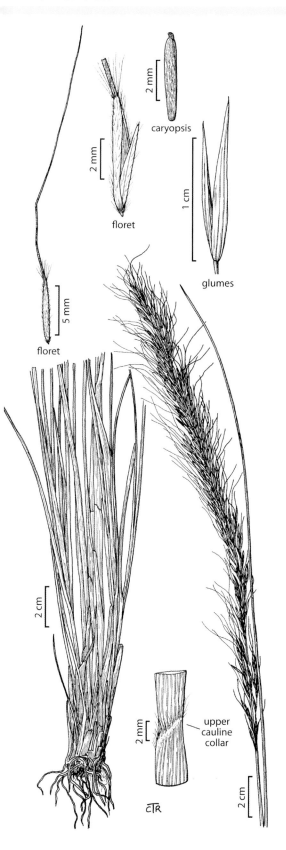

caryopsis

floret

glumes

floret

upper
cauline
collar

CTR

8. *Achnatherum robustum* (Vasey) Barkworth (sleepygrass). A large cespitose perennial growing to over 2 m tall. Most likely occurs at higher elevations in the Trans-Pecos area of the state. Plants contain a narcotic that produces torpor in horses and cattle who ingest it, but it is not lethal (Allred 2005). The toxin is probably produced by an endophytic fungus infesting the plant. The compound responsible for the poison is diacetone alcohol (Epstein, Gerber, and Karler 1964). The plant has the potential to be used as a native ornamental in arid regions of the state with cold winters.

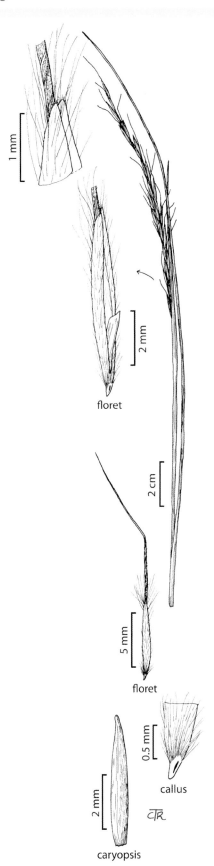

floret

floret

callus

caryopsis

CTR

9. *Achnatherum scribneri* (Vasey) Barkworth (Scribner's needlegrass). Strongly cespitose perennial. The sharp callus is distinctive. Just extending into Texas in the mountains of the Trans-Pecos, where it grows in pinyon-juniper and ponderosa pine woodlands.

Aegilops L.
(Poöideae: Triticeae)

Annuals. Inflorescence a bilateral spike usually with 1–3 rudimentary spikelets below. Disarticulation either at the base of the spikes or in the rachis; spikelets falling attached to the internodes above and below. Spikelets solitary at the nodes with 2–8 florets. Glumes oblong to nearly rectangular, rounded on the back, coriaceous and becoming indurate at maturity, truncate, dentate, or 1–5-awned. Lemmas rounded on the back, toothed, 1–3-awned. Paleas chartaceous, 2-keeled, keels ciliated. Basic chromosome number, $x = 7$. Photosynthetic pathway, C_3. Represented in Texas by a single species. This genus is sometimes included with *Triticum*.

(Gould 1975b; Hatch, Gandhi, and Brown 1990)

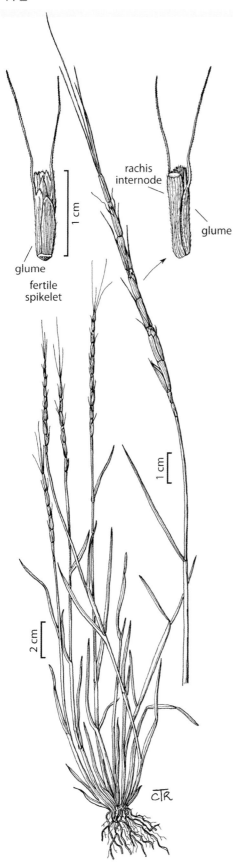

glume

fertile
spikelet

rachis
internode

glume

1 cm

1 cm

2 cm

CTR

1. *Aegilops cylindrica* Host (jointed goatgrass). Annual. Culms to about 50 cm tall, usually with numerous tillers. Spikes up to 15 cm long. Spikelets cylindric, 3–5 florets. Glumes and lemmas awned. A troublesome weed introduced from the Mediterranean region. Typically found in disturbed areas or winter wheat fields.

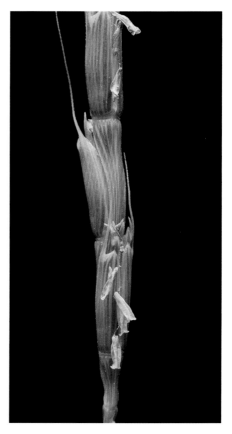

Agropyron Gaertn.
(Poöideae: Triticeae)

Perennial, densely cespitose, occasionally rhizomatous. Inflorescence a bilateral spike with spikelets 1 per node. Spikelets divergent or pectinate on a very thick and tough rachis, with 3–10 florets; disarticulation is above the glumes below each floret. Glumes strongly keeled to the base, tapering to an acuminate or to short awn tip in which the veins converge. Lemmas keeled, tipped like that of the glumes. Paleas membranous, about as long as the lemmas, usually adhering to the caryopsis. Basic chromosome number, $x = 7$. Photosynthetic pathway, C_3. Represented in Texas by a single species. Species previously included in this genus by Gould (1975b) have been moved to other genera, such as *Elymus, Leymus, Pascopyrum, Pseudoroegneria,* and *Thinopyrum*.

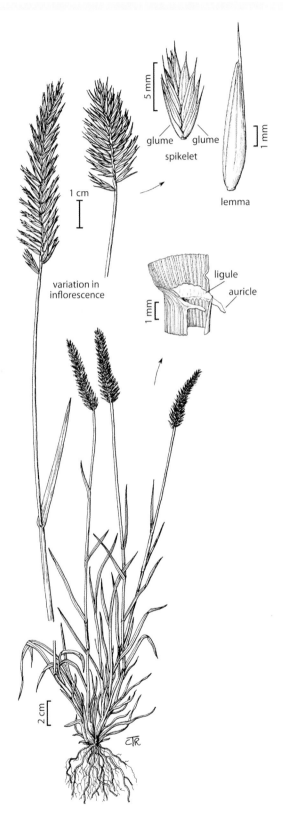

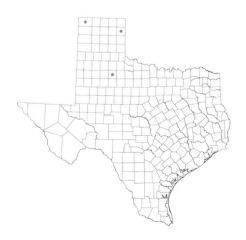

1. *Agropyron cristatum* (L.) Gaertn. (crested wheatgrass). Erect perennial up to 1 m tall. Spikes up to 15 cm long. Spikelets typically diverging at angles of 30°–95° with 3–8 florets. An easily established, introduced forage species used extensively in revegetation of depleted rangelands and along roadsides throughout the western United States. Numerous cultivars are available. Expected in the Panhandle region of the state.

Agrostis L.
(Poöideae: Poeae)

Low to moderately tall annuals and perennials with slender culms, flat or involute blades, and open or contracted panicles. Spikelets small, 1-flowered, disarticulating above the glumes. Rachilla usually not hairy or extended beyond the insertion of the floret. Glumes thin, lanceolate, acute to acuminate, nearly equal, the lower usually 1-nerved, the upper 1- or 3-nerved. Lemma thin, broad, 3- or 5-nerved, acute to obtuse or truncate at the apex, glabrous or with a tuft of hair at the base, awnless or awned from the middle or below. Palea hyaline, usually small or absent. Basic chromosome number, $x = 7$. Photosynthetic pathway, C_3. Represented in Texas by 7 species.

1. Paleas at least $^2/_5$ as long as lemmas.
   2. Panicles contracted, branches appressed.
      3. Stolons present, rhizomes absent ....7. *A. stolonifera*
      3. Stolons absent, plant cespitose or rhizomes sometimes present ....2. *A. exarata*
   2. Panicles open, branches spreading.
      4. Stolons present, rhizomes absent...7. *A. stolonifera*
      4. Stolons absent, rhizomes present.. 3. *A. gigantea*
1. Paleas absent or less than $^2/_5$ as long as lemmas.
      5. Lemmas awned.
         6. Plants annual; awn flexible, anthers 1 ....1. *A. elliottiana*
         6. Plants perennial; awn straight; anthers 3.
            7. Panicle strongly contracted, spikelike.....................................2. *A. exarata*
            7. Panicle open, not spikelike.
               8. Panicles diffuse, nearly as wide as long, entire panicle disarticulating and becoming a tumbleweed, leaves mostly basal .... 6. *A. scabra*
               8. Panicle open, not nearly as wide as long, not becoming a tumbleweed, leaves basal and cauline ....5. *A. perennans*
      5. Lemmas awnless.
         9. Panicles contracted, branches appressed.........................2. *A. exarata*
         9. Panicles open or diffuse, branches descending.
            10. Lemmas <1.2 mm long ....4. *A. hyemalis*
            10. Lemmas >1.2 mm long.
               11. Panicles diffuse, nearly as wide as long, entire panicle disarticulating and becoming a tumbleweed, leaves mostly basal.. 6. *A. scabra*
               11. Panicle open, not nearly as wide as long, not becoming a tumbleweed, leaves basal and cauline............. 5. *A. perennans*

(Harvey 2007a)

1. *Agrostis elliottiana* Schult. (Elliott's bent, annual ticklegrass). An annual <0.5 m tall. Basal leaf blades typically withered by anthesis. Differs from A. *scabra* and A. *hyemalis* in its flexible awn and single anther. Poor livestock and wildlife values (Hatch, Schuster, and Drawe 1999).

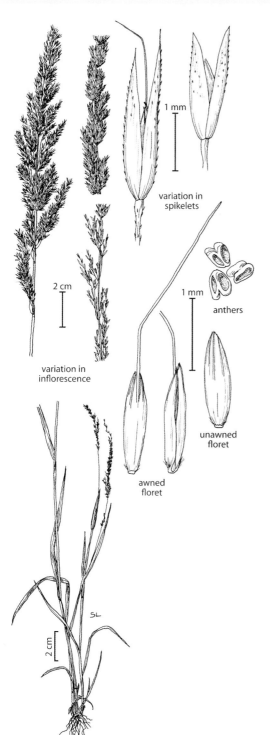

1 mm

variation in
spikelets

2 cm

1 mm

anthers

variation in
inflorescence

unawned
floret

awned
floret

2 cm

SL

2. *Agrostis exarata* Trin. (spike bent). A native perennial typically found in moist, open ground. Occurs only in the far western part of Texas, typically at higher altitudes. An excellent forage species (Allred 2005).

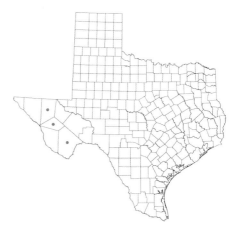

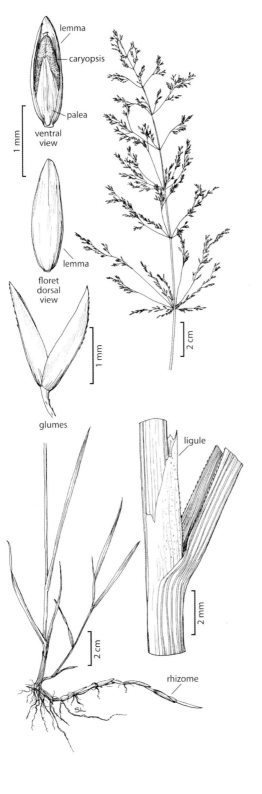

3. *Agrostis gigantea* Roth (redtop bent).
Rhizomatous perennial that can grow to
>1 m tall. Panicle is typically more open
than that in *A. stolonifera*. A weedy species
of roadsides, ditches, and fields but can also
be an important soil stabilizer. An excellent
forage species (Allred 2005). Expected in west
Texas along the New Mexico border.

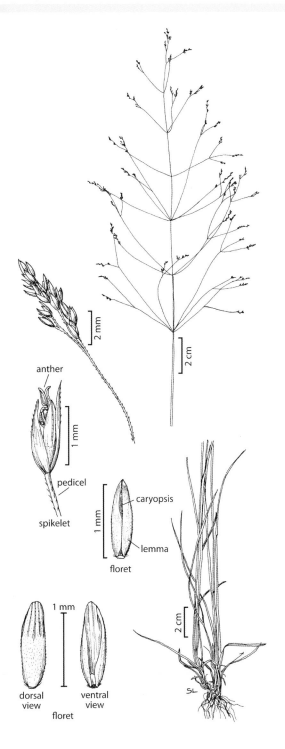

4. *Agrostis hyemalis* (Walter) Britton, Sterns, & Poggenb. (winter bent, fly-away grass, ticklegrass). A weak perennial or annual. Diffuse panicle often detaches at the base and becomes a tumbleweed. Common along roadsides. It differs from the closely related *A. scabra* in its smaller spikelets and anthers, conspicuous culm leaves, and clustered spikelets. Poor livestock and wildlife values (Hatch, Schuster, and Drawe 1999).

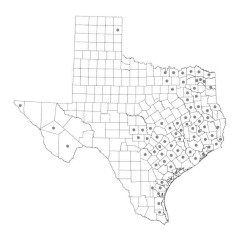

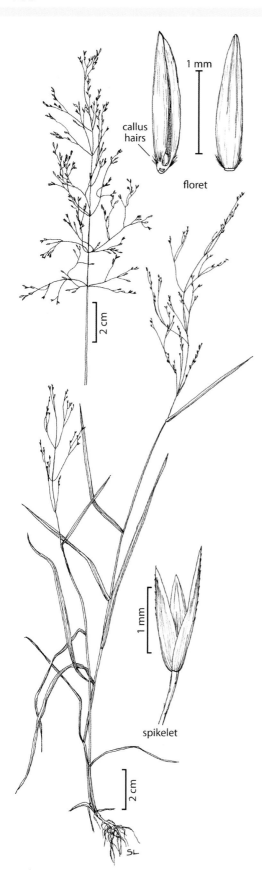

callus
hairs

floret

spikelet

5. *Agrostis perennans* (Walter) Tuck. (autumn bent, upland bent). Perennial found along roadsides, in fields, and on stream banks. Grows in shadier and moister sites than *A. scabra*. Common in the eastern part of the state.

6. *Agrostis scabra* Willd. (ticklegrass, rough bent). A weak perennial or occasionally an annual. Diffuse panicle frequently detaches at the base and becomes a tumbleweed. Grows in a variety of habitats and frequently is one of the first plants to appear on burned-over sites within its range.

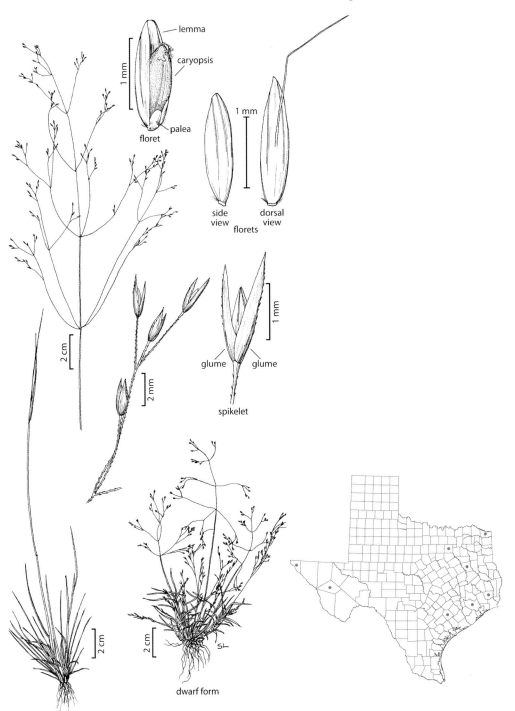

lemma

caryopsis

palea

floret

1 mm

side view

dorsal view

florets

1 mm

glume   glume

spikelet

2 cm

2 mm

2 cm

2 cm

dwarf form

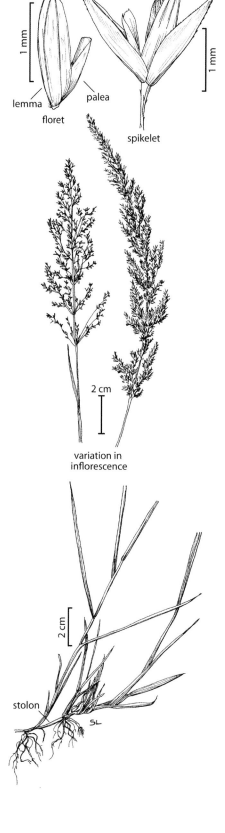

lemma
floret
palea

spikelet

1 mm

1 mm

variation in
inflorescence

2 cm

stolon
SL

2 cm

7. *Agrostis stolonifera* L. (creeping bent).
A stoloniferous perennial that can form
large mats. It is often found in areas that are
temporarily flooded, as well as along roadsides
and ditches. Differs from *A. gigantea* by having
stolons, lacking rhizomes, and having a more
closed inflorescence. Poor livestock and
wildlife values (Hatch, Schuster, and Drawe
1999). Reported to be highly palatable but
not abundant enough to be important (Powell
1994).

Aira L.
(Poöideae: Poeae)

Delicate, tufted annuals with thin, subfiliform, mostly basal leaves. Inflorescence an open or contracted panicle of small, 2-flowered spikelets. Rachis disarticulating above the glumes and between the florets, not prolonged behind the upper floret. Glumes about equal, longer than the lemmas, thin, lanceolate, 1- or obscurely 3-nerved. Lemmas firm, rounded on the back, awned from below the middle with a usually geniculate and twisted, hairlike awn, tapering to 2 slender teeth or setae at the tip. Awn of lower lemma sometimes lacking or reduced. Palea thin, shorter than the lemma. Basic chromosome number, $x = 7$. Photosynthetic pathway, C_3. Represented in Texas by a single species and variety.

(Wipff 2007a)

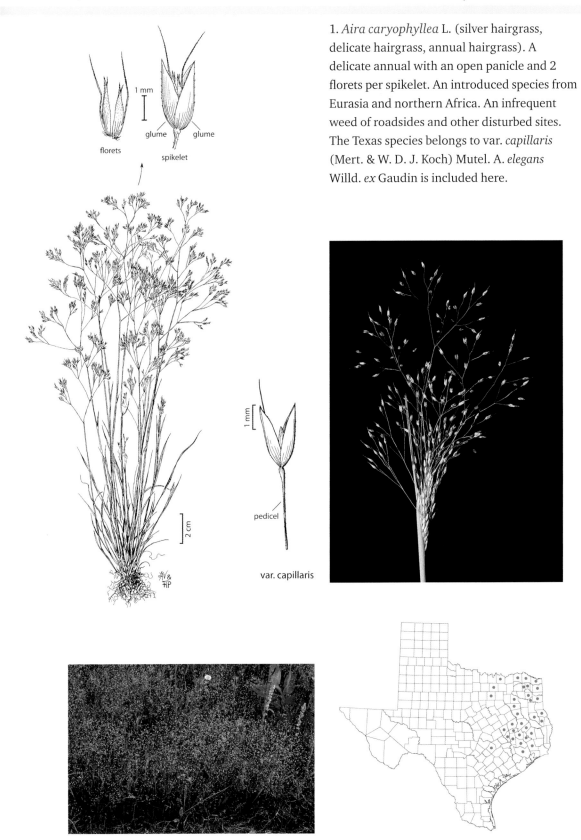

florets

glume   glume

spikelet

1 mm

pedicel

var. capillaris

2 cm

1. *Aira caryophyllea* L. (silver hairgrass, delicate hairgrass, annual hairgrass). A delicate annual with an open panicle and 2 florets per spikelet. An introduced species from Eurasia and northern Africa. An infrequent weed of roadsides and other disturbed sites. The Texas species belongs to var. *capillaris* (Mert. & W. D. J. Koch) Mutel. A. *elegans* Willd. *ex* Gaudin is included here.

Allolepis Soderstrom and Decker
(Chloridoideae: Cyndonteae)

Dioecious perennial with or without rhizomes and with long, creeping stolons. Culms decumbent at base, 30–60 cm tall. Blades flat, firm, glabrous or scabrous, with bicellular microhairs not sunken in the epidermis. Inflorescence a contracted panicle of large, 4–8-flowered, unisexual spikelets. Pistillate spikelets about twice as large as staminate ones. Glumes unequal, acute. Lemmas of pistillate spikelets broad, coriaceous, laterally compressed, acute at the apex, with 3 strong nerves and indistinct intermediate nerves, the margins thin and erose. Paleas of the pistillate spikelets broad and bowed out below, narrow above, keeled, the 2 keels with narrow, erose, or toothed wings. Caryopsis tightly enclosed by the base of the palea. Lemmas of the staminate spikelets thin, 3-nerved. Palea of the staminate spikelets not bowed out and lacking winged keels. Basic chromosome number, $x = 10$. Photosynthetic pathway, C_4. A monotypic genus found only in the Big Bend region of Texas and adjacent areas of Mexico. Represented in Texas by a single species.

(Wipff 2003a)

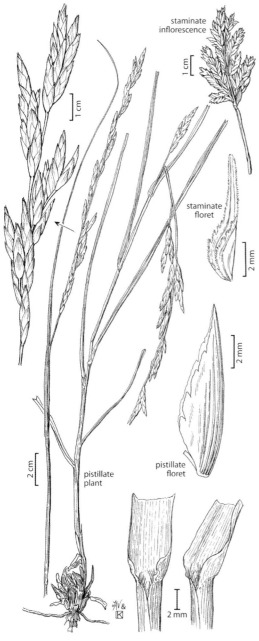

staminate
inflorescence

staminate
floret

pistillate
plant

pistillate
floret

1. *Allolepis texana* (Vasey) Soderstrom &
H. F. Decker (false saltgrass). A stoloniferous
perennial of sandy or silty soils. Resembles
Distichlis spicata (L.) Greene, but *A. texana*
does not produce rhizomes or occur on
alkaline soils.

Alopecurus L.
(Poöideae: Poeae)

Cespitose annuals and perennials, a few rhizomatous, with flat blades and dense, cylindric, spikelike panicles of 1-flowered spikelets. Disarticulation below the glumes, the spikelets falling entire. Glumes equal, awnless, usually united at the margins below, ciliate on the keels. Lemma about as long as the glumes, firm, 5-nerved, obtuse, awned from below the middle. Margins of the lemma usually united near the base. Palea and lodicules typically absent. Basic chromosome number, $x = 7$. Photosynthetic pathway, C_3. The genus closely resembles *Phleum*, but *Alopecurus* has awned lemmas. Represented in Texas by 2 species.

1. Spikelets 4.5–7.5 mm long; glumes keeled, glabrous except for the veins; lemma apex acute...2. A. *myosuroides*
1. Spikelets 2.1–3.1 mm long; glumes not keeled, pubescent or rarely completely glabrous; lemma apex obtuse ....1. A. *carolinianus*

(Crins 2007)

1. *Alopecurus carolinianus* Walter (Carolina foxtail, tufted foxtail). A native annual found in wet areas, ditches, and other moist habitats. It is also a weed of rice fields. Fairly common in the eastern part of the state.

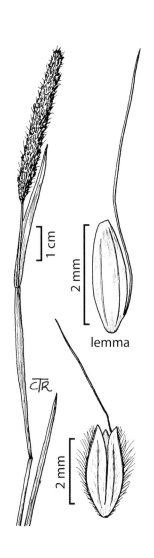

lemma

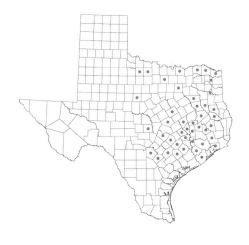

2. *Alopecurus myosuroides* Huds. (blackgrass, mouse foxtail, slimspike foxtail, slender meadow foxtail). A tufted annual introduced from Eurasia that is a weed of winter wheat fields but generally does not spread outside cultivation. Rare in Texas and perhaps not a constant member of the flora. Reported for Texas in Jones, Wipff, and Montgomery (1997); however, it is not included in Crins (2007) for the state. Closest collections have been in south-central New Mexico and northeastern Louisiana.

lemma

spikelet

Amelichloa Arriaga & Barkworth
(Poöideae: Stipeae)

Cespitose perennial with erect culms and mostly basal leaves. Cleistogenes frequently present in lower leaf sheaths. Auricles absent, and ligules ciliate. Leaf blades often stiff, involute, and sharp-pointed. Inflorescence a narrow open or loosely contracted panicle with few spikelets. Spikelets have 1 floret that disarticulates above the glumes and beneath the floret. Glumes exceed the floret and are 1–5-veined. Floret is fusiform and terete. Lemmas variously pubescent, and callus is blunt. Lemma awns once- or twice-geniculate, persistent. Paleas about as long as the lemmas, flat. Hairy, 2-veined, veins ending at or near the apices. Basic chromosome number, $x = 7$. Photosynthetic pathway, C_3. Represented in Texas by a single introduced species. Genus not included in Gould (1975b); Hatch, Gandhi, and Brown (1990); or Jones, Wipff, and Montgomery (1997).

(Arriaga 2007)

1. *Amelichloa clandestina* (Hack.) Arriaga & Barkworth (Mexican needlegrass). A cespitose perennial species with knotty rhizomes at the base of the culms. An introduced species that has become accidentally established in the central part of the state along roadsides and pastures. Potentially could become noxious because cattle will not graze the sharp-pointed leaf blades. Goats, however, will consume it.

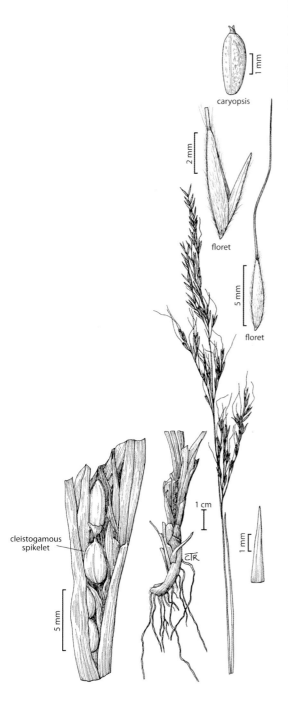

caryopsis

floret

floret

cleistogamous spikelet

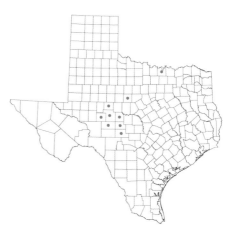

Andropogon L.
(Panicoideae: Andropogoneae)

Cespitose perennials with usually stiffly erect culms, rounded or flattened and keeled sheaths, and flat or folded blades. Ligule membranous. Flowering culm much-branched and "broomlike" in some species, unbranched or little-branched above the base in others. Each culm or branch terminated by an inflorescence of 2 to several racemose branches. Sessile spikelet of a pair fertile. Pediceled spikelet well developed, rudimentary, or absent. Disarticulation in the rachis, the sessile spikelets falling attached to the associated pedicel and section of the rachis. Glumes large, firm, awnless. Lemmas of the sterile and fertile florets membranous, lemma of the fertile floret awned or awnless. Basic chromosome numbers, $x = 9$ and 10. Photosynthetic pathway, C_4. Represented in Texas by 6 species and 5 varieties.

1. Pedicellate spikelets well developed, usually staminate.
    2. Sessile spikelet with awn 8–25 mm long; rhizomes sometimes present, the internodes usually <2 cm long...1. *A. gerardii*
    2. Sessile spikelet with awn <11 mm long; rhizomes always present, the internodes >2 cm long.. 4. *A. hallii*
1. Pedicellate spikelets (excluding terminal ones) not well developed, greatly reduced or absent.
    3. Sessile spikelets 5–8 mm long.
        4. Stamens 3; plants of East or South Texas .... 5. *A. ternarius*
        4. Stamens 1; plants of West Texas (Brewster County)..... see *Schizachyrium spadiceum*
    3. Sessile spikelets <5 mm long.
        5. Flowering culms profusely branched and rebranched,
           broomlike .... 2. *A. glomeratus*
        5. Flowering culm moderately rebranched, not broomlike.
            6. Plants usually <1 m tall; peduncles usually 5–31 mm long; rames usually 2.8–4.2 mm long; rames exserted or not, subtending sheath prominent and highly inflated .... 3. *A. gyrans*
            6. Plants usually >1 m tall; peduncles usually 4–6 mm long; rames usually 1.7–2.8 mm long; rames usually partially enclosed, subtending sheath slender and only slightly inflated ....6. *A. virginicus*

(Campbell 2003)

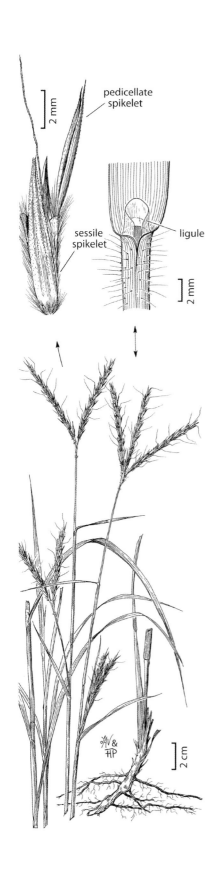

pedicellate
spikelet

2 mm

sessile
spikelet

ligule

2 mm

2 cm

1. *Andropogon gerardii* Vitman (big bluestem). A tall, rhizomatous perennial, clump-forming species of drier soil. A dominant plant of the tallgrass prairie throughout North America. Much more common in Texas than the distribution map indicates. It is used in restoration, as an ornamental, and for erosion control. Listed as poor for wildlife and good for livestock (Hatch and Pluhar 1993) and has good livestock and fair wildlife values (Hatch, Schuster, and Drawe 1999).

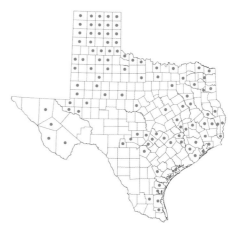

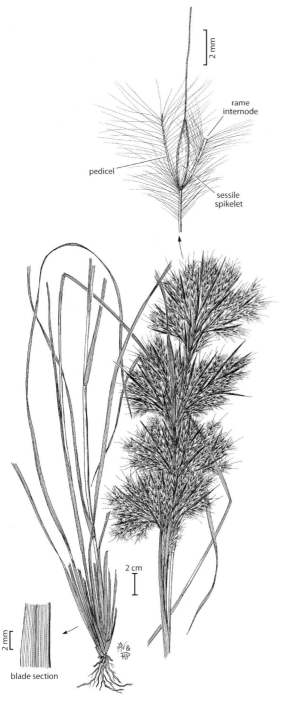

2. *Andropogon glomeratus* (Walter) Britton, Sterns, & Poggenb. (bushy bluestem, bushybeard bluestem, bushy beardgrass). A cespitose perennial growing up to 2.5 m tall. A very common and easily recognizable roadside species of seepy areas, ditches, bogs, and other wet areas. Five varieties have been named, but none are recognized here. This species has considerable potential as an ornamental (Allred 2005). It has poor livestock and wildlife values (Hatch, Schuster, and Drawe 1999; Powell 1994).

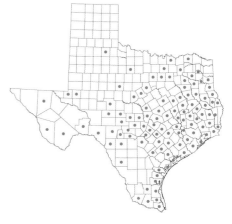

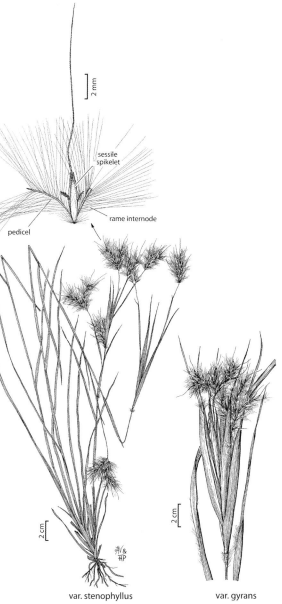

var. stenophyllus   var. gyrans

3. *Andropogon gyrans* Ashe (Elliott's bluestem, Elliott's beardgrass). A cespitose, native perennial of varied habitats. Erroneously referred to as *A. elliottii*. Two varieties are recognized: var. *gyrans*; and var. *stenophyllus* (Hack.) C. S. Campb. They can be distinguished by the following characters:

1. Rames usually hidden in the upper, inflated
   leaf sheaths; plants of well-drained
   soils .... var. *gyrans*
1. Rames usually exposed at maturity; plants
   of wet habitats .... var. *stenophyllus*

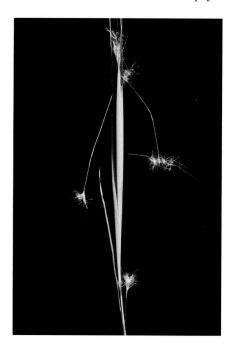

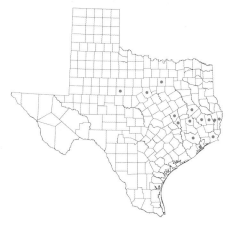

pedicellate
spikelet

sessile
spikelet

2 mm

rame

peduncle

2 cm

ligule

2 mm

ʒAV &
ℋP

4. *Andropogon hallii* Hack. (sand bluestem, Hall's bluestem). A tall, showy perennial species of sandy soils, with extensive rhizomes. It can hybridize with *A. gerardii* and has sometimes been included as a subspecies, subsp. *hallii* (Hack.) Wipff.

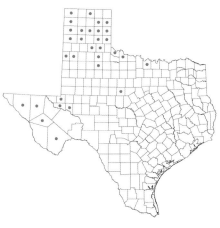

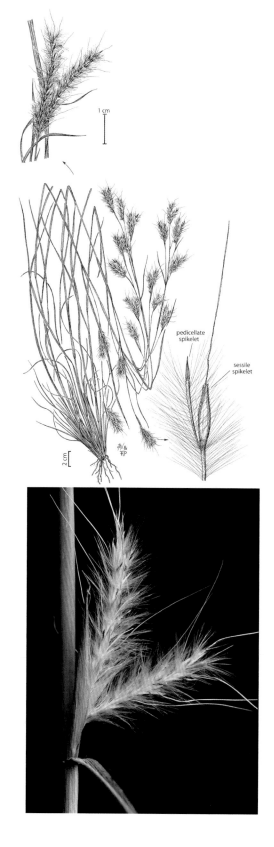

5. *Andropogon ternarius* Michx. (split bluestem, splitbeard beard-grass, silvery beardgrass, feather beardgrass). A cespitose, native perennial common in the eastern half of the state. It is used to stabilize poor and sandy soils and as an ornamental. The Texas species belongs to the typical variety, var. *ternarius*. Listed as poor for wildlife and livestock (Hatch and Pluhar 1993).

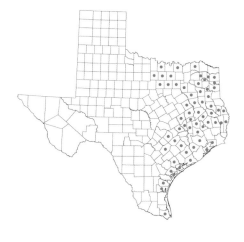

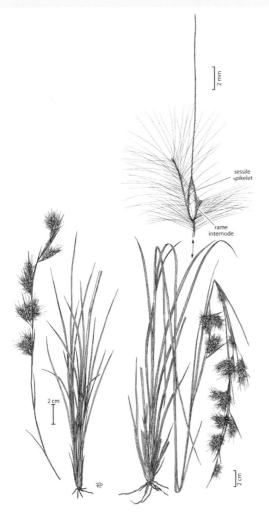

6. *Andropogon virginicus* L. (broomsedge bluestem, yellowsedge bluestem, Virginia bluestem, broomsedge). A cespitose, native perennial, widespread and common in East Texas. Grows on well-drained and wet soils; can be weedy and invade disturbed sites. Considerable variation in degree of pubescence, amount of glaucousness, and size of rames and peduncles. Listed as poor for wildlife and livestock (Hatch and Pluhar 1993) but does provide nesting habitat for ground-nesting birds (Hatch, Schuster, and Drawe 1999). Telfair (2006) listed it as important for wildlife. Two varieties are found in Texas: var. *virginicus*; and var. *glaucus* Hack. They can be distinguished by the following characters:

1. Leaves bluish-green, glaucous.... var. *glaucus*
1. Leaves green, sometimes glaucous .... var. *virginicus*

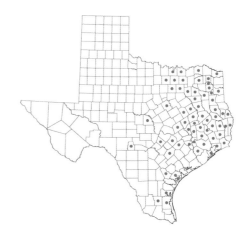

Anisantha K. Koch
(Poöideae: Bromeae)

Annuals. Sheath margins connate. Ligules membranous. Auricles absent. Inflorescence a simple panicle, sometimes racemose when immature. Disarticulation above the glumes. Lower glume 1–3-nerved. Lemmas long and narrow, 10–30 mm long and 0.5–1.5 mm wide, 5-nerved or more, acute to acuminate, apical teeth usually >2 mm long; awned from between the lemma teeth, awn usually 10–60 mm long. Basic chromosome number, $x = 7$. Photosynthetic pathway, C_3. Represented in Texas by 4 species. Some authors include this as *Bromus* sect. *Genea* (Gould 1975b; Hatch, Gandhi, and Brown 1990; Jones, Wipff, and Montgomery 1997).

1. Lemmas 20–35 mm long .... 1. *A. diandrus*
1. Lemmas <20 mm long.
    2. Panicle branches shorter than the spikelets .... 2. A. *rubens*
    2. Panicle branches longer than the spikelets.
        3. Lemmas 14–20 mm long .... 3. *A. sterilis*
        3. Lemmas 9–12 mm long .... 4. *A. tectorum*

(Pavlick and Anderton 2007)

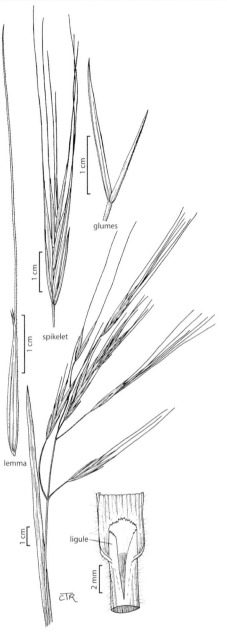

1. *Anisantha diandrus* (Roth) Tutin (great brome, ripgut brome). Annual introduced from Europe that has become a weed in disturbed areas, roadsides, and waste places. The sharp, stiff awns can cause injury to the eyes, nose, and underbelly of grazing animals.

2. *Anisantha rubens* (L.) Nevski (foxtail chess, foxtail brome, red brome). Annual introduced from Europe. Primarily a weed of disturbed sites and waste areas. The red heads are distinctive (Allred 2005).

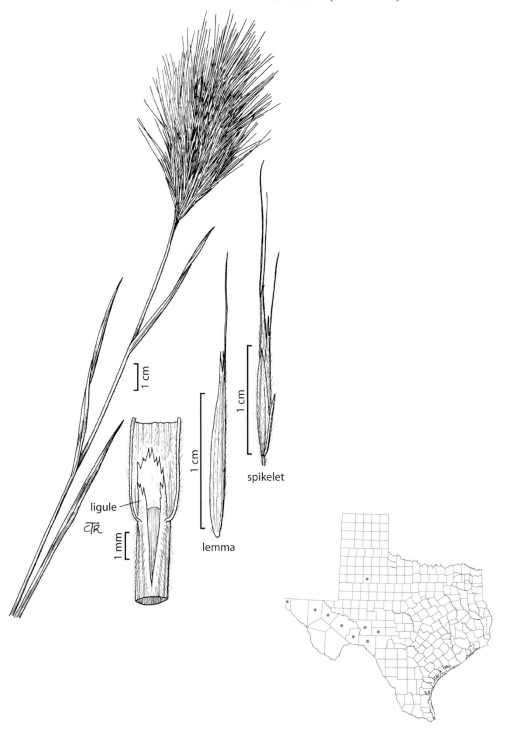

ligule

lemma

spikelet

glumes

1 cm

1 cm

upper
glume

lower
glume

lemma

spikelet

5 cm

CTR

3. *Anisantha sterilis* (L.) Nevski (barren brome). Annual introduced from Europe. Rare in Texas but widespread in the eastern and western United States. A weed of waste places, roadsides, and other disturbed sites.

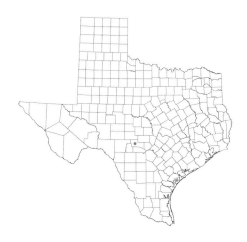

4. *Anisantha tectorum* (L.) Nevski (cheatgrass, downy chess). A widespread annual from Europe. Tends to begin growth in winter or early spring and provides good-quality forage until emergence of seed heads (Allred 2005). An aggressive and troublesome weed throughout most of the United States.

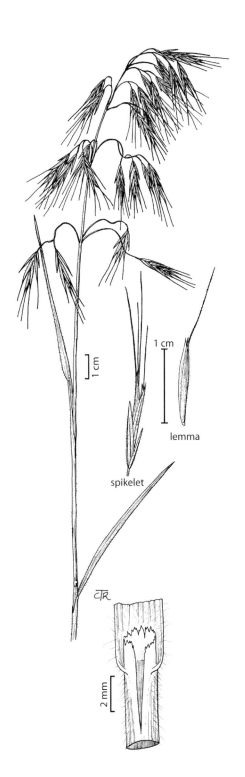

1 cm

1 cm

lemma

spikelet

2 mm

CTR

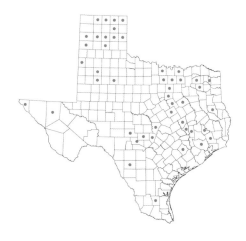

Anthenantia P. Beauv.
(Panicoideae: Paniceae)

Slender, erect perennials with short rhizomes and firm, flat blades. Spikelets conspicuously villous, short-pediceled in small, loosely contracted panicles, disarticulating below the glume. First glume absent, the second glume 5-nerved, obovate, rounded at the apex, densely villous. Lemma of the sterile floret villous, 5-nerved, about as long as the second glume. Lemma of the fertile floret 3-nerved, cartilaginous, dark brown, much narrower than the second glume, with pale, thin, flat margins. Palea narrow, the 2 nerves nearly parallel. Basic chromosome number, $x = 10$. Photosynthetic pathway, C_4. Represented in Texas by 2 species.

1. Lower leaves not auriculate; fertile lemma and palea dark reddish-brown to nearly black .... 1. *A. rufa*
1. Lower leaves auriculate; fertile lemma and palea brown .... 2. *A. villosa*

(Kral 2004; Wipff 2003b)

1. *Anthenantia rufa* (Elliott) Schult. (purple silkyscale). Erect perennial from knotty rhizomes. Plants generally reddish-brown to purple, spikelets with spreading hairs. Generally a species of wetter areas and bog edges. Kral (2004) reported that this species is possibly in East Texas, but he had not seen any specimens.

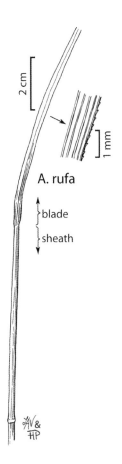

A. rufa

blade

sheath

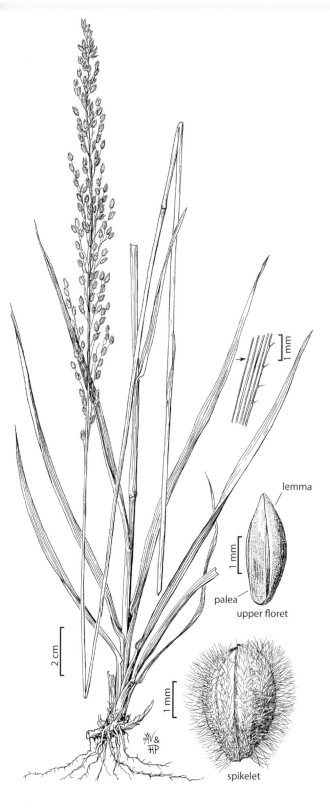

2. *Anthenantia villosa* (Michx.) P. Beauv. (green silkyscale). Erect perennial from scaly rhizomes. Spikelets typically greenish with appressed hairs. Found mostly in coastal prairies and marshes and pine flatwoods. Kral (2004) has separated out another species, *A. texana* Kral, which has hairy adaxial leaf surfaces and reddish longitudinal bands on the glume and upper lemma. Genetic study is needed to determine if they are indeed 2 species. Diggs et al. (2006) recognized *A. texana*. Poor livestock and wildlife values (Hatch, Schuster, and Drawe 1999).

Anthoxanthum L.
(Poöideae: Poeae)

Annuals or perennials with fragrant herbage due to the presence of coumarin. Sheaths open with overlapping margins; ligules membranous, short; auricles absent. Inflorescence a contracted, often spicate panicle. Spikelets with 1 large fertile floret and 2 sterile reduced florets below. Disarticulation above the glumes, the fertile and reduced florets falling together. Glumes unequal, broad, thin, acute or mucronate; the lower 1-nerved and shorter than the upper, the upper 3-nerved and longer than the florets. Lemmas of the reduced florets hairy, awned from the back, notched or toothed at the apex. Lemmas of the fertile floret broad, rounded, firm, shiny brown, awnless. Palea present and 1-nerved in the fertile floret, absent from the reduced florets. Basic chromosome number, $x = 5$. Photosynthetic pathway, C_3. Represented in Texas by 2 species. Genus not included in Gould (1975b) or Hatch, Gandhi, and Brown (1990).

The fragrance emitted by the members of this genus is from the compound coumarin. This substance is an anticoagulant that is the main ingredient in the drug coumadin (Allred and Barkworth 2007). Schouten and Veldkamp (1985) recommended merging this genus with *Hierochloë*. Indigenous peoples use the plants for incense, weaving, hair and skin tonic, analgesic, and insecticide.

1. Plants annual; ligules 1–3 mm long; blades 1–5 mm wide; panicles 1–4 cm long .... 1. A. *aristatum*
1. Plants perennial; ligules 2–7 mm long; blades 3–10 mm wide; panicles 3–14 cm long .... 2. A. *odoratum*

(Allred and Barkworth 2007)

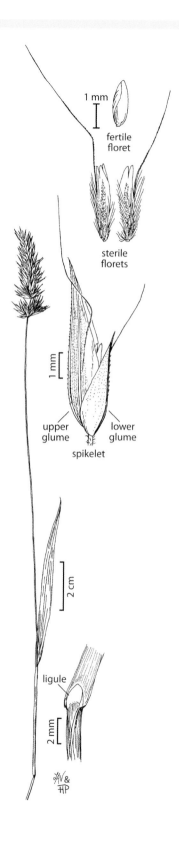

1. *Anthoxanthum aristatum* Boiss. (vernalgrass). Annual introduced from Europe. Not widespread in North America but occasionally found in open and disturbed sites. Rare in Texas and perhaps not a constant member of the state's flora. The Texas species belongs to the typical subspecies, subsp. *aristatum*.

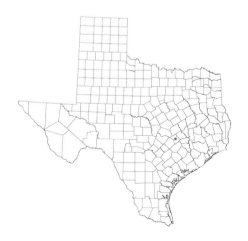

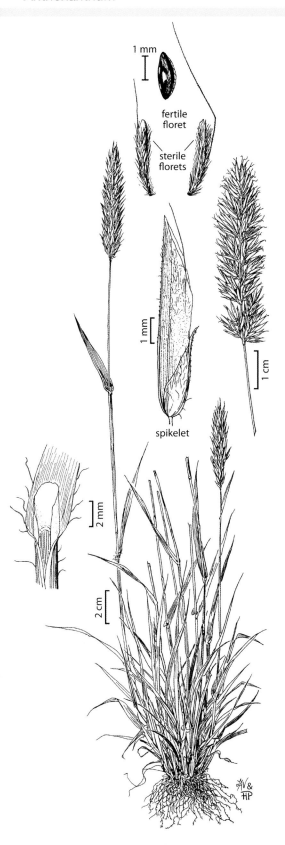

fertile
floret

sterile
florets

spikelet

2. *Anthoxanthum odoratum* L. (sweet vernalgrass). Perennial introduced from Europe. Mostly found on the east and west coasts of North America and more common than *A. aristatum*. Rare in Texas and perhaps not a constant member of the state's flora. Plants contain the potentially poisonous compound coumarin, which gives them a strong, sweet odor when wilting or drying.

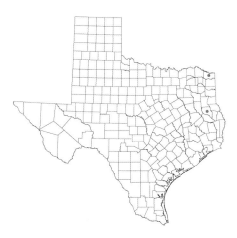

Apera Adans.
(Poöideae: Poeae)

Tufted annuals with weak culms, flat blades, and open or contracted, densely flowered panicles. Spikelets small, 1-flowered, disarticulating above the glumes. Rachilla prolonged along back of the floret as a naked bristle. Glumes membranous, lanceolate, acuminate at the tip, the lower 1-nerved and usually slightly shorter than the 3-nerved upper. Lemma firm, rounded on the back, indistinctly 5-nerved, puberulent at the base, awned from near the tip with a straight or flexuous awn. Palea well developed, 2-nerved, as long as the lemma or slightly shorter. Mature caryopsis narrowly oblong, tightly enclosed between the firm lemma and palea. Basic chromosome number, $x = 7$. Photosynthetic pathway, C_3. Represented in Texas by a single species. Genus not included in Gould (1975b) or Hatch, Gandhi, and Brown (1990).

1. *Apera spica-venti* (L.) P. Beauv. (common windgrass, loose silkybent). Annual with culms up to 1 m tall. Sheaths and spikelets often purplish. Lemma awned, awns 5–12 mm long. A European introduction occasionally found in disturbed areas. Perhaps not a constant member of the Texas flora (Allred 2007a).

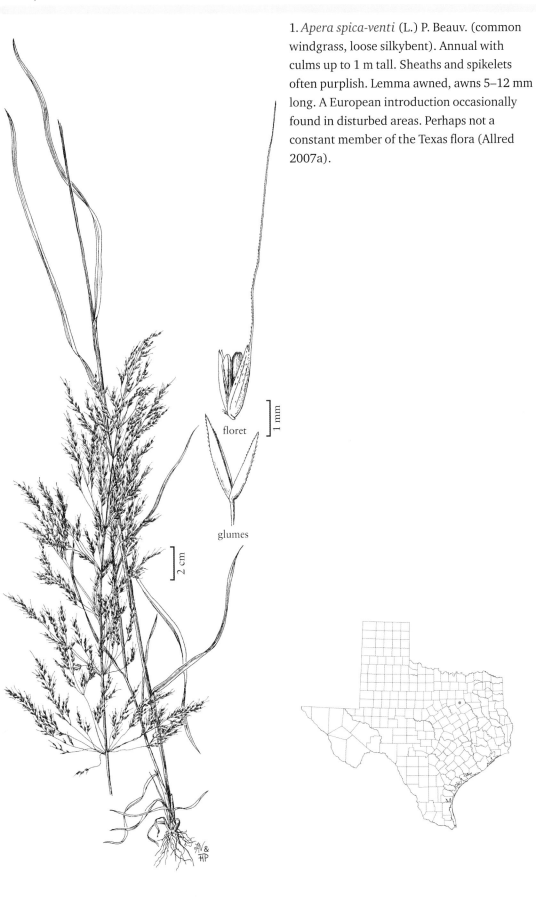

floret

1 mm

glumes

2 cm

Aristida L.
(Aristidoideae: Aristideae)

Annuals or perennials, cespitose, lacking rhizomes or stolons. Inflorescence an open or contracted panicle of usually large, 1-flowered spikelets. Disarticulation above the glumes. Glumes thin, lanceolate, with a strong central nerve and occasionally 2 lateral veins. Lemma indurate, terete, 3-nerved, with a hard, sharp-pointed callus at the base, tapering gradually or less frequently abruptly to an awn column bearing usually 3 stiff awns, the lateral awns partially or totally reduced in a few species (sometimes interpreted as a single awn separated into 3 parts). Caryopses long and slender, permanently enclosed in the firm lemma. Basic chromosome number, $x = 11$. Photosynthetic pathway, C_4. Represented in Texas by 18 species and 13 varieties.

1. Plants annual.
  2. Lower glumes 3–7-veined.
    3. Lateral awns at least ¾ as long as central awn .... 11. *A. oligantha*
    3. Lateral awns no more than ½ as long as central awn, sometimes
      absent ....16. *A. ramosissima*
  2. Lower glumes 1–2-, rarely 3-, veined.
    4. Central awns spirally coiled at base.
      5. Lateral awns erect, <4 mm long .... 5. *A. dichotoma*
      5. Lateral awns spreading, >5 mm long...3. *A. basiramea*
    4. Central awns straight to curved at the base, but not spirally coiled.
      6. Lateral awns ⅓ or less as long as central awns.
        7. Awns flattened at base ....1. *A. adscensionis*
        7. Awns terete at base.
          8. Lemmas 2.5–10 mm long; central awn divergent,
            straight .... 10. *A. longespica*
          8. Lemmas 8–22 mm long; central awn with a distinct semicircular curve at
            the base..16. *A. ramosissima*
      6. Lateral awns well developed, at least ½ as long as central awns.
        9. Junction of the lemma and awns evident; awns
          disarticulating ....4. *A. desmantha*
        9. Junction of lemma and awns obscure, awns persistent.
          10. Awns flattened at base...1. *A. adscensionis*
          10. Awns terete at base... 10. *A. longespica*
1. Plants perennial.
  11. Lateral awns reduced, ⅓ or less as long as central awns.
    12. Lateral awns <3 mm long or absent.
      13. Central awns reflexed at maturity; lemma tips often twisted.
        14. Blades flat or folded, 1–2 mm wide .... 17. *A. schiedeana*
        14. Blades tightly involute, 0.5 mm wide............................... 7. *A. gypsophila*
      13. Central awns straight or arched; lemma tip not twisted................. 18. *A. ternipes*
    12. Lateral awns 3–20 mm or more.
      15. Anther 0.8–1.0 mm long.

16. Spikelet divergent and pedicels with axillary pulvini...............8. *A. havardii*

16. Spikelets appressed and pedicels without axillary pulvini ...6. *A. divaricata*

15. Anthers 1.2–3.0 mm long...18. *A. ternipes*

11. Lateral awns well developed, at least ½ as long as or longer than the central awns.

  17. Rachis nodes and leaf sheaths lanose or floccose, sheaths sometimes glabrous .... 9. *A. lanosa*

  17. Rachis nodes glabrous, scabrous, or with straight hairs; leaf sheaths sometimes floccose.

    18. Lower primary panicle branches, at least, divergent and with axillary pulvini.

      19. Panicles narrow and contracted, usually only the lower 1 or 2 branches spreading and with pulvini .... 15. *A. purpurea*

      19. Almost all panicle branches spreading and with pulvini.

        20. Anther 0.8–1.0 mm long.

          21. Spikelet divergent and pedicels with axillary pulvini ....
....8. *A. havardii*

          21. Spikelets appressed and pedicels without axillary pulvini..............
.... 6. *A. divaricata*

        20. Anthers 1.2–3.0 mm long.

          22. Base of blades with scattered hairs 1.5–3.0 mm long on the adaxial (top) surface ....18. *A. ternipes*

          22. Base of blades glabrous on the adaxial surface, or if with hairs, hairs <0.5 mm long.

            23. Glumes reddish; awns ascending to divaricate, 8–140 mm long... 15. *A. purpurea*

            23. Glumes brownish; awns spreading, 5–15 mm long...................
.... 13. *A. pansa*

    18. Lower panicle branches appressed and without axillary pulvini.

      24. Lower glumes usually ⅓ to ¾ as long as upper glumes .... 15. *A. purpurea*

      24. Lower glumes usually more than ¾ as long as upper glumes.

        25. Lemma apices (tips) twisted for 3–6 mm .... 2. *A. arizonica*

        25. Lemma apices straight or only slightly twisted.

          26. Lower glumes prominently 2-keeled .... 12. *A. palustris*

          26. Lower glumes 1-keeled .... 14. *A. purpurascens*

(Allred 2003a)

2 mm

glumes

lemma

2 cm

1. *Aristida adscensionis* L. (sixweeks threeawn). An annual species of waste places, roadsides, and other disturbed sites, often in sandy soils. Local environmental conditions cause extreme variation in plant height, longevity, and panicle and awn lengths.

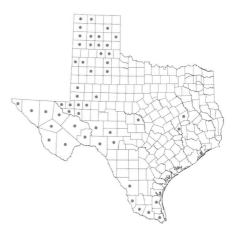

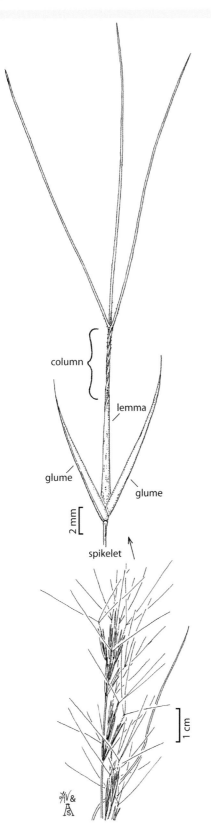

2. *Aristida arizonica* Vasey (Arizona threeawn). A cespitose, or occasionally rhizomatous, perennial. It generally grows in pine, pine-oak, or pinyon-juniper woodlands. It is sometime mistaken for *A. purpurea* var. *nealleyi*, but A. *arizonica* has flat, curly leaf blades and longer awns.

column

lemma

glume

glume

2 mm

spikelet

1 cm

3. *Aristida basiramea* Engelm. *ex* Vasey
(forked threeawn). An annual plant with erect
culms that branch at most nodes. Typically
found in sandy soils in open waste areas.
It is closely related to *A. dichotoma*, but *A. basiramea* has longer lateral awns. Relatively
rare in the state.

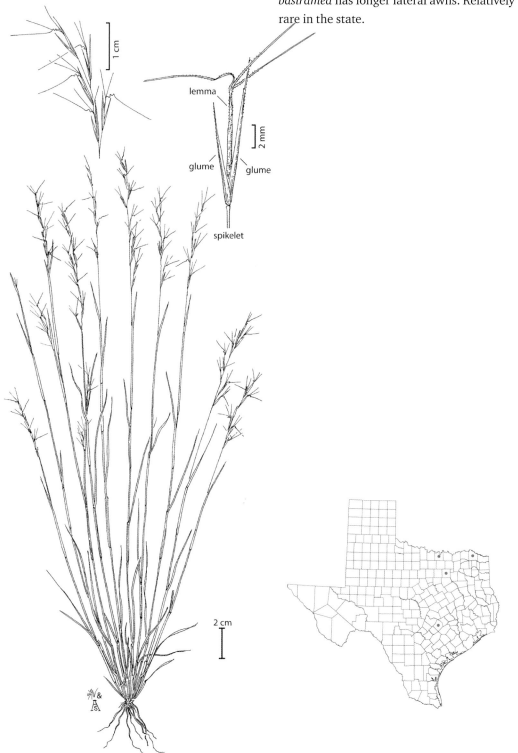

lemma

glume   glume

spikelet

4. *Aristida desmantha* Trin. and Rupr.
(curly threeawn, western threeawn, western
tripleawn grass). An annual species found
in waste places and sandy areas. As one of
the vernacular names implies, the awns are
characteristically curved to strongly arched
near their bases. Fairly common in the eastern
half of the state. Listed as poor for livestock
and wildlife (Hatch, Schuster, and Drawe
1999).

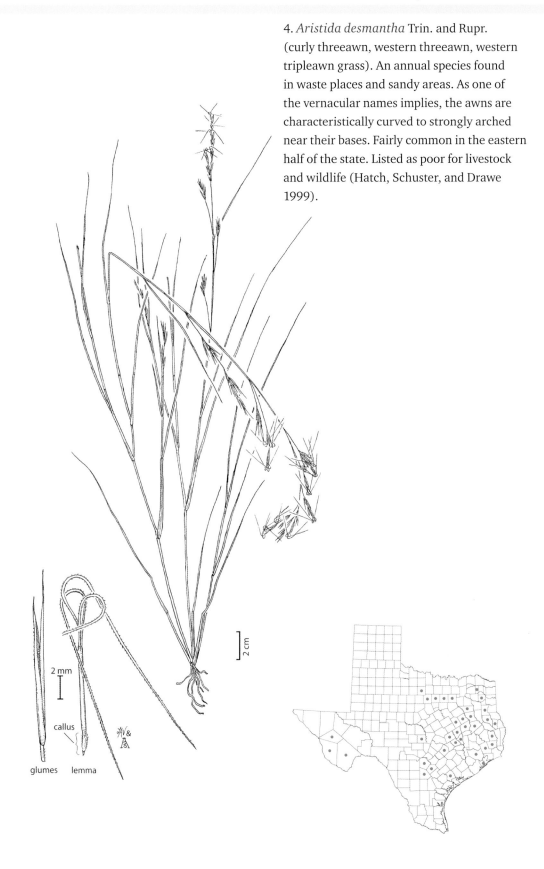

2 cm

2 mm

callus

glumes   lemma

5. *Aristida dichotoma* Michx. (churchmouse threeawn, poverty grass, pigbutt threeawn). An annual with erect culms that branch at most nodes. Typically found on sandy soils in open waste places and pinewoods. The Texas species belongs to the typical variety, var. *dichotoma*.

lemma

spikelet

var. dichotoma

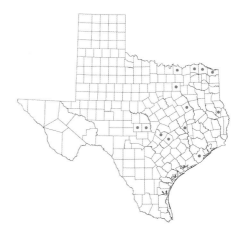

6. *Aristida divaricata* Humb. & Bonpl. *ex*
Willd. (poverty grass, poverty threeawn). A
cespitose perennial with erect or sprawling
culms. It occurs mostly in pinyon-juniper
woodlands and desert grasslands. Closely
related to *A. havardii*, but *A. divaricata*
typically has more secondary branching
and more pedicels without axillary pulvini,
resulting in a more compact inflorescence with
appressed spikelets.

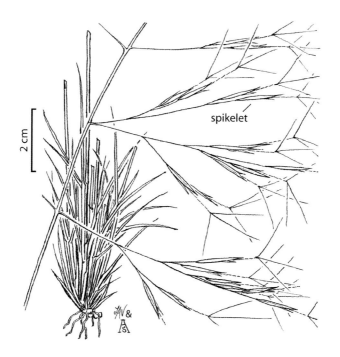

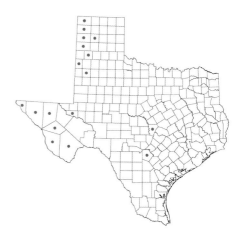

7. *Aristida gypsophila* Beetle (gypsum threeawn). A cespitose perennial with erect, unbranched culms. A species of thorn scrub communities typically found on limestone or gypsum hills. Closely related to *A. pansa*, but *A. gypsophila* has greatly reduced lateral awns and is usually taller. This species is not included in Gould (1975b) or Hatch, Gandhi, and Brown (1990). Species is of limited forage value (Powell 1994).

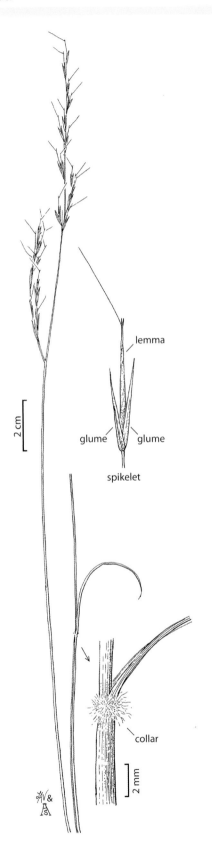

lemma

glume | glume

spikelet

collar

8. *Aristida havardii* Vasey (Havard's threeawn). A cespitose perennial often with culms tightly clustered into distinct clumps. It occurs in desert grasslands and pinyon-juniper woodlands, typically on sandy soils. Closely related to *A. divaricata*, but *A. havardii* usually lacks secondary branches and has pedicels with axillary pulvini, which results in a more open inflorescence with spreading spikelets. Reported as A. *barbata* Fourn. in Gould (1975b) and Hatch, Gandhi, and Brown (1990).

spikelet

2 cm

glumes   lemma

2 mm

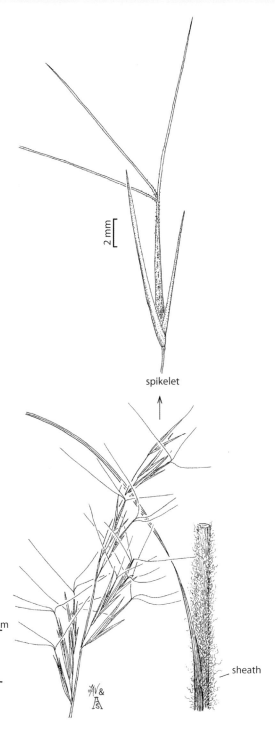

2 mm

spikelet

1 cm

sheath

9. *Aristida lanosa* Muhl. *ex* Elliott (woolly threeawn, woollysheath threeawn, woolly tripleawn). A loosely cespitose perennial with erect, unbranched culms. The vernacular name alludes to the densely pubescent leaf sheaths. It is found on sandy soils in dry fields, uplands, and pine-oak woods. Poor forage values (Hatch, Schuster, and Drawe 1999).

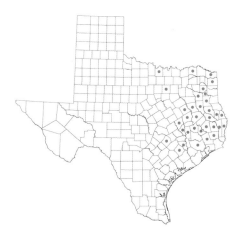

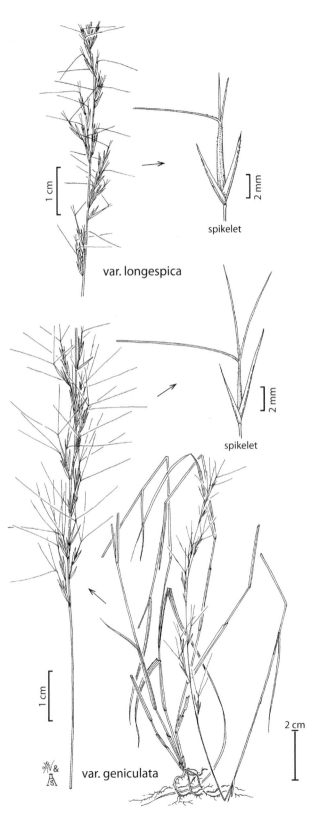

var. longespica

spikelet

var. geniculata

10. *Aristida longespica* Poir. (slimspike threeawn). An annual, often with spreading to sometimes prostrate culms. It is found along roadsides, sandy fields, and woodland openings. Poor forage values (Hatch, Schuster, and Drawe 1999). Two varieties occur in Texas: var. *geniculata* (Raf.) Fernald (Kearney's threeawn, plains threeawn); and the typical variety, var. *longespica*. They can be distinguished by the following characters:

1. Central awns 8–30 mm long, lateral awns 6–20 mm long...........................var. *geniculata*
1. Central awns 1–15 mm long and/or lateral awns usually 0–5 mm long ....var. *longespica*

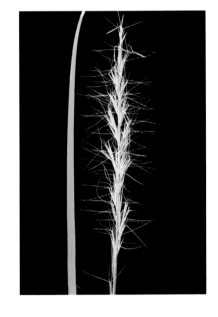

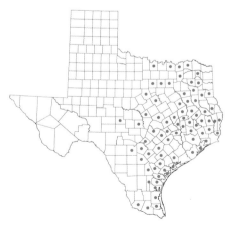

11. *Aristida oligantha* Michx. (oldfield threeawn, prairie threeawn, few-flowered threeawn). An annual with highly branched culms. It occurs in fallow or abandoned fields, along roadsides and railroads, and in other waste places and is most common in sandy soils. Diggs et al. (2006) reported that this species has allelopathic effects on other plants and is often found on harvester ant mounds. Listed as poor for wildlife and livestock (Hatch and Pluhar 1993).

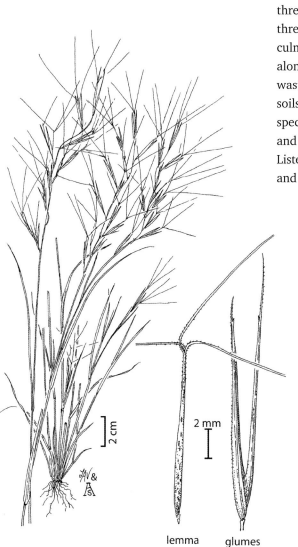

lemma   glumes

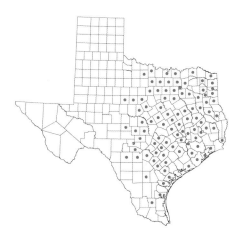

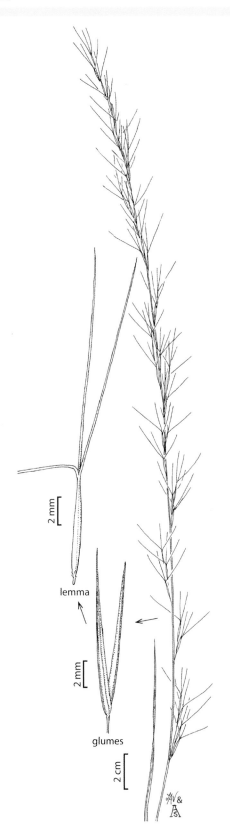

12. *Aristida palustris* (Chapm.) Vasey (longleaf threeawn). A cespitose perennial with hard, knotty bases. Similar to *A. lanosa* but with glabrous leaf sheaths. A distinctive species that grows in wet areas such as bogs, pitcher plant savannahs, and wet flatwoods and prairies.

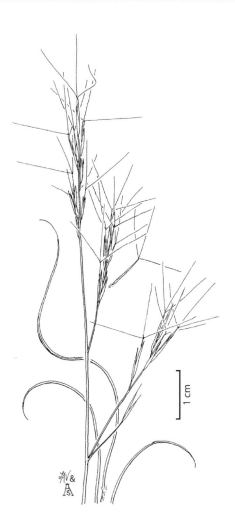

13. *Aristida pansa* Wooton & Standl. (Wooton's threeawn). A cespitose perennial with erect, unbranched culms. Most common in desert shrub communities but does extend into juniper woodlands. It differs from the closely related *A. gypsophila* by having 3 well-developed awns rather than 1.

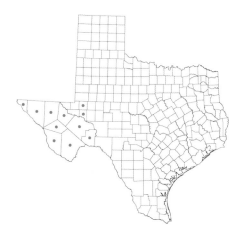

14. *Aristida purpurascens* Poir. (arrowfeather threeawn, arrowgrass). A cespitose perennial with knotty but non-rhizomatous culms. Typically found in mesic places in sandy or sometimes clayey soils. Two varieties occur in Texas: the typical variety, var. *purpurascens*; and var. *virgata* (Trin.) Allred (Trinius three-awn). Gould (1975b) and Hatch, Gandhi, and Brown (1990) listed var. *virgata* as a separate species (*A. virgata* Trin.). The varieties can be distinguished by the following character:

1. Central awn about as twice as thick at the base as the lateral awns..............var. *virgata*
1. Central awn about the same thickness as the lateral awns .... var. *purpurascens*

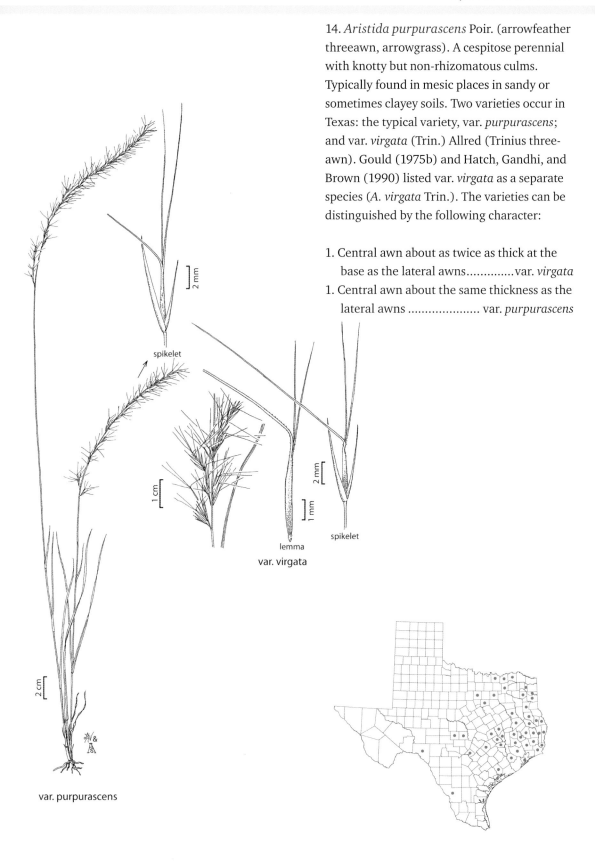

spikelet

var. virgata

lemma

spikelet

var. purpurascens

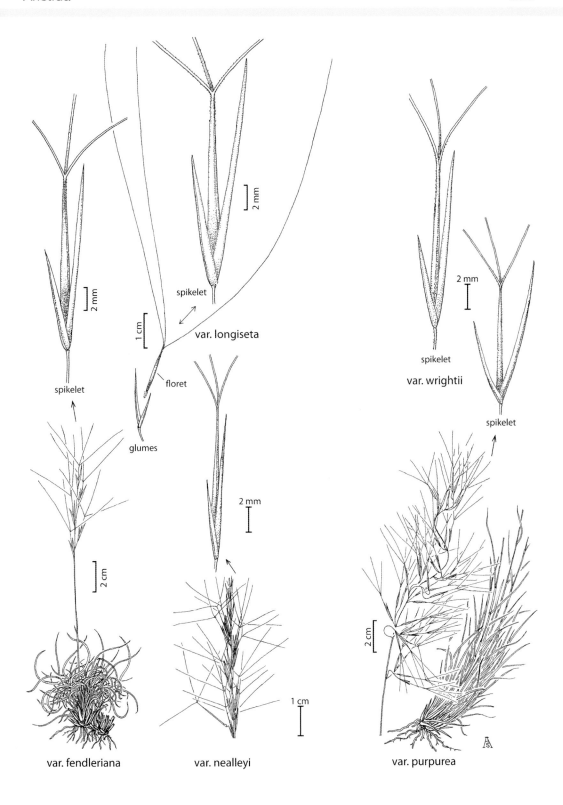

var. longiseta

spikelet

floret

glumes

spikelet

2 mm

spikelet

var. wrightii

spikelet

spikelet

2 mm

2 mm

2 mm

2 cm

2 mm

1 cm

2 cm

1 cm

var. fendleriana

var. nealleyi

var. purpurea

15. *Aristida purpurea* Nutt. (purple threeawn, purple needlegrass). A densely cespitose perennial with erect, unbranched culms, panicles open or contracted. A widespread and variable species. *A. brownii* Warnock is included here. Listed as poor for wildlife and fair or poor for livestock (Hatch and Pluhar 1993). Numerous integrating varieties have been named, 5 of which occur in Texas: var. *fendleriana* (Steud.) Vasey (Fendler's threeawn); var. *longiseta* (Steud.) Vasey (red threeawn, dogtown-grass, longawned threeawn); var. *nealleyi* (Vasey) Allred; var. *purpurea*; and var. *wrightii* (Nash) Allred (Wright's threeawn, Wright's tripleawn grass). Sometimes they can be distinguished by the following characters:

1. Awns 35–140 mm long.
    2. Lemma apices 0.1–0.3 mm wide; awns 35–60 mm long; upper glumes shorter than 16 mm .... var. *purpurea*
    2. Lemma apices 0.3–0.8 mm wide; awns 40–140 mm long; upper glumes 14–25 mm long .... var. *longiseta*
1. Awns 5–35 mm long.
    3. Lemma apices 0.1–0.3 mm wide; awns 0.1–0.3 mm wide at base.
        4. Panicle branches and pedicels drooping .... var. *purpurea*
        4. Panicle branches and pedicels ascending .... var. *nealleyi*
    3. Lemma apices 0.2–0.3 mm wide; awns 0.2–0.3 mm wide at the base.
        5. Panicle branches and pedicels lax or drooping .... var. *purpurea*
        5. Panicle branches and pedicels erect and straight.
            6. Panicle usually 15–30 mm long; blades 4–10 cm long.............. var. *fendleriana*
            6. Panicle usually 15–30 mm long; blades usually 1–25 cm long................... var. *wrightii*

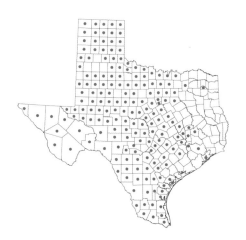

16. *Aristida ramosissima* Engelm. *ex* A. Gray (S-curve threeawn). An annual with short, wiry culms. It grows in waste places such as abandoned fields and roadsides. As the vernacular name implies, there is a distinctive semicircular bend or kink in the central awn. Relatively rare in Texas.

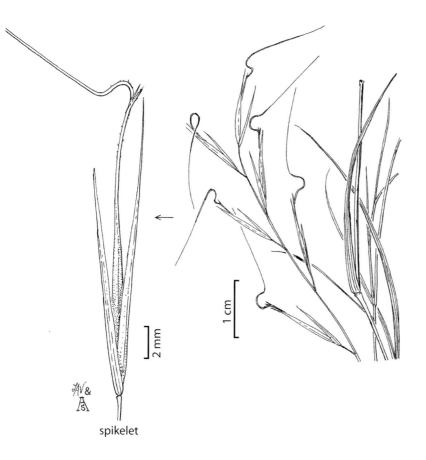

spikelet

17. *Aristida schiedeana* Trin. & Rupr. (single threeawn, singleawn aristida, beggarstick threeawn). A cespitose perennial with long, erect, wiry culms. It occurs primarily in pinyon-juniper and ponderosa pine woodlands. As the vernacular name implies, this species has lateral awns that are very short or lacking altogether. Reported as the best threeawn forage species in the United States (Powell 1994). Texas plants belong to var. *orcuttiana* (Vasey) Allred & Valdes-Reyna (Orcutt's threeawn), which has glabrous lower glumes, collars, and throats. Gould (1975b) and Hatch, Gandhi, and Brown (1990) listed this as a separate species (*A. orcuttiana* Vasey).

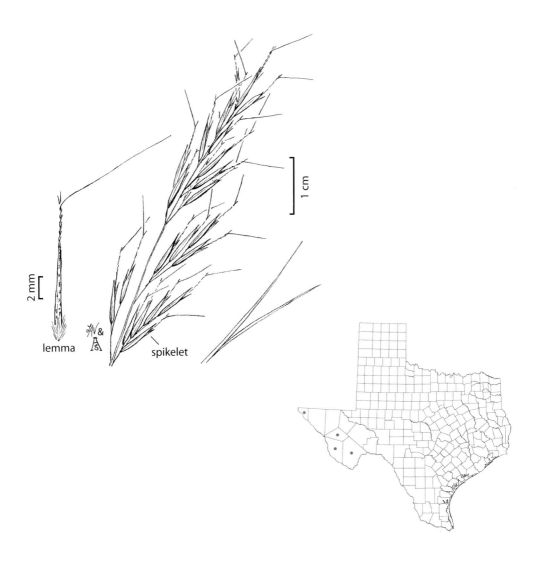

2 mm

1 cm

lemma

spikelet

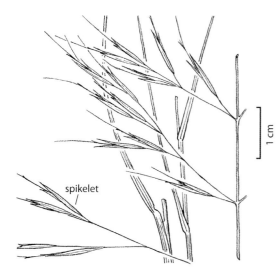

spikelet

var. ternipes

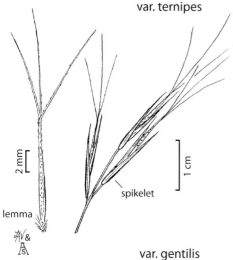

2 mm

lemma

var. gentilis

18. *Aristida ternipes* Cav. (spidergrass, hook threeawn). A cespitose perennial with unbranched, wiry culms. It occurs on dry, rocky slopes and along roadsides. Fair forage value until the inflorescence is developed (Powell 1994). Two varieties are recognized: var. *ternipes*; and var. *gentilis* (Henrard) Allred (spidergrass). They can be distinguished by the following character:

1. Lateral awns >2 mm long .... var. *gentilis*
1. Lateral awns <2 mm long .... var. *ternipes*

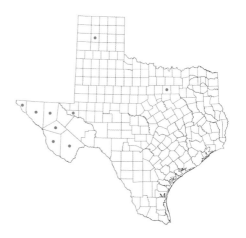

Arthraxon P. Beauv.
(Panicoideae: Andropogoneae)

Low, creeping, or decumbent annuals with broad, flat, mostly short, cordate, clasping blades. Inflorescence of few (rarely 1) to several short, sparingly rebranched, spicate or racemose branches, these scattered or digitate at the culm apex. Pediceled spikelet of a pair completely reduced and absent, the pedicels also usually absent except near the base of the inflorescence. Rachis disarticulating at the base of the sessile spikelet. Glumes subequal, the first firm, acute, the second acute or acuminate with membranous margins. Lemma of the fertile floret hyaline, entire or minutely notched, usually awned from the back. Basic chromosome number, $x = 9$. Photosynthetic pathway, C_4. Represented in Texas by a single species. Genus not included in Gould (1975b) or Hatch, Gandhi, and Brown (1990).

(Thieret 2003a)

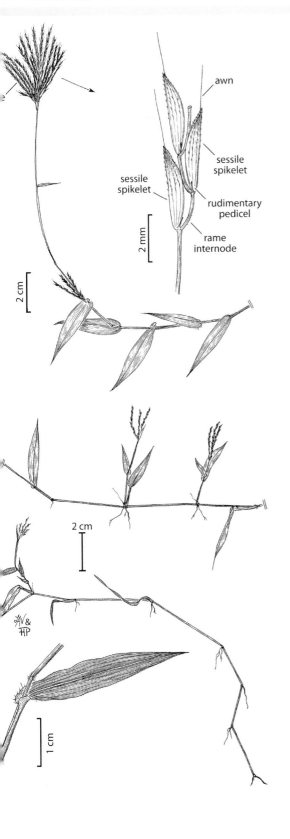

awn

sessile
spikelet

sessile
spikelet

rudimentary
pedicel

rame
internode

2 mm

2 cm

2 cm

1 cm

♂AV &
ℍP

1. *Arthraxon hispidus* (Thunb.) Makino (jointhead, small carpgrass). Low annual with weak culms often rooting at lowermost nodes. Pedicel greatly reduced and difficult to discern. Introduced from Asia. Found along roadsides, ditches, and abandoned fields. The Texas species belongs to the typical variety, var. *hispidus*. Diggs et al. (2006) cite this species as used for medicinal purposes in China and a source of yellow dye.

Arundinaria Michx.
(Bambusoideae: Bambuseae)

Small, shrubby to large, woody perennials with culms from stout, creeping rhizomes. Leaves of the main culms and larger branches with thin, papery sheaths and reduced or rudimentary blades. Leaves of the branch tips with firm sheaths; broad, flat blades; and a petiole-like constriction between the sheath and blade. Sheaths with a few stiff hairs or bristles on either side at the apex and a short, firm, erose to ciliate membrane across the collar. Ligule a firm, short, membranous rim. Spikelets produced at intervals of 5–10 years in small panicles or racemes on branchlets of the leafy shoots and also on special flowering shoots arising from the rhizomes. Spikelets (in Texas species) large, many-flowered, mostly 3–7 cm long, borne on slender pedicels. Disarticulation above the glumes and between the florets. Glumes similar to the lemmas but shorter, the first sometimes minute or absent. Lemmas large, thin, many-nerved, gradually tapering to an acute, acuminate, or short-awned apex. Palea large, 2-nerved, and 2-keeled. Stamens and lodicules 3. Stigmas usually 3, the style solitary. Caryopsis narrowly elliptic, terete, about 1 cm long. Basic chromosome number, $x = 12$. Photosynthetic pathway, C_3. Represented in Texas by 2 native species and 8 additional exotic ornamentals (Jones, Wipff, and Montgomery 1997).

1. Culm internodes sulcate; culm leaves deciduous .... 1. *A. gigantea*
1. Culm internodes usually terete; leaves persistent or slowly deciduous .... 2. *A. tecta*

(Clark and Triplett 2007)

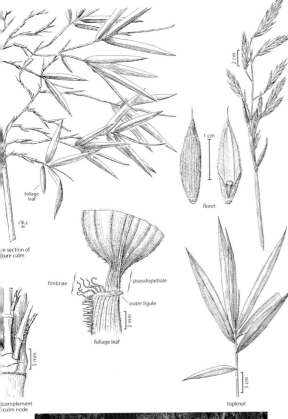

foliage leaf

e section of
ture culm

fimbriae

pseudopetiole

outer ligule

2 mm

foliage leaf

5 mm

complement
culm node

1 cm

floret

5 cm

topknot

2 cm

1. *Arundinaria gigantea* (Walter) Muhl. (giant cane, river cane, switch cane, southern cane). Rhizomatous perennial with culms to 8 m tall and 3 cm wide. Rhizomes without air chambers. Typically forms large clusters or colonies, but overgrazing, farming, and burning have greatly reduced the abundance of this species. Texas plants belong to the typical subspecies (subsp. *gigantea*). Most common member of the genus in Texas. It is found in mesic pine flatwoods, in marshy areas, and along stream banks, but it prefers drier sites than A. *tecta*. Provides cover for wildlife and regrowth is highly palatable (Hatch, Schuster, and Drawe 1999).

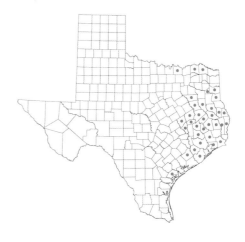

2. *Arundinaria tecta* (Walter) Muhl. (switch cane). Rhizomatous perennial with culms to 2.5 m tall and 2 cm wide. Rhizomes with air chambers. Least common member of the genus in Texas. It is found in swampy areas, in wet woods, and along stream banks. It prefers wetter sites than *A. gigantea*. This species is sometimes treated as a variety or subspecies under *A. gigantea*. Not included in Hatch, Gandhi, and Brown (1990); Jones, Wipff, and Montgomery (1997); or Diggs et al. (2006).

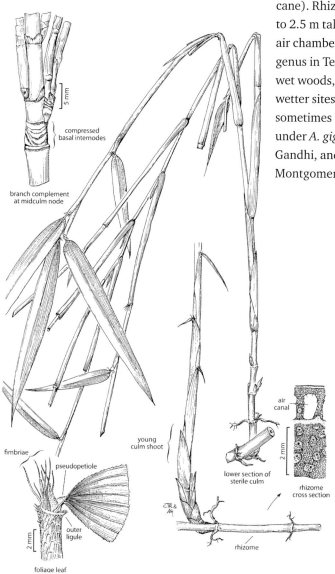

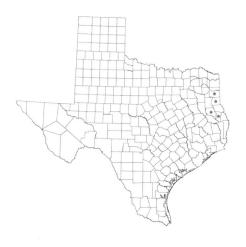

Arundo L.
(Arundinoideae: Arundineae)

Tall, stout, rhizomatous perennials with broad blades and large, plumose panicles. Ligule a short, lacerate, ciliate membrane. Spikelets with several florets, these successively smaller from the basal one upward. Rachilla glabrous, disarticulating above the glumes and between the florets. Glumes large, somewhat unequal, thin, 3-nerved, tapering to a slender point. Lemmas thin, 3-nerved, villous on the lower half, tapering to a point or short awn, the lateral nerves often extended as short teeth. Palea thin, 2-keeled. Basic chromosome number, $x = 12$. Photosynthetic pathway, C_3. Represented in Texas by a single species.

(Allred 2003b)

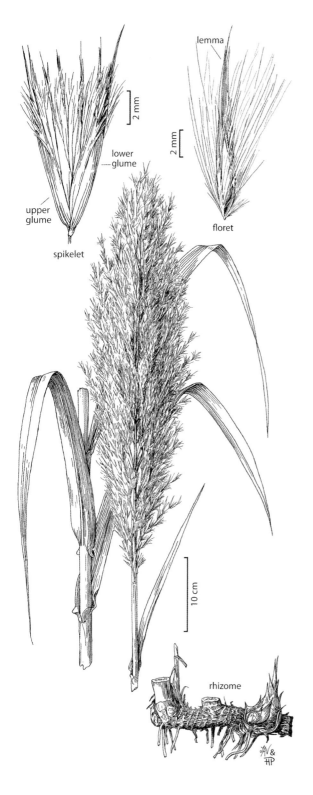

1. *Arundo donax* L. (giant cane, giantreed). A rhizomatous perennial with culms up to 10 m tall that are topped by a large, whitish inflorescence. Clones typically grow in large, conspicuous clumps. This species is usually found in ditches, around culverts, and along roadsides, rivers, and streams. It is an aggressive weed in Texas, especially in the Rio Grande Valley, but it does have potential as biofuel for cellulosic ethanol production. Not a bamboo, but many people refer to it as one. The culms are used for many of the same purposes as those of bamboo. Palatable to livestock and wildlife and provides nesting habitat and cover for wildlife (Hatch, Schuster, and Drawe 1999). Reeds for woodwind instruments are made from dried *Arundo* culms.

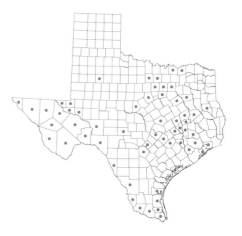

Avena L.
(Poöideae: Poeae)

In Texas, annuals with moderately tall, weak culms, broad, flat blades, and panicles of large, pendulous, usually 2–3-flowered spikelets on slender pedicels. The inflorescence of depauperate plants may be reduced to a raceme of a few spikelets or even a single spikelet. Disarticulation above the glumes and between the florets. Glumes about equal, broad, thin, several-nerved, longer than the lower floret and often exceeding the upper. Lemmas rounded on the back, 5–7-nerved, tough and indurate, often hairy on the callus, usually bearing a stout, geniculate awn from below the notch of a bifid apex. In the cultivated oats, A. *fatua*, the awn is absent or reduced. Palea 2-keeled, shorter than the lemma. Basic chromosome number, $x = 7$. Photosynthetic pathway, C_3. Represented in Texas by 2 species and 2 typical varieties.

1. Callus glabrous; florets not disarticulating from the glumes, attached to the plant at maturity .... 2. *A. sativa*
1. Callus bearded; florets disarticulating above the glumes .... 1. *A. fatua*

(Baum 2007)

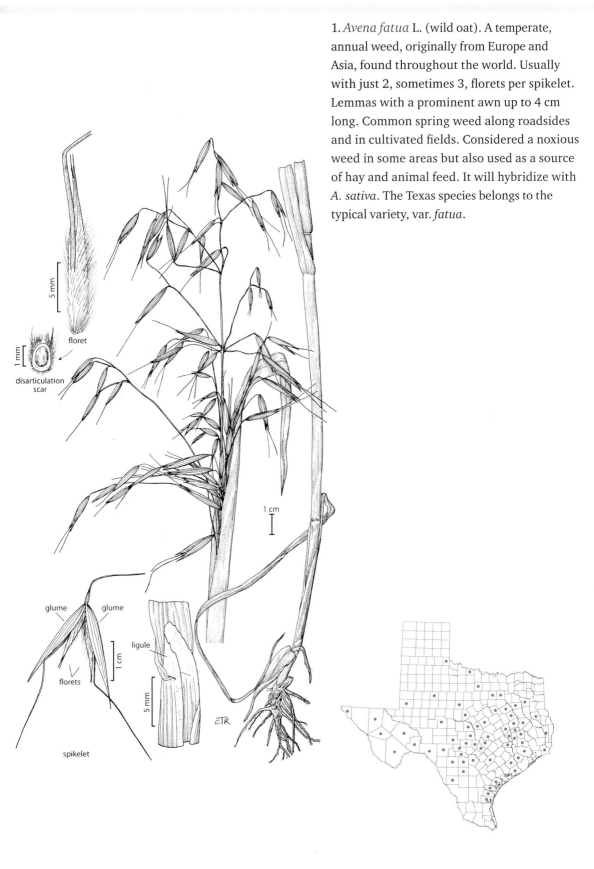

1. *Avena fatua* L. (wild oat). A temperate, annual weed, originally from Europe and Asia, found throughout the world. Usually with just 2, sometimes 3, florets per spikelet. Lemmas with a prominent awn up to 4 cm long. Common spring weed along roadsides and in cultivated fields. Considered a noxious weed in some areas but also used as a source of hay and animal feed. It will hybridize with *A. sativa*. The Texas species belongs to the typical variety, var. *fatua*.

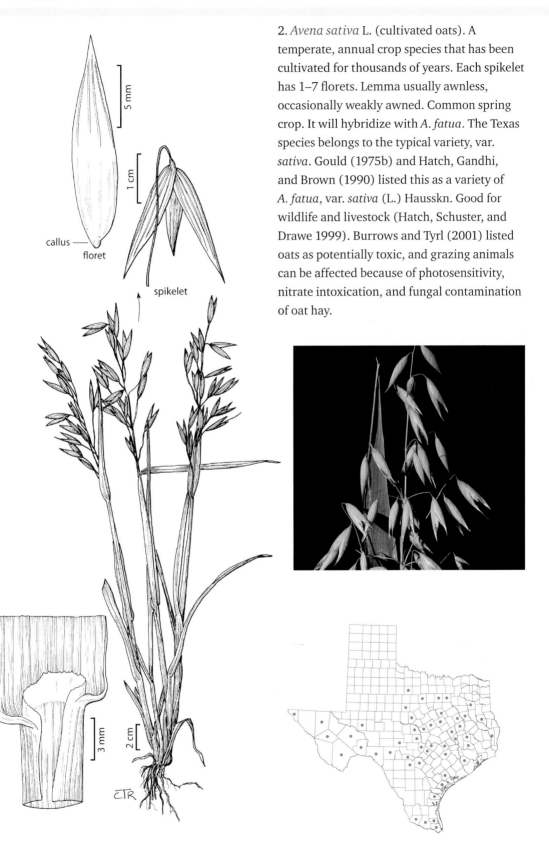

callus

floret

spikelet

2. *Avena sativa* L. (cultivated oats). A temperate, annual crop species that has been cultivated for thousands of years. Each spikelet has 1–7 florets. Lemma usually awnless, occasionally weakly awned. Common spring crop. It will hybridize with *A. fatua*. The Texas species belongs to the typical variety, var. *sativa*. Gould (1975b) and Hatch, Gandhi, and Brown (1990) listed this as a variety of *A. fatua*, var. *sativa* (L.) Hausskn. Good for wildlife and livestock (Hatch, Schuster, and Drawe 1999). Burrows and Tyrl (2001) listed oats as potentially toxic, and grazing animals can be affected because of photosensitivity, nitrate intoxication, and fungal contamination of oat hay.

Axonopus P. Beauv.
(Panicoideae: Paniceae)

Stoloniferous or cespitose perennials, with flat or folded, rounded or pointed blades. Inflorescence with 2 to several slender spicate branches, the spikelets solitary at the nodes and rather widely spaced in 2 rows on one side of a flattened rachis. Spikelets with the rounded back of the lemma of the fertile floret oriented away from the branch rachis. Disarticulation below the spikelet. First glume absent, second glume and lemma of the sterile floret about equal, narrowly ovate or oblong, often pointed beyond the fertile lemma. Lemma and palea of the fertile floret indurate, oblong, glabrous, usually obtuse at the apex. Basic chromosome number, $x = 10$. Photosynthetic pathway, C_4. Represented in Texas by 3 species.

1. Spikelets >3.5 mm long; upper glumes glabrous...3. *A. furcatus*
1. Spikelets <3.5 mm long; upper glumes sparsely pilose.
    2. Upper glumes and lower lemmas extending beyond the upper florets .... 1. *A. compressus*
    2. Upper glumes and lower lemmas not or only scarcely extending beyond the upper
       florets ....2. *A. fissifolius*

(Barkworth 2003a)

1. *Axonopus compressus* (Sw.) P. Beauv.
(broadleaf carpetgrass). Stoloniferous
perennial, sometimes rhizomatous as well.
Occasionally used as a lawn or pasture
grass but escapes and becomes a weed.
Most common in disturbed, moist sites. Not
included in Hatch, Gandhi, and Brown (1990).

upper glume   1 mm   lower lemma
                       upper floret
spikelet
2 cm

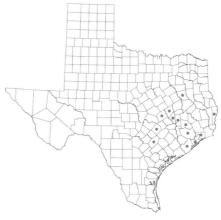

2. *Axonopus fissifolius* (Raddi) Kuhlm.
(common carpetgrass). Tufted or stoloniferous
perennial. Stolons often with pilose nodes.
Sometimes used as a lawn or pasture grass but
escapes and becomes a weedy species. Most
common in disturbed, moist sites. *Axonopus
affinis* Chase is included here. Listed as fair
for wildlife and livestock (Hatch and Pluhar
1993).

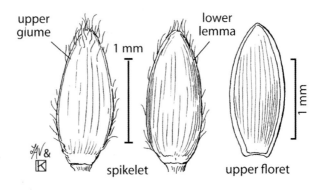

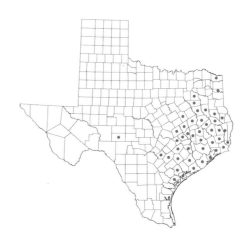

3. *Axonopus furcatus* (Flüggé) Hitchc. (big carpetgrass, flat crabgrass). Stoloniferous perennial found mostly in wet areas such as pond margins, ditches, marshes, and riverbanks. Not cultivated. Probably the least common of the 3 species in the state.

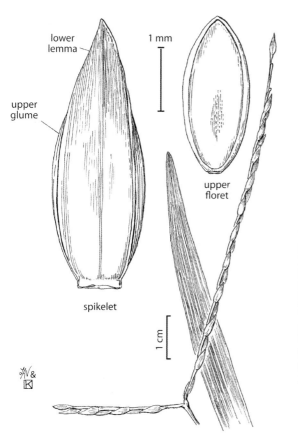

lower lemma

upper glume

1 mm

upper floret

spikelet

1 cm

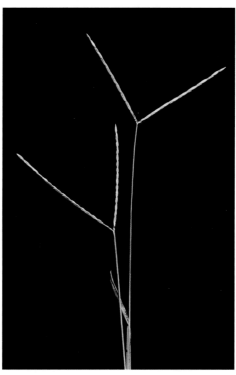

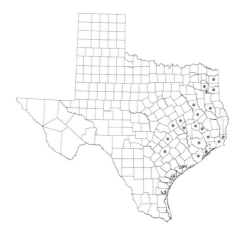

Bambusa Schreb.
(Bambusoideae: Bambuseae)

Perennials. Culms woody, lignified, and usually freely branching above and with a highly developed rhizome system. Culm and branch leaves different. Culm leaves usually bladeless or with rapidly deteriorating blade. Branch leaves with a petiole-like constriction between the sheath and blade. Auricles usually present. Synflorescences usually spicate. Pseudospikelets 1–5 cm long with 3–12 florets. Disarticulation above the glumes and between the florets. Glumes several. Lemmas ovate, acute, unawned. Paleas not exceeding the lemmas, 2-keeled. Anthers 6. Basic chromosome number, $x = 12$. Photosynthetic pathway, C_3. Of the 6 species reported from Texas (Jones, Wipff, and Montgomery 1997), only 2 (*B. multiplex* and *B. arundinacea*) are generally reported to have escaped cultivation after being planted as ornamentals (Stapleton 2007). *Bambusa beecheyana, B. oldhamii, B. textilis, B. tuldoides*, and *B. ventricosa* are only considered cultivated as ornamentals (see checklist). Neither Gould (1975b) nor Hatch, Gandhi, and Brown (1990) listed any *Bambusa*.

1. Culm sheath auricles well developed, to 5 cm long .... 1. *B. arundinacea*
1. Culm sheath auricles absent or poorly developed .... 2. *B. multiplex*

(Stapleton 2007)

1. *Bambusa arundinacea* (Retz.) Willd. (giant bamboo). Rhizomatous perennial, forming dense clumps with thorny tangled branches. Culms to 20 m tall and 20 cm thick. Internodes sometimes nearly solid. Found as an ornamental, occasionally escaped or around old nurseries and abandoned homesteads.

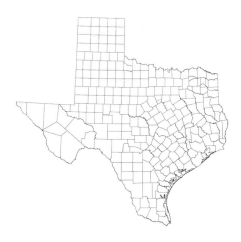

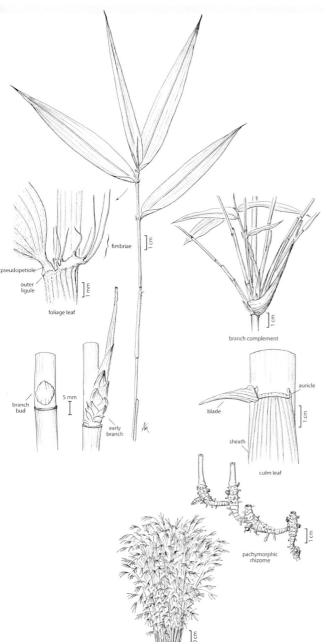

fimbriae

1 cm

pseudopetiole

outer ligule

1 mm

foliage leaf

branch complement

1 cm

branch bud

5 mm

early branch

auricle

blade

sheath

1 cm

culm leaf

pachymorphic rhizome

1 cm

20 cm

2. *Bambusa multiplex* (Lour.) Raeusch. *ex* Schult. and Schult. F. (hedge bamboo). Densely clumped, rhizomatous perennial. Culms usually <7 m tall and 2.5 cm thick. Many leaves on each branchlet and shorter height make this species an excellent hedge plant. Some variants have both striped leaves and culms. This species occurs as an ornamental, occasionally escaped or persisting around abandoned homesteads.

Blepharidachne Hack.
(Chloridoideae: Cynodonteae)

Low, tufted perennial or annual. Culms with widely spaced internodes and spur shoots with short, congested internodes. Prophyll with 2 aristate appendages. Panicles short, congested, few-flowered, not or only slightly exserted above the subtending leaves. Sheaths open. Ligule of minute hairs or absent. Spikelets 4-flowered, first and second florets sterile or staminate, third floret fertile (perfect or pistillate), and fourth floret reduced to a 3-awned rudiment. Glumes about equal, thin, 1-nerved, acute or acuminate. Disarticulation above the glumes but not between the florets. Lemmas 3-nerved, 3-lobed, and 3-awned at the apex, the awns conspicuously plumose. Basic chromosome number, $x = 7$. Photosynthetic pathway, C_4. Represented in Texas by a single species.

(Valdes-Reyna 2003a)

1. *Blepharidachne bigelovii* (S. Watson) Hack.
(Bigelow's desertgrass). An inconspicuous
perennial species of desert flats, mesas, and
limestone hills in the Trans-Pecos region of
Texas. Not often collected, and similar in
appearance to *Dasyochloa pulchella* (Kunth)
Willd. *ex* Rydb. Gould (1975b) cited this
species only from Mexico and Texas, but Allred
(2005) reported it from southern New Mexico
as well.

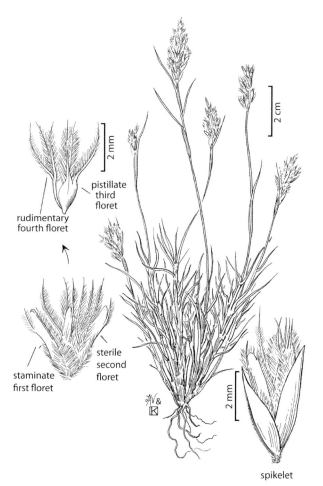

pistillate
third
floret

rudimentary
fourth floret

staminate
first floret

sterile
second
floret

2 cm

2 mm

spikelet

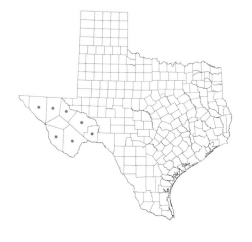

Blepharoneuron Nash
(Chloridoideae: Cynodonteae)

Perennial, cespitose, with culms 20–70 cm tall. Leaves mostly in basal tuft. Sheaths open; ligule membranous, short, rounded, appearing as a continuation of the sheath margins; auricles absent. Inflorescence a loosely contracted panicle. Spikelets with 1 floret, grayish-green, borne on slender pedicels, pedicels glandular. Disarticulation above the glumes. Glumes 2, broad, subequal, 1-veined. Lemma firmer than the glumes and slightly longer, rounded on the back, 3-veined, the veins usually densely ciliate-pubescent, the apex acute, occasionally mucronate. Palea slightly exceeding lemma in length, pubescent on the 2 veins. Basic chromosome number, $x = 8$. Photosynthetic pathway, C_4. Represented in Texas by a single species.

(Peterson and Annable 2003)

1. *Blepharoneuron tricholepis* (Torr.) Nash.
(pine dropseed, hairy dropseed). A densely
cespitose perennial with erect culms <1 m tall.
Inflorescences are loosely contracted panicles.
Veins of the lemma densely pubescent.
Generally found in dry pine-oak woodlands
or dry mountain meadows. Gould (1975b)
reported most collections from the Chisos
Mountains at medium to high elevations,
and Powell (1994) found it to be common on
medium to upper elevations on slopes on and
near the 3 highest peaks of the Trans-Pecos.
Considered highly palatable to all classes of
livestock (Powell 1994).

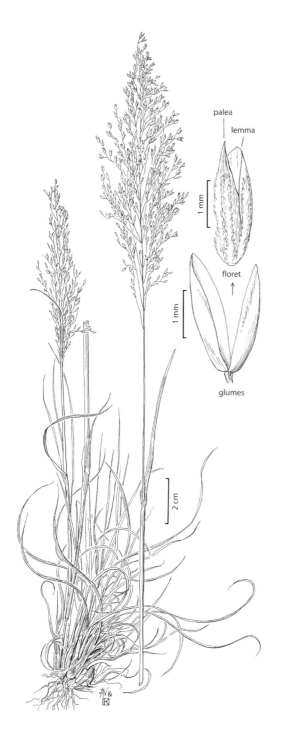

Bothriochloa Kuntze
(Panicoideae: Andropogoneae)

Cespitose perennials with erect or decumbent-spreading culms and flat blades. Inflorescence
a terminal panicle, the spikelets on few to several spicate primary branches, these sparingly
rebranched in a few species. Pedicels and upper rachis internodes with a central groove or
membranous area. Sessile spikelets of a pair fertile and awned, more or less triangular in outline,
the lower glume dorsally flattened, the upper glume with a median keel. Pediceled spikelet
staminate or sterile, usually well developed. Disarticulation in the rachis, the sessile spikelet
falling attached to a pedicel and section of the rachis. Basic chromosome number, $x = 10$.
Photosynthetic pathway, C_4. Represented in Texas by 12 species, 1 subspecies, and 2 varieties.

1. Pedicellate spikelets about equal in length to the sessile spikelets.
  2. Sessile spikelets >5 mm long .... 12. *B. wrightii*
  2. Sessile spikelets <5 mm long.
    3. Rachises longer than the branches .... 3. *B. bladhii*
    3. Rachises shorter than the branches.
      4. Lower glumes of sessile spikelets pitted .... 10. *B. pertusa*
      4. Lower glumes of sessile spikelets not pitted .... 7. *B. ischaemum*
1. Pedicellate spikelets shorter than sessile spikelets.
      5. Sessile spikelets <4.5 mm long, lemmas awnless or awn <17 mm long.
        6. Lemma of sessile spikelets awnless or awn <6 mm long .... 5. *B. exaristata*
        6. Lemma of sessile spikelets with awn 7–17 mm long.
          7. Hairs below sessile spikelets ¼ as long as spikelets, sparse .... 3. *B. bladhii*
          7. Hairs below sessile spikelets ½ as long as or longer than spikelets, abundant.
            8. Glumes acute; leaves mostly basal; culms usually <2 mm
            thick .... 8. *B. laguroides*
            8. Glumes blunt; leaves cauline; culms usually >2 mm
            thick .... 9. *B. longipaniculata*
      5. Sessile spikelets >4.5 mm long, lemmas awned, awn 18–35 mm long.
        9. Rachises usually <5 cm long; branches 2–9.
          10. Culm nodes densely pubescent, hairs spreading, 3–7 mm
          long .... 11. *B. springfieldii*
          10. Culm nodes glabrous or puberulent, hairs ascending, always <3 mm long.
            11. Lower branches of the inflorescences rebranched; leaves mostly
            cauline, blades usually >2 mm wide .... 6. *B. hybrida*
            11. Lower branches of inflorescences not rebranched; leaves mostly
            basal, blades usually <2 mm wide .... 4. *B. edwardsiana*
        9. Rachises usually >5 cm long; branches >9.
          12. Culms >1.3 m tall, glaucous below nodes; nodes with
          spreading hairs, hairs 2–6 mm long .... 1. *B. alta*
          12. Culms usually <1.2 m tall, not glaucous below nodes; nodes
          with ascending hairs, hairs <3 mm long .... 2. *B. barbinodis*

(Allred 2003c)

1. *Bothriochloa alta* (Hitchc.) Henrard (tall bluestem). Tall (up to 2.5 m) perennial found in swales, ditches, and areas where moisture accumulates. Appears as a robust form of *B. barbinodis* (Allred 2005). It is relatively scarce.

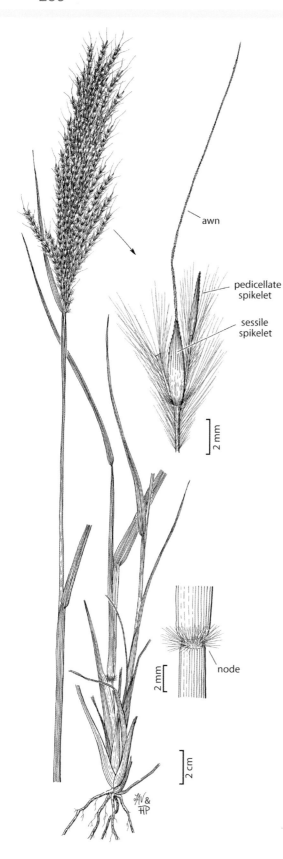

awn

pedicellate spikelet

sessile spikelet

2 mm

2 mm node

2 cm

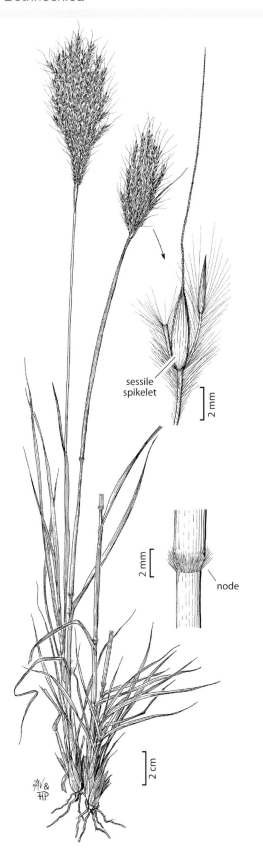

sessile
spikelet

2 mm

2 mm

node

2 cm

JAV &
HP

2. *Bothriochloa barbinodis* (Lag.) Herter (cane bluestem, pinhole bluestem). Widespread perennial. Shorter than *B. alta*, taller than and with shorter nodal hairs than *B. springfieldii*, and lacking the glaucous leaves of *B. wrightii*. It is sometimes used as an ornamental. No varieties are recognized. Found along roadsides, drainways, and gravelly slopes. Listed as poor for wildlife and fair for livestock (Hatch and Pluhar 1993). Good forage species according to Powell (1994).

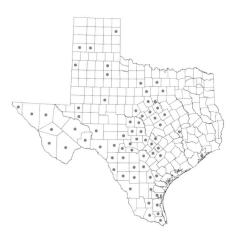

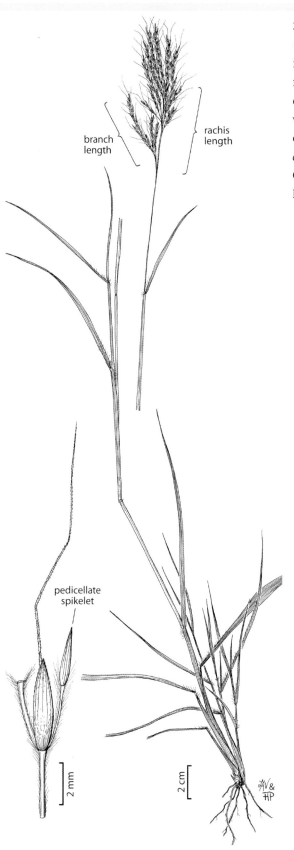

branch
length

rachis
length

pedicellate
spikelet

2 mm

2 cm

3. *Bothriochloa bladhii* (Retz.) S. T. Blake (Australian bluestem). A perennial species introduced from Asia and Africa for livestock forage that has escaped and become established. It occurs along roadsides, in waste areas and other disturbed sites, and occasionally on rangelands. *Bothriochloa caucasica* (Trin.) C. E. Hubb. is included here. Good forage according to Hatch, Schuster, and Drawe (1999).

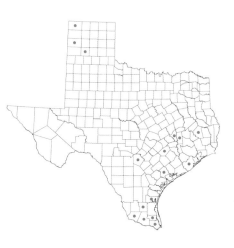

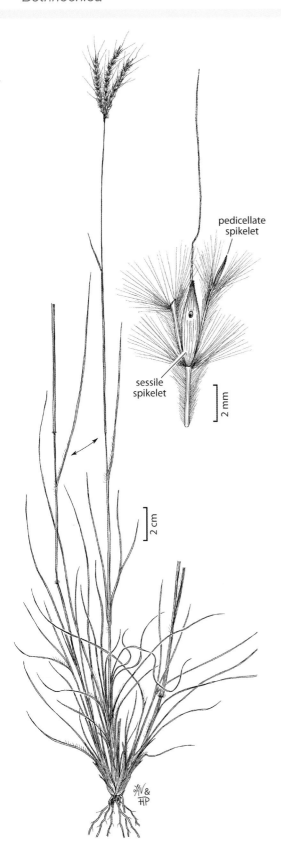

4. *Bothriochloa edwardsiana* (Gould) Parodi (Merrill's bluestem). Slender, short (<65 cm), erect perennial. Found in rocky grasslands of the Edwards Plateau on calcareous soils. Differs from the closely related *B. hybrida* by its lower stature, more cauline leaves, and narrower blades.

5. *Bothriochloa exaristata* (Nash) Henrard (awnless bluestem). Perennial up to 1.5 m tall. Sheaths with a white, powdery bloom. Generally found in heavy clay soils in fields and along roadsides in coastal prairies. More restricted in distribution and less abundant than *B. longipaniculata*.

pedicellate spikelet

2 mm

2 cm

6. *Bothriochloa hybrida* (Gould) Gould
(hybrid bluestem). A stiffly erect, moderately
branched perennial. It occurs in rangelands,
pastures, roadsides, and disturbed areas on
mainly calcareous soils. It has more basal
leaves, wider blades, and a more robust habit
than the closely related *B. edwardsiana*.
Perhaps this species originated as a hybrid
between *B. edwardsiana* and *B. laguroides*
subsp. *torreyana*. Fair forage according to
Hatch, Schuster, and Drawe (1999).

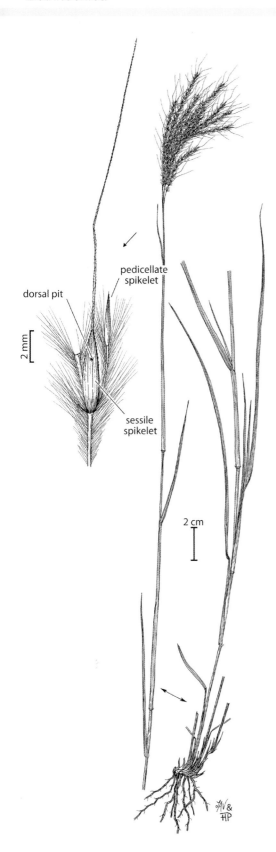

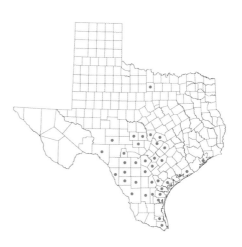

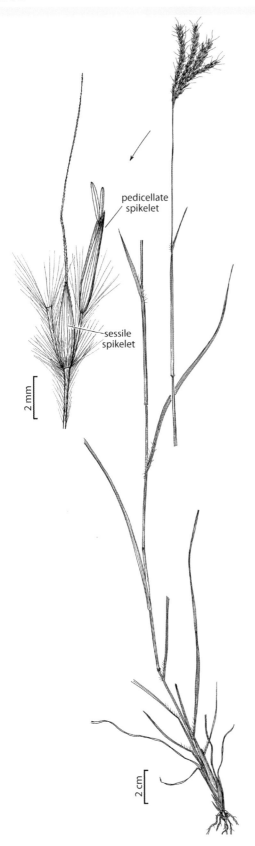

pedicellate spikelet

sessile spikelet

2 mm

2 cm

7. *Bothriochloa ischaemum* (L.) Keng (King Ranch bluestem, KR bluestem). Plants typically tufted but becoming stoloniferous and/or rhizomatous with frequent mowing or close grazing. Introduced from Europe and Asia as a forage species and for erosion control, but now escaped and established throughout most of the state. Found in seeded pastures, along roadsides, and in other moderately disturbed sites. One of the most common roadside grasses throughout much of the state. Listed as fair for wildlife and livestock (Hatch and Pluhar 1993). Powell (1994) reported it as highly palatable to livestock. Two varieties have been identified: var. *ischaemum*; and var. *songarica* (Rupr. *ex* Fisch. & C. A. Mey.) Celarier & J. R. Harlan. They can be distinguished by the following character:

1. Nodes glabrous .... var. *ischaemum*
1. Nodes pubescent .... var. *songarica*

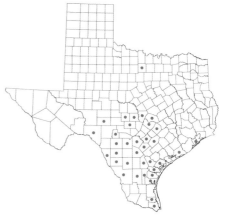

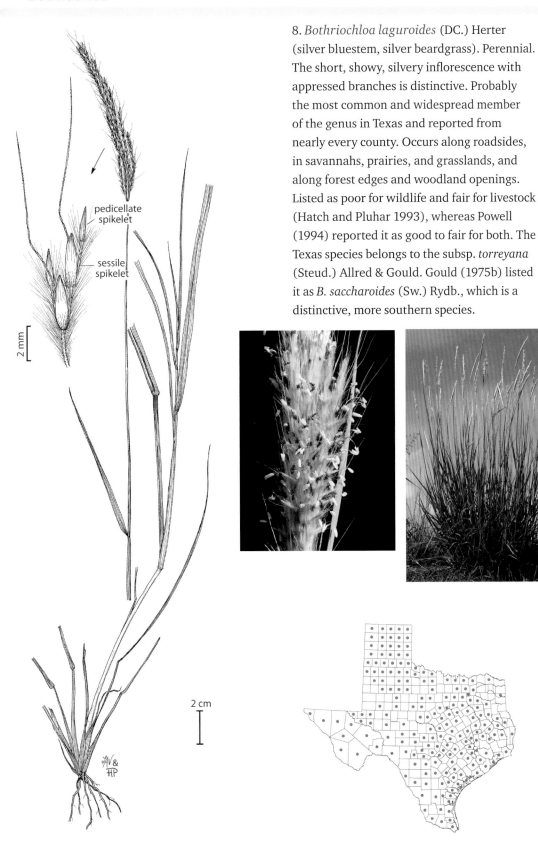

pedicellate
spikelet

sessile
spikelet

2 mm

2 cm

8. *Bothriochloa laguroides* (DC.) Herter (silver bluestem, silver beardgrass). Perennial. The short, showy, silvery inflorescence with appressed branches is distinctive. Probably the most common and widespread member of the genus in Texas and reported from nearly every county. Occurs along roadsides, in savannahs, prairies, and grasslands, and along forest edges and woodland openings. Listed as poor for wildlife and fair for livestock (Hatch and Pluhar 1993), whereas Powell (1994) reported it as good to fair for both. The Texas species belongs to the subsp. *torreyana* (Steud.) Allred & Gould. Gould (1975b) listed it as *B. saccharoides* (Sw.) Rydb., which is a distinctive, more southern species.

9. *Bothriochloa longipaniculata* (Gould) Allred & Gould (longspike silver bluestem). Tall (up to 2 m), robust perennial. Typically grows at low elevations in the Texas Gulf Coastal Plains. Most frequently found in heavy clay soils. Gould (1975b) listed this as a variety of *B. saccharoides* (var. *longipaniculata* Gould). Good for wildlife according to Hatch, Schuster, and Drawe (1999).

rame internode

2 mm

groove

pedicellate spikelet

2 mm

sessile spikelet

2 cm

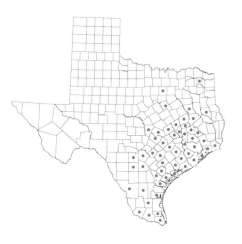

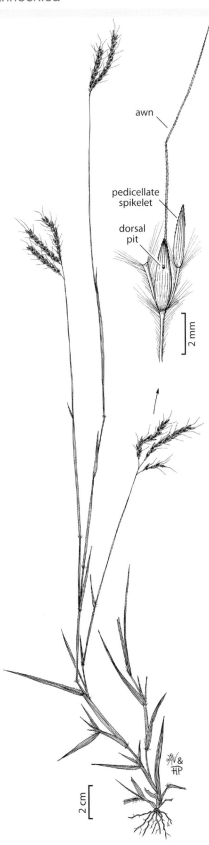

awn

pedicellate
spikelet

dorsal
pit

2 mm

10. *Bothriochloa pertusa* (L.) A. Camus (pitted bluestem). Tufted or stoloniferous species introduced as a forage plant. The pitting is not distinct or consistent. Escaped and established in moist, disturbed sites of lower elevations.

2 cm

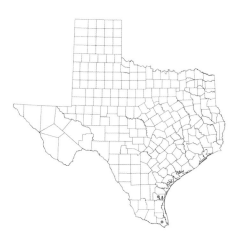

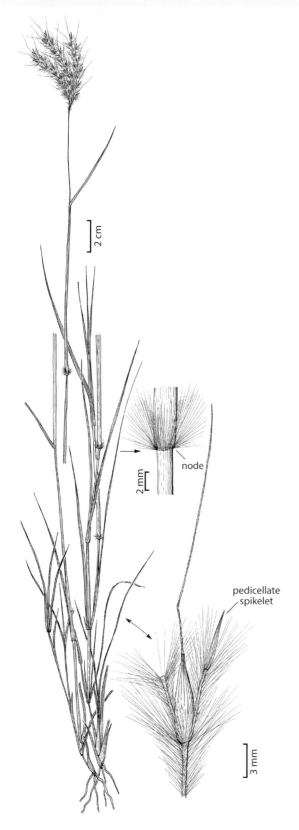

11. *Bothriochloa springfieldii* (Gould) Parodi (Springfield bluestem). Erect, unbranched, perennial. Nodes prominently bearded, hairs spreading, white and up to 7 mm long. Species is found in rangelands, plains, and uplands and prefers sandy areas. It has lower stature, longer nodal hairs, and fewer panicle branches than *B. barbinodes*. It has wider, nonciliate leaf blades and pubescent nodes, which distinguish it from B. *edwardsiana*. It is highly palatable to livestock (Powell 1994).

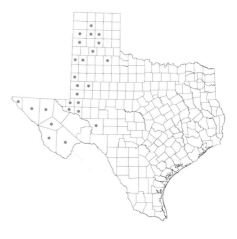

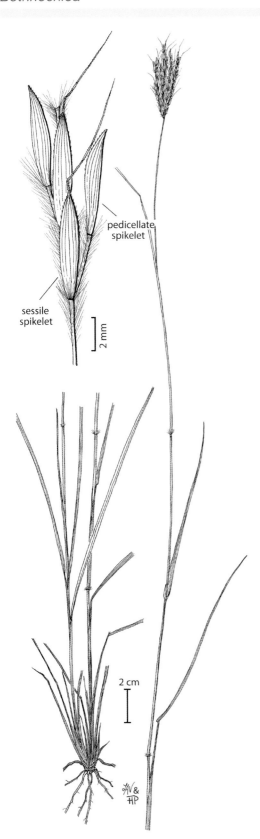

pedicellate spikelet

sessile spikelet

2 mm

2 cm

12. *Bothriochloa wrightii* (Hack.) Henrard (Wright's bluestem). Relatively short (<70 cm), sparsely branched, perennial. Pedicellate spikelets largest in the genus and foliage glaucous. Found in West Texas in grasslands and shrubby areas in the pine-oak woodlands of mid-elevations. Perhaps not a constant member of Texas flora since last collected in the United States in 1930 and in New Mexico in 1904 (Allred 2003c, 2005). Not included in Gould (1975b) or Hatch, Gandhi, and Brown (1990).

Bouteloua Lag.
(Chloridoideae: Cynodonteae)

Annuals and perennials of diverse habit, some cespitose, others with rhizomes or stolons. Leaves mostly basal, with flat or folded blades. Ligule usually a ring of hairs. Inflorescences of several to numerous short, spicate branches, these closely or distantly spaced along the main axis and bearing relatively few (1–15) large, nonpectinate, sessile spikelets in 2 rows along the margins of an angular or flattened rachis. Disarticulation at the base of the branch rachis. Spikelets with 1 fertile floret and 1–2 rudimentary florets above. Glumes lanceolate, 1-nerved, unequal to nearly equal, awnless or short-awned. Lemmas membranous, 3-nerved, the midnerve often extending into an awn, the lateral veins occasionally short-awned. Paleas membranous, the 2 veins occasionally awn-tipped. Basic chromosome number, $x = 10$. Photosynthetic pathway, C_4. Represented in Texas by 7 species and 5 varieties. Members previously included in *Bouteloua* subgenus *Chondrosum* have been elevated to generic status (*Chondrosum*).

1. Inflorescences spicate branches, usually 20–50 or more per culm.
  2. Branches bearing mostly 1 spikelet ....6. *B. uniflora*
  2. Branches bearing 2–9 spikelets.
    3. Leaf blades (at least some) >2.5 mm wide; ligule 0.3–0.5 mm long .... 3. *B. curtipendula*
    3. Leaf blades involute, <2.5 mm wide; ligule 1.0–1.5 mm long .... 7. *B. warnockii*
1. Inflorescences spicate branches, <20 per culm.
    4. Plants annual ....1. *B. aristidoides*
    4. Plants perennial.
      5. Glumes hairy, the hairs not confined to the midvein.
        6. Glumes and rachis branches densely and conspicuously hairy .... 2. *B. chondrosioides*
        6. Glumes and rachis branches sparsely and inconspicuously hairy ....5. *B. rigidiseta*
      5. Glumes glabrous or with short hairs on the midvein.
        7. Lower glume 4–6 mm long, nearly as long and wide as the upper
          glume .... 4. *B. repens*
        7. Lower glume 2–3 mm long, much shorter and narrower than upper
          glume .... 7. *B. warnockii*

(Gould 1975b; Wipff 2003c)

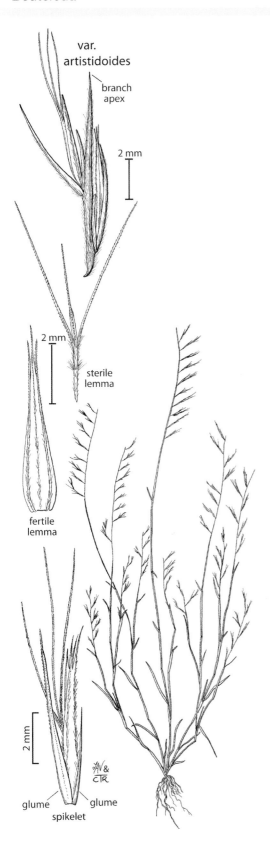

var. artistidoides

branch apex

2 mm

2 mm

sterile lemma

fertile lemma

2 mm

glume   glume

spikelet

1. *Bouteloua aristidoides* (Kunth) Griseb. (needle grama, sixweeks grama). Short-lived annual. Sheaths usually much shorter than the internodes. Found along washes and in gravelly soils; dry, rocky slopes; and graded roadsides. Diggs et al. (2006) reported that the inflorescence branches get caught in sheep's wool, thus reducing its value, and of little forage value (Powell 1994). The Texas species belongs to the typical variety, var. *aristidoides*.

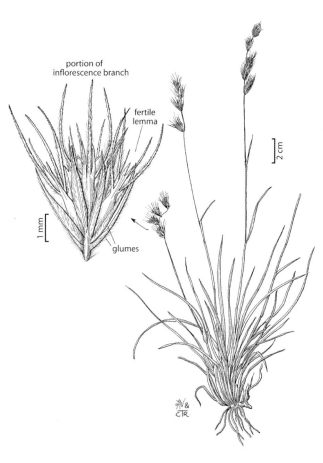

portion of
inflorescence branch

fertile
lemma

2 cm

1 mm

glumes

2. *Bouteloua chondrosoides* (Kunth) Benth. *ex* S. Watson (sprucetop grama, woolly spiked grama, Harvard's grama). Cespitose perennial with a firm crown but without stolons or rhizomes. Glumes and rachis branches densely and conspicuously hairy. Grows on dry, rocky slopes and grassy plateaus. An excellent forage where abundant (Powell 1994).

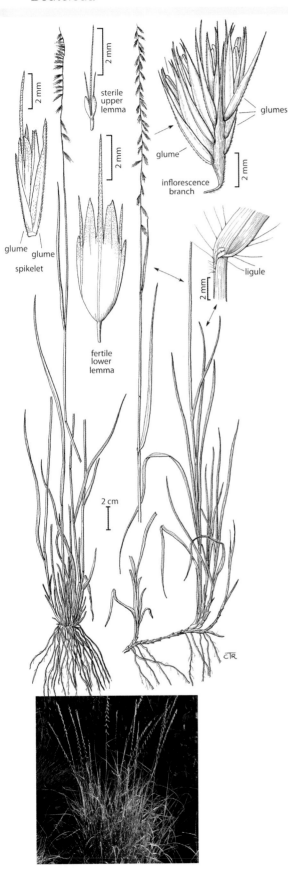

2 mm

sterile upper lemma

2 mm

glume

inflorescence branch

glumes

2 mm

glume glume

spikelet

fertile lower lemma

ligule

2 mm

2 cm

CTR

3. *Bouteloua curtipendula* (Michx.) Torr. (sideoats grama). A cespitose perennial with or without rhizomes. Base of leaf margins usually with white, papillose-based hairs. Widely distributed species abundant in most grasslands in the state. This is the Texas state grass. Listed as good for wildlife and livestock (Hatch and Pluhar 1993). Powell (1994) described this species as one of the best forage species in the Trans-Pecos. Telfair (2006) listed the seeds as especially important for wildlife. Two varieties occur within the state: var. *curtipendula*; and var. *caespitosa* Gould & Kapadia. They can be distinguished by the following character:

1. Plants with long rhizomes......var. *curtipendula*
1. Plants without long rhizomes, sometimes from a knotty base .... var. *caespitosa*

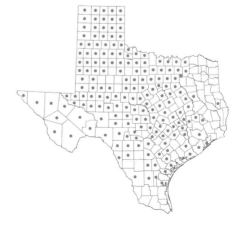

4. *Bouteloua repens* (Kunth) Scribn. & Merr.
(slender grama, large-mesquite grama). Weak,
cespitose perennial. Lower glume nearly as
long and wide as the upper glume. Widespread
in South Texas in open areas on numerous soil
types.

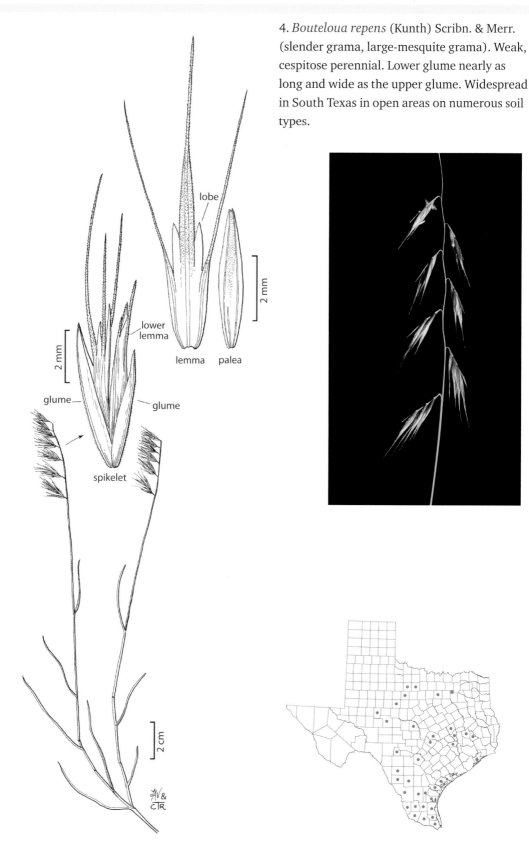

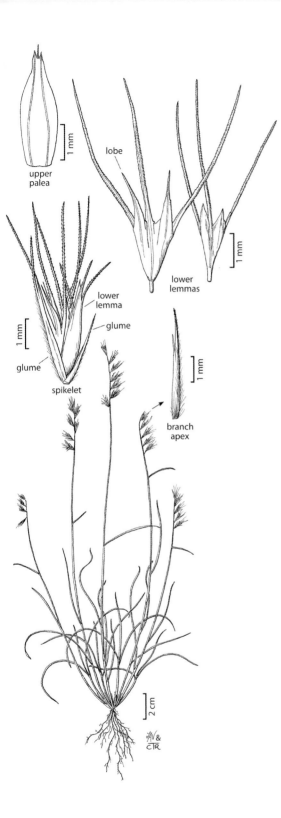

upper palea

lobe

lower lemmas

lower lemma

glume

glume

spikelet

branch apex

5. *Bouteloua rigidiseta* (Steud.) Hitchc. (Texas grama, mesquitegrass). Cespitose perennial forming small, distinct clumps. Glumes and rachis branches sparsely and inconspicuously hairy. Grows in grasslands, open woods, and roadsides on clayey or sandy clay soils. One of the earliest warm-season grasses to flower (Diggs et al. 2006). Listed as poor for wildlife and livestock (Hatch and Pluhar 1993). The Texas species belongs to the typical variety, var. *rigidiseta*.

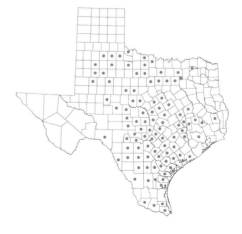

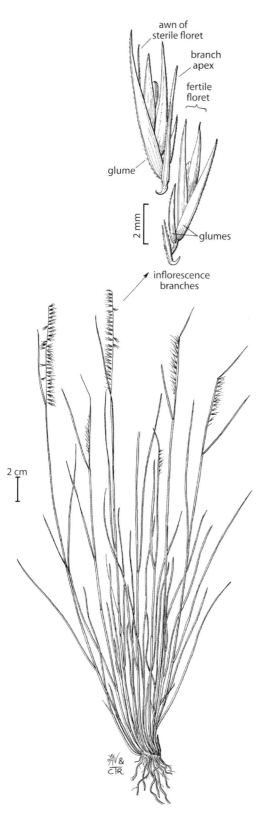

awn of
sterile floret

branch
apex

fertile
floret

glume

2 mm

glumes

inflorescence
branches

2 cm

6. *Bouteloua uniflora* Vasey (Nealley's grama, one-flowered grama). A cespitose perennial lacking both stolons and rhizomes. Most panicle branches support only 1 spikelet. Found in fertile, limestone-derived soils. Only in Texas, northern Mexico, and 1 collection in Utah. The Texas species belongs to the typical variety, var. *uniflora*.

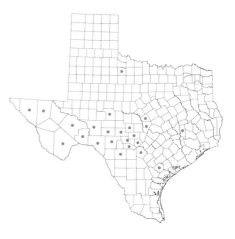

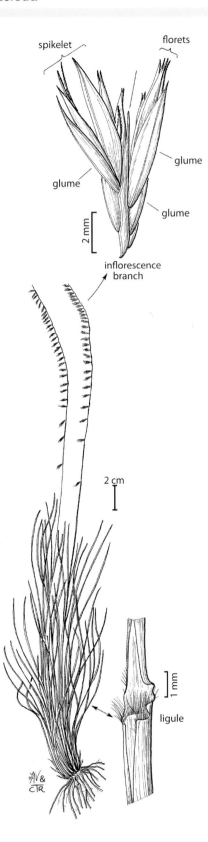

spikelet

florets

glume

glume

glume

2 mm

inflorescence
branch

2 cm

1 mm

ligule

7. *Bouteloua warnockii* Gould & Kapadia
(Warnock's grama). A cespitose perennial
lacking both stolons and rhizomes. Restricted
in distribution to limestone ledges, dry slopes
below limestone outcrops, or gypsum in
the Trans-Pecos region of the state. Listed
as a good forage species (Powell 1994).
Reported to hybridize with *B. curtipendula* var.
caespitosa.

Brachyelytrum P. Beauv.
(Poöideae: Brachyelytreae)

Slender, erect perennial with short, knotty rhizomes. Blades flat and folded. Inflorescence a short, few-flowered panicle. Spikelets with 1 floret, disarticulation above the glumes. Rachilla extended beyond the floret as a bristle. Lower glume absent or minute; upper glume short, narrowly lanceolate, occasionally awned. Lemma about 1 cm long, firm, narrow, rounded on the back, 5-veined, extending into an awn 1–3 cm long. Basic chromosome number, $x = 11$. Photosynthetic pathway, C_3. Represented in Texas by a single introduced species.

The tribe Brachyelytreae also has been placed in the Bambusoideae because of the presence of arm and fusoid cells and broad seedling leaves.

(Stephenson and Saarela 2007)

1. *Brachyelytrum erectum* (Schreb.) P. Beauv. (southern shorthusk, bearded shorthusk, long-awned woodgrass). Rhizomatous perennial with culms up to 1 m tall; nodes densely pubescent. Spikelet terete or dorsally compressed. Lemma awned, up to 20 mm long. Plants are expected in moist woods and forests of far East Texas. The Texas species belongs to the typical variety, var. *erectum*.

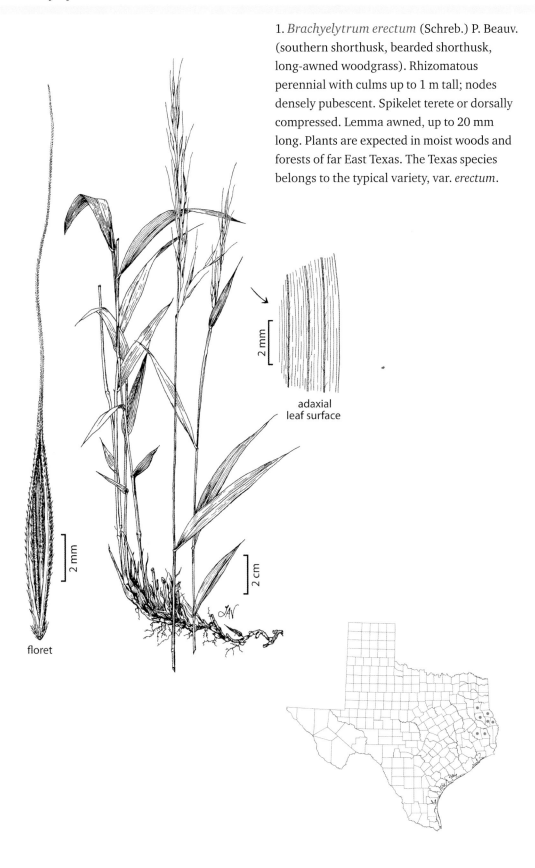

2 mm

adaxial
leaf surface

2 mm

2 cm

floret

Briza L.
(Poöideae: Poeae)

Tufted annuals and perennials with usually showy open panicles of several-flowered, awnless spikelets. Florets crowded and horizontally spreading. Glumes subequal, broad, thin and papery, rounded on the back, 3–9-nerved. Lemmas similar to the glumes, broader than long, usually with 7–9 veins, these distinct or indistinct, and a broadly rounded apex. Palea short to nearly as long as the lemma. Basic chromosome number, $x = 7$. Photosynthetic pathway, C_3. Represented in Texas by 3 species, 2 of which are grown as ornamentals (Gould 1975b; Jones, Wipff, and Montgomery 1997).

1. Plants perennial with short rhizomes .... 2. *B. media*
1. Plants annual without rhizomes.
    2. Spikelets 10–20 mm long .... 1. *B. maxima*
    2. Spikelets 2–7 mm long .... 3. *B. minor*

(Snow 2007)

1. *Briza maxima* L. (big quackgrass). Annual with culms to 80 cm long. Panicles open, few-flowered. Spikelets 10–20 mm long, oval to elliptic. A native of the Mediterranean region. No locations are indicated on the distribution map because the species apparently does not persist outside of cultivation in Texas. It has been reported from the Piney Woods, Cross Timbers and Prairies, and Gulf Prairies and Marshes. Poor for wildlife and livestock according to Hatch, Schuster, and Drawe (1999).

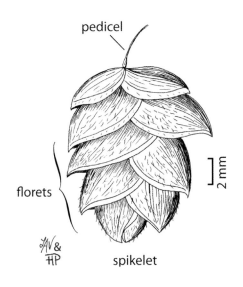

2. *Briza media* L. (perennial quackgrass). Perennial species with rhizomes. Panicles open, about as long as wide. Spikelets 4.0–5.5 mm long. Palea strongly V-shaped in cross section. A European native introduced as an ornamental. No locations are indicated on the distribution map because the species apparently does not persist outside of cultivation in Texas. Not included in Gould (1975b) or Hatch, Gandhi, and Brown (1990).

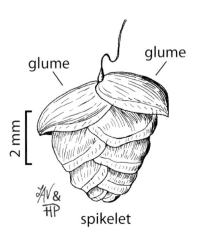

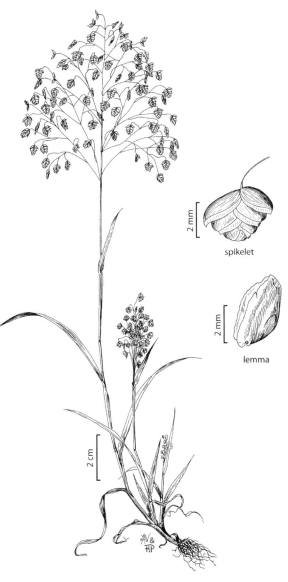

3. *Briza minor* L. (little quakinggrass). Delicate annual native to the Mediterranean region. Panicle open, about as wide as long. Spikelets triangular in shape, inflated, papery. Grows in numerous habitats from swamps to open woodlands and roadsides. Poor for wildlife and livestock according to Hatch, Schuster, and Drawe (1999).

spikelet

lemma

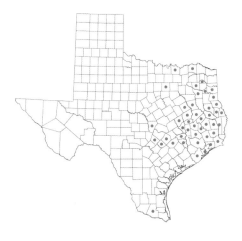

Bromopsis (Dumort.) Fourr.
(Poöideae: Bromeae)

Perennials and annuals, cespitose or rhizomatous. Sheath margins connate. Auricles absent. Ligule membranous. Inflorescence a large, open panicle. Disarticulation above the glumes and between the florets. Spikelets large, with numerous florets, slightly laterally compressed to rounded on the back, florets closely overlapping. Glumes unequal, acute or blunt, rarely awned, shorter than the lowest floret, lower with fewer veins than the upper. Lemmas 5–9-nerved, bifid at the apex, awned from between the teeth. Palea adnate to the caryopsis. Basic chromosome number, $x = 7$. Photosynthetic pathway, C_3. Represented in Texas by 8 species. Sometimes included in *Bromus* as section *Bromopsis* (Gould 1975b; Hatch, Gandhi, and Brown 1990; Jones, Wipff, and Montgomery 1997).

1. Plants rhizomatous ....3. *B. inermis*
1. Plants not rhizomatous.
  2. Plants annual .... 8. *B. texensis*
  2. Plants perennial.
    3. Blades of cauline leaves with a midvein that is not narrowed below junction with sheath... 1. *B. anomalus*
    3. Blades of cauline leaves with a midvein that is narrowed below junction with sheath.
      4. Glumes usually pubescent, rarely glabrous.
        5. Lemmas with awns up to 3 mm long ....5. *B. porteri*
        5. Lemmas with awns 3 mm or longer.. 6. *B. pubescens*
      4. Glumes usually glabrous, occasionally pubescent.
        6. Lemmas with pubescent margins and backs, sometimes nearly glabrous .... 4. *B. lanatipes*
        6. Lemmas with densely hirsute or pilose margins.
          7. Lemmas with all backs glabrous... 2. *B. ciliatus*
          7. Lemmas, at least the upper ones in a spikelet, pubescent ....7. *B. richardsonii*

(Pavlick and Anderton 2007)

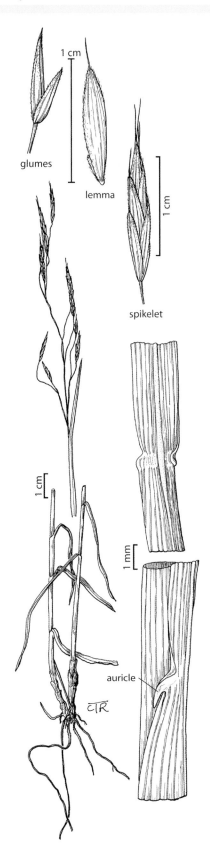

glumes

lemma

spikelet

1 cm

1 cm

1 cm

1 mm

auricle

CTR

1. *Bromopsis anomalus* (F. Rupr. *ex* E. Fourn.) Holub. (nodding brome, Mexican brome). Nonrhizomatous perennial. Panicles 10–20 cm long, nodding. Differs from the closely related B. *porteri* by the presence of auricles and leaves with midribs that narrow just below the collar. Grows on rocky slopes in the Trans-Pecos region, and one report from Bexar County.

2. *Bromopsis ciliatus* (Michx.) Holub. (fringed brome). Nonrhizomatous perennial that can grow to 1.5 m tall. Panicles 10–20 cm long, open, nodding. Found in mesic areas of the western mountains in meadows, thickets, woods, and stream banks. Not included in Gould (1975b) or Hatch, Gandhi, and Brown (1990).

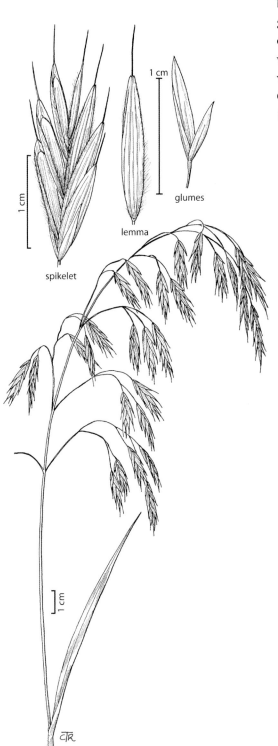

1 cm

glumes

lemma

1 cm

spikelet

1 cm

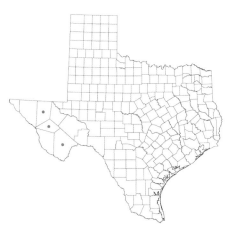

3. *Bromopsis inermis* (Leyss.) Holub. (smooth brome). Rhizomatous perennial. Panicle 10–20 cm long, erect, branches ascending to appressed. Leaf blades flat with a distinctive M or W imprint (most obvious on fresh material). A native of Eurasia introduced in the United States as a species to stabilize roadsides or "improve" pastures or hay meadows. Excellent forage species, especially for horses. Not included in Gould (1975b) or Hatch, Gandhi, and Brown (1990).

lemmas

1 cm

upper glume   lower glume

spikelet

CTR

ligule

auricle

5 cm

5 mm

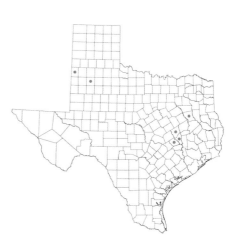

4. *Bromopsis lanatipes* (Shear) Holub. (woolly brome). A nonrhizomatous perennial. Panicles 1–25 cm long, open, nodding, branches ascending to spreading. Generally found in mesic sites in forests of the mountainous regions of the Trans-Pecos and one report from Deaf Smith County.

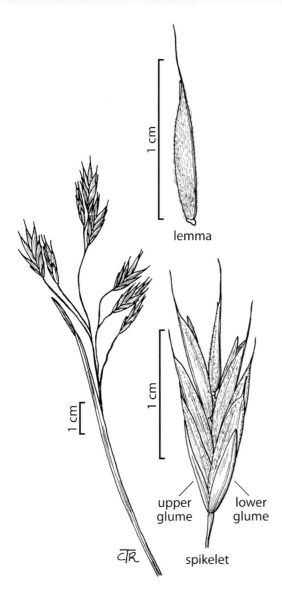

1 cm

lemma

1 cm

1 cm

upper glume

lower glume

CTR

spikelet

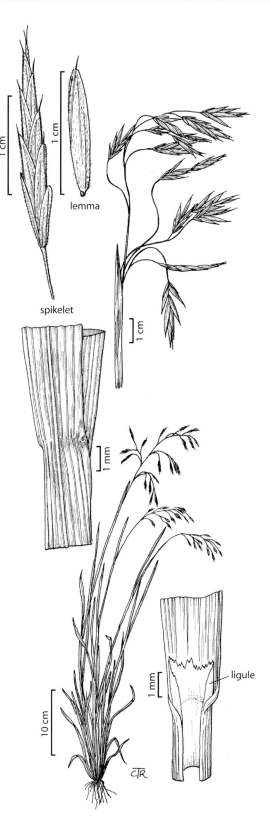

lemma

spikelet

ligule

5. *Bromopsis porteri* (J. M. Coult.) Holub (nodding brome). A nonrhizomatous perennial growing to about 1 m tall. Panicles 7–20 cm long, open, often 1-sided, branches slender, often recurved and flexuous. Grows in the mountainous regions of the Trans-Pecos along forest openings, meadows, grassy slopes, and other mesic sites. It is closely related to *B. anomalus*, but it differs in lacking auricles and not having leaf blades with narrowed midribs just below the collar. Not included in Gould (1975b) or Hatch, Gandhi, and Brown (1990).

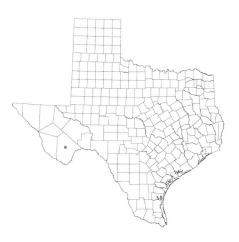

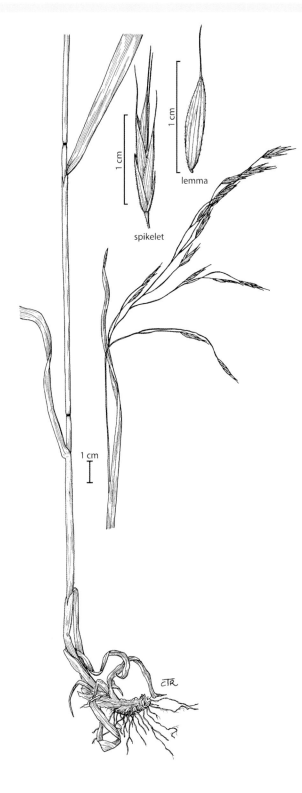

1 cm

lemma

spikelet

1 cm

6. *Bromopsis pubescens* (Muhl. *ex* Willd.) Holub (Canada brome, hairy woodland brome). A perennial growing to 1.5 m tall. Panicles 10–25 cm long, open and nodding, branches usually spreading and the lower ones often reflexed. Found in moist woods and along stream banks throughout much of East Texas. Specimens with some 5-veined lemmas, usually referred to as *B. nottowayanus*, are included here.

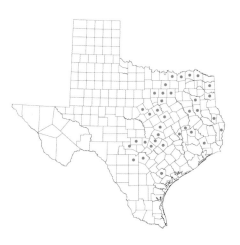

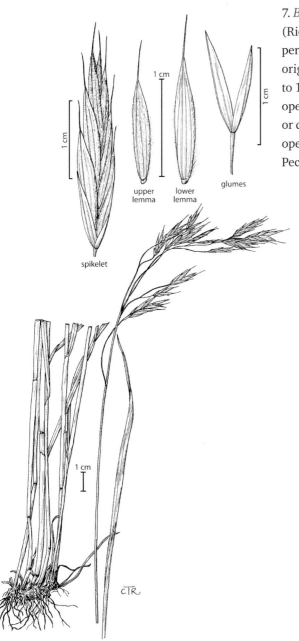

1 cm

upper
lemma

lower
lemma

glumes

spikelet

1 cm

CTR

7. *Bromopsis richardsonii* (Link) Holub. (Richardson's brome). Nonrhizomatous perennial. Mature plants, however, do originate from a knotty crown. It can grow up to 1.5 m tall. Panicles 10–25 cm long, closed or open, nodding, branches appressed, ascending or drooping, filiform. Found in meadows and open woods in the mountains of the Trans-Pecos region.

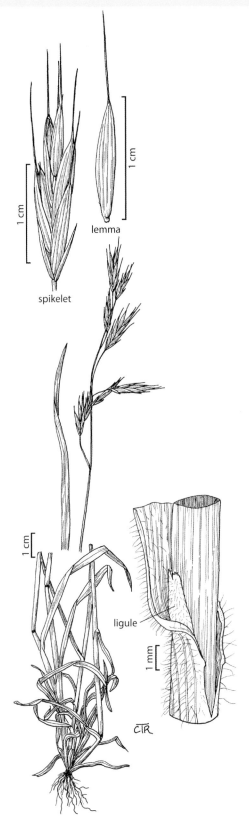

spikelet

lemma

ligule

C̄TR

8. *Bromopsis texensis* (Shear) Holub. (Texas brome). An annual species. Leaf blades uniformly pubescent with short, soft hairs on both sides. Panicles 10–15 cm with erect or, more often, spreading branches, erect or drooping. Restricted to thickets and oak mottes in Central and South Texas. Listed as a Texas endemic (Gould 1975b, as *Bromus texensis* Shear), but it also occurs in Mexico. A relatively uncommon species, although it may be locally abundant in the spring. Not listed as rare by Jones, Wipff, and Montgomery (1997) or Poole et al. (2007). Fair to poor forage in the spring and some seed for birds (Hatch, Schuster, and Drawe 1999).

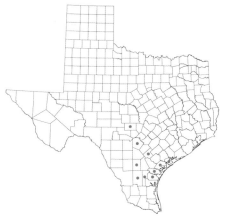

Bromus L.
(Poöideae: Bromeae)

Annuals. Leaf blades and sheaths usually densely pubescent. Sheaths connate for most of their length; ligules membranous, erose; auricles absent. Inflorescence a terminal, simple, panicle, rarely a raceme (most common in underdeveloped plants), branches short and appressed to long and flexuous. Disarticulation above the glumes and between the florets. Spikelets large, usually with 5 to many florets, terete to very slightly laterally compressed. Glumes subequal, shorter than the lowest lemma, lower usually shorter and narrower than the upper. Lemmas broad, only slightly laterally compressed, if at all, 5- or more-nerved, awnless or awned from an entire or bifid apex. Awns straight or only slightly geniculate at maturity. Palea nearly as long as the lemma, attached to the caryopsis. Basic chromosome number, $x = 7$. Photosynthetic pathway, C_3. Represented in Texas by 7 species and 2 subspecies.

1. Rachilla internodes visible at maturity .... 7. *B. secalinus*
1. Rachilla internodes obscure at maturity.
    2. Lemma awn originating <1.5 mm from apex, not twisted at base.
        3. Panicle branches shorter than spikelets .... 3. *B. hordeaceus*
        3. Panicle branches, at least some, longer than the spikelets.
            4. Lemmas <8 mm long, margins rolled .... 6. *B. racemosus*
            4. Lemmas >8 mm long, margins bluntly angled .... 2. *B. commutatus*
    2. Lemma awn originating >1.5 mm from the apex, often twisted at base.
        5. Panicles erect at maturity; branches shorter than the spikelets.
            6. Lemmas densely pubescent evenly over the back .... 5. *B. lanceolatus*
            6. Lemmas glabrous with some hairs near the apices .... 3. *B. hordeaceus*
        5. Panicle drooping at maturity; branches, at least some, longer than the spikelets.
            7. Lower glumes 7–10 mm long; upper glume 8–12 mm long .... 1. *B. arenarius*
            7. Lower glumes 4–7 mm long; upper glumes 5–8 mm long .... 4. *B. japonicus*

(Pavlick and Anderton 2007)

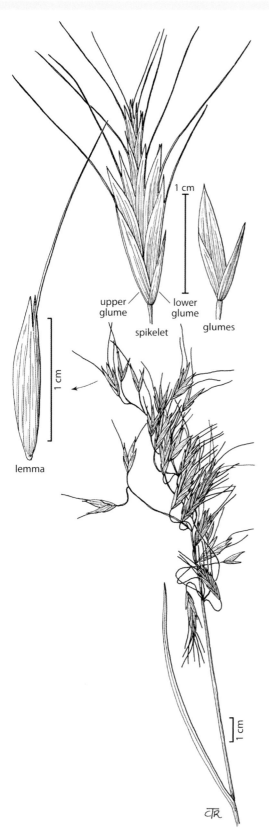

upper
glume
lower
glume

spikelet

glumes

lemma

1 cm

1 cm

1 cm

1. *Bromus arenarius* Labill. (Australian brome). An annual introduced from Australia. Similar to the common *B. japonicus*, but *B. arenarius* has larger glumes and lemmas that are typically pilose. Found in sandy, disturbed sites or waste areas. Not listed in Gould (1975b); Hatch, Gandhi, and Brown (1990); or Jones, Wipff, and Montgomery (1997). Rare in East Texas.

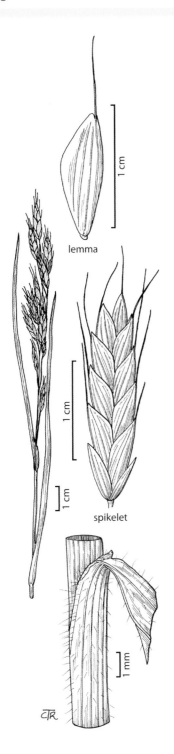

lemma

1 cm

spikelet

1 cm

1 cm

1 mm

CTR

2. *Bromus commutatus* Schard. (meadow brome, hairy chess). An annual, weedy introduction from Europe. Sheaths densely hairy with stiff hairs. Leaf blades pilose. Panicle usually erect. A common weedy species of roadsides, disturbed sites, and fallow fields. Gould (1975b); Hatch, Gandhi, and Brown (1990); and Diggs et al. (2006) included this in *B. secalinus* L.

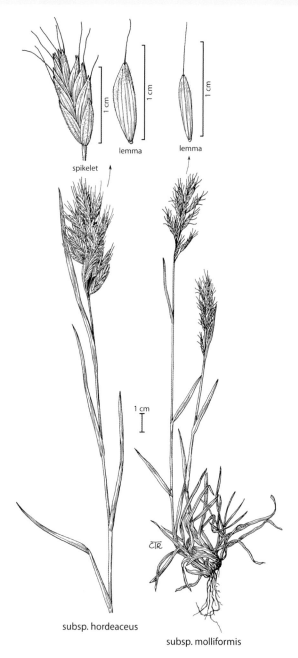

lemma

lemma

spikelet

subsp. hordeaceus

subsp. molliformis

3. *Bromus hordeaceus* L. (lopgrass). An annual, weedy, introduced species from Europe. Panicles relatively short and squatty. Pubescence on glumes and lemmas generally conspicuous. A weedy plant found in waste places, roadsides, and fields. Gould (1975b) and Hatch, Gandhi, and Brown (1990) refer to this species as B. *mollis* L. Two weak subspecies are sometimes recognized in Texas: subsp. *hordeaceus*; and subsp. *molliformis* (J. Lloyd *ex* Billot) Maire & Weiller. They can be distinguished by the following characters:

1. Awns flattened at base, divaricate at maturity .... subsp. *molliformis*
1. Awns terete at base, straight or slightly curved outward at maturity............ subsp. *hordeaceus*

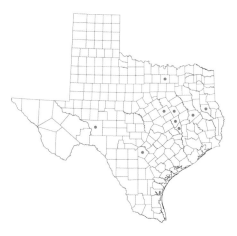

4. *Bromus japonicus* Thunb. (Japanese brome). An annual weedy species from Europe and Asia. Similar to *B. arenarius* but *B. japonicus* generally has shorter glumes and lemmas, and less pubescent spikelets. A common weed of disturbed sites and waste areas. Commonly found growing with *B. secalinus*. The lemmas of *B. japonicus* are softer, more prominently nerved, and have longer apical teeth than those of *B. secalinus* (Gould 1975b). Poor for wildlife and livestock (Hatch, Schuster, and Drawe 1999).

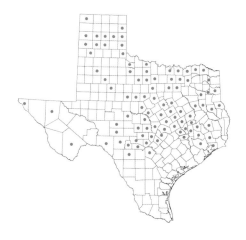

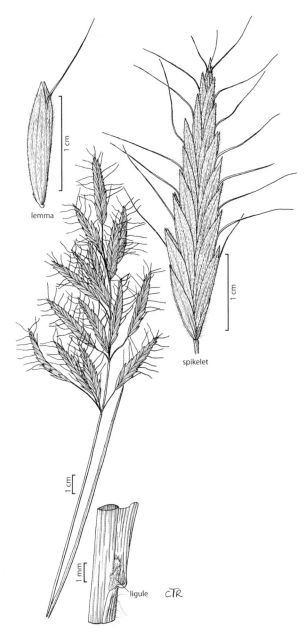

lemma

spikelet

1 cm

1 cm

1 mm

ligule

CTR

5. *Bromus lanceolatus* Roth (lanceolate brome). An annual introduced from southern Europe. Characterized by its short pedicels, long spikelets, and evenly pilose glumes and lemmas. It is a weedy species of waste areas and disturbed sites. This species is included in Gould (1975b) and Hatch, Gandhi, and Brown (1990) as *B. macrostachys*. Diggs et al. (2006) stated that the Texas specimens more closely resemble *B. alopercurus* (= *B. caroli-henrici*).

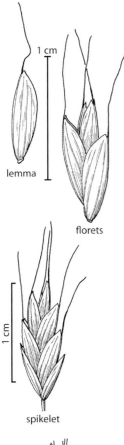

1 cm

lemma

florets

1 cm

spikelet

1 cm

CTR

6. *Bromus racemosus* L. (smooth brome). An annual introduced from western Europe and the Baltic region. This species differs from *B. secalinus* by not having inrolled lemma margins at maturity, and floret bases and rachilla internodes are concealed at maturity. A weed of disturbed sites and waste places. Uncommon in Texas. Gould (1975b) and Jones, Wipff, and Montgomery (1997) included this species in *B. secalinus*.

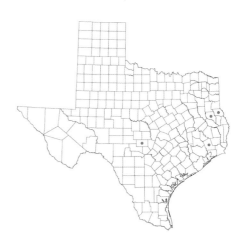

side view

floret dorsal view

ventral view

1 cm

caryopsis

2 mm

1 cm

upper glume   lower glume

spikelet

1 cm

5 cm

7. *Bromus secalinus* L. (rye brome). An annual introduced from Europe. It differs from the closely related *B. racemosus* by having inrolled lemma margins, visible floret bases, and rachilla internodes at maturity. A common weed in disturbed sites, waste areas, and roadsides. Poor for wildlife and livestock (Hatch, Schuster, and Drawe 1999).

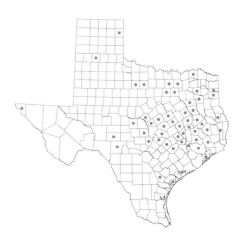

Buchloë Engelm.
(Chloridoideae: Cynodonteae)

Perennials, low, mat-forming, stoloniferous. Sheaths open; ligule a fringe of short hairs; auricles absent. Staminate and pistillate spikelets in separate inflorescences, usually on different plants (dioecious) but not infrequently on the same plant (monoecious). Staminate spikelets with 2 florets, sessile, and closely crowded on 1–4 short, spicate inflorescence branches, these well exserted above the leafy portion of the plant. Pistillate spikelets with 1 fertile floret, in deciduous, capitate, burlike clusters of 2–4, these present in the leafy portion of the plant and usually partially included in expanded leaf sheaths. Rachis and lower ⅔ of upper glume of pistillate spikelets thickened, indurate. Lemma of pistillate spikelet membranous, 3-nerved, usually glabrous and awnless. Basic chromosome number, $x = 10$. Photosynthetic pathway, C_4. Represented in Texas by the only member of the genus.

(Snow 2003a)

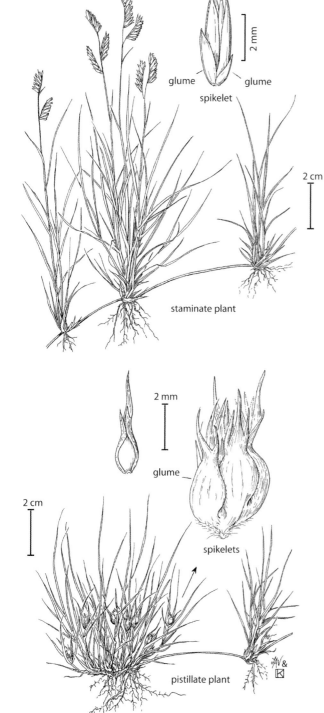

glume   glume

spikelet

2 mm

2 cm

staminate plant

2 mm

glume

spikelets

2 cm

pistillate plant

1. *Buchloë dactyloides* (Nutt.) Engelm. (buffalograss). A low, mat-forming perennial with long stolons. The species has both male and female plants (dioecious); occasionally both male and female flowers are found on the same plant (monoecious). A common species that is a valuable forage plant that withstands heavy grazing. It is frequently used as a low-water, low-maintenance turfgrass species. It is often confused with *Hilaria belangeri*, which has pilose nodes. Some authors combine this genus with *Bouteloua*. Listed as fair for wildlife and good for livestock (Hatch and Pluhar 1993).

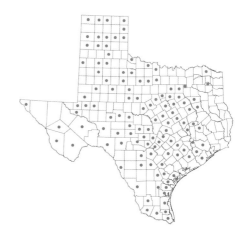

Calamovilfa (A. Gray) Hack
(Chloridoideae: Cynodonteae)

Tall, stout perennials, with short or widely spreading rhizomes and large, open or contracted panicles of small spikelets. Spikelets with 1 floret; the rachilla disarticulating above the glumes, not prolonged beyond the floret. Glumes firm, unequal, 1-nerved, acute, the second nearly as long as the lemma, the first shorter. Lemma firm, 1-nerved, rounded on the back, glabrous or pubescent, bearded on the callus. Palea usually greatly reduced. Basic chromosome number, $x = 10$. Photosynthetic pathway, C_4. Represented in Texas by a single species.

(Thieret 2003b)

glumes

floret

2 mm

2 mm

2 cm

1. *Calamovilfa gigantea* (Nutt.) Scribn. and Merr. (big sandreed, giant sandreed). A perennial with long, shiny rhizomes. Culms can grow to 2.5 m tall, with blades to 12 mm wide and almost 1 m long. The plant is found on sand dunes, riverbanks, and floodplains. Plants are too coarse to be of much value for livestock but are important as a soil binder (Powell 1994).

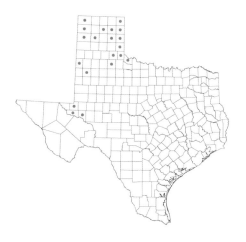

Cathestecum J. Presl
(Chloridoideae: Cynodonteae)

Low, mostly stoloniferous annuals and perennials with tufted leaves and small, spicate, few-flowered inflorescences. Spikelets in subsessile clusters of 3, the terminal (central) one perfect, the 2 lower (lateral) ones usually staminate or sterile. Spikelet clusters few to several, scattered on the short culm rachis and deciduous as a whole. Terminal spikelet of each cluster with 1 perfect floret below and 1 or more reduced florets above. Glumes unequal, usually hairy, the first short, thin, nerveless or 1-nerved, and the second about as long as the lemma, 1-nerved. Lemmas typically 3-nerved, the nerves extending into awns at the broadly notched and 2-lobed apex. Palea well developed, the nerves usually extending into setae. Basic chromosome number, $x = 10$. Photosynthetic pathway, C_4. Represented in Texas by a single species.

(Gould and Shaw 1983)

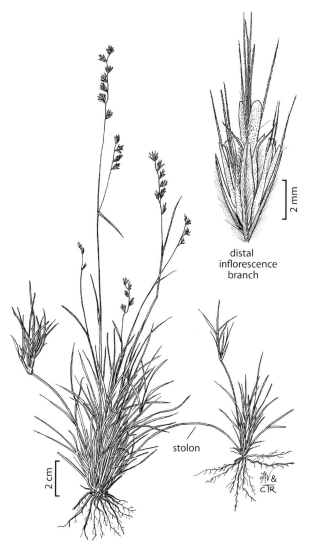

distal
inflorescence
branch

2 mm

stolon

2 cm

1. *Cathestecum erectum* (Vasey) Hack. (false grama). Short (<40 cm) perennial that can be either dioecious or monoecious. Stoloniferous, stolons thin and strongly arching. Found on dry hills in the Big Bend region of the state. Some advocate placing this genus in *Bouteloua*.

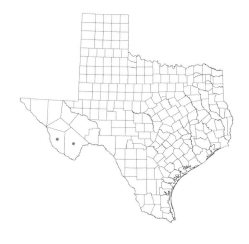

Cenchrus L.
(Panicoideae: Paniceae)

Annuals and perennials, mostly with weak, geniculate-decumbent culms and soft, flat blades but a few with tall, coarse, stiffly erect culms. Ligule a fringe of hairs. Spikelets enclosed in fascicles, these sessile or subsessile on a short, stout rachis. Fascicles of bristles and/or flattened spines (modified branches) fused together at least at the base. Bristles and spines usually retrorsely barbed. Spikelets several in each fascicle, this readily disarticulating at maturity. Glumes thin, membranous, unequal. Lower glume 1- or 3-nerved, the upper 1–7-nerved. Lemma of the sterile floret thin, 1–7-nerved, equaling or exceeding the upper glume. Palea of the sterile floret about equaling the lemma. Lemma of the fertile floret thin, membranous, 5- or 7-nerved, tapering to a slender, usually acuminate tip, the margins not inrolled. Caryopsis elliptic to ovoid, dorsally flattened. Basic chromosome number, $x = 17$. Photosynthetic pathway, C_4. Represented in Texas by 5 species. The sharp spines can cause painful wounds, get caught in clothing and pet fur, and are a general nuisance. After a single encounter, most learn and remember this genus. All species are commonly referred to as "stickers."

1. Fascicles burlike; some bristles flattened at the base.
    2. Fascicles with 1 whorl of fused, flattened inner bristles, subtended by 5–25 terete outer bristles.
        3. Rachis internodes <2mm long; most outer bristles equaling or exceeding the inner, flattened bristles ....1. *C. brownii*
        3. Rachis internodes >2 mm long; most outer bristles about .5 as long as the inner, flattened bristles ....2. *C. echinatus*
    2. Fascicles with more than 1 whorl of flattened inner bristles, sometimes subtended by terete outer bristles.
        4. Inner bristles usually <1 mm wide at base; fascicles with 45–75 bristles.......................
        .... 3. *C. longispinus*
        4. Inner bristles usually 1–3 mm wide at base; fascicles with 8–43 bristles .... 5. *C. spinifex*
1. Fascicles not burlike; all bristles terete .... 4. *C. myosuroides*

(Stieber and Wipff 2003)

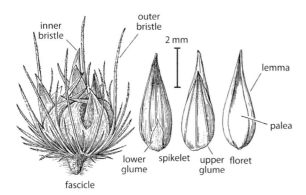

inner bristle
outer bristle
2 mm
lemma
palea
lower glume
spikelet
upper glume
floret
fascicle

1. *Cenchrus brownii* Roem. and Schult. (slimbristle sandbur, green sandbur). An annual with erect or decumbent culms that can grow to nearly 1 m in length. Sheaths are slightly compressed. Panicles 4–15 cm long. A native species most frequently found in sandy waste places of the Southeast. The single Texas collection in Val Verde County may have been a chance introduction, and this species may not be a persistent member of Texas flora. Not included in Gould (1975b) or Hatch, Gandhi, and Brown (1990).

2 cm

inner bristle

2 mm

lemma

palea

outer bristle

fascicle

lower glume

upper glume

floret

spikelet

2 cm

2 cm

2. *Cenchrus echinatus* L. (southern sandbur, hedgehog grass). An annual with geniculate culms to nearly 1 m long. Rooting may occur at the lower nodes. Sheaths shorter than to about equal to the internodes, compressed. Panicles 2–12 cm long. A common plant of the Coastal Plains and Piney Woods. Typically found in sandy waste places and disturbed sites. Poor for wildlife and livestock (Hatch, Schuster, and Drawe 1999).

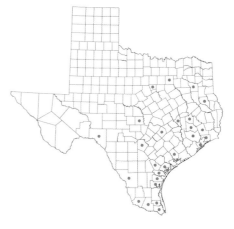

3. *Cenchrus longispinus* (Hack.) Fernald (mat sandbur, long-spine sandbur, innocent-weed). An annual, tufted species with many branches arising from the base. Sheaths strongly compressed-keeled. Panicles 1–10 cm long. Most common in the High Plains and expected elsewhere throughout the state. A plant of sandy fields, roadsides, and disturbed or waste places. Often confused with *C. spinifex*, but *C. longispinus* has longer spikelets, more bristles overall, narrower inner bristles, and outer bristles that are usually terete. No forage value (Powell 1994).

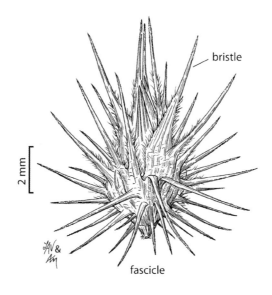

bristle

2 mm

fascicle

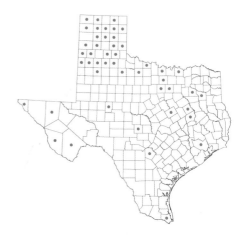

4. *Cenchrus myosuroides* Kunth (big sandbur). A perennial that can grow to nearly 2 m tall. Sheaths compressed. Panicles 4–25 cm long. Found along roadsides, disturbed sites, and waste places. Only long-lived perennial *Cenchrus* species in Texas. Listed as fair for wildlife and good for livestock (Hatch and Pluhar 1993; Powell 1994); poor for wildlife and fair for livestock along the Texas Coastal Prairies and Marshes (Hatch, Schuster, and Drawe 1999).

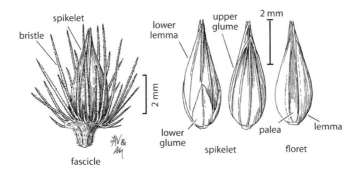

5. *Cenchrus spinifex* Cav. (common sandbur, grassbur, coastal sandbur). Annual or rarely a very short-lived perennial. Culms tufted and reaching 1 m tall. Sheaths compressed. Panicles 3–8 cm long. Abundant and widespread throughout the state. Most commonly confused with *C. longispinus*, but *C. spinifex* has shorter spikelets, fewer bristles overall, wider inner bristles, and outer bristles that are usually flattened. Poor for wildlife and livestock (Hatch, Schuster, and Drawe 1999; Powell 1994, as *C. insertus* M. A. Curtis).

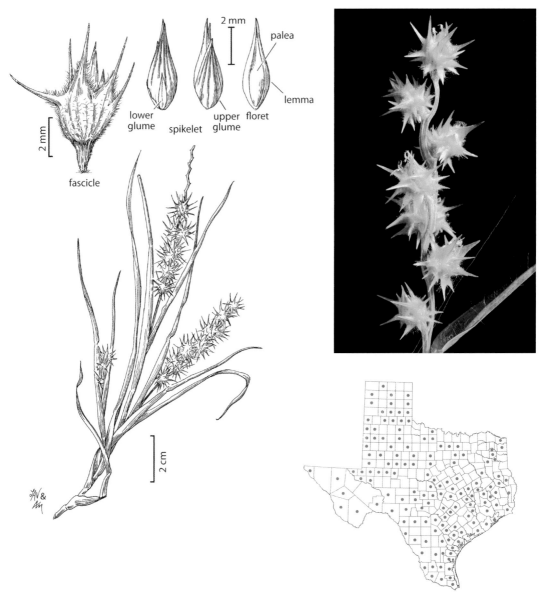

2 mm

palea

lemma

lower glume   spikelet   upper glume   floret

2 mm

fascicle

2 mm

2 cm

Ceratochloa P. Beauv.
(Poöideae: Bromeae)

Annuals or biennials or infrequently perennials, cespitose. Sheaths connate for most of their length; ligules prominent, membranous, erose; auricles absent. Inflorescence a simple, terminal panicle (rarely a raceme in very immature plants). Disarticulation above the glumes and between the florets. Spikelets relatively large, strongly laterally compressed (strongly flattened), with 3–9 florets. Glumes subequal, strongly keeled, shorter than the lowermost lemma. Lemmas strongly flattened, many-nerved, awned from a bifid or entire apex or awnless. Palea nearly as long as the lemma. Basic chromosome number, $x = 7$. Photosynthetic pathway, C_3. Represented in Texas by 4 species and 2 varieties.

A segregate of *Bromus* based on strongly laterally compressed spikelets and keeled lemmas. Included in Gould (1975b); Hatch, Gandhi, and Brown (1990); and Jones, Wipff, and Montgomery (1997) as *Bromus* section *Ceratochloa*.

1. Upper glume about as long as the lowest lemma in spikelet, marginal hairs (if present) longer than hairs elsewhere on lemma .... 1. *C. arizonica*
1. Upper glume shorter than the lowest lemma in spikelet, marginal hairs similar in length to those elsewhere on lemma.
  2. Lemmas and sheath throats glabrous.. 4. *C. polyantha*
  2. Lemmas and sheath throats with hairs.
    3. Lemmas 9–13-veined; veins riblike .... 3. *C. cathartica*
    3. Lemmas 7–9-veined; veins not riblike .... 2. *C. carinata*

(Pavlick and Anderton 2007)

1. *Ceratochloa arizonica* (Shears) Holub. (Arizona brome). A tufted annual. Panicles 12–25 cm long, contracted or somewhat open. Spikelets with 4–8 florets, florets spreading at maturity. Anthers about 0.5 mm long, suggesting that most seed production is from cleistogamy. Found mostly in disturbed areas, along roadsides, in waste places, and on field margins. Gould (1975b) and Hatch, Gandhi, and Brown (1990) reported the species (as *Bromus arizonicus*) as occurring within the Trans-Pecos region of the state, but the species was not reported for Texas in Pavlick and Anderton (2007). They show it as occurring in the southern counties of New Mexico that border Texas. So the species should be expected in the western part of the state.

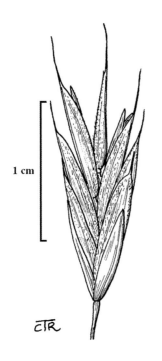

1 cm

CTR

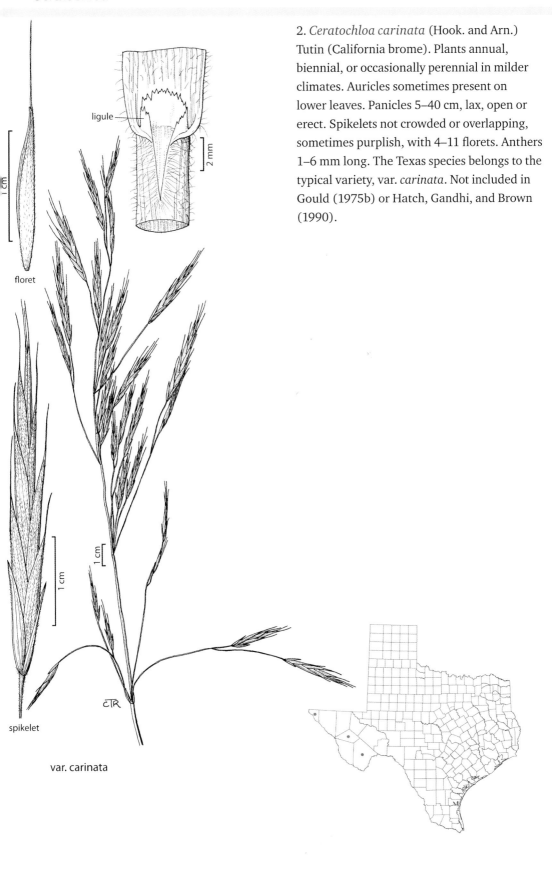

ligule

2 mm

floret

1 cm

spikelet

1 cm

1 cm

CTR

var. carinata

2. *Ceratochloa carinata* (Hook. and Arn.) Tutin (California brome). Plants annual, biennial, or occasionally perennial in milder climates. Auricles sometimes present on lower leaves. Panicles 5–40 cm, lax, open or erect. Spikelets not crowded or overlapping, sometimes purplish, with 4–11 florets. Anthers 1–6 mm long. The Texas species belongs to the typical variety, var. *carinata*. Not included in Gould (1975b) or Hatch, Gandhi, and Brown (1990).

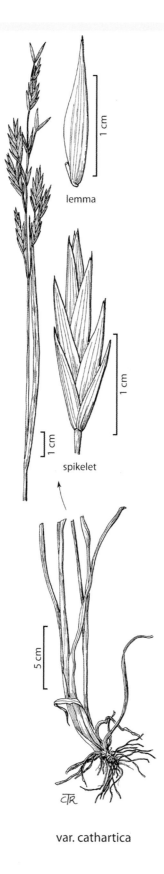

lemma

spikelet

var. cathartica

3. *Ceratochloa cathartica* (Vahl.) Herter (rescue grass, rescue brome, Schraders-grass). Annual or biennial. Panicles 8–30 cm long, open, erect or nodding. Spikelets not crowded or overlapping, with 6–12 florets. Some florets cleistogamous. Common species and one of the first to begin growth in late fall–early spring. A South American species introduced as a forage crop that is now common throughout much of North America. Hatch and Pluhar (1993) listed it as fair for wildlife and good for livestock, as *Bromus unioloides* (Willd.) HBK. Powell (1994) reported it as fair forage, as *Bromus catharticus* Vahl. The Texas species belongs to the typical variety, var. *cathartica*.

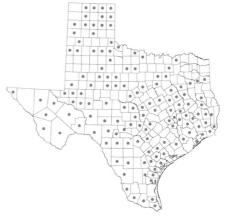

lemma

1 cm

1 cm

1 cm

2 mm

spikelet

CTR

4. *Ceratochloa polyantha* (Scribn.) Tzvelev (Colorado brome, Great Basin brome). A cespitose perennial species with culms up to 1.2 m tall. Panicles 10–20 cm long, open or contracted. Spikelets not crowded or overlapping, with 4–11 florets. Included in *C. carinata* (as *Bromus carinatus*) in Jones, Wipff, and Montgomery (1997).

Chasmanthium Link
(Centothecoideae: Centotheceae)

Moderately tall perennials, some rhizomatous, with broad, flat blades. Ligule a fringe of hairs
or a hyaline, ciliate membrane. Inflorescence open or contracted, the spikelets long- or short-
pediceled. Spikelets 2- to many-flowered, the lower 1–4 sterile. Disarticulation above the
glumes and between the florets. Glumes subequal, shorter than the lemmas, acute to acuminate,
3–7-nerved, laterally compressed and keeled, the keel serrulate. Lemmas 5–15-nerved,
compressed-keeled, the keel serrate or ciliate. Palea about as large as the lemma, bowed out at
the base, 2-keeled, the keels serrate-winged. Flowers perfect, with a single stamen and 2 fleshy,
cuneate, lobed-truncate, 2–4-nerved lodicules. Ovary glabrous, with a single style and 2 plumose
stigmas. Caryopsis ovate to elliptic, laterally compressed, with an embryo less than ½ the length
of the grain. Basic chromosome number, $x = 12$. Photosynthetic pathway, C_3. Represented in
Texas by 3 species.

1. Inflorescence open, the branches drooping; pedicels 10–30 mm long; spikelets 20–50 mm
   long ....1. *C. latifolium*
1. Inflorescence contracted, branches erect or ascending; pedicels 1–5 mm long; spikelets
   5–18 mm long.
    2. Collars and sheaths pilose; culms up to 3.5 mm thick at the nodes; fertile lemmas
      7–9-veined, usually curved .... 3. *C. sessiliflorum*
    2. Collars and sheaths glabrous; culms to 1 mm thick at the nodes; fertile lemmas 3–7-veined,
      usually straight..2. *C. laxum*

(Sánchez-Ken and Clark 2003)

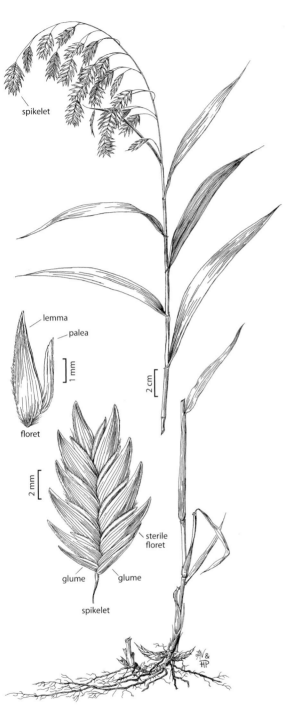

spikelet

lemma
palea

1 mm

floret

2 cm

2 mm

sterile
floret

glume   glume

spikelet

1. *Chasmanthium latifolium* (Michx.) H. O. Yates (broadleaf chasmanthium, inland seaoats, creek-oats, Indian woodoats). A rhizomatous perennial with culms to 1.5 m tall. Leafy for 80% of its height. Spikelets with 6–25 florets, lower 1–3 sterile, fertile ones diverging at about 45°. Widespread and abundant species in woodlands, in forest openings, and along stream banks. Popular as an ornamental. Sensitive to grazing; birds and rodents eat the seeds (Hatch, Schuster, and Drawe 1999).

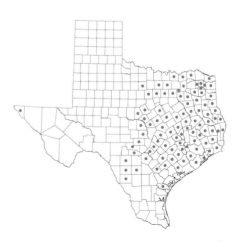

2. *Chasmanthium laxum* (L.) H. O. Yates
(slender chasmanthium, slender woodoats). A
rhizomatous perennial with culms to 1.3 m tall.
Leafy for about 50% of its height. Spikelets
with 3–7 florets, lower 1 or less frequently,
2, sterile. Fertile florets diverging up to
about 45°. Found in woodlands, forests, and
meadows throughout East Texas. Fair for
wildlife and fair to poor forage for livestock
(Hatch, Schuster, and Drawe 1999).

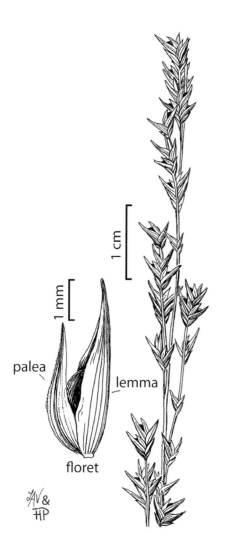

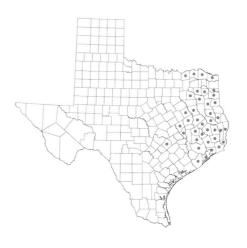

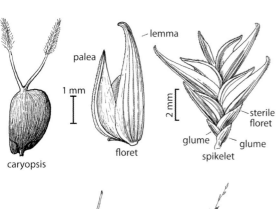

caryopsis

palea

1 mm

lemma

floret

glume glume

spikelet

sterile floret

2 mm

3. *Chasmanthium sessiliflorum* (Poir.) H. O. Yates (long-leaf chasmanthium, narrow-leaf chasmanthium, long-leaf woodoats). A cespitose or rhizomatous perennial with culms to 1.5 m tall. Leafy for about 40% of its height. Spikelets with 4–8 florets, lower 1–2 sterile, fertile florets diverge up to 80°. Found in woods, swamps, and forest openings throughout East Texas. Jones, Wipff, and Montgomery (1997) listed it as a variety of *C. laxum*.

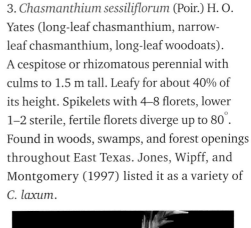

2 cm

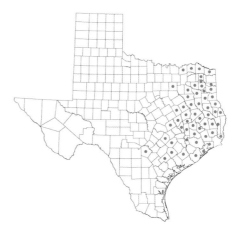

Chloris Sweet
(Chloridoideae: Cynodonteae)

Annuals and perennials with flowering culms from erect or decumbent and stoloniferous bases. Lower portion of culms often flattened, and laterally compressed. Sheaths sharply keeled. Inflorescence with few to several persistent, spicate branches, these mostly digitate or clustered at the culm apex but loosely distributed along the main axis in a few species. Spikelets with a single perfect floret and 1–3 staminate or rudimentary florets above. Disarticulation above the glumes. Glumes thin, lanceolate, 1-nerved (except in cleistogamous, subterranean spikelets), strongly unequal to nearly equal. Lemmas broadly to narrowly ovate or oblong, awnless or more frequently awned from a minutely bifid apex, 1–5-nerved, but usually with 3 strong veins, glabrous or ciliate-pubescent or puberulent on the veins. Palea well developed, strongly 2-nerved. Basic chromosome number, $x = 10$. Photosynthetic pathway, C_4. Represented in Texas by 12 species and 2 varieties.

Gould (1975b) and Hatch, Gandhi, and Brown (1990) included *C. chloridea*; it has since been moved to the genus *Enteropogon*. Contrary to Hatch, Gandhi, and Brown, *C. polydactyla* (L.) Sw. (a synonym of *C. elata* Desv.) does not occur in Texas.

1. Spikelets with 3 or more florets, the lowest perfect, others staminate or sterile, uppermost floret sometimes completely reduced and only a swollen rachilla present.
    2. Terminal florets strongly inflated .... 2. *C. barbata*
    2. Terminal florets not inflated, or only so at the apices.
        3. Spikelets 5–7 per centimeter of inflorescence branch .... 8. *C. pilosa*
        3. Spikelets 10–14 per centimeter of inflorescence branch.
            4. Plants annual; third florets shorter than subtending rachilla segments ....12. *C. virgata*
            4. Plants perennial; third florets as long as or longer than subtending rachilla segment.
                5. Lowest lemmas >2.7 mm long...3. *C. canterae*
                5. Lowest lemmas <2.7 mm long.
                    6. Panicles with 4–7 branches; branches to 8 cm long; awns of the first sterile floret up to 1.5 mm long ....4. *C. ciliata*
                    6. Panicles with up to 30 branches; branches to 20 cm long; awns of the first sterile floret up to 4 mm long..7. *C. gayana*
1. Spikelets with 2 florets, lowest perfect and upper staminate or sterile.
    7. Lemma of upper florets lobed for ⅓ of length; lobes sometimes awned .... 6. *C. divaricata*
    7. Lemma of upper florets lobed, unlobed, or lobed less than ¼ of length; lobes, if present, ... unawned.
        8. Lowest lemmas unawned, mucronate, or short-awned, the awns <1.5 mm long.
            9. Plants annual or short-lived perennial; culms 30–200 cm long .... 8. *C. pilosa*
            9. Plants perennial; culms usually 50 cm or less.
                10. Upper florets inflated, about as long as wide ....5. *C. cucullata*
                10. Upper florets not inflated, at least twice as long as wide ....9. *C. submutica*
        8. Lowest lemmas always awned, awns 1–11 mm long.

11. Lowest lemmas with a prominent groove on each side .... 8. *C. pilosa*

11. Lowest lemmas without a prominent groove on each side.

    12. Panicle branches in 2 or more distinct whorls.

        13. Panicle branches spikelet-bearing to the base ....11. *C. verticillata*

        13. Panicle branches naked on the basal 2–5 cm............................. 10. *C. texensis*

    12. Panicle branches usually in a terminal, digitate cluster; occasionally a second, incomplete whorl develops.

        14. Apical margins of lower lemmas with longer hairs up to 1.5 mm long.

            15. Plants annual12. *C. virgata*

            15. Plants perennial7. *C. gayana*

        14. Apical margins of lower lemmas glabrous or with appressed hairs <1.5 mm long.

            16. Panicle branches naked on the basal 2–5 cm..................... 10. *C. texensis*

            16. Panicle branches naked on the lower 2 cm or with spikelets to the base.

                17. Second florets 0.1–0.5 mm wide.......................1. *C. andropogonoides*

                17. Second florets 0.5–1.0 mm wide ....7. *C. gayana*

(Barkworth 2003b)

Note: *Chloris verticillata*, *C. cucullata*, and *C. andropogonoides* hybridize and intergrade in South and Central Texas. Large, mixed, hybrid populations are not uncommon and exhibit varying degrees of similarity to both parents. *Chloris subdolichostachya* Mull., found in Gould (1975b) and Hatch, Gandhi, and Brown (1990), is a common hybrid (Barkworth 2003b).

1. *Chloris andropogonoides* E. Fourn. (slimspike windmill grass). Perennial, sometimes exhibiting short stolons. Culms relatively short, not exceeding 50 cm. Ligules 0.5–0.8 mm long. Panicle branches 6–13, usually digitate, occasionally with a poorly developed second whorl. Entire inflorescence disarticulates, panicle falling intact. Lemmas with awns 2–5 mm long. This taxon grows along roadsides, lawns, pastures, and prairie relicts in Central, South, and coastal Texas. Poor forage according to Hatch, Schuster, and Drawe (1999).

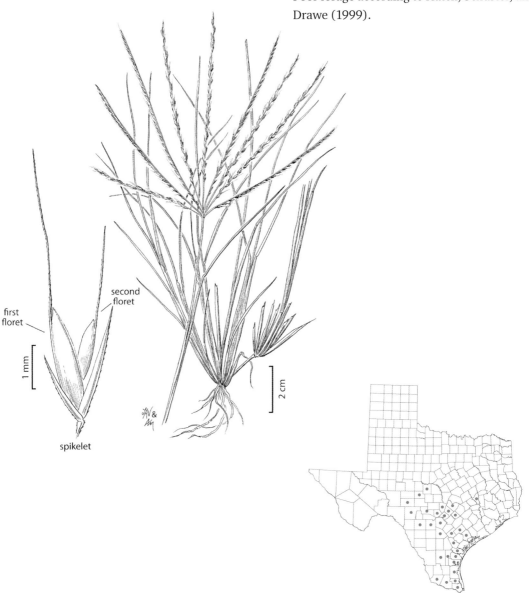

first floret

second floret

1 mm

spikelet

2 cm

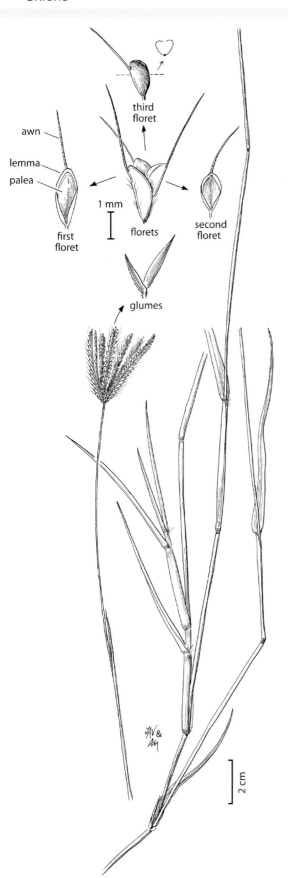

awn
lemma
palea
first
floret
third
floret
florets
second
floret
glumes

1 mm

2 cm

2. *Chloris barbata* Sw. (swollen windmillgrass). An annual with erect or decumbent culms that root at the lower nodes. Ligules 0.3–0.5 mm long. Panicle branches 7–15, erect. Second and third florets strongly inflated. Disarticulation below the glumes. Tends to be a weedy species found along beaches in coastal areas, field margins, and waste areas. Included in Gould (1975b) and Hatch, Gandhi, and Brown (1990) as *C. inflata* Link.

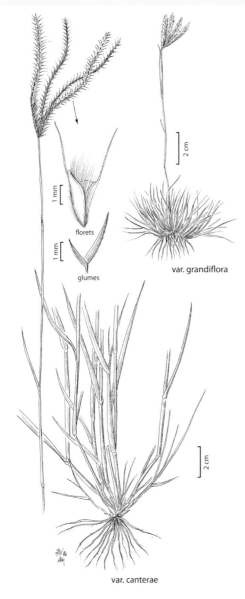

florets

glumes

var. grandiflora

var. canterae

3. *Chloris canterae* Arechav. (Paraguayan windmillgrass). A cespitose perennial with culms to about 1 m tall. Ligules about 0.5 mm long. Panicle branches 2–9, curved, erect. Disarticulation below the glumes. An introduced species from South America. An invasive species of roadsides and ditches that is spreading throughout the more humid areas of the state. It is very common in Brazos County. The specific epithet is spelled *canterai* by Gould (1975b); Hatch, Gandhi, and Brown (1990); and Jones, Wipff, and Montgomery (1997). Two varieties occur in Texas: var. *canterae* is found in more mesic areas; and var. *grandiflora* (Roseng. and Izag.) D. E. Anderson occurs on drier sites. They can be distinguished by the following characters:

1. Leaves mostly cauline; panicle branches 4–14 cm long ....var. *canterae*
1. Leaves mostly basal; panicle branches 3–6 cm long............................... var. *grandiflora*

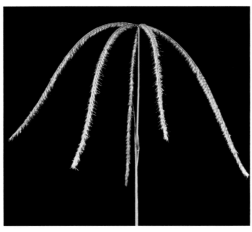

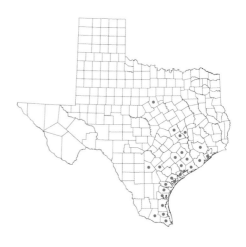

first
floret

florets

glumes

1 mm

1 mm

2 cm

4. *Chloris ciliata* Sw. (fringed windmillgrass, fringed chloris). A cespitose perennial with culms to about 1 m tall. Ligules absent or 0.3–0.4 mm long. Panicle branches 2–7, digitate, ascending or, less frequently, spreading. Disarticulation below the glumes. This species is a native to grasslands of the Gulf Coast and is also found along roadsides and in ditches.

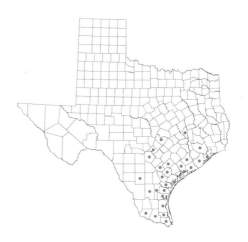

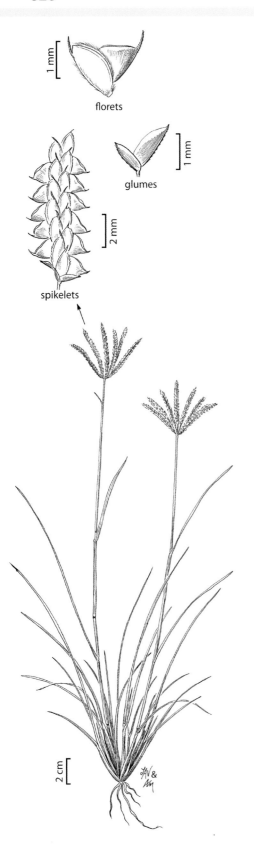

florets

glumes

spikelets

2 cm

5. *Chloris cucullata* Bisch. (hooded windmillgrass, hooded fingergrass, crowfoot). A perennial with erect culms up to 60 cm or longer. Ligules 0.7–1.0 mm long. Panicle branches 1–20, in several closely spaced whorls. Disarticulation below the glumes. A common weed of roadsides and waste places throughout most of Texas. Also found in prairies, where it is considered a fair forage species (Hatch and Pluhar 1993). Most commonly found on sandy soils.

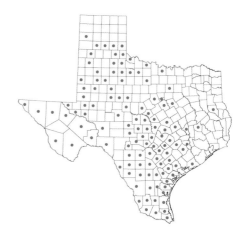

6. *Chloris divaricata* R. Br. (spreading windmill grass). A perennial, sometimes with stolons, culms only to 50 cm long. Ligules ciliolate. Panicle branches 3–9, spreading. Spikelets appressed to branches. Lemma awns up to 9 mm long. Disarticulation below the glumes. Introduced from Australia; rare; mostly found in the Gulf Coast region.

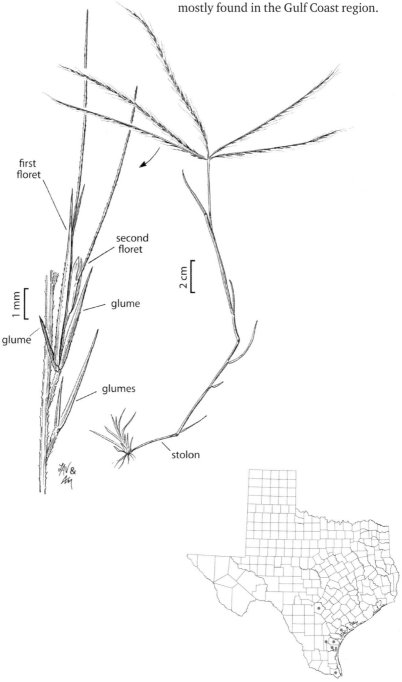

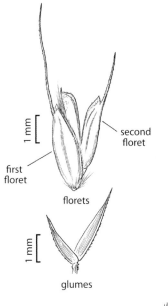

first
floret

second
floret

florets

glumes

7. *Chloris gayana* Kunth (Rhodesgrass). A stoloniferous perennial with erect culms up to 3 m tall. Ligules ciliate. Panicle branches 9–30, usually spreading. Disarticulation below the glumes. An introduction from southeastern Africa, now established as a forage species particularly in the Gulf Coast and South Texas regions. It has commonly escaped along roadsides. Good forage producer (Hatch, Schuster, and Drawe 1999).

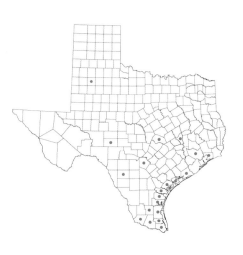

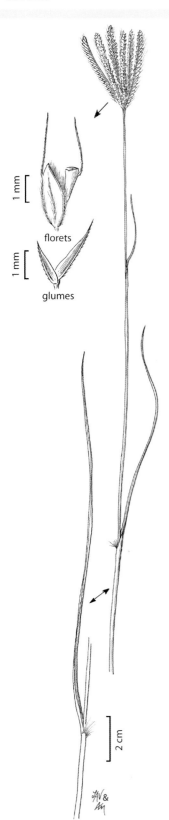

florets

glumes

1 mm

1 mm

2 cm

8. *Chloris pilosa* Schumach. (African windmillgrass). A short-lived perennial that can produce stolons, or occasionally an annual. Culms erect or more frequently decumbent, up to 2 m tall. Panicle branches 5–9, erect. An African species that is sometimes planted for forage. Once collected in Kleberg County but may not be a persistent member of the state grass flora.

9. *Chloris submutica* Kunth (Mexican windmillgrass). A perennial, occasionally with stolons. Culms up to 75 cm tall. Ligules membranous. Panicle branches 5–17, in 1–3 closely spaced whorls, erect when young, spreading and reflexed at maturity. Disarticulation below the glumes. Relatively restricted species; reported only from Jeff Davis County in Texas. Not included in Gould (1975b).

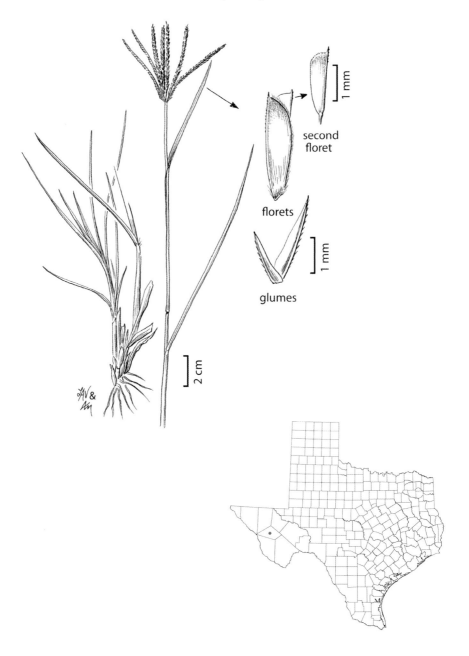

second floret

florets

glumes

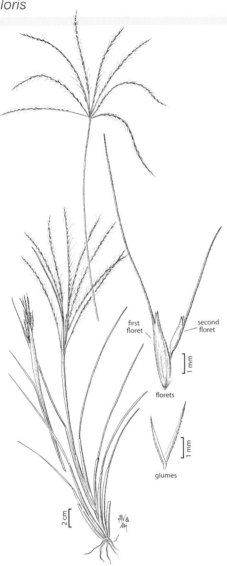

first floret

second floret

florets

] 1 mm

] 1 mm

glumes

2 cm [

10. *Chloris texensis* Nash (Texas windmillgrass). Cespitose perennial with culms to 45 cm tall. Ligules membranous. Panicle branches 8–10, usually digitate, rarely with a poorly developed second whorl, divergent with maturity. Disarticulation below the glumes. Reported as a rare Texas endemic (Gould 1975b). Listed as a Category V species (watch-listed) by the Texas Organization for Endangered Species (TOES) (Jones, Wipff, and Montgomery 1997). Poole et al. (2007, 34) reported the global rank of the species as "G2 = 6–20 occurrences known globally; imperiled and very vulnerable to extinction throughout its range"; and state rank as "S2 = 6–20 known occurrences in Texas; imperiled in the state because of rarity; very vulnerable to extirpation from the state." Poor forage according to Hatch, Schuster, and Drawe (1999). The occurrence in Jeff Davis County is questionable.

11. *Chloris verticillata* Nutt. (tumble windmillgrass). A short (<50 cm) perennial that sometimes roots at the lower nodes. Ligules 0.5–1.5 mm long. Panicle branches 10–16, in several distinct, well-separated whorls and a solitary vertical terminal branch. Entire inflorescence disarticulates, panicle falling intact. A common weed of roadsides, waste areas, lawns, and disturbed sites. An indicator of overutilized rangeland (Diggs et al. 2006) and of poor value for livestock or wildlife (Hatch and Pluhar 1993; Powell 1994).

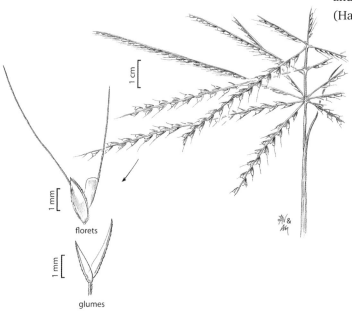

1 cm

1 mm

florets

1 mm

glumes

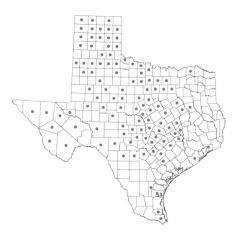

12. *Chloris virgata* Sw. (showy chloris,
feather windmillgrass, feather fingergrass).
A perennial, occasionally with stolons and
culms up to 1 m tall. Ligules up to 4 mm long.
Panicle branches 4–20, erect, sometimes wavy.
Lemmas with long distal hairs. Disarticulation
below the glumes. A widespread weed of many
habitats, but common along road rights-of-
way. Poor forage according to Hatch, Schuster,
and Drawe (1999), but Powell (1994) reported
it as palatable to livestock.

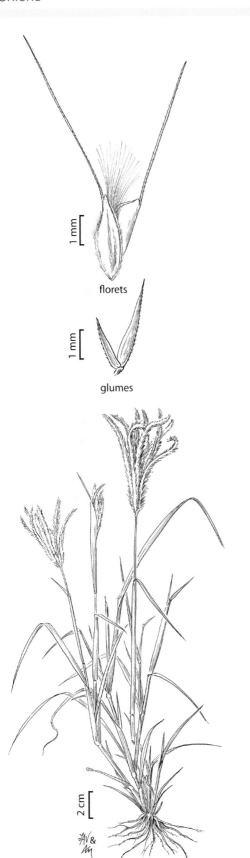

florets

glumes

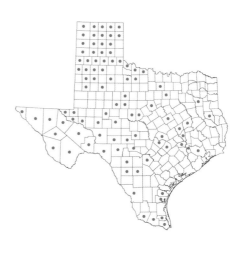

Chondrosum Desv.
(Chloridoideae: Cynodonteae)

Annuals and perennials of diverse habit, some cespitose, others with rhizomes or stolons. Ligule usually a ring of hairs. Inflorescence of 1 to numerous short, spicate branches, these closely or distantly spaced along the main axis and bearing numerous (20–80), small, closely spaced, pectinate, sessile spikelets in 2 rows along the margins of an angular or flattened rachis. Disarticulation above the glumes, branches persistent. Spikelets with 1 fertile floret and 2–3 reduced or rudimentary florets above. Glumes lanceolate, 1-nerved, unequal to nearly equal, awnless or short-awned. Lemmas membranous, 3-nerved, the midnerve often extending into an awn, the lateral veins occasionally short-awned. Palea membranous, the 2 veins occasionally awn-tipped. Basic chromosome number, $x = 10$. Photosynthetic pathway, C_4. Represented in Texas by 9 species, 2 subspecies, and 4 varieties.

A segregate of *Bouteloua*, where many authors still place it as a subgenus, including Gould (1975b); Hatch, Gandhi, and Brown (1990); and Jones, Wipff, and Montgomery (1997).

1. Upper glumes at least on some spikelets with papillose-based hairs.
    2. Panicle branches extending beyond the base of terminal spikelets .... 5. *C. hirsutum*
    2. Panicle branches terminating in a spikelet .... 4. *C. gracile*
1. Upper glumes without papillose-based hairs.
    3. Lower cauline internodes woolly-pubescent .... 3. *C. eriopodum*
    3. Lower cauline internodes glabrous, occasionally pubescent immediately below the nodes.
        4. Lemma of perfect spikelets terminating in membranous lobes.
            5. Palea of perfect spikelets awned, awns 1–2 mm long; panicle branches 2–20.
                6. Lowest lemmas glabrous, awns 3–4 m long; spikelets 6–20 per inflorescence
                branch .... 6. *C. kayi*
                6. Lowest lemmas pilose, awns 0.5–3.0 mm long; spikelets 20–50 per
                inflorescence branch .... 1. *C. barbatum*
            5. Palea of perfect spikelets not awned, or veins excurrent to <1 mm long.
                7. Plants annual .... 8. *C. simplex*
                7. Plants perennial.
                    8. Culms with 2–3 nodes; lower palea bilobed, veins sometimes
                    excurrent .... 4. *C. gracile*
                    8. Culms with 4–5 nodes; lower palea acute, veins not excurrent.
                        9. Lower culm internodes with a thick, white, chalky bloom at least above;
                        plants rhizomatous, growing on gypsum soils .... 2. *C. brevisetum*
                        9. Lower culm internodes without a bloom; plants cespitose, growing on
                        limestone soils .... 7. *C. ramosa*
        4. Lemma of perfect spikelets not terminating in membranous lobes .... 9. *C. trifidum*

(Wipff 2003c)

Note: *Chondrosum parryi* Fourn. is an annual or short-lived perennial found in the New Mexico counties along the border with West Texas. It should be expected in El Paso, Hudspeth, Culberson, and Reeves counties. It would probably key out close to *C. gracile*, but it is usually an annual and has panicle branch axes with papillose-based hairs.

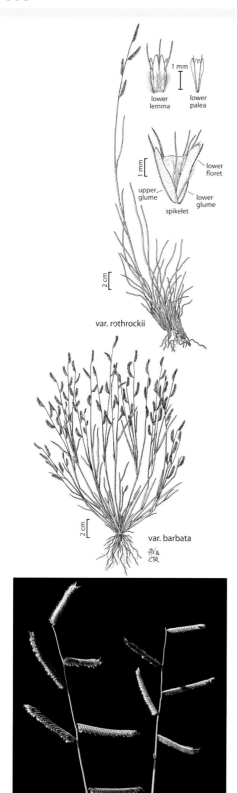

var. rothrockii

var. barbata

1. *Chondrosum barbatum* (Lag.) W. Clayton (sixweeks grama, Rothrock's grama). An annual or rarely a short-lived perennial, usually tufted but occasionally with stolons. Panicle with 2–11 branches. Branches persistent, straight or arcuate, axes terminating in a well-developed spikelet. Disarticulation above the glumes. Spikelets pectinate. Lemmas with 3 awns. Lower paleas with 2 awns. First rudiment with 3 awns, second rudiment awnless. Usually found in sandy soils on dry grasslands, roadsides, waste areas, and disturbed sites. Abundant enough after summer rains to provide some forage value (Powell 1994). Two varieties are recognized: var. *barbatum*, the more common one in Texas; and var. *rothrockii* (Vasey) R. B. Shaw. They can be distinguished by the following characters:

1. Plants annual; culms
    geniculate .... var. *barbatum*
1. Plants perennial; culms erect ......var. *rothrockii*

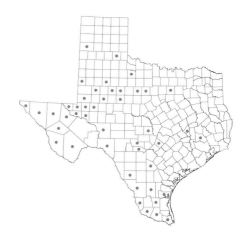

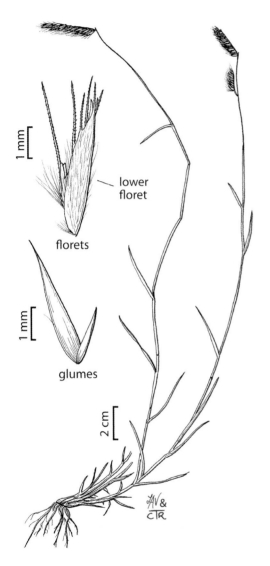

2. *Chondrosum brevisetum* (Vasey) W. Clayton (gypsum grama). A cespitose perennial, sometimes with scaly rhizomes. Culms to 40 cm tall, hard and knotty at the base. Distal portion of lower internodes have a thick, white, chalky bloom. Panicles with 1–4 braches. Branches persistent, straight to slightly arcuate, mostly appressed, terminating in a reduced spikelet. Disarticulation above the glumes. Locally abundant in gypsum soils in the Trans-Pecos region and of "modest" forage value (Powell 1994).

3. *Chondrosum eriopodum* Torr. (black grama, woollyfoot grama). A perennial with short rhizomes and long, densely woolly-pubescent stolons. Panicles with 2–8 branches. Branches persistent, densely woolly-pubescent at the base. Disarticulation above the glumes. A species of grasslands, shrublands, and foothills. Highly palatable species that persists only where shrubs and cacti protect it from overgrazing. The woolly internodes and stolons are distinctive. Listed as good for wildlife and livestock (Hatch and Pluhar 1993). Powell (1994) described it as highly palatable for livestock.

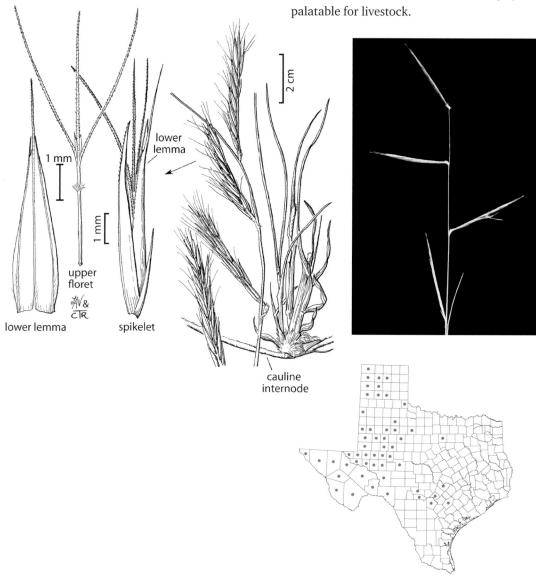

lower lemma

1 mm

lower lemma

1 mm

upper floret

spikelet

2 cm

cauline internode

4. *Chondrosum gracile* Willd. *ex* Kunth (blue grama, eyelash grass). A perennial, sometimes with stout rhizomes, generally forming large turf mats. Panicles with 1–3 or rarely up to 6 branches. Branches persistent, arcuate, without papillose-based hairs. Disarticulation above the glumes. A dominant species of mixed and shortgrass prairies. Sometimes used as a native ornamental. Listed as good for wildlife and livestock (Hatch and Pluhar 1993). Provides abundant forage in the Trans-Pecos (Powell 1994).

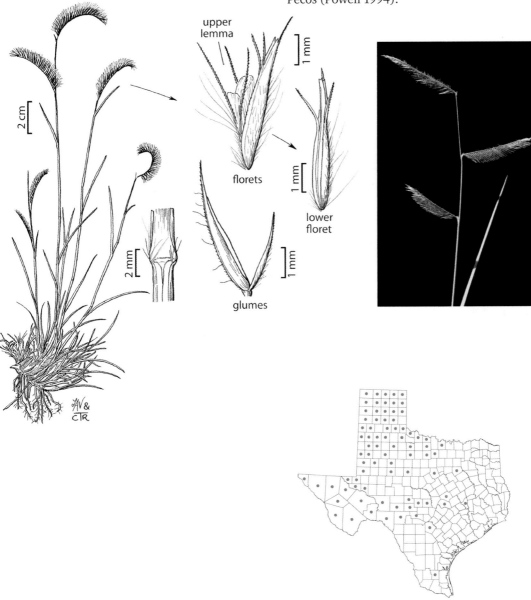

upper lemma

florets

lower floret

glumes

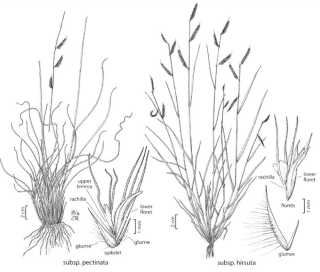

upper
lemma

rachilla

glume
spikelet

subsp. pectinata

rachilla

lower
floret

florets

glume

lower
floret

glumes

subsp. hirsuta

5. *Chondrosum hirsutum* (Lag.) Kunth
(hairy grama, tall grama). A perennial,
occasionally with stolons. Panicles 1–6
branches, sometimes digitate. Branches
persistent, straight, axes extend up to 1 cm
beyond the terminal spikelet, with papillose-
based hairs. Disarticulation above the glumes.
A widespread species of numerous habitats,
often in rocky areas or clay soils. Listed as
fair for wildlife and livestock (Hatch and
Pluhar 1993; Powell 1994). Two subspecies
are recognized: subsp. *hirsutum*; and subsp.
pectinatum (Featherly) R. B. Shaw. Gould
(1975b) and Hatch, Gandhi, and Brown
(1990) consider the latter as a separate
species (*Bouteloua pectinata* Featherly).
The subspecies can be distinguished by the
following characters:

1. Rachilla internodes subtending second
   florets without an apical tuft of hairs;
   culms usually decumbent and
   branched ....subsp. *hirsutum*
1. Rachilla internodes subtending second
   florets with an apical tuft of hairs; culms
   usually erect and
   unbranched .... subsp. *pectinatum*

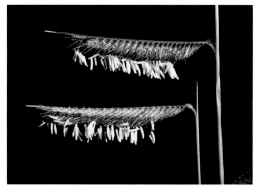

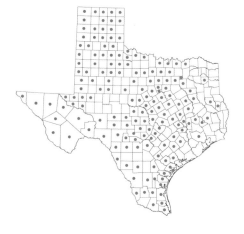

6. *Chondrosum kayi* (Warnock) W. Clayton (Kay's grama). A cespitose perennial lacking both stolons and rhizomes. Panicles with 7–20 branches. Branches persistent, terminating in a spikelet. Spikelets pectinate. Lemmas with 3 awns. Lower paleas reduced to 2 awns. Second florets reduced to 1–3 awns. A Texas endemic known only from the mountainous limestone terrain along the Rio Grande in the Big Bend region. It is listed as a Category V species (watch-listed) by the Texas Organization for Endangered Species (TOES). It is related to C. *trifidum* but differs in having stouter, erect culms and more numerous and longer inflorescence branches; and the lemma awns are usually shorter, whereas the lemma body is longer (Gould 1975b).

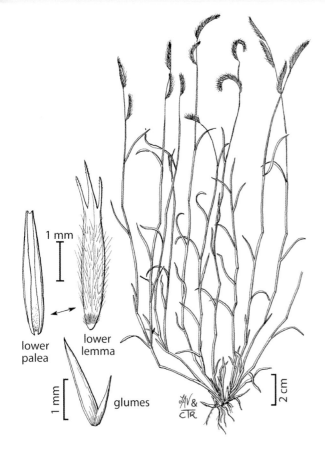

1 mm

lower
palea

lower
lemma

1 mm

glumes

2 cm

7. *Chondrosum ramosa* (Scribn. *ex* Vasey) R. B. Shaw (chino grama). A densely cespitose perennial with hard, knotty bases and culms to 60 cm tall. Distal portion of lower internodes without a thick, white, chalky bloom. Panicles with 1–4 branches. Branches persistent, ascending to widely spreading, becoming arcuate, terminating in a reduced spikelet. Disarticulation above the glumes. Locally abundant on rocky limestone slopes and flats in the Trans-Pecos region. Gould (1975b) and Hatch, Gandhi, and Brown (1990) included this species under *Bouteloua breviseta*. Powell (1994) stated that this species has forage potential.

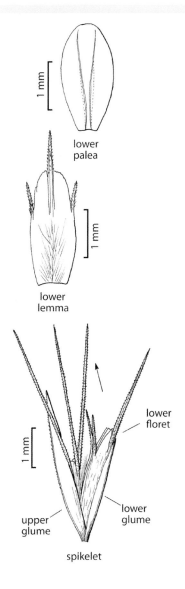

lower
palea

lower
lemma

lower
floret

upper
glume

lower
glume

spikelet

8. *Chondrosum simplex* (Lag.) Kunth (mat grama). An annual with decumbent, short (<35 cm), and rarely branched culms. Panicles usually with 1 branch at the culm apex, rarely 2–4. Branches persistent, straight, arcuate, or nearly circular, terminating in a reduced spikelet. Disarticulation above the glumes. Spikelets pectinate. Found in dry plains, rocky slopes, and disturbed areas.

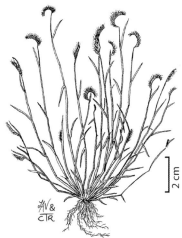

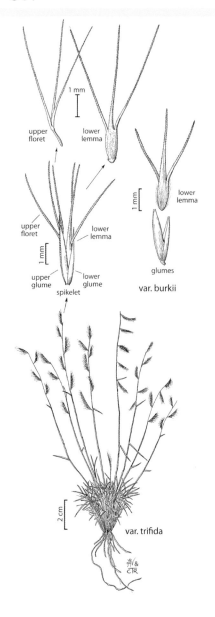

var. burkii

var. trifida

9. *Chondrosum trifidum* (Thurb.) W. Clayton (red grama). A cespitose perennial, although older plants occasionally have rhizomes. Panicles 2–7 branches, more or less evenly distributed. Branches straight to slightly arcuate, axes terminating in a spikelet. Disarticulation above the glumes. Spikelets appressed to pectinate, reddish-purple. Lower lemmas with 3 awns. It generally grows on rocky hillsides, dry plains, and shrublands. *Chondrosum trifidum* is related to *C. kayi* but differs in having more delicate, geniculate culms and fewer and shorter inflorescence branches; and the lemma awns are usually longer, whereas the lemma body is shorter (Gould 1975b). Listed as poor for wildlife and livestock (Hatch and Pluhar 1993). Powell (1994) saw it as fair forage during early-season growth. Two varieties are recognized: var. *trifidum*; and var. *burkii* (Scribn. *ex* S. Watson) R. B. Shaw. They can be distinguished by the following characters:

1. Lower lemmas glabrous or sparsely appressed pubescent along both sides of the veins; awns (3.2–4.0–6.6 mm long ....var. *trifidum*
1. Lower lemmas densely appressed pubescent; awns 2.2–4.5 mm long .... var. *burkii*

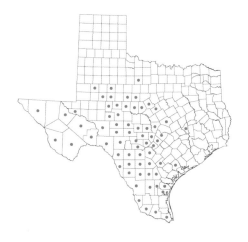

Chrysopogon Trin.
(Panicoideae: Andropogoneae)

Cespitose annuals and perennials with flat blades and few-flowered panicles of large spikelets borne at the tips of slender, spreading branches. Rames usually in threes, 1 sessile and perfect and 2 pediceled and reduced. Glumes of the fertile spikelet firm and hard, rounded on the back, awnless, with a hard, sharp-pointed, hairy callus at the base. Lemmas of the sterile and fertile florets thin, membranous, the latter with a stout, usually geniculate awn. Basic chromosome number, $x = 10$. Photosynthetic pathway, C_4. Represented in Texas by 2 species.

1. Plants annual; upper lemmas of sessile spikelets awned, awns up to 16 cm long .... 1. *C. pauciflorus*
1. Plants perennial; upper lemmas of sessile spikelets with a mucro or awn not more than 5 mm long .... 2. *C. zizanioides*

(Hall and Thieret 2003)

1. *Chrysopogon pauciflorus* (Chapm.)
Benth. *ex* Vasey (Florida rhaphis). An annual
up to 1 m tall. Panicle 20–30 cm long, open
with divergent branches. Rames consisting
of 3 spikelets. Upper lemma awned, awns
10–16 cm long. It occurs in pine flatwoods,
fields, marshes, and disturbed sites. Not
included in Gould (1975b).

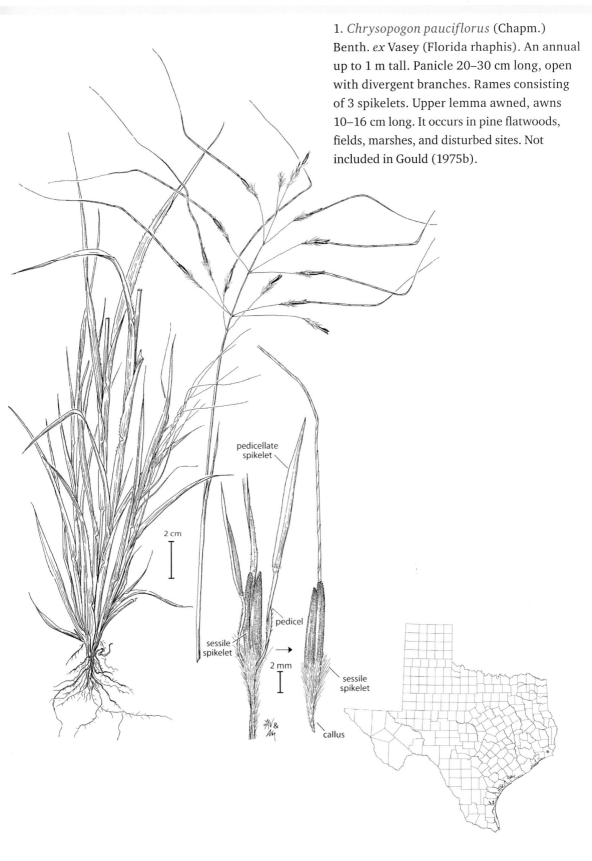

2. *Chrysopogon zizanioides* (L.) Roberty
(vetiver). A cespitose perennial with culms up
to 3 m tall. Panicles 15–35 cm long. Branches
numerous and ascending. Rames usually in
pairs, terminal one consisting of triplets. Upper
lemmas with a mucro or short awn to 2 mm
long. An introduced species from Asia. It does
not spread vegetatively, and many cultivars
have no or very low seed set, so it does not
spread aggressively. It is used for flood
control, water-quality improvement, pollution
control, land and mining rehabilitation, and
handicrafts. Oils from the roots are used as a
perfume. Numerous biocidal effects have been
identified. Not included in Gould (1975b) or
Hatch, Gandhi, and Brown (1990).

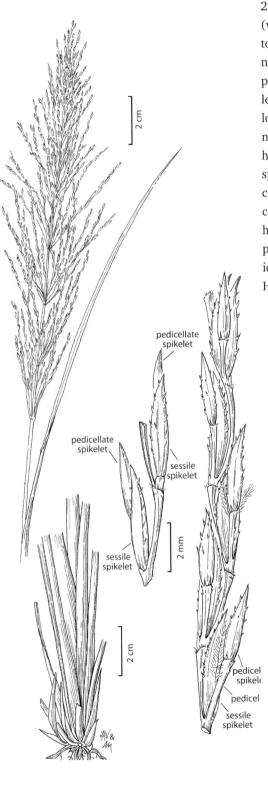

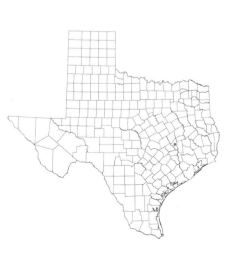

Cinna L.
(Poöideae: Poeae)

Tall perennials with slender culms; thin, flat blades; and open or contracted panicles. Spikelets 1-flowered, disarticulating below the glumes, the rachilla usually prolonged behind the palea as a small, glabrous or scabrous stub or bristle. Glumes about equal or the lower somewhat shorter, lanceolate, 1–3-nerved, acute at the apex. Lemma similar to the glumes, 3–5-nerved, the midnerve usually projecting as a short, straight awn just below the acute or narrowly rounded and notched tip. Palea slightly shorter than the lemma, 2- or apparently 1-nerved, with a single keel and with the veins close together. Stamens 1 or 2. Basic chromosome number, $x = 7$. Photosynthetic pathway, C_3. Represented in Texas by a single species.

(Brandenburg 2007a)

1. *Cinna arundinacea* L. (stout woodgrass). Perennial. Culms up to 2 m tall with a bulblike base. Inflorescences are large, loose panicles. Spikelets with a pronounced stipe and extension of the rachilla. Lemma with a short awn. Palea 1-veined. Only 1 stamen present in the perfect florets. Species of swamps and wet forests, found along streams and rivers and in floodplains. Uncommon.

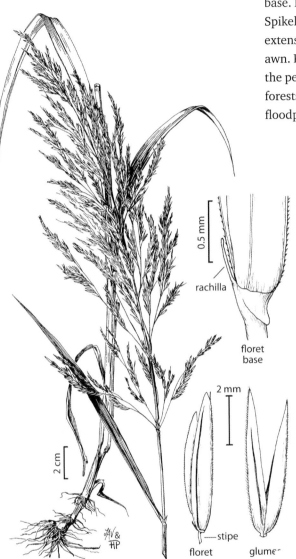

0.5 mm

rachilla

floret base

2 mm

stipe

floret   glume

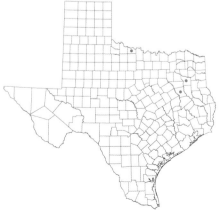

Coelorachis Brongn.
(Panicoideae: Andropogoneae)

Perennials with erect, usually much-branched culms and numerous cylindric, spicate racemes. Spikelets awnless, in pairs on a thickened, readily disarticulating rachis. Sessile spikelet of a pair perfect, the pediceled spikelet staminate, both more or less sunken in the corky rachis. First glume of the sessile spikelet thick and firm, rounded at the apex, smooth or variously pitted, covering the remainder of the spikelet. Lemmas of the sterile and fertile florets membranous and hyaline. Pediceled spikelet reduced, sterile, the pedicel short and thick. Basic chromosome number, $x = 9$. Photosynthetic pathway, C_4. Represented in Texas by 2 species.

1. Culms and sheaths terete .... 1. *C. cylindrica*
1. Culms and sheaths compressed-keeled... 2. *C. rugosa*

(Allen 2003a)

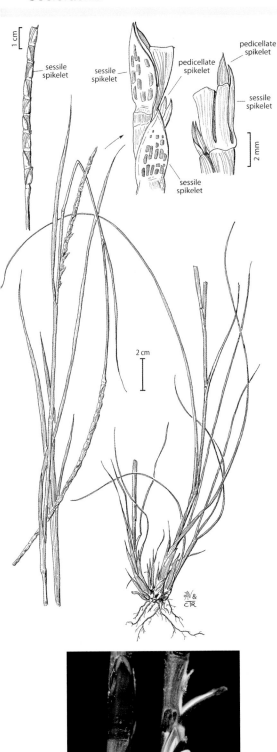

sessile spikelet

sessile spikelet

pedicellate spikelet

pedicellate spikelet

sessile spikelet

sessile spikelet

1 cm

2 mm

2 cm

1. *Coelorachis cylindrica* (Michx.) Nash (Carolina jointgrass). A perennial with short rhizomes, culms to 1.2 m tall. Sheaths terete. Rames often purple. Lower glumes with circular pits on the sides, the central regions usually developing rectangular pits at maturity. Found in tallgrass prairies, in forest openings or along margins, and along roadsides. Reported by Diggs et al. (2006) as a significant member of sandy and mima mound prairies. Palatable but rated as poor because of low abundance (Hatch, Schuster, and Drawe 1999).

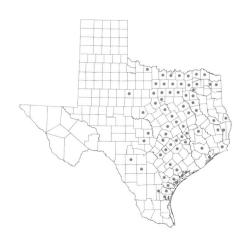

2. *Coelorachis rugosa* (Nutt.) Nash (wrinkled jointgrass). A cespitose perennial, culms to 1.2 m tall. Sheaths compressed-keeled. Lower glumes distinctively transverse-rugose. Found in pine flatwoods, savannahs, bogs, and wet prairies.

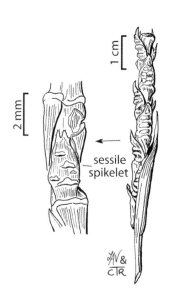

sessile
spikelet

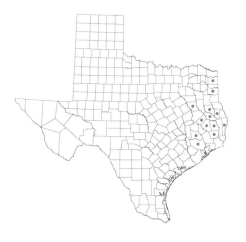

Coix L.
(Panicoideae: Andropogoneae)

Tall, coarse annual or perennial grasses, with thick culms; broad, flat blades; and unisexual spikelets. Spikelets on spicate branches, the staminate in twos or threes on a continuous rachis above the pistillate, the latter enclosed in hard, bony, beadlike involucres of modified bracts. Staminate spikelets 2-flowered, with thin, obscurely nerved glumes and hyaline lemma and palea. Pistillate spikelets 3 in each involucre, 1 fertile and 2 sterile. Glumes of the fertile floret hyaline below, firmer at the pointed tip. Lemma and palea of fertile floret hyaline. Basic chromosome number, $x = 10$. Photosynthetic pathway, C_4. Represented in Texas by a single species.

(Thieret 2003c)

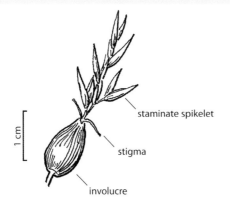

staminate spikelet

1 cm

stigma

involucre

1. *Coix lacryma-jobi* L. (Job's tears). This species is widespread in tropical regions of the world. It is a tall, coarse annual or perennial and has been occasionally cultivated in the warmer parts of the United States. Although frequent and often weedy in the American tropics, this grass is believed to be native only to the Old World. The hard, bony, fruiting involucres have long been used as beads, especially for rosaries, in tropical countries. The involucres can be white, blue, pink, straw, gray, brown, or black.

2 cm

Cortaderia Stapf
(Danthonioideae: Danthonieae)

Large, cespitose perennials with culms in dense clumps. Leaves densely clustered at the base of tall, slender, floriferous shoots. Leaf blades long and narrow, tough and fibrous, with rough, serrulate margins. Ligule absent (in Texas species). Plants dioecious, the pistillate spikelets plumose, in showy, silvery panicles 25–100 cm long, and the staminate spikelets glabrous, in large but nonshowy panicles. Disarticulation above the glumes and between the florets, the rachilla internodes jointed, the lower part glabrous, the upper bearded, forming a stipe for the floret. Spikelets 2–4-flowered, with narrow, papery, 1-nerved glumes and lemmas, the latter tapering to a narrow point or delicate awn. Lemmas of the pistillate spikelets with long, spreading hairs on the back and base. Chromosome number, $x = 12$. Photosynthetic pathway, C_4. Represented in Texas by 2 species.

1. Sheaths hairy; panicles elevated well above the foliage .... 1. *C. jubata*
1. Sheaths glabrous or sparsely hairy; panicles elevated only slightly above the
   foliage.. 2. *C. selloana*

(Allred 2003d)

1. *Cortaderia jubata* (Lemoine *ex* Carrièe) Stapf (purple pampasgrass). Coarse perennial with culms to 7 m tall. Panicles high above the foliage, deep violet when young. Spikelets are all pistillate. An ornamental that has escaped and is a serious weed in California. Jones, Wipff, and Montgomery (1997) listed it in Texas.

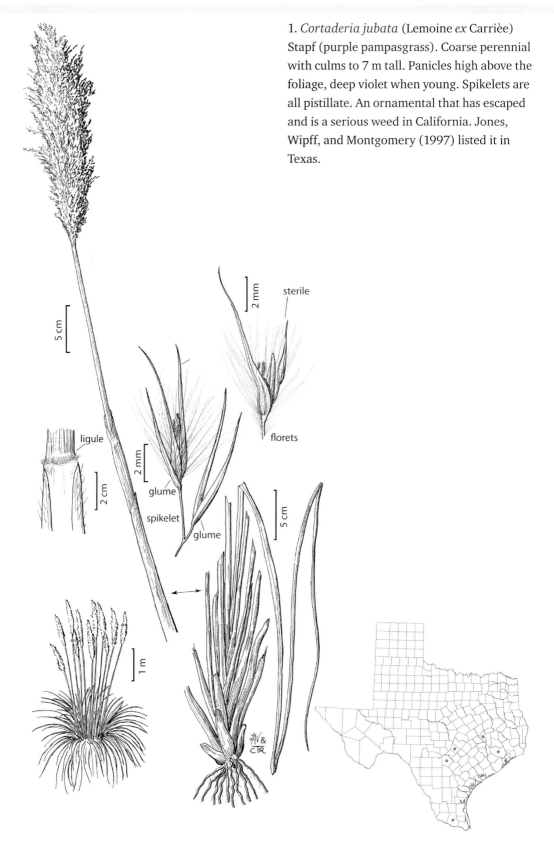

2. *Cortaderia selloana* (Schult. and Schult.f.) Asch. & Graebn. (pampasgrass). Coarse perennial with culms to 4 m tall. Dioecious or sometimes monoecious. Panicles are only slightly above the foliage, purplish when young, becoming white at maturity. The species is widely introduced in the warmer parts of the world. In the southern United States the female plants of pampasgrass, with their large, plumose panicles, commonly are grown as lawn ornamentals. One of the most common and easily recognized ornamentals in Texas. It probably occurs in every county in Texas but is rarely collected.

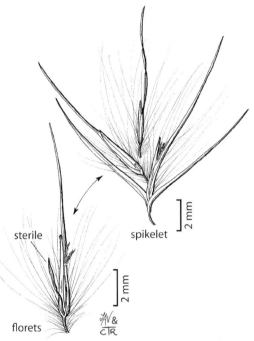

sterile

spikelet

2 mm

florets

2 mm

JAV & CTR

Cottea Kunth
(Chloridoideae: Pappophoreae)

Tufted perennial with culms 30–70 cm tall. Leaves usually pilose, the blades flat or folded. Ligule a ring of hairs. Inflorescence a narrow but open panicle with rather stout, short, erect-spreading branches. Spikelets large, with 6–10 florets, the upper ones reduced. Disarticulation above the glumes and between the florets. Glumes subequal, about as long as the lemmas, broadly lanceolate, with 7–13 fine nerves, the midnerve sometimes continued as a short awn. Lemmas broad, irregularly lobed and cleft, hairy below, with 9–13 strong nerves, these extended into awns of varying lengths. Palea broad, slightly longer than the body of the lemma. Basic chromosome number, $x = 10$. Photosynthetic pathway, C_4. Represented in Texas by the only member of the genus.

(Reeder 2003a)

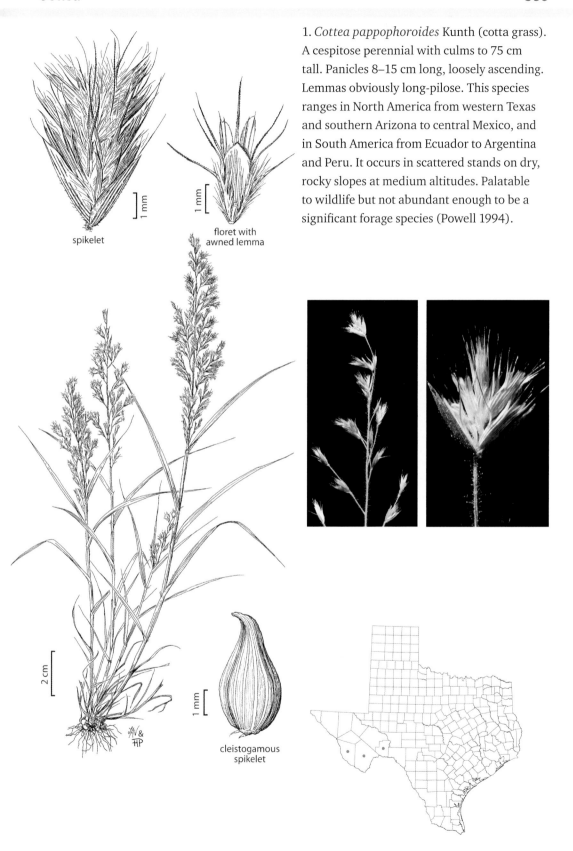

spikelet

floret with
awned lemma

cleistogamous
spikelet

1. *Cottea pappophoroides* Kunth (cotta grass). A cespitose perennial with culms to 75 cm tall. Panicles 8–15 cm long, loosely ascending. Lemmas obviously long-pilose. This species ranges in North America from western Texas and southern Arizona to central Mexico, and in South America from Ecuador to Argentina and Peru. It occurs in scattered stands on dry, rocky slopes at medium altitudes. Palatable to wildlife but not abundant enough to be a significant forage species (Powell 1994).

Critesion Raf.
(Poöideae: Triticeae)

Annual or cespitose perennials. Inflorescence a spicate raceme with 3 spikelets at a node (triad). Rachis usually fragmenting at maturity, disarticulating above each node, the short internode falling with the triad of spikelets. Central spikelet usually sessile, with 1 floret and a bristlelike extension of the rachilla. Glumes side by side, usually awnlike and separate to the base. Lemmas firm, rounded on the back, dorsally compressed, and terminating in a short awn. Palea slightly shorter than the lemma and usually adnate to the caryopsis. Lateral spikelets typically smaller than the central one, staminate, sterile, or more frequently reduced to 3 awnlike structures. Basic chromosome number, $x = 7$. Photosynthetic pathway, C_3. Represented in Texas by 5 species and 5 subspecies. A segregate of *Hordeum*, where it was placed by Gould (1975b); Hatch, Gandhi, and Brown (1990); and Jones, Wipff, and Montgomery (1997).

1. Plants perennial.
    2. Glumes 15–85 mm long, usually strongly divergent......................................2. *C. jubatum*
    2. Glumes 7–19 mm long, usually erect or ascending.............................1. *C. brachyantherum*
1. Plants annual.
      3. Auricles to 8 mm long, prominent..4. *C. murinum*
      3. Auricles absent or to 0.3 mm long; obscure.
        4. Glumes bent, strongly divergent..2. *C. jubatum*
        4. Glumes straight, ascending, or only slightly divergent.
          5. Lemmas of lateral spikelets with awns 3–8 mm long ....3. *C. marinum*
          5. Lemmas of lateral spikelets awnless or awns <3 mm long................5. *C. pusillum*

(Bothmer, Baden, and Jacobsen 2007)

1. *Critesion brachyantherum* (S. Nevski)
Barkworth & Dewey (meadow barley, northern
barley). A cespitose perennial that can grow
to nearly 1 m tall. Spikes 3.0–8.5 cm long.
Glumes ascending or slightly divergent at
maturity. Lemmas with a 3–15 mm long
awn. The Texas species belongs to the typical
subspecies, subsp. *brachyantherum*. Neither
Gould (1975b) nor Bothmer, Baden, and
Jacobsen (2007) included it in the state;
however, Jones, Wipff, and Montgomery
(1997) listed it. The species is also found in
New Mexico (Allred 2005). Expected in the
Guadalupe and Davis mountains and perhaps
in the western counties of the Panhandle
region.

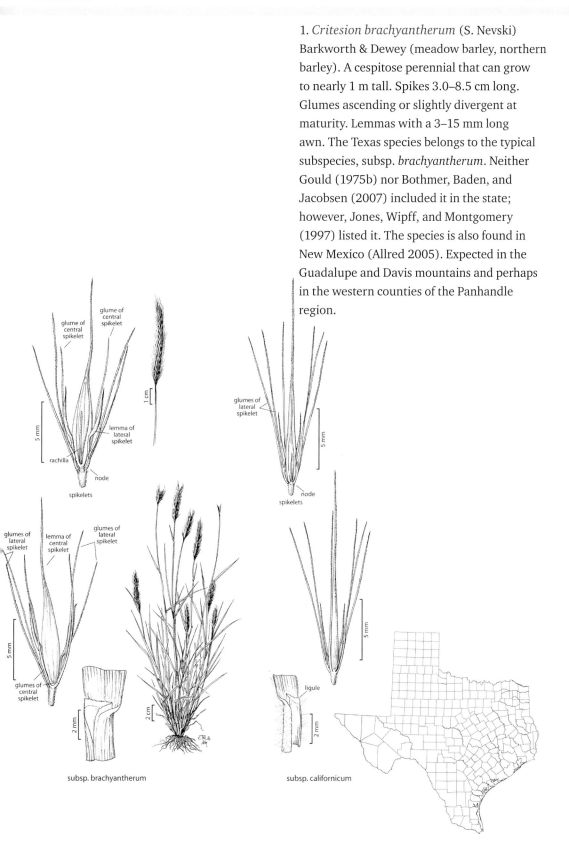

subsp. brachyantherum

subsp. californicum

2. *Critesion jubatum* (L.) S. Nevski (foxtail barley, squirreltail barley, squirreltail grass, skunktail grass). Perennial, sometimes appearing annual when young or flowering the first season. Spikes 3–15 cm long, whitish-green when immature but becoming purplish, very soft and showy. Glumes strongly divergent when mature. Lemmas with awns 10–90 mm long. Generally a weedy species of saline areas, ditches, lake margins, and other disturbed sites. Very attractive and showy plant, sometimes grown as an ornamental. The Texas species belongs to the typical subspecies, subsp. *jubatum*. Of poor grazing value (Powell 1994), and awns can sometimes cause mechanical injuries to livestock.

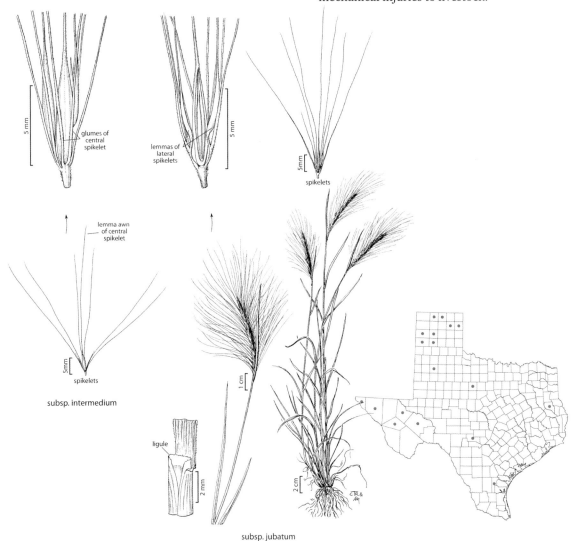

glumes of central spikelet

lemmas of lateral spikelets

5 mm

5 mm

5mm

spikelets

lemma awn of central spikelet

5mm

spikelets

subsp. intermedium

ligule

2 mm

1 cm

2 cm

subsp. jubatum

3. *Critesion marinum* (Huds.) Á. Löve
(Mediterranean barley, sea barley). An annual
introduced from Eurasia. Spikes 1–7 cm long,
5–10 mm wide, dense, greenish or purplish
on the awns. Glumes straight to divergent.
Lemmas with awn 5–20 mm long. Rachises
break up at maturity. A weed of disturbed
habitats. Common in Brazos County.
Not included in Gould (1975b); Hatch,
Gandhi, and Brown (1990); Jones, Wipff,
and Montgomery (1997); or Diggs et al.
(2006). The Texas species belongs to subsp.
gussoneanum (Parl.) Barkworth and D. R.
Dewey.

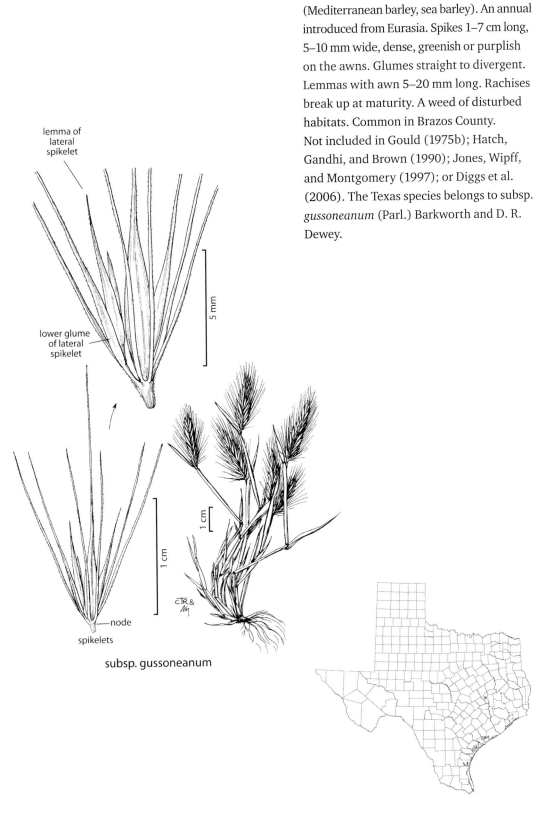

lemma of
lateral
spikelet

5 mm

lower glume
of lateral
spikelet

1 cm

1 cm

node

spikelets

subsp. gussoneanum

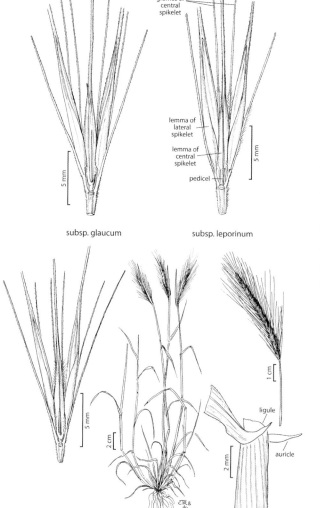

subsp. glaucum   subsp. leporinum

glumes of central spikelet

lemma of lateral spikelet

lemma of central spikelet

pedicel

5 mm

ligule

auricle

subsp. murinum

4. *Critesion murinum* (L.) Á. Löve (smooth barley, mouse barley, hare barley). A Eurasian annual introduced as a common weed into areas of human disturbance. The only *Critesion* species in Texas with well-developed auricles up to 8 mm long. Spikes 3–8 cm long, 7–16 mm wide, pale green to distinctively reddish. Rachises disarticulate at maturity. Lemmas awned, 20–40 mm long. Found in disturbed sites. Because of its size, it might provide slightly better forage in early stages then other barleys (Powell 1994). Two subspecies are recognized in Texas: subsp. *leporinum* (J. Link) G. Arcangeli; and subsp. *glaucum* (Steud.) W. A. Weber. They can be distinguished by the following character:

1. Lemmas of the central floret much shorter than those of the lateral florets ....subsp. *leporinum*
1. Lemmas of the central floret about equal in length to those of the lateral florets .... subsp. *glaucum*

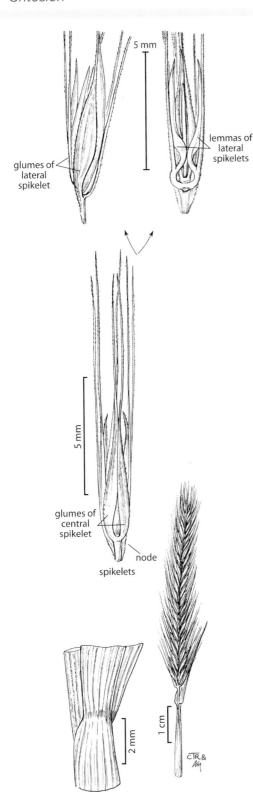

5 mm

lemmas of
lateral
spikelets

glumes of
lateral
spikelet

5 mm

glumes of
central
spikelet

node

spikelets

2 mm

1 cm

CTR &
AM

5. *Critesion pusillum* (Nutt.) Á. Löve (little barley, mouse barley). An annual growing to 60 cm tall. Spike 2–9 cm long, 3–7 mm wide, often enclosed in uppermost sheath at maturity. Glumes straight, not diverging. Lemmas usually short-awned. A widespread weedy plant of overgrazed pastures, roadsides, and waste places. One of the first plants to "green up" in early spring. Listed as poor for wildlife and livestock (Hatch and Pluhar 1993, as *Hordeum pusillum* Nutt.).

Ctenium Panz.
(Chloridoideae: Cynodonteae)

Perennials, mostly with tall, slender culms. Inflorescences a short terminal, spikelike panicle of 1–3 strongly pectinate branches, appears as a branch that has assumed a terminal position. Branch usually strongly curved. Spikelets several-flowered, but with 1 perfect floret, sessile in 2 rows on one side of the branch (unilateral). Lowest glume small, thin, 1-veined. Upper glume firm, 3- or 4-veined, about as long as the fertile lemma, bearing a stout, divergent awn from the middle. Lemmas thin, 3-veined, pubescent in the lateral veins, with a stout, divergent awn borne dorsally just below the tip. Lowermost 2 florets sterile, the third floret fertile, the upper 1–3 florets sterile and successively smaller. Palea of perfect floret about as long as lemma. Basic chromosome number, $x = 10$. Photosynthetic pathway, C_4. Represented in Texas by a single species just recently reported for the state (Diggs et al. 2006).

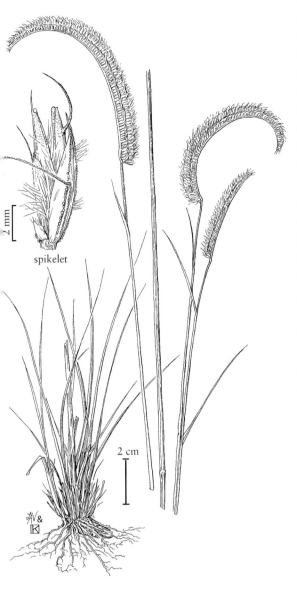

spikelet

2 cm

1. *Ctenium aromaticum* (Walter) Alph. Wood (toothache grass). Strongly cespitose perennial with erect culms to 1.5 m tall. Panicles with 1 branch; branches 5–15 cm long, curved, axes extending beyond the spikelets. Spikelets 8–11 mm long. Upper glume bidentate with a row of glands on either side of midrib, awn to 4 mm long, strongly divergent. Perfect lemmas 4–5 mm long, pilose on the lateral veins, awned from a bifid apex; awn 3–4 mm long, straight or divergent. A common species of the pine flatwoods, prairies, savannahs, and pitcher plant bogs of the southeastern United States. Found in Newton County across the boundary from Louisiana, where populations were known to occur (Diggs et al. 2006). A spicy, tingling, or numbing sensation occurs when tasting the freshly dug roots, thus the vernacular name. Inflorescences, which curl, twist, or corkscrew, sometimes used in floral arrangements. Species has potential as an ornamental. It furnishes fair forage after fires (Barkworth 2003c).

Cynodon Rich.
(Chloridoideae: Cynodonteae)

Low, mostly mat-forming stoloniferous and/or rhizomatous perennials. Culms much-branched, mostly with short internodes. Leaf blades flat, short, narrow, soft, and succulent. Ligule a fringe of hairs. Inflorescence with 2 to several slender, spicate branches, these digitately arranged at the culm apex. Spikelets sessile in 2 rows on a narrow, somewhat triangular branched rachis. Spikelets with 1 fertile floret, the rachilla prolonged beyond the palea as a bristle and occasionally bearing a rudimentary lemma. Glumes slightly unequal, lanceolate, awnless, 1-nerved, the upper nearly as long as the lemma. Lemma firm, laterally compressed, awnless, 3-nerved, usually puberulent on the midnerve. Palea narrow, 2-nerved, as long as the lemma. Basic chromosome number, $x = 9$. Photosynthetic pathway, C_4. Represented in Texas by 5 species and 2 varieties. *Cynodon dactylon* is the only species listed by Gould (1975b) and Hatch, Gandhi, and Brown (1997).

1. Glumes less than ⅓ the length of the spikelet, 0.1–0.6 mm long .... 4. *C. plectostachyus*
1. Glumes more than ½ the length of the spikelet, 1.1–2.6 mm long.
  2. Panicle branches 0.7–2.1 cm long .... 5. *C. transvaalensis*
  2. Panicle branches >2 cm long.
    3. Plants stoloniferous and rhizomatous; stolons not woody; culms <50 cm .... 2. *C. dactylon*
    3. Plants stoloniferous, stolons woody, lying flat on the ground; culms can be >50 cm long.
      4. Lemma keels usually pubescent; panicle branches usually in 2–5 whorls, stiff, red or purplish; culms woody .... 1. *C. aethiopicus*
      4. Lemma keels usually glabrous; panicle branches usually in 1 whorl, lax, greenish; culm not woody... 3. *C. nlemfuënsis*

(Barkworth 2003d)

1. *Cynodon aethiopicus* Clayton and J. R. Harland (Ethiopian dogstooth grass). A stoloniferous perennial, stolons and culms becoming woody. Panicles with 3–20 branches. Branches in 2–5 whorls, stiff, reddish or purplish. An East African introduction. Known to be growing along the canal bank in the Santa Ana National Wildlife Refuge in Hidalgo County.

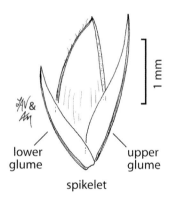

lower glume   upper glume

spikelet

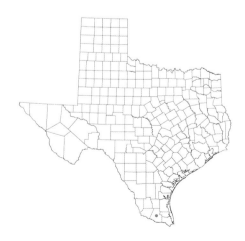

1 mm

spikelet

panicle
branch

stolon

2 cm

rhizome

var. dactylon

2. *Cynodon dactylon* (L.) Pers. (Bermudagrass). A stoloniferous and usually rhizomatous species. Panicles with 2–9 branches. Branches in a single whorl, ascending to only slightly divergent. A successful introduction for lawns, golf courses, and pastures and as hay. A large number of cultivars available. Reported by Diggs et al. (2006) to cause hay fever in humans and photosensitivity, staggers syndrome, and acute respiratory distress syndrome in livestock. Probably occurs in every county in the state, but collectors frequently ignore weeds and cultivars. Listed as poor for wildlife and good for livestock (Hatch and Pluhar 1993). The Texas species belongs to the typical variety, var. *dactylon*.

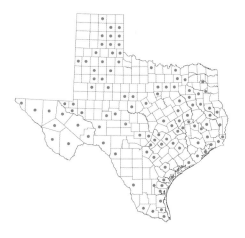

3. *Cynodon nlemfuënsis* Vanderyst (African Bermudagrass). A stoloniferous perennial, stolons becoming woody, culms not woody. Panicles with 4–13 branches. Branches usually in 1 whorl, lax, green. An introduced forage species from Africa. Collected in Cameron and Kenedy counties. The Texas species belongs to the typical variety, var. *nlemfuënsis*.

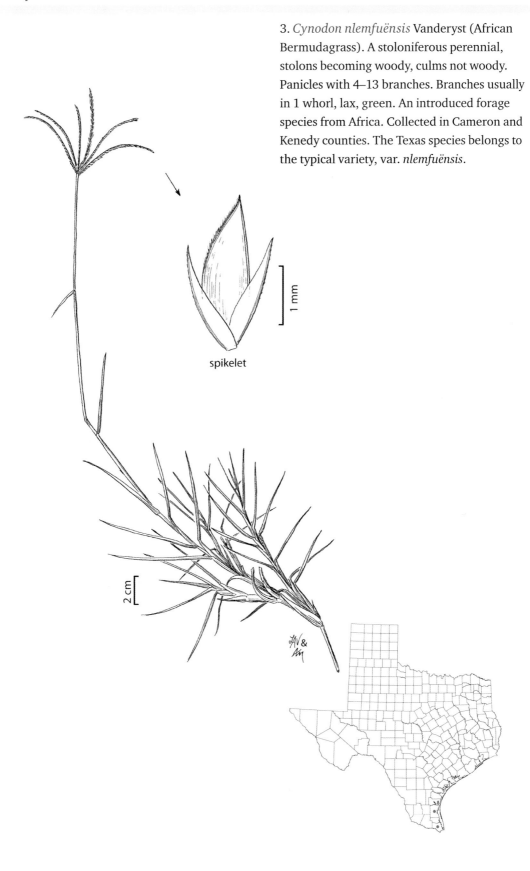

spikelet

1 mm

2 cm

4. *Cynodon plectostachyus* (K. Schum.) Pilg. (stargrass). A perennial with arching stolons. Panicles with 6–20 branches. Branches in 2–7 closely spaced whorls. An introduction from Africa as a forage species. It is not frost tolerant. Rare in Texas. Not included in Jones, Wipff, and Montgomery (1997).

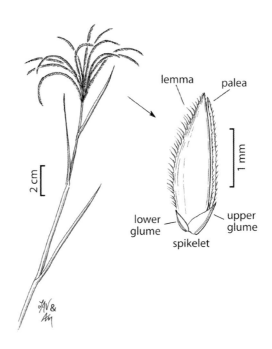

5. *Cynodon transvaalensis* Burtt Davey (African dogstooth grass). Plants stoloniferous and rhizomatous perennials. Panicles with 1–4 branches. Branches in a single whorl, reflexed at maturity. An introduction from Africa as a lawn grass but not a great success. Perhaps not a constant member of the Texas flora. Listed for the state by Jones, Wipff, and Montgomery (1997).

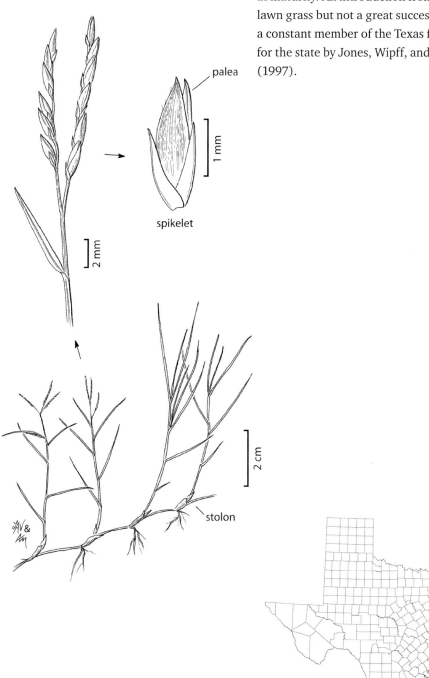

palea

1 mm

spikelet

2 mm

2 cm

stolon

Cynosurus L.
(Poöideae: Poeae)

Tufted annuals and perennials, with narrow, flat leaf blades and subcapitate or spicate panicles. Spikelets of 2 kinds, fertile and sterile, these together in dense clusters, the fertile ones sessile, the sterile ones short-pediceled, almost concealing the fertile ones. Fertile spikelets 1–5-flowered, disarticulating above the narrow glumes. Lemmas of the fertile floret rounded on the back, inconspicuously 5-veined, terminating in a short awn. Sterile spikelets with 2 narrow, 1-veined glumes and several narrow, 1-nerved lemmas on a stiff, continuous rachis. Basic chromosome number, $x = 7$. Photosynthetic pathway, C_3. Represented in Texas by a single species.

1. *Cynosurus echinatus* L. (bristly dogtail, rough dogtail, spiny dogtail grass). An annual introduced from southern Europe. Unusual in having 2 different types of spikelets. It is usually found in Mediterranean climates. A single collection in Texas, and it may not be a constant member of the Texas flora. Not included in Gould (1975b); Hatch, Gandhi, and Brown (1990); or Jones, Wipff, and Montgomery (1997).

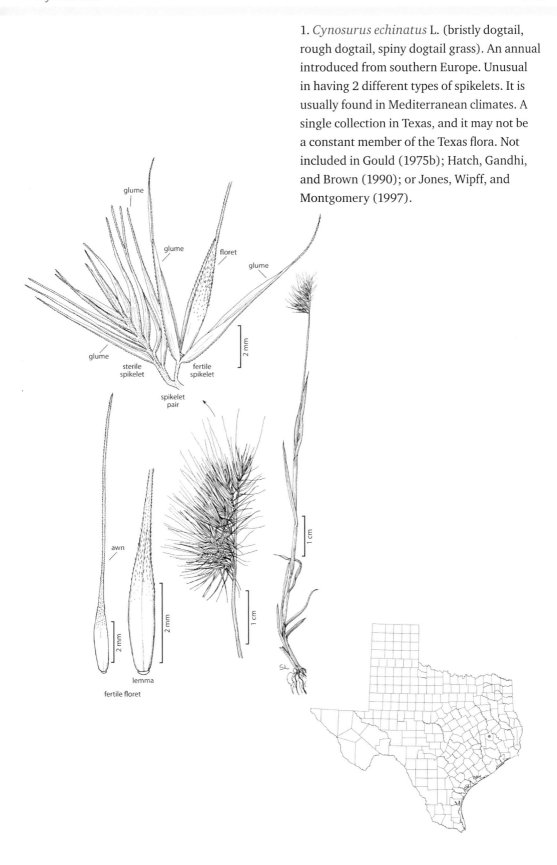

Dactylis L.
(Poöideae: Poeae)

Perennial, erect with densely clumped culms. Sheaths distinctly flattened or keeled, open; ligules membranous, usually erose at the tip; auricles absent. Inflorescence a 1-sided panicle. Spikelets with 2–5 fertile florets, laterally flattened, crowded in dense asymmetrical clusters at the tips of stiff, erect or spreading, inflorescence branches. Disarticulation above the glumes and between the florets. Glumes unequal, keeled, hispid-ciliate on the keel, 1–3-nerved, acute to acuminate or ending in a short awn. Lemmas keeled, awnless or short-awned, 5-nerved, hispid-ciliate on the keel. Palea well developed, short-ciliate on the keels. Basic chromosome number, $x = 7$. Photosynthetic pathway, C_3. Represented in Texas by the only species in the genus.

(Allred 2007b)

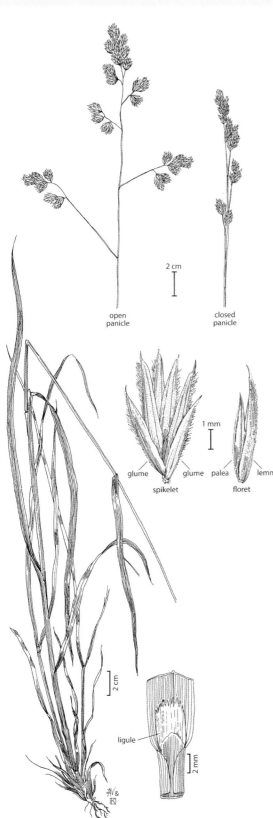

2 cm

open
panicle

closed
panicle

1 mm

glume   glume   palea   lemma

spikelet   floret

2 cm

ligule

2 mm

1. *Dactylis glomerata* L. (orchardgrass).
Perennial with culms 2 m long or longer.
Leaf blades with a prominent midrib and
white margins. Panicles much-reduced, lower
branch spreading, upper branches appressed.
Spikelets clustered at the end of branches.
An introduced pasture species and excellent
forage for most kinds and classes of livestock.
It escapes and becomes established in low,
moist places and is often used in stabilization
of disturbed sites.

Dactyloctenium Willd.
(Chloridoideae: Cynodonteae)

Annuals or short-lived perennials with thick, weak culms that are often decumbent and root at the lower nodes. Sheaths open; ligule ciliate with a membranous base; auricles absent. Inflorescence of usually 2 to several digitately arranged, unilateral spicate branches. Spikelets closely placed and pectinate in 2 rows on one side of a short, stout rachis, this projecting as a point beyond the insertion of the uppermost spikelet. Spikelets with 2 to several florets, laterally compressed. Disarticulation often between the glumes, the lower remaining on the rachis. Glumes 2, keeled, subequal, 1-nerved, the lower awnless, the upper mucronate or with a short, stout awn. Lemmas firm, broad, 3-nerved, abruptly narrowing to a beaked, usually short-awned tip. Palea well developed, about as long as the lemma. Basic chromosome number, $x = 10$. Photosynthetic pathway, C_4. Represented in Texas by a single species.

(Hatch 2003a)

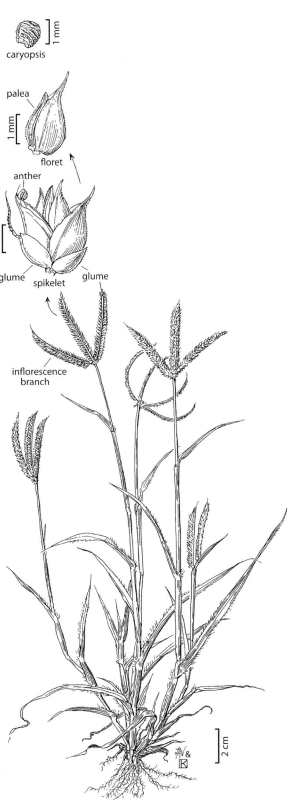

caryopsis

palea

floret

anther

glume spikelet glume

inflorescence branch

1. *Dactyloctenium aegyptium* (L.) Willd. (Durban crowfoot, crowfoot, Egyptian crowfoot). A species introduced from Africa, where it is sometimes used for food and drink. It is usually an annual but occasionally is a short-lived, stoloniferous perennial. It has a distinctive compressed culm. Panicle with 2–8 branches. Branches extending beyond the last spikelet for several millimeters. Seeds have been used as food in time of famine. Used in Africa and Asia as a lawn grass and is an important pasture grass (Diggs et al. 2006). Potentially toxic due to presence of cyanide (Burrows and Tyrl 2001). A weedy species of disturbed sites in Texas. Palatable to livestock but poor value due to limited productivity (Hatch, Schuster, and Drawe 1999).

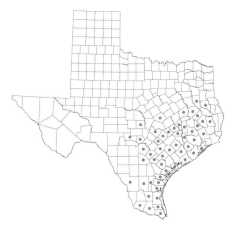

Danthonia DC.
(Danthonioideae: Danthonieae)

Low to moderately tall cespitose perennials with few-flowered panicles. Spikelets several-flowered, disarticulating above the glumes and between the florets. Glumes about equal, 1–5-nerved, much longer than the lemmas. Lemmas rounded and hairy on the back, indistinctly several-nerved, with a well-developed callus and a 2-toothed apex, the midnerve diverging as a stout, flat, twisted, geniculate awn at the base of the apical cleft. Palea broad, well developed. Basic chromosome number, $x = 6$. Photosynthetic pathway, C_3. Represented in Texas by 2 species.

The flattened awn above the point of insertion with the lemma is distinctive for the genus.

1. Lemmas sparsely pubescent with hairs of uniform length; awns 10–17 mm long; sheaths usually glabrous or inconspicuously pubescent...2. *D. spicata*
1. Lemmas villous with long hairs on margins and short hairs on back; awns 5–10 mm; sheaths usually villous and conspicuously pubescent .... 1. *D. sericea*

(Gould 1975b; Darbyshire 2003)

1. *Danthonia sericea* Nutt. (downy oatgrass, silky wildoatgrass). A perennial plant up to 1.2 m tall. Culms not disarticulating at maturity. Inflorescence with 5–30 spikelets. Branches usually erect or ascending. It is found generally in open woods on dry, sandy sites. Rare; only collection from Bowie County.

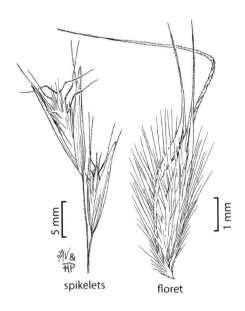

spikelets   floret

spikelet

2 mm

florets

2. *Danthonia spicata* (L.) P. Beauv ex Roem. & Schult. (poverty oatgrass, curly oatgrass). A perennial with culms rarely up to 1 m tall. Culms disarticulate at the nodes at maturity. Inflorescence with 5–18 spikelets. Branches stiff, appressed to ascending. Generally found in open, sandy woods.

Dasyochloa Willd. *ex* Rydb.
(Chloridoideae: Cynodonteae)

Perennial, stoloniferous. Culms composed of elongated internodes topped by a fascicle of leaves, the fascicle finally touching the ground and rooting. Sheaths open; ligules a short, ciliated ring of hairs; auricles absent. Inflorescences and small, compact panicle, capitate, usually not exceeding the blades of the fascicle, white-woolly. Spikelets with 5–15 florets. Glumes 2, subequal, 1-nerved, often tinged with purple, acuminate, awn-tipped. Lemmas deeply cleft, long-pilose on the lower half, often purple-tinged, awn from the bifid apex and barely exceeding the lobes in length. Palea about ½ as long as lemma, long-pilose below, veins extending into awns. Basic chromosome number, $x = 8$. Photosynthetic pathway, C_4. Represented in Texas by the only species in the genus.

(Valdes-Reyna 2003b)

1. *Dasyochloa pulchella* (Kunth) Willd. *ex* Rydb. (fluffgrass). Low, tufted, stoloniferous perennial. Lemmas densely pubescent, midvein extending into short awn. Paleas densely pilose below. A plant of arid, western regions of the state. It has been included in *Erioneuron* and *Tridens* but most closely resembles *Munroa*. Gould (1975a, 1975b) includes it in *Erioneuron*.

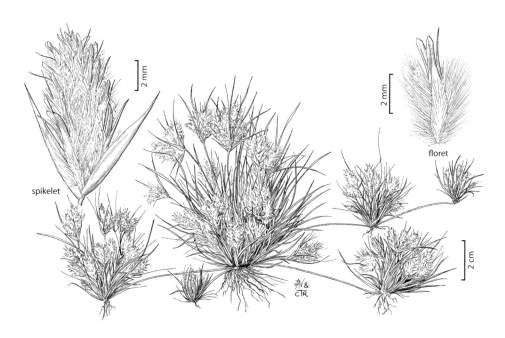

spikelet

floret

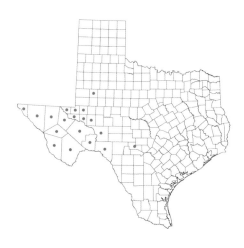

Desmazeria Dumort.
(Poöideae: Poeae)

Annuals, low and tufted with narrow, contracted panicles or spicate racemes. Leaves glabrous, basal, and cauline, blades flat or convolute when dry. Auricles absent. Ligules membranous, lacerate at the apex. Spikelets 4–25-flowered, awnless, short-pediceled, disarticulation above the glumes and between the florets. Glumes unequal to subequal, shorter than to almost as long as the lemmas, 1–5-veined; lower glume lanceolate, upper glume elliptic or oblong. Lemmas 5-nerved, broad, nearly terete, blunt at apex. Lemma and palea indurate and tightly enclosing the caryopsis at maturity. Basic chromosome number, $x = 7$. Photosynthetic pathway, C_3. Represented in Texas by a single species and variety.

(Tucker 2007)

1. *Desmazeria rigida* (L.) Tutin. (fern grass).
A low, tufted annual found along roadsides,
railroads, livestock pens, field margins, and
lawns. An introduced species from southern
Europe. It has been referred to as *Scleropoa
rigida* (L.) Griseb. and *Catapodium rigidum*
(L.) C. E. Hubb. (Gould 1975b). The Texas
species belongs to the typical subspecies,
subsp. *rigida*.

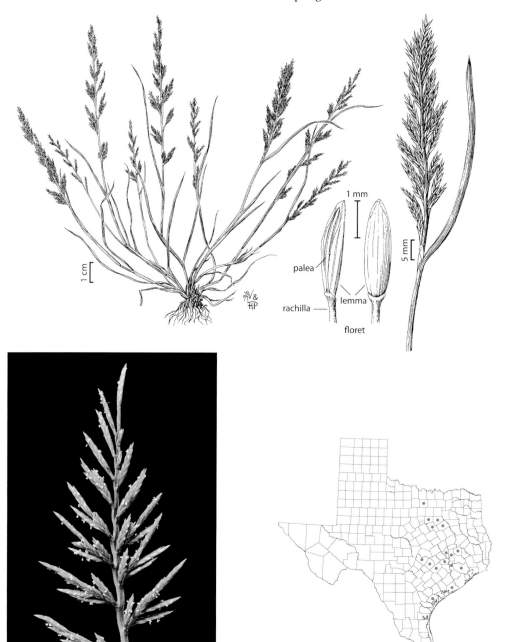

Diarrhena P. Beauv.
(Poöideae: Diarrheneae)

Slender, erect perennials. Leaf blades long, flat, 1–2 cm broad. Inflorescence a narrow, erect or drooping, few-flowered panicle. Spikelets mostly 3–5-flowered, disarticulating above the glumes and between the florets. Glumes unequal, shorter than the lemmas, the first 1-nerved, the second 3- or 5-nerved. Lemmas firm, tapering to a point, 3-nerved, the nerves converging at the tip. Palea broad, obtuse, strongly 2-nerved. Stamens 2 or 3, infrequently 1. Caryopsis large, hard, shiny, turgid, beaked above, at maturity conspicuously exserted from between the spreading lemma and palea. Basic chromosome number, $x = 10$. Photosynthetic pathway, C_3. Represented in Texas by 2 species.

1. Calluses pubescent on all but the most mature lemmas; mature fruit tapering to a blunt beak, 1.3–1.8 mm long .... 1. *D. americana*
1. Calluses glabrous on all mature lemmas; mature fruit tapering to a distinctive bottlenose-shaped beak, 1.8–2.5 mm long .... 2. *D. obovata*

(Brandenburg 2007b)

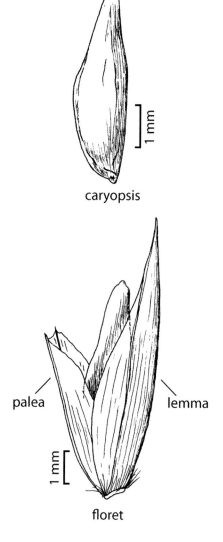

caryopsis

palea

lemma

floret

1 mm

1 mm

1. *Diarrhena americana* P. Beauv. (American beakgrain). A perennial with long, scaly rhizomes. Auricles pubescent. Mature fruit tapering into a blunt beak. A species restricted to the United States, where it grows in moist woods and pine-hardwood forests. Included in Jones, Wipff, and Montgomery (1997) as a synonym of *D. obovata*. Hatch, Gandhi, and Brown (1990) listed it from east Texas. Digg et al. (1999) reported it from eastern Oklahoma but unknown in Texas.

2. *Diarrhena obovata* (Gleason) Brandenburg (obovate beakgrain, hairy beakgrain). A perennial with long, scaly rhizomes. Auricles glabrous or pubescent. Mature fruit tapering into a distinct bottlenose-shaped beak. Usually found in rich woodlands. Not included in Gould (1975b) or Hatch, Gandhi, and Brown (1990). Perhaps not a constant member of the Texas flora.

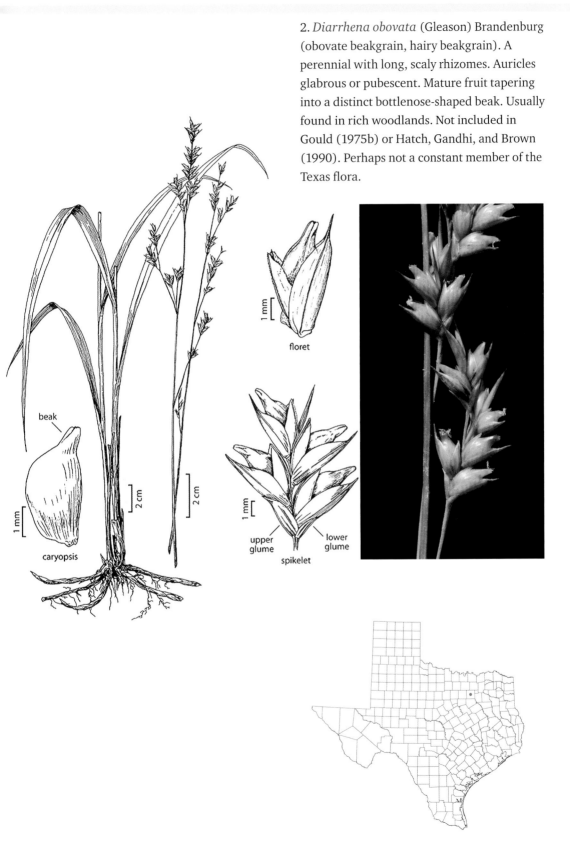

floret

beak

caryopsis

upper glume

lower glume

spikelet

Dichanthelium (Hitchc. and Chase) Gould
(Panicoideae: Paniceae)

Perennials, typically tufted and usually forming early in the growing season a basal rosette of shorter, broader blades than those of the culms. Ligule usually a ring of hairs, infrequently a short membrane or absent; pseudoligule sometimes present immediately behind the true ligule, hairs 1–5 mm long. Secondary branching after first culm growth and elongation, often resulting in densely clustered fascicles of much-reduced leaves and inflorescences. Inflorescences typically a small panicle with spreading or occasionally contracted branches, the spikelets short- or long-pediceled. Disarticulation below the glumes. Glumes both present, but the lowermost often greatly reduced. Lower floret usually neuter but staminate in a few species. Upper floret perfect, with a shiny, glabrous, coriaceous lemma and palea. The lemma tightly inrolled over the palea. Basic chromosome number, $x = 9$. Photosynthetic pathway, C_3. Represented in Texas by 26 species and 25 subspecies.

Jones, Wipff, and Montgomery (1997) do not separate this genus from *Panicum*.

Members of this genus have 2 distinct flowering periods. Primary panicles are typically produced in April until June or July; these are usually open with at least some chasmogamy occurring. Secondary panicles are produced from late May–early June until fall. These are usually closed, contained in subtending leaf sheaths, and cleistogamous. Seed set is often more pronounced in the fall than in the spring.

A difficult and often confusing genus. As is usually the case, if the genus is easy to determine, the species will be difficult. This complex affords grass identifiers an opportunity for hours of fun and frustration.

1. Basal leaf blades similar in shape to those of the lower cauline leaves; culms branching from near the base in the fall.
    2. Blades soft, 3–12 mm wide, usually ciliate.
        3. Spikelets with papillose-based hairs.. 11. *D. laxiflorum*
        3. Spikelets pubescent, but hairs not papillose-based .... 24. *D. strigosum*
    2. Blades stiff, 1–5 mm wide, not ciliate.
        4. Upper glumes and lower lemmas forming a beak; spikelets 3.2–4.3 mm long 7. *D. depauperatum*
        4. Upper glumes and lower lemmas not forming a beak; spikelets 2.0–3.4 mm long 12. *D. linearifolium*
1. Basal leaf blades forming a rosette, or basal leaves absent; culms branching from the midculm nodes in the fall.
    5. Bases of culms hard, cormlike; basal leaves absent.
        6. Lower glumes not encircling the pedicels, subadjacent to the upper glumes .... 17. *D. pedicellatum*
        6. Lower glumes encircling the pedicels, attached about 0.2 mm below the upper glume .... 14. *D. nodatum*

5. Bases of the culms not cormlike; basal rosettes present.

    7. Blades cordate, with white, cartilaginous margins; spikelets 1.0–1.8 mm long.

        8. Cauline blades >10 cm long, >14 mm wide, veins evident; panicles less than ½ as wide as long .... 18. *D. polyanthes*

        8. Cauline blades <10 cm long, <14 mm wide, veins obscure; panicles more than ½ as wide as long .... 23. *D. sphaerocarpon*

    7. Blades not cordate, margins usually not white and cartilaginous; spikelets >1.9 mm long.

        9. Lower glumes attached about 0.2 mm below the upper, the bases ..... clasping the pedicels.

            10. Blades 2–7 cm long; spikelets planoconvex in side view 19. *D. portoricense*

            10. Blades 4–16 cm long; spikelets biconvex in side view.

                11. Culms densely villous; nodes densely bearded 6. *D. consanguineum*

                11. Culms and nodes glabrous or variously pubescent but not villous 1. *D. aciculare*

        9. Lower glumes attached immediately below the upper, the bases not clasping the pedicels.

            12. Sheaths mottled with pale spots and constricted at the apex.

                13. Nodes often swollen, densely bearded above a viscid glabrous ring; blades densely soft pubescent .... 22. *D. scoparium*

                13. Nodes not swollen, glabrous or sparsely pubescent; blades glabrous or sparsely pubescent.

                    14. Cauline blades <15 mm wide, tips attenuate, involute .... 21. *D. scabriusculum*

                    14. Cauline blades >15 mm wide, tips acuminate, flat .... 4. *D. clandestinum*

            12. Sheaths not mottled with pale spots or constricted at the apex Key A

Key A

1. Ligules a ciliated membrane; plants usually rhizomatous.

    2. Ligules 0.3 mm or shorter; all lower florets sterile .... 5. *D. commutatum*

    2. Ligules 0.4–1.0 mm long; some lower florets staminate.

        3. Nodes glabrous or slightly bearded; spikelets <4 mm long .... 10. *D. latifolium*

        3. Nodes densely, retrorsely bearded; spikelets >3.8 mm long .... 3. *D. boscii*

1. Ligules of hairs; plants not rhizomatous.

      4. Upper glumes usually with an orange or purple spot at base, veins prominent; spikelets 2.5–4.3 mm long.

        5. Nodes glabrous or sparsely pubescent; abaxial leaf blade surface not glabrous or variously pubescent but not velvety .... 15. *D. oligosanthes*

5. Nodes densely bearded with spreading hairs; abaxial leaf blade surface velvety pubescent.

    6. Spikelets 3.7–4.3 mm long; ligules 2–5 mm long; adaxial leaf blade surface glabrous or variously pubescent, but not velvety......................20. *D. ravenelii*

    6. Spikelets 2.5–3.2 mm long; ligules 0.5–1.0 mm long; adaxial leaf blade surface velvety pubescent13. *D. malacophyllum*

4. Upper glumes without an orange or purple spot at base, veins not prominent; spikelets 0.8–3.0 mm long.

    7. Ligules 1–5 mm long, or the culms with hairs of 2 lengths (long hairs and also puberulent).

        8. Spikelets 0.8–1.1 mm long; culms <1 mm thick ....26. *D. wrightianum*

        8. Spikelets 1.1–3.0 mm long; culms usually >1 mm thick.

            9. Sheaths glabrous or with hairs up to 3 mm long; spikelets 1.1–2.1 mm long 2. *D. acuminatum*

            9. Sheaths with hairs to 4 mm long; spikelets 1.8–3.0 mm long 16. *D. ovale*

    7. Ligules absent or <2 mm long; culms glabrous or with hairs of only one length.

        10. Culms usually more than 1 mm thick; plants up to 1 m tall   8. *D. dichotomum*

        10. Culms usually <1 mm thick; plants only up to 0.5 m tall.

            11. Culms usually reclining; leaf blades without white margin; ligules usually >1 mm long .... 9. *D. ensifolium*

            11. Culms usually erect; leaf blades with white margins; ligules usually <1 mm long .... 25. *D. tenue*

(Freckmann and Lelong 2003a)

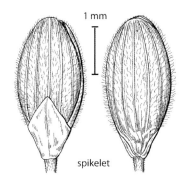

subsp. angustifolium

1. *Dichanthelium aciculare* (Desv. *ex* Poir.) Gould & C. A. Clark (narrow-leaved panicgrass, needleleaf rosettegrass). Cespitose perennial with poorly developed basal rosettes. Culms up to 75 cm tall, erect. Sheaths shorter than the internodes and often with papillose-based hairs. Spikelets 1.7–3.6 mm long. Lower glumes clasping, about 0.5 mm below the upper glumes. Found in open, sandy areas. Common in East Texas. Two subspecies have been recognized: subsp. *aciculare*; and subsp. *angustifolium* (Elliott) Freckmann & Lelong. They can be distinguished by the following character:

1. Spikelets 1.7–2.3 mm long.......subsp. *aciculare*
1. Spikelets 2.4–3.6 mm
    long .... subsp. *angustifolium*

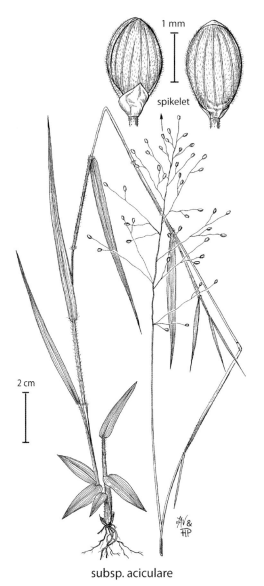

subsp. aciculare

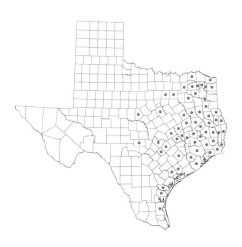

1 mm

2 cm

2. *Dichanthelium acuminatum* (Sw.) Gould & C. A. Clark (hairy panicgrass). Cespitose perennial, usually with a well-developed basal rosette. Culms up to but rarely over 1 m tall. Typically with pubescent sheaths, blades, nodes, and/or internodes; glabrous forms common as well. Common species throughout the state in varied habitats. Occurs from dry to wet sites, sandy to clayey soils, and open to shaded areas. According to Freckmann and Lelong (2003) this is the "most polymorphic and troublesome species in the genus." They recognize 10 subspecies that hybridize and integrate, not only among themselves but with other members of the genus as well. Four of the 10 subspecies have been reported for Texas: subsp. *acuminatum*; subsp. *lindheimeri* (Nash) Freckmann & Lelong; subsp. *longiligulatum* (Nash) Freckmann & Lelong; and subsp. *spretum* (Schult.) Freckmann & Lelong. They can sometimes be distinguished by the following characters:

1. Lower portion of the culms and lower sheaths usually glabrous or sparsely pubescent.
    2. Primary panicles congested, >2 times as long as wide .... subsp. *spretum*
    2. Primary panicles open, <2 times as long as wide.
        3. Blades green or purple, margins not ciliate at the base ....
        .... subsp. *longiligulatum*
        3. Blades often yellowish-green, margins usually with papillose-based cilia at base .... subsp. *lindheimeri*
1. Lower portion of the culms and lower sheaths usually densely pubescent or puberulent................... subsp. *acuminatum*

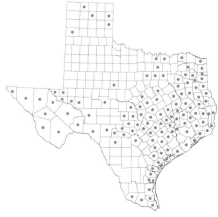

palea
lemma

lower
glume

upper floret

spikelet

node

2 mm

3. *Dichanthelium boscii* (Poir.) Gould & C. A. Clark (Bosc's panicgrass). Perennial. Basal rosette well developed from knotty rhizomes. Lower florets usually staminate. Close to *D. latifolium* but differs in having more bearded nodes and more pubescent spikelets. Generally found in partially open, dry oak-hickory forests. Fairly common species. Not included in Gould (1975b).

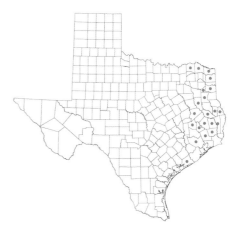

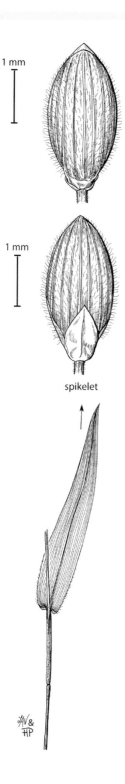

1 mm

1 mm

spikelet

4. *Dichanthelium clandestinum* (L.) Gould (deer-tongue panic, deer-tongue grass). Perennial plants found in large clumps with thick rhizomes. Basal rosettes well developed and distinct. Leaf blades wider, blunter, and flatter than in D. *scabriusculum*. Stiff hairs on the sheath can be irritating to the skin. Infrequent in wet, sandy, partially shaded woodlands in East Texas.

5. *Dichanthelium commutatum* (Schult.) Gould (variable panicgrass, variable rosettegrass). Perennial either cespitose or from knotty rhizomes; basal rosettes well developed. Lower florets sterile. Fairly common in East Texas on sandy soils in partially open to deeply shaded woodlands. Four integrating subspecies are found in Texas: subsp. *ashei* (T. G. Pearson *ex* Ashe) Freckmann & Lelong; subsp. *commutatum*; subsp. *equilaterale* (Scribn.) Freckmann & Lelong; and subsp. *joorii* (Vasey) Freckmann & Lelong. They can be distinguished by the following characters:

1. Culms densely puberulent; culm leaf blades symmetrical at base; rosette leaf blades <3 cm long and <6 mm wide .... subsp. *ashei*
1. Culms glabrous or sparsely puberulent; culm leaf blades asymmetrical at base; rosette leaf blades sometimes >4 cm long and 10 mm wide.
　2. Lower glumes about .5 length of spikelet ....subsp. *equilaterale*
　2. Lower glumes about .25 length of spikelet.
　　3. Blades strongly asymmetrical; spikelets >2.9 mm long; lower lemmas pointed.............................. subsp. *joorii*
　　3. Blades almost symmetrical; spikelets <2.9 mm long; lower lemmas blunt ....subsp. *commutatum*

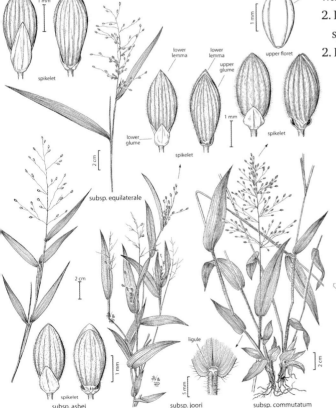

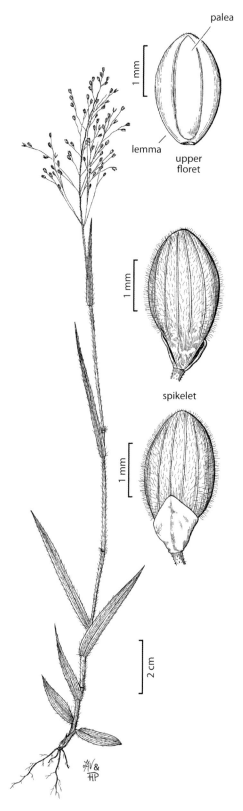

palea

1 mm

lemma

upper
floret

1 mm

spikelet

1 mm

2 cm

JV &
HP

6. *Dichanthelium consanguineum* (Kunth)
Gould & C. A. Clark (blood rosettegrass,
Kunth's panicgrass). Cespitose perennial
with poorly developed basal rosettes. Culms
erect and to about 0.5 m tall. Nodes densely
bearded. Internodes densely villous. Sheaths
and both sides of blades villous. Spikelets
1.4–1.8 mm long. Found in sandy woods
and pine flatlands. Gould (1975a) included
this species in *D. angustifolium*, which in this
treatment and in Jones, Wipff, and Montgomery
(1997, as *Panicum*) is considered a subspecies of
D. aciculare.

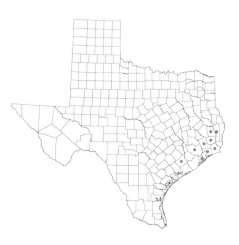

1 mm

spikelet

2 cm

2 cm

HP

7. *Dichanthelium depauperatum* (Muhl.) Gould (starved panicgrass, starved rosettegrass). Cespitose perennial with poorly differentiated basal rosette. Culms >50 cm long. Nodes bearded; internodes usually pubescent. Sheaths longer than the internodes. Blades narrow, 4 mm wide or less. Spikelets 3.2–4.3 mm long. Scattered along forest edges and in woodland openings, primarily in sandy areas.

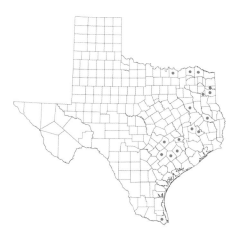

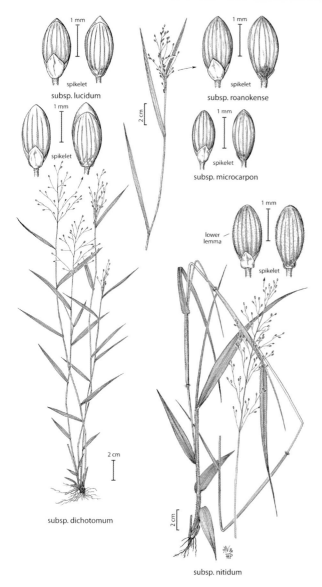

subsp. lucidum

subsp. roanokense

spikelet

subsp. microcarpon

lower lemma

spikelet

subsp. dichotomum

subsp. nitidum

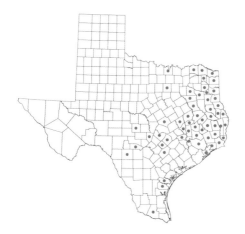

8. *Dichanthelium dichotomum* (L.) Gould (forked panicgrass, cypress rosettegrass). Cespitose perennial from a knotty crown with well-developed and differentiated basal rosettes. Culms up to 1 m long, more or less erect. Uppermost leaf sheaths often with whitish glandular spots between the veins. A variable species from numerous and varied habitats; found from dry to wet areas, from sandy to clayey soils, and in shaded to open sites. Freckmann and Lelong (2003a) provided the following statement about *D. dichotomum*: "It is a polymorphic and ubiquitous species, with many of its intergrading subspecies exhibiting traits of other widespread and variable species . . . which grow at the same site." This complex provides additional entertainment for grass identifiers with free time on their hands. Eight subspecies have been recognized; 5 occur within Texas: subsp. *dichotomum*; subsp. *lucidum* (Ashe) Freckmann & Lelong; subsp. *microcarpon* (Muhl. *ex* Elliott) Freckmann & Lelong; subsp. *nitidum* (Lam.) Freckmann & Lelong; and subsp. *roanokense* (Ashe) Freckmann & Lelong. Diggs et al. (2006) also included subsp. *yadkinense* in Texas. The subspecies can sometimes be distinguished by the following characters:

1. Lower nodes hairy.
     2. Spikelets 1.5–1.8 mm
          long .... subsp. *microcarpon*
     2. Spikelets 1.8–2.5 mm long.
          3. Spikelets pubescent; midculm blades
               usually 7–14 mm wide ..... subsp. *nitidum*
          3. Spikelets glabrous; midculm blades 5–7
               mm wide .... subsp. *dichotomum*
1. Lower nodes glabrous.
          4. Culms weak, often flattened,
               sprawling .... subsp. *lucidum*
          4. Culms erect, terete.
               5. Blades usually spreading;
                    spikelets 1.8–2.3 mm
                    long .... subsp. *dichotomum*
               5. Blades usually ascending;
                    spikelets 1.5–1.8 mm
                    long .... subsp. *roanokense*

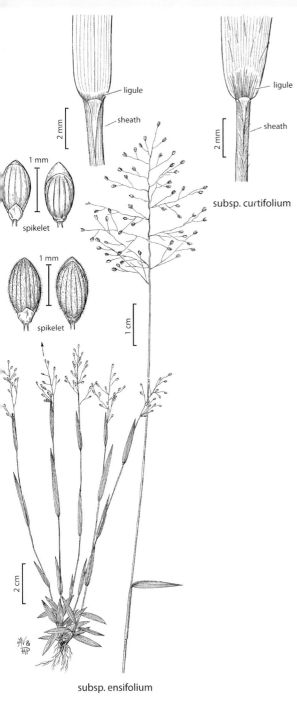

ligule

sheath

2 mm

1 mm

spikelet

1 mm

spikelet

1 cm

2 cm

subsp. ensifolium

ligule

sheath

2 mm

subsp. curtifolium

9. *Dichanthelium ensifolium* (Baldwin *ex* Elliott) Gould (sword-leaf panicgrass, sword-leaf rosettegrass). Cespitose perennial with well-differentiated basal rosette. Culms relatively short, rarely exceeding 40 cm in length. Sheaths much shorter than the internodes. Blades only up to 3.5 cm long. Spikelets only 1.2–1.5 mm long. Generally found in wet, sandy areas; often found in the presence of *Sphagnum* moss (Gould 1975b). Generally smaller than the closely related *D. tenue*. Two subspecies have been recognized: subsp. *curtifolium* (Nash) Freckmann & Lelong; and subsp. *ensifolium*. They can be distinguished by the following characters:

1. Cauline sheaths glabrous; ligules 0.2–1.0 mm long......................................subsp. *ensifolium*
1. Cauline sheaths sparsely pilose; ligules 1–2 mm long................ subsp. *curtifolium*

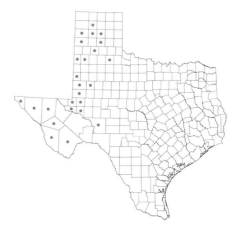

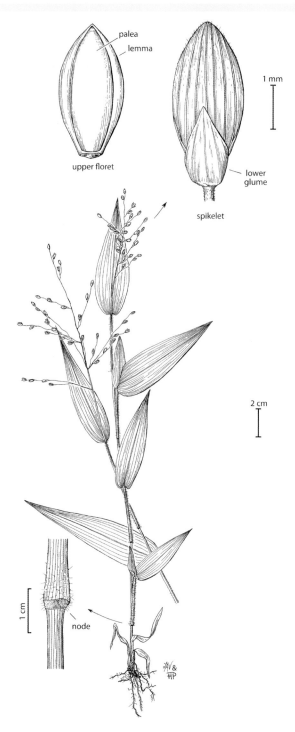

palea
lemma

upper floret

1 mm

lower glume

spikelet

node

2 cm

1 cm

10. *Dichanthelium latifolium* (L.) Harvill. (broadleaf panicgrass, broadleaf rosettegrass). Perennial with a well-developed basal rosette from knotty rhizomes. Lower florets staminate. Close to *D. boscii* but differs in less pubescent nodes and spikelets. Reported only from Harrison County in East Texas. Not included in Gould (1975b).

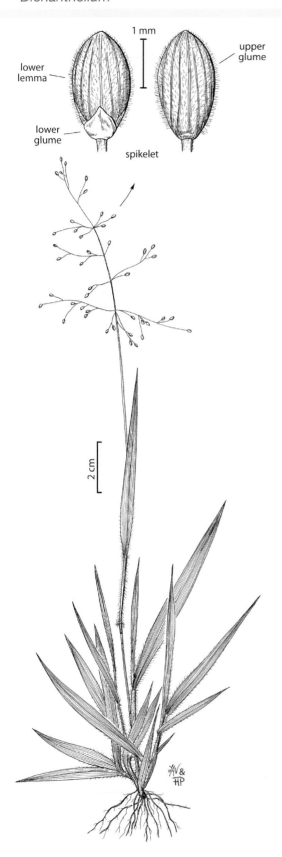

lower lemma

upper glume

lower glume

spikelet

11. *Dichanthelium laxiflorum* (Lam.) Gould (soft-tufted panicgrass, open-flower rosettegrass). Densely tufted perennial with basal rosette poorly differentiated from upper cauline leaves. Culms generally >50 cm long, erect or radiating from the large tuft of basal leaves. Nodes bearded; internodes glabrous. Spikelet 1.7–2.3 mm long. Relatively common species of deciduous woods and the Gulf Coast Prairies and Marshes.

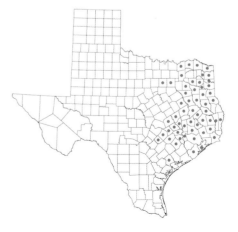

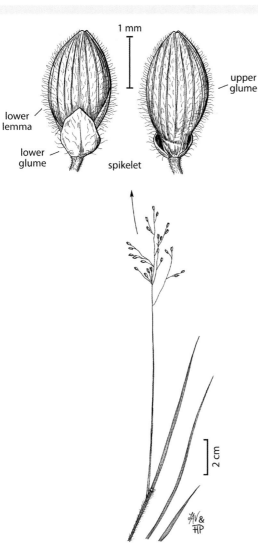

1 mm

upper
glume

lower
lemma

lower
glume

spikelet

2 cm

12. *Dichanthelium linearifolium* (Scribn.)
Gould (linear-leaved panicgrass, slim-leaf
rosettegrass). Cespitose perennial with poorly
differentiated basal rosettes. Culms up to 50 cm
long, drooping to less frequently erect. Sheaths
longer than the internodes. Blades erect to
ascending. Spikelets 2.0–3.2 mm long. Fairly
common in sandy woods, open areas, prairies,
and rocky outcrops.

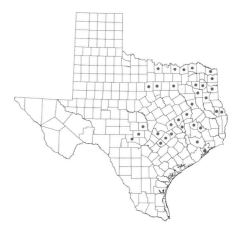

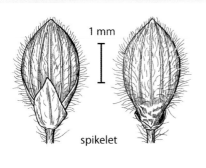

1 mm

spikelet

13. *Dichanthelium malacophyllum* (Nash) Gould (soft-leaved panicgrass, soft-leaved rosettegrass). Cespitose perennial with a well-differentiated basal rosette. Nodes densely bearded; internodes puberulent and with papillose-based, soft hairs. Blades velvety-pubescent on both sides. Found in sandy or rocky woodlands; more common on clayey soils.

2 cm

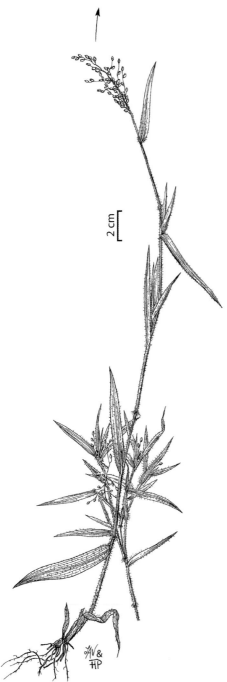

14. *Dichanthelium nodatum* (Hitchc. & Chase) Gould (Sarita panicgrass, Sarita rosettegrass). Perennial without a basal rosette, culms from a hard, cormlike base. Similar to *D. pedicellatum* but differs in having lower glumes that encircle the pedicel and are separated from the upper glume; thicker, shorter, and blunter leaf blades with cilia to above the middle. Reported as endemic to Texas (Gould 1975b) but also reported for Mexico (Diggs et al. 2006). Not listed as rare by Jones, Wipff, and Montgomery (1997) or Poole et al. (2007). Found primarily on sandy grasslands in South Texas and the Gulf Coast areas. Good for livestock but low productivity, large seeds for birds, and utilized by deer in the spring (Hatch, Schuster, and Drawe 1999).

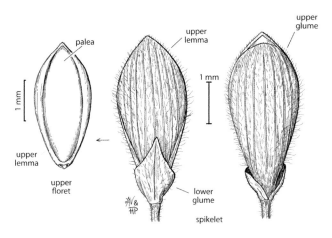

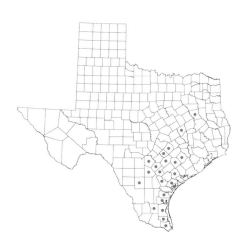

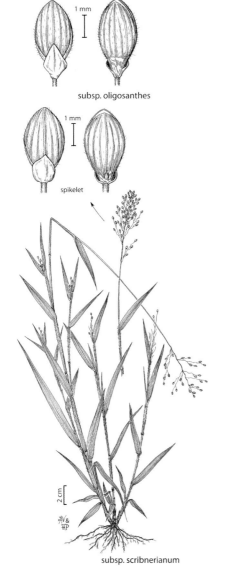

subsp. oligosanthes

spikelet

subsp. scribnerianum

15. *Dichanthelium oligosanthes* (Schult.) Gould (few-flowered panicgrass, Heller's rosettegrass). Cespitose perennial with well-developed and distinctive basal rosettes. Upper glumes strongly veined and with an orange or purple basal area. Grows in diverse habitats, including sandy, loamy, or clayey soils in open, partial, or full shade. Fairly common and reported from all regions of the state. Listed as fair for wildlife and livestock (Hatch and Pluhar 1993). Forage for deer and seeds are especially important to wildlife (Telfair 2006). Two subspecies are recognized: subsp. *oligosanthes*; and subsp. *scribnerianum* (Scribner's panicgrass). The latter is more common than the former. They can be distinguished by the following characters:

1. Blades 10 times longer than wide, partially involute; ligules 2–3 mm long .... subsp. *oligosanthes*
1. Blades less than 10 times longer than wide, usually flat; ligules 1.0–1.5 mm long .... subsp. *scribnerianum*

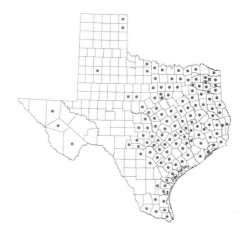

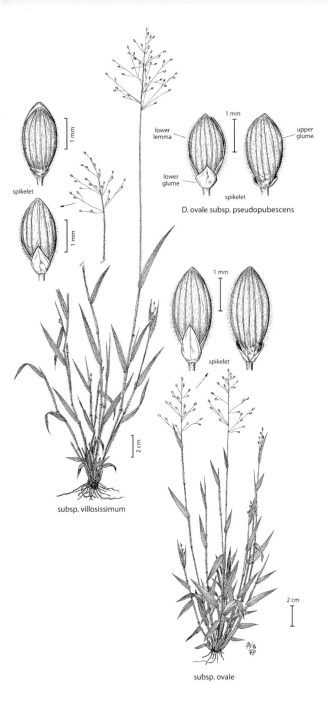

spikelet

1 mm

1 mm

spikelet

subsp. villosissimum

2 cm

subsp. ovale

2 cm

lower
lemma

lower
glume

1 mm

upper
glume

spikelet

D. ovale subsp. pseudopubescens

1 mm

spikelet

16. *Dichanthelium ovale* (Elliott) Gould &
C. A. Clark (stiff-leaved panicgrass, egg-leaved
rosettegrass). A cespitose perennial with a
well-differentiated basal rosette. Blades up to 8
cm long, usually more or less erect and ciliate.
Nodes and internodes, especially the lower,
with long-appressed or spreading hairs. Plants
of dry, open, sandy, and rocky areas. Three
subspecies are recognized: subsp. *ovale*; subsp.
pseudopubescens (Nash) Freckmann & Lelong;
and subsp. *villosissimum* (Nash) Freckmann &
Lelong. Diggs et al. (2006) also included subsp.
praecocius (Hitchc. and Chase) Freckmann &
Lelong in Texas. The subspecies can sometimes
be distinguished by these characters:

1. Lower sheaths and culm internodes with
   ascending or appressed, nonpapillose-based
   hairs, longest hairs <4 mm long or nearly
   glabrous.
   2. Spikelets 2.5–3.0 mm long; basal blades
      with long hairs on or near the margins and
      bases ....subsp. *ovale*
   2. Spikelets 2.1–2.6 mm long; basal
      blades usually without long hairs on or
      near the margins and bases ....
      .... subsp. *pseudopubescens*
1. Lower sheaths and culm internodes with
   spreading, sometimes retrorse, papillose-
   based hairs, some hairs >4 mm long ....
   ....subsp. *villosissimum*

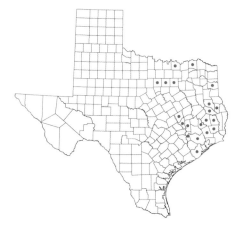

17. *Dichanthelium pedicellatum* (Vasey) Gould (corm-based panicgrass, cedar panicgrass). Perennial. Basal rosette absent and culms arising from a hard, cormlike base. Closely related to *D. nodatum* but differs in having the lower glume almost adjacent to the upper glume and not encircling the pedicel; thinner, longer, and more acuminate leaf blades with cilia only at the base. Found on limestone outcrops and dry oak woodlands in the Edwards Plateau, Cross Timbers and Prairies, and Blackland Prairies.

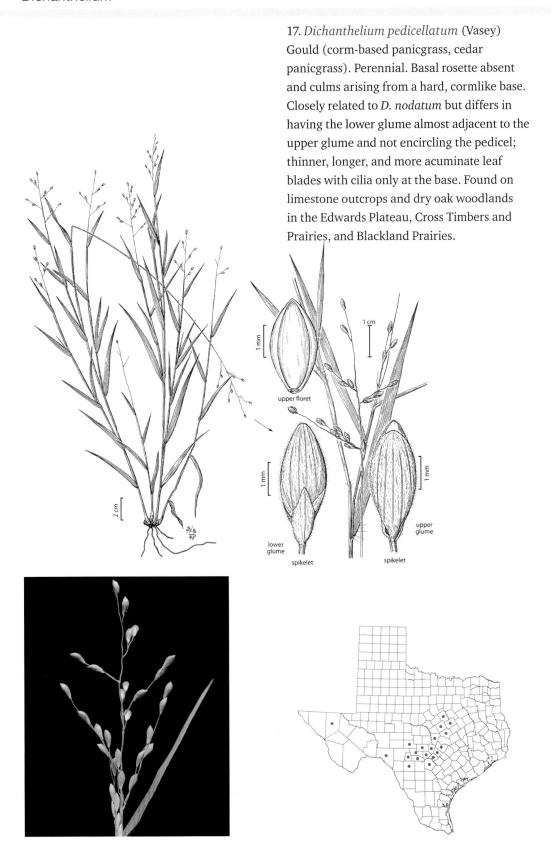

18. *Dichanthelium polyanthes* (Schult.) Mohlenbr. (many-flowered panicgrass, many-flowered rosettegrass, leafy rosettegrass). Cespitose perennial with few culms per tuft. Basal rosette well differentiated. Culms up to 1 m long, erect. Blades ciliate with papillose-based hairs, margins whitish. Found in shaded areas, along stream banks, and in ditches. Gould (1975b) listed it as a variety of *D. sphaerocarpon*.

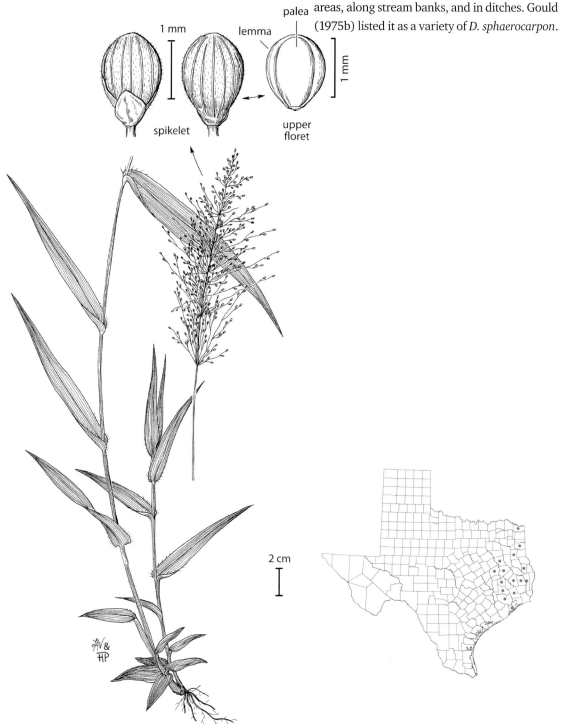

1 mm

spikelet

19. *Dichanthelium portoricense* (Desv. *ex* Ham.) B. F. Hansen & Wunderlin (hemlock witchgrass, blunt-glumed panicgrass). Cespitose perennial with well-developed and well-differentiated basal rosette. Culms up to 50 cm long. Blade bases often ciliate with papillose-based hairs, margins whitish. Spikelets 1.5–2.6 mm long, planoconvex in side view. Found in sandy areas, woodlands, savannahs, and coastal dune regions. Not included in Gould (1975b). Freckman & Lelong (2003a) recognize two subspecies; the typical one, subsp. *portoricense*, and subsp. *patulum* (Scribn. & Merr.) Freckman & Lelong. However, none are recognized here.

2 cm

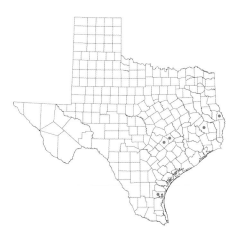

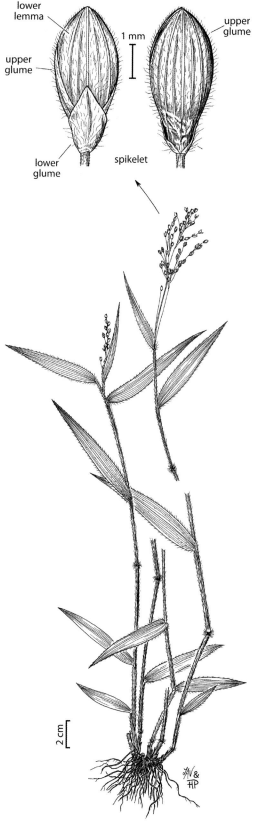

20. *Dichanthelium ravenelii* (Scribn. & Merr.) Gould (Ravenel's panicgrass). Cespitose perennial with a well-developed and distinctive basal rosette. Nodes densely bearded with spreading hairs above a glabrous ring; internodes usually puberulent with longer hairs also present. Upper glumes shorter than the spikelets with a purplish area at the base. Grows in dry, sandy areas in East Texas. Gould (1975b) included *D. boscii* here.

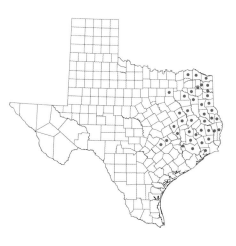

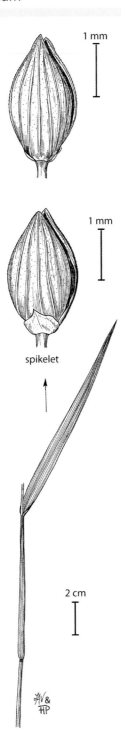

spikelet

21. *Dichanthelium scabriusculum* (Elliott) Gould & C. A. Clark (tall-swamp panicgrass, woolly rosettegrass). Perennial. Plants grow in large clumps from thick rhizomes. Culms can reach 1.5 m tall. Basal rosettes are well developed. Rachis and inflorescence branches usually mottled. Spikelets often purplish. Leaf blades narrower, more pointed, and involute than in D. *clandestinum*. Found in wet, sandy, open sites like shores, river and stream banks, and bogs. Not included in Gould (1975b).

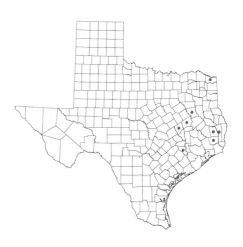

22. *Dichanthelium scoparium* (Lam.) Gould (velvety panicgrass). Perennial. Found in small clumps with thick rhizomes. Basal rosette well developed and distinctive. Plants can grow to 1.5 m tall. Culm nodes densely bearded over a viscid, glabrous ring. Internodes velvety-pubescent. Panicle branches often with purplish viscid spots. Scattered in wet, sandy, open, or disturbed sites in pine woods and post oak savannahs.

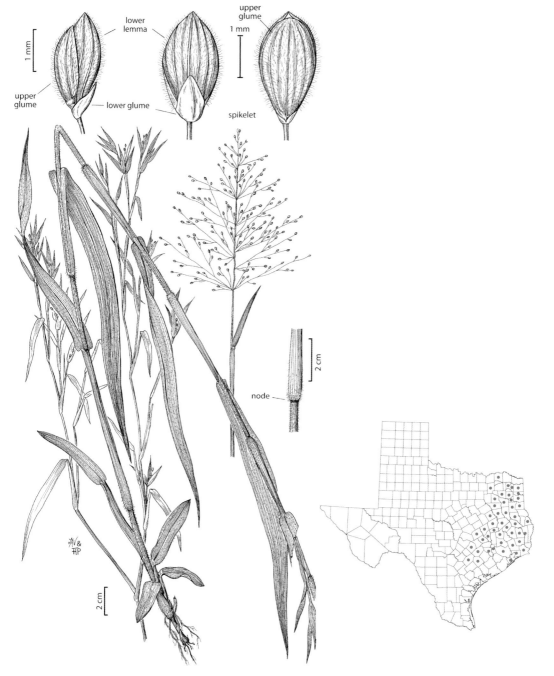

upper glume

lower lemma

lower glume

upper glume

spikelet

1 mm

node

2 cm

2 cm

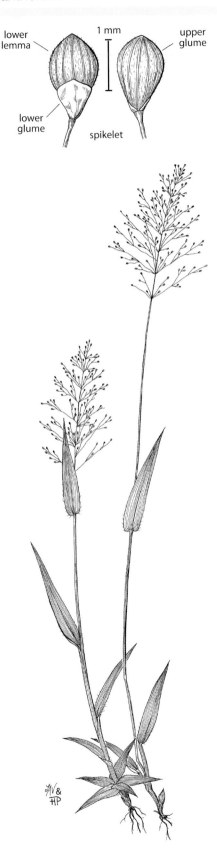

lower lemma

1 mm

upper glume

lower glume

spikelet

23. *Dichanthelium sphaerocarpon* (Elliott) Gould (round-fruited panicgrass, round-seeded rosettegrass). Cespitose perennial with well-differentiated and well-developed basal rosette. Culms to 50 cm long. Blades with papillose-based cilia and whitish margins. Spikelets 1.4–1.8 mm long. Found in dry, open areas and along roadsides. Good for livestock but low productivity; large seeds for birds; and utilized by deer in the spring (Hatch, Schuster, and Drawe 1999).

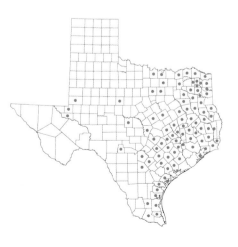

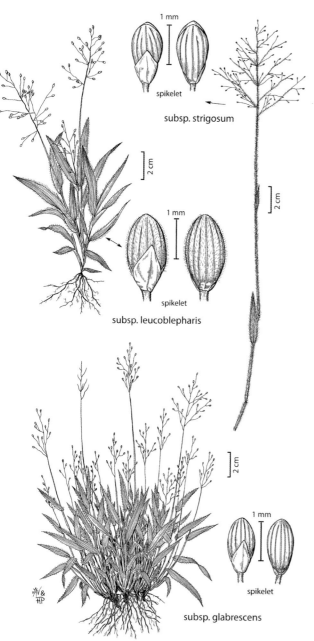

1 mm

spikelet

subsp. strigosum

2 cm

2 cm

1 mm

spikelet

subsp. leucoblepharis

2 cm

1 mm

spikelet

subsp. glabrescens

24. *Dichanthelium strigosum* (Muhl. *ex* Elliott) Freckmann (cushion-tufted panicgrass). Densely tufted perennial with culms up to 45 cm long. Leaf blade margins with prominent papillose-based hairs. Spikelets 1.1–2.1 mm long. Most common in sandy, open woods and boggy areas. Not included in Gould (1975b). Three recognized subspecies occur in Texas: subsp. *glabrescens* (Griseb.) Freckmann & Lelong; subsp. *leucoblepharis* (Trin.) Freckmann & Lelong; and subsp. *strigosum*. They can be distinguished by the following characters:

1. Spikelets pubescent, 1.6–2.1 mm long; lower glume about ½ as long as the spikelet ....subsp. *leucoblepharis*
1. Spikelets glabrous, 1.1–1.8 mm long; lower glume about ⅓ as long as the spikelet.
   2. Blades pilose ....subsp. *strigosum*
   2. Blades glabrous............subsp. *glabrescens*

25. *Dichanthelium tenue* (Muhl.) Freckmann & Lelong (slender panicgrass). Cespitose perennial with well-developed and well-differentiated basal rosettes. Culms up to 55 cm long. Nodes glabrous. Sheaths much shorter than the internodes. Spikelets 1.3–1.7 mm long. Found in sandy forest openings, flatwoods, and disturbed sites. Generally larger than the closely related *D. ensifolium*. Not listed for Texas in Gould (1975b) or Jones, Wipff, and Montgomery (1997).

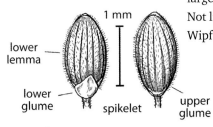

lower lemma

lower glume

spikelet

upper glume

1 mm

2 cm

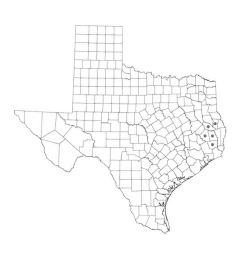

26. *Dichanthelium wrightianum* (Scribn.)
Freckmann (Wright's panicgrass). Weakly
cespitose perennial with conspicuous basal
rosette. Culms usually 50 cm or less. Upper
panicle branches and pedicels sometimes
viscid. Generally found in wet, sandy areas
such as pine flatwoods and margins of streams
and ponds. Rare in East Texas (reported only
from Tyler, Polk, and Hardin counties). This
species is not included in Gould (1975b) or
Jones, Wipff, and Montgomery (1997).

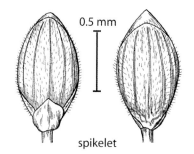

0.5 mm

spikelet

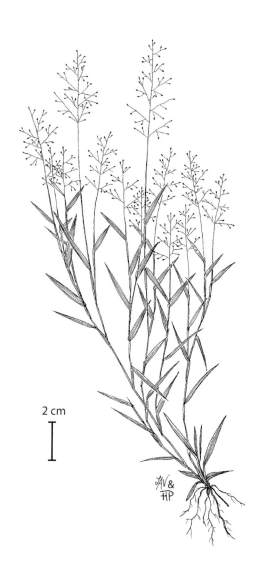

2 cm

Dichanthium Willemet
(Panicoideae: Andropogoneae)

Low to moderately tall perennials, mostly cespitose but some with extensive creeping stolons. Inflorescence as in *Bothriochloa*, but pedicels and internodes of the rachis flat or rounded, without a groove or membranous central area, and lower pair of spikelets of the inflorescence branches usually both sterile and awnless. Basic chromosome number, $x = 10$. Photosynthetic pathway, C_4. Represented in Texas by 3 species and 1 subspecies.

1. Pedicellate spikelet sterile; plants without stolons...3. *D. sericeum*
1. Pedicellate spikelet staminate; plants usually stoloniferous.
   2. Rame bases pilose .... 2. *D. aristatum*
   2. Rame bases glabrous.. 1. *D. annulatum*

(Barkworth 2003e; Gould 1975b)

1. *Dichanthium annulatum* (Forssk.) Stapf (ringed dichanthium, Kleberg bluestem). Stoloniferous perennial. Erect portion of culms generally >60 cm long. Nodes glabrous or short-puberulent. Lower glumes sparsely pubescent below. Introduced as a forage species but of relatively low quality for both livestock and wildlife. Found along roadsides, ditches, disturbed areas, and pastures. Poor for livestock and wildlife but good cover for birds and deer (Hatch, Schuster, and Drawe 1999).

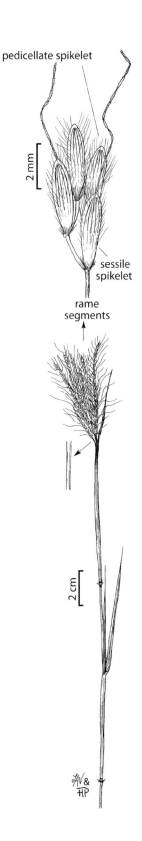

pedicellate spikelet

2 mm

sessile spikelet

rame segments

2 cm

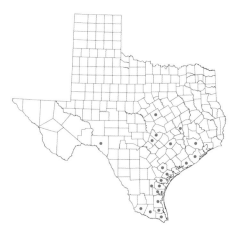

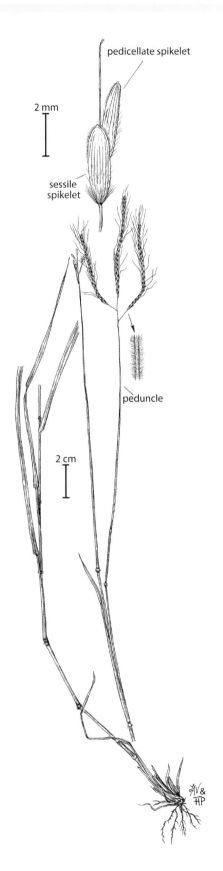

pedicellate spikelet

2 mm

sessile
spikelet

peduncle

2 cm

2. *Dichanthium aristatum* (Poir.) C. E. Hubb.
(awned dichanthium, Angleton bluestem).
Stoloniferous perennial, stolons sometimes
as long as 2 m. Erect portion of culms usually
about 35 cm tall. Nodes glabrous or densely
puberulent. Lower margins of lower glumes
ciliate. Sometimes used as a lawn species
but introduced as a forage grass from Asia.
Established as a weed of roadsides, ditches,
and other disturbed areas. Poor forage for
livestock and wildlife but good cover for birds
and deer (Hatch, Schuster, and Drawe 1999).

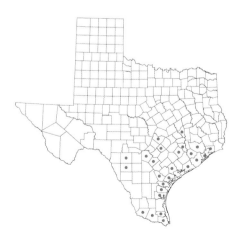

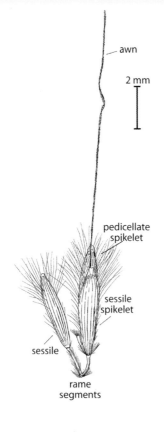

awn

2 mm

pedicellate
spikelet

sessile
spikelet

sessile

rame
segments

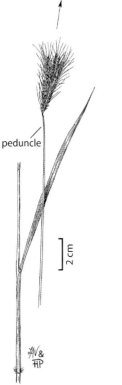

peduncle

2 cm

JAV &
HP

3. *Dichanthium sericeum* (R. Br.) Camus
(silky bluestem, Queensland bluegrass).
Cespitose perennial. Culms up to 1.2 m tall.
Nodes densely pilose. Lower glume of the
sessile spikelets with an arch of hairs below the
apex. Introduced as a potential forage species
from Australia. Reported to be fair to poor
as a forage species for livestock and wildlife
but good cover for birds and deer (Hatch,
Schuster, and Drawe 1999). The Texas species
belongs to the typical subspecies, subsp.
sericeum.

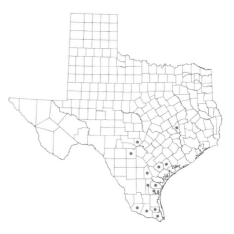

Digitaria Haller
(Panicoideae: Paniceae)

Annuals and perennials, with erect or decumbent-spreading, stoloniferous culms. Blades mostly thin and flat. Inflorescence a panicle with few to numerous slender, spikelike, racemose branches, these unbranched or sparingly branched near the base. Spikelets slightly planoconvex; solitary, paired, or in groups of 3–5; subsessile or short-pediceled in 2 rows on a 3-angled, often winged rachis. Disarticulation below the glumes. Lower glume minute or absent. Upper glume well developed but usually shorter than the lemma of the sterile floret. Veins of glumes and lemma of the sterile floret glabrous, puberulent, or long-ciliate. Lemma of the fertile floret relatively narrow, acute or acuminate, firm and cartilaginous, but not hard, the margins thin, flat, not inrolled over the palea. Basic chromosome number, $x = 9$. Photosynthetic pathway, C_4. Represented in Texas by 15 species and 3 varieties.

The witchgrasses were once separated into the genus *Leptoloma*, and the cottontops were placed in *Trichachne*.

Digitaria velutina (Forssk.) P. Beauv., a plant included on the noxious weed list by the U.S. Department of Agriculture, has erroneously been reported from the state (Wipff 2003d).

All species are considered highly palatable to livestock (Powell 1994).

1. Inflorescences open panicles; spikelets solitary.
    2. Spikelets >3.5 mm long .... 1. *D. arenicola*
    2. Spikelets <3.3 mm long.
        3. Lower lemmas 7-veined .... 5. *D. cognata*
        3. Lower lemmas 5-veined .... 12. *D. pubiflora*
1. Inflorescence panicles of spikelike branches; spikelets in groups of 2–5.
        4. Spikelets paired; pedicels not adnate to the panicle branches.
            5. Upper lemmas brown, becoming dark brown with maturity.
                6. Ligules ciliate, 0.1–1.5 mm long; spikelets including pubescence 1.3–3.1 mm long.
                    7. Lower glumes 0.3–1.0 mm long... 7. *D. hitchcockii*
                    7. Lower glumes absent or 0.1 mm long .... 6. *D. filiformis*
                6. Ligules not ciliate, 1–6 mm long; spikelets including pubescence 3.7–7.5 mm long.
                    8. Terminal pedicels of primary branches 7–20 mm long.........11. *D. patens*
                    8. Terminal pedicels of primary branches 1.5–6.0 mm long.
                        9. Lower lemmas pubescent between the veins and on the margins .... 8. *D. insularis*
                        9. Lower lemmas pubescent on the margins, sometimes on the lateral veins.. 3. *D. californica*
            5. Upper lemmas variously colored but not brown and not turning a dark brown with maturity.

10. Panicle branches not winged; or if winged, the wings less than ½ as wide as midrib .... 14. *D. texana*

10. Panicle branches winged; wings more than ½ as wide as the midribs.

    11. Plants perennial, stoloniferous, and/or rhizomatous ....10. *D. milanjiana*

    11. Plants annual; neither stoloniferous nor rhizomatous.

        12. Lateral veins of the lower lemmas scabrous throughout or on the lower ⅔.. 13. *D. sanguinalis*

        12. Lateral veins of the lower lemmas smooth throughout or only scabrous on the lower ⅓.

            13. Lower lemma of the lower spikelet of each pair 7-veined; the lateral 2 on each side crowded toward the margins, the 3 central veins equally spaced .... 2. *D. bicornis*

            13. Lower lemma of the lower spikelet of each pair 5–7-veined, the 2 or 3 veins on each side crowded together near the margins and well separated from the midrib ....4. *D. ciliaris*

4. Spikelets in groups of 3–5 on the middle portion of the primary branches.

        14. Primary panicle branches with winged margins, wings at least ½ as wide as midrib.

            15. Plants always with axillary panicles in the lower leaf sheaths; spikelets 1.7–2.3 mm long .... 9. *D. ischaemum*

            15. Plants without axillary panicles; spikelets 1.2–1.7 mm long .... 15. *D. violascens*

        14. Primary panicle branches without winged margins, or if winged, wings less than ½ as wide as midrib.. 6. *D. filiformis*

(Wipff 2003d)

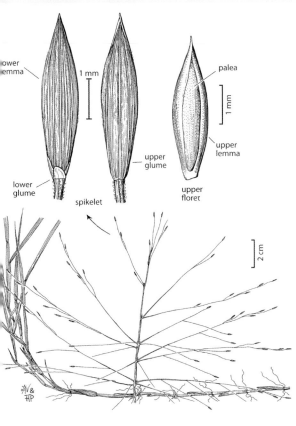

1. *Digitaria arenicola* (Swallen) Beetle (sand witchgrass). Perennial with long, creeping rhizomes. Panicles simple, open with divergent branches. Upper lemmas dark brown. Only found in deep sands along the Gulf Coast. Gould (1975b) included this as a variety of *D. cognata* (as *Leptoloma cognatum* var. *arenicola*). Fair forage for livestock but poor for wildlife (Hatch, Schuster, and Drawe 1999).

2. *Digitaria bicornis* (Lam.) Roem. & Schult.
(Asian crabgrass). Annual with decumbent-
spreading culms, sometimes reported as
stoloniferous. Lower sheaths with papillose-
based hairs. Panicles of 3–8 spikelike primary
branches. Upper lemmas yellow, light gray, or
sometimes light brown at maturity. Contrary
to the vernacular name, it is considered to be
a native species. A relatively common plant of
sandy, coastal areas.

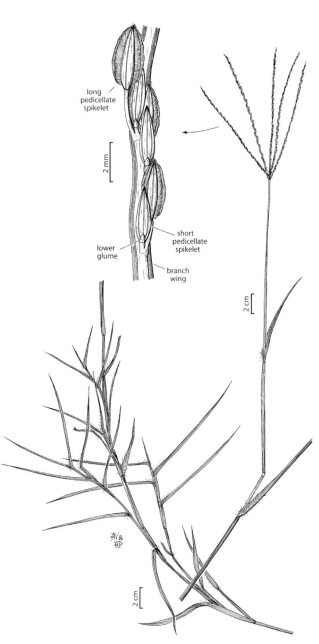

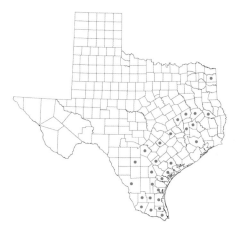

upper
glume   spikelet

lower
lemma

lower
glume

spikelet

3. *Digitaria californica* (Benth.) Henrard
(California cottontop, Arizona cottontop).
Cespitose perennial. Basal sheaths villous.
Panicles with 4–10, usually appressed,
spikelike primary branches. Lower lemmas
densely pubescent; upper lemmas becoming
dark brown. Found along roadsides, open
grasslands, and well-drained sites. Listed as
fair for wildlife and good for livestock (Hatch
and Pluhar 1993).

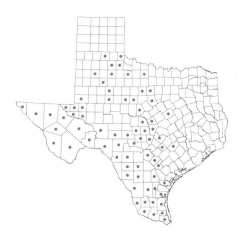

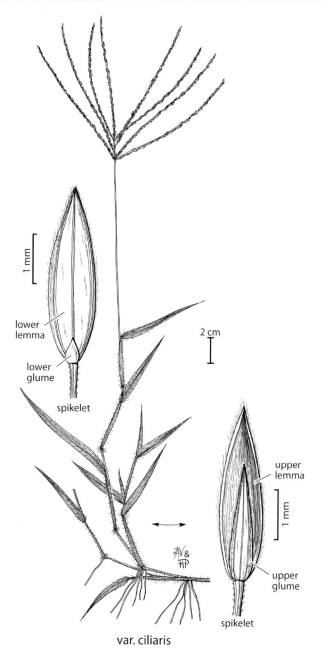

lower
lemma

1 mm

lower
glume

spikelet

2 cm

upper
lemma

1 mm

upper
glume

spikelet

var. ciliaris

4. *Digitaria ciliaris* (Retz.) Koeler (southern crabgrass). Annual with decumbent culms that root at the nodes. Sheaths with papillose-based hairs. Panicles of 2–10 spikelike primary branches. Lower lemmas with hairs mostly along margins. Upper lemmas yellow, gray, or tan, becoming brown and purple-tinged with age. A common species of disturbed sites throughout the state. The Texas species belongs to the typical variety, var. *ciliaris*.

5. *Digitaria cognata* (Schult.) Pilg. (fall witchgrass, Carolina crabgrass). Cespitose perennial without rhizomes. Panicle simple, open, with divergent branches, usually disarticulating as a whole and acting as a tumbleweed. Upper lemmas dark brown. Found in dry, sandy soils throughout most of the state east of the 100th meridian; rarer to the west. Telfair (2006) listed the seeds as especially important for wildlife. Diggs et al. (2006) listed 2 subspecies, but none are recognized here. Reported as fair for wildlife and livestock (Hatch and Pluhar 1993).

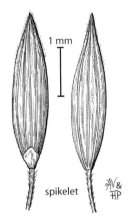

1 mm

spikelet

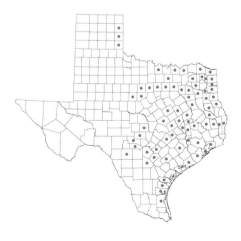

6. *Digitaria filiformis* (L.) Koeler (slender crabgrass, slender fingergrass). Annual or short-lived perennial. Panicle with spikelike primary branches. Upper lemmas dark brown. Found in sandy, open, often disturbed areas in the eastern third of the state. There are 2 varieties found in Texas: the typical one, var. *filiformis*; and var. *villosa* (Walter) Fernald. They can be distinguished by the following characters:

1. Spikelets 1–2 mm long; panicles 3–13 cm long; culms 10–80 cm long .... var. *filiformis*
1. Spikelets 2–3 mm long; panicles 10–25 cm long; culms 75–150 cm long ....var. *villosa*

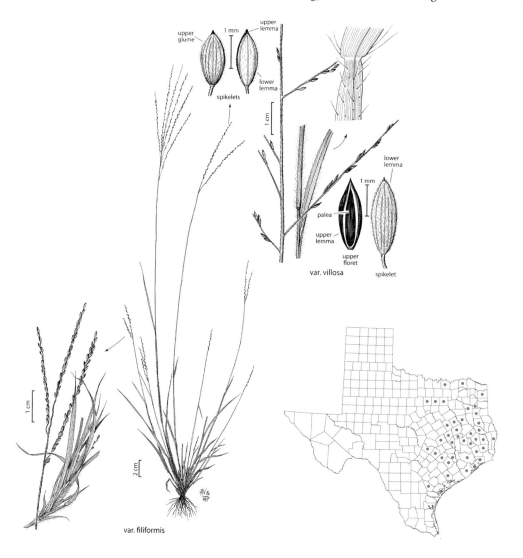

var. filiformis

upper
lfloret

spikelet

7. *Digitaria hitchcockii* (Chase) Stuck.
(shortleaf cottontop). Perennial with short,
knotty rhizomes. Basal sheaths tomentose.
Panicle with 3–6 appressed branches. Upper
lemmas dark brown at maturity. A rare species
of dry, open, rocky slopes of West Texas.

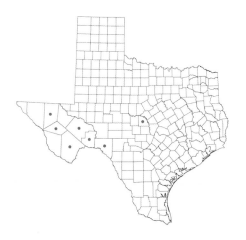

1 mm

palea

lemma

upper floret

upper glume

lower lemma   lower glume

2 cm

2 cm

8. *Digitaria insularis* (L.) Mez *ex* Ekman (sourgrass, sour cottontop). Perennial with knotty base created by short rhizomes. Villous basal sheaths and blades. Panicles with numerous spikelike primary branches. Lower lemmas with long, golden brown hairs. Upper lemmas dark brown at maturity. Frequent along ditches and other low, wet areas. Poor for livestock and wildlife (Hatch, Schuster, and Drawe 1999). Powell (1994) reported that the vernacular name comes from an unpleasant odor, much like rotten lemons, emitted from crushed foliage.

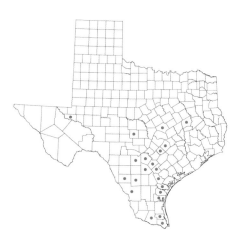

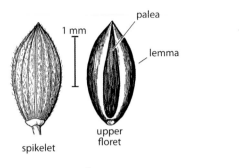

palea

lemma

1 mm

upper floret

spikelet

9. *Digitaria ischaemum* (Schreb.) Muhl. (smooth crabgrass). Annual, generally decumbent and rooting at the lower nodes. Basal sheaths glabrous or slightly pubescent. Panicles with 2–7 spikelike primary branches. Upper lemmas dark brown at maturity. A Eurasian introduction that has become weedy. Found in open woods and along roadsides, lawns, fields, and other disturbed areas.

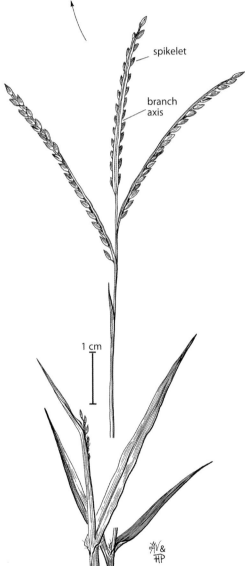

spikelet

branch axis

1 cm

10. *Digitaria milanjiana* (Rendle) Stapf
(perennial crabgrass). Perennial with rhizomes
and/or stolons. Panicles of 2–18 spikelike
primary branches. Upper lemmas gray to
tan at maturity. Not reported from Texas in
Gould (1975b), Turner et al. (2003), or Wipff
(2003d), but listed for the state in Jones,
Wipff, and Montgomery (1997). Perhaps not a
constant member of the Texas flora.

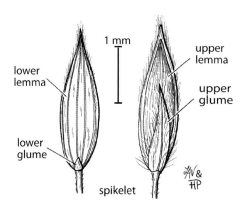

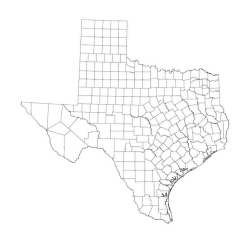

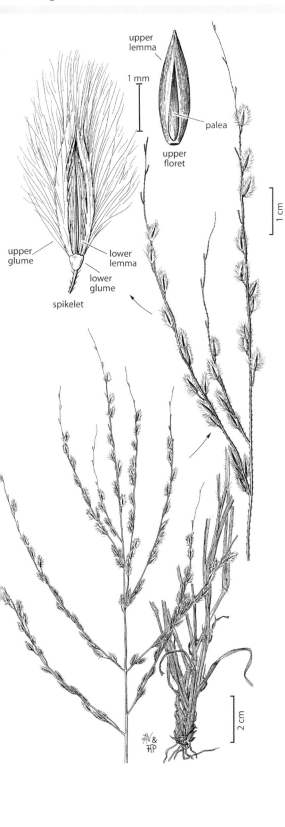

upper lemma

1 mm

palea

upper floret

upper glume

lower lemma

lower glume

spikelet

1 cm

2 cm

11. *Digitaria patens* (Swallen) Henrard (Texas cottontop). Cespitose perennial. Basal sheaths villous. Panicles with 4–10 spikelike primary branches, slightly ascending to divergent at maturity. Lower lemmas densely villous. Upper lemmas dark brown at maturity. Generally found in disturbed sites or on dry, sandy soils. Fair to good forage for livestock but poor for wildlife (Hatch, Schuster, and Drawe 1999).

spikelet

1 mm

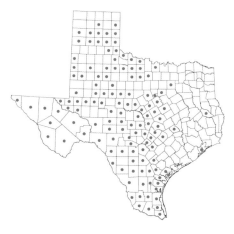

12. *Digitaria pubiflora* (Vasey) Wipff (western witchgrass). Perennial with or without rhizomes. Panicle open, simple, with divergent branches. Upper lemmas dark brown. Found in dry, rocky, or sandy soils over most of the state, but most common in the western and northern parts. Listed as a subspecies (subsp. *pubiflora*) of *D. cognata* in Jones, Wipff, and Montgomery (1997), Turner et al.(2003), and Diggs et al. (2006).

2 cm

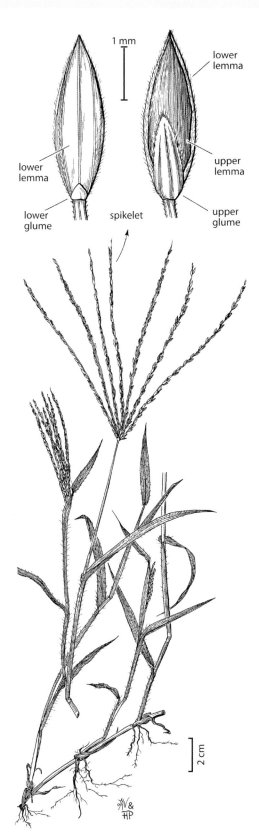

1 mm

lower
lemma

lower
lemma

lower
glume

spikelet

upper
lemma

upper
glume

2 cm

13. *Digitaria sanguinalis* (L.) Scop. (hairy
crabgrass, common crabgrass). Annual with
decumbent culms that root at the nodes.
Sheaths keeled and with papillose-based hairs.
Panicles with 3–15 spikelike primary branches.
Upper lemmas yellow or gray, often with
purple margins when young, becoming brown
at maturity. A European introduction that is a
common weed of waste places and disturbed
sites. Poor for livestock and wildlife (Hatch,
Schuster, and Drawe 1999).

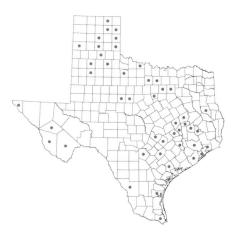

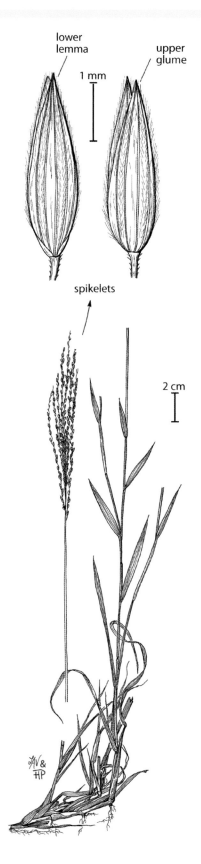

lower
lemma

1 mm

upper
glume

spikelets

2 cm

14. *Digitaria texana* Hitchc. (Texas crabgrass,
Texas fingergrass). Perennial, sometimes
decumbent and rooting at lower nodes. Basal
sheaths villous. Panicles of 5–10 primary
spikelike branches. Upper lemmas purple at
maturity. Found in sandy oak woodlands and
prairies. Listed as a Texas endemic by Diggs et
al. (2006); however, Wipff (2003d) reported
it from southwest Florida as well. The Brazos
County record has been questioned, and it
could be the result of mislabeling (Webster
and Hatch 1990). Not listed as rare by Jones,
Wipff, and Montgomery (1997) or Poole et al.
(2007). Poor to fair forage for livestock and
poor wildlife values (Hatch, Schuster, and
Drawe 1999).

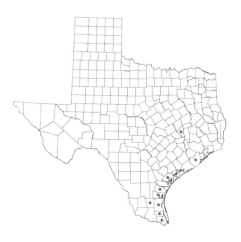

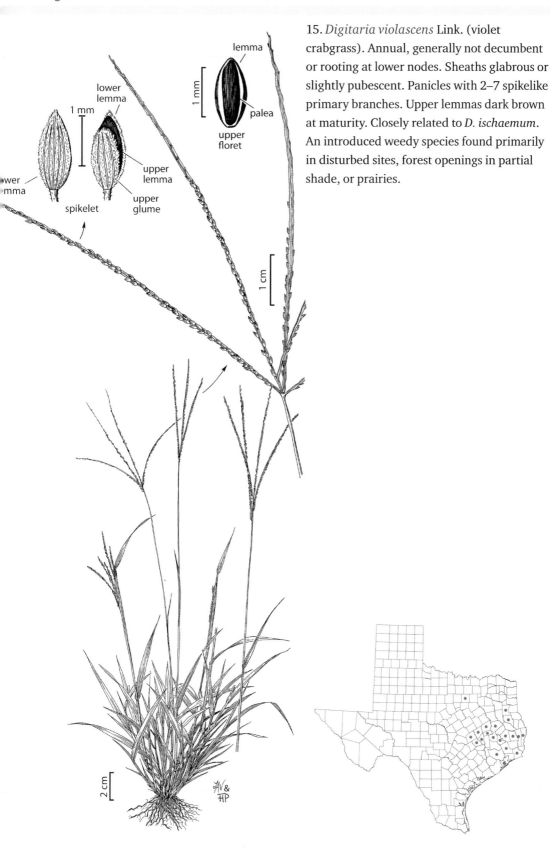

lemma

1 mm

palea

upper
floret

lower
lemma

1 mm

wer
mma

spikelet

upper
lemma

upper
glume

1 cm

2 cm

AV &
HP

15. *Digitaria violascens* Link. (violet crabgrass). Annual, generally not decumbent or rooting at lower nodes. Sheaths glabrous or slightly pubescent. Panicles with 2–7 spikelike primary branches. Upper lemmas dark brown at maturity. Closely related to *D. ischaemum*. An introduced weedy species found primarily in disturbed sites, forest openings in partial shade, or prairies.

Distichlis Raf.
(Chloridoideae: Cynodonteae)

Low, dioecious (rarely monoecious) perennials with stout rhizomes and firm, glabrous culms and leaves. Culms with many nodes and internodes, these usually completely covered by the distichous and closely overlapping leaf sheaths. Ligule a short, fringed membrane. Spikelets large, several-flowered, laterally flattened, in short, contracted panicles or racemes. Disarticulation above the glumes and between the florets. Glumes unequal, acute, 3–5-nerved. Lemmas broad, indistinctly 9–11-nerved, awnless, laterally compressed, those of the pistillate spikelets coriaceous, thicker than those of the staminate spikelets. Palea large, strongly 2-keeled. Basic chromosome number, $x = 10$. Photosynthetic pathway, C_4. Represented in Texas by a single species.

(Barkworth 2003f)

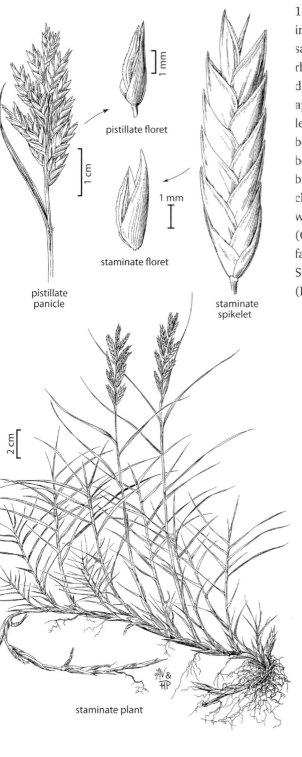

pistillate floret

1 cm

1 mm

staminate floret

pistillate
panicle

staminate
spikelet

staminate plant

1. *Distichlis spicata* (L.) Greene (saltgrass, inland saltgrass, desert saltgrass, coastal saltgrass, alkali grass, spike grass). A highly rhizomatous perennial of saline soils. Plants dioecious; pistillate and staminate individuals appear similar except that the staminate lemmas are thinner and the paleas less bowed. Some accounts have differentiated between maritime and inland populations, but inland plants show a variable series of characteristics resulting from growing in a wide range of edaphic and climatic conditions (Gould 1975b). Listed as poor for wildlife and fair for livestock (Hatch and Pluhar 1993). Some useful forage is produced in saline areas (Powell 1994).

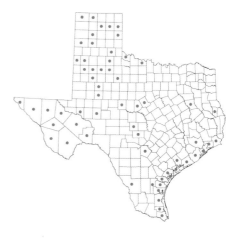

Echinochloa P. Beauv.
(Panicoideae: Paniceae)

Coarse annuals and perennials, usually with weak, succulent culms and broad, flat blades. Ligule a ring of hairs or absent. Inflorescence a contracted or moderately open panicle, with few to numerous, simple or rebranched, densely flowered branches. Spikelets subsessile, in irregular fascicles or regular rows, disarticulating below the glumes. Glumes and lemma of sterile floret usually with stout spicules and long or short hairs, these often glandular-pustulate at the base. Lower glume well developed but much shorter than the upper, acute or slightly awned. Upper glume and lemma of the sterile floret about equal, acute, short-awned or with a long flexuous awn. Lemma of the fertile floret indurate, smooth and shiny, with inrolled margins and usually an abruptly pointed apex. Palea of the fertile floret similar to the lemma in texture, broad but narrowing to a pointed tip that is free from the lemma margins. Basic chromosome number, $x = 9$. Photosynthetic pathway, C_4. Represented in Texas by 7 species and 6 varieties.

A genus of widespread, weedy species; however, Telfair (2006) listed the seeds as especially important for wildlife.

1. Ligule of stiff hairs present on lower leaves .... 6. *E. polystachya*
1. Ligule absent on all leaves.
    2. Spikelets near base of panicle not disarticulating; upper lemmas exposed at maturity .... 4. *E. esculenta*
    2. Spikelets disarticulating at maturity; upper lemma not exposed at maturity.
        3. Panicle branches unbranched, usually 2 cm or shorter .... 1. *E. colona*
        3. Panicle branches rebranched, sometimes inconspicuously so, usually >2 cm long.
            4. Lower sheaths hispid or hirsute .... 7. *E. walteri*
            4. Lower sheaths glabrous.
                5. Upper lemmas broadly ovate or elliptic, sometimes with a line of minute hairs across the base of an early-withering tip.
                    6. Apex of upper lemmas narrowly acute to acuminate, gradually transitioning into a stiff, membranous, usually mucronate tip, minute line of hairs below tip absent .... 5. *E. muricata*
                    6. Apex of upper lemmas rounded or broadly acute, abruptly transitioning into a membranous tip, a line of hairs present at the base of tip .... 2. *E. crus-galli*
                5. Upper lemmas narrowly ovate to elliptical, never with a line of hairs below the tip.
                    7. Spikelets 2.5–3.4 mm long; lower lemmas unawned or with awns 3–15 mm long, curved.. 3. *E. crus-pavonis*
                    7. Spikelets 3–5 mm long; lower lemmas usually awned, 8–25 mm long, straight .... 7. *E. walteri*

(Gould 1975b; Michael 2003)

Note: Reports of *E. paludigena* Wiegand and *E. frumentacea* Link. in Texas are apparently based on misidentifications.

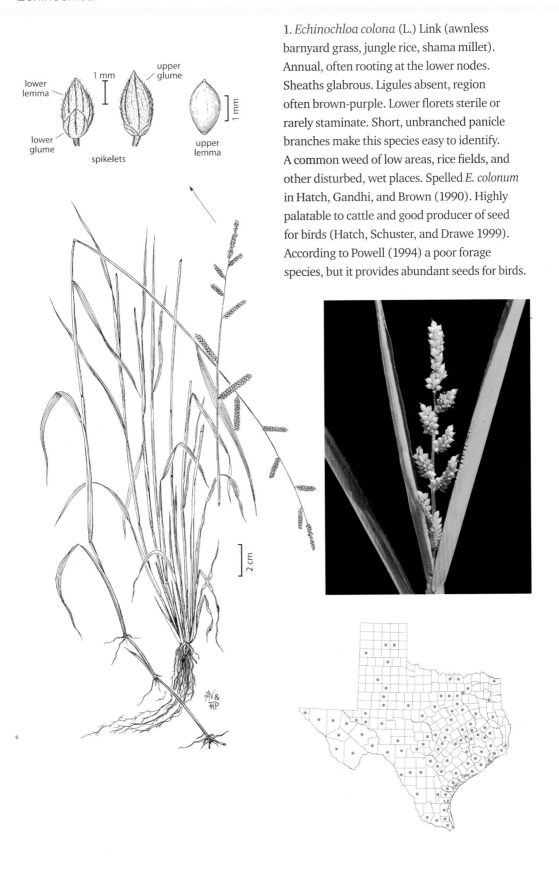

lower
lemma

1 mm

upper
glume

lower
glume

1 mm

spikelets

upper
lemma

2 cm

JAV &
HP

1. *Echinochloa colona* (L.) Link (awnless barnyard grass, jungle rice, shama millet). Annual, often rooting at the lower nodes. Sheaths glabrous. Ligules absent, region often brown-purple. Lower florets sterile or rarely staminate. Short, unbranched panicle branches make this species easy to identify. A common weed of low areas, rice fields, and other disturbed, wet places. Spelled *E. colonum* in Hatch, Gandhi, and Brown (1990). Highly palatable to cattle and good producer of seed for birds (Hatch, Schuster, and Drawe 1999). According to Powell (1994) a poor forage species, but it provides abundant seeds for birds.

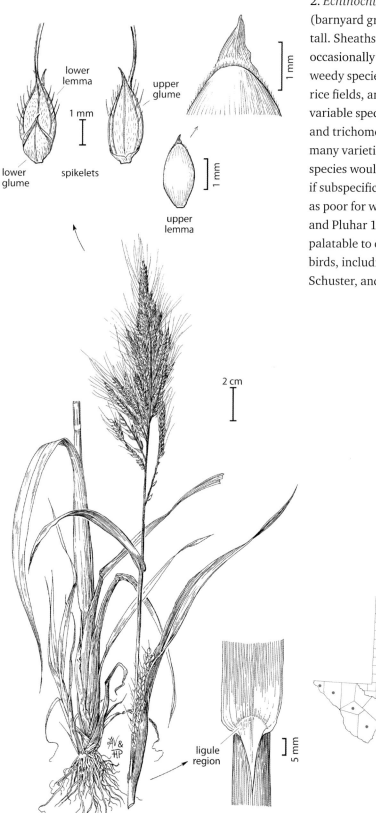

lower lemma

1 mm

lower glume

spikelets

upper glume

1 mm

upper lemma

1 mm

2 cm

ligule region

5 mm

2. *Echinochloa crus-galli* (L.) P. Beauv. (barnyard grass). Annual with culms to 2 m tall. Sheaths glabrous. Ligules absent, region occasionally pubescent. Lower florets sterile. A weedy species of disturbed sites, waste places, rice fields, and other wet places. Extremely variable species in awn length, pubescence, and trichome abundance and length, and many varieties have been named. The Texas species would belong to the typical variety if subspecific taxa were recognized. Listed as poor for wildlife and livestock (Hatch and Pluhar 1993). However, listed as highly palatable to cattle and important for attracting birds, including game birds and ducks (Hatch, Schuster, and Drawe 1999).

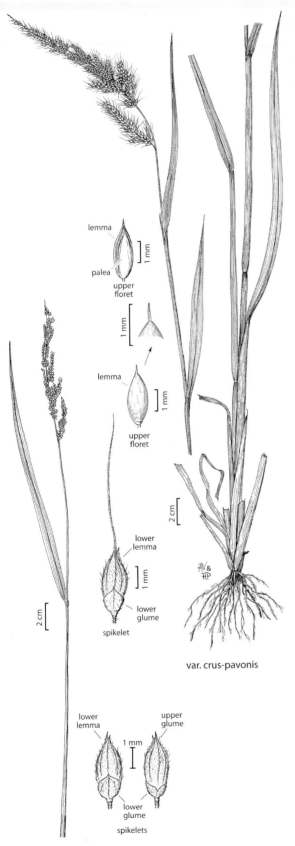

spikelet

var. crus-pavonis

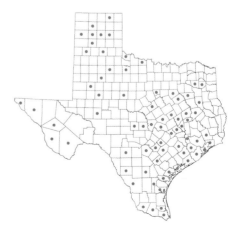

var. macra

3. *Echinochloa crus-pavonis* (Kunth.) Schult. (Gulf Coast barnyardgrass, Gulf cockspur grass). Annual or occasionally short-lived perennial. Culms to 1.5 m tall. Nodes glabrous. Sheaths glabrous, often purplish. Ligules absent. Lower florets sterile. Found in ditches, disturbed areas, marshes, and standing water. According to Gould (1975b) probably the most common and widespread *Echinochloa* in the state. Listed as highly palatable to cattle, and are important for attracting birds, including game birds and ducks (Hatch, Schuster, and Drawe 1999). There are 2 varieties: var. *crus-pavonis*; and var. *macra* (Wiegand) Gould. The former is not listed in Gould (1975b). Gould and Jones, Wipff, and Montgomery (1997) both spell the latter as *macera*, and it is the most common variety in the state. The varieties can be distinguished by the following characters:

1. Panicles drooping; palea of lower florets well developed...................var. *crus-pavonis*
1. Panicles erect; palea of lower florets absent or rudimentary ....var. *macra*

4. *Echinochloa esculenta* (A. Braun)
H. Scholtz (Japanese millet). Annual with
culms to 2 m tall. Sheaths glabrous. Ligules
absent, regions sometimes pubescent. Lower
florets sterile. A cultivated species used for
fodder, grain, or birdseed. The Texas record
was probably introduced from birdseed.
Jones, Wipff, and Montgomery (1997) listed
this as *E. frumentacea* Link, a closely related
cultivated plant from India. Perhaps not a
constant member of the Texas flora.

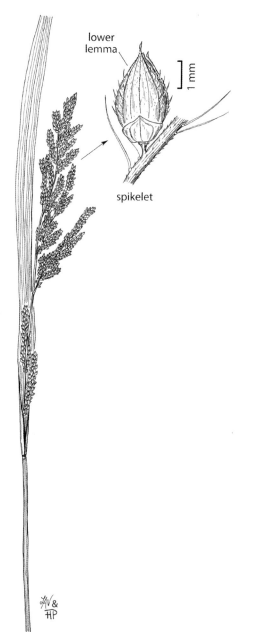

lower
lemma

1 mm

spikelet

5. *Echinochloa muricata* (P. Beauv.) Fernald (American barnyard-grass, cockspur, rough barnyardgrass). Annual with culms to 1.5 m long, rooting at the lower nodes. Sheaths glabrous. Ligules absent. Lower florets sterile. Found in wet areas on both disturbed and undisturbed sites; typically not a weed of rice fields. Listed as highly palatable to cattle, and are important for attracting birds, including game birds and ducks (Hatch, Schuster, and Drawe 1999). Two varieties occur in Texas: var. *microstachya* Wiegand; and var. *muricata*. They can be distinguished by the following characters:

1. Spikelets <3.8 mm long; lower lemmas if awned, awn <10 mm long ....var. *microstachya*
1. Spikelets >3.5 mm long; lower lemmas with awns 6–16 mm long .... var. *muricata*

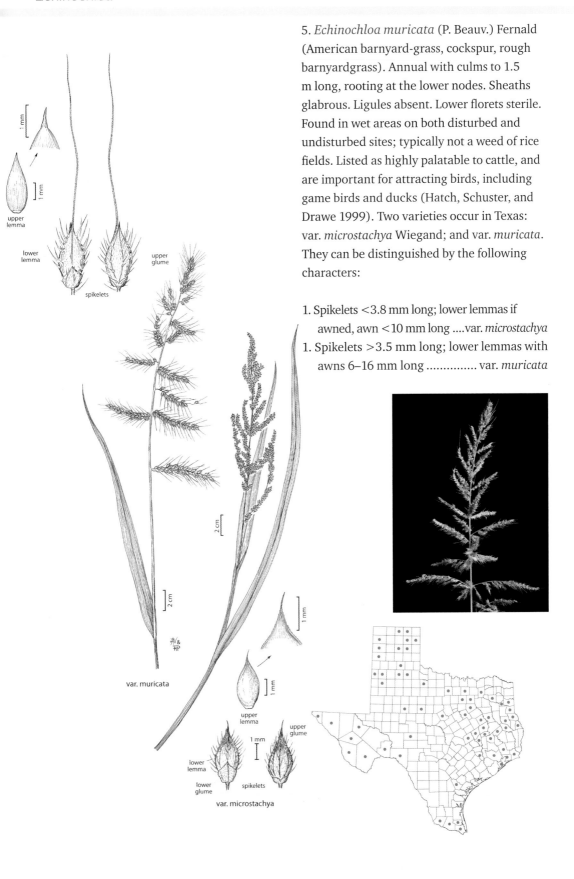

6. *Echinochloa polystachya* (Kunth.) Hitchc. (creeping river grass). Perennial with culms up to 2 m long and 1 cm thick. Ligule present on lower leaves. Lower florets staminate. Found in coastal marshes, ditches, and low places with standing water. Listed as highly palatable to cattle, and important for attracting birds, including game birds and ducks (Hatch, Schuster, and Drawe 1999). Two varieties occur in Texas: var. *polystachya*; and var. *spectabilis* (Nees *ex* Trin.) Mart. Crov. They can be distinguished by the following character:

1. Culms and leaf sheaths glabrous ....var. *polystachya*
1. Culm nodes and leaf sheaths pubescent .... var. *spectabilis*

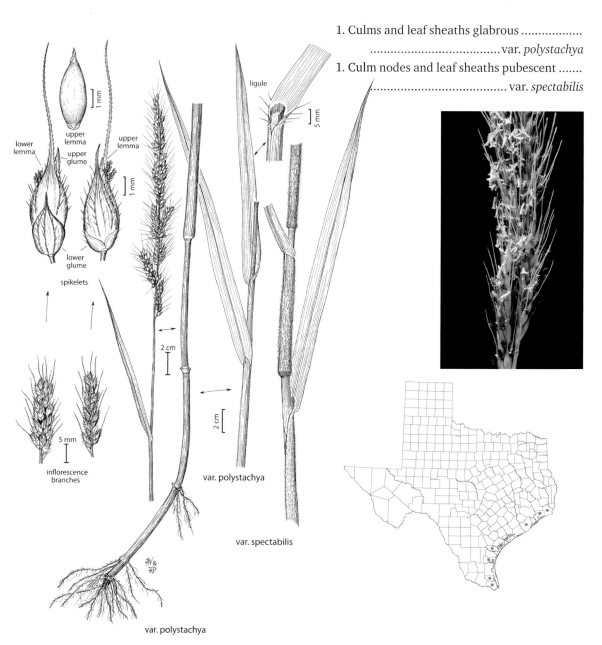

upper
lemma

lower
lemma
upper
lemma
upper
glume

1 mm

1 mm

lower
glume

spikelets

ligule

5 mm

2 cm

2 cm

5 mm

inflorescence
branches

var. polystachya

var. spectabilis

var. polystachya

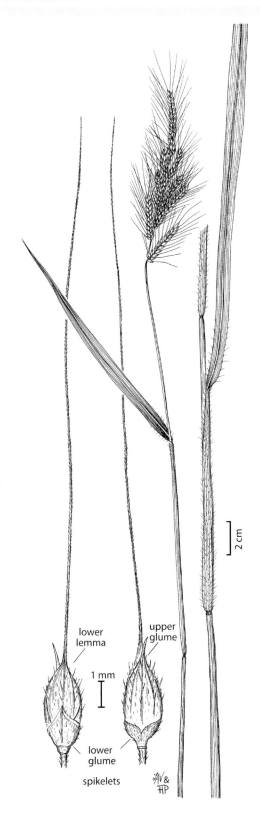

7. *Echinochloa walteri* (Pursh) A. Heller
(coast barnyardgrass, long-awn cockspur,
coast cockspur). Annual with culms up to 2 m
long and 2.5 cm thick. Nodes pilose to villous,
upper sometimes glabrous. Ligules absent.
Sheaths usually hispid. Lower florets sterile.
Generally found in wet areas such as ditches,
standing water, and brackish marshes. Listed
as highly palatable to cattle, and important
for attracting birds, including game birds and
ducks (Hatch, Schuster, and Drawe 1999).

lower
lemma

upper
glume

1 mm

lower
glume

spikelets

2 cm

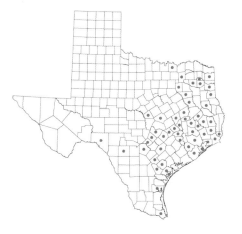

Ehrharta Thunb.
(Ehrhartoideae: Ehrharteae)

Perennial, in Texas species, cespitose and occasionally with rhizomes. Culms up to 2 m tall, erect, glabrous. Auricles present, ciliate. Ligules membranous, lacerated, glabrous. Blades flat or involute, surfaces glabrous, margins hairy and wavy; sometimes disarticulating from the sheath. Inflorescence a raceme or panicle, branches usually spreading. Disarticulation above the glumes but not between the florets. Spikelets single, pediceled, terete or laterally compressed, with 3 florets, lower 2 sterile, distal floret fertile. Glumes 5–7-veined. Sterile florets consisting only of indurate lemmas. Perfect lemmas indurate, 5–7-veined, keeled, unawned. Basic chromosome number, $x = 12$. Photosynthetic pathway, C_3. Represented in Texas by a single species.

(Barkworth 2007b)

sterile   fertile floret

sterile

2 mm

florets

spikelet

1 cm

1. *Ehrharta calycina* Sm. (perennial veldtgrass). A large cespitose perennial, occasionally with scaly rhizomes. Spikelets often purplish and with 2 sterile florets below the fertile one. A species from Africa introduced around the world as a potential forage species, but it does not withstand heavy grazing well. Not collected since 1950, and perhaps not a constant member of the Texas flora.

Eleusine Gaertn.
(Chloridoideae: Cynodonteae)

Low, spreading annuals with thick, weak culms and soft, flat or folded, succulent leaf blades. Ligule a short, lacerate membrane. Spikelet sessile in 2 rows on 2 (occasionally 1) to several branches digitately arranged at the culm apex. One or 2 branches frequently developed below the apical cluster. Spikelets 3- to several-flowered, disarticulating above the glumes and between the florets. Glumes firm, acute, unequal, the lower short, 1-nerved, the upper 3–7-nerved. Lemmas acute, awnless or mucronate, broadly keeled, 3-nerved, the lateral veins very close to the midnerve. Palea shorter than the lemma. Grain plump, with minutely transversely rugose seed loosely enclosed in a thin pericarp. Basic chromosome number, $x = 9$. Photosynthetic pathway, C_4. Represented in Texas by 2 species.

1. Panicles with 1–3 branches, attached in a single digitate cluster......................2. *E. tristachya*
1. Panicle with 4–20 branches, 1, or rarely 2, branches attached below the terminal, digitate
    cluster...1. *E. indica*

(Hilu 2003)

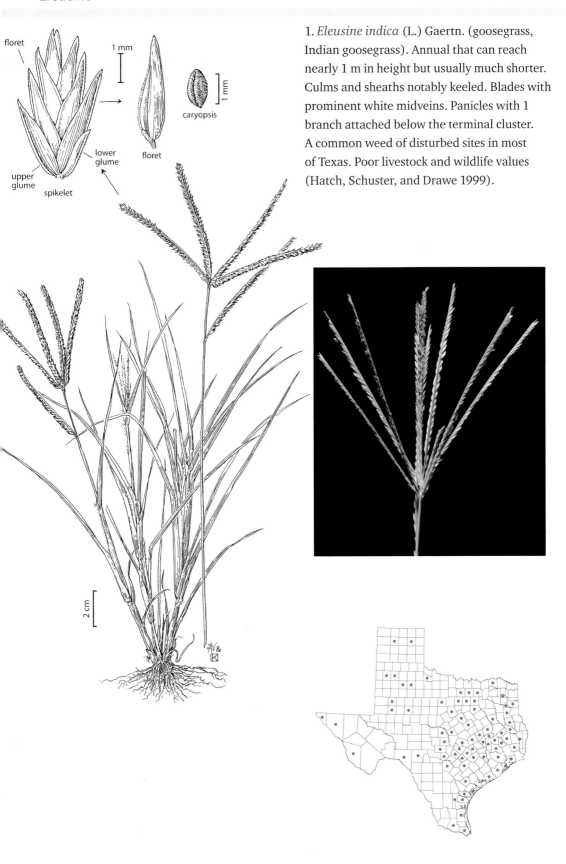

floret

1 mm

caryopsis

1 mm

floret

lower glume

upper glume

spikelet

2 cm

1. *Eleusine indica* (L.) Gaertn. (goosegrass, Indian goosegrass). Annual that can reach nearly 1 m in height but usually much shorter. Culms and sheaths notably keeled. Blades with prominent white midveins. Panicles with 1 branch attached below the terminal cluster. A common weed of disturbed sites in most of Texas. Poor livestock and wildlife values (Hatch, Schuster, and Drawe 1999).

2. *Eleusine tristachya* (Lam.) Lam. (threespike goosegrass). Annual with compressed culms <0.5 m long. Panicles with 1–3 branches terminal and no branch(es) below. A tropical American introduction from ballast dumps. Specimen from Brazos County is from university research area and probably does not persist (Diggs et al. 2006). Species is probably not a constant member of the Texas flora. Not listed in Gould (1975b); Hatch, Gandhi, and Brown (1990); or Jones, Wipff, and Montgomery (1997).

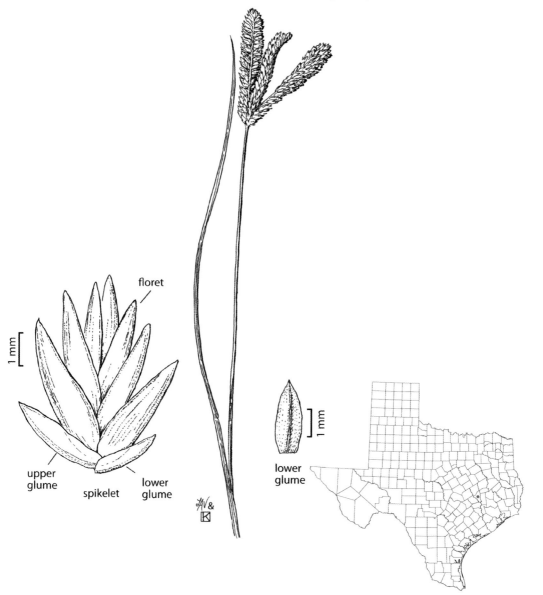

floret

1 mm

upper glume

spikelet

lower glume

lower glume

1 mm

Elionurus Humb. & Bonpl. *ex* Willd.
(Panicoideae: Andropogoneae)

Cespitose perennials with slender, moderately tall culms and narrow, flat or involute blades. Inflorescence a cylindric, spikelike raceme, with disarticulation in the rachis. Spikelets awnless, paired, the pediceled spikelets staminate, similar in size and appearance to the sessile, perfect ones. First glume of the perfect spikelet firm and moderately coriaceous, the second glume thinner. Lemmas of the sterile and fertile florets thin and hyaline, the paleas absent. Basic chromosome number, $x = 10$. Photosynthetic pathway, C_4. Sometimes spelled *Elyonurus* (Gould 1975b). Represented in Texas by 2 species.

1. Lower glumes densely pilose; culms hirsute below the nodes ....1. *E. barbiculmis*
1. Lower glumes glabrous or nearly so; culms glabrous throughout.................. 2. *E. tripsacoides*

(Barkworth 2003g)

1. *Elionurus barbiculmis* Hack. (woolspike
balsamscale). Cespitose perennial with
densely hirsute culms immediately below the
nodes. Rame internodes villous. Lower glumes
densely hirsute. Found on mesas, rocky slopes,
and canyons, usually at elevations of about
1,200 ft (365 m). Palatable to livestock but not
abundant (Powell 1994).

sessile spikelet

pedicellate
spikelets

3 mm

sessile
spikelet

sessile
spikelet

1 cm

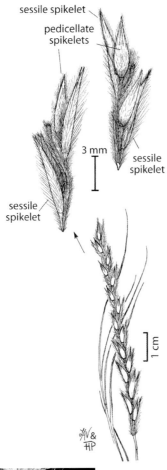

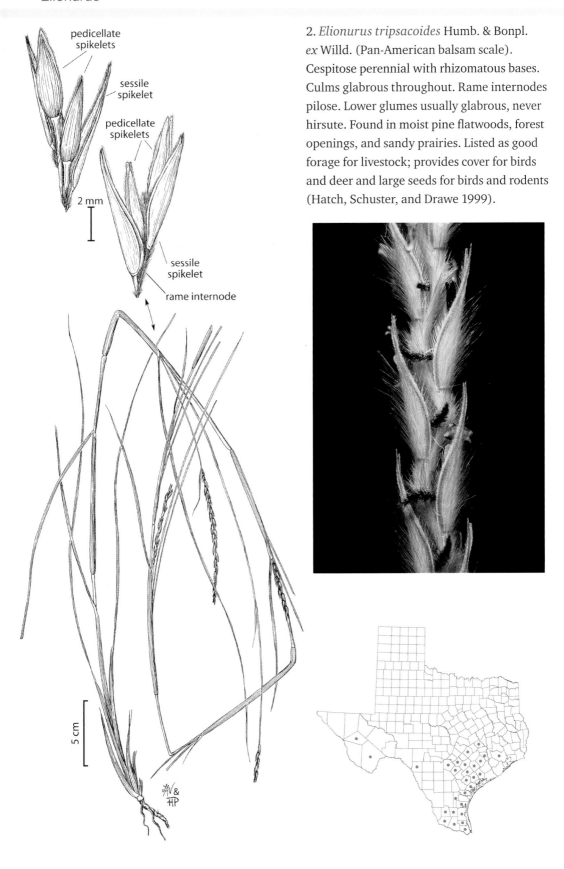

pedicellate
spikelets

sessile
spikelet

pedicellate
spikelets

2 mm

sessile
spikelet

rame internode

5 cm

2. *Elionurus tripsacoides* Humb. & Bonpl.
ex Willd. (Pan-American balsam scale).
Cespitose perennial with rhizomatous bases.
Culms glabrous throughout. Rame internodes
pilose. Lower glumes usually glabrous, never
hirsute. Found in moist pine flatwoods, forest
openings, and sandy prairies. Listed as good
forage for livestock; provides cover for birds
and deer and large seeds for birds and rodents
(Hatch, Schuster, and Drawe 1999).

Elymus L.
(Poöideae: Triticeae)

Perennials, occasionally with rhizomes. Inflorescence a bilateral spike or spicate raceme. Spikelets 1–4 per node, generally ascending, with 2–7 florets. Glumes lanceolate to subulate, rarely lacking. Lemmas longer than the glumes, rounded on the back, acute to long-awned at the tip, awn straight or curved outward. Palea well developed. Basic chromosome number, $x = 7$. Photosynthetic pathway, C_3. Species once included in *Agropyron* and *Sitanion* are now included in *Elymus*. Represented in Texas by 14 species, 1 subspecies, and 5 varieties.

1. Spikelets 1 at all or most nodes; glumes with flat, nonindurate bases.
  2. Anthers 3–7 mm long.
    3. Lemmas, at least some, with strongly divergent awns .... 1. *E. arizonicus*
    3. Lemmas awnless, or if awned, awns straight or flexuous .... 9. *E. repens*
  2. Anthers 0.7–3.0 mm long .... 12. *E. trachycaulus*
1. Spikelets 2–5 at all or most nodes; glumes usually with terete, indurate bases.
    4. Rachises disarticulating at maturity .... 4. *E. elymoides*
    4. Rachises not disarticulating at maturity.
      5. Glume bodies with 0–1(–2) veins, linear or tapering from the base, 0.1–0.6 mm wide.
        6. Spikelets (6–)9–15(–22) mm long excluding awns, 2–5 florets per spikelet; lemma awns outcurved at maturity .... 6. *E. interruptus*
        6. Spikelets 18–40 mm long excluding awns, 3–8 florets per spikelet; lemma awns straight or slightly curved at maturity.
          7. Anthers 2.5–4.0 mm long; lemmas hirsute, scabrous or thinly strigose; spike internodes without green lateral bands .... 8. *E. pringlei*
          7. Anthers 4.5–6.0 mm long; lemmas glabrous, smooth; spike internodes with green lateral bands .... 11. *E. texensis*
      5. Glume bodies with 2–5(–8) veins, widening or parallel above the base, 0.2–2.3 mm wide.
        8. Glume bases flat and veined, or if terete, then indurate and without veins for <1 mm.
          9. Spikes erect to slightly nodding, internodes 5–14 mm long; nodes usually exposed .... 6. *E. interruptus*
          9. Spikes nodding to pendent; internodes 2–12 mm long; nodes usually concealed .... 2. *E. canadensis*
        8. Glume bases terete, indurate, without veins for 0.5–4.0 mm
      10. Glumes persistent; spikes nodding.
        11. Blade adaxial surface densely villous, dark glossy green .... 13. *E. villosus*
        11. Blade adaxial surface glabrous to scabrous, dull green .... 10. *E. riparius*
      10. Glumes disarticulating; spikes erect.
        12. Spikes <2.5 cm wide including awns; glume awns 0–10(–15) cm long.
          13. Lemma awns 5–20 mm long at midspike; lower leaf blades not markedly smaller or less persistent than upper blades .... 14. *E. virginicus*

13. Lemma awns 0.5–4.0 mm long at midspike; lower leaf blades shorter, narrower, and senescing earlier than those of the upper ....
....3. *E. curvatus*

12. Spikes >2.2 cm wide including awns; glume awns (10–)15–30 mm ..... long.

14. Internodes 4–7 mm long; blades usually dark glossy green under the glaucous bloom, auricles 2–3 mm long .... 7. *E. macgregorii*

14. Internodes 3–5 mm long; blades usually dull green, with or .... without glaucous bloom; auricles 0–2 mm long .... 5. *E. glabriflorus*

(Barkworth, Campbell, and Salomon 2007)

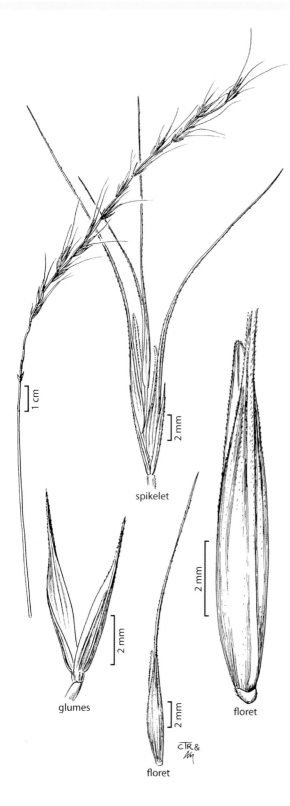

1. *Elymus arizonicus* (Scrib. & J. G. Sm.)
Gould (Arizona wheatgrass). Cespitose
perennial with erect culms to 1 m long. Nodes
exposed. Auricles 0–1 mm long. Ligules
to 1 mm long on basal leaves. Distinctive
drooping spike and solitary spikelets. Grows
in canyons and on rocky slopes. Jones,
Wipff ,and Montgomery (1997) listed this as
Pseudoroegneria arizonica (Scrib. & J. G. Sm.)
Á. Löve. Hatch, Gandhi, and Brown (1990)
listed it as *Elytrigia arizonica* (Scrib. & J. G.
Sm.) D. R. Dewey.

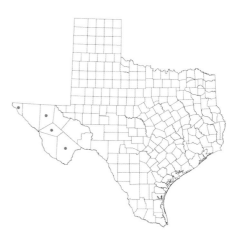

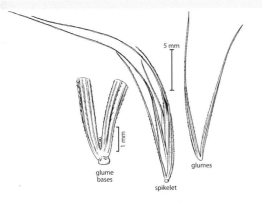

var. brachystachys

glume bases

glumes

spikelet

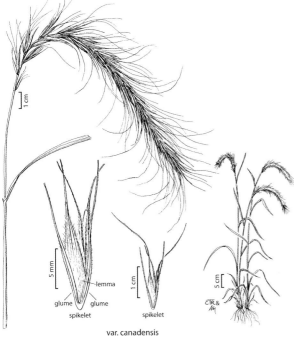

lemma

glume glume

spikelet

spikelet

var. canadensis

2. *Elymus canadensis* L. (Great Plains wildrye, Canada wildrye, nodding wildrye). Plants perennial, rarely with short rhizomes, usually cespitose, often glaucous. Culms to 1.8 m long, erect. Nodes usually concealed by sheaths. Auricles 1.5–4.0 mm long, brown or purplish-black. Ligules to 2 mm long. Anthesis begins in May. Common species found on sandy to gravelly soils in grasslands, ditches, and roadsides; in thickets and woodlands near streams; and on stream banks. Listed as fair for wildlife and good for livestock (Hatch and Pluhar 1993). Two varieties occur in Texas: var. *brachystachys* (Scribn. & C. R. Ball) Farw.; and the typical one, var. *canadensis*. They can be distinguished by the following characters:

1. Lemmas usually villous or hispid; spikes nodding or pendent; internodes often strongly glaucous .... var. *canadensis*
1. Lemmas usually glabrous or sometimes hirsute; spikes nodding or sometimes erect; internodes not strongly glaucous ....var. *brachystachys*

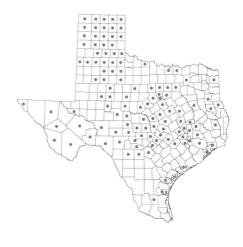

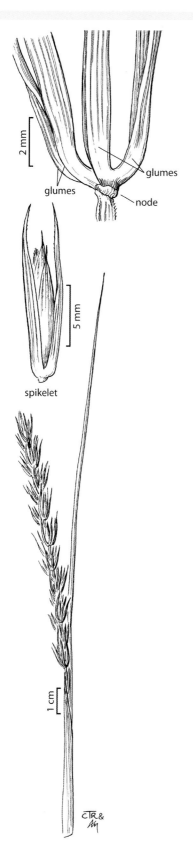

2 mm

glumes
glumes
node

5 mm

spikelet

1 cm

CTR&
AM

3. *Elymus curvatus* Piper (awnless wildrye).
Cespitose perennial with erect, often glaucous
culms. Culms to 1.1 m long. Auricles 0–1 mm
long. Ligules <1 mm long. Anthesis begins
in late June. Found on damp soils and in
bottomlands, open forests, grasslands, and
disturbed sites. Reported only from Cottle
County.

4. *Elymus elymoides* (Raf.) Swezey (squirreltail). Cespitose, often glaucous, perennial. Culms to only 65 cm long, erect or slightly decumbent. Nodes mostly concealed. Auricles 0–1 mm long, purplish. Ligules <1 mm long. Anthesis beginning in late May. A common and widespread species mostly west of the 100th meridian in Texas. Grows generally in dry, open, rocky grasslands, woods, thickets, and disturbed sites. The Texas species belongs to the subsp. *brevifolius* (J. M. Sm.) Barkworth.

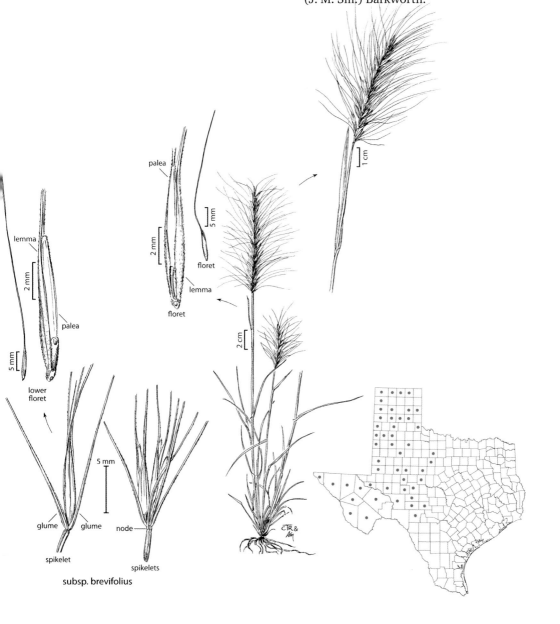

subsp. brevifolius

5. *Elymus glabriflorus* (Vasey *ex* L. H. Dewey) Scribn. & C. R. Ball (southeastern wildrye). Cespitose perennial with usually erect culms to 1.4 m long. Nodes mostly concealed. Ligules <1 mm long. Auricles 0–2 mm long, usually purplish-brown. Anthesis begins in mid-June. Found on moist to dry soils in open woods, thickets, grasslands, roadsides, and old fields. Diggs et al. (2006) included it in *E. virginicus*.

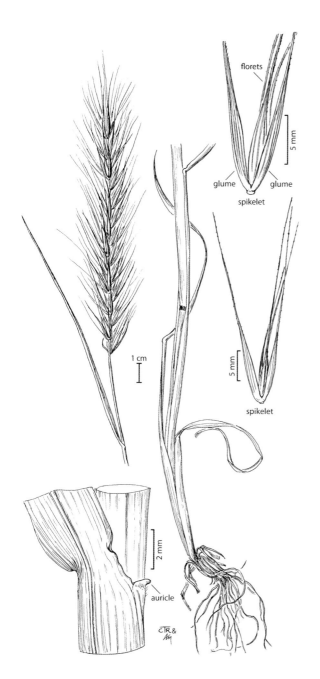

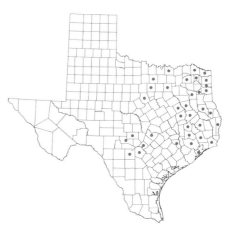

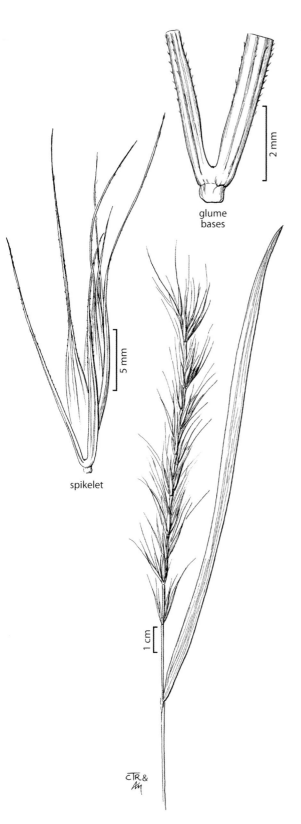

glume
bases

2 mm

5 mm

spikelet

1 cm

CTR &
AM

6. *Elymus interruptus* Buckley (southwestern wildrye). Cespitose and sometimes glaucous perennial. Culms erect or slightly decumbent to 1.2 m long. Nodes usually exposed. Auricles 0–2 mm long, pale or reddish-brown. Ligules to 1 mm long. Anthesis begins in May. Found in rocky soils in canyons, woodlands, and thickets.

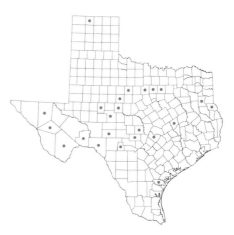

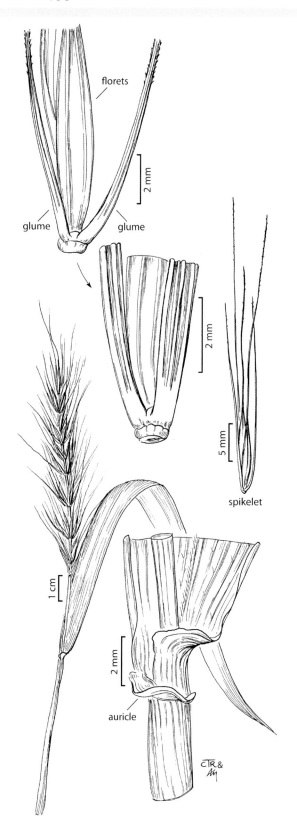

florets

2 mm

glume   glume

spikelet

1 cm

2 mm

auricle

7. *Elymus macgregorii* R. Brooks & J. J. N.
Campb. (early wildrye). Cespitose perennial
with usually erect culms to 1.2 m long. Nodes
mostly exposed. Auricles 2–3 mm long, usually
purplish-black, sometimes light brown. Ligules
<1 mm long. Anthesis usually begins in mid-
May. Found in moist, deep soils in woods and
thickets; most abundant east of the 100th
meridian. Diggs et al. (2006) included it
within *E. virginicus*.

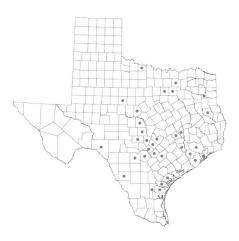

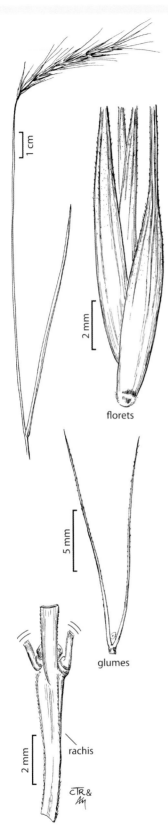

florets

glumes

rachis

8. *Elymus pringlei* Scribn. & Merr. (Mexican wildrye). Cespitose perennial with erect culms to 1.1 m long. Nodes usually exposed. Auricles about 1 mm long, pale or brownish. Ligules about 1 mm long. Anthesis usually begins in May. Found in canyons and on hillsides and in woodlands at elevations of 1,500–2,300 ft (450–700 m). Not yet collected in Texas, but it has been found across the border in Mexico near Big Bend National Park. It is expected to occur on the Texas side in that area.

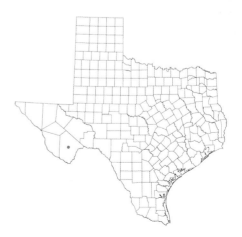

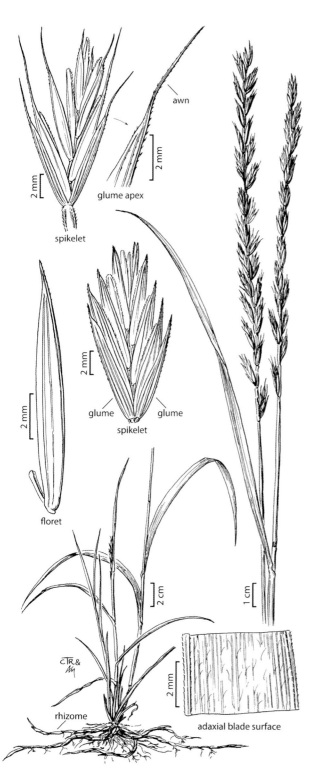

awn

2 mm

glume apex

2 mm

spikelet

2 mm

2 mm

glume   glume

spikelet

2 mm

floret

2 cm

1 cm

CTR & M

2 mm

rhizome

adaxial blade surface

9. *Elymus repens* (L.) Gould (couchgrass, quackgrass). Strongly rhizomatous perennial with culms to 1 m long. Auricles 0.3–1.0 mm long. Ligules 0.2–1.5 mm long. An introduced species that grows well in disturbed sites and spreads by the rhizomes as well as seed. Considered to be a good forage species. Reported only from Bandera County. Diggs et al. (2006) tentatively reported the species, based on National Park Service citations, from the Big Thicket National Preserve.

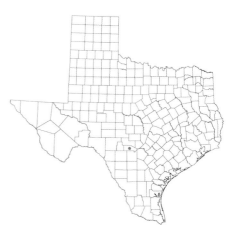

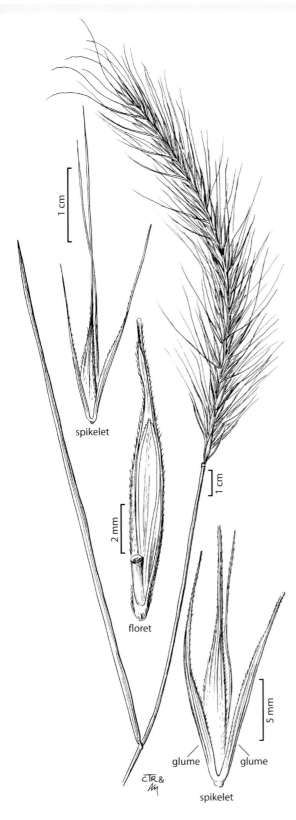

spikelet

floret

1 cm

2 mm

1 cm

glume ╲   ╱ glume

5 mm

spikelet

CTR &
AM

10. *Elymus riparius* Wiegand (eastern riverbank wildrye). Cespitose, often glaucous perennial. Culms to 1.6 m long, erect or rooting at the lower nodes. Auricles 0–2 mm long. Ligules <1 mm long. Anthesis begins in late June. Usually grows along streams and ditches on sandy, alluvial soils. The report from Uvalde County is far west of the normal distribution of the species.

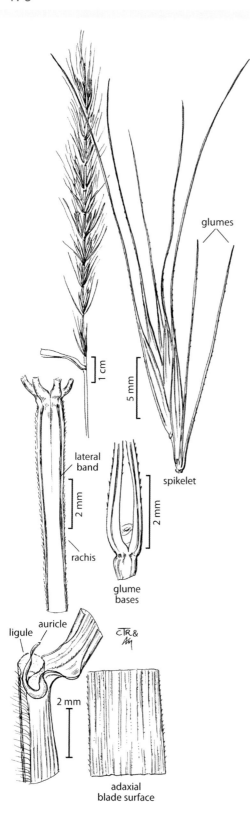

11. *Elymus texensis* J. J. N. Campb. (Texas wildrye). Cespitose, often glaucous perennial. Culms to 1.1 m long, erect. Nodes mostly exposed. Auricles to about 2 mm long, pale to purplish-brown. Ligules 1–2 mm long. Anthesis in May. Known from calcareous bluffs and hills in juniper woods and grassy areas of the Edwards Plateau. Collected in Uvalde, Gillespie, and Burnet counties. Not listed in Gould (1975b); Hatch, Gandhi, and Brown (1990); Jones, Wipff, and Montgomery (1997); or Turner et al. (2003).

12. *Elymus trachycaulus* (Link.) Gould.
(slender wheatgrass). Perennial, sometimes
with rhizomes. Culms to 1.5 m long, erect.
Nodes usually exposed. Auricles absent or
to 1 mm long. Ligules 0.2–0.8 mm long. A
widespread and common species, but rare in
Texas. The Texas species belongs to the typical
variety, var. *trachycaulus*.

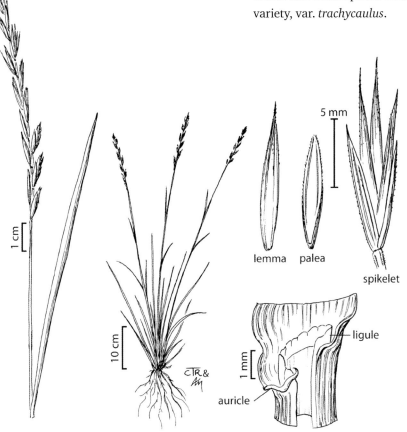

lemma   palea

spikelet

ligule

auricle

subsp. trachycaulus

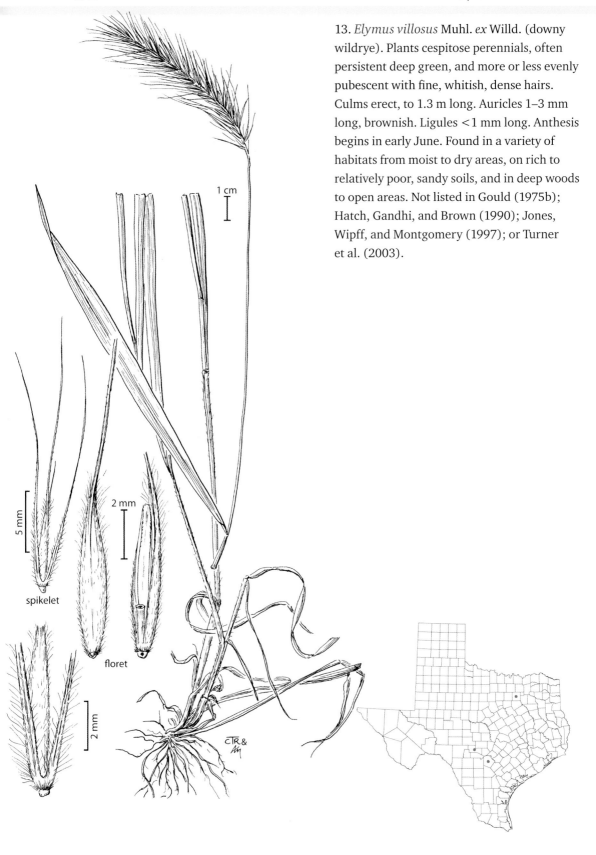

13. *Elymus villosus* Muhl. *ex* Willd. (downy wildrye). Plants cespitose perennials, often persistent deep green, and more or less evenly pubescent with fine, whitish, dense hairs. Culms erect, to 1.3 m long. Auricles 1–3 mm long, brownish. Ligules <1 mm long. Anthesis begins in early June. Found in a variety of habitats from moist to dry areas, on rich to relatively poor, sandy soils, and in deep woods to open areas. Not listed in Gould (1975b); Hatch, Gandhi, and Brown (1990); Jones, Wipff, and Montgomery (1997); or Turner et al. (2003).

1 cm

5 mm

2 mm

spikelet

floret

2 mm

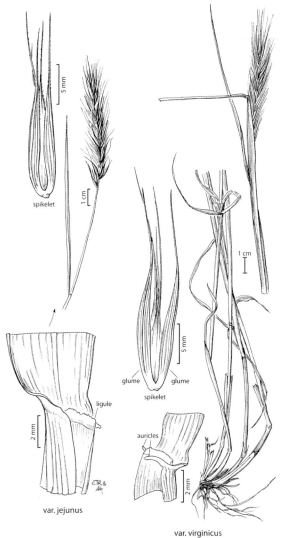

spikelet

5 mm

1 cm

ligule

2 mm

var. jejunus

glume   glume

spikelet

5 mm

auricles

2 mm

var. virginicus

1 cm

14. *Elymus virginicus* L. (Virginia wildrye). Cespitose perennial with erect culms to 1.3 m long. Nodes concealed or exposed. Auricles absent or to 1.8 mm long, brown. Ligules <1 mm long. Anthesis usually begins in mid-June. Common species in fertile soils, grasslands, woodlands, and disturbed areas. Hatch and Pluhar (1993) listed it as fair for wildlife and good for livestock. Two varieties are found in Texas: var. *jejunus* (Ramaley) Bush; and the typical one, var. *virginicus*. They can be distinguished by the following characters:

1. Spikes partially sheathed; glumes 1.0–2.3 mm wide, bowed in the basal 2–4 mm; plants not glaucous, yellowish-brown or purplish at maturity .... var. *virginicus*
1. Spikes mostly exserted; glumes 0.5–1.8 mm wide, bowed in the basal 1–2 mm; plants usually glaucous, yellowish or reddish-brown at maturity .... var. *jejunus*

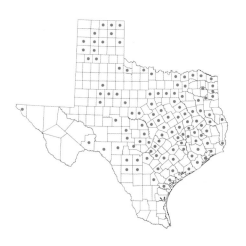

Enneapogon Desv. *ex* P. Beauv.
(Chloridoideae: Pappophoreae)

Low, tufted perennials with narrow, often spikelike panicles. Ligule a ring of hairs. Spikelets several-flowered, the upper florets reduced. Disarticulation above the glumes and tardily between the florets. Glumes subequal, lanceolate, with 5 to numerous veins. Lemmas broad, much shorter than the glumes, firm, rounded on the back, strongly 9-nerved and with 9 equal, plumose awns. Palea slightly longer than the body of the lemma. Basic chromosome number, $x = 10$. Photosynthetic pathway, C_4. Represented in Texas by a single species.

(Reeder 2003b)

1. *Enneapogon desvauxii* P. Beauv. (nineawn pappusgrass). Perennial, generally <50 cm long. Sheaths and blades pubescent. Panicles spikelike and grayish or lead-colored. Only the lowest florets perfect. Lemmas with 9 veins, each terminating in an awn up to 4 mm long. Cleistogamous spikelets in lower sheaths, typically larger than the aerial spikelets, awnless or with very short awns. Infrequently found on dry, rocky slopes. Not an important forage species (Powell 1994).

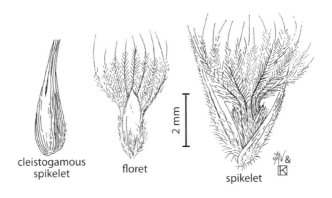

cleistogamous
spikelet   floret   spikelet

2 mm

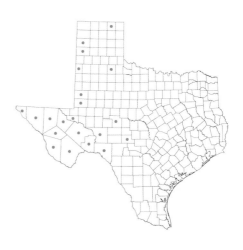

Enteropogon Nees
(Chloridoideae: Cynodonteae)

Plants perennial. Texas species has rhizomes. Culms erect, up to 1 m tall. Sheaths open, ligules membranous, and blades flat. Inflorescence a panicle, 3–30 nondisarticulating branches; branches digitate or racemose. Spikelets single, imbricate, 2–6 florets, lower florets bisexual and fertile, remaining florets progressively reduced, usually the uppermost rudimentary or sterile. Glumes shorter than the lower lemmas. Lemmas stiff, 3-veined, ridged over the veins, apices acute or bidentate, usually awned from between the teeth. Basic chromosome number, $x = 10$. Photosynthetic pathway, C_4. Represented in Texas by a single species.

(Barkworth 2003h)

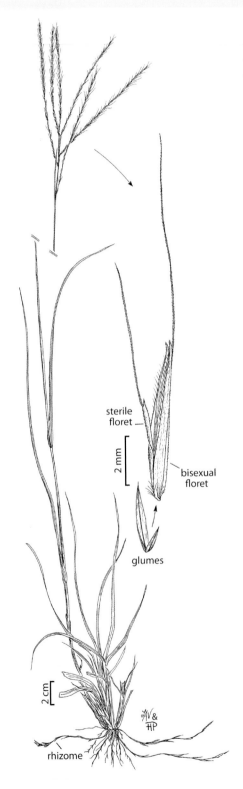

sterile
floret

2 mm

bisexual
floret

glumes

2 cm

rhizome

1. *Enteropogon chlorideus* (J. Presl) Clayton (buryseed umbrellagrass). A perennial cespitose grass with very slender and delicate rhizomes, each rhizome terminating in a cleistogamous spikelet. Similar in this characteristic to *Amphicarpum* (peanut grass) of the Southeast and Atlantic coast. This genus is sometimes included in *Chloris*, but it differs from *Chloris* in its dorsal compression, indurate lemmas, and conspicuous ridges over the midvein. Poor forage (Hatch, Schuster, and Drawe 1999).

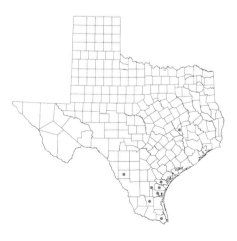

Eragrostis Wolf
(Chloridoideae: Cynodonteae)

Cespitose annuals and perennials, a few with rhizomes. Inflorescence an open (infrequently contracted) panicle. Spikelets 3- to many-flowered, awnless. Glumes 2, unequal, 1–3-nerved, shorter than the lemmas. Lemmas 3-nerved, acute or acuminate at the apex, keeled or rounded on the back. Palea strongly 2-nerved and usually 2-keeled, often ciliolate on the keels, as long or nearly as long as the lemma. Glumes, lemmas, and mature caryopsis usually early deciduous, the paleas persistent on the rachilla. Basic chromosome number, $x = 10$. Photosynthetic pathway, C_4. Represented in Texas by 32 species, 2 subspecies, and 5 varieties. *Eragrostis polytricha* Nees (= E. *trichocolea* Hack. & Arechav.) has been reported, but not confirmed, for the state (Diggs et al. 2006). Telfair (2006) listed this genus as important for wildlife.

1. Plants annual.
  2. Palea keels prominently ciliate.
    3. Spikelets 1.0–3.6 mm long, with 4–12 florets.
      4. Anthers 2; pedicels mostly shorter than the spikelets, straight .... 6. *E. ciliaris*
      4. Anthers 3; pedicels as long as or longer than the spikelets, curved ....2. *E. amabilis*
    3. Spikelets 5–20 mm long, with 10–40 florets ....5. *E. cilianensis*
  2. Palea keels not ciliate.
    5. Plants mat-forming; basal portion of the culms prostrate and rooting at the nodes.
      6. Spikelets bisexual; anthers 2 .... 12. *E. hypnoides*
      6. Spikelets unisexual; plants monoecious; anthers 3 .... 23. *E. reptans*
    5. Plants not mat-forming; basal portion of the culms not prostrate nor rooting at the nodes.
      7. Ligules membranous .... 14. *E. japonica*
      7. Ligules with a membranous base and ciliate above, cilia often longer than the basal portion.
        8. Caryopses with a ventral groove.
          9. Spikelets 1.5–5.0 mm long, with 2–7 florets; pedicels divergent ....4. *E. capillaris*
          9. Spikelets 4–11 mm long, with 5–15 florets; pedicels appressed to slightly divergent ....17. *E. mexicana*
        8. Caryopses without a ventral groove.
          10. Plants without glandular pits or bands.
            11. Lemmas 1.6–3.0 mm long; caryopses white or light brown; plant cultivated, sometimes escaping .... 31. *E. tef*
            11. Lemmas 1.0–2.2 mm long; caryopses brownish; plants not cultivated.
              12. Panicle branches single or paired at lower 2 nodes; lemmas conspicuously veined; lower glumes at least ½ as long as the lowest lemma ....20. *E. pectinacea*
              12. Panicle branches whorled at lower 2 nodes; lemmas faintly

veined; lower glumes less than ½ as long as the lowest lemma...21. *E. pilosa*

10. Plants with glandular pits or bands somewhere (below the nodes, sheaths, blades, rachises, panicle branches, pedicels, keels of the lemmas, and\or palea) .... Key A

Key A

1. Spikelets 0.5–1.4 mm wide; pedicels 1–10 mm long .... 21. *E. pilosa*
1. Spikelets 1.1–4.0 mm wide; pedicels 0.2–4.0 mm long.
  2. Lemmas 2.0–2.8 mm long, with 1–3 pits along the keels; disarticulation below the florets, the rachillas persistent; anthers yellow5. *E. cilianensis*
  2. Lemmas 1.4–1.8 mm long; rarely with 1or 2 pits on keels; disarticulation below the lemmas, both the palea and rachilla persistent; anthers reddish-brown.
    3. Panicle nodes with glandular tissue forming a ring or band, area shiny or yellowish; anthers 3; blade margins without glands; pedicels without glands ....3. *E. barrelieri*
    3. Panicle nodes without rings or bands below; anthers 2; blade margins sometimes with glands; pedicels usually with glandular bands..18. *E. minor*
1. Plants perennial.
  13. Paleas with lower portion forming a wing on each side.............................29. *E. superba*
  13. Paleas without lower portion forming a wing.
    14. Plants rhizomatous; paleas falling with lemmas and caryopses.
      15. Sheaths, blades, and culms not viscid or glandular; lemmas leathery 27. *E. spectabilis*
      15. Sheaths, blades, and culms viscid and sometimes glandular; lemmas membranous.
        16. Pedicels 0.2–1.2 mm long; lemmas 1.5–2.2 mm long........ 7. *E. curtipendicellata*
        16. Pedicels 1.5–12 mm long; lemmas 1.1–1.4 mm long....................26. *E. silveana*
    19. Plants not rhizomatous; paleas often persistent.
      17. Panicles dense, spikelike; spikelets with 2–3 florets .... 28. *E. spicata*
      17. Panicles not spikelike; spikelets with 1–45 florets.
        18. Caryopses with a grooved adaxial surface ....Key B
        18. Caryopses not grooved on the adaxial surface.................................Key C

Key B

1. Caryopses dorsally compressed, translucent, light brown.
  2. Lemmas 1.8–3.0 mm long; panicles 16–40 cm long; ligules 0.6–1.3 mm long....... 8. *E. curvula*
  2. Lemmas 1.4–1.7 mm long; panicles 6–18 cm long; ligules 0.3–0.5 mm long . 15. *E. lehmanniana*
1. Caryopses laterally compressed, terete or only slightly dorsally compressed, opaque, reddish-brown.
  3. Lemmas strongly keeled, with conspicuous lateral nerves, often greenish.
    4. Pedicels with a glandular band; culms with a glandular band below the node ....
    ....30. *E. swallenii*
    4. Pedicels and culms without glandular bands.
      5. Upper glumes generally longer than the lower lemmas; caryopses 0.8–1.3 mm long .... 32. *E. trichodes*
      5. Upper glumes shorter than the lower lemmas; caryopses 0.6–0.8 mm long ....
      .... 19. *E. palmeri*

3. Lemmas sometimes only weakly keeled, with inconspicuous veins.

    6. Lemmas 1.2–1.8 mm long; culms up to 70 cm long ....16. *E. lugens*

    6. Lemmas 1.6–3 mm long; culms up to 120 cm long.

        7. Spikelets greenish with a purple tinge; sheaths densely hirsute with papillose-based hairs on the collar, back, and base....................................... 11. *E. hirsuta*

        7. Spikelets olive or lead-colored; sheaths never with papillose-based hairs, sometimes pubescent.

            8. Lemmas 1.6–2.2 mm long 13. .... *E. intermedia*

            8. Lemmas 2–3 mm long.

                9. Caryopses 0.8–1.6 mm long.. 10. *E. erosa*

                9. Caryopses 0.6–0.8 mm long.. 19. *E. palmeri*

Key C

1. Anthers 2.

    2. Panicles 14–45 cm wide, open, diffuse; primary branches lax.

        3. Spikelets with appressed pedicels; paleas not persistent......................... 22. *E. refracta*

        3. Spikelets with divergent pedicels; paleas persistent .... 9. *E. elliottii*

    2. Panicles 2–17 cm wide, contracted to open; primary branches stiff ......24. *E. secundiflora*

1. Anthers 3.

    4. Primary panicle branches not rebranched; spikelets on each branch sessile or subsessile; pedicels <0.4 mm long ....25. *E. sessilispica*

    4. Primary panicle branches usually rebranched; spikelet usually pedicellate; pedicels >0.4 mm long.

        5. Spikelets 1.3–2.0 mm long, with 1–3 florets .... 1. *E. airoides*

        5. Spikelets 2–19 mm long, with 2–22 florets.

            6. Spikelets 2.0–4.5 mm long.

                7. Blades 25–60 cm long; lemmas 1.6–2.4 mm long; sheaths densely hirsute, with papillose-based hairs on base, collar, and back .... 11. *E. hirsuta*

                7. Blades 4–22 cm long; lemmas 1.2–1.8 mm long; sheaths variously pubescent but never with papillose-based hairs..16. *E. lugens*

            6. Spikelets 4–19 mm long.

                8. Lemmas 1.8–3.0 mm long; ligules 0.6–1.3 mm long............... 8. *E. curvula*

                8. Lemmas 1.4–1.7 mm long; ligules 0.3–0.5 mm long .... 15. *E. lehmanniana*

(Peterson 2003a)

1. *Eragrostis airoides* Nees (darnel lovegrass).
Cespitose perennial without rhizomes. Glands
absent; glumes deciduous. An introduction
from South America that is found only
in Brazos County along roadsides and in
disturbed areas.

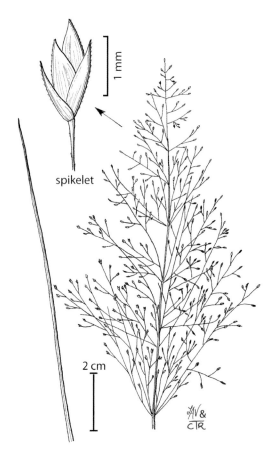

spikelet

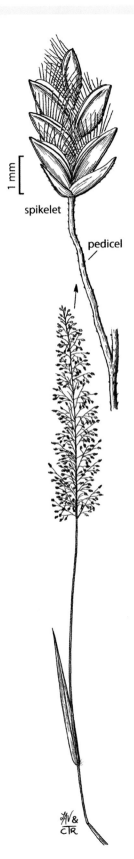

1 mm

spikelet

pedicel

2. *Eragrostis amabilis* (L.) Wight & Arn *ex* Nees (Japanese lovegrass). An annual, introduced species. Paleas with ciliated keels; glumes persistent; glands absent. Listed as present in the state by Jones, Wipff, and Montgomery (1997) but not by Hatch, Gandhi, and Brown (1990); Peterson (2003a); or Turner et al. (2003). Gould (1975b) and Diggs et al. (2006) suggested that it might be here.

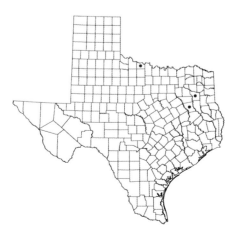

3. *Eragrostis barrelieri* (Mediterranean lovegrass). Tufted annual. Culms with a distinctive ring of glandular tissue below the nodes, rings often shiny or yellowish. Paleas persistent. An introduction from Europe found along roadways and other disturbed sites, especially railroad yards. Poor livestock and wildlife values (Hatch, Schuster, and Drawe 1999).

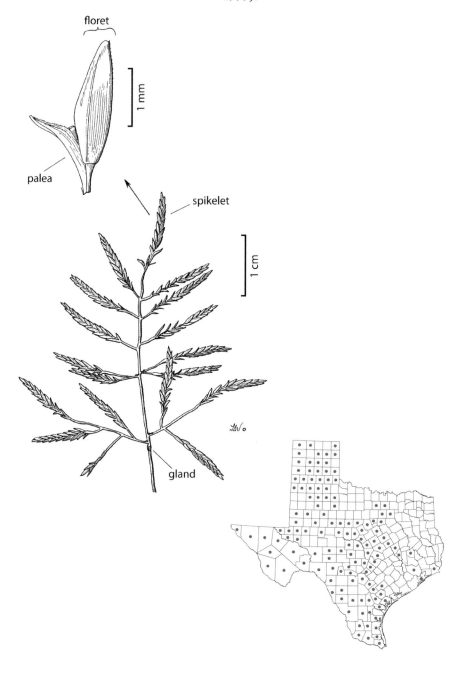

4. *Eragrostis capillaris* (L.) Nees (lacegrass, tiny lovegrass). An annual with short culms (<60 cm). Glands absent; paleas persistent. Typically a ruderal species as well as found along sandy banks of streams and rivers. Poor livestock and wildlife values (Hatch, Schuster, and Drawe 1999).

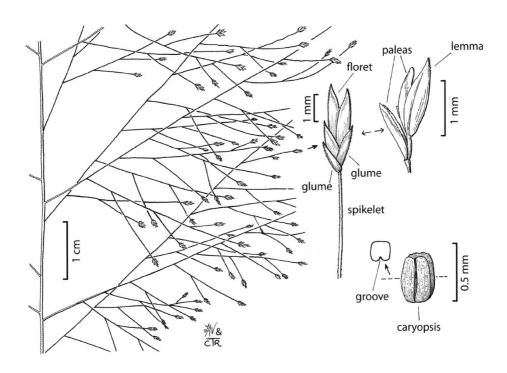

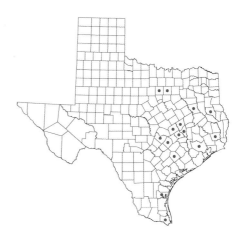

gland

palea

palea

lemma

floret

1 mm

1 mm

spikelet

5. *Eragrostis cilianensis* (All.) Vignolo *ex* Janch. (stinkgrass). Tufted annual. Glands sometimes present below the nodes, on the sheath, blades, glumes, and lemmas. Rachillas persistent. A European introduction that is now a ruderal species. Fresh plants have an unpleasant odor equated to "crushed cockroaches" and reported to be toxic to livestock (Diggs et al. 2006). Poor livestock and wildlife values (Hatch, Schuster, and Drawe 1999; Powell 1994).

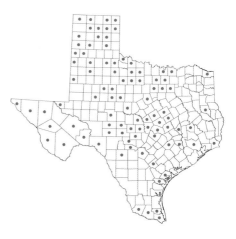

6. *Eragrostis ciliaris* (L.) R. Br. (gophertail lovegrass). An annual, introduced species. Paleas with prominent ciliated keels. Glumes persistent; glands absent. Naturalized in the United States as a weedy species. Found along roadsides, in waste places, and sometimes in saline habitats. Prefers sandy soils. Poor livestock and wildlife values (Hatch, Schuster, and Drawe 1999). The Texas species belongs to the typical variety, var. *ciliaris*.

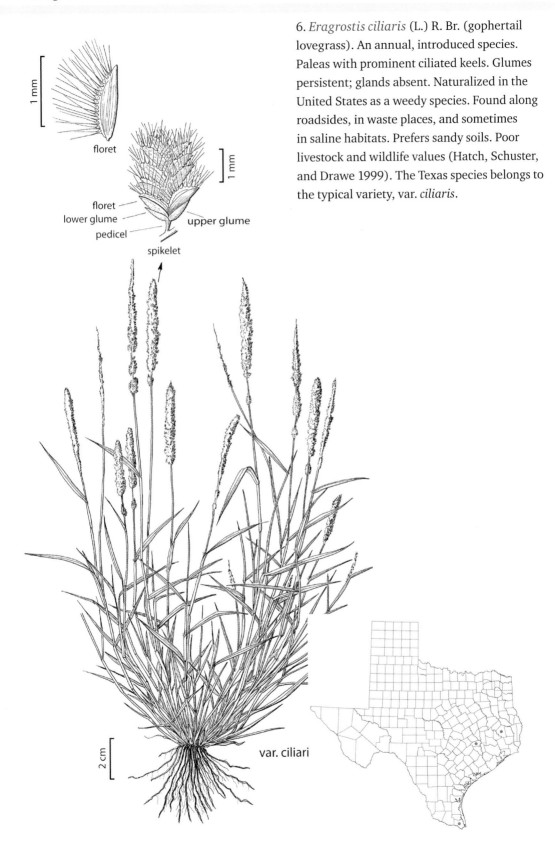

floret

floret
lower glume
pedicel
upper glume
spikelet

var. ciliari

7. *Eragrostis curtipendicellata* Buckley
(gummy lovegrass, short-stalked lovegrass).
Perennial with short, knotty rhizomes. Foliage
and inflorescences viscid. Glumes persistent.
A common species easily identified by its
"stickiness." Found along roadsides, oak
woodland openings, and field margins. Poor
livestock and wildlife values (Hatch, Schuster,
and Drawe 1999; Powell 1994).

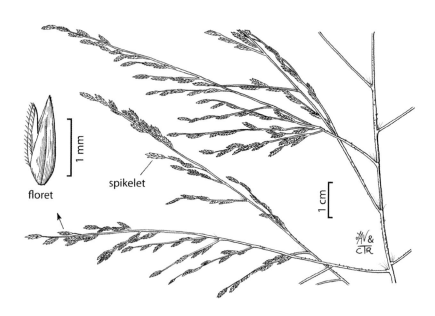

floret

spikelet

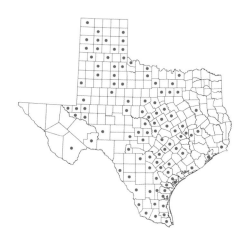

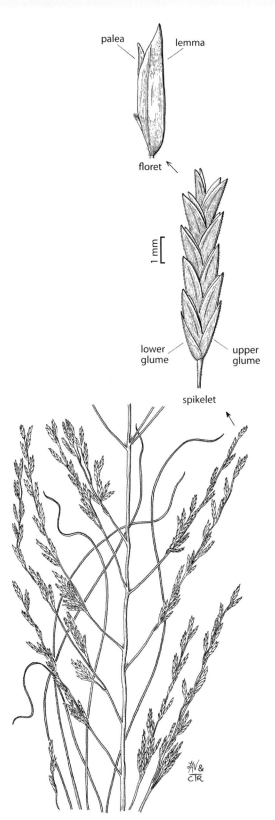

palea   lemma

floret

1 mm

lower glume   upper glume

spikelet

8. *Eragrostis curvula* (Schrad.) Nees (weeping lovegrass). A cespitose perennial with culms to 1.5 m long. Rachillas persistent after disarticulation. Glands absent. An African introduction frequently used in revegetation efforts because it establishes easily, but it often escapes. Found in waste areas, along roadsides, and sometimes in woodlands and forests. Hard stubble from closely grazed plants can cause mechanical injury to cattle ("tender-foot") (Diggs et al. 2006). Listed as poor for wildlife and fair for livestock (Hatch and Pluhar 1993). Fair forage value for livestock according to Powell (1994).

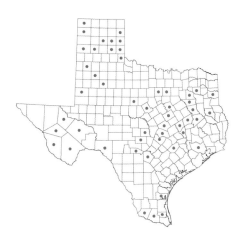

9. *Eragrostis elliottii* S. Watson (Elliott's lovegrass). Cespitose perennial without rhizomes. Glands absent; paleas persistent. Found in sandy pine forests, oak woodlands through the Coastal Plains, and less frequently in the Cross Timbers and Prairies. Prefers sandy soils.

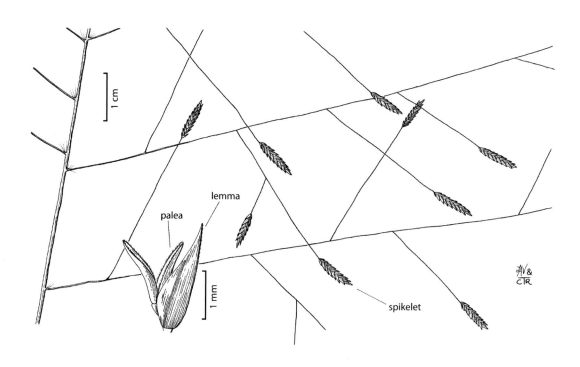

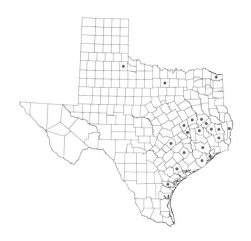

10. *Eragrostis erosa* Scribn. *ex* Beal
(Chihuahua lovegrass). Perennial without
rhizomes. Glands absent; paleas persistent.
Found on rocky slopes in pinyon-juniper
woodlands of West Texas. Probably a good
forage species (Powell 1994).

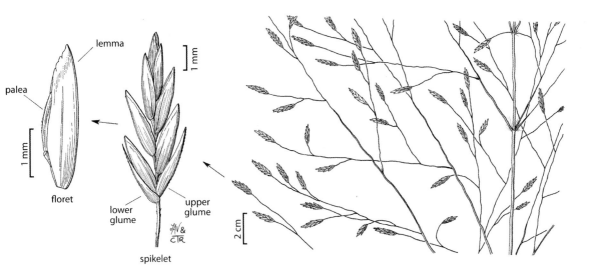

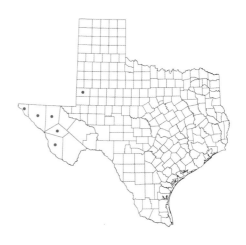

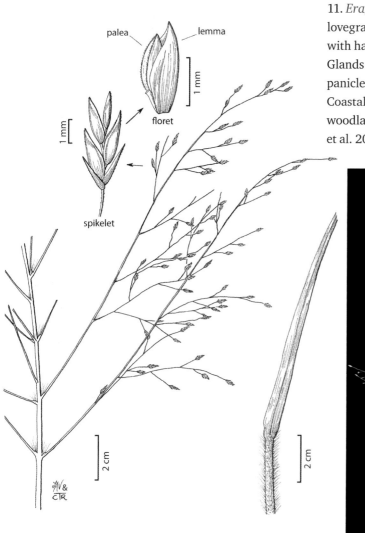

palea   lemma

floret

spikelet

11. *Eragrostis hirsuta* (Michx.) Nees (bigtop lovegrass, stout lovegrass). Cespitose perennial with hardened but not rhizomatous bases. Glands absent; paleas persistent. Large open panicles are indicative. Generally found in the Coastal Plains, longleaf pine stands, or oak woodlands. Also found in river bottoms (Diggs et al. 2006).

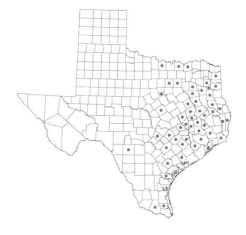

12. *Eragrostis hypnoides* (Lam.) Britton, Sterns, & Poggenb. (teel or teal lovegrass, smooth creeping lovegrass). An annual, mat-forming plant, rooting at the lower nodes, stoloniferous. Paleas persistent after disarticulation; glands absent. Found along the muddy or sandy shores of lakes, ponds, or rivers; also in wet disturbed sites. Ducks eat the spikelets (Hatch, Schuster, and Drawe 1999).

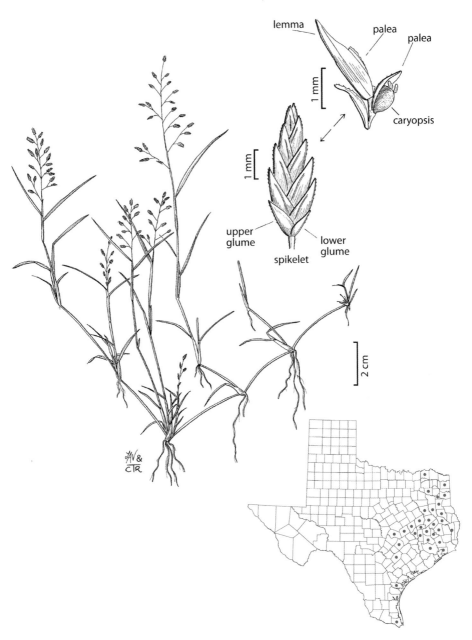

13. *Eragrostis intermedia* Hitchc. (plains lovegrass). Cespitose perennial without rhizomes. Glands absent; paleas persistent. Found in clayey, sandy, or rocky soils along roadsides, waste areas, or grasslands. Closely related to the more common *E. lugens* but has differing flowering period and caryopses with a prominent adaxial groove. Reported by Hatch and Pluhar (1993) and Powell (1994) to be fair or poor for wildlife and good for livestock.

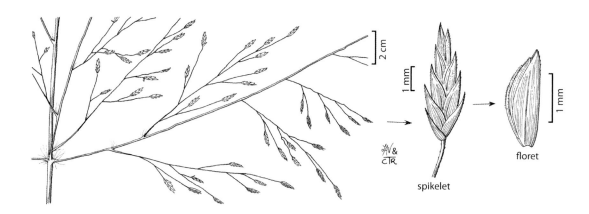

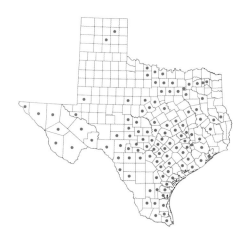

14. *Eragrostis japonica* (Thunb.) Trin. (pond lovegrass, Japanese lovegrass). An annual with culms to 1 m long. Rachilla and glumes persistent; glands absent. An introduced species found in sandy soils along rivers and streams. *Eragrostis glomerata* (Walter) L. H. Dewey of Hatch, Gandhi, and Brown (1990); Jones, Wipff, and Montgomery (1997); and Turner et al. (2003).

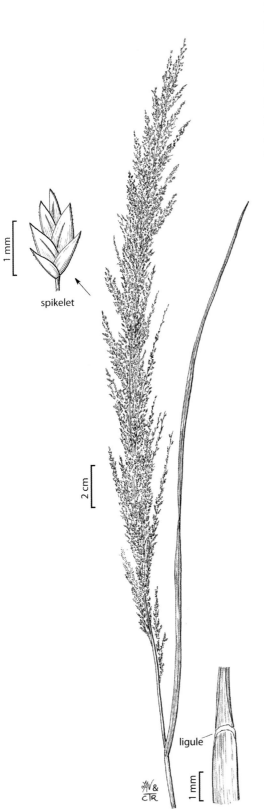

spikelet

ligule

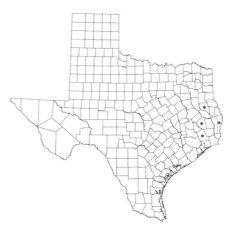

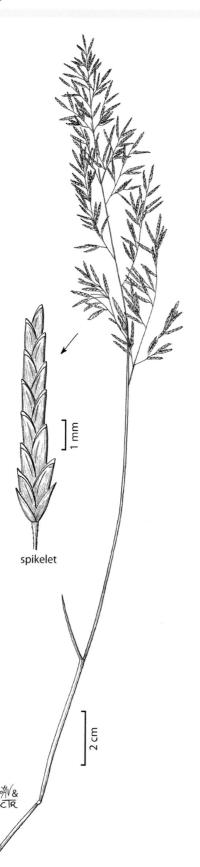

spikelet

1 mm

2 cm

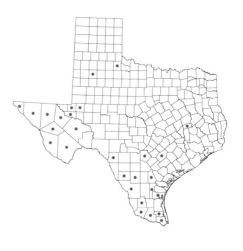

15. *Eragrostis lehmanniana* Nees (Lehmann's lovegrass). A cespitose perennial with culms to 80 cm long. Paleas persistent after disarticulation; glands absent. An African introduction used in erosion control. Found in dry, sandy soils along roadsides and in other disturbed sites. Good forage species (Powell 1994).

16. *Eragrostis lugens* Nees (mourning lovegrass). Cespitose perennial without rhizomes. Glands absent; paleas persistent. Found on sand dunes and riverbanks. Closely related to *E. intermedia*. Fair forage (Hatch, Schuster, and Drawe 1999).

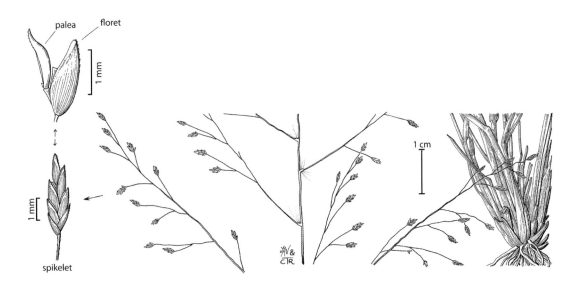

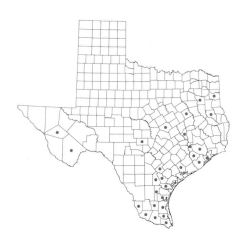

17. *Eragrostis mexicana* (Hornem.) Link (Mexican lovegrass). An annual that can grow to 1.3 m tall. Culms sometimes with a ring of glandular depressions below the nodes. Sheaths sometimes with glandular pits. Paleas persistent. A weedy species of roadsides, fields, and other disturbed sites. Fair forage value (Powell 1994). The Texas species belongs to the typical subspecies, subsp. *mexicana*.

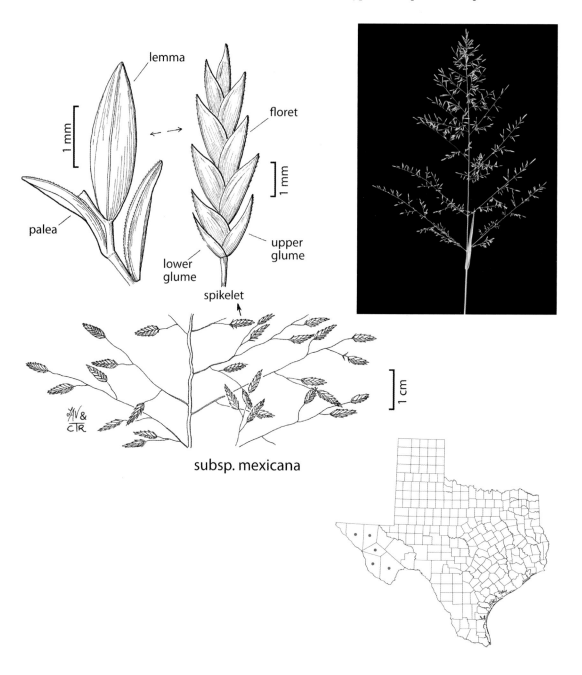

lemma

floret

1 mm

1 mm

palea

upper glume

lower glume

spikelet

1 cm

subsp. mexicana

18. *Eragrostis minor* Host (little lovegrass). Tufted annual. Glands sometimes forming a ring below nodes and on pedicel; rachillas persistent, but not paleas. An introduction from Europe that is found along roadsides and disturbed sites, especially around railroad yards.

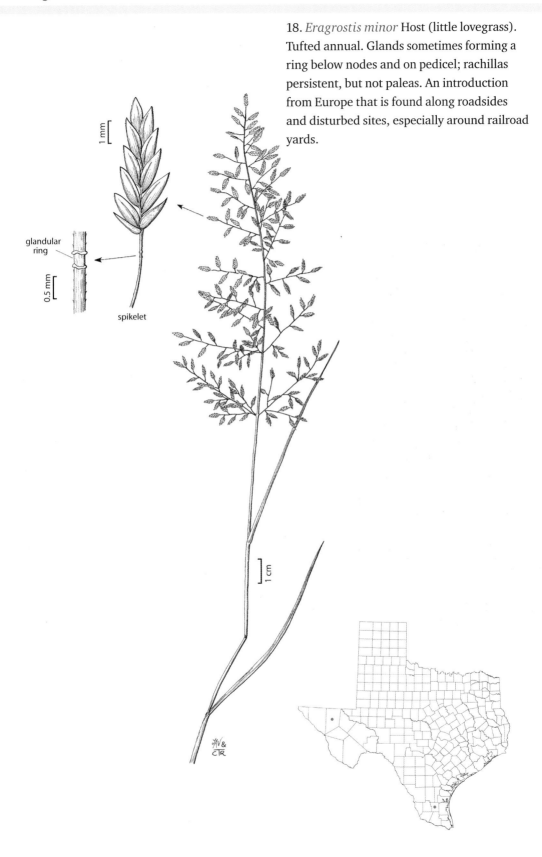

1 mm

glandular ring

0.5 mm

spikelet

1 cm

JAV & CTR

19. *Eragrostis palmeri* S. Watson (Rio Grande lovegrass). Cespitose perennial without rhizomes. Glands absent; paleas persistent. Found on rocky slopes and hills growing with pinyon pine and mesquite. Not common.

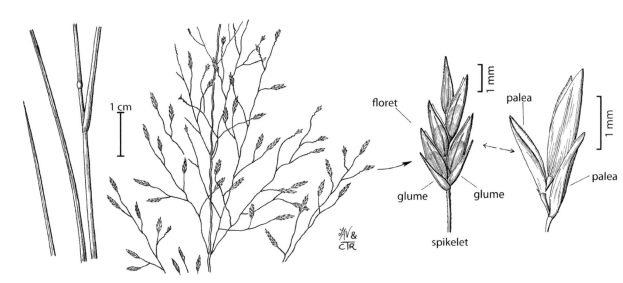

floret

palea

glume   glume

palea

spikelet

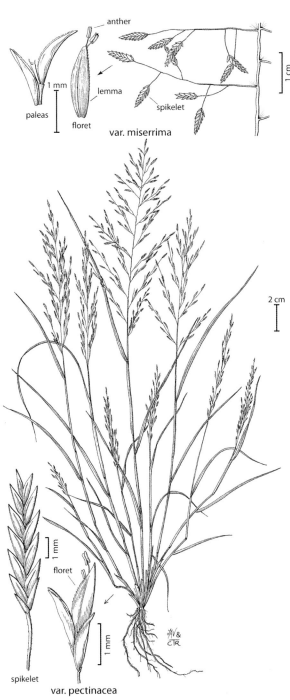

anther

lemma

paleas

floret

spikelet

var. miserrima

1 mm

1 cm

20. *Eragrostis pectinacea* (Michx.) Nees (tufted lovegrass). Tufted annual. Glands absent; paleas persistent. A native ruderal species found along roadsides, in fields, and in gardens. Two varieties occur in Texas: var. *miserrima* (E. Fourn.) Reeder [E. *tephrosanthos* Schult. in Powell (1994)]; and the typical one, var. *pectinacea*. They can be distinguished by the following character:

1. Pedicels appressed ....var. *pectinacea*
1. Pedicels divergent ....var. *miserrima*

2 cm

spikelet

floret

1 mm

1 mm

var. pectinacea

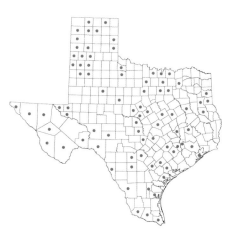

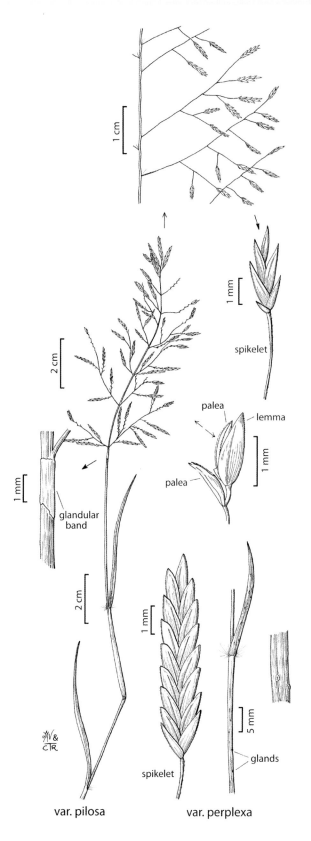

var. pilosa   var. perplexa

21. *Eragrostis pilosa* (L.) P. Beauv. (India lovegrass). An annual occasionally with glands on the culms, sheaths, blade midribs, to sometimes scattered over the entire plant. Paleas eventually fall, and rachillas persistent. A Eurasian introduction that is a common weed of roadsides, gardens, fields, and other disturbed sites. Two varieties occur in Texas: var. *perplexa* (L. H. Harv.) S. D. Koch; and the typical variety, var. *pilosa*. They can be distinguished by the following characters:

1. Glandular pits scattered over the entire plant; lemmas 1.8–2.0 mm long ....var. *perplexa*
1. Glandular pits absent or a few on the culms; lemmas 1.2–1.8 mm long ....var. *pilosa*

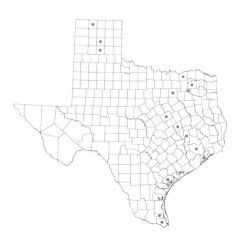

22. *Eragrostis refracta* (Muhl.) Scribn.
(coastal lovegrass). Cespitose perennial
without rhizomes. Glands absent; panicles
very open; glumes persistent. Found on
the Coastal Plains in sandy pine flatwoods,
savannahs, marshes, and grasslands.

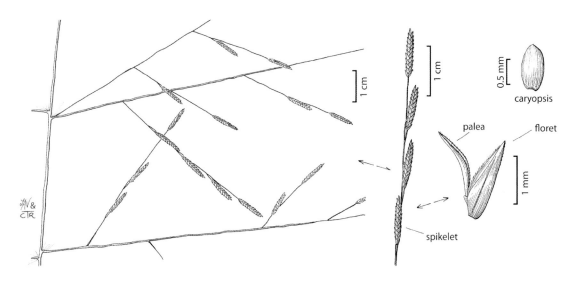

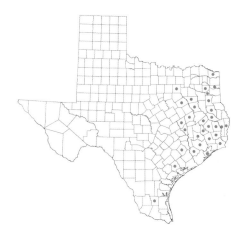

23. *Eragrostis reptans* (Michx.) Nees (creeping lovegrass). An annual, dioecious, mat-forming plant. Male and female plants similar. Paleas persistent on pistillate plants; glands absent. Found in wet sandy, clayey, or gravelly soils along lakes, ponds, and rivers. Listed as *Neeragrostis reptans* (Michx.) Nicora in Gould (1975b) and Hatch, Gandhi, and Brown (1990).

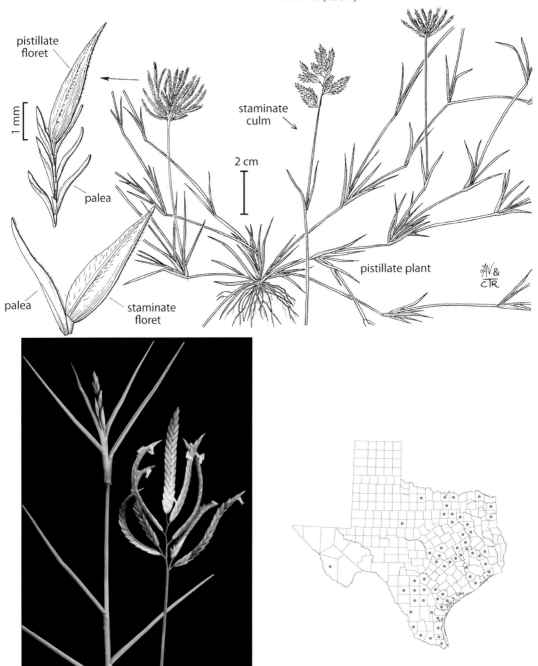

24. *Eragrostis secundiflora* J. Presl (red lovegrass). Cespitose perennial without rhizomes. Glands absent. Florets falling intact before the glumes. Found in sandy areas, dunes, beaches, grasslands, and disturbed sites. Listed as poor for wildlife and livestock (Hatch and Pluhar 1993). The Texas species belongs to subsp. *oxylepis* (Torr.) S. D. Koch.

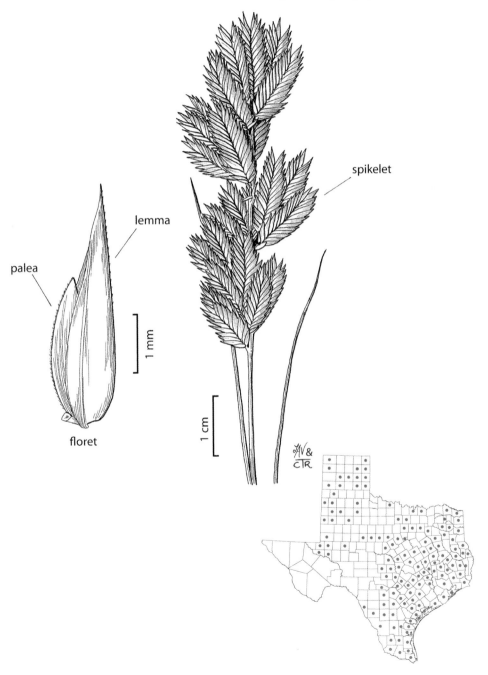

palea

lemma

1 mm

floret

spikelet

1 cm

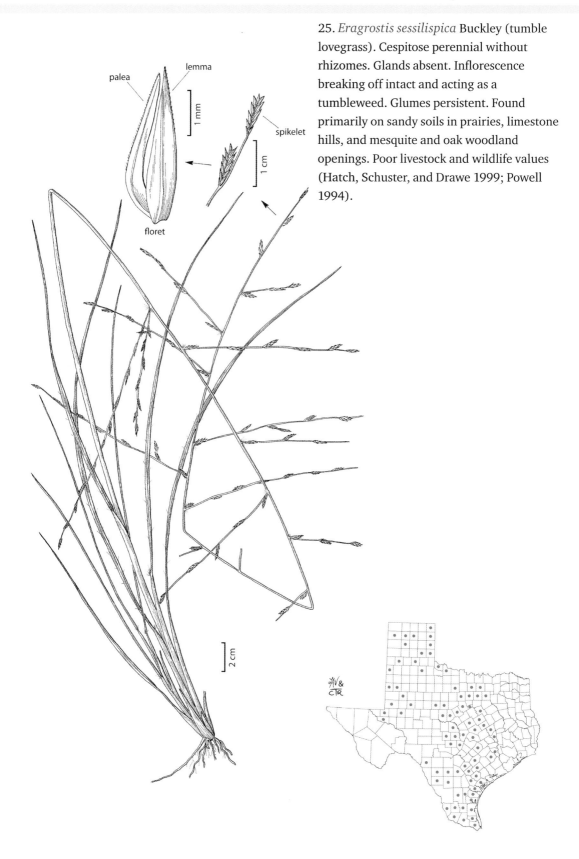

palea   lemma

1 mm

spikelet

1 cm

floret

2 cm

25. *Eragrostis sessilispica* Buckley (tumble lovegrass). Cespitose perennial without rhizomes. Glands absent. Inflorescence breaking off intact and acting as a tumbleweed. Glumes persistent. Found primarily on sandy soils in prairies, limestone hills, and mesquite and oak woodland openings. Poor livestock and wildlife values (Hatch, Schuster, and Drawe 1999; Powell 1994).

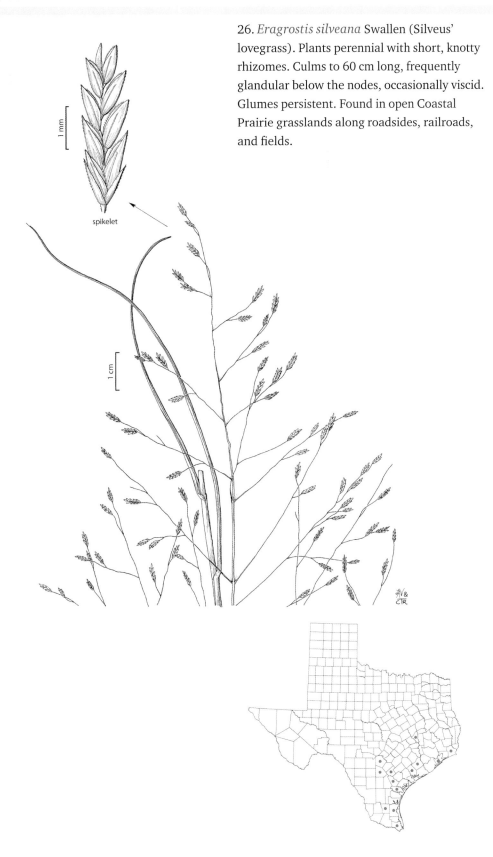

spikelet

26. *Eragrostis silveana* Swallen (Silveus' lovegrass). Plants perennial with short, knotty rhizomes. Culms to 60 cm long, frequently glandular below the nodes, occasionally viscid. Glumes persistent. Found in open Coastal Prairie grasslands along roadsides, railroads, and fields.

27. *Eragrostis spectabilis* (Pursh) Steud.
(purple lovegrass). Perennial with short,
knotty rhizomes. Glands absent; glumes
persistent. Found along roadsides and in
abandoned fields, openings in hardwood
forests, mesquite-acacia shrublands, and
shortgrass steppes. It is sometimes used as an
ornamental. Poor livestock and wildlife values
(Hatch, Schuster, and Drawe 1999).

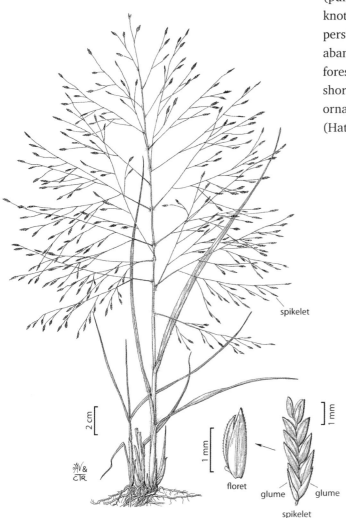

spikelet

2 cm

1 mm

floret

1 mm

glume   glume

spikelet

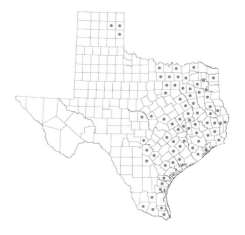

28. *Eragrostis spicata* Vasey (spike lovegrass). Cespitose perennial without rhizomes. One of the few members of the genus with a very dense, spikelike panicle. Glands absent; glumes persistent. Found in moist areas in prairies, mesquite-acacia shrublands, and post oak savannahs.

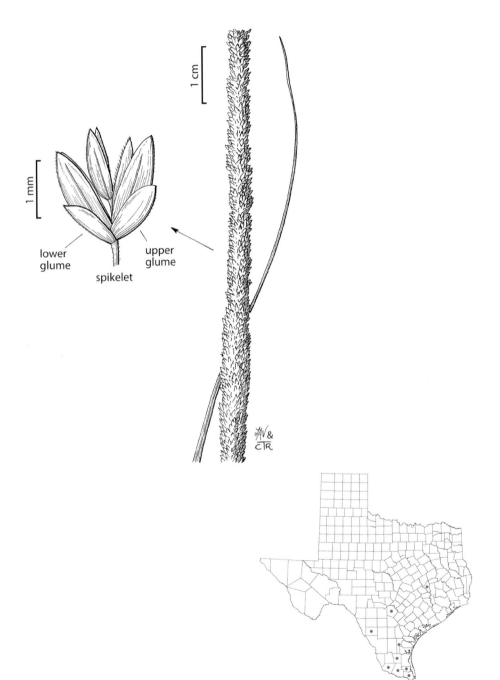

1 cm

1 mm

lower
glume

upper
glume

spikelet

29. *Eragrostis superba* Peyr. (Wilmann's
lovegrass, sawtooth lovegrass). Perennial.
Glands absent; leaf margins sharply scabrous;
glumes persistent. Introduced from Africa
for feeding trials, erosion control, and
revegetation. Found along roadsides and in
mesquite-acacia and oak-juniper woodlands.
Species has large, flattened spikelets (similar
to those of *Chasmanthium*) and could be used
as an ornamental (Diggs et al. 2006).

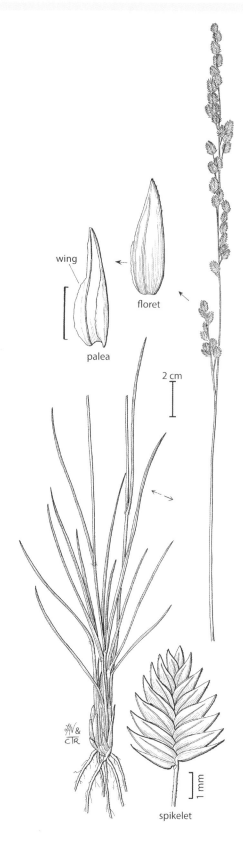

wing

floret

palea

2 cm

1 mm

spikelet

30. *Eragrostis swallenii* Hitchc. (Swallen's lovegrass). Cespitose perennial without rhizomes. Culms to 70 cm long with glandular bands below the nodes. Paleas persistent. Found in coastal grasslands with *Andropogon* and *Spartina*.

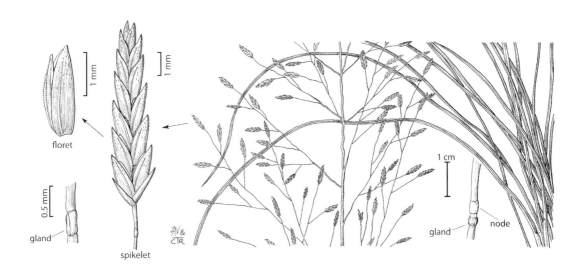

floret

gland

spikelet

1 mm

0.5 mm

1 mm

1 cm

gland

node

31. *Eragrostis tef* (Zucc.) Trotter (teff). A cultivated, tufted annual. Glands absent. Caryopses falling before the glumes and lemmas; paleas persistent. Occasionally escapes from cultivation and occurs along field margins and roadsides. Used as grain and cattle fodder in Ethiopia (Barkworth et al. 2003).

caryopsis

spikelet

32. *Eragrostis trichodes* (Nutt.) Alph. Wood (sand lovegrass). Perennial without rhizomes. Glands absent; paleas persistent. Found in sandy grasslands, open sandy woods, along roadsides, and frequently in mixed oak savannahs. Sometimes used as an ornamental. Listed as poor for wildlife and good for livestock (Hatch and Pluhar 1993).

spikelet

floret

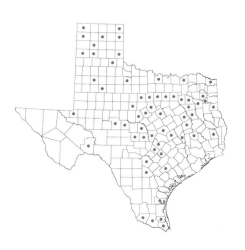

Eremochloa Büse
(Panicoideae: Andropogoneae)

Perennials with compressed, spikelike racemes. Pediceled spikelet greatly reduced, in some species represented only by the stiffly erect pedicel. Sessile spikelets dorsally compressed, awnless, imbricated along one side of, but not sunken in, the slender rachis. Basic chromosome number, $x = 9$. Photosynthetic pathway, C_4. Represented in Texas by a single species.

(Thieret 2003d)

1. *Eremochloa ophiuroides* (Munro) Hack.
(centipedegrass). A low, stoloniferous
perennial that forms a dense turf. Sheaths
sharply keeled. Inflorescence a spikelike
raceme. It has been introduced with
considerable success as a lawn grass
throughout the southeastern part of Texas.
Escapes and also occurs along roadsides,
parking lots, and other disturbed areas.

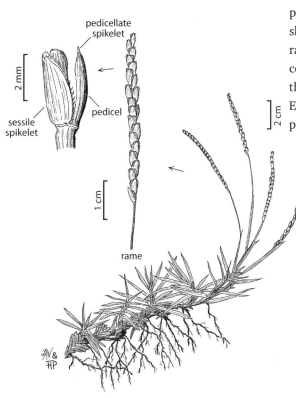

pedicellate
spikelet

2 mm

sessile
spikelet

pedicel

1 cm

rame

2 cm

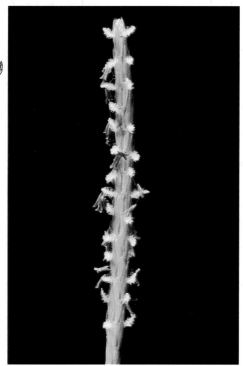

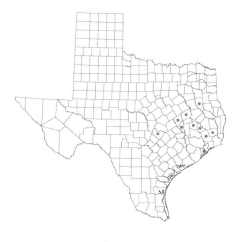

Eriochloa Kunth
(Panicoideae: Paniceae)

Cespitose annuals and perennials. Sheaths open; ligule a ciliated membrane; auricles absent. Inflorescence a loosely contracted panicle, the spikelets subsessile or short-pediceled on unbranched or sparingly rebranched primary branches. Disarticulation below the glumes. Lower glume reduced and fused with the rachis node to form a cup or disk. Upper glume and lemma of the sterile floret about equal, usually scabrous, hispid or hirsute, acute or more commonly acuminate at the apex. Lemma of the fertile floret indurate, glabrous, finely rugose, with slightly inrolled margins, apiculate or short-awned at the apex. Basic chromosome number, $x = 9$. Photosynthetic pathway, C_4. Represented in Texas by 6 species and 2 varieties.

1. Spikelets solitary at the middle of the panicle branches, sometimes in pedicellate pairs near the base.
    2. Lemma of upper florets unawned or rarely with a short awn, awn 0.1–0.2 mm long; plants perennial.. 6. *E. sericea*
    2. Lemma of upper florets always awned, awn 0.4–1.1 mm long; plants annual ...... 2. *E. contracta*
1. Spikelets pedicellate paired or triplets at the middle of panicle branches, sometimes solitary distally.
    3. Plants annual.. 1. *E. acuminata*
    3. Plants perennial.
        4. Plants with rhizomes or stolons.
            5. Plants with stolons .... 3. *E. polystachya*
            5. Plants with rhizomes .... 5. *E. punctata*
        4. Plants cespitose without rhizomes or stolons ....4. *E. pseudoacrotricha*

(Shaw, Webster, and Bern 2003)

upper glume

upper glume

1 mm

spikelet

spikelet

var. minor

1 cm

2 cm

var. acuminata

1. *Eriochloa acuminata* (J. Presl) Kunth (southwestern cupgrass). Annual, sometimes rooting at the lower nodes. Panicle loosely contracted. Lower glumes absent; lower palea absent; upper lemma with short awn 0.1–0.3 mm long. A ruderal species of roadsides, ditches, fields, and other waste places. Two varieties occur in Texas: the typical one, var. *acuminata*; and var. *minor* (Vasey) R. B. Shaw. They can be distinguished by the following characters:

1. Spikelets 4–6 mm long, long-acuminate or tapering to a short awn .... var. *acuminata*
1. Spikelets 3.8–4.0 mm long; acute .... var. *minor*

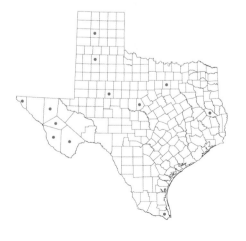

2. *Eriochloa contracta* Hitchc. (prairie cupgrass). Annual, sometimes rooting at the lower nodes. Panicles closed. Lower glumes absent; lower palea absent; upper lemma awned. Found in fields, ditches, and other disturbed sites. The species is becoming a troublesome weed in Kansas. Poor livestock and wildlife values (Hatch, Schuster, and Drawe 1999).

1 cm

ligule

lemma

palea

upper glume

1 mm

1 mm

upper floret

spikelet

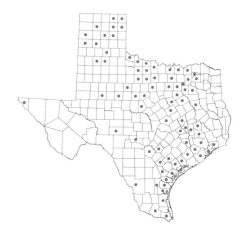

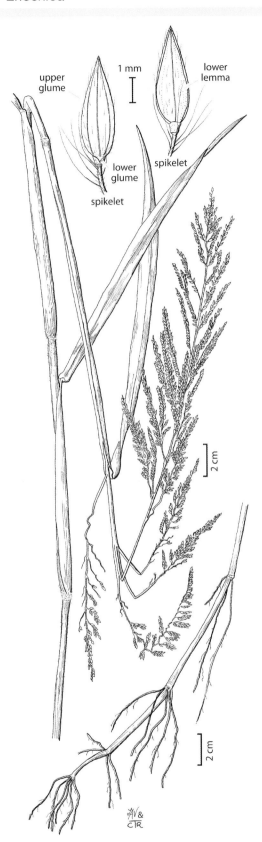

3. *Eriochloa polystachya* Kunth (Caribbean cupgrass). Stoloniferous perennial. Lower glumes present as an extension of the callus. Lower paleas present or less frequently absent. Upper lemma mucronate. Introduced as a forage species, now escaped. Prefers wet, clayey sites.

upper
glume

spikelet

4. *Eriochloa pseudoacrotricha* (Stapf
ex Thell.) J. M . Black (vernal cupgrass).
Cespitose perennial without rhizomes or
stolons. Panicles with loosely appressed
branches. Lower glumes absent; lower paleas
absent; upper lemmas awned. An Australian
introduction found in South Texas.

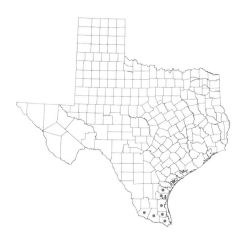

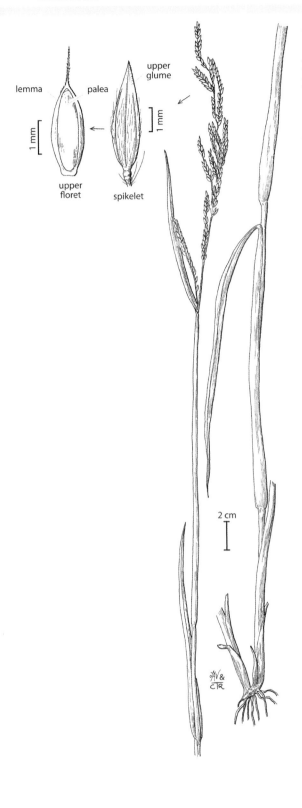

5. *Eriochloa punctata* (L.) Desv. *ex* Ham. (Louisiana cupgrass, everlasting grass). Rhizomatous perennial with culms to 1.5 m long. Panicles contracted. Lower glumes absent; lower palea absent; upper lemma awned. Found in coastal marshes and along watercourses, wet swales, and ditches throughout the Coastal Plains.

6. *Eriochloa sericea* (Scheele) Munro *ex* Vasey (Texas cupgrass). Cespitose perennial with short rhizomes, producing a somewhat knotty base. Panicles closed. Lower glume absent; lower palea absent; upper lemma awnless or very short-awned (0.1–0.2 mm). Prefers clayey soils in native grasslands, prairies, and oak-juniper woodlands. This species and *Nassella leucotricha* are the earliest native, perennial grasses to "green up" and provide abundant spring forage. Listed as fair for wildlife and good for livestock (Hatch and Pluhar 1993).

Erioneuron Nash
(Chloridoideae: Cynodonteae)

Low, tufted perennials with narrow, often involute, cartilaginous-margined leaf blades. Sheath open; ligule a ring of hairs; auricles absent. Inflorescence a short, usually capitate raceme or panicle of several-flowered spikelets, these disarticulating above the glumes and between the florets. Glumes large, membranous, subequal, 1-nerved. Lemmas broad, rounded on the back, 3-nerved, conspicuously long-hairy along the veins at least below, the midnerve short-awned, each lateral nerve often prolonged as a short mucro. Palea slightly shorter than the lemma, ciliate on the keels, long-hairy on the lower part between the veins. Basic chromosome number, $x = 8$. Photosynthetic pathway, C_4. Represented in Texas by 3 species and 1 variety.

A segregate of *Tridens*.

1. Lemmas entire or with teeth to 0.5 mm long; both glumes shorter than the lowest floret; ligules 2–4 mm long...3. *E. pilosum*
1. Lemmas 2-lobed, lobes 1.0–2.5 mm long; upper glumes equaling or exceeding the lowest floret; ligules <1 mm long.
    2. Lemma lateral veins forming mucros to 1 mm long...2. *E. nealleyi*
    2. Lemma lateral veins not forming mucros..1. *E. avenaceum*

(Valdés-Reyna 2003c)

1. *Erioneuron avenaceum* (Kunth)
Tateoka (large-flowered tridens, shortleaf
woollygrass). Perennial. Occasionally
stoloniferous. Ligules to 0.5 mm long. Found
on limestone hills and rocky outcrops. The
Texas species belongs to the typical variety,
var. *avenaceum*.

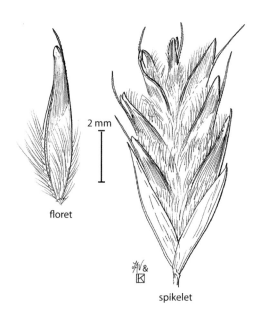

floret

2 mm

spikelet

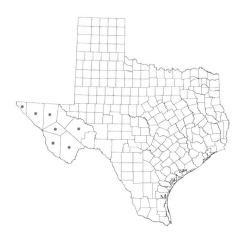

2. *Erioneuron nealleyi* (Vasey) Tateoka
(Nealley's erioneuron, Nealley's
woollygrass). Tufted perennial. Ligules
0.2–0.6 mm long. Found on limestone hills
and rocky outcrops. Gould (1975b) listed
this as a variety of *E. avenaceum*, var. *nealleyi*
(Vasey) Gould.

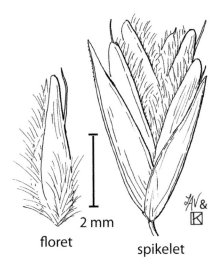

floret   2 mm   spikelet

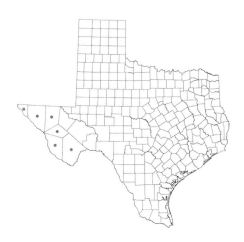

3. *Erioneuron pilosum* (Buckley) Nash (hairy tridens, hairy woollygrass). Tufted perennial. Ligules 2–4 mm long. Widespread species found on a variety of habitats from dry, rocky hills to open grasslands to pinyon-juniper woodlands. Also found in disturbed habitats such as roadsides and waste areas. Listed as poor for wildlife and livestock (Hatch and Pluhar 1993).

1 cm

floret

2 mm

spikelet

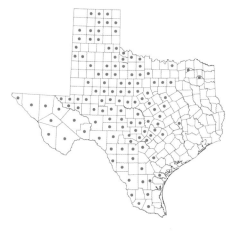

Eustachys Desv.
(Chloridoideae: Cynodonteae)

Perennials, cespitose, rhizomatous, or stoloniferous. Culms to 1.5 m tall, erect or decumbent, flattened. Auricles absent. Ligules short-ciliate, to 0.5 mm long. Blades typically flat or folded. Inflorescences are panicles of several to many nondisarticulating, spikelike branches. Branches digitately arranged with spikelets in 2 rows on the abaxial sides of the branches. Spikelets single, laterally compressed, with 2–3 florets. Lowest florets bisexual; upper florets staminate or completely reduced. Disarticulation above the glumes. Glumes short, 1-veined. Lower lemmas 3-veined, unawned, with a mucro, or awned (1.5 mm). Some authors include *Eustachys* in *Chloris*. Basic chromosome number, $x = 10$. Photosynthetic pathway, C_4. Represented in Texas by 4 species. *Eustachys distichophylla* (Lag.) Nees had been reported from Texas, but the report was based on misidentified material of *E. retusa* (Diggs et al. 2006).

This genus has characteristic blunt leaf blades.

1. Keels of the lowest lemma in each spikelet glabrous .... 4. *E. retusa*
1. Keels of the lowest lemma in each spikelet pubescent.
    2. Spikelets 1.5–2.5 mm long.
        3. Lowest lemma in each spikelet dark brown, lateral veins with appressed hairs <0.5 mm
             long .... 3. *E. petraea*
        3. Lowest lemma in each spikelet yellowish- to reddish-brown, lateral veins with spreading
             hairs >0.5 mm long ....1. *E. caribaea*
    2. Spikelets 2.6–3.7 mm long .... 2. *E. neglecta*

(Aulbach 2003)

1. *Eustachys caribaea* (Spreng.) Herter
(chickenfoot grass, Caribbean fingergrass).
Tufted culms to 70 cm long, erect. Lower
lemmas yellowish- to reddish-brown at
maturity, lateral veins and keels with
appressed spreading hairs to 1 mm long,
mucronate. An introduction from South
America found growing along roadsides.

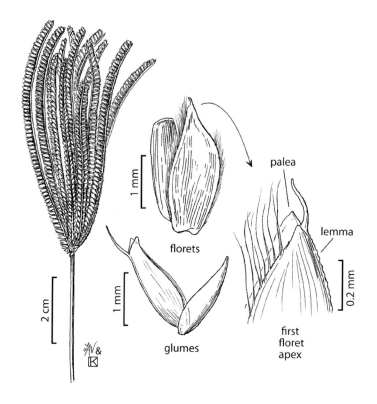

florets

glumes

palea

lemma

first
floret
apex

2. *Eustachys neglecta* (Nash) Nash (four-spike fingergrass). Tufted perennial with culms to 1.2 m long. Lower lemmas tan to light brown at maturity, lateral veins and keels with appressed hairs to 0.7 mm long, awned, awn 0.4–0.6 mm long. Apparently found only in Brazos County along roadsides.

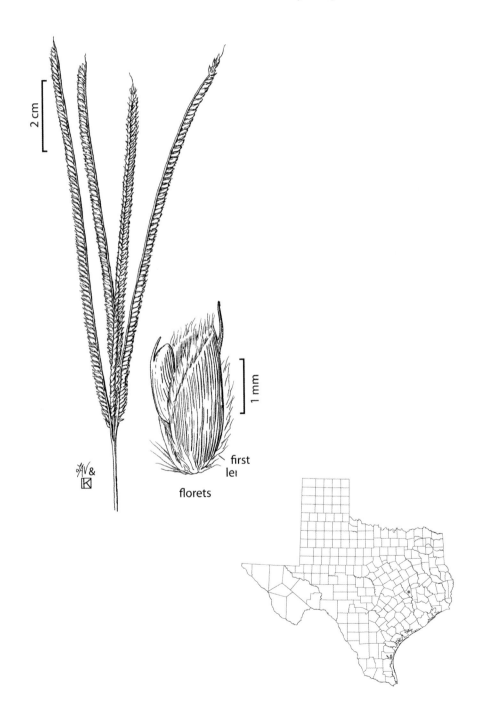

2 cm

1 mm

first
lei

florets

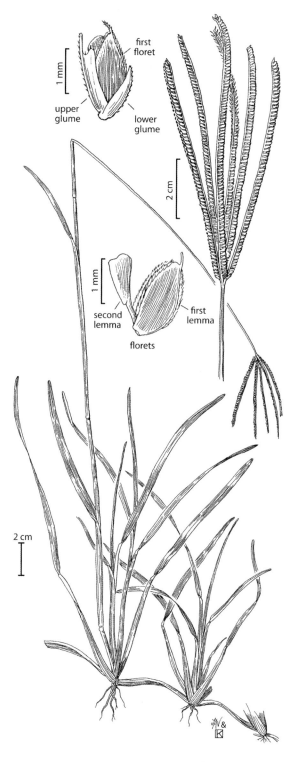

3. *Eustachys petraea* (Sw.) Desv. (pinewoods
fingergrass, stiff-leaf fingergrass).
Stoloniferous perennial with culms to 1 m
long. Lowest lemma dark brown at maturity,
mucronate, with appressed hairs on the keels
and lateral veins. Found in open, sandy areas,
along roadsides, and in brackish marshes. Poor
livestock and wildlife values (Hatch, Schuster,
and Drawe 1999).

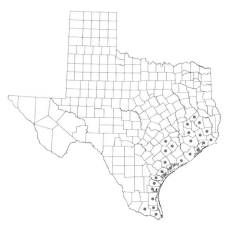

4. *Eustachys retusa* (Lag.) Kunth (Argentine fingergrass). Perennial. Culms to 80 cm tall, rooting at the lower nodes. Lower lemmas have lateral veins with spreading, white hairs that are 1–2 mm long; apices mucronate. An introduction from South America that now grows along roadsides, sandy fields, and waste areas.

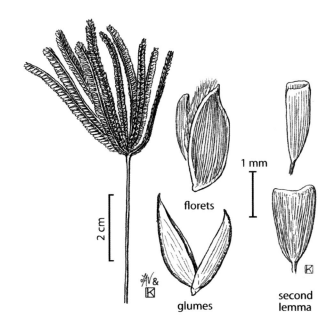

florets

glumes

1 mm

second lemma

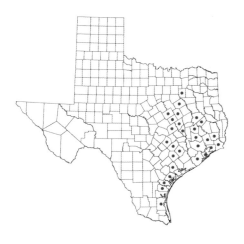

Festuca L.
(Poöideae: Poeae)

Tufted perennials with usually thin, flat or narrow and involute blades, and 3- to several-flowered spikelets in open or contracted panicles. Disarticulation above the glumes and between the florets. Glumes narrow, unequal, acute or acuminate, 1–3-nerved. Lemmas thin or firm, rounded on the back, usually 5–7-nerved, awned from a narrow, entire or minutely bifid apex, or awnless. Palea free from the caryopsis. Stamens 3. Basic chromosome number, $x = 7$. Photosynthetic pathway, C_3. Represented in Texas by 6 species.

1. Blades usually flat or loosely convolute or conduplicate, some more than 3 mm wide.
   2. Ligules 2–9 mm long ....2. *F. ligulata*
   2. Ligules to 1.5 (-2.0) mm long.
      3. Lemmas 3–5 mm long; unawned.
         4. Inflorescence branches usually reflexed at maturity; spikelets not or only slightly imbricate.. 5. *F. subverticillata*
         4. Inflorescence branches ascending to spreading at maturity; spikelets closely imbricate .... 3. *F. paradoxa*
      3. Lemmas 5–12 mm long, unawned or with awns to 2 mm long.................... 6. *F. versuta*
1. Blades usually conduplicate or folded and <2.5 mm wide.
   5. Ligules 2–9 mm long; spikelets with 2–3 florets ....2. *F. ligulata*
   5. Ligules to 1.5(–2.0) mm long.
      6. Culms densely tufted, usually erect at base; basal sheaths brown or gray, not shredding into fibers; blades usually glaucous; rhizomes absent ..... 1. *F. arizonica*
      6. Culms not densely tufted, usually decumbent at base; basal sheaths reddish-brown to purple; with age shedding into fibers; blades not glaucous; rhizomes usually present ....4. *F. rubra*

(Darbyshire and Pavlick 2007; Gould 1975b)

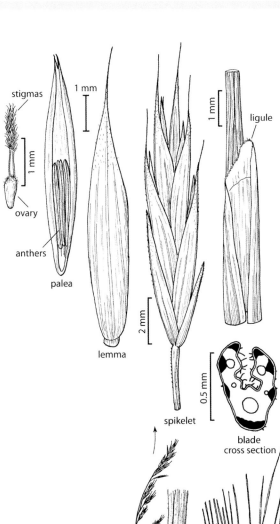

stigmas

ovary

anthers

palea

lemma

1 mm

1 mm

2 mm

0.5 mm

spikelet

blade
cross section

ligule

1 mm

1. *Festuca arizonica* Vasey (Arizona fescue).
Perennial without rhizomes. Sheaths closed
for less than ½ their length, not shredding.
Panicles contracted or loosely open, erect;
1–2 branches per node. Lemmas usually with
an awn to 3 mm long. Found in high mountain
meadows and open montane forests. Provides
forage and cover for wildlife (Powell 1994).

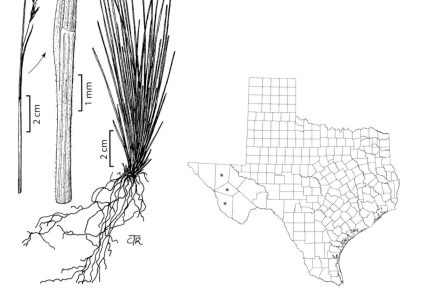

2 cm

1 mm

2 cm

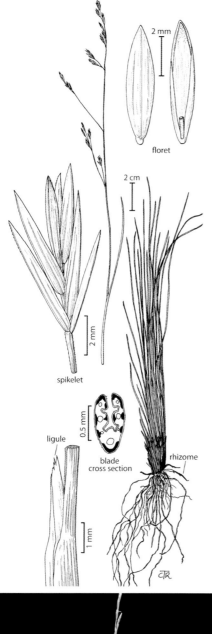

floret

spikelet

ligule

blade
cross section

rhizome

2. *Festuca ligulata* Swallen (Guadalupe fescue). Perennial with short rhizomes. Sheaths closed for less than ⅓ their length, not shredding into fibers. Panicles open or loosely contracted, 1–2 branches per node. Spikelets borne near branch tips. The long ligules (2–9 mm) are distinctive. Found on moist, shady slopes in the West Texas mountains. It is federally listed as an endangered species.

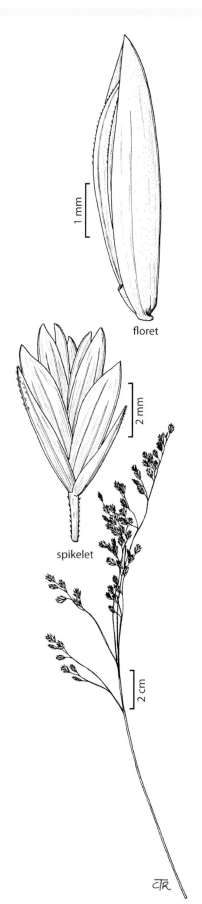

floret

spikelet

3. *Festuca paradoxa* Desv. (cluster fescue). Perennial without rhizomes. Sheaths closed for less than 1/3 their length, shredding into fibers. Panicles open to loosely closed, erect, 1–2 branches per node. Spikelets imbricate, clustered toward the tips of the branches. Only in far northeastern Texas on prairies; in open woods; and on low, open ground.

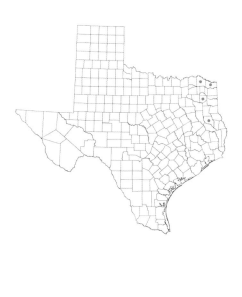

stigmas

1 mm

ovary

lemma

2 mm

1 mm

anther

2 mm

upper glume   lower glume

spikelet

2 cm

ligule

1 mm

CTR

4. *Festuca rubra* L. (red fescue). Perennial, usually rhizomatous, sometimes stoloniferous as well. Sheaths closed for about ¾ their length, shredding into fibers. Panicles usually loosely contracted or sometimes open, 1–3 branches per node, erect. Spikelets toward the tips of the short branches. Found in moist shaded areas of the highest peaks in the Davis Mountains. About 10 subspecies have been identified, but none are recognized here.

5. *Festuca subverticillata* (Pers.) E. B. Alexeev (nodding fescue). Perennial without rhizomes. Sheaths closed for less than 1/3 their length, shredding into fibers. Panicles open, drooping, 1–3 branches per node. Spikelets toward the branch tips. Only a few records for the eastern part of the state. Found in mixed hardwood forests. Listed as *F. obtusa* Bieler in Gould (1975b).

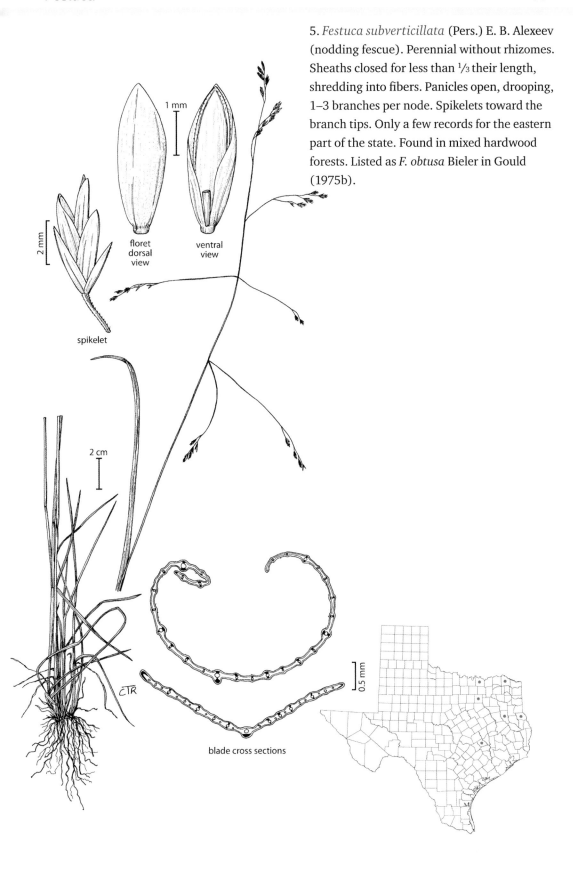

floret dorsal view

ventral view

spikelet

blade cross sections

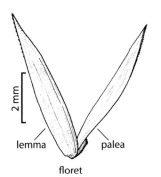

lemma   palea

floret

upper   lower
glume   glume

spikelet

2 cm

ℂℝ

6. *Festuca versuta* Beal (Texas fescue).
Perennial without rhizomes. Sheaths closed for
less than $1/3$ their length, shredding into fibers.
Panicles open, erect, with usually 2 branches
per node. Spikelets toward the branch tips.
Relatively rare species found in moist, shady
areas on rocky slopes in wooded areas.

Gastridium P. Beauv.
(Poöideae: Poeae)

Low annuals with weak culms, flat blades, membranous ligules, and cylindric, tightly contracted panicles. Spikelets 1-flowered, disarticulating above the glumes. Rachilla prolonged behind the palea as a minute bristle. Glumes long, somewhat unequal, lanceolate or acuminate, 1-nerved, firm and shiny, awnless. Lemma much shorter than the glumes, broad and blunt, rounded on the back, 5-nerved, awnless or with a delicate awn from below the minutely toothed apex. Palea about as long as the lemma. Basic chromosome number, $x = 7$. Photosynthetic pathway, C_3. Represented in Texas by a single species.

(Wipff 2007b)

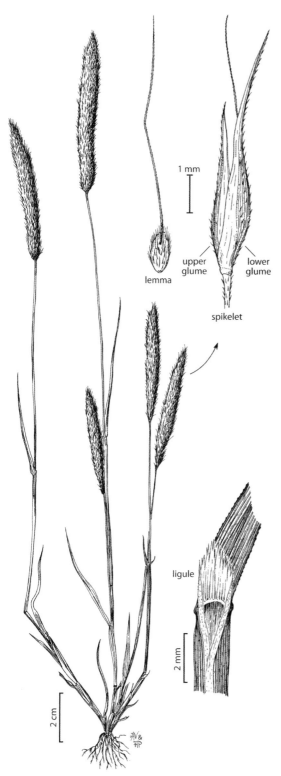

1 mm

lemma

upper
glume

lower
glume

spikelet

ligule

2 mm

2 cm

AV &
HP

1. *Gastridium phleoides* (Nees and Meyen)
C. E. Hubb. (nitgrass). An annual with
weak culms. Panicles pale green, branches
appressed. Rachillas prolonged, densely
pubescent. Lemmas $1/5$ to $1/2$ the length of
glumes, long-awned. A native of Asia and
Africa, it is now a widely distributed weed of
dry, disturbed sites. Gould (1975b) listed this
as G. *ventricosum* (Gouan) Schinz &Thell.
Wipff (2007b) cited a known collection from
West Texas in 1884 by Nealley; and Gould
(1975b) reported the species from the Trans-
Pecos region. Perhaps not a constant member
of the Texas flora.

Glyceria R. Br.
(Poöideae: Meliceae)

Perennials, strongly rhizomatous. Sheaths connate nearly to the top; ligules prominent, membranous, erose to deeply lacerate, sometimes fused in the front; auricles absent. Inflorescence in Texas species usually a large, open panicle, occasionally with flexuous branches. Spikelets with 3 to many florets, often purplish. Disarticulation above the glumes and between the florets. Glumes shorter than the lowermost lemmas, 1-nerved, awnless. Lemmas broad, firm, often erose, prominently 5- or more-nerved. Palea large, broad, sometimes longer than the lemma. Basic chromosome number, $x = 10$. Photosynthetic pathway, C_3. Represented in Texas by 4 species and 2 varieties. Species can sometimes be poisonous to livestock due to the presence of dhurrin (Diggs et al. 2006).

1. Spikelets laterally compressed, length 1–4 times the width; palea keels not winged.
  2. Lemma tips almost flat; anthers 3; veins of one or both glumes usually extending to the apices .... 1. *G. grandis*
  2. Lemma tips bow-shaped; anthers 2; veins of both glumes terminating below the apices .... 4. *G. striata*
1. Spikelets terete or cylindric, length >5 times the width; palea keels usually winged.
  3. Pedicels 1–6 mm long; plants to 80 cm tall .... 2. *G. notata*
  3. Pedicels 0.5–1.7 mm long; plants usually over 80 cm tall .... 3. *G. septentrionalis*

(Barkworth and Anderton 2007)

1. *Glyceria grandis* S. Watson (American mannagrass). Perennial with erect or decumbent culms that root at the lower nodes. Panicles open, branches lax, drooping. Found on banks and in water of streams, rivers, ponds, and wet meadows. Only one report from Sabine County. Not reported by Gould (1975b) or Jones, Wipff, and Montgomery (1997).

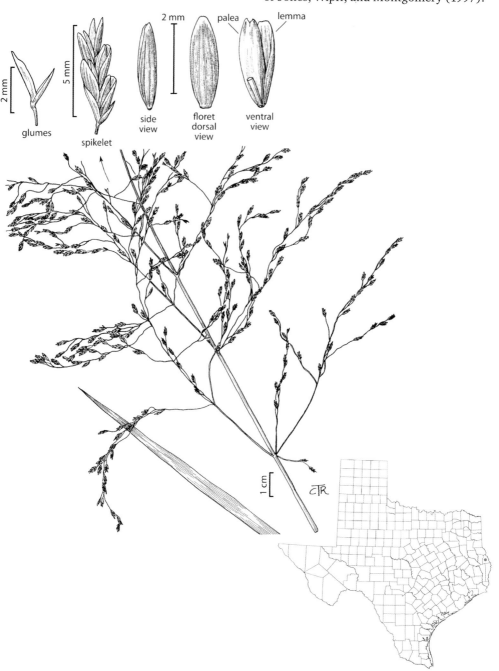

glumes

spikelet

2 mm

side view

floret dorsal view

palea

lemma

ventral view

1 cm

CTR

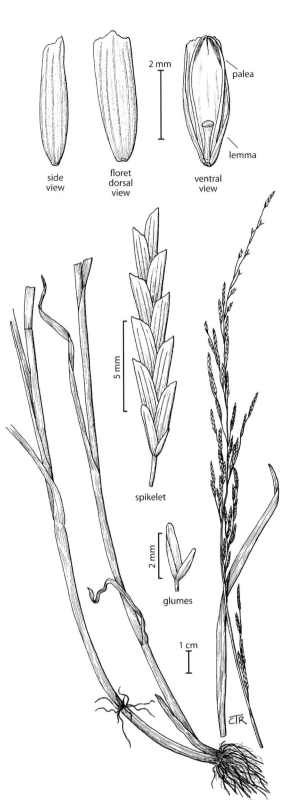

side
view

floret
dorsal
view

ventral
view

palea

lemma

2 mm

spikelet

5 mm

glumes

2 mm

1 cm

2. *Glyceria notata* Chevall. (marked glyceria). Perennial with culms <80 cm long that frequently root at the lowermost nodes. Panicles spreading to loosely closed. A Eurasian introduction that has been reported from Waller County. Perhaps not a constant member of the Texas flora.

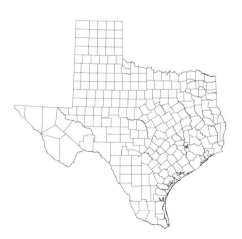

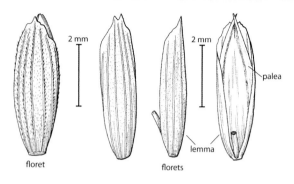

floret

var. arkansana

palea

lemma

florets

var. septentrionalis

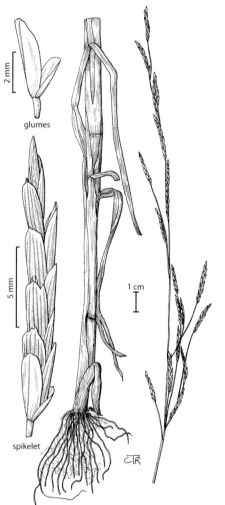

glumes

spikelet

ctr

3. *Glyceria septentrionalis* Hitchc. (northern mannagrass, eastern mannagrass, floating mannagrass). Perennial, culms frequently rooting at lower nodes. Panicles usually contracted, sometimes lower branches lax. Found along the margins of streams, ponds, and lakes. Two varieties have been recognized: var. *arkansana* (Fernald) Steyerm. and Kučera; and var. *septentrionalis*. Gould (1975b) and others listed var. arkansana as a separate species (*G. arkansana* Fernald). The varieties can be distinguished by the following character:

1. Lemmas hirsute over the veins .. var. *arkansana*
1. Lemmas scabrous............ var. *septentrionalis*

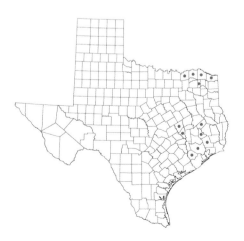

4. *Glyceria striata* (Lam.) Hitchc. (ridged mannagrass, fowl mannagrass, nerved mannagrass). Perennial, sometimes rooting at the lower nodes. Panicles open, nodding, branches lax and drooping at maturity. Found in swamps, bogs, and marshes and around ponds and other wet places. Plants may produce cyanide, but poisoning is rare (Allred 2005). Palatable to livestock (Powell 1994).

2 mm

dorsal view   ventral view

floret

1 cm

1 mm

floret

1 mm

glumes

2 mm

spikelet

Gymnopogon P. Beauv.
(Chloridoideae: Cynodonteae)

Perennials with stiff, erect, many-noded culms. Base of plant often with short, knotty rhizomes. Leaves stiff and firm with rounded, overlapping sheaths and short, flat, stiff, usually spreading blades. Spikelets subsessile, rather widely spaced in 2 rows on slender, spreading primary inflorescence branches, these distributed on the upper portion of the culm. Spikelets 1–3-flowered, the rachilla prolonged behind the terminal fertile floret as a slender stipe bearing a rudimentary floret. Disarticulation above the glumes. Glumes narrow, nearly equal, 1-nerved, acuminate, the second exceeding the lemmas in length. Lemmas narrow, rounded on the back, 3-nerved, usually bearing a delicate awn from a minutely bifid apex. Basic chromosome number, $x = 10$. Photosynthetic pathway, C_4. Represented in Texas by 2 species.

1. Plants with elongated rhizomes; lemmas with an awn 0.8–3.0 mm long........2. *G. brevifolius*
1. Plants with short, knotty rhizomes; lemmas with an awn 1–12 mm long ....1. *G. ambiguus*

(Smith 2003)

1. *Gymnopogon ambiguus* (Michx.) Britton, Sterns, and Poggenb. (bearded skeletongrass, beardgrass). A coarse perennial with a knotty, rhizomatous base. Panicle branches naked for less than ⅓ their length. It grows in sandy soils in woodlands of East Texas and is the more common of the 2 species. Gould (1975b) did not recognize this as a distinct species.

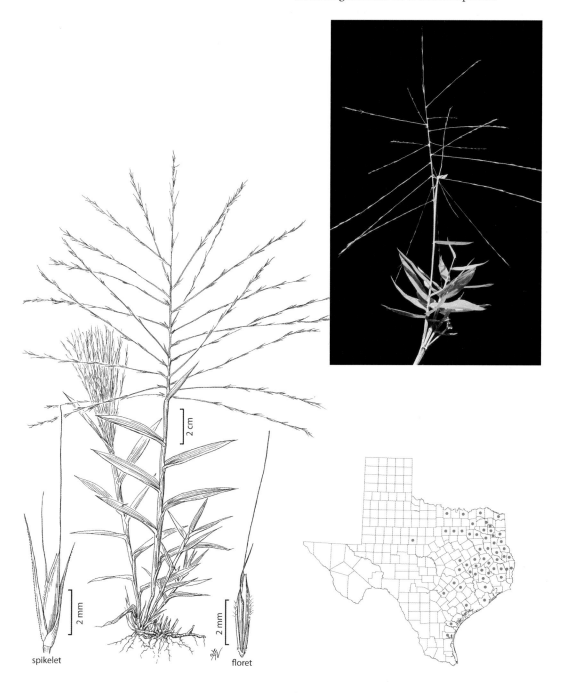

2 cm

2 mm

2 mm

spikelet

floret

2. *Gymnopogon brevifolius* Trin. (shortleaf skeletongrass). Plants with elongated rhizomes. Panicle branches naked for at least ⅓ their length. Generally found in partial shade in sandy, pine flatwoods. Relatively rare in Texas.

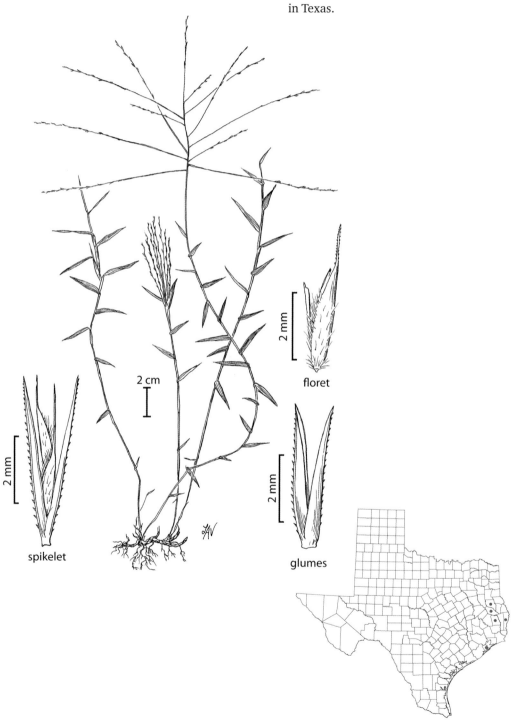

2 mm

floret

2 cm

2 mm

spikelet

2 mm

glumes

Hackelochloa Kuntze
(Panicoideae: Andropogoneae)

Much-branched annual with erect or decumbent-spreading culms. Blades flat and thin.
Spikelets in short, spikelike racemes. Racemes subtended by, and often partially enclosed in,
expanded leaf sheaths. Spikelets in pairs of 1 sessile and 1 pediceled, on a stout, short rachis.
Disarticulation at the nodes of the rachis. Sessile spikelet fertile, awnless, globose, the first
glume thick and rounded, coarsely rugose or alveolate. Lemmas of the sterile and fertile florets
membranous, hyaline. Pediceled spikelets ovate-lanceolate, with thin, flat glumes. Pedicels
fused on one side of the rachis. Basic chromosome number, $x = 7$. Photosynthetic pathway, C_4.
Represented in Texas by the only member of the genus.

Thieret (2003e)

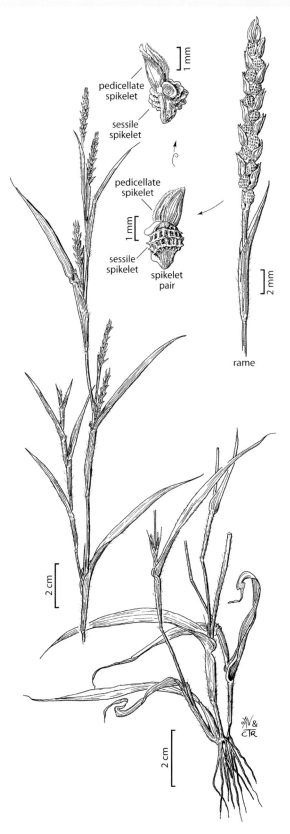

1. *Hackelochloa granularis* (L.) Kuntze (pitscale grass). This species is widely distributed and weedy in tropical regions of the world but apparently introduced in the Americas. Rare in East Texas and perhaps not a constant member of the Texas flora, although scattered throughout the Gulf Coast region of the southern United States. Some authors have included it in the genus *Mnesithea*. Not included in Gould (1975a, 1975b); Hatch, Gandhi, and Brown (1990); or Jones, Wipff, and Montgomery (1997).

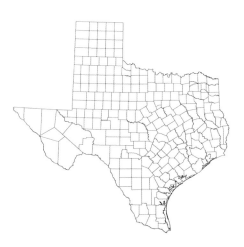

Hainardia Greuter
(Poöideae: Poeae)

Annual plants. Culm internodes solid. Uppermost sheaths partially enclosing the inflorescences. Ligules membranous. Blades flat or convolute. Inflorescences single, cylindric spike. Spikelets embedded in the rachis, dorsiventrally compressed, 1 or 2 florets; if second present, it is reduced and sterile. Lower glumes absent from all but the terminal spikelets. Upper glumes firm, rigid, 3–9-veined, unawned. Lower lemmas membranous, 3–5-veined, unawned. Basic chromosome number, $x = 13$. Photosynthetic pathway, C_4. Represented in Texas by the only member of the genus.

(Smith 2007)

1. *Hainardia cylindrica* (Willd.) Greuter (hardgrass, thintail). Annual with culms up to 50 cm long. Upper sheaths typically inflated and enclosing the lower part of the inflorescence. An introduced European species that occurs in coastal salt marshes and alkaline soils. Previously, this species was included in *Monerma* and *Rottboellia*.

Hemarthria R. Br.
(Panicoideae: Andropogoneae)

Perennials, usually cespitose, rhizomatous or occasionally rooting at the lowest nodes. Culms to 1.5 m tall, usually branched above the base. Sheaths mostly glabrous. Ligules membranous and ciliate. Blades linear to linear-lanceolate. Inflorescence terminal and axillary, with 1 or rarely 2 flattened rames borne on a common peduncle. Spikelets partially embedded in the rame axes. Disarticulation in the rames, sometimes tardily so. Spikelets in pairs, one sessile and one pedicellate, dorsally compressed. Sessile spikelets with 2 florets, lower reduced, upper perfect. Calluses blunt. Lower glumes coriaceous. Upper glumes equaling the lower glumes in length, chartaceous to membranous, sometimes awned. Lower florets reduced to hyaline lemmas. Upper floret bisexual, lemmas unawned. Pedicellate spikelet similar to sessile ones, staminate or sterile. Basic chromosome numbers, $x = 9$ and 10. Photosynthetic pathway, C_4. Represented in Texas by a single species.

(Allen 2003b)

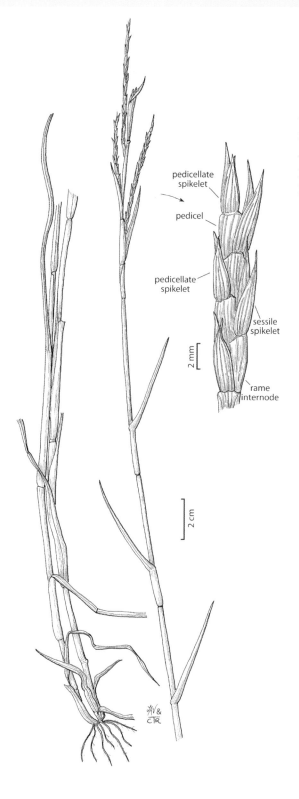

1. *Hemarthria altissima* (Poir.) Stapf & C. E. Hubbard (limpograss). Perennial with rhizomes and occasionally stolons. Culms up to 1.5 m long, flattened. Pedicel flattened, about as wide as or wider than the upper spikelet. An introduction from the Mediterranean region. Previously included in *Manisuris* and *Rottboellia* (Allen 2003b). Good livestock and poor wildlife values (Hatch, Schuster, and Drawe 1999).

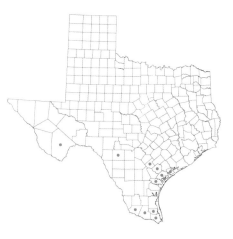

Hesperostipa (Elias) Barkworth
(Poöideae: Stipeae)

Cespitose perennials. Inflorescence a terminal, open or contracted, panicle of 1-flowered spikelets. Disarticulation above the glumes and below the florets. Glumes long and narrow, tapering from the base to a hairlike tip, longer than the lemmas. Lemmas indurate, margins overlapping at maturity, the upper portion fused into a ciliate crown, terminating into a stout, persistent, geniculate awn. Paleas equal to the lemma in length, pubescent, coriaceous. Basic chromosome number, $x = 11$. Photosynthetic pathway, C_3. Represented in Texas by 2 species and 2 subspecies. A segregate of *Stipa*.

1. Awns pilose.. 2. *H. neomexicana*
1. Awns scabrous .... 1. *H. comata*

(Barkworth 2007c)

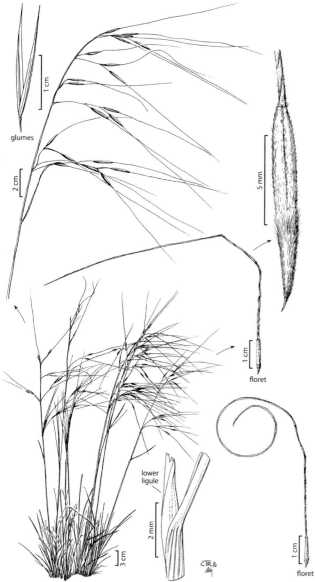

glumes

floret

lower
ligule

subsp. intermedia

subsp. comata

1. *Hesperostipa comata* (Trin. & Rupr.) Barkworth (needle-and-thread). Tufted perennial. Ligules to 7 mm long. Awns 65–225 mm long. Found exclusively west of the 100th meridian in mid- and shortgrass areas and pinyon-juniper woodlands. Of only fair to good forage value for livestock, and relatively poor for wildlife (Powell 1994). Two subspecies occur in the state: the typical one, subsp. *comata*; and subsp. *intermedia* (Scribn. & Tweedy) Barkworth. They can be distinguished by the following characters:

1. Terminal awn segment 40–120 mm long; lower nodes and panicles partially concealed by sheaths .... subsp. *comata*
1. Terminal awn segment 30–80 mm long; lower nodes exposed; panicles fully exserted .... subsp. *intermedia*

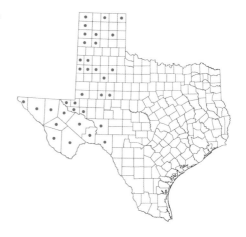

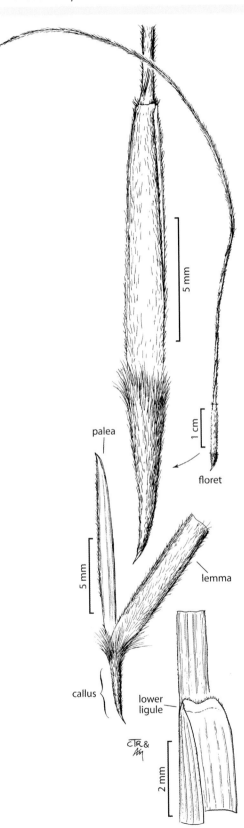

palea

lemma

callus

lower
ligule

CTR&
AM

5 mm

1 cm

floret

5 mm

2 mm

2. *Heterostipa neomexicana* (Thurb.) Barkworth (New Mexico needlegrass). Tufted perennial. Ligules to 3 mm long. Awns 120–220 mm long. Found exclusively west of the 100th meridian in rocky areas in grasslands and oak and pinyon woodlands. Of only fair to good forage value for livestock, and relatively poor for wildlife. The sharp callus can cause injury to animals' eyes, tongues, and ears (Powell 1994).

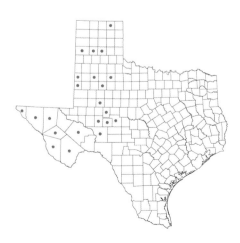

Heteropogon Pers.
(Panicoideae: Andropogoneae)

Tufted annuals and perennials. Blades flat or keeled and folded on the midnerve. Culms usually branched at the upper nodes, terminating in unilateral spicate racemes. Spikelets in pairs, 1 sessile and the other short-pediceled. Sessile spikelets, except for the lowermost, fertile and long-awned, with a sharp-pointed, bearded callus. Both the sessile and pediceled spikelets staminate or sterile (and awnless) on the lower portion of the raceme. Disarticulation in the rachilla, at the base of the callus of the fertile spikelets. Glumes of the fertile spikelet about equal, the first thick, indurate, and dark-colored at maturity, enclosing the second glume. Glumes of the sterile or staminate spikelets thin, the first broad, green, faintly many-nerved. Lemmas of the sterile and fertile florets membranous, the latter with a stout geniculate and twisted awn. Mature fertile spikelet superficially similar to the floret of stipoid grasses in appearance. Basic chromosome number, $x = 10$. Photosynthetic pathway, C_4. Represented in Texas by 2 species.

1. Plants perennial; glumes of pedicellate spikelets without glandular pits .... 1. *H. contortus*
1. Plants annual, glumes of pedicellate spikelets with glandular pits .... 2. *H. melanocarpus*

(Barkworth 2003i)

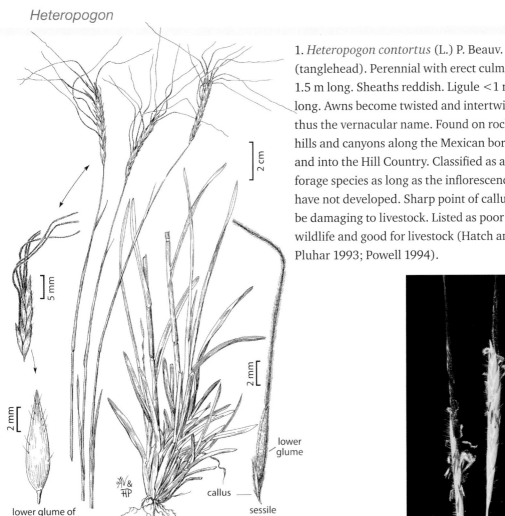

lower glume of
pedicellate spikelet

callus

lower
glume

sessile
spikelet

1. *Heteropogon contortus* (L.) P. Beauv. (tanglehead). Perennial with erect culms to 1.5 m long. Sheaths reddish. Ligule <1 mm long. Awns become twisted and intertwined, thus the vernacular name. Found on rocky hills and canyons along the Mexican border and into the Hill Country. Classified as a good forage species as long as the inflorescences have not developed. Sharp point of callus can be damaging to livestock. Listed as poor for wildlife and good for livestock (Hatch and Pluhar 1993; Powell 1994).

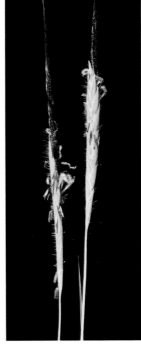

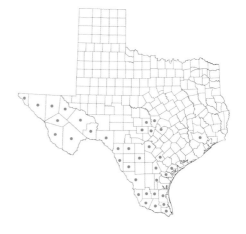

2. *Heteropogon melanocarpus* (Ell.) Benth. (sweet tanglehead). Annual to 2 m tall. Culms often with prop roots. Sheaths with a row of glands along the keel. Ligules 2–4 mm long. Found in open disturbed areas and sandy roadsides. Reported only from Waller County. Fresh plants smell like citronella oil. Perhaps not a constant member of the Texas flora.

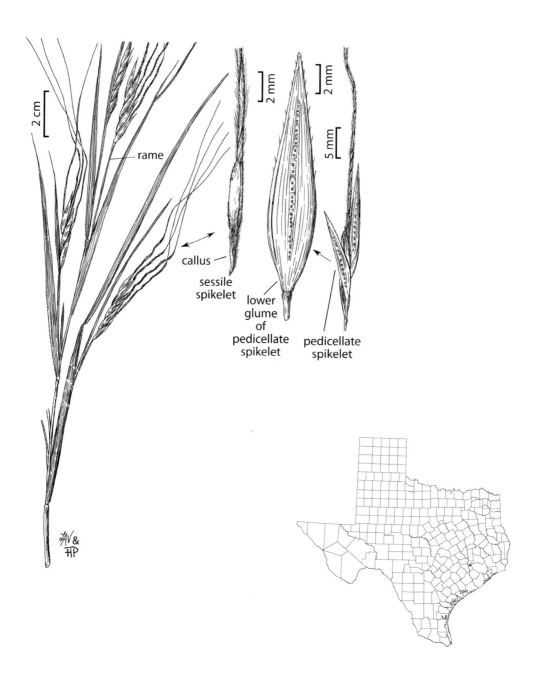

rame

callus

sessile
spikelet

lower
glume
of
pedicellate
spikelet

pedicellate
spikelet

2 cm

2 mm

2 mm

5 mm

Hilaria Kunth.
(Chloridoideae: Cynodonteae)

Perennials, mostly stoloniferous. Inflorescence a slender, dense, bilateral spike, the spikelets in clusters of 3 at each node of a zigzag rachis, the clusters deciduous as a whole. Spikelets of the cluster dissimilar, the 2 lateral ones 2-flowered, staminate, and the central one 1-flowered, perfect. Glumes firm, flat, usually asymmetrical, bearing an awn on one side from about the middle. Lemmas thin, 3-nerved, awned or awnless. Palea about as large as the lemma and similar in texture. Basic chromosome number, $x = 9$. Photosynthetic pathway, C_4. Represented in Texas by 2 species and 2 varieties. Fungal infections of the plants can result in ergot alkaloid poisoning in livestock.

Rhizomatous perennials historically included in this genus are now found in *Pleuraphis*.

1. Glumes pale to purplish, awned from below midlength .... 1. *H. belangeri*
1. Glumes gray to dark brown, awned from above midlength .... 2. *H. swallenii*

(Barkworth 2003j)

1. *Hilaria belangeri* (Steud.) Nash (curly mesquite). Perennial, usually stoloniferous, with culms to 35 cm long. Glumes of lateral spikelets with scattered glands at the base or glands absent. Found on mesas and plains of South Texas, Edwards Plateau, and High Plains. Listed as poor for wildlife and fair for livestock (Hatch and Pluhar 1993). Powell (1994) called it a significant forage species. Two varieties occur in Texas: the typical variety, var. *belangeri*; and var. *longifolia* (Vasey) Hitchc. They can be distinguished by the following characters:

1. Stolons present; ligules about 1.0–1.5 mm long ....var. *belangeri*
1. Stolons absent; ligules about 2.0–2.5 mm long .... var. *longifolia*

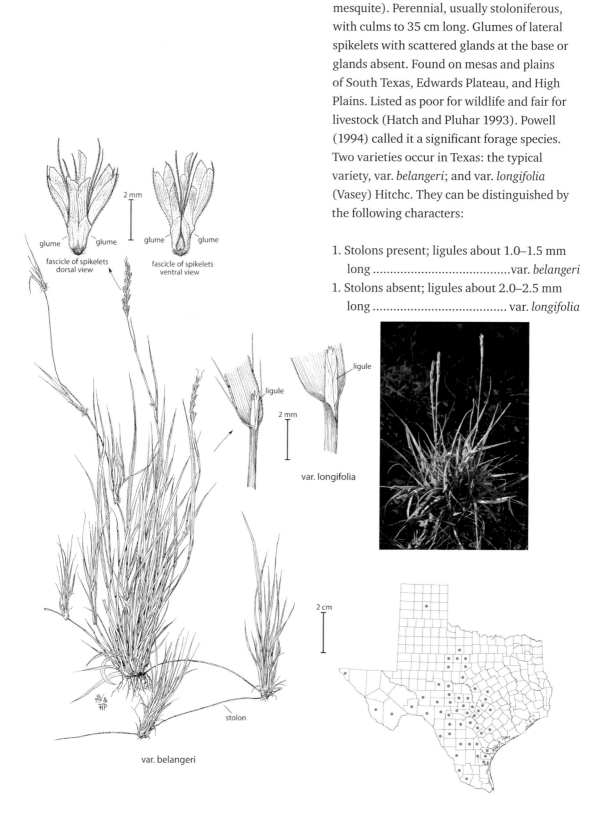

2 mm

glume   glume
fascicle of spikelets
dorsal view

glume   glume
fascicle of spikelets
ventral view

ligule

2 mm

ligule

var. longifolia

2 cm

stolon

var. belangeri

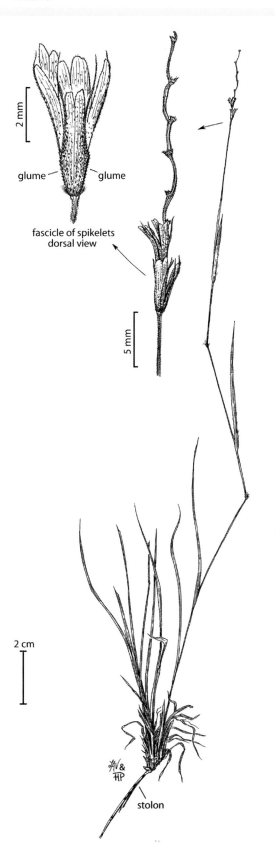

2 mm

glume — — glume

fascicle of spikelets
dorsal view

5 mm

2 cm

AV &
HP

stolon

2. *Hilaria swallenii* Cory (Swallen's curly mesquite). Stoloniferous perennial with culms to 35 mm long. Lateral glumes evenly and densely spotted with glands. Found on plains and rocky mesas of West Texas. Most common *Hilaria* in the Trans-Pecos and produces good forage throughout the year (Powell 1994).

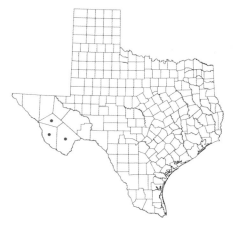

Holcus L.
(Poöideae: Poeae)

Perennial in Texas, with velvety pubescent herbage. Culms weak and rather succulent. Sheaths open to the base; ligules a short, fringed membrane, often puberulent; auricles absent. Inflorescence a contracted panicle. Spikelets with 2 florets, the lower floret perfect, the upper floret smaller, staminate or sterile. Disarticulation below the glumes. Glumes large, thin, subequal, longer than the florets, the lower 1-nerved, the upper 3-nerved. Lemma of the lower floret faintly 3–5-nerved, awnless, blunt at the apex. Lemma of the upper floret with a short, often hooked awn near the apex. Palea of the perfect floret as long as or slightly longer than the lemmas. Basic chromosome number, $x = 7$. Photosynthetic pathway, C_3. Represented in Texas by a single species.

(Standley 2007a)

glumes

caryopsis

florets

2 mm

ligule

2 mm

2 cm

1. *Holcus lanatus* L. (velvetgrass, creeping softgrass, Yorkshire fog). A perennial species that can grow to 1 m in height. Internodes, sheaths, and blades covered with a soft, dense pubescence. Lemmas with a bifid apex, with an awn up to 2 mm that becomes twisted and hooks at maturity. An early European introduction that occurs in marshes, disturbed sites, and waste places. Reported to cause inflammation of mucous membranes of the mouth of grazing animals (Diggs et al. 2006).

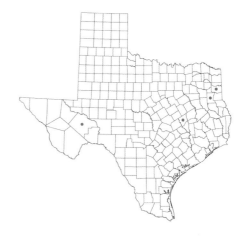

Hopia Zuloaga & Morrone
(Panicoideae: Paniceae)

Perennial with short rhizomes and long stolons. Nodes swollen and villous, especially on the stolons. Ligules papery; blades linear-lanceolate. Inflorescences contracted panicles. Panicle branches spikelike, appressed, 1-sided. Spikelets borne in pairs, ellipsoid to obovoid. Lower glumes nearly as long as the spikelet, 5–7-veined. Upper glumes 7–11-veined, blunt. Lower florets staminate; lower paleas well developed. Upper florets indurate, smooth. Basic chromosome number, $x = 10$. Photosynthetic pathway, C_4. Represented in Texas by the only member of the genus.

(Zuloaga et al. 2007)

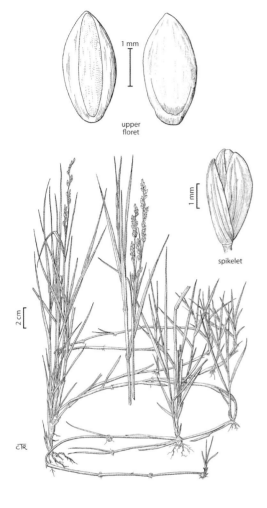

upper
floret

spikelet

1. *Hopia obtusa* (Kunth) Zuloaga & Morrone (vine mesquite). Strongly stoloniferous perennial with swollen and villous nodes. Stolons sometimes meters long and draping over small shrubs. Found on clayey to sandy banks, ditches, and moist depressions and swales on rangelands. A recent segregate of *Panicum*. Palatable species that can withstand heavy grazing by livestock, fair for wildlife, and used by native Americans for food (Powell 1994).

Hordeum L.
(Poöideae: Triticeae)

Annual. Inflorescence a spicate raceme with 3 spikelets at a node (triad), all 3 fertile and sessile on a strong rachis. Spikelets each with 1 perfect floret. Glumes side by side, narrowly lanceolate to awnlike. Lemma rounded on the back, conspicuously awned. Palea shorter than the lemma and adnate to the caryopsis. Basic chromosome number, $x = 7$. Photosynthetic pathway, C_3. Represented in Texas by a single species and subspecies. Other species, typically included in this genus, with lateral spikelets reduced or absent, are in *Critesion*.

(Bothmer, Baden, and Jacobsen 2007)

1. *Hordeum vulgare* L. (barley). Annual that can grow to 1.5 m in height. Auricles well developed. Inflorescence a spike, 3 spikelets per node, and each producing seed. Rachises not disarticulating at maturity. Lemmas typically long-awned. A native of Eurasia commonly cultivated in all temperate areas of the world. Frequently escapes from cultivation and is found in abandoned fields, along roadsides, and in disturbed sites. The Texas species belongs to the typical subspecies, subsp. *vulgare*. Other members of the genus have been moved to *Critesion*. Good forage for livestock and fair wildlife values (Hatch, Schuster, and Drawe 1999).

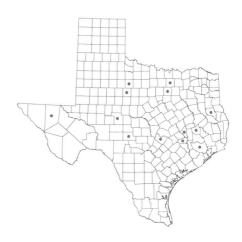

Hyparrhenia Anderss. ex Stapf
(Panicoideae: Andropogoneae)

Perennials, mostly tall, coarse, and robust. Flowering culms terminating in a series of short branches, each bearing a panicle of 2 flowering branchlets, these separate or on a common peduncle and subtended by an expanded, spathelike leaf sheath. Spikelets in pairs of 1 sessile and 1 pediceled, the lowermost pairs sessile, awnless, staminate or neuter, the uppermost pairs as in *Andropogon*, with the sessile spikelet bisexual and awned and the pediceled spikelet staminate or neuter and usually awnless. In some species only 1 of the sessile spikelets of each branch is fertile. Disarticulation usually in the rachis, the spikelets falling singly or in clusters. Glumes of the sessile spikelet large, firm, flat or rounded, the first exceeding the second in length and thickness. Lemma of the sterile floret and lemma and palea of the fertile floret thin and membranous, the palea minute or absent. Lemma of the fertile floret usually with a stout, geniculate awn. Basic chromosome number, $x = 10$. Photosynthetic pathway, C_4. Represented in Texas by 2 species.

1. Spikelets with whitish to dark yellow hairs ....*H. hirta*
1. Spikelets with reddish hairs .... 2. *H. rufa*

(Barkworth 2003k)

1. *Hyparrhenia hirta* (Nees) Stapf (thatching grass). Cespitose perennial with short rhizomes. Lemmas of sessile spikelets awned, awn to 3.5 cm long. An African introduction typically found on stony or rocky soils. Used for thatching. Cultivated and perhaps not a constant member of the Texas flora.

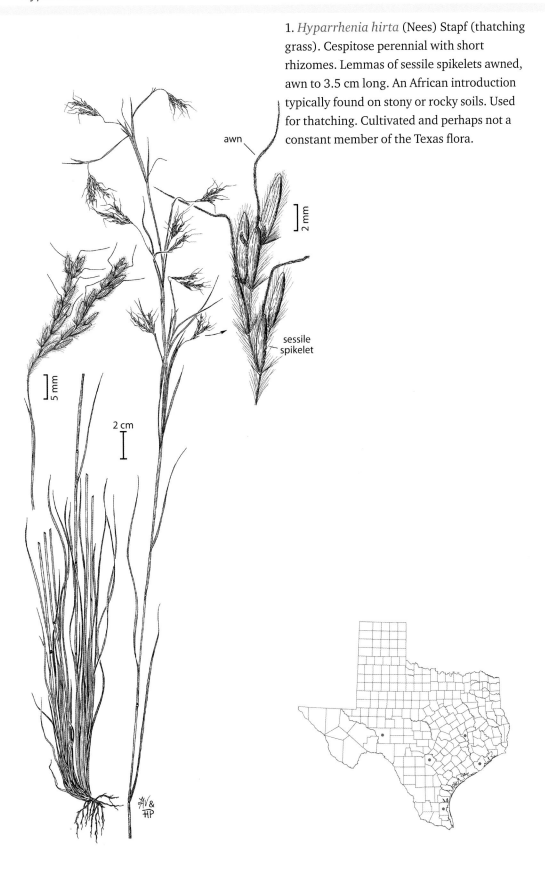

awn

2 mm

sessile spikelet

5 mm

2 cm

2. *Hyparrhenia rufa* (Nees) Stapf (jaragua grass). Cespitose perennial with short rhizomes. Lemmas of sessile spikelets awned, awns 2–3 cm long. An Eastern Hemisphere introduction that grows in ditches, roadsides, swamps, and flatwoods. Reported only from Brazos County, and perhaps not a constant member of the Texas flora.

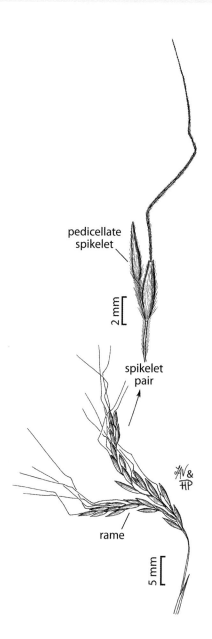

pedicellate spikelet

2 mm

spikelet pair

rame

5 mm

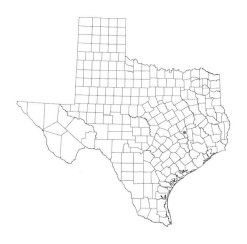

Imperata Cirillo
(Panicoideae: Andropogoneae)

Coarse, mostly tall perennials with stout, creeping rhizomes. Inflorescence a dense, contracted, narrowly oblong or slender and spikelike panicle, the spikelets more or less obscured by silky hairs. Spikelets all alike, awnless, with a tuft of long, silky hair at the base. Disarticulation below the spikelet, the long-hairy callus falling with the spikelet. Glumes subequal, thin, several-nerved. Lemmas of the sterile and fertile florets hyaline, and palea of the fertile floret membranous or hyaline. Basic chromosome number, $x = 10$. Photosynthetic pathway, C_4. Represented in Texas by 2 species.

1. Stamens 2, filaments not dilated at base...2. *I. cylindrica*
1. Stamens 1, filament dilated at base ....1. *I. brevifolia*

(Gabel 2003)

1. *Imperata brevifolia* Vasey (satintail).
A large cespitose perennial with culms to
1.3 m long. Panicles with lower branches
diverging. Found in semiarid regions. It has
not been recently collected and perhaps is not
a constant member of the Texas flora except as
an ornamental.

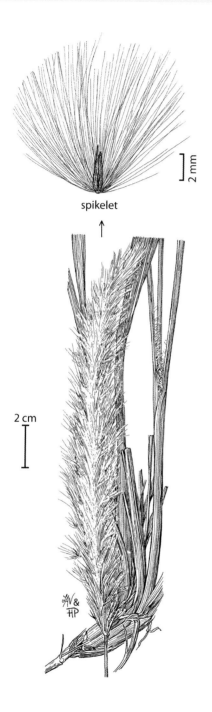

spikelet

2 mm

2 cm

2. *Imperata cylindrica* (L.) P. Beauv.
(cogongrass, bladygrass, cottongrass,
satintail). A large cespitose perennial with
rhizomes. Plants can grow to over 2 m in
height. Panicles narrowly cylindric. Listed as
a noxious weed by the U.S. Department of
Agriculture. Cultivar most common in Texas
is "Red Baron" with reddish foliage. Found
along roadsides and in other disturbed areas.
Reported only from Brazos County.

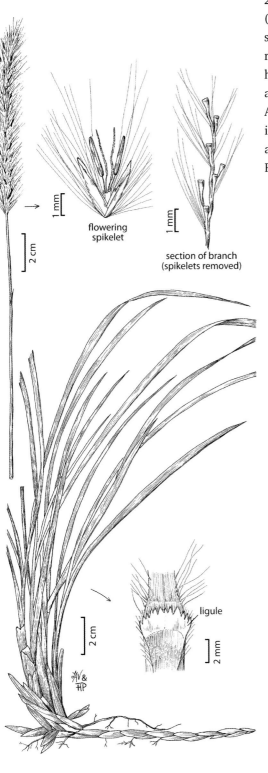

flowering
spikelet

section of branch
(spikelets removed)

ligule

Ixophorus Schltdl.
(Panicoideae: Paniceae)

Perennial, occasionally with succulent culms. Culms to 1.5 m tall and 1 cm thick. Ligules to 1.5 mm long, ciliated membrane. Leaves with flat blades, midrib often white. Inflorescence an open, terminal panicle. Each pedicel with a terete bristle that can be up to 1.2 cm long. Spikelets dorsally compressed, with 2 florets. Lower floret staminate, upper floret perfect. Lower glumes less than ½ as long as the upper glume. Upper lemma awnless. Stigmas bright red. Disarticulation below the glumes. Basic chromosome number, x = unknown. Photosynthetic pathway, C_4. Represented in Texas by a single species. Genus not included in Gould (1975b); Hatch, Gandhi, and Brown (1990); or Jones, Wipff, and Montgomery (1997), but mapped in Turner et al. (2003).

(Hiser 2003)

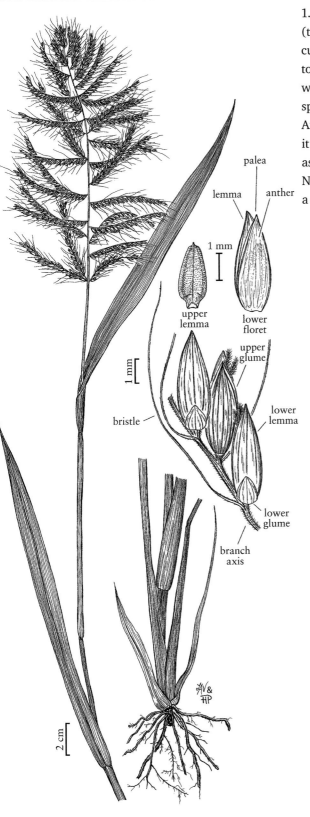

1. *Ixophorus unisetus* (J. Presl) Schltdl. (turkey grass, crane grass). Perennial with culms to 1.5 m long and 1 cm thick. Blades to 60 cm long and 2 cm wide. Panicle open, with 4–50 branches. Spikelets 3–4 mm long. A species of Mexico, Central America, and South America. Collected in Kleberg County, where it was being experimentally grown and tested as a potential forage species for South Texas. Not known to have persisted, thus may not be a constant member of the Texas flora.

Koeleria Pers.
(Poöideae: Poeae)

Perennial, cespitose. Leaf blades frequently corkscrewing at maturity. Sheaths open to the base; ligules short, membranous; auricles absent. Inflorescence usually a contracted, spicate panicle, except at anthesis when the panicle branches are extended. Spikelets usually 2–4-flowered, shiny, strongly flattened, rachilla usually extended beyond the uppermost floret as a slender bristle or bearing a vestigial floret. Disarticulation above the glumes and between the florets. Glumes large, thin, acute, unequal, the lower shorter and 1-nerved, the upper longer, broader, and 3–5-nerved. Lemmas thin, shining, the lowermost usually slightly longer than the glumes, awnless or rarely awned from a bifid apex. Palea large, scarious, and colorless. Basic chromosome number, $x = 7$. Photosynthetic pathway, C_3. Represented in Texas by a single species.

(Standley 2007b)

palea, lemma, glumes, floret, rachis, 1 mm, 2 mm, 2 cm

1. *Koeleria macrantha* (Ledeb.) Schult. (prairie junegrass). A loosely cespitose perennial. Culms glabrous, pubescent below the inflorescence and near the nodes. Leaf blades usually twisted. Inflorescences densely contracted except at anthesis when branches spread, interrupted at base, golden. Lemmas shiny, midvein sometimes extended into a short (1 mm) awn. Found in grasslands, dry prairies, and woodlands. A high-quality forage for livestock and wildlife (Powell 1994).

Lachnagrostis Trin.
(Poöideae: Poeae)

Annual or short-lived perennial, occasionally with rhizomes. Culms up to 80 cm tall, erect or geniculate. Sheaths open; auricles absent; ligules membranous; blades flat or folded. Inflorescences are panicles, erect or lax, open, usually more than ½ as wide as long. Branches naked at base, and spikelets situated near the tips. Spikelets with 1 or rarely 2 florets, laterally compressed; rachilla markedly prolonged beyond the base of the floret. Disarticulation above the glumes, in perennials the inflorescences detaching, but persistent in annuals. Glumes exceeding the florets, unawned. Lemmas usually pubescent, apices often dentate, usually awned. Paleas ½ to as long as lemmas, hyaline, 2-veined, extending into mucro. Basic chromosome number, $x = 7$. Photosynthetic pathway, C_3. Represented in Texas by a single species.

(Harvey 2007b)

1. *Lachnagrostis filiformis* (G. Forst.) Trin. (Pacific bent). Cespitose perennial, sometimes rhizomatous. Ligules 2–8 mm long, lacerated. Spikelets green to yellowish and tinged with purple. This is a Southern Hemisphere species introduced into disturbed sites in Texas. Last collection in Texas was in 1902, so perhaps not a constant member of the state's flora. The genus has frequently been included in *Agrostis* but differs in a suite of characters.

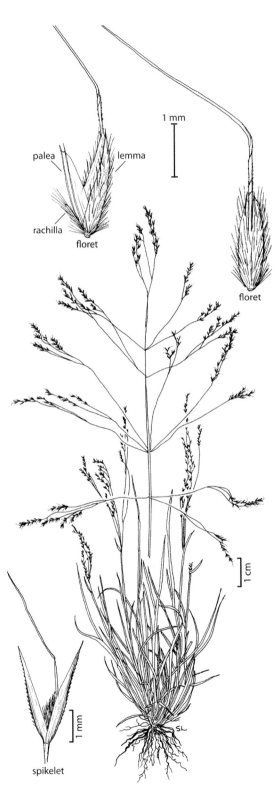

palea   lemma

rachilla

floret

1 mm

floret

SL

spikelet

1 mm

1 cm

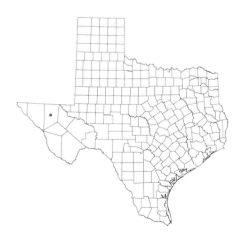

Lamarckia Moench.
(Poöideae: Poeae)

Low, short-lived annual with weak culms, soft, flat blades, and contracted panicles of fascicled spikelets. Terminal spikelet of each fascicle fertile, those below sterile, the fascicles falling entire. Fertile spikelet with a single perfect floret on a slender rachilla and a rudimentary floret borne above on a long bristlelike stipe. Lemmas of both the fertile and reduced florets with a delicate awn 5–10 mm long. Sterile spikelets mostly with 3–6 empty florets, the lemmas well developed, broad, awnless, scarious above. Glumes of the fertile and sterile spikelets similar, narrow, pointed. Basic chromosome number, $x = 7$. Photosynthetic pathway, C_3. Represented in Texas by the only member of the genus.

(Clark 2007)

1. *Lamarckia aurea* (L.) Moench. (goldentop). This adventive is a cool-season annual weed of disturbed soils. It is a very attractive plant that has potential as an ornamental, but it does become weedy. No report of this species in Gould (1975b) or Clark (2007); however, listed by Jones, Wipff, and Montgomery (1997) to be in the state. Perhaps not a constant member of the Texas flora.

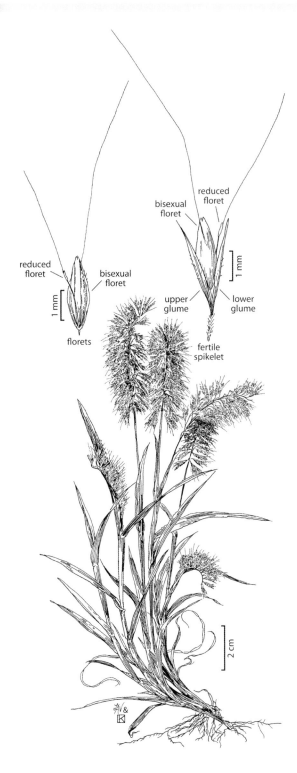

reduced floret

bisexual floret

bisexual floret

reduced floret

florets

upper glume

lower glume

fertile spikelet

1 mm

2 cm

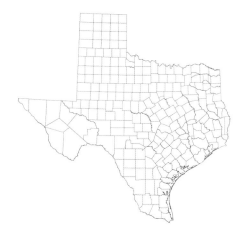

Leersia Sw.
(Ehrhartoideae: Oryzeae)

Rhizomatous perennials with flat blades and open panicles. Ligule a short, firm membrane, often continued laterally as short sheath auricles. Spikelets 1-flowered, strongly compressed laterally, awnless, subsessile and crowded at the branch tips. Disarticulation below the spikelet. Glumes absent. Lemma firm or indurate, boat-shaped, 5-nerved, tightly enclosing the margins of a firm, narrow, usually 3-nerved palea. Stamen number varying from 1 to 6. Basic chromosome number, $x = 12$. Photosynthetic pathway, C_3. Represented in Texas by 5 species.

1. Panicle branches with spikelets nearly to base .... 1. *L. hexandra*
1. Panicle branches bare of spikelets on lower 1.5–4.0 cm.
   2. Spikelets 3–4 mm broad .... 2. *L. lenticularis*
   2. Spikelets 2 mm or less broad.
      3. Spikelets 4.0–5.5 mm long and 1.5–2.0 mm broad .... 4. *L. oryzoides*
      3. Spikelets 1.3–3.5 mm long and 1 mm or less broad.
         4. Spikelets 2.2–3.5 mm long, finely hispid; culms decumbent; rhizomes well
            developed .... 5. *L. virginica*
         4. Spikelets 1.3–2.0 mm long, glabrous; culms erect; rhizomes not
            developed .... 3. *L. monandra*

(Gould 1975b; Pyrah 2007)

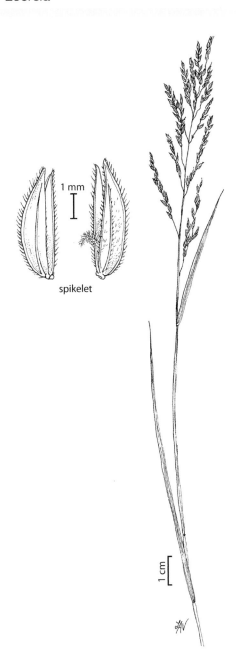

spikelet

1. *Leersia hexandra* Sw. (clubhead cutgrass, southern cutgrass). Perennial with elongated, but not scaly, rhizomes. Ligules 1–3 mm long. Panicle branches 1–2 per node, spikelet-bearing to near the base. Lemmas and paleas ciliate on the keels. Found in wet soils along streams, ditches, swales, swamps, and lakes. Often grows in shallow water. "Palatable to livestock but of low forage value due to limited abundance; provides seed for waterfowl" (Hatch, Schuster, and Drawe 1999, 184).

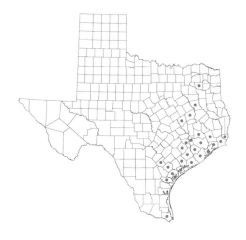

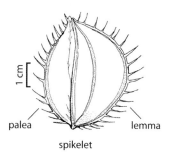

palea   lemma

spikelet

2. *Leersia lenticularis* Michx. (catchfly grass, oatmeal grass). Perennial with elongated, scaly rhizomes. Ligules 0.5–1.5 mm long. Panicle branches 1 or rarely 2 per node, lacking spikelets on the lower $^1/_3$ Lemmas and paleas ciliate on the keels. Found in wet, marshy soils along streams, lakeshores, and swales.

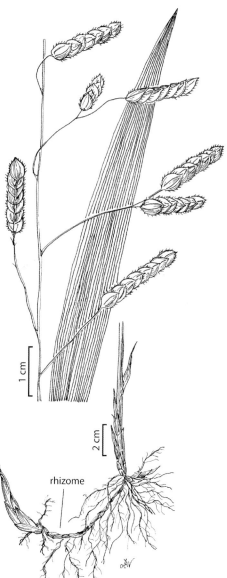

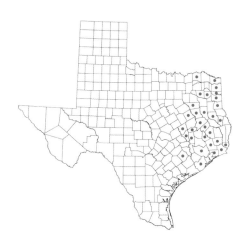

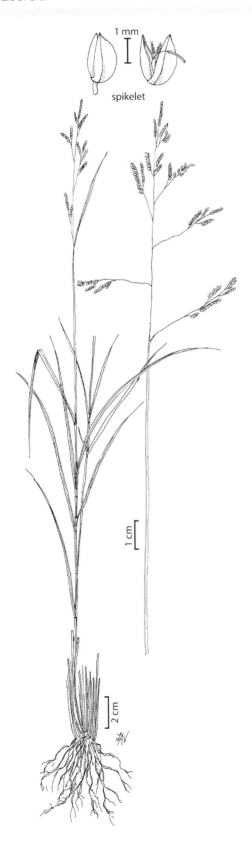

1 mm

spikelet

1 cm

2 cm

3. *Leersia monandra* Sw. (bunch cutgrass, canyongrass, whitegrass). Cespitose perennial, rhizomes lacking. Ligules 1.5–4.0 mm long. Panicle branches 1 per node, with spikelets on distal ¹/₃. Lemmas and paleas glabrous. Found in wet ditches, in swales, along lakeshores, and in shaded woodland sites. Sometimes sold as an ornamental. Poor livestock and wildlife values (Hatch, Schuster, and Drawe 1999).

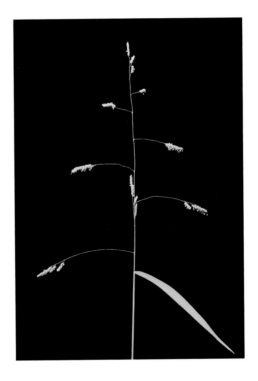

4. *Leersia oryzoides* (L.) Sw. (rice cutgrass).
Perennial with elongated, but not scaly,
rhizomes. Ligules 0.5–1.0 mm long. Panicle
branches with 2 or more per lower node, 1
on upper nodes; lower ⅓ naked of spikelets.
Lemmas and paleas ciliate on the keels.
Cleistogenes found enclosed in upper leaf
sheaths. Found mostly in saturated soils along
streams, rivers, ponds, lakes, and swales.
Often grows in shallow water. Seeds provide
nourishment to ducks and other waterfowl
(Gould 1975b).

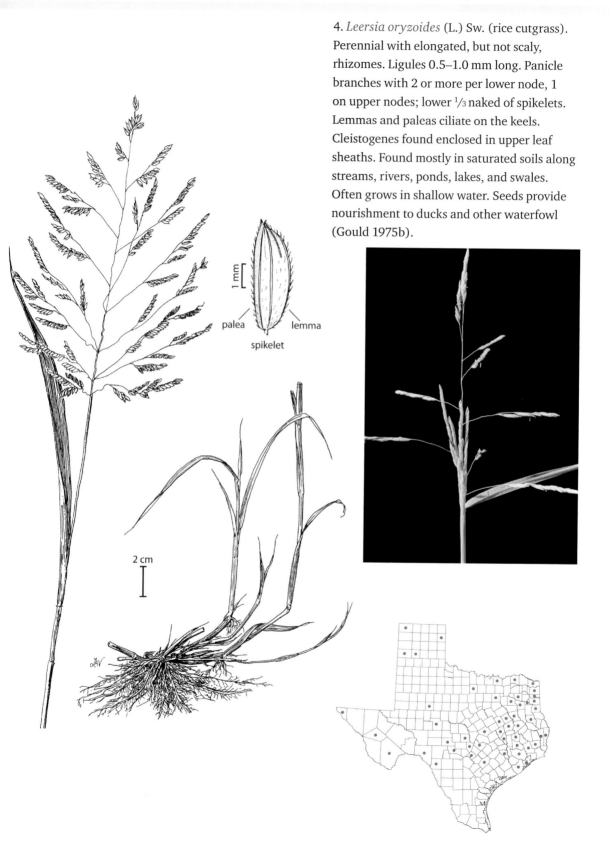

palea   lemma

spikelet

1 mm

2 cm

spikelet

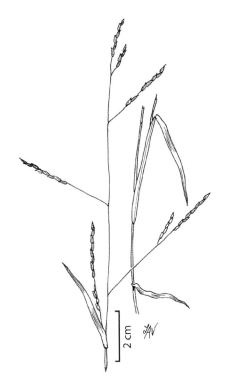

5. *Leersia virginica* Willd. (whitegrass, white cutgrass). Perennial with short, scaly rhizomes. Ligules 1–3 mm long. Panicle branches 1 per node, with spikelets lacking on the lower ⅓. Lemmas and paleas ciliate or sometimes glabrous on keels. Found in low, wet sites along streams and lakes and in marshes, ditches, and swales.

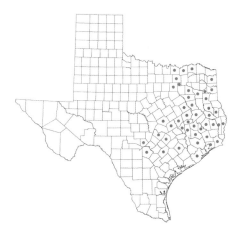

Leptochloa P. Beauv.
(Chloridoideae: Cynodonteae)

Annuals and perennials, cespitose. Sheaths open; ligule a ciliate membrane; auricles absent. Inflorescence a panicle, with few to numerous unbranched primary branches distributed along the upper portion of the culm or clustered near the tip. Spikelets with 2 to several florets, overlapping and closely spaced on the branches. Disarticulation above the glumes and between the florets. Glumes 2, thin, 1-nerved or the second occasionally 3-nerved, acute, awnless or with a mucronate tip. Lemmas 3-nerved, frequently puberulent on the veins, apex acute to obtuse or notched, awnless, awned, or with a mucro. Palea well developed, and frequently puberulent on the veins. Caryopsis apex 2-horned. Basic chromosome number, $x = 10$. Photosynthetic pathway, C_4. Represented in Texas by 8 species and 4 subspecies. Telfair (2006) listed the genus as important for wildlife.

1. Panicle branches digitate or subdigitate.
    2. Lemma apices obtuse to truncate, often emarginated .... 2. *L. dubia*
    2. Lemma apices acute.
        3. Panicle branches digitate .... 1. *L. chloridiformis*
        3. Panicle branches subdigitate .... 7. *L. virgata*
1. Panicle branches racemose.
        4. Ligules 2–8 mm long, attenuate .... 3. *L. fusca*
        4. Ligules 0.3–5.0 mm long, truncate to obtuse.
            5. Sheaths with papillose-based hairs .... 5. *L. panicea*
            5. Sheaths glabrous or if hairy, hairs not papillose-based.
                6. Panicles with 25–150 branches.
                    7. Lemmas 2.4–3.0 mm long; spikelets 4–5 mm long .... 6. *L. panicoides*
                    7. Lemmas 1.2–2.4 mm long; spikelets 2.8–3.4 mm long .... 4. *L. nealleyi*
                6. Panicles with 2–25 branches.
                    8. Plants perennial.
                        9. Lemmas 4–5 mm long, apices obtuse to truncate, often emarginated; secondary panicles concealed in lower sheaths .... 2. *L dubia*
                        9. Lemmas 1.5–3.6 mm long; secondary panicles not in lower sheaths .... 7. *L. virgata*
                    8. Plants annual.
                        10. Panicles 2–17 cm long, with 5–23 branches .... 8. *L. viscida*
                        10. Panicles 20–35 cm long, with 20–90 branches .... 6. *L. panicoides*

(Snow 2003b; Gould 1975b)

1. *Leptochloa chloridiformis* (Hack.) Parodi (Argentine sprangletop). Stout perennial with erect culms to 1.5 m tall. Ligules 1–2 mm long, a lacerated and ciliated membrane. Panicle branches digitate, 5–20. Lemmas acute to obtuse, minutely emarginated, with a mucro. Introduced from South America and reported only from Cameron County from dry, disturbed sites. Perhaps not a permanent member of the Texas flora.

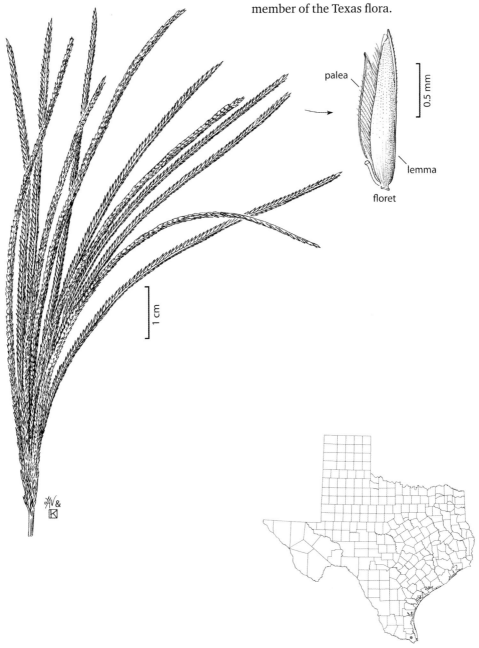

palea

0.5 mm

lemma

floret

1 cm

2. *Leptochloa dubia* (Kunth) Nees (green sprangletop, Texas crowfoot). Tufted perennial without rhizomes or stolons. Ligules a fringed membrane about 0.5 mm long. Panicle branches 2–15. Lemmas truncate and usually emarginated. Cleistogamous spikelets developed in the axis of the lower sheaths. Found in grasslands and rocky and sandy soils. Forage value reported as fair for wildlife and good for livestock (Hatch and Pluhar 1993); but it can accumulate hydrogen cyanide in new growth after rains (Diggs et al. 2006). Powell (1994) reported it as an excellent forage species in the Trans-Pecos.

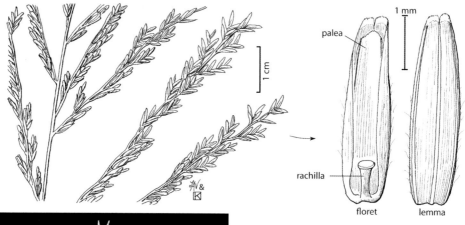

1 cm

1 mm

palea

rachilla

floret

lemma

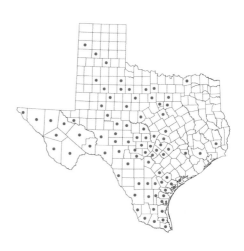

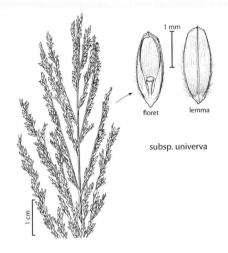

floret   lemma

subsp. univerva

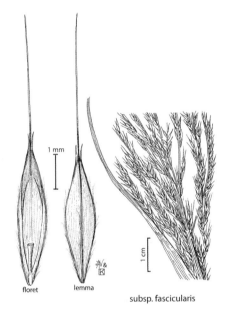

floret   lemma

subsp. fascicularis

3. *Leptochloa fusca* (L.) Kunth (bearded sprangletop, Mexican sprangletop). Annuals or weak perennials with somewhat succulent culms. Ligules 2–8 mm long, membranous, sometimes with lateral lobes appearing as auricles. Panicle branches 3–35. Lemmas sometimes with a dark spot at the base, emarginated to bifid, unawned, mucronate, or awned, awns to 3 mm long. Found along streams, lakeshores, and swales and in brackish marshes along the coast. Good forage and fair for wildlife (Hatch, Schuster, and Drawe 1999). Two subspecies are found in Texas: subsp. *fascicularis* (Lam.) N. Snow; and subsp. *uninervia* (J. Presl) N. Snow. They can be distinguished by the following characters:

1. Uppermost leaf blades exceeding the panicles; panicles partly enclosed by upper sheaths; mature lemmas with a dark spot on the base ....subsp. *fascicularis*
1. Uppermost leaf blade exceeded by the panicles; panicles exserted from upper sheaths; mature lemmas lacking a dark spot....................................subsp. *uninervia*

Gould (1975b) and Jones, Wipff, and Montgomery (1997) recognize these subspecies as distinct species.

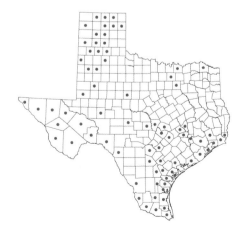

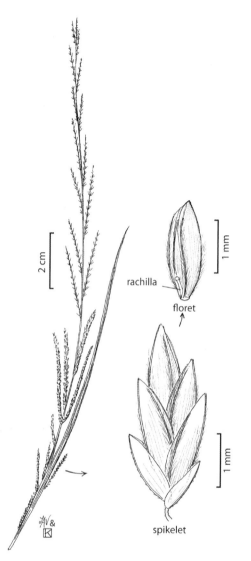

rachilla

floret

spikelet

2 cm

1 mm

1 mm

4. *Leptochloa nealleyi* Vasey (Nealley's sprangletop). Annual with culms to 2.5 m long. Ligules 1.5–3.0 mm long, truncate, erose. Panicle branches 25–75, appressed. Lemmas acute, awnless. Found along the coast, often in saline sites; also in wet, muddy soils of marshes, swales, and rivers. Poor livestock and fair wildlife values (Hatch, Schuster, and Drawe 1999).

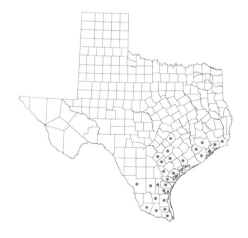

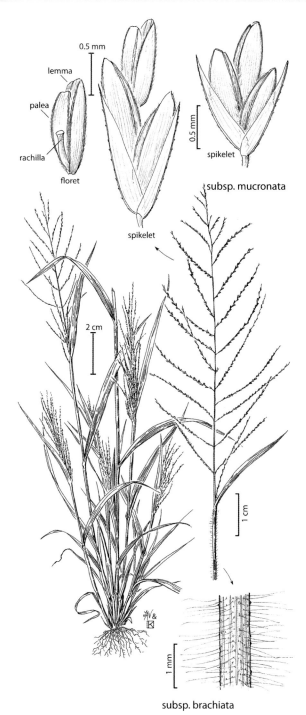

0.5 mm

lemma

palea

rachilla

floret

spikelet

0.5 mm

spikelet

subsp. mucronata

2 cm

1 cm

1 mm

subsp. brachiata

5. *Leptochloa panicea* (Retz.) Ohwi (red sprangletop, Mississippi sprangletop). Annual with erect culms to 1.5 m long. Ligules 0.5–3.0 mm long. Panicle branches 3–100. Lemmas acute to obtuse. A weedy species of disturbed sites such as roadsides, gardens, and abandoned fields.

Two subsp. are found in Texas: subsp. *brachiata* (Steud.) N. Snow; and subsp. *mucronata* (Michx.) Nowack. Subsp. *brachiata* is considered a noxious weed by the U.S. Department of Agriculture. Gould (1975b) recognized *L. filiformis* (Lam.) P. Beauv., which is now considered a synonym of subsp. *brachiata*. Jones, Wipff, and Montgomery (1997) recognized *L. mucronata* (Michx.) Kunth, which is now subsp. *mucronata*. The subspecies can be distinguished by the following characters:

1. Glumes exceeding the florets; lemmas 0.9–1.2 mm long.....................subsp. *mucronata*
1. Glumes not, or only slightly, exceeding the florets; lemmas 1.3–1.7 mm long ....
....subsp. *brachiata*

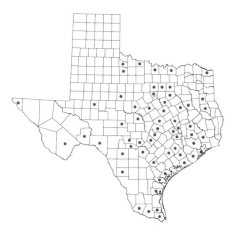

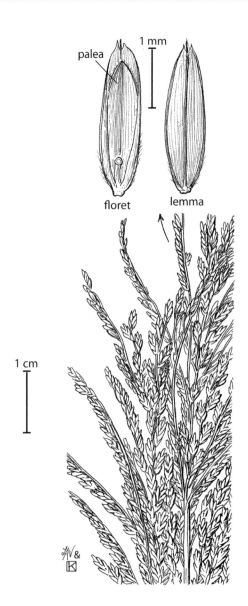

6. *Leptochloa panicoides* (J. Presl) Hitchc. (Amazon sprangletop). Annual with culms to 1.2 m long. Ligules membranous, 2–4 mm long. Panicle branches 20–90, erect. Lemmas acute, unawned or sometimes mucronate. Found on muddy shores of lakes, in swamps, and in swales. Relatively rare.

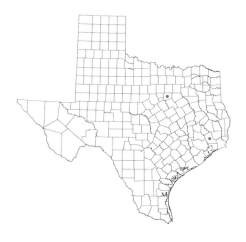

7. *Leptochloa virgata* (L.) P. Beauv. (tropical sprangletop). Erect perennial with culms to 2 m tall. Ligules a minute (0.3–1.0 mm) fringed membrane. Panicle branches 9–25. Lemmas acute, unawned, mucronate or awned, awns to 11 mm long. Found in moist Coastal Prairies areas along swales, ditches, and marshes.

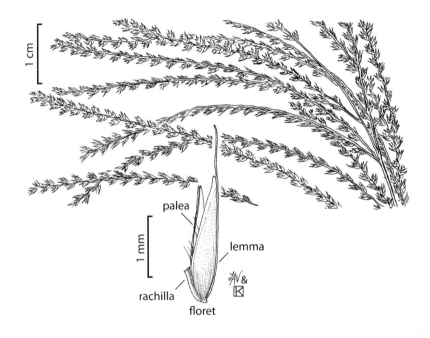

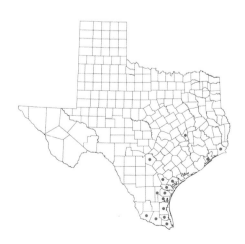

8. *Leptochloa viscida* (Scribn.) Beal (Sonoran sprangletop). Annual with often prostrate culms to 60 m long. Ligules membranous, 1.0–2.5 mm long, truncate. Panicles with 5–23 branches. Lemmas acute, mucronate. A ruderal species of waste sites. Reported only from El Paso and Jeff Davis counties.

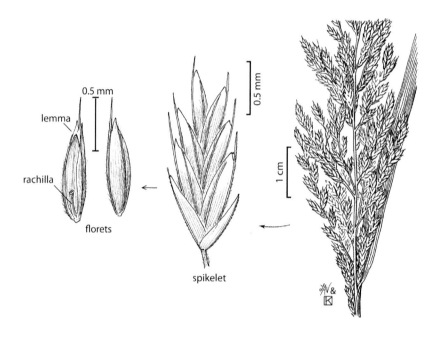

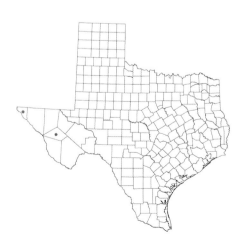

Leymus Hochst.
(Poöideae: Triticeae)

Perennials, cespitose or occasionally with rhizomes. Inflorescence a bilateral spike or spicate raceme. Disarticulation above the glumes and between the florets. Spikelets 2 or more per node, sessile, with 2–7 florets. Glumes shorter than the lemmas, narrowly lanceolate to awl-shaped, stiff. Lemmas acute to short-awned. Palea shorter than the lemma. Basic chromosome number, $x = 7$. Photosynthetic pathway, C_3. Represented in Texas by 2 species.

A segregate of *Elymus*.

1. Lemmas densely villous; culms 1–3 mm thick .... 2. *L. triticoides*
1. Lemmas glabrous or with a few scattered hairs; culms 2.5–12 mm thick............. 1. *L. arenarius*

(Barkworth 2007d)

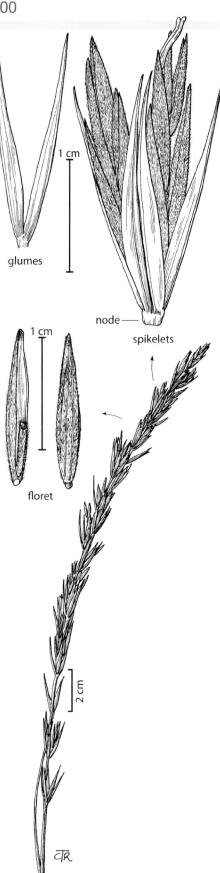

1 cm

glumes

node —

spikelets

1 cm

floret

2 cm

CTR

1. *Leymus arenarius* (L.) Hochst. (European dunegrass). Rhizomatous perennial, strongly glaucous. Ligules 0.2–2.5 mm long. Lemmas densely villous. A European introduction grown as an ornamental in Texas (Jones, Wipff, and Montgomery 1997).

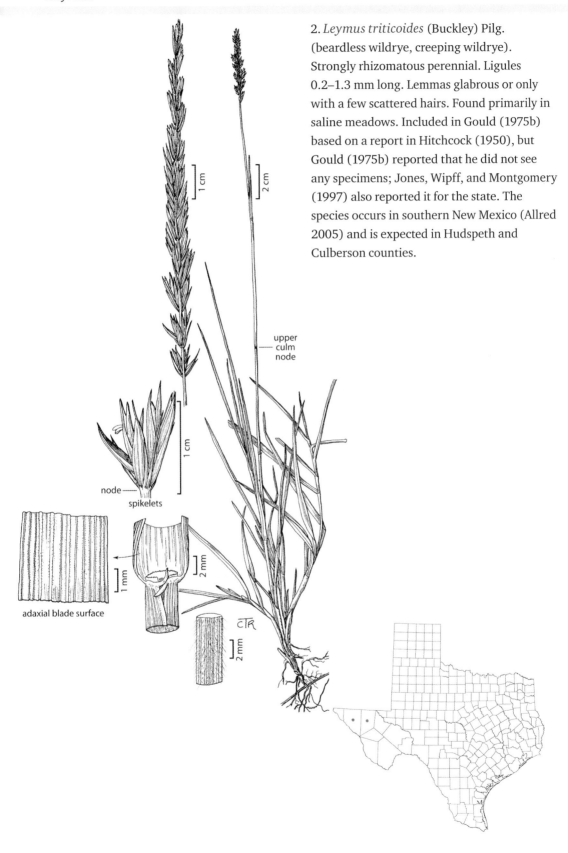

2. *Leymus triticoides* (Buckley) Pilg.
(beardless wildrye, creeping wildrye).
Strongly rhizomatous perennial. Ligules
0.2–1.3 mm long. Lemmas glabrous or only
with a few scattered hairs. Found primarily in
saline meadows. Included in Gould (1975b)
based on a report in Hitchcock (1950), but
Gould (1975b) reported that he did not see
any specimens; Jones, Wipff, and Montgomery
(1997) also reported it for the state. The
species occurs in southern New Mexico (Allred
2005) and is expected in Hudspeth and
Culberson counties.

1 cm

2 cm

upper
culm
node

1 cm

node

spikelets

1 mm

2 mm

CTR

2 mm

adaxial blade surface

Limnodea L.
(Poöideae: Poeae)

Low, short-lived annual, with soft culms; broad, thin, flat blades; and a loosely contracted panicle. Spikelets 1-flowered, disarticulation below the glumes, the spikelets falling entire. Rachilla extended behind the palea as a fine bristle. Glumes about equal, lanceolate, firm, gradually narrowing to an acute tip. Lemma membranous, glabrous, nerveless, rounded on the back, 2-toothed at the apex, and bearing a slender, bent awn between the teeth. Palea membranous, shorter than the lemma, 2-keeled. Basic chromosome number, $x = 7$. Photosynthetic pathway, C_3. A monotypic genus.

(Snow 2007b)

1. *Limnodea arkansana* (Nutt.) L. H. Dewey
(Ozarkgrass). A tufted annual to 60 cm tall.
Ligules 1–2 mm long. Lemmas awned, awns
5–15 mm long. A common weedy species of
open woodlands, stream banks, and brushy
grasslands. Poor livestock and wildlife values
(Hatch, Schuster, and Drawe 1999).

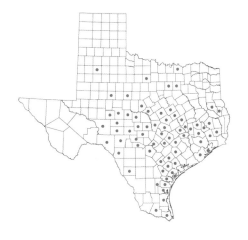

Lolium L.
(Poöideae: Poeae)

Annuals and short-lived perennials with usually succulent culms and flat or folded blades. Inflorescence a spike of several-flowered spikelets, these borne solitary and oriented edgewise at the nodes of a continuous rachis. Lower glume absent except on the terminal spikelet. Upper glume usually large, broad, several-nerved, awnless. Lemmas 5–9-nerved, rounded on the back, awnless or awned from a usually broad apex. Palea large. Basic chromosome number, $x = 7$. Photosynthetic pathway, C_3. Represented in Texas by 4 species and 1 subspecies.

1. Plants perennial.
    2. Plants long-lived perennial, with 2–10 florets per spikelet; lemmas unawned or awned, awns to 8 mm long ....2. *L. perenne*
    2. Plants short-lived perennial, with 10–20 florets per spikelet; lemmas usually awned, awns to 15 mm long .... 1. *L. multiflorum*
1. Plants annual.
    3. Lemmas usually awnless; spikelets sunken into the rachises........................3. *L. rigidum*
    3. Lemmas usually long-awned; spikelet not sunken into the rachises.
        4. Glumes ¼ to ½ the length of the spikelet.. 1. *L. multiflorum*
        4. Glumes ¾ to as long as the spikelet .... 4. *L. temulentum*

(Terrell 2007a)

1. *Lolium multiflorum* Lam. (annual ryegrass, Italian ryegrass). Annual or short-lived perennial with culms to 1.5 m long. Glumes ¼ to ½ as long as the florets. Used as a cover crop, temporary lawn grass, restoration species, and soil enhancer. Found escaped in disturbed sites. Gould (1975b) and others combined this with *L. perenne*. Potentially poisonous from an endophytic fungus that produces alkaloid toxins.

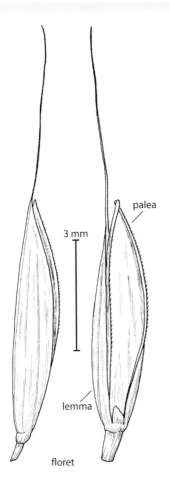

palea

3 mm

lemma

floret

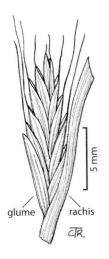

glume   rachis

5 mm

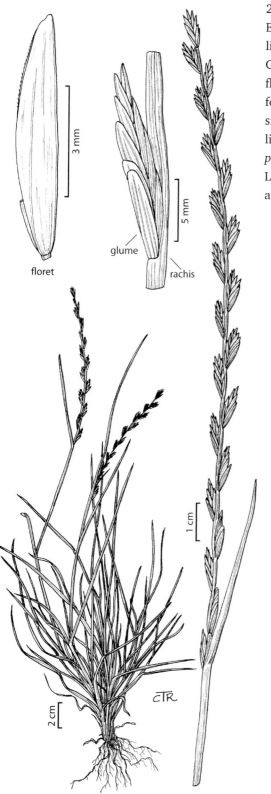

floret

glume

rachis

2. *Lolium perenne* L. (perennial wildrye, English wildrye, perennial ryegrass). Long-lived perennial with culms to about 1 m long. Glumes from ¼ to about as long as the lowest floret. Used as a lawn grass, for forage, and for revegetation. Found escaped in disturbed sites. Jones, Wipff, and Montgomery (1997) listed 2 varieties, var. *aristatum* Willd. and var. *perenne*; however, none are recognized here. Listed as fair for wildlife and livestock (Hatch and Pluhar 1993).

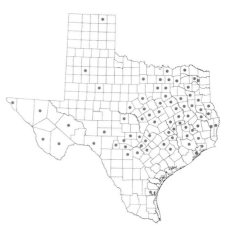

3. *Lolium rigidum* Gaudin (stiff ryegrass).
Annual with culms to 70 cm long. Glumes ¾
to as long as or longer than the florets. A weed
of disturbed sites, roadsides, and waste places.
Reported to be toxic. One reported from
Walker County.

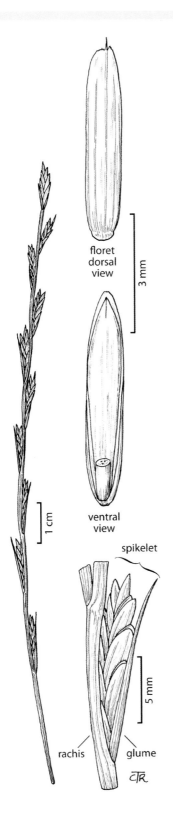

floret
dorsal
view

3 mm

ventral
view

1 cm

spikelet

5 mm

rachis   glume

CTR

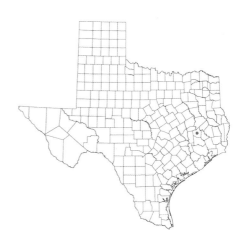

4. *Lolium temulentum* L. (darnel, darnel ryegrass, poison darnel). Annual with culms to 1.2 m long. Glumes as long as or longer than the florets. Typically a weed of grain fields. Escaped and found in disturbed areas, along roadsides, and in waste places. The Texas species belongs to the typical subspecies, subsp. *temulentum*. This species is thought to be the weed tares mentioned in the Bible. An endophytic fungus that produces toxic alkaloids can infect the caryopses. Grazing animals can be poisoned by consuming infected plants. Historically, the fungus poisoned flour when this species was an abundant weed in *Triticum* fields.

palea

5 mm

floret

lemma

5 mm

glume   rachis

spikelet

1 cm

CTR

subsp. temulentum

Luziola Juss.
(Ehrhartoideae: Oryzeae)

Low, rather delicate perennials, some plants possibly annual, with weak, often submerged culms and small panicles of 1-flowered, unisexual spikelets. Ligule large, thin, membranous. Pistillate and staminate spikelets in the same inflorescence or more often in separate inflorescences on the same plant, the staminate usually terminal, the pistillate at the middle and upper culm nodes. Disarticulation below the spikelet. Glumes absent. Lemma and paleas about equal, thin, several- to many-nerved. Caryopsis globose or oblong, smooth or striate, free from the lemma and paleas. Stamens usually 6. Basic chromosome number, $x = 12$. Photosynthetic pathway, C_3. Represented in Texas by 2 species.

1. Culms prostrate, usually immersed; pistillate inflorescences mostly included in the sheath, only stigmas exposed .... 1. *L. fluitans*
1. Culms erect or suberect, usually emergent; pistillate inflorescences exserted ....2. *L. peruviana*

(Terrell 2007b)

1. *Luziola fluitans* (Michx.) Terrell & H. Rob. (silverleaf grass, southern watergrass). Stoloniferous perennial with submerged culms. Ligules to 2 mm long. Stigmas 3–6 mm long. Found in lakes and streams. The Texas species belongs to the typical variety, var. *fluitans*. Listed as *Hydrochloa caroliniensis* P. Beauv. in Gould (1975b) and Hatch, Gandhi, and Brown (1990). A significant wildlife food.

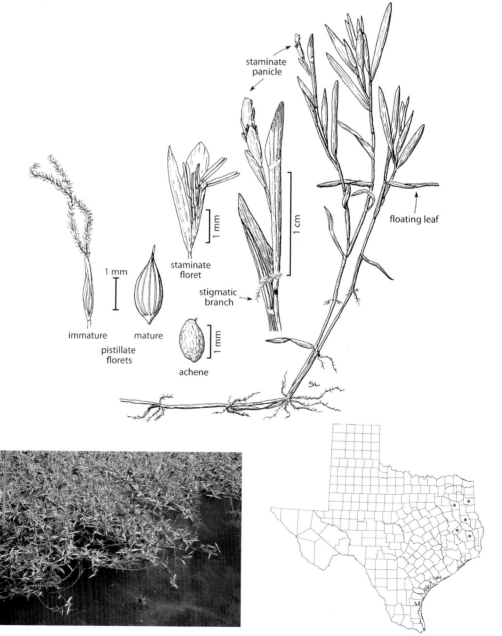

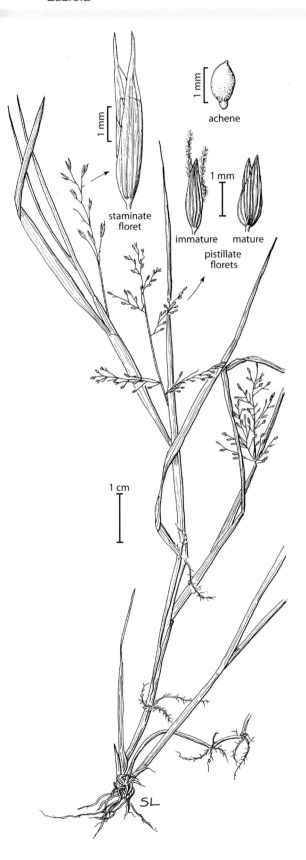

staminate
floret

achene

immature   mature

pistillate
florets

2. *Luziola peruviana* J. F. Gmel. (Peruvian watergrass). Stoloniferous perennial with mostly emergent culms. Ligules to 5 mm long. Stigmas 2–3 mm long. Found in wet places and shallow water along streams and lakes. Not listed in Gould (1975b); Hatch, Gandhi, and Brown (1990); or Jones, Wipff, and Montgomery (1997).

Lycurus Kunth
(Chloridoideae: Cynodonteae)

Tufted perennials with dense, spikelike panicles of 1-flowered spikelets. Sheaths laterally compressed and sharply keeled. Blades narrow, flat, or folded, usually with a whitish midnerve and margins. Inflorescence mostly 3–8 cm long and 5–8 mm thick. Spikelets short-pediceled, deciduous in pairs together with the pedicels, the lowermost spikelet of the pair often sterile or staminate. Glumes shorter than the lemma, the lower 2- or 3-nerved and with 2 or 3 short awns, the upper similar but 1-nerved and 1-awned. Palea similar in texture to the lemma and about as long, enclosed by the lemma only at the base. Basic chromosome number, $x = 10$. Photosynthetic pathway, C_4. Represented in Texas by 2 species.

1. Distal portion of leaf terminating with a fragile, awnlike tip 4–7 mm long; ligules generally in 1 piece, 3–12 mm long, elongate, acute to acuminate; culms erect .... 2. *L. setosus*
1. Distal portion of leaf acute and the midnerve extending beyond the end of the leaf as a short mucro or bristle 1–3 mm long; ligules 1.5–3.0 mm long with narrow triangular lobes to each side, 1.5–3.0 mm long; culms ascending to erect, often geniculate at base .... 1. *L. phleoides*

(Reeder 2003)

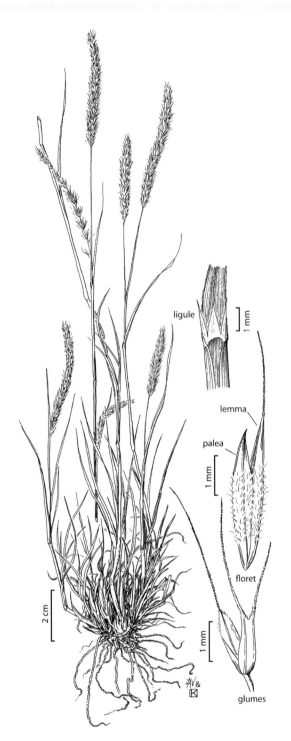

ligule

lemma

palea

floret

glumes

2 cm

1 mm

1 mm

1 mm

1. *Lycurus phleoides* Kunth (wolftail, common wolfstail). Cespitose perennial, sometimes with geniculate culms. Ligules 1.5–3.0 mm long, often with lobes extending from each side. Found on desert grasslands, open slopes, and rocky hills. Powell (1994) thought this species should be palatable to livestock.

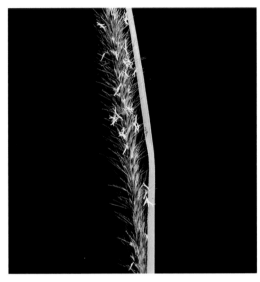

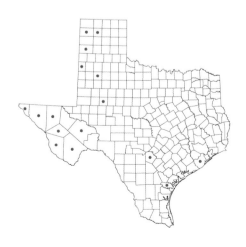

2. *Lycurus setosus* (Nutt.) C. Reeder (bristly wolfstail, wolftail). Densely cespitose, erect perennial. Ligules 3–12 mm long, hyaline, acuminate. Found on open desert grasslands, rocky slopes, and mesas. Gould (1975b) and Hatch, Gandhi, and Brown (1990) do not include this species. According to Powell (1994) this species is an important forage grass for cattle and perhaps for wildlife.

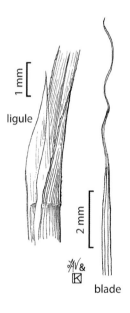

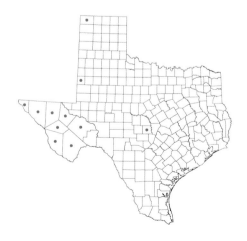

Megathyrsus
(Panicoideae: Paniceae)

Cespitose perennial with short, thick rhizomes. Culms erect or geniculate, root at lower nodes. Sheaths usually shorter than the internodes; collars densely pubescent; ligules of hairs; blades flat. Panicles up to 65 cm long, highly branched, branches usually spikelike. Spikelets up to 4.5 mm long, usually glabrous. Lower glumes about ⅓ as long as spikelet, 1–3-veined. Upper glumes about as long as spikelet, 5-veined. Lower florets staminate. Upper florets perfect, transversely rugose. Basic chromosome number, $x = 9$. Photosynthetic pathway, C_4. Represented in Texas by a single species.

(Simon and Jacobs 2003)

1. *Megathyrsus maximus* (Jacq.) B. K.
Simon & S. W. L. Jacobs (Guinea grass).
Rhizomatous perennial with culms up to
2.5 m tall and 1 cm thick. Sheaths typically
shorter than the internodes; collars densely
pubescent; blades flat and up to 1 m long.
An introduced species from Africa that is
sometimes cultivated as a forage crop and
escapes. This species has been included in
Panicum and *Urochloa* (Gould 1975b; Hatch,
Gandhi, and Brown 1990; Jones, Wipff, and
Montgomery 1997).

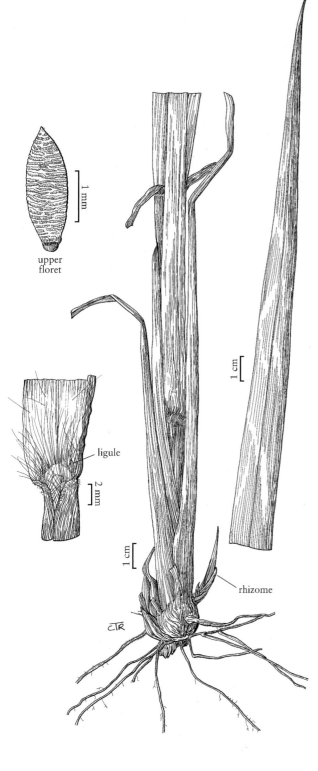

upper
floret

ligule

rhizome

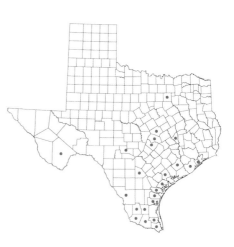

Melica L.
(Poöideae: Meliceae)

Perennials, cespitose. Sheaths connate; ligules membranous, erose-lacerate; auricles absent. Inflorescence paniculate, branches short, appressed, few-flowered. Spikelets with 2–5 florets, reflexed on short, pubescent pedicels. Disarticulation below the glumes, each spikelet at maturity falling entire, or above the glumes. Glumes subequal. Lemma with 5 prominent and several obscure veins. Palea about ½ to ⅔ as long as lemma. Basic chromosome number, $x = 9$. Photosynthetic pathway, C_3. Represented in Texas by 5 species and 2 varieties.

Plants that disarticulate above the glumes and typically produce corms are sometimes included in the genus *Bromelica*.

1. Spikelets disarticulate below the glumes; pedicels bent sharply just below the spikelet.
  2. Lemmas with twisted hairs .... 2. *M. montezumae*
  2. Lemmas glabrous or scabrous.
    3. Rudiments acute to acuminate, resembling the perfect florets .... 5. *M. porteri*
    3. Rudiments clublike, not resembling the perfect florets.
      4. Rudiments at an angle to the rachilla; panicle branches with 2–5 spikelets ....
      .... 3. *M. mutica*
      4. Rudiments parallel to rachilla; panicle branches with 5–20 spikelets ...... 4. *M. nitens*
1. Spikelets above the glumes; pedicels more or less straight .... 1. *M. bulbosa*

(Barkworth 2007e)

lemma

palea

2 mm

floret

1 cm

corm

1. *Melica bulbosa* Geyer *ex* Porter & J. M. Coult. (oniongrass). Rhizomatous perennial with culms producing corms. Rudiments resembling the perfect florets. Found in open woodlands, on dry slopes, and along streams. Reported only from Jeff Davis and Sutton counties. Also known as *Bromelica bulbosa* (Geyer *ex* Porter and J. M. Coult.) W. A. Weber.

2. *Melica montezumae* Piper (Montezuma melic). Cespitose perennial without rhizomes or corms. Rudiments clublike, not resembling perfect florets. Found in shady habitats of the West Texas mountains. Excellent forage for livestock where abundant (Powell 1994).

floret

rudiment

pedicel

spikelet

2 mm

2 mm

2 cm

3. *Melica mutica* Walter (two-flower melic, narrow melic). Perennial with short rhizomes; culms not producing corms. Rudiments clublike and at a sharp angle to rachilla. Found in forest openings in sandy soils, woods, and thickets, mainly in East and Central Texas.

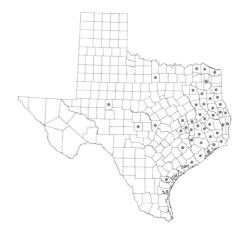

4. *Melica nitens* (Scribn.) Nutt. *ex* Piper
(three-flower melic, tall melic). Perennial with
short rhizomes; culms not producing corms.
Rudiments clublike and not resembling perfect
florets. Found in open woods and on moist
slopes, rocky grasslands, and canyon bottoms.
Palatable to most grazing animals (Powell
1994).

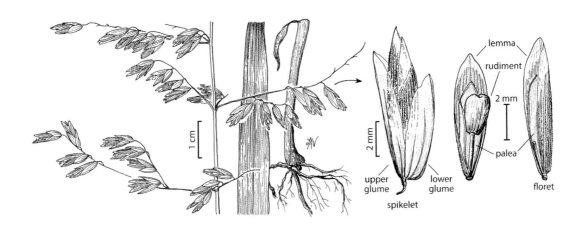

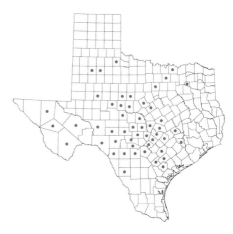

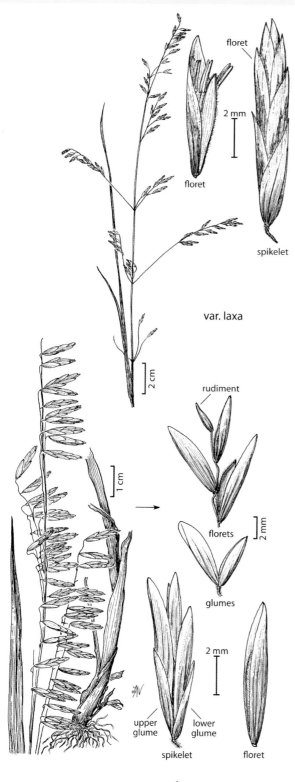

floret

2 mm

floret

spikelet

var. laxa

2 cm

rudiment

1 cm

florets

2 mm

glumes

2 mm

upper
glume

lower
glume

spikelet

floret

var. porteri

5. *Melica porteri* Scribn. (Porter's melic).
Perennial with short rhizomes; culms not
producing corms. Rudiments resembling
perfect florets. Found in open woods and on
rocky slopes, often near streams. Palatable
to livestock and wildlife but rarely abundant
enough to be of major importance (Powell
1994). Two varieties occur in Texas: var. *laxa*
W. Boyle; and the typical one, var. *porteri*.
They can be distinguished by the following
characters:

1. Glumes purplish-tinged; panicle branches
     flexible.. var. *laxa*
1. Glumes green or pale; panicle branches
     straight .... var. *porteri*

Melinis P. Beauv.
(Panicoideae: Paniceae)

Annuals and perennials, mostly with erect, moderately tall culms. Ligule a ring of hairs. Blades thin and flat. Inflorescence an open or contracted much-branched panicle, the spikelets on slender pedicels. Spikelets laterally compressed, disarticulating below the glumes. First glume minute to about ⅓ as long as the second. Second glume and lemma of the lower floret about equal, usually silky-villous with fine hairs, tapering to a notched, usually awned apex. Lower floret usually staminate, with a well-developed palea. Lemma of the fertile floret much shorter than the second glume and lemma of the lower floret, slender, membranous, glabrous, narrowly rounded at the apex, the margins thin and not inrolled over the palea. Basic chromosome number, $x = 9$. Photosynthetic pathway, C_4. Represented in Texas by a single species.

Sometimes included in the genus *Rhynchelytrum* as *R. repens* (Willd.) C. E. Hubb. (Gould 1975b; Hatch, Gandhi, and Brown 1990).

(Wipff 2003e)

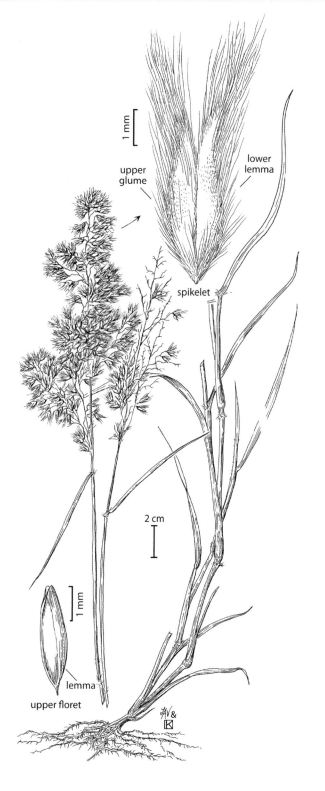

upper
glume

lower
lemma

1 mm

spikelet

2 cm

1 mm

lemma

upper floret

1. *Melinis repens* (Willd.) Zizka (red natal grass). Annual or short-lived perennial, sometimes rooting at the lower nodes. Glumes white, rose, or dark purple. A weedy species found along roadsides and other disturbed areas. Sometimes grown as an ornamental. Poor livestock and wildlife values (Hatch, Schuster, and Drawe 1999). The Texas species belongs to subsp. *repens*.

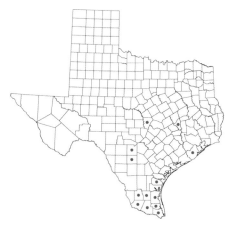

Microchloa R. Br.
(Chloridoideae: Cynodonteae)

Tufted annuals and perennials with slender culms and densely clumped, filiform leaves. Inflorescence a slender, curved, unilateral spike, with 1-flowered spikelets closely imbricated in 2 rows on one side of a narrow, flattened rachis. Disarticulation usually between the glumes. Glumes firm, lanceolate, subequal, 1-nerved, acute. Lemma thin and membranous, slightly shorter than the glumes, awnless, acute, with a midnerve and 2 short lateral nerves, the latter not always apparent. Palea similar to the lemma in texture but slightly shorter. Caryopsis oval, flattened, reddish-brown. Basic chromosome numbers, $x = 10$ and 12. Photosynthetic pathway, C_4. Represented in Texas by a single species.

(Sánchez 2003)

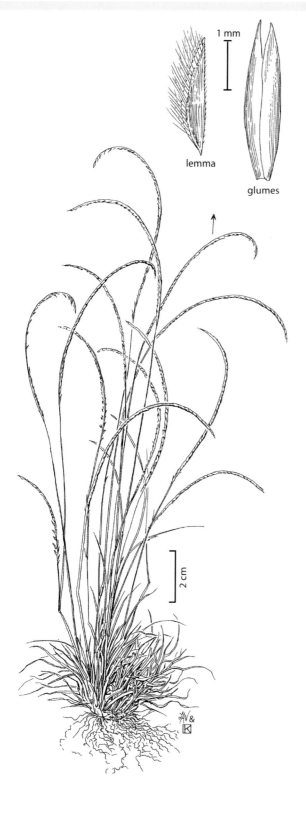

1 mm

lemma

glumes

1. *Microchloa kunthii* Desv. (Kunth's smallgrass). Small perennial forming dense tufts. Panicle with 1 curved, unilateral, spicate branch. Found on granite outcrops and rocky slopes in the Big Bend region. Not included in Gould (1975b) or Hatch, Gandhi, and Brown (1990). Of limited forage value according to Powell (1994).

2 cm

Microstegium Nees
(Panicoideae: Andropogoneae)

Annuals with weak, usually decumbent-spreading culms. Inflorescence of few to several short, spreading, racemose branches, these widely spaced or subdigitate on the main culm. Spikelets all alike and fertile, in pairs of 1 sessile and 1 pediceled or both unequally pediceled. Disarticulation first in the nodes of the rachis and secondarily at the base of the spikelets. Glumes equal, usually firm but thin. Lemma of the sterile floret reduced or absent. Lemma of the fertile floret hyaline, awnless or awned from an entire or notched apex. Palea of fertile floret small or absent. Lodicules 2, cuneate. Stamens 3 or 2, the anthers small. Basic chromosome number, $x = 10$. Photosynthetic pathway, C_4. Represented in Texas by a single species.

(Thieret 2003f)

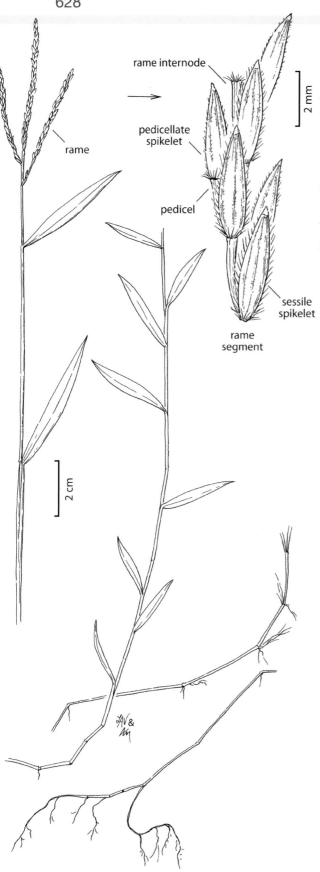

rame

rame internode

pedicellate
spikelet

pedicel

sessile
spikelet

rame
segment

2 mm

2 cm

1. *Microstegium vimineum* (Trin.) A. Camus
(Nepalese browntop, flexible sasa grass,
eulalia, Japanese grass, Japanese stilt grass).
Annual with culms to 1 m tall, lower culms
prostrate and root at the nodes, forming
dense monotypic stands. Found in bottomland
forests, along stream banks, and in wetland
areas but occurs in disturbed sites as well.
Often confused with *Leersia virginica* Willd.
(Thieret 2003f). Can become a troublesome
weed of moist, shaded habitats. Not included
in Gould (1975b).

Miscanthus Anderss.
(Panicoideae: Andropogoneae)

Tall, coarse perennials, with long, flat blades and flabellate, silky-pubescent panicles of numerous, crowded, closely flowered branches. Spikelets all alike and perfect, unequally pediceled in pairs along a continuous rachis, with a tuft of long, silky hair at the base of the glumes and on the callus. Disarticulation below the callus. Glumes subequal, mostly thin but firm, awnless. Lemma of the sterile floret membranous, awnless, slightly shorter than the glumes. Lemma of the fertile floret membranous, hyaline, awnless or with a geniculate and twisted awn from a bifid, toothed apex. Palea of the fertile floret small, membranous. Basic chromosome number, $x = 9$. Photosynthetic pathway, C_4. Represented in Texas by a single ornamental species. Jones, Wipff, and Montgomery (1997) reported *M. transmorrisonensis* B. Hayata as being cultivated in Texas as well; however, Barkworth (2003l) did not report this species in the United States.

(Barkworth 2003l)

1. *Miscanthus sinensis* Anderss. (silvergrass, eulalia, Chinese silvergrass, plumegrass). Cespitose perennial forming large clumps with short rhizomes. Culms to 2 m tall. A native of Asia, now a widely used ornamental grass. About 40 cultivars are available.

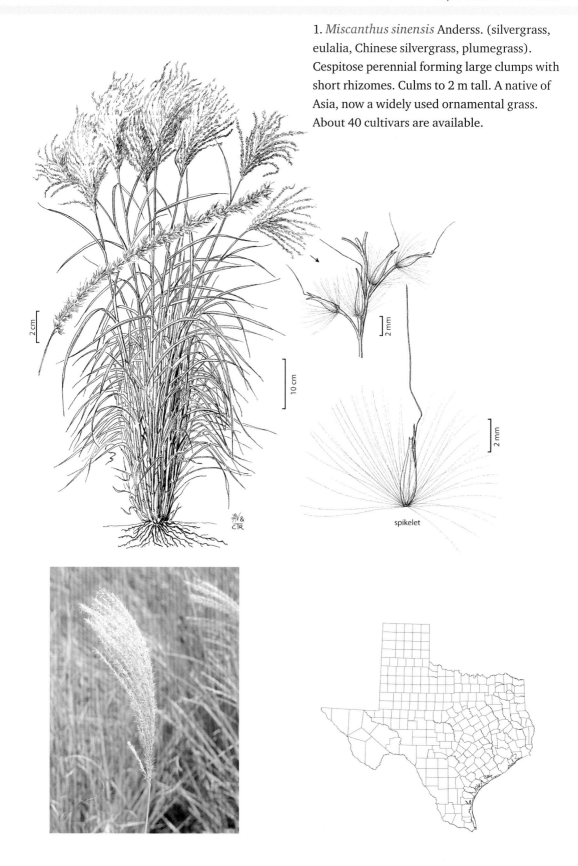

spikelet

Monanthochloë Engelm.
(Chloridoideae: Cynodonteae)

Low, mat-forming, dioecious perennials with wiry, decumbent, much-branched culms and tufted leaves, these with blades mostly <1 cm long. Spikelets 3–5-flowered, only the lower fertile, borne in the axils of fascicled leaves terminating the short, erect branches. Glumes absent. Lemmas several-nerved, those of the pistillate spikelets like the leaf blades in texture. Palea narrow, 2-nerved, enfolding the caryopsis. Basic chromosome number, $x = 10$. Photosynthetic pathway, C_4. Represented in Texas by a single species.

(Thieret 2003g)

1. *Monanthochloë littoralis* Engelm.
(shoregrass). Mat-forming perennial
with long, wiry stolons. Plants dioecious.
Inflorescence composed of a single spikelet
enclosed in upper leaf sheaths. Found in
moist, sandy, saline soils along the Gulf Coast.
Poor livestock and wildlife values according to
Hatch, Schuster, and Drawe (1999).

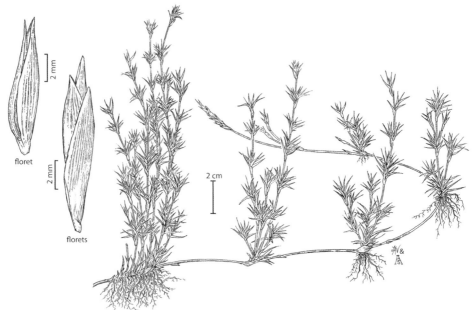

floret

florets

2 mm

2 mm

2 cm

Moorochloa Veldcamp
(Panicoideae: Paniceae)

Annuals, frequently rooting at the lower nodes. Leaves cauline; sheaths open; ligules membranous with a ciliate fringe. Inflorescences terminal; branches erect to ascending. Disarticulation below the glumes and between the florets. Spikelets single, dorsally compressed, convex, in 2 rows, with 2 florets. Lower florets sterile or staminate; upper florets perfect. Upper glumes and lower lemmas villous and 3–5-veined. Upper lemmas indurate, smooth, shiny, 5–7-veined. Basic chromosome number, $x = 9$. Photosynthetic pathway, C_4. Represented in Texas by a single species. *Bracharia* was once the name for this genus and included many of the species now included in *Urochloa*. It differs from *Urochloa* in its smooth, rounded upper floret and from *Panicum* in its disarticulating upper floret.

This species is not listed in Gould (1975b); Hatch, Gandhi, and Brown (1990); or Jones, Wipff, and Montgomery (1997).

Veldkamp (2004)

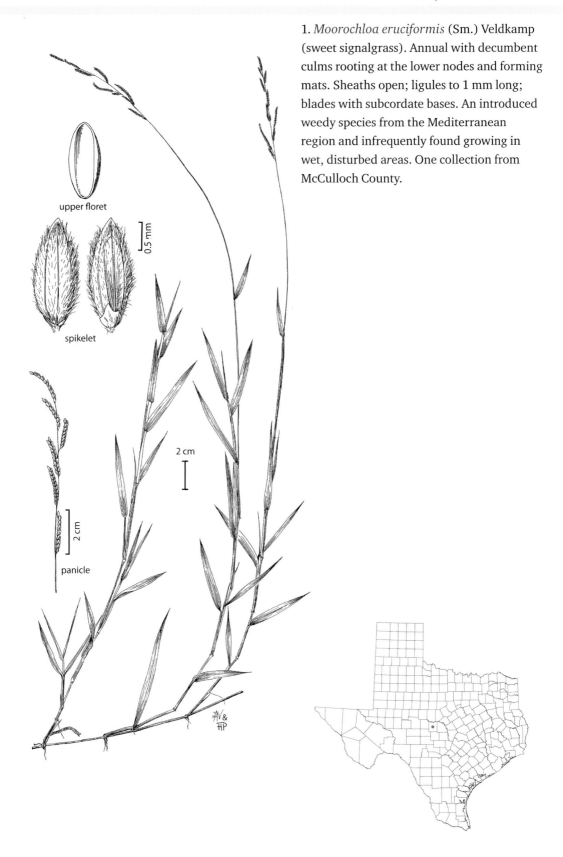

upper floret

0.5 mm

spikelet

2 cm

panicle

2 cm

1. *Moorochloa eruciformis* (Sm.) Veldkamp (sweet signalgrass). Annual with decumbent culms rooting at the lower nodes and forming mats. Sheaths open; ligules to 1 mm long; blades with subcordate bases. An introduced weedy species from the Mediterranean region and infrequently found growing in wet, disturbed areas. One collection from McCulloch County.

Muhlenbergia Schreb.
(Chloridoideae: Cynodonteae)

Plants of diverse habitats, from delicate, tufted annuals to large, coarse, cespitose perennials. Several species with creeping rhizomes. Sheaths open; ligules usually a prominent membrane; auricles absent. Inflorescence an open or contracted panicle, spikelike in a few species. Spikelets typically 1-flowered, a second or third floret occasionally produced. Disarticulation above the glumes. Glumes usually 1-nerved or nerveless, occasionally 3-nerved, mostly shorter than the lemma, obtuse, acute, acuminate or short-awned. Lemma as firm as, or firmer than, the glumes, 3-nerved (the veins indistinct in some species), with a short, usually bearded callus at the base and a single flexuous awn at the apex, less frequently mucronate or awnless. Palea well developed, shorter than or about equaling the lemma. Caryopsis usually not falling free from the lemma and palea. Basic chromosome number, $x = 10$. Photosynthetic pathway, C_4. Represented in Texas by 49 species, 3 varieties, and 1 hybrid.

Jones, Wipff, and Montgomery (1997) reported that M. *dumosa* Scribn. *ex* Vasey is grown as an ornamental within the state; however, Peterson (2003b) has no record for Texas. It is not included here. Also, M. x *involuta* Swallen (seep muhly, canyon muhly) is a hybrid between M. *lindheimeri* and M. *reverchonii*. It is fairly common along the eastern edge of the Edwards Plateau (Bandera, Bennett, Bexar, Blanco, Comal, Hays, Kendall, Travis, and Williamson counties) (Turner et al. 2003; Diggs et al. 2006).

1. Plants annual.
    2. Lemmas unawned or if mucronate, mucro only to 1 mm long.
        3. Glumes strigose, hairs 0.1–0.3 mm long.
            4. Sheaths and internodes strigose; lemmas 1.3–2.0 mm long; mucronate .... 43. *M. texana*
            4. Sheaths and internodes glabrous; lemmas 0.8–1.5 mm long; not mucronate ....23. *M. minutissima*
        3. Glumes glabrous.
            5. Panicles contracted; culms often rooting at lower nodes .... 16. *M. filiformis*
            5. Panicles open or diffuse; culms not rooting at lower nodes.
                6. Ligules with lateral lobes ....17. *M. fragilis*
                6. Ligules without lateral lobes .... 46. *M. uniflora*
    2. Lemmas awned.
            7. Upper glumes 2- or 3-veined, each ending as a tooth at the apex .... 8. *M. crispiseta*
            7. Upper glumes 1-veined or veinless.
                8. Lower glumes 2-veined, bifid; disarticulation at the base of the pedicels.
                    9. Glumes about as long as the lemmas; lemmas 2.5–4.5 mm long; awns 6–15 mm long ....10. *M. depauperata*
                    9. Glumes to ⅔ as long as the lemmas; lemmas 3.5–6.0 mm long; awns 10–20 mm long ....5. *M. brevis*

8. Lower glumes 1-veined, not bifid (if present); disarticulation above the glumes.

    10. Lemma awns 1–5 mm long.

        11. Glumes 0.1–0.3 mm long; ligules up to 0.5 mm long...34. *M. schreberi*

        11. Glumes 0.8–1.8 mm long; ligules 1.0–2.5 mm long.

            12. Paleas minutely appressed, pubescent on the lower ½... 43. *M. texana*

            12. Paleas glabrous .... 13. *M. eludens*

    10. Lemma awns 10–35 mm long.

        13. Spikelets paired; lower glumes of lower spikelets <0.8 mm long; lower glumes of upper spikelets to 8 mm long .... 11. *M. diversiglumis*

        13. Spikelets not paired; lower glumes of all spikelets similar .... 42. *M. tenuifolia*

1. Plants perennial.

  14. Plants with scaly, creeping rhizomes.

    15. Panicles open.

      16. Lemmas awned, awns up to 20 mm long...28. *M. pungens*

      16. Lemmas unawned or mucronate.

        17. Ligules hyaline with lateral lobes; blade margins and midveins white, thick ....2. *M. arenacea*

        17. Ligules ciliate, without lateral lobes; blade margins and midveins greenish, thin ... .4. *M. asperifolia*

    15. Panicles contracted.

      18. Blades <2.7 mm wide, flat, folded, or involute at maturity.

        19. Lemmas awned, awns 1–25 mm long.

          20. Lemma awns usually <4(–6) mm long...............................20. *M. glauca*

          20. Lemma awns 4–25 mm long.

            21. Ligules with lateral lobes to 3 mm long; leaves with dark brown spots. ....25. *M. pauciflora*

            21. Ligules without lateral lobes; leaves without dark brown spots............ ....26. *M. polycaulis*

        19. Lemmas unawned, mucronate, or if awned, awn <1 mm long.

          22. Lemmas and paleas pubescent, hairs 0.4–1.2 mm long.

            23. Glumes acuminate; blades usually flat ....20. *M. glauca*

            23. Glumes acute; blades tightly involute.

              24. Lemmas 2.6–4.0 mm long; glumes nearly as long as lemmas..44. *M. thurberi*

              24. Lemmas 1.4–2.5 mm long; glumes less than ⅔ as long as ..... lemmas ....48. *M. villiflora*

          22. Lemmas and paleas glabrous or with hairs <0.3 mm long.

            25. Glumes more than ½ as long as the lemmas ....30. *M. repens*

25. Glumes less than ½ as long as the lemmas ....
....47. *M. utilis*
18. Blades 2–15 mm wide; flat at maturity.
    26. Glumes awned, 3–8 mm long (including the awn).......... 29. *M. racemosa*
    26. Glumes unawned or if awned, awn <4 mm long.
        27. Lemmas usually completely glabrous .... 19. *M. glabrifloris*
        27. Lemmas with hairs, these sometimes restricted to callus.
            28. Lemma bases with hairs about as long as the florets .... 1. *M. andina*
            28. Lemma bases glabrous or with hairs shorter than the floret.
                29. Axillary panicles often present, always partly included in
                    sheaths; culm internodes smooth, shiny.
                    30. Ligules 0.2–0.6 mm long; glumes 1.4–2.0 mm long, up to
                        ⅔ as long as the lemmas .... 6. *M. bushii*
                    30. Ligules 0.7–1.7 mm long; glumes 2–4 mm long, about as
                        long as the lemmas ....*M. frondosa*
                29. Axillary panicles absent, or if present, always exserted on
                    elongated peduncles.
                    31. Glumes much shorter than the lemmas, acute.
                        32. Internodes glabrous; lemmas unawned or awn <1 mm
                        long.. 37. *M. sobolifera*
                        32. Internodes pubescent; lemmas awned, awn to 12 mm
                        long.. 41. *M. tenuiflora*
                    31. Glumes nearly as long as the lemmas, acuminate.
                        33. Ligules 0.4–1.0 mm; panicles dense 22. *M. mexicana*
                        33. Ligules 1.0–2.5 mm; panicles not dense
                          .... 40. *M.sylvatica*
14. Plants without rhizomes.
  34. Upper glumes usually 3-veined and 3-toothed.
    35. Upper glumes as long as or longer than the lemmas........................ 39. *M. straminea*
    35. Upper glumes ½ to ⅔ as long as the lemmas ....24. *M. montana*
  34. Upper glumes usually 1-veined, or if more, veined not toothed.
    36. Panicles loosely contracted, open or diffuse; usually naked basally
      37. Lemmas unawned or awn to 4(–6) mm long.
        38. Basal sheaths flattened, usually keeled ....14. *M. emersleyi*
        38. Basal sheaths rounded, never keeled.
          39. Culms 10–60 cm long, decumbent.
            40. Blades arcuate, 0.3–0.9 mm wide .... 45. *M. torreyi*
            40. Blades not arcuate, 1–2 mm wide .... 3. *M. arenicola*
          39. Culms 40–160 cm long; erect.
            41. Panicles about as long as wide, not diffuse........... 31. *M. reverchonii*
            41. Panicles longer than wide, diffuse.
                42. Glumes more than ½ as long as lemmas .... 15. *M. expansa*
                42. Glumes less than ½ as long as lemmas................. 7. *M. capillaris*

37. Lemmas awned, awn 6–35 mm long.
    43. Plants conspicuously branched, bushy; culms wiry ....27. *M. porteri*
    43. Plants typical bunchgrasses.
        44. Basal sheaths laterally compressed..14. *M. emersleyi*
        44. Basal sheaths terete.
            45. Glumes not awned or mucronate; spikelets purple.
                46. Panicles 8–40 cm wide; in East Texas ....7. *M. capillaris*
                46. Panicles 2–12 cm wide; in West Texas .... 33. *M. rigida*
            45. Glume awned or mucronate .... Key A
36. Panicles narrow, 0.2–5.0 cm wide; spikelet-bearing to the base.
    47. Panicles spikelike.
        48. Culms decumbent, rooting at lower nodes, usually 3–20 cm long.. 16. *M. filiformis*
        48. Culms stiffly erect, not rooting at lower nodes, usually 20–150 cm long.
            49. Basal sheaths rounded; panicles >15 cm long .... 32. *M. rigens*
            49. Basal sheaths keeled; panicles 4–16 cm long.
                50. Ligules <1 mm long; paleas glabrous ....9. *M. cuspidata*
                50. Ligules >1 mm long; paleas pubescent .... 49. *M. wrightii*
    47. Panicles not spikelike .... Key B

Key A

1. Lemmas shiny, smooth; blades tightly involute, 0.2–1.2 mm wide ....36. *M. setifolia*
1. Lemmas usually scabrous; blade flat or involute, 1–4 mm wide.
    2. Panicles about as wide as long, not diffuse .... 31. *M. reverchonii*
    2. Panicles narrower than long, diffuse.
        3. Lemmas without setaceous teeth, or teeth <1 mm long ....7. *M. capillaris*
        3. Lemma with setaceous teeth 1–5 mm long.. 35. *M. sericea*

Key B

1. Ligules 10–45 mm long; glumes as long as or longer than the florets .... 21. *M. lindheimeri*
1. Ligules 0.2–10 mm long; glumes shorter than the florets.
    2. Lemma awns 0.5–6.0 mm long.
        3. Blades 10–60 mm long ....12. *M. dubia*
        3. Blades to 10 cm long ....34. *M. schreberi*
    2. Lemma awns 6–40 mm long.
        4. Glumes 0.3–1.0 mm long, not awned .... 38. *M. spiciformis*
        4. Glumes 1.2–3.5 mm long, usually awn-tipped.
            5. Ligules with lateral lobes; sheaths and blades usually with dark brown spots;
                lemmas and paleas almost glabrous .... 25. *M. pauciflora*
            5. Ligules without lateral lobes; sheaths and blades without dark brown spots;
                lemmas and paleas pubescent.

6. Anthers 0.9–1.5 mm long, yellowish; panicles 7–20 cm long ..... 42. *M. tenuifolia*
6. Anthers 1.5–2.0 mm long, orange; panicles 2–13 cm long ....26. *M. polycaulis*

(Peterson 2003b)

1. *Muhlenbergia andina* (Nutt.) Hitchc.
(foxtail muhly). Rhizomatous perennial
with culms to about 80 cm long. Ligules
membranous, 0.5–1.5 mm long. Panicles
dense; branches appressed. Rare; collected
only once in Hudspeth County. Found along
stream banks and in wet thickets.

glumes

floret

1 mm

1 mm

2 cm

1 cm

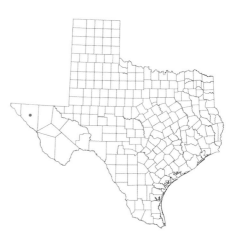

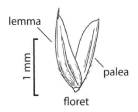

lemma

1 mm

palea

floret

1 mm

glumes

2. *Muhlenbergia arenacea* (Buckley) Hitchc. (ear muhly, sand muhly). Rhizomatous perennial. Culms to 40 cm long, decumbent, somewhat compressed-keeled. Ligules 0.5–2.0 mm long, hyaline, with lateral lobes. Panicles open. Found on flats, plains, washes, depressions, and open grasslands. Species is of limited forage value (Powell 1994).

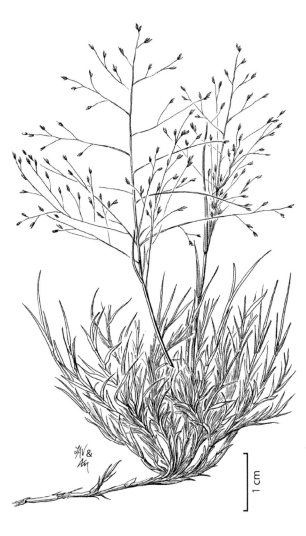

1 cm

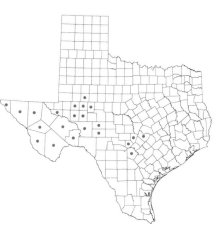

3. *Muhlenbergia arenicola* Buckley (sand muhly). Perennial, cespitose, forming a distinct "ring." Ligules 2–9 mm long, often with lateral lobes. Panicles diffuse. Found on sandy mesas, limestone outcrops, and in open desert grasslands. Closely related to *M. torreyi*, but *M. arenicola* is more robust. This is the more common of the 2 species. Reported to have fair forage value (Powell 1994).

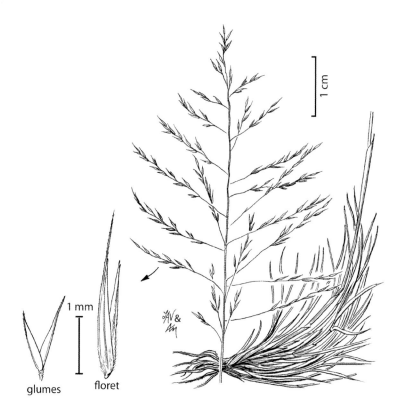

glumes   floret

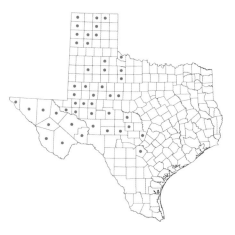

glumes   1 mm   floret

4. *Muhlenbergia asperifolia* (Nees and Meyen *ex* Trin.) Parodi (scratchgrass, alkali muhly, rough-leafed muhly). Rhizomatous perennial, occasionally stoloniferous as well. Plants often take on a bushy appearance due to branching at the upper nodes. Culms usually around 60 cm, occasionally taller, compressed-keeled. Ligules 0.2–1.0 mm long, membranous, without lateral lobes. Panicles open. Spikelets often with 2 florets. A smut often infects the caryopses. Found in moist alkaline soils, along playas, and in sandy washes.

2 cm

rhizome

5. *Muhlenbergia brevis* C. O. Goodd. (short muhly). Tufted annual. Culms to 20 cm long, erect or sprawling. Sheaths keeled. Ligules 1–3 mm long, membranous, sometimes with lateral lobes. Panicles contracted, not long-exserted. Found on rocky slopes, rock outcrops, and pinyon-juniper and pine-oak woodlands. Less frequent than the closely related *M. depauperata.*

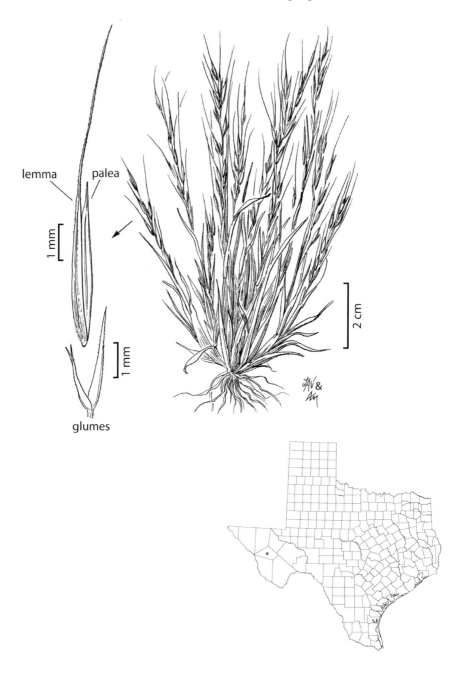

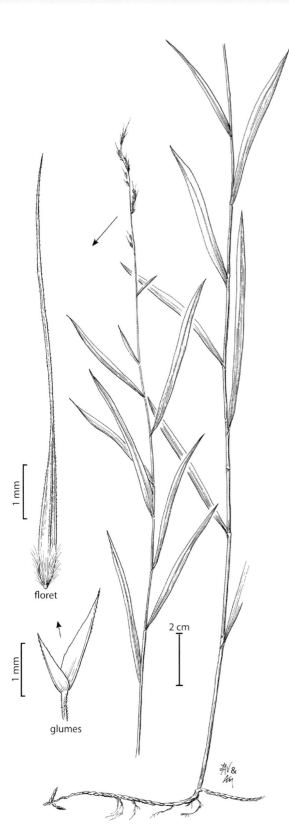

6. *Muhlenbergia bushii* R. W. Pohl (nodding muhly). Perennial with long, creeping, coarse rhizomes. Culms to 1.1 m tall, erect early then becoming much-branched above at maturity, giving the plant a "bush" appearance. Ligules membranous, 0.2–0.6 mm long. Panicles dense, axillary and terminal; branches appressed. Found in moist woodlands and along streams, often in clay soils. Infrequent in Texas.

1 mm

floret

1 mm

glumes

2 cm

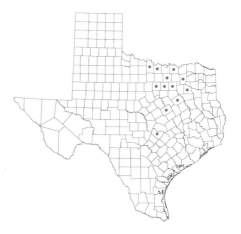

7. *Muhlenbergia capillaris* (Lam.) Trin. (hairyawn muhly, longawned hairgrass, slender muhly, pink muhly). Cespitose perennial without rhizomes. Culms to 1.5 m long, erect. Sheaths become fibrous with age but never coil. Ligules 2–10 mm long, membranous, decurrent. Panicles open, diffuse, reddish, often contained in subtending sheaths. Found in rocky or clayey soils in open woodlands. Frequently grown as an ornamental. Poor livestock and wildlife values (Hatch, Schuster, and Drawe 1999).

2 cm

1 mm

floret

glumes

8. *Muhlenbergia crispiseta* Hitchc. (Mexicali muhly). Tufted annual. Culms to 16 mm long. Ligules 1–2 cm long, membranous. Panicles loosely opened, long-exserted. Found on rock outcrops, in rocky drainages in pine-oak and pinyon-juniper woodlands. A Mexican species with a disjunct population in Brewster County. Not included in Gould (1975b) or Hatch, Gandhi, and Brown (1990).

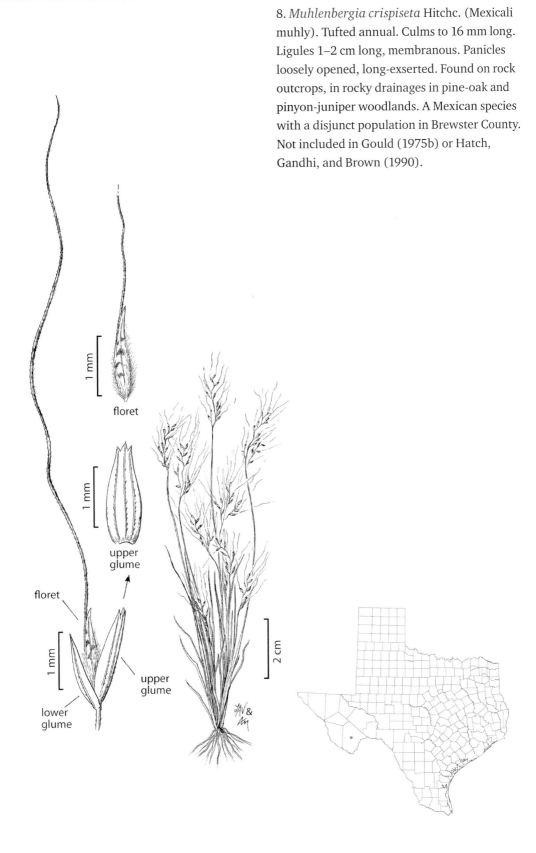

floret

upper glume

floret

upper glume

lower glume

1 mm

1 mm

1 mm

2 cm

9. *Muhlenbergia cuspidata* (Torr. *ex* Hook.) Rydb. (plains muhly). Perennial with knotty base, without rhizomes. Culms to 60 cm tall, compressed, erect. Panicles spikelike. Found in limestone rock outcrops, sandy drainages, and rocky slopes. This species not listed in Gould (1975b); Hatch, Gandhi, and Brown (1990); or Jones, Wipff, and Montgomery (1997).

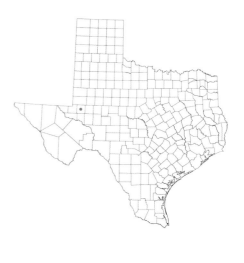

10. *Muhlenbergia depauperata* Scribn.
(sixweeks muhly). Tufted annual. Culms
to 15 cm long. Sheaths spirally coiled with
age, keeled. Ligules 1.4–2.5 mm long,
membranous, with lateral lobes. Panicles
contracted, primarily in the leaves, not long-
exserted. Found in rocky flats, outcrops,
bedrock, and desert grasslands. This species is
more frequent than the closely related *M. brevis*.

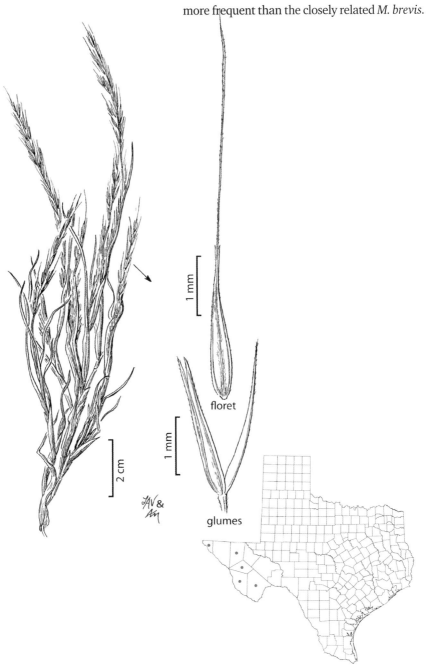

floret

1 mm

1 mm

2 cm

glumes

11. *Muhlenbergia diversiglumis* Trin.
(mixedglume muhly). Sprawling annual.
Culms decumbent, rooting at the nodes.
Ligules 0.5–0.8 mm long, membranous.
Panicles open. A Mexican introduction recently
collected in Galveston County. Not included
in Gould (1975b); Hatch, Gandhi, and Brown
(1990); or Jones, Wipff, and Montgomery
(1997).

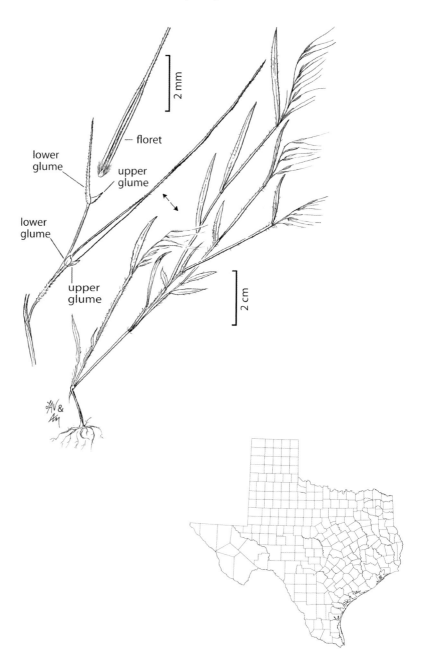

lower
glume

upper
glume

lower
glume

upper
glume

floret

2 mm

2 cm

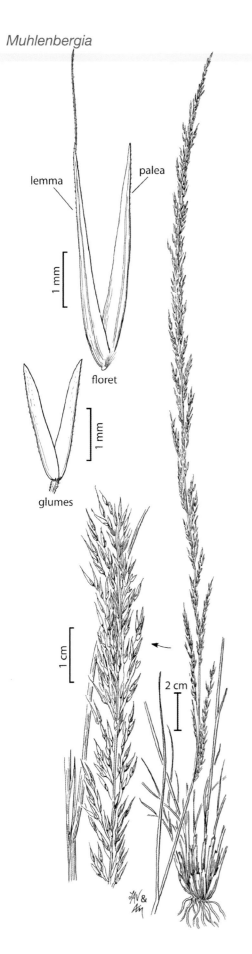

lemma

palea

1 mm

floret

glumes

1 mm

1 cm

2 cm

12. *Muhlenbergia dubia* E. Fourn. (pine muhly, pineland muhly). Perennial, densely cespitose without rhizomes. Culms to 1 m long, erect. Ligules 4–10 mm long, membranous, firm. Panicles contracted, gray-green. Found on steep slopes, on rock outcrops, and along draws. Powell (1994) thought it should be good forage where abundant.

13. *Muhlenbergia eludens* C. Reeder
(gravelbar muhly). Tufted annual. Culms to
40 cm long, slender, erect. Sheaths keeled.
Ligules 1.5–2.5 mm long, membranous.
Panicles open. Found along roadsides, in
gullies and washes, and on rocky slopes.

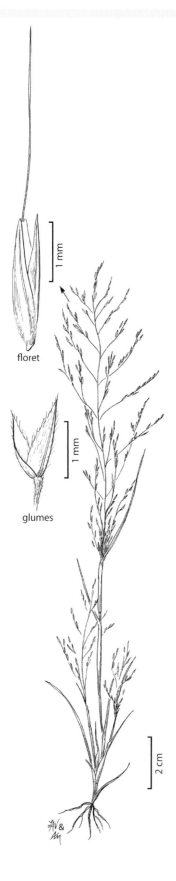

floret

glumes

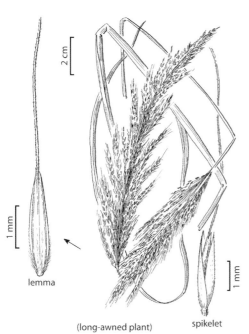

1 mm

lemma

(long-awned plant)

spikelet

1 mm

14. *Muhlenbergia emersleyi* Vasey (bullgrass). Perennial, not rhizomatous. Culms to 1.5 m long, stout, erect, compressed-keeled. Ligules 10–25 mm long, membranous. Panicles loosely contracted to open, light purple to light brownish. Found on rocky slopes and cliffs and in canyons, washes, and arroyos. Coarse and unpalatable according to Powell (1994).

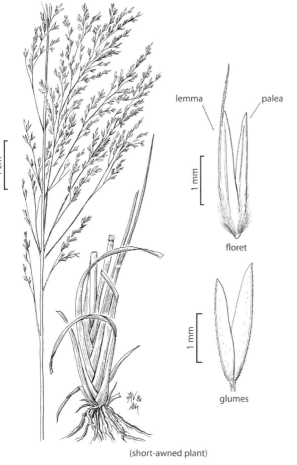

lemma   palea

1 mm

floret

1 mm

glumes

(short-awned plant)

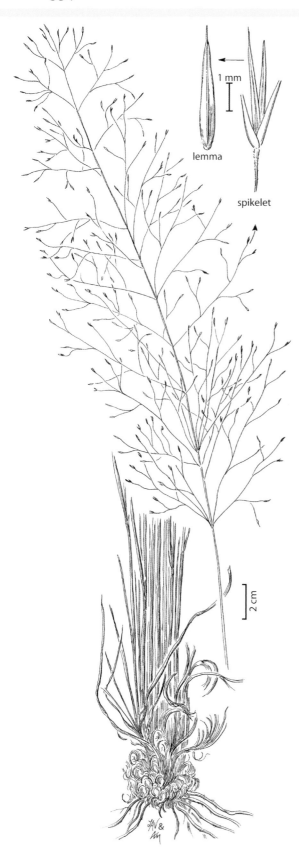

lemma

spikelet

1 mm

2 cm

15. *Muhlenbergia expansa* (Poir.) Trin. (savannah hairgrass, cutover muhly). Perennial, not rhizomatous. Culms to 1.5 m long, erect. Sheaths becoming fibrous and spirally coiled at the base when mature. Ligules 2–10 mm long, membranous, decurrent. Panicles open. Found in continually wet soils, pitcher plant bogs, and pine flatwoods. Apparently rare in far East Texas.

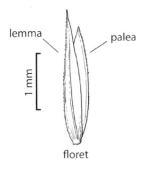

lemma   palea

1 mm

floret

1 mm

glumes

16. *Muhlenbergia filiformis* (Thurb. *ex* S. Watson) Rydb. (pull-up muhly). Tufted annual. Culms to 30 cm long, often rooting at lower nodes. Ligules 1–3 mm long, membranous. Panicles spikelike, interrupted near the base, long-exserted. Found in open moist areas and along streams. Apparently rare in Texas; reported only from Brazos County. Not included in Gould (1975b); Hatch, Gandhi, and Brown (1990); or Jones, Wipff, and Montgomery (1997).

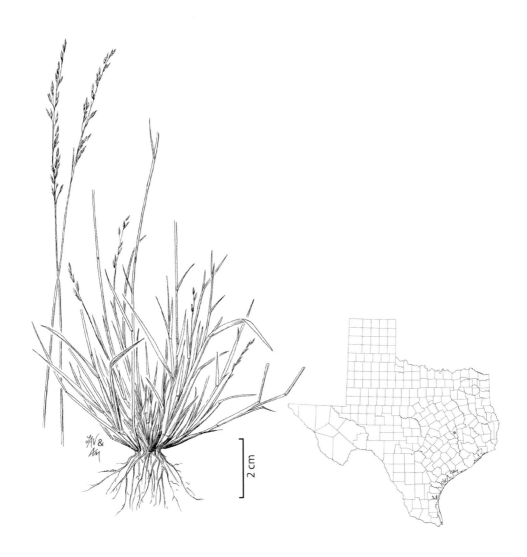

2 cm

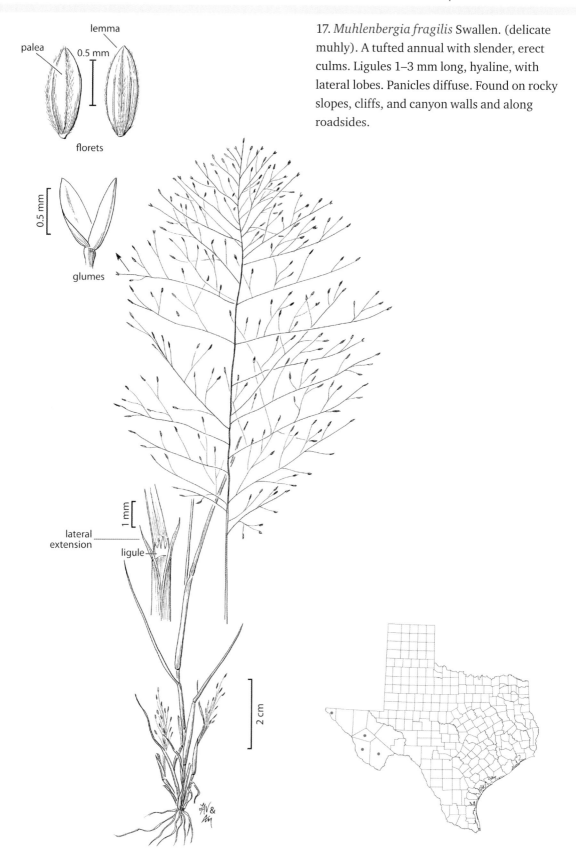

palea
lemma
0.5 mm
florets

0.5 mm
glumes

lateral
extension
1 mm
ligule

2 cm

17. *Muhlenbergia fragilis* Swallen. (delicate muhly). A tufted annual with slender, erect culms. Ligules 1–3 mm long, hyaline, with lateral lobes. Panicles diffuse. Found on rocky slopes, cliffs, and canyon walls and along roadsides.

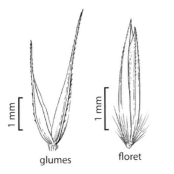

glumes floret

18. *Muhlenbergia frondosa* (Poir.) Fernald (wirestem muhly). Perennial with coarse, creeping, and scaly rhizomes. Culms to 1.1 m long, much-branched above, giving the plants a "bush" appearance. Panicles terminal and axillary; branches appressed or slightly diverging. Infrequent; found in forest openings and disturbed sites in woodlands.

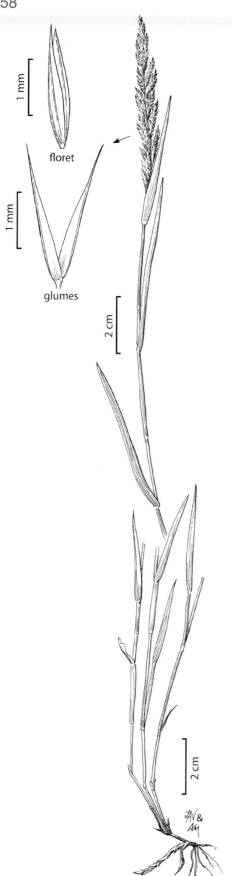

floret

glumes

19. *Muhlenbergia glabrifloris* Scribn. (inland muhly). Perennial with long, coarse, scaly rhizomes. Ligules 0.5–1.5 mm long, membranous. Panicles terminal and axillary; branches appressed. Found in moist oak woodlands. Rare; reported only from Dallas County. Gould (1975b) and Hatch, Gandhi, and Brown (1990) spell the specific epithet *glabriflora*.

20. *Muhlenbergia glauca* (Nees) B. D. Jacks. (desert muhly). Rhizomatous perennial with erect or decumbent culms, to 60 cm long. Ligules 0.5–2.0 mm long, membranous. Panicles contracted. Found on rocky slopes, canyon walls, cliffs, and rock outcrops.

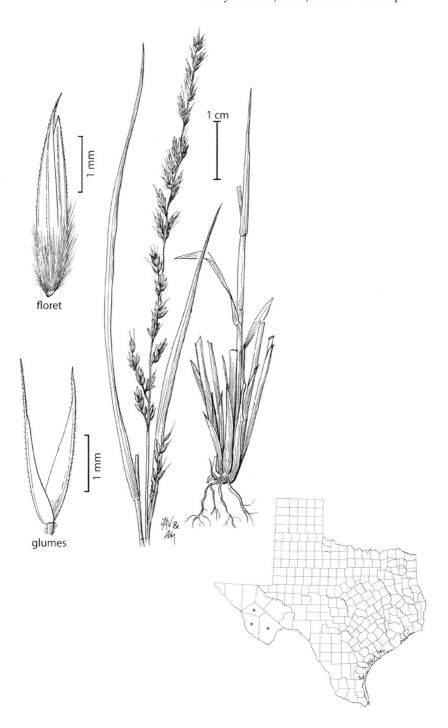

floret

1 mm

1 cm

glumes

1 mm

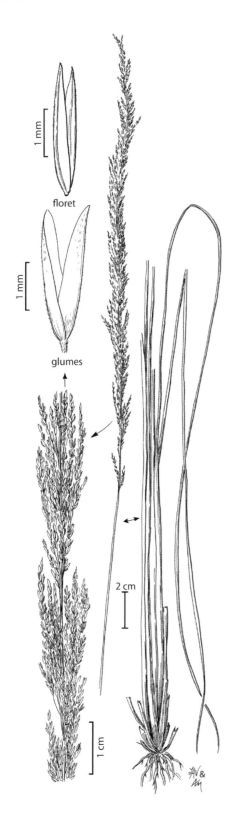

floret

glumes

1 mm

1 cm

2 cm

21. *Muhlenbergia lindheimeri* Hitchc. (Lindheimer's muhly). Cespitose perennial without rhizomes. Culms to 1.5 m long, stout, erect, compressed-keeled. Ligules 10–34 mm long, firm and brown basally, membranous distally. Panicles contracted, purplish. Grown as an ornamental although relatively infrequent in the wild. Found in sandy draws and open areas.

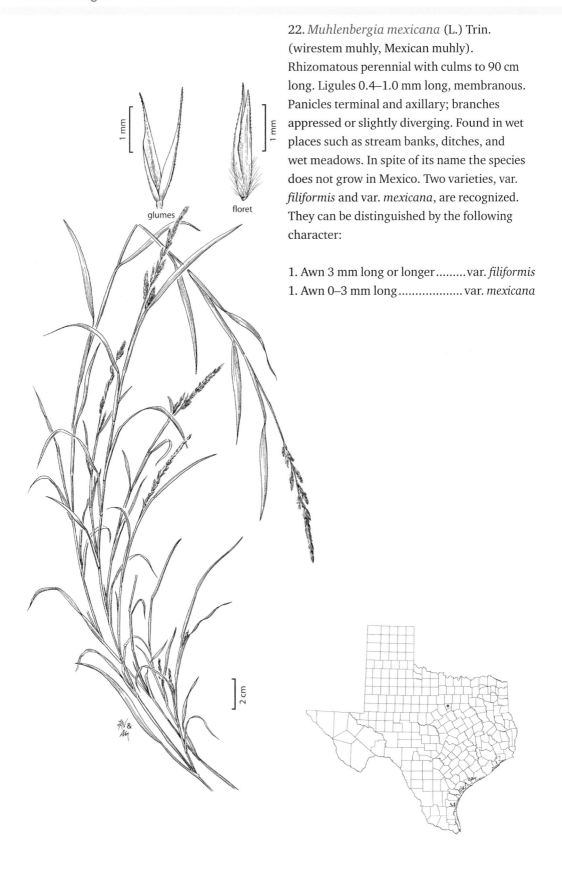

glumes

floret

1 mm

1 mm

2 cm

22. *Muhlenbergia mexicana* (L.) Trin. (wirestem muhly, Mexican muhly). Rhizomatous perennial with culms to 90 cm long. Ligules 0.4–1.0 mm long, membranous. Panicles terminal and axillary; branches appressed or slightly diverging. Found in wet places such as stream banks, ditches, and wet meadows. In spite of its name the species does not grow in Mexico. Two varieties, var. *filiformis* and var. *mexicana*, are recognized. They can be distinguished by the following character:

1. Awn 3 mm long or longer.........var. *filiformis*
1. Awn 0–3 mm long..................var. *mexicana*

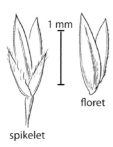

spikelet   floret   1 mm

23. *Muhlenbergia minutissima* (Steud.) Swallen (annual muhly, least muhly, sixweeks muhly). Delicate annual. Culms to 40 cm long, slender, erect. Ligules 1.0–2.5 mm long, hyaline, sometimes with lateral lobes. Panicles open. Found in sandy open areas, in gravelly drainages, on flats, along roadsides, and on rocky slopes.

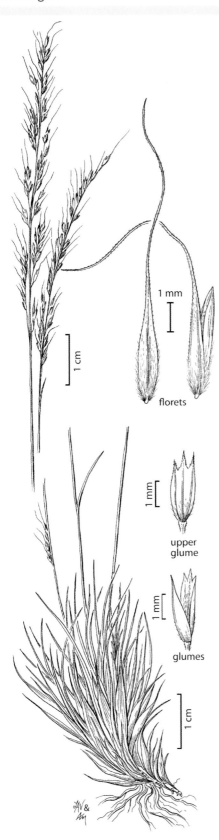

1 cm

1 mm

florets

1 mm

upper
glume

1 mm

glumes

1 cm

24. *Muhlenbergia montana* (Nutt.) Hitchc. (mountain muhly). Cespitose perennial without rhizomes. Culms to 80 cm long, erect. Ligules 4–20 mm long, membranous, acuminate. Panicles with slightly appressed to divergent branches, not dense. Found in the mountains of West Texas on rocky slopes and in open pine forests, dry meadows, and open grasslands. Powell (1994) reported that the species is palatable and good forage, mainly for wildlife.

25. *Muhlenbergia pauciflora* Buckley (New Mexico muhly). Perennial with a hard, knotty base. Culms decumbent and rooting at lower nodes. Ligules 1–5 mm long, with lateral lobes that are 1–3 mm longer than the central portion. Panicles loosely contracted. Found on rocky slopes, cliffs, canyons, and rock outcrops. Palatable and sometimes abundant enough to be of some forage value (Powell 1994).

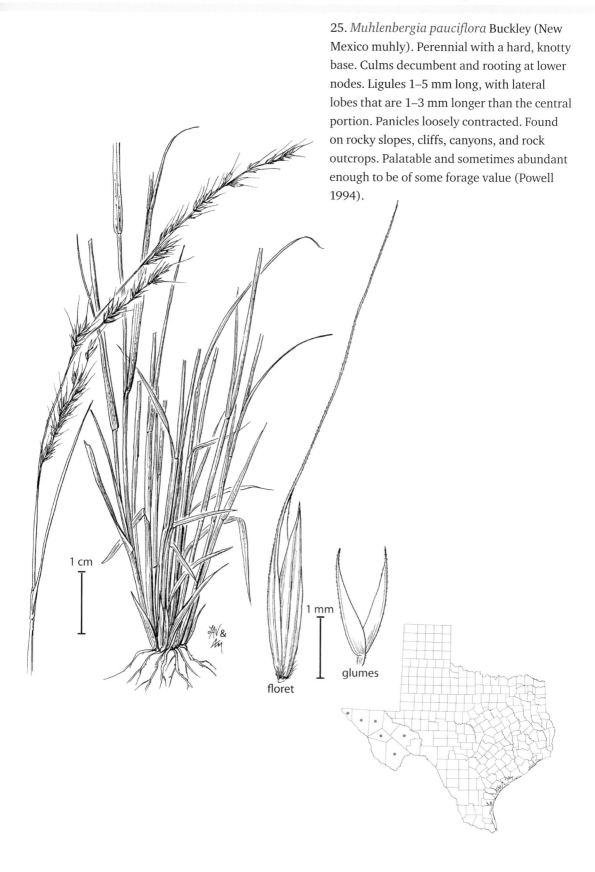

1 cm

1 mm

floret

glumes

26. *Muhlenbergia polycaulis* Scribn. (cliff muhly, many-stemmed muhly). Perennial, occasionally with rhizomes. Ligules 0.5–2.5 mm long, membranous. Panicles contracted. Found in open areas on steep rocky slopes, cliffs, canyon walls, and rock outcrops. Evidently palatable to livestock but restricted in availability and abundance (Powell 1994).

1 cm

glumes

1 mm

floret

glumes

palea

lemma

floret

2 cm

AV &
AM

27. *Muhlenbergia porteri* Scribn. *ex* Beal.
(bush muhly, mesquitegrass, bush-grass).
Perennial with a knotty base, not rhizomatous.
Culms to 1 m long, wiry, freely branching,
"bushy." Ligules 1.0–2.5 mm long with lateral
lobes. Panicles open, purplish. Grows among
boulders, on rocky slopes, in dry arroyos, and
on desert flats and grasslands. Almost always
growing in the protection of shrubs. Listed as
poor for wildlife and good for livestock (Hatch
and Pluhar 1993). Reported as excellent forage
for livestock (Powell 1994).

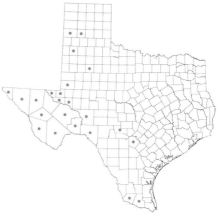

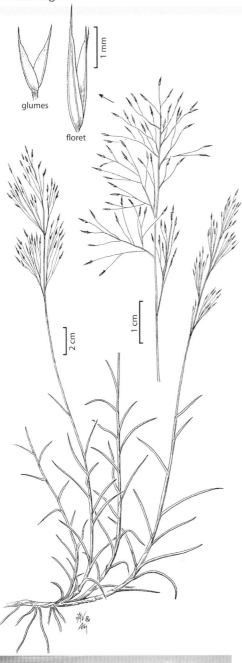

glumes

floret

28. *Muhlenbergia pungens* Thurb. *ex* A. Gray (pungent muhly, sandhill muhly). Perennial with scaly, creeping rhizomes. Ligules 0.2–1.0 mm long, with lateral lobes. Panicles more or less open. Found on loose sand near dunes, desert flats, and desert shrub and open woodland communities.

29. *Muhlenbergia racemosa* (Michx.)
Britton, Sterns & Poggenb. (green muhly,
marsh muhly, satingrass). Rhizomatous
perennial with erect culms to 1.1 m tall.
Ligules 0.5–1.5 mm long, membranous.
Panicles dense, contracted. Infrequent species
found along ditches and seasonally wet
prairies.

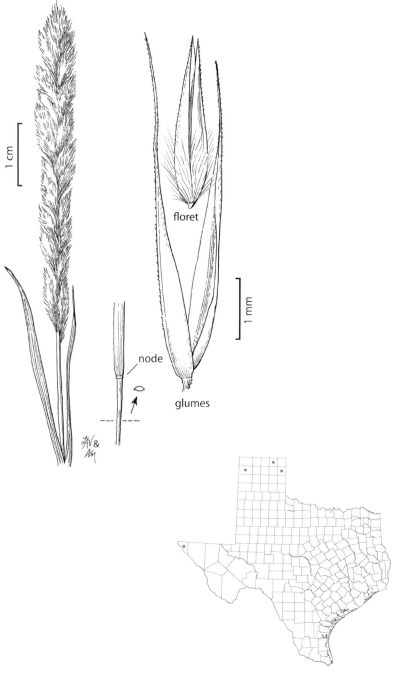

1 cm

floret

1 mm

node

glumes

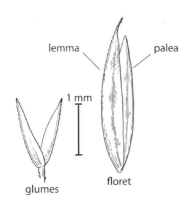

lemma

palea

1 mm

glumes

floret

30. *Muhlenbergia repens* (J. Presl) Hitchc. (creeping muhly, red muhly). Perennial with hard, shiny, creeping rhizomes. Culms to 40 cm long, forming dense mats. Ligules 0.1–1.0 mm long, membranous, often split down the middle. Panicles contracted, not dense, often included in subtending sheaths. Found in desert grasslands, forming rather large stands. Can form an attractive lawn, and it should be palatable to livestock and wildlife (Powell 1994).

1 cm

31. *Muhlenbergia reverchonii* Vasey and Scribn. (seep muhly, Reverchon's muhly). Cespitose perennial without rhizomes. Culms to 80 cm, erect. Ligules 2–9 mm long, firmer near the base. Panicles loosely contracted to open. Found on rocky slopes, flats, and rock outcrops. Resembles *M. capillaris* but has shiny and smooth lemmas (Diggs et al. 2006).

lemma

palea

1 mm

glumes

floret

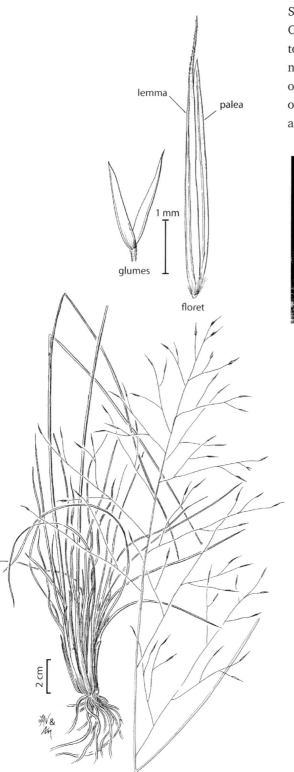

2 cm

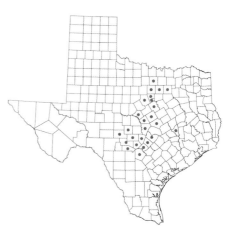

32. *Muhlenbergia rigens* (Benth.) Hitchc.
(deer muhly, deergrass). Cespitose perennial
without rhizomes. Culms to 1.5 m long, stout,
erect. Ligules 0.5–3.0 mm long, firm. Panicles
spikelike, green-gray, strong appressed
branches. Found in sandy washes, canyon
bottoms, sandy slopes, and frequently along
small streams. Used as an ornamental. Plants
are too coarse to be of value as forage (Powell
1994).

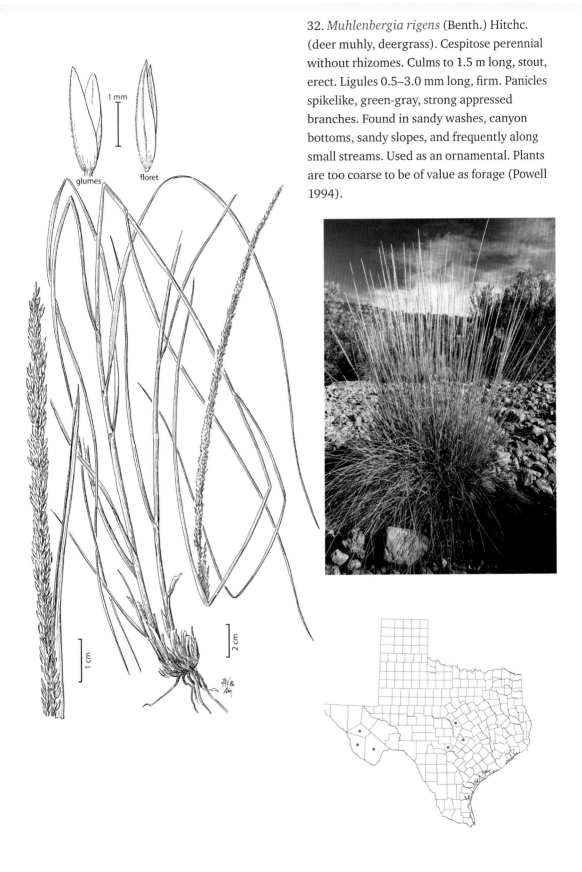

1 mm

glumes   floret

1 cm

2 cm

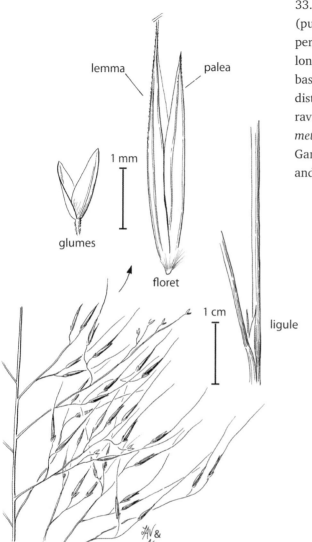

lemma   palea

1 mm

glumes

floret

1 cm

ligule

33. *Muhlenbergia rigida* (Kunth) Trin. (purple muhly, stiff muhly). Cespitose perennial without rhizomes. Culms to 1 m long, erect. Ligules 1–15 mm long, firmer basally. Panicles loosely contracted to open, distinctively purple. Found on rocky slopes and ravines. Grown as an ornmental. *Muhlenbergia metcalfei* M. E. Jones of Gould (1975b); Hatch, Gandhi, and Brown (1990); and Jones, Wipff, and Montgomery (1997) belongs here.

34. *Muhlenbergia schreberi* J. F. Gmel.
(nimblewill, satingrass, Schreber's muhly).
Perennial with weak, slender culms, often
stoloniferous and rooting at lower nodes.
Ligules 0.2–0.5 mm long, membranous.
Panicles with appressed branches. Found
in dry woods and prairies, on rocky slopes,
along stream banks, in river bottoms, and in
disturbed areas. Poor livestock and wildlife
values (Hatch, Schuster, and Drawe 1999).

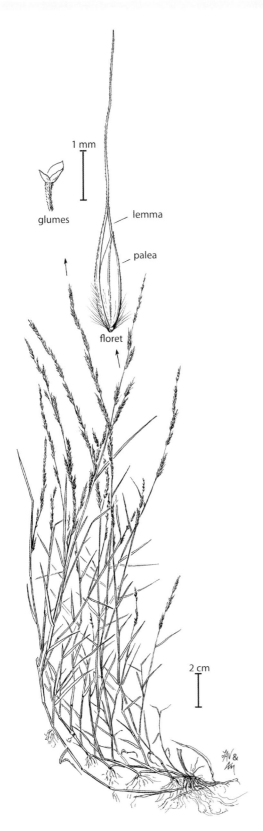

1 mm

glumes

lemma

palea

floret

2 cm

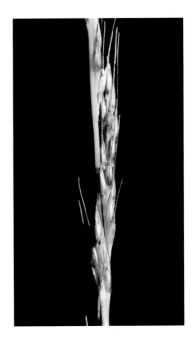

35. *Muhlenbergia sericea* (Michx.) P. M.
Peterson (dune muhly, purple muhly).
Cespitose perennial without rhizomes. Culms
to 1.7 m long, erect. Sheaths becoming fibrous
at the base of the plant, but not spirally coiled.
Ligules 4–8 mm long. Panicles open, diffuse,
purplish. Found in sandy maritime habitats.
Grown as an ornamental. Included in Gould
(1975b); Hatch, Gandhi, and Brown (1990);
and Jones, Wipff, and Montgomery (1997) as
M. filipes M. A. Curtis.

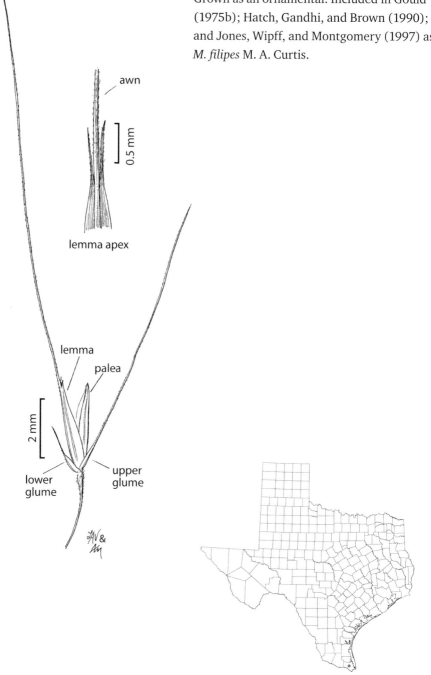

36. *Muhlenbergia setifolia* Vasey (curlyleaf muhly, bristle-leafed muhly). Perennial without rhizomes. Culms to 80 cm long. Ligules 4–7 mm long, membranous. Blades characteristically curly. Panicles loosely contracted. Found on rocky slopes, in dry arroyos, and on brushy flats. Typically grows in the protection of shrubs or cacti.

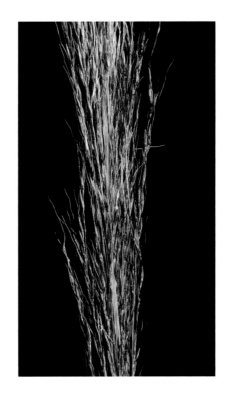

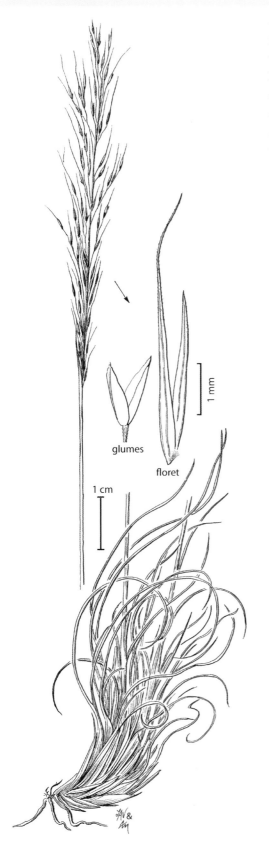

glumes

floret

1 mm

1 cm

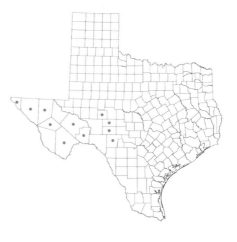

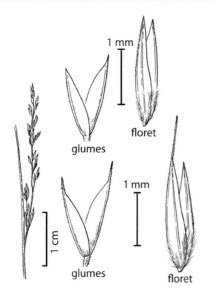

1 mm

glumes

floret

1 mm

glumes

floret

1 cm

37. *Muhlenbergia sobolifera* (Michx. *ex* Willd.) Trin. (rock muhly, rock dropseed). Rhizomatous perennial with erect culms to 1 m long. Ligules 0.3–1.0 mm long, membranous. Panicles terminal and axillary, narrow; branches appressed, little overlap. Found in partial shade, along rock outcrops, and on rocky slopes. No varieties are recognized here.

2 cm

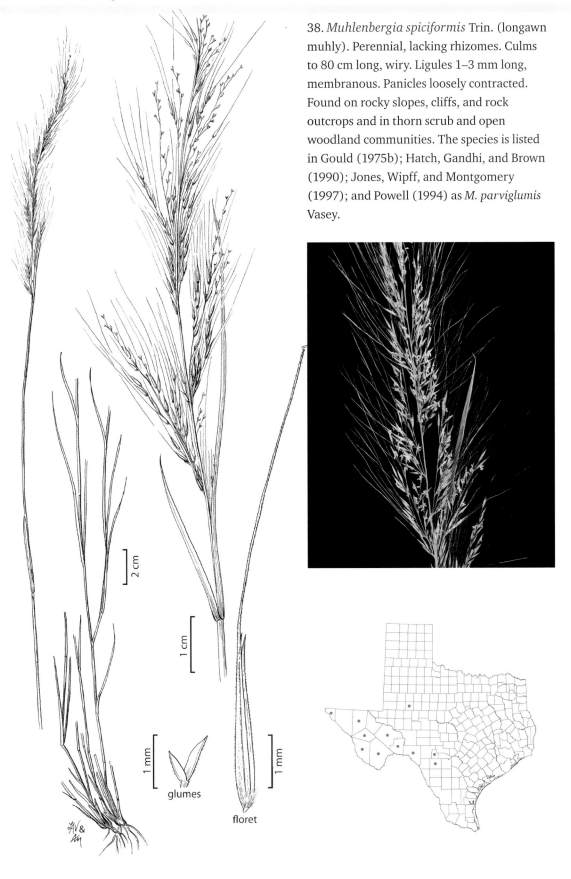

38. *Muhlenbergia spiciformis* Trin. (longawn muhly). Perennial, lacking rhizomes. Culms to 80 cm long, wiry. Ligules 1–3 mm long, membranous. Panicles loosely contracted. Found on rocky slopes, cliffs, and rock outcrops and in thorn scrub and open woodland communities. The species is listed in Gould (1975b); Hatch, Gandhi, and Brown (1990); Jones, Wipff, and Montgomery (1997); and Powell (1994) as *M. parviglumis* Vasey.

2 cm

1 cm

1 mm

glumes

1 mm

floret

39. *Muhlenbergia straminea* Hitchc.
(screwleaf muhly). Cespitose perennial
without rhizomes. Culms to 70 cm long, erect.
Leaves "corkscrewed." Ligules 10–20 mm long,
membranous. Panicles with loosely appressed
branches. Found on rocky slopes, canyon
bottoms, and open pine forests. Not included
in Gould (1975b); Hatch, Gandhi, and Brown
(1990); or Jones, Wipff, and Montgomery
(1997).

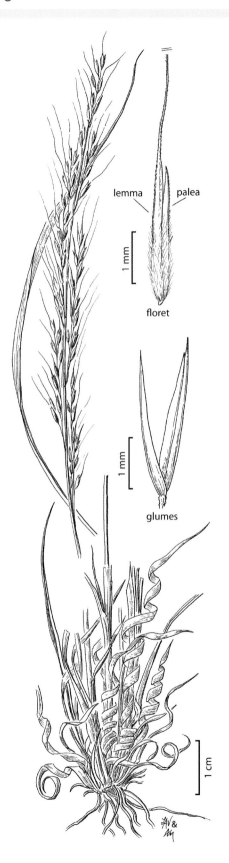

lemma   palea

1 mm

floret

1 mm

glumes

1 cm

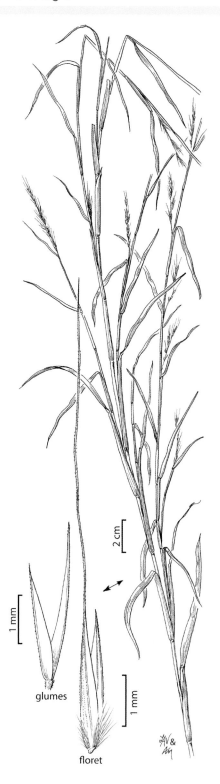

glumes

floret

2 cm

1 mm

1 mm

40. *Muhlenbergia sylvatica* (Torr.) Torr. *ex* A. Gray (forest muhly, woodland muhly). Rhizomatous perennial with erect culms to 1.1 mm long. Ligules 1.0–2.5 mm long, membranous. Panicles terminal and axillary; branches ascending to closely appressed. Infrequent; found in forest openings and along shaded stream banks.

41. *Muhlenbergia tenuiflora* (Willd.) Britton, Sterns, & Poggenb. (slimflowered muhly). Perennial with scaly rhizomes. Culms to 1.2 m long, erect. Ligules 0.4–12 mm long, membranous. Panicles usually terminal, branches appressed. Found on sandy to rocky slopes, canyon bottoms, and mixed hardwood forests. The species is not listed in Gould (1975b) or Hatch, Gandhi, and Brown (1990).

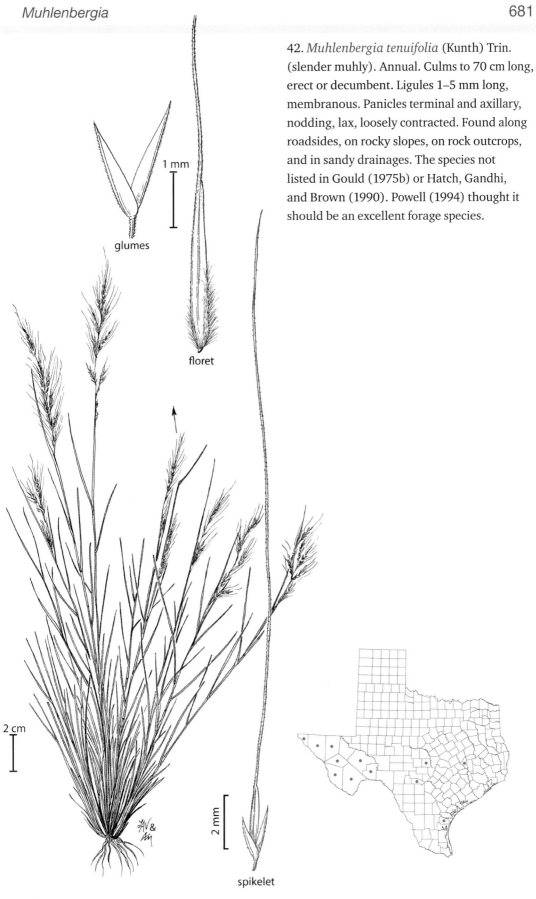

42. *Muhlenbergia tenuifolia* (Kunth) Trin. (slender muhly). Annual. Culms to 70 cm long, erect or decumbent. Ligules 1–5 mm long, membranous. Panicles terminal and axillary, nodding, lax, loosely contracted. Found along roadsides, on rocky slopes, on rock outcrops, and in sandy drainages. The species not listed in Gould (1975b) or Hatch, Gandhi, and Brown (1990). Powell (1994) thought it should be an excellent forage species.

glumes

1 mm

floret

2 cm

2 mm

spikelet

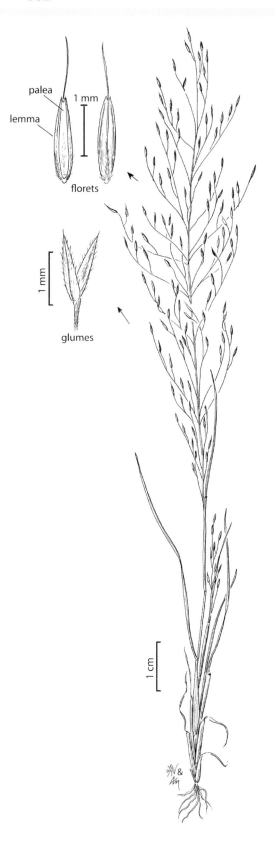

43. *Muhlenbergia texana* Buckley (Texas muhly). Tufted annual. Culms to 35 cm long, slender, erect. Ligules 0.9–2.5 mm long, hyaline. Panicles loosely opened. Found on open slopes, on rock outcrops, and in sandy drainages. Gould (1975b); Hatch, Gandhi, and Brown (1990); and Turner et al. (2003) listed this as a synonym of *M. minutissima*.

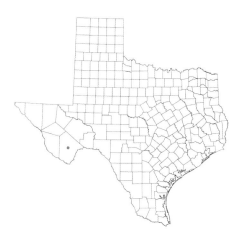

44. *Muhlenbergia thurberi* (Scribn.) Rydb.
(Thurber's muhly). Rhizomatous perennial.
Culms to 35 cm long, clumped, erect or
decumbent. Ligules 0.9–1.2 mm long,
membranous. Panicles contracted, not dense,
sometimes interrupted at the base. Found in
moist areas near canyon cliffs and rock ledges.
This species not listed in Gould (1975b);
Hatch, Gandhi, and Brown (1990); or Jones,
Wipff, and Montgomery (1997).

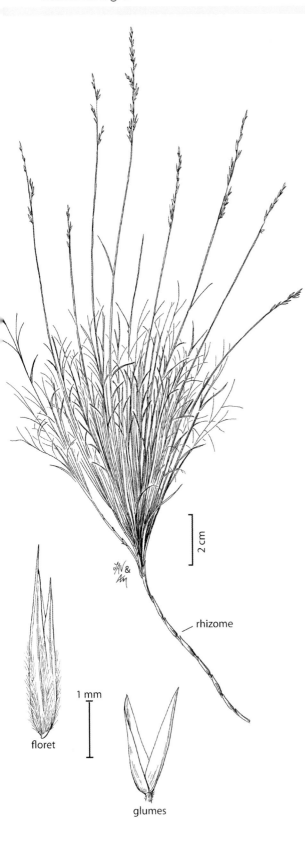

2 cm

floret

1 mm

glumes

rhizome

45. *Muhlenbergia torreyi* (Kunth) Hitchc. *ex* Bush (ring muhly, ringgrass). Perennial, rhizomes absent. As the base of the plant grows, the center dies out, leaving a distinct "ring." Ligules 2–7 mm, often with lateral lobes, membranous. Panicles diffuse. Found in desert grasslands, on sandy mesas, and on rocky slopes. An overabundance of this plant is indicative of overgrazed areas. Closely related to *M. arenicola*, but *M. torreyi* is more diminutive in all features. *Muhlenbergia torreyi* is less frequent than *M. arenicola* in Texas. When green, it is palatable to cattle, but it is a low forage producer (Powell 1994).

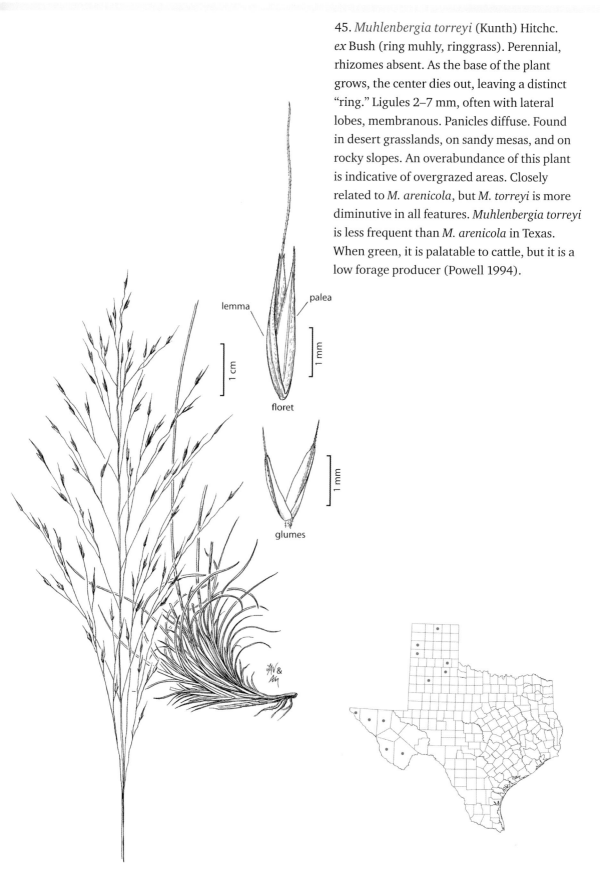

lemma

palea

floret

1 mm

1 cm

glumes

1 mm

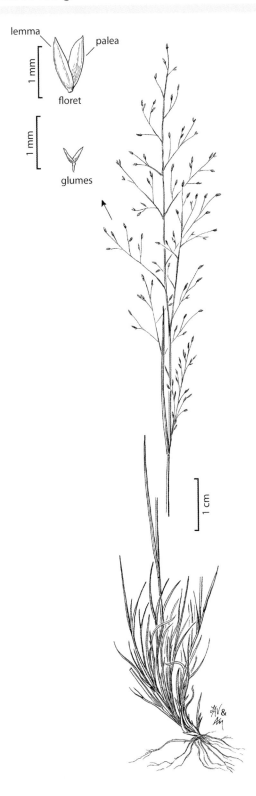

lemma

palea

1 mm

floret

1 mm

glumes

1 cm

46. *Muhlenbergia uniflora* (Muhl.) Fernald (bog muhly). Loosely matted perennial. Culms to 45 cm long, compressed-keeled. Ligules 0.5–1.5 mm long, membranous, without lateral lobes. Panicles diffuse. A native of eastern North America and probably introduced into Texas. Perhaps not a constant member of the Texas flora. Not included in Gould (1975b); Hatch, Gandhi, and Brown (1990); or Jones, Wipff, and Montgomery (1997).

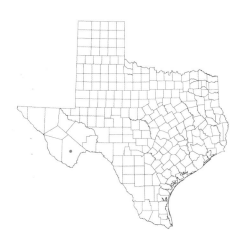

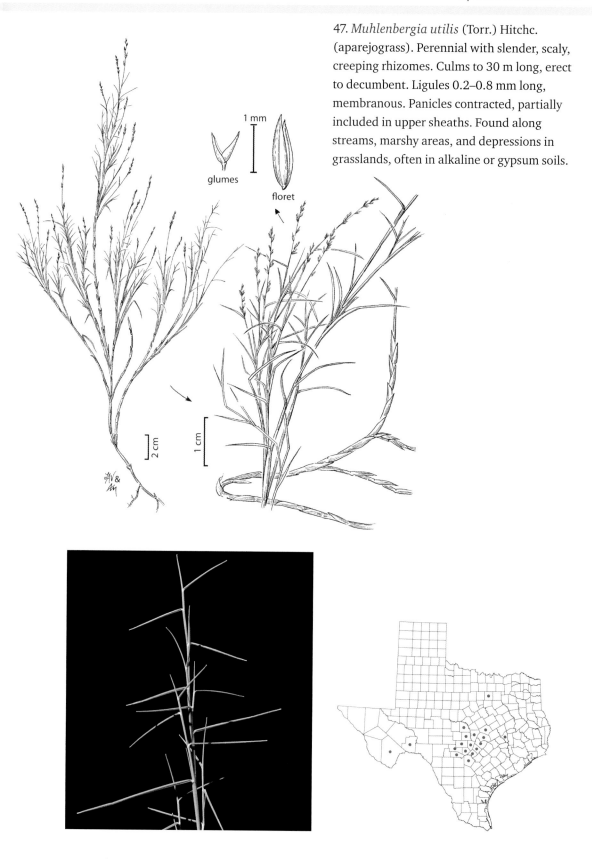

47. *Muhlenbergia utilis* (Torr.) Hitchc. (aparejograss). Perennial with slender, scaly, creeping rhizomes. Culms to 30 m long, erect to decumbent. Ligules 0.2–0.8 mm long, membranous. Panicles contracted, partially included in upper sheaths. Found along streams, marshy areas, and depressions in grasslands, often in alkaline or gypsum soils.

glumes

1 mm

floret

2 cm

1 cm

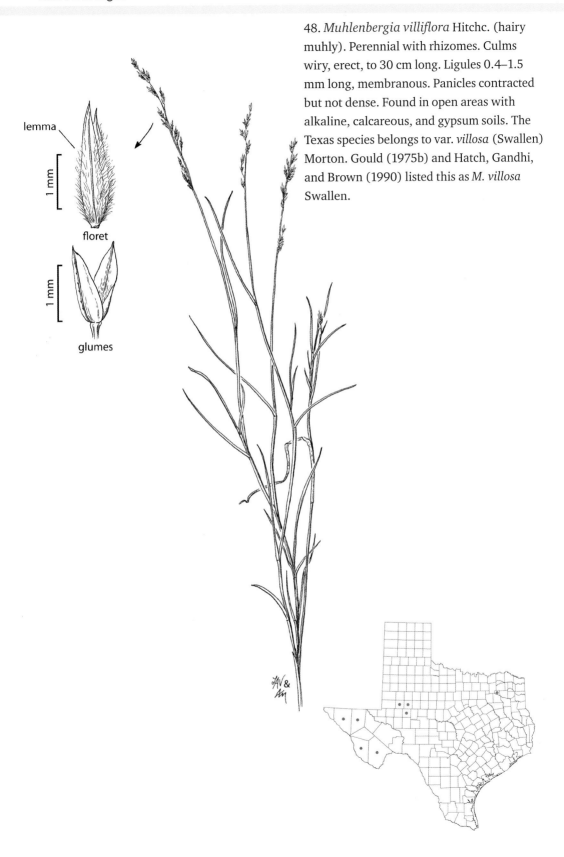

lemma

1 mm

floret

1 mm

glumes

48. *Muhlenbergia villiflora* Hitchc. (hairy muhly). Perennial with rhizomes. Culms wiry, erect, to 30 cm long. Ligules 0.4–1.5 mm long, membranous. Panicles contracted but not dense. Found in open areas with alkaline, calcareous, and gypsum soils. The Texas species belongs to var. *villosa* (Swallen) Morton. Gould (1975b) and Hatch, Gandhi, and Brown (1990) listed this as *M. villosa* Swallen.

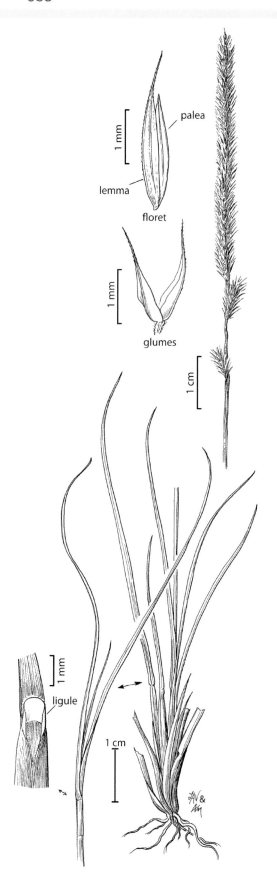

49. *Muhlenbergia wrightii* Vasey *ex* J. M. Coult. (spike muhly). Perennial, rhizomes lacking. Culms to 60 cm long, compressed, erect. Ligules 1–3 mm long, membranous. Panicles spikelike, dense. Found on rocky slopes and grasslands. Apparently rare and reported only from Jeff Davis County, but it is abundant throughout New Mexico (Allred 2005). Perhaps it has been overlooked in Texas. Gould (1975b); Hatch, Gandhi, and Brown (1990); Jones, Wipff, and Montgomery (1997); and Turner et al. (2003) did not include this species.

Munroa Torr.
(Chloridoideae: Cynodonteae)

Low, tufted annuals with spreading, much-branched culms. Leaves short, the blades stiff, pungent, and flat. Inflorescences of small, subsessile clusters of spikelets almost hidden in fascicles of leaves at the branch tips. Spikelets few-flowered, the lower ones of a cluster 3- or 4-flowered, the upper ones 2- or 3-flowered. Typically lower florets pistillate or perfect, distal florets sterile and rudimentary. Disarticulation above the glumes and sometimes between the florets, or beneath the leaves subtending the branches. Glumes of the lower spikelets subequal, narrow, acute, 1-nerved, slightly shorter than the lowermost lemma. Glumes of the upper spikelets unequal, the lower reduced or even absent. Lemmas prominently 3-nerved, those of the lower spikelets coriaceous, usually with tufts of hairs on the margins near the middle, gradually narrowing above, abruptly mucronate or short-awned from a slender, spreading tip. Palea narrow, membranous. Basic chromosome number, $x = 8$. Photosynthetic pathway, C_4.

In the original publication of the genus, the name was misspelled *Monroa*, and some suggest that this should be the correct spelling for the genus (Jones, Wipff, and Montgomery 1997). Represented in Texas by a single species.

(Valdes-Reyna 2003d)

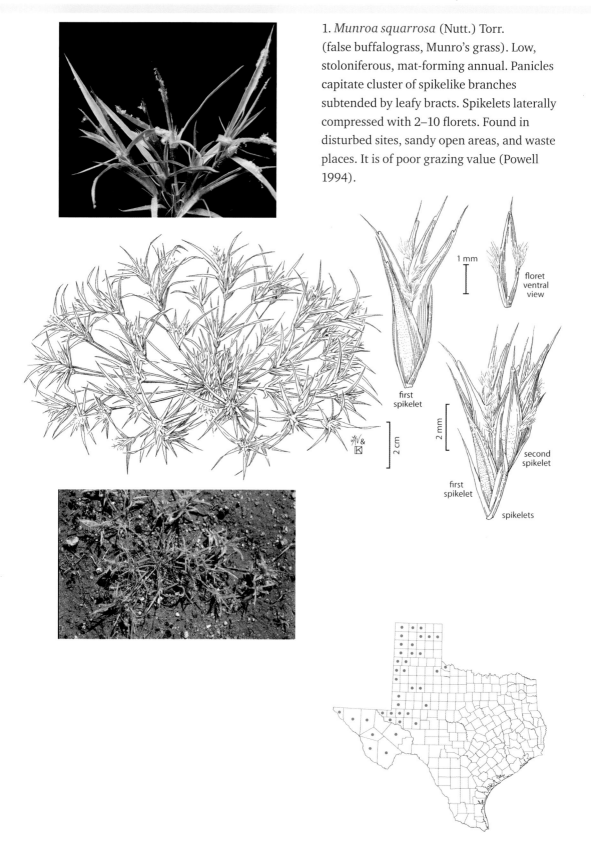

1. *Munroa squarrosa* (Nutt.) Torr.
(false buffalograss, Munro's grass). Low,
stoloniferous, mat-forming annual. Panicles
capitate cluster of spikelike branches
subtended by leafy bracts. Spikelets laterally
compressed with 2–10 florets. Found in
disturbed sites, sandy open areas, and waste
places. It is of poor grazing value (Powell
1994).

1 mm

floret
ventral
view

first
spikelet

2 mm

second
spikelet

first
spikelet

spikelets

2 cm

Nassella E. Desv.
(Poöideae: Stipeae)

Cespitose perennials. Inflorescence a terminal contracted or occasionally an open panicle of 1-flowered spikelets. Glumes longer than the floret, tapering to an acuminate tip, frequently purplish at the base before drying. Lemmas strongly convolute, coriaceous, with a crown at the apex, terminating into a persistent, usually twisted and geniculate awn. Paleas up to $\frac{1}{3}$ the length of the lemmas, glabrous, and without veins. Basic chromosome number, $x = 11$. Photosynthetic pathway, C_3. Represented in Texas by 3 species. A segregate of *Stipa*.

1. Florets 1.5–3.0 mm long ....2. *N. tenuissima*
1. Florets 3.4–13 mm long.
    2. Florets 6–13 mm long; lemmas constricted below the crown; awns 40–90 mm long.. 1. *N. leucotricha*
    2. Florets 3.4–5.5 mm long; lemmas not constricted below the crown; awns 19–32 mm long..3. *N. viridula*

(Barkworth 2007f)

1. *Nassella leucotricha* (Trin and Rupr.)
R. W. Pohl (Texas wintergrass, Texas nassella,
Texas needlegrass, speargrass, Texas
tussockgrass). Perennial with culms to 1.2 m
long. Ligules 0.2–1.2 mm long, membranous,
sometimes longest on the sides. Crowns
0.8–2.0 mm long with 1–2 mm hairs. Species
increases on grasslands and prairies with
disturbance; also found on roadsides and
other moderately disturbed sites. Provides
abundant and important spring forage. Sharp
callus can do mechanical damage to grazing
animals. Reported in Gould (1975b) and
Hatch, Gandhi, and Brown (1990) as *Stipa
leucotricha*. Listed as fair for wildlife and good
to fair for livestock (Hatch and Pluhar 1993, as
S. leucotricha Trin. and Rupr.; Powell 1994).
However, in the Coastal Prairies and Marshes
the forage is considered high value in the
spring for cattle and wildlife (Hatch, Schuster,
and Drawe 1999).

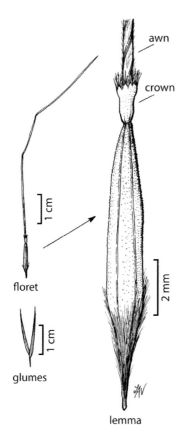

awn

crown

1 cm

floret

2 mm

glumes

1 cm

lemma

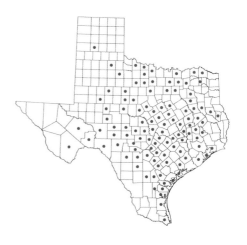

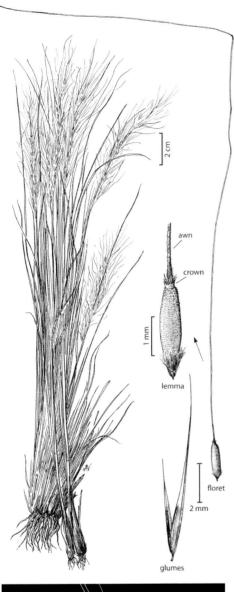

2. *Nassella tenuissima* (Trin.) Barkworth (fineleaved needlegrass, fineleaved nassella). Cespitose perennial with erect, slender culms to 1 m long. Ligules 1–5 mm long, acute. Crowns 0.1–0.2 mm long, hairs <0.5 mm long. Locally abundant on rocky slopes, desert grasslands, and oak-pine woodlands. An attractive species frequently used as an ornamental, but it does escape and can be an aggressive weed on disturbed sites. Reported in Gould (1975b) and Hatch, Gandhi, and Brown (1990) as *Stipa tenuissima*.

3. *Nassella viridula* (Trin.) Barkworth (green
needlegrass, green nassella). Cespitose perennial
with erect culms to 1.2 m long. Ligules 0.2–1.2
mm long, membranous. Crowns 0.4–0.5 mm
long, hairs 0.50–0.75 mm long. Found on
grasslands and open woodlands, typically
on sandy soils. Reported from San Saba and
Washington counties. If established, these
represent disjunct populations from the closest
known plants in northern New Mexico and
the panhandle of Oklahoma. Not included in
Gould (1975b); Hatch, Gandhi, and Brown
(1990); or Jones, Wipff, and Montgomery
(1997).

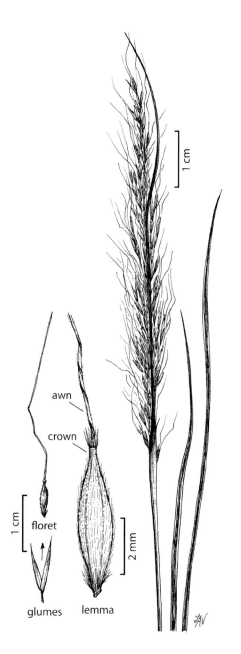

Oplismenus P. Beauv.
(Panicoideae: Paniceae)

Low, usually prostrate and creeping annuals and perennials with culms freely branching below and short, thin, flat leaf blades. Inflorescences small, few-flowered, with spikelets subsessile and crowded on 2 to several short, distant, spicate branches. Disarticulation below the glumes. Glumes about equal, awned from a minutely notched apex, the awn of the first glume considerably longer than that of the second. Lemma of sterile floret about as long as the second glume, mucronate or short-awned. Lemma of fertile floret smooth, indurate, elliptic, acute at the apex, the margins enclosing the palea. Basic chromosome number, $x = 9$. Photosynthetic pathway, C_3. Represented in Texas by a single species.

(Wipff 2003f)

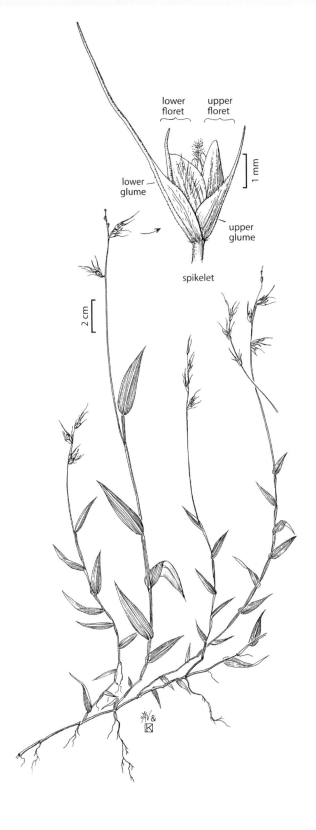

lower floret upper floret

lower glume

upper glume

spikelet

2 cm

1 mm

1. *Oplismenus hirtellus* (L.) P. Beauv. (basketgrass, bristle basketgrass, long-leaf basketgrass). Perennial that often roots at the lower nodes. Flower culms to 25 cm long. Ligules a ciliate membrane, 0.4–1.6 mm long. Panicles with 2–10 short, primary branches. Awns are viscid. Found primarily in moist, shady areas. Poor livestock and wildlife values (Hatch, Schuster, and Drawe 1999). The Texas species belongs to the subsp. *setarius* (Lam.) Mez *ex* Ekman.

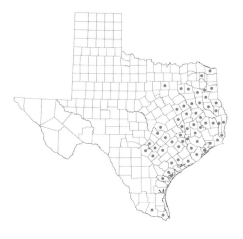

Oryza L.
(Ehrhartoideae: Oryzeae)

Annuals and a few perennials, with herbaceous culms; flat, lanceolate blades; and large, apparently 1-flowered spikelets borne in open or somewhat contracted panicles. Ligules membranous. Spikelets laterally flattened, with a thick, firm, 5-nerved, awnless or awned lemma and a large, 2-nerved palea of similar texture tightly enclosing the caryopsis. At the base of the lemma (in *O. sativa*) are 2 short, pointed bracts, these superficially resembling glumes but more correctly interpreted as being reduced lemmas of rudimentary florets. Disarticulation below the spikelet. Stamens usually 6, lodicules 2. Basic chromosome number, $x = 12$. Photosynthetic pathway, C_3. Represented in Texas by a single species.

(Barkworth and Terrell 2007)

palea lemma

upper glume   lower glume   callus

sterile floret   sterile floret

spikelet

ligule

SL

1. *Oryza sativa* L. (rice, Asian rice, cultivated rice, red rice). Plants succulent, annual or occasionally perennial. Culms to 2 m long, erect, rooting at the lower nodes. Ligules 10–30 mm long, acute. Panicles nodding, loosely open. Cultivated and occasionally escapes. This species, along with *Zea mays* and *Triticum aestivum,* provides the primary carbohydrate needs of humans. Considered good forage and seed for livestock and waterfowl (Hatch, Schuster, and Drawe 1999).

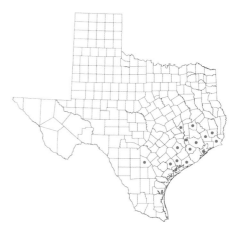

Panicum L.
(Panicoideae: Paniceae)

Annuals and perennials of extremely diverse habit. Ligule a membrane or a ring of hairs. Inflorescence an open or contracted panicle. Disarticulation below the glumes. Glumes usually both present, the lower commonly short. Lowermost floret sterile or occasionally staminate, with a lemma similar to the glumes in texture and usually as long as, or slightly longer than, the upper glume. Lemma of the upper floret shiny and glabrous in Texas species, smooth, firm or indurate, tightly clasping the palea with thick, inrolled margins. Palea of the upper floret like the lemma in texture. Basic chromosome numbers, $x = 9$ and 10. Photosynthetic pathways, C_3, C_4. Represented in Texas by 24 species and 16 subspecies.

1. Panicle branches 1-sided; longest pedicels <2 mm long.
    2. Lower florets staminate; upper lemmas thin, only clasping paleas at the base...14. *P. hemitomon*
    2. Lower florets sterile; upper lemmas indurate, clasping the paleas throughout their length.
        3. Glumes and lower lemmas without keeled midveins; pedicels with hairs near the apices .... 21. *P. tenerum*
        3. Glumes and lower lemmas with keeled midveins; pedicels glabrous.
            4. Plants rhizomatous; culms and sheaths not compressed-keeled .... 2. *P. anceps*
            4. Plants cespitose; culms and sheaths compressed-keeled .... 20. *P. rigidulum*
1. Panicle branches usually not 1-sided; longest pedicels 2–20 mm long.
        5. Upper glumes and lower lemmas warty-tuberculate.
            6. Lower lemmas verrucose with warts; spikelets 1.7–2.2 mm long ....
            ....23. *P. verrucosum*
            6. Lower lemmas tuberculate; spikelets 3.2–4.0 mm long ...... 5. *P. brachyanthum*
        5. Upper glumes and lower lemmas not warty-tuberculate.
            7. Upper florets transversely rugose; culms strongly compressed....................
            ....see *Zuloagaea bulbosa*
            7. Upper florets not transversely rugose; culms not compressed.
                8. Plants with rhizomes about 1 cm thick and with large, pubescent, scalelike leaves.. 3. *P. antidotale*
                8. Plants without rhizomes or with rhizomes <1 cm thick, and small, glabrous, scalelike leaves.
                    9. Plants perennial, rhizomatous; lower floret staminate.
                        10. Lower glumes 0.5–1.5 mm long, less than ½ as long as spikelet; lower paleas not hastate-lobed.
                            11. Plants with long, scaly rhizomes; lower glumes faintly veined; upper florets widest at or above the middle, apices rounded.....
                            .... 19. *P. repens*
                            11. Plants with short, knotty rhizomes; lower glumes prominently veined; upper florets widest below the middle, apices slightly beaked... 8. *P. coloratum*

10. Lower glumes 1.8–4.0 mm long; more than ½ as long as spikelet; lower paleas hastate-lobed.

    12. Panicles contracted; plants glabrous throughout .... 1. *P. amarum*

    12. Panicles open; plants pilose at least at base of blades ....24. *P. virgatum*

9. Plants annual or perennial, usually without rhizomes; lower florets sterile ....Key A

Key A

1. Lower glumes less than ⅓ as long as spikelet; sheaths compressed; plants succulent or spongy ....9. *P. dichotomiflorum*

1. Lower glumes more than ⅓ as long as spikelet; sheaths rounded; plants not succulent.

  2. Spikelets 4.0–6.5 mm long.

    3. Plants annual; upper florets 2.0–2.5 mm wide; lower paleas truncate to bilobed; upper glumes and lower lemmas slightly exceeding the upper florets in length..................... ....17. *P. miliaceum*

    3. Plants perennial; upper florets 1.0–1.1 mm wide; lower paleas acute; upper glumes and lower florets exceeding the upper floret by 3–4 mm........................... 7. *P. capillariodes*

  2. Spikelets 1.0–4.2 mm long.

    4. Plants perennial.

      5. Lower panicle branches whorled.

        6. Sheaths without prickly hairs causing skin irritation; lower paleas 1.3–1.7 mm long... 15. *P. hirsutum*

        6. Sheaths without prickly hairs; lower paleas 1.4–2.2 mm long; panicle becomes a tumbleweed.. 4. *P. bergii*

      5. Lower panicle branches solitary.

        7. Blades glabrous and glaucous on the adaxial surface................. 13. *P. hallii*

        7. Blades sparsely to densely hirsute on adaxial surface.

          8. Blades with a prominent white midrib; spikelets 2.1–2.9 mm long ....10. *P. diffusum*

          8. Blades without a prominent white midrib; spikelets 2.6–3.4 mm long ....12. *P. ghiesbreghtii*

    4. Plants annual.

      9. Blades lanceolate, 4–6 times longer than wide........... 22. *P. trichoides*

      9. Blades linear, >10 times longer than wide.

        10. Panicles >2 times longer than wide........................... 11. *P. flexile*

        10. Panicle <1.5 times as long as wide.

          11. Upper glumes and lower lemmas with prominent veins; upper glumes more than ½ as long as spikelet ....16. *P. hirticaule*

          11. Upper glumes and lower lemmas faintly veined; upper glumes less than ½ as long as spikelet.

12. Panicles usually more than ½ the total length of plant, becoming a tumbleweed at maturity .... 6. *P. capillare*

12. Panicles less than ½ the total length of plant, usually not becoming a tumbleweed .... 18. *P. philadelphicum*

(Freckmann and Lelong 2003b; Gould 1975b)

subsp. amarulum

spikelet

spikelet

subsp. amarum

1. *Panicum amarum* Elliott (bitter panicum, bitter beachgrass). Rhizomatous perennial. Rhizomes stout, glabrous and glaucous. Culms to 2.5 cm thick, erect or decumbent. Ligules 1–5 mm long. Panicles more or less contracted, 10–80 cm long. Found in coastal sand dunes and wet, sandy soil. Palatable to cattle (Hatch, Schuster, and Drawe 1999). Two subspecies occur in Texas: subsp. *amarulum* (Hitchc. and Chase) Freckmann & Lelong; and the typical one, subsp. *amarum*. They can sometimes be distinguished by the following characters:

1. Culms often bunched and decumbent; lower glumes with 3–5 veins...... subsp. *amarulum*
1. Culms usually solitary; lower glumes with 7–9 veins .... subsp. *amarum*

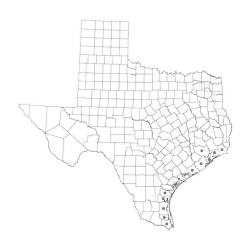

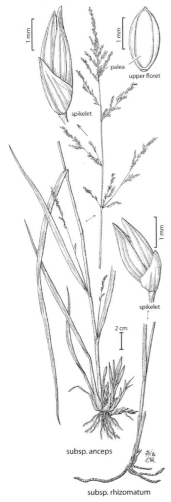

palea

upper floret

spikelet

spikelet

2 cm

subsp. anceps

subsp. rhizomatum

2. *Panicum anceps* Michx. (beaked panicgrass, beaked panicum). Rhizomatous perennial. Culms to 1.3 m long. Sheaths laterally compressed. Ligules membranous, <0.5 mm long. Panicles 1–40 cm long, open. Found in forests or shaded, grassy areas, usually in sandy, low, and moist places. Listed as fair for wildlife and good for livestock (Hatch and Pluhar 1993). Telfair (2006) listed the seeds and rhizomes as especially important for wildlife. Two subspecies occur in the state: the typical one, subsp. *anceps*; and subsp. *rhizomatum* (Hitchc. & Chase) Freckmann & Lelong. They can be distinguished by the following characters:

1. Rhizomes short and stout; spikelets 2.7–3.9 mm long .... subsp. *anceps*
1. Rhizomes long and slender; spikelets 2.3–2.8 mm long..................subsp. *rhizomatum*

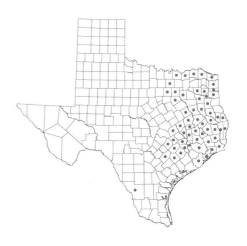

1 mm

spikelet

3. *Panicum antidotale* Retz. (blue panicgrass, blue panicum). Perennial with thick (1 cm), knotted, pubescent, and scaly rhizomes. Culms to 3 m long, often compressed, hard, almost woody. Ligules a fringed membrane, 0.5–1.0 mm long. Panicles 10–45 cm long, open to somewhat contracted. A species introduced from India and reseeded as a forage species. Escaped and established in open, disturbed areas and fields. Listed as good for wildlife and livestock but can be poisonous under some circumstances (Hatch and Pluhar 1993; Powell 1994).

2 cm

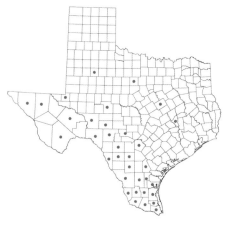

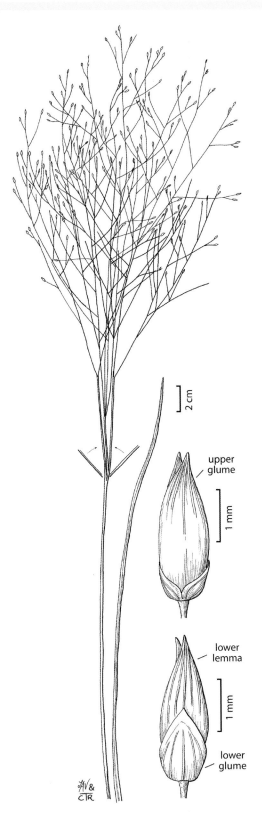

4. *Panicum bergii* Arechav. (Berg's witchgrass). Strongly cespitose perennial. Culms to 1.4 m long, stout, erect. Ligules a ciliated membrane to 3 mm long. Panicles 10–40 cm long, open, breaking at maturity and becoming a tumbleweed. A South American introduction now found in ditches, field margins, shallow depressions in grasslands, and occasionally flooded sites. Gould (1975b) and Hatch, Gandhi, and Brown (1990) listed this as *P. pilomayense* Hack.

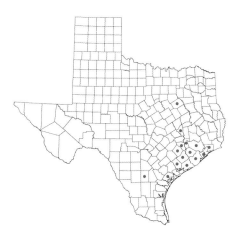

5. *Panicum brachyanthum* Steud. (prairie
panicgrass, pimple panicum). Annual with
spreading culms. Culms wiry, often with
purple streaks and dots, rooting at the lower
nodes. Ligules a minute (0.3 mm long) ciliated
membrane. Panicles 4–17 cm long, nearly as
wide as long. Spikelets densely tuberculate,
with hairs arising from wartlike bases. Found
in dry areas such as prairies, woodland
borders, and roadsides. The other species with
wart spikelets, P. *verrucosum*, prefers wetter
sites.

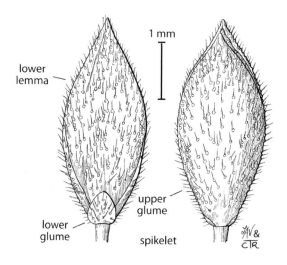

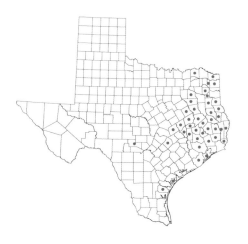

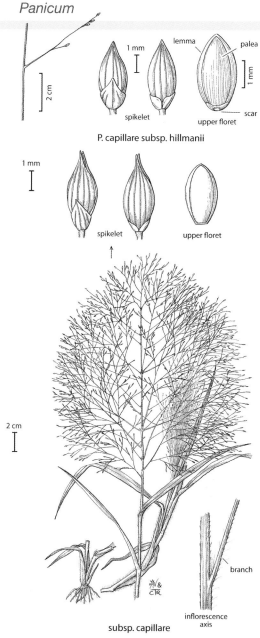

P. capillare subsp. hillmanii

subsp. capillare

6. *Panicum capillare* L. (witchgrass, common witchgrass). Annual, more or less covered with papillose-based hairs, often purple or bluish. Culms to 1.3 m tall. Ligules a ciliated membrane, 0.5–2.0 mm long. Panicles 13–50 cm long, open, disarticulating at the base and becoming a tumbleweed. A weed of disturbed sites such as gardens, roadsides, fields, and waste places. Poor livestock and wildlife values (Hatch, Schuster, and Drawe 1999). Telfair (2006) listed the seeds as especially important for wildlife. Jones, Wipff, and Montgomery (1997) reported 4 varieties from the state, but now only 2 subspecies are recognized: subsp. *capillare*; and subsp. *hillmanii* (Chase) Freckmann and Lelong. Gould (1975b); Hatch, Gandhi, and Brown (1990); and Jones, Wipff and Montgomery listed the latter as a separate species. The subspecies can be distinguished by the following characters:

1. Lower paleas absent; pedicels and secondary branches strongly divergent...... subsp. *capillare*
1. Lower paleas present; pedicels and secondary branches appressed to narrowly divergent ....subsp. *hillmanii*

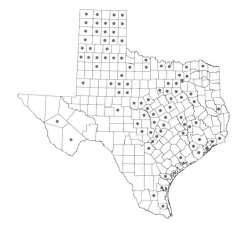

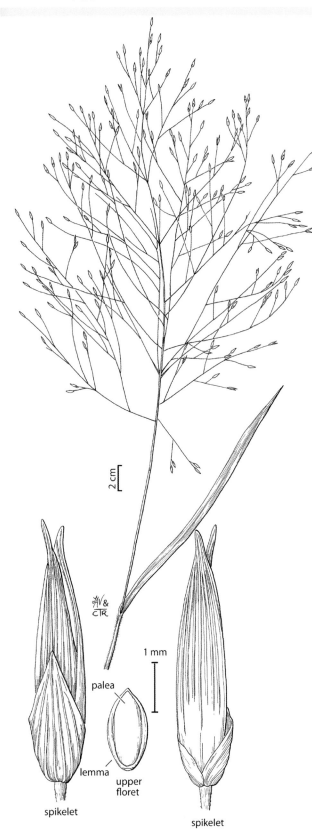

7. *Panicum capillarioides* Vasey (long-beaked witchgrass, southern witchgrass, slender panicgrass). Perennial with a knotty basal crown. Culms to 75 cm long, often bent at the nodes. Ligules a ciliated membrane <1 mm long. Panicles 15–30 cm long, open. Upper glumes and lower lemmas extended 3–4 mm past the upper floret. Found in sandy grasslands and around oak mottes. Poor livestock and wildlife values (Hatch, Schuster, and Drawe 1999).

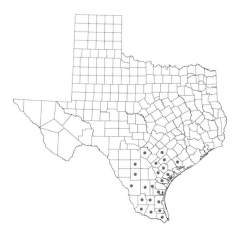

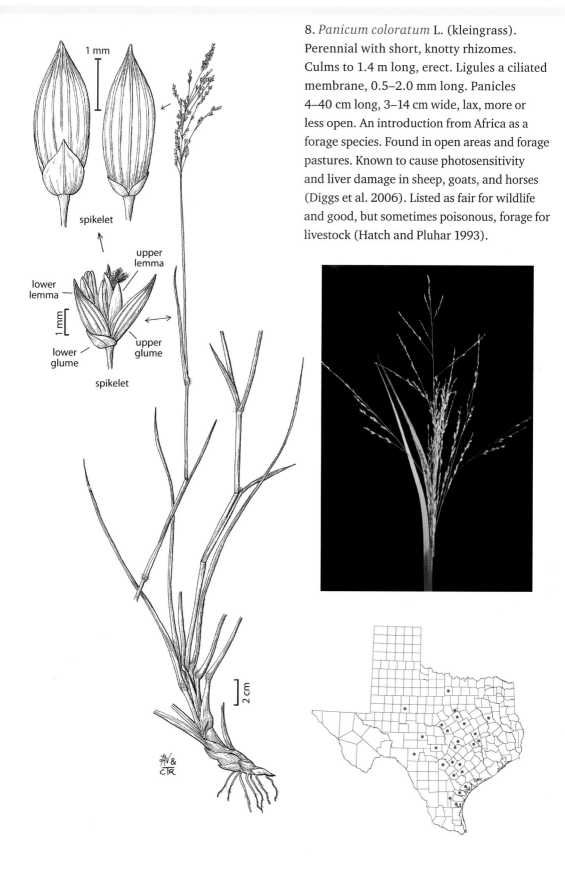

1 mm

spikelet

upper
lemma

lower
lemma

1 mm

lower
glume

upper
glume

spikelet

2 cm

8. *Panicum coloratum* L. (kleingrass). Perennial with short, knotty rhizomes. Culms to 1.4 m long, erect. Ligules a ciliated membrane, 0.5–2.0 mm long. Panicles 4–40 cm long, 3–14 cm wide, lax, more or less open. An introduction from Africa as a forage species. Found in open areas and forage pastures. Known to cause photosensitivity and liver damage in sheep, goats, and horses (Diggs et al. 2006). Listed as fair for wildlife and good, but sometimes poisonous, forage for livestock (Hatch and Pluhar 1993).

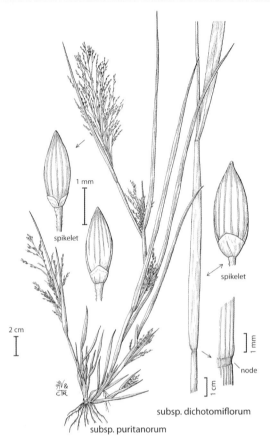

1 mm

spikelet

spikelet

2 cm

node

1 cm

1 mm

𝒥𝒱 &
CTR

subsp. dichotomiflorum

subsp. puritanorum

9. *Panicum dichotomiflorum* Michx. (fall panicum, spreading witchgrass). Coarse annual with erect or decumbent culms to 2 m long, succulent or spongy. Ligules a ciliated membrane <1 mm long. Panicles 4–40 cm long, diffuse. Found in wet, disturbed areas such as fallow fields, roadsides, ditches, and receding shorelines and sometimes in shallow water. Fair livestock forage and poor to fair seed producer (Hatch, Schuster, and Drawe 1999). Telfair (2006) listed the seeds as especially important for wildlife. The Texas species belongs to the typical subspecies, subsp. *dichotomiflorum*. Can cause photosensitivity (Burrows and Tyrl 2001).

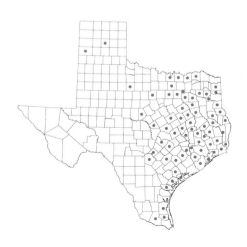

10. *Panicum diffusum* Sw. (spreading witchgrass). Perennial with short rhizomes. Culms to 1 m long, spreading. Ligules a fringe of hairs to 4 mm long. Panicles 3–35 cm long; open. Found along riverbanks, roadways, ditches, swales, and disturbed sites. Poor livestock and wildlife values (Hatch, Schuster, and Drawe 1999).

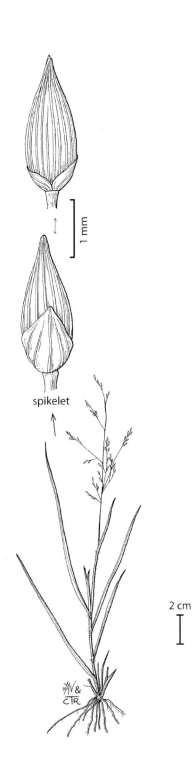

1 mm

spikelet

2 cm

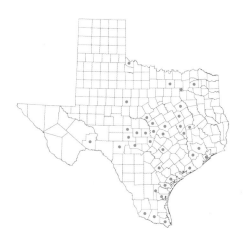

11. *Panicum flexile* (Gatt.) Scribn. (wiry
witchgrass). Annual. Culms to 75 cm long.
Ligules a fringe of hairs, 0.5–1.5 mm long.
Panicles 5–45 cm long, open. Found primarily
on limestone-derived soils. This species is
listed as P. *flexicaule* in Turner et al. (2003).
Rare in Texas; generally found in the eastern
United States.

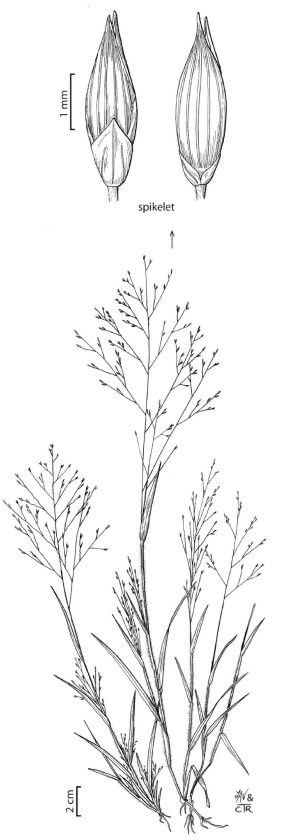

spikelet

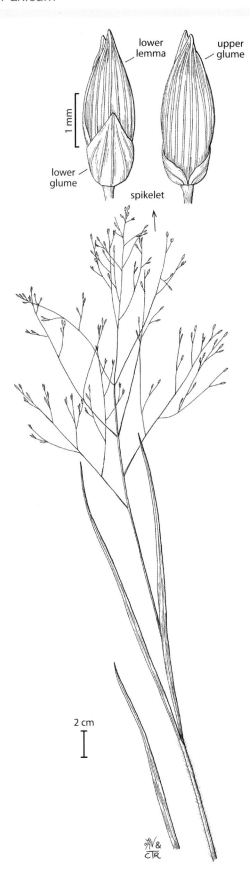

lower lemma

upper glume

1 mm

lower glume

spikelet

2 cm

12. *Panicum ghiesbreghtii* Fourn. (Ghiesbreght's witchgrass, Ghiesbreght's panicum). Cespitose perennial. Culms to 1.2 m long. Ligules a ciliated membrane to 5 mm long. Panicles 7–35 cm long, open. Found in low-lying areas, moist ground, and thickets. Restricted to the southernmost counties.

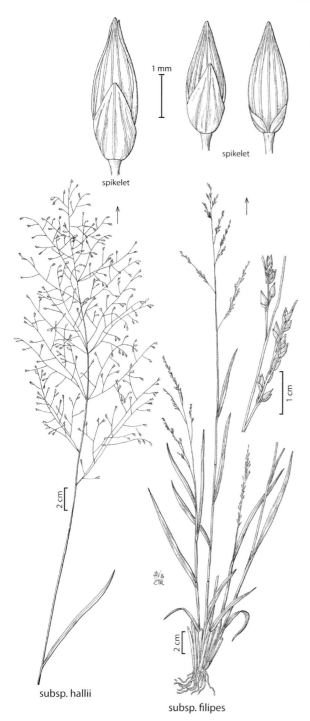

1 mm

spikelet

spikelet

subsp. hallii

subsp. filipes

13. *Panicum hallii* Vasey (Hall's panicum, Hall's witchgrass). Cespitose perennial with culms to 1 m tall. Ligules a ciliated membrane, 0.6–2.0 mm long. Leaf blades in the typical subspecies characteristically curl upon drying. Panicles 7–30 cm long, more or less open. Common species of grasslands, rangelands, pastures, roadsides, oak and pine savannahs, and desert regions. Listed as fair for wildlife and fair for livestock (Hatch & Pluhar 1993; Powell 1994). Telfair (2006) listed the seeds as especially important for wildlife. Two subspecies are recognized: subsp. *filipes* (Scribn.) Freckmann & Lelong; and the typical one, subsp. *hallii*. The latter grows on more arid sites than does the former. They can be distinguished by the following characters:

1. Spikelets 3.0–4.2 mm long; leaves curl at maturity....................................subsp. *hallii*
1. Spikelets 2.1–3.0 mm long; leaves do not curl at maturity........................subsp. *filipes*

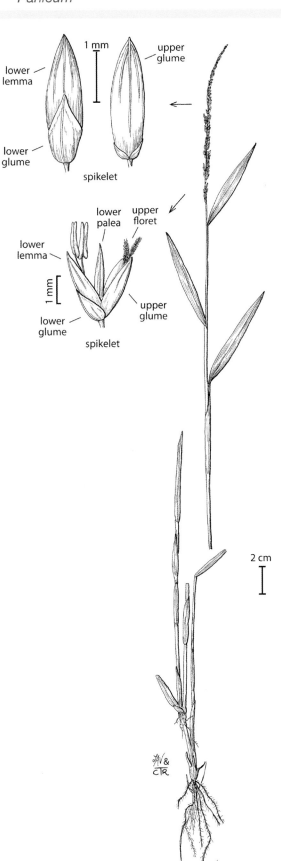

lower lemma

1 mm

upper glume

lower glume

spikelet

lower palea

upper floret

lower lemma

1 mm

lower glume

upper glume

spikelet

2 cm

14. *Panicum hemitomon* Schult. (maidencane, Simpson's grass). Perennial, aquatic or semiaquatic, forming extensive colonies from rhizomes. Culms to 2 m long, rooting at lower nodes if submerged. Ligules <1 mm long. Panicles 10–30 cm long, narrow, closed. Found in water or wet soils in marshes and swamps and along streams, canals, ditches, and lakes. Provides good forage (Hatch, Schuster, and Drawe 1999).

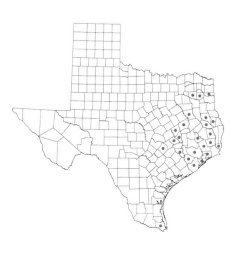

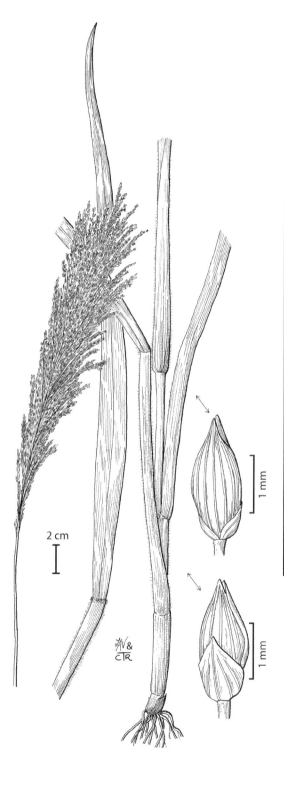

2 cm

15. *Panicum hirsutum* Sw. (giant witchgrass, hairy panicum). Perennial, forming large clumps with short rhizomes. Culms to 3 m tall. Sheaths covered with papillose-based hairs that penetrate and irritate the skin when plants handled. Ligules a ciliated membrane to 5 mm long. Panicles 25–45 cm long, closed to slightly open. Localized in the Brownsville area, often in partial shade along riverbanks or ditches.

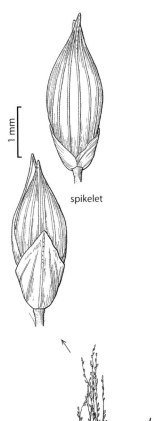

1 mm

spikelet

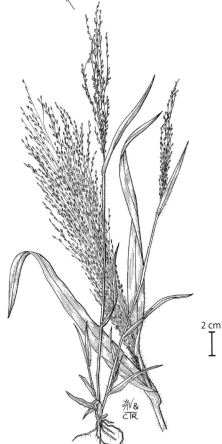

2 cm

16. *Panicum hirticaule* J. Presl (roughstalked witchgrass). Annual with culms to 1 m long. Ligules a fringe of hairs 0.9–3.5 mm long. Panicles 9–30 cm long, sometimes nodding, more or less open. Found in open areas along roadsides, waste places, and disturbed sites. Species is of little forage value (Powell 1994). The Texas species belongs to the typical subspecies, subsp. *hirticaule*.

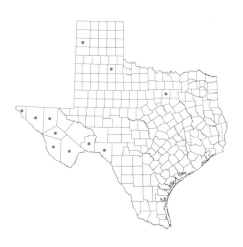

subsp. miliaceum

17. *Panicum miliaceum* L. (broomcorn millet, proso millet, hog millet, panic millet, Russian millet, common millet). Coarse annual with culms to 2 m long. Ligules a ciliated membrane, 1.5–3.0 mm long. Panicles 6–20 cm long, more or less contracted. Cultivated in Texas for bird seed or a crop to attract and sustain game birds. Occasionally escapes and is found along roadsides and cornfields. Seeds are sometimes fed to hogs and used for human consumption (Powell 1994). The Texas species belongs to the typical subspecies, subsp. *miliaceum*. This species has the lowest water requirements of any cereal grain (Freckmann & Lelong 2003b).

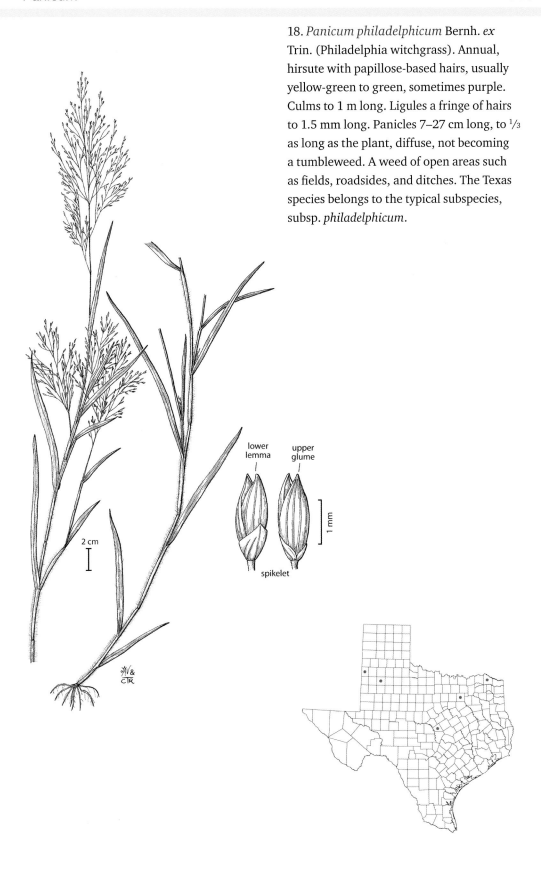

18. *Panicum philadelphicum* Bernh. *ex* Trin. (Philadelphia witchgrass). Annual, hirsute with papillose-based hairs, usually yellow-green to green, sometimes purple. Culms to 1 m long. Ligules a fringe of hairs to 1.5 mm long. Panicles 7–27 cm long, to ⅓ as long as the plant, diffuse, not becoming a tumbleweed. A weed of open areas such as fields, roadsides, and ditches. The Texas species belongs to the typical subspecies, subsp. *philadelphicum*.

lower lemma

upper glume

1 mm

spikelet

2 cm

spikelet

spikelet

19. *Panicum repens* L. (torpedo grass). Rhizomatous perennial that forms large colonies. Culms to 90 cm long, erect. Ligules ciliated membrane to 1 mm long. Panicles 3–24 cm long, open. Lower glumes truncate to subacute, less than $^2/_5$ as long as spikelet. Found primarily on coastal sands and shores of lakes and ponds.

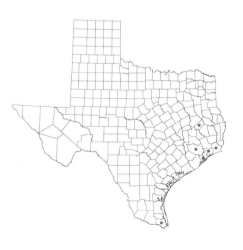

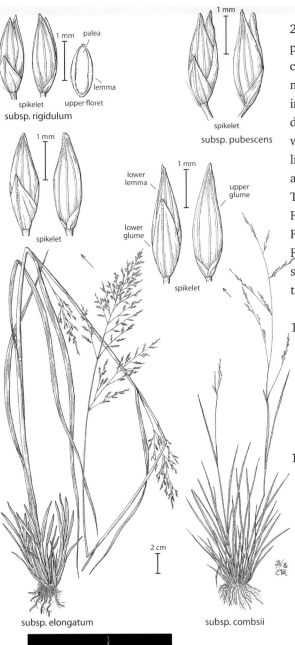

spikelet
upper floret
subsp. rigidulum

spikelet
subsp. pubescens

lower
lemma
lower
glume
spikelet

upper
glume
spikelet

2 cm

subsp. elongatum

subsp. combsii

20. *Panicum rigidulum* Bosc *ex* Nees (redtop panicum). Perennial. Culms to 1.5 m long, stout, compressed. Ligules a ciliated membrane to 3 mm long. Panicles 9–40 cm long, open. Found in swamps, moist woodlands and grasslands, ditches, swales, drainage areas, floodplains, and wet flatwoods. Rarely occurs on dry sites. Fair for livestock and good for wildlife (Hatch, Schuster, and Drawe 1999). Four subspecies occur in Texas: subsp. *combsii* (Scribn. and C. R. Ball) Freckmann & Lelong; subsp. *elongatum* (Scribn.) Freckmann & Lelong; subsp. *pubescens* (Vasey) Freckmann & Lelong; and the typical subspecies, subsp. *rigidulum*. They can be distinguished by the following characters:

1. Blades mostly flat and glabrous; ligules membranous, 0.3–1.0 mm long.
    2. Spikelets 1.6–2.5 mm long, >0.6 mm wide, green or purplish .... subsp. *rigidulum*
    2. Spikelets 2.4–3.0 mm long, usually <0.6 mm wide, stipitate, purplish .... subsp. *elongatum*
1. Blades mostly folded or involute; ligules ciliated membrane, 0.5–3.0 mm long.
    3. Spikelets 2.0–2.7 mm long, often obliquely set on the pedicel ....subsp. *pubescens*
    3. Spikelets 2.6–2.8 mm long, erect on the pedicels.................. subsp. *combsii*

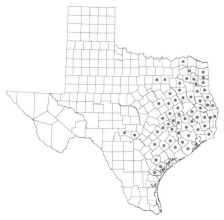

21. *Panicum tenerum* Beyr. *ex* Trin. (blue-joint panicgrass, southeastern panic). Perennial with short, knotted rhizomes. Culms to 1 m long, erect. Ligules a minute ciliated membrane. Panicles 2–12 cm long, somewhat contracted. Found in wet or mesic sandy soils, depressions, swales, ditches, pine savannahs, marshes, and bogs and along pond margins.

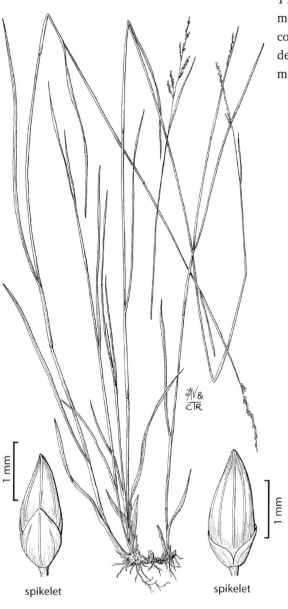

2 cm

1 mm

spikelet

1 mm

spikelet

22. *Panicum trichoides* Sw. (small-flowered panicgrass, tropical panicgrass). Annual with culms to 1 m long, erect or creeping, and rooting at the lower nodes. Ligules a ciliated membrane <1 mm long. Panicles 4–25 cm long, diffuse. A recent tropical American introduction; found as a weed in Travis and Cameron counties.

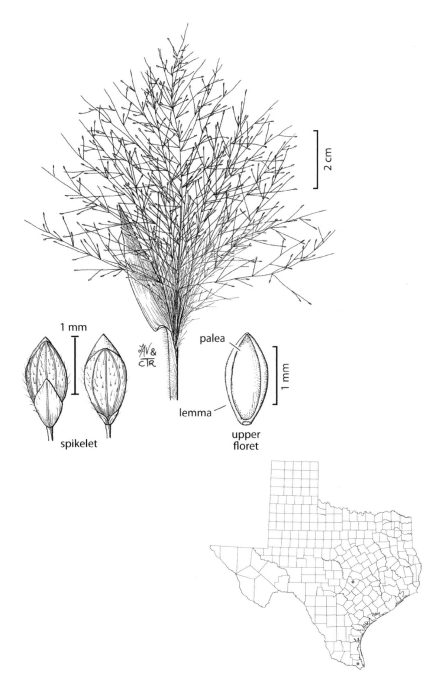

2 cm

1 mm

spikelet

palea

1 mm

lemma

upper floret

23. *Panicum verrucosum* Muhl. (warty panicgrass). Annual with sprawling culms to 1.5 m long, wiry, often with purple dots and streaks, rooting at the lower nodes. Ligules a ciliated membrane 0.2–0.5 mm long. Panicles 5–30 cm long, as wide as long. Spikelets covered in wartlike glands. Found in open, moist or wet sands around swamps, marshes, lakes, roadside ditches, and swales. The other species with warty spikelets, *P. brachyanthum*, is found on drier sites.

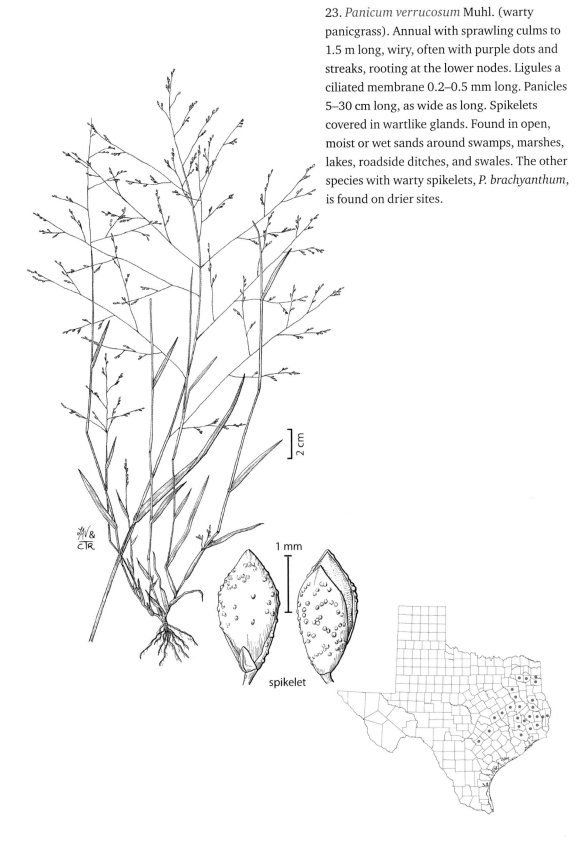

2 cm

1 mm

spikelet

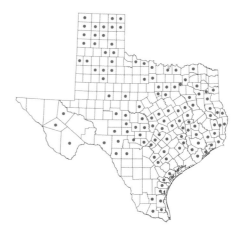

24. *Panicum virgatum* L. (switchgrass). Rhizomatous perennial. Culms to 3 m tall, solitary or forming large clumps. Ligules a fringed membrane, 1.5–6.0 mm long. Panicles 10–55 cm long, open. A widespread and abundant species. Found in tallgrass prairies, dry slopes, and open oak or pine woodlands and along shores, riverbanks, moist lowlands, swales, and ditches. Cultivars are being championed as a major source of cellulosic biomass for biofuels (ethanol) production. Listed as fair for wildlife and good for livestock (Hatch and Pluhar 1993; Powell 1994). Telfair (2006) listed the seeds as especially important for wildlife.

Pappophorum Schreb.
(Chloridoideae: Pappophoreae)

Erect, cespitose perennials with slender, contracted, usually spikelike panicles of bristly spikelets. Ligule a ring of hairs. Blades long, narrow, flat or folded. Spikelets 3–6-flowered, only the lower 1–3 fertile, disarticulating above the glumes, the florets falling together. Glumes subequal, thin and membranous, 1-nerved. Lemmas firm, rounded on the back, indistinctly many-nerved, the nerves extending into 11 or more unequal, glabrous or scabrous awns. Palea about as long as the body of the lemma. Basic chromosome number, $x = 10$. Photosynthetic pathway, C_4. Represented in Texas by 2 species.

1. Panicles pinkish or purplish at maturity; perfect florets, usually 2; lemma body of lowermost floret 3–4 mm long.. 1. *P. bicolor*
1. Panicles white or tawny; perfect florets, usually 1; lemma body of lowermost floret 2–3 mm long .... 2. *P. vaginatum*

(Gould 1975b; Reeder 2003c)

1. *Pappophorum bicolor* E. Fourn. (pink pappusgrass). Perennial with erect culms to 1 m long. Panicles contracted but with a few slightly spreading branches. Awns <1.5 times as long as the lemma body, erect. Found in open grasslands, valleys, and roadsides. Occurs only in Texas and Mexico. Listed as poor for wildlife and fair for livestock (Hatch and Pluhar 1993; Powell 1994).

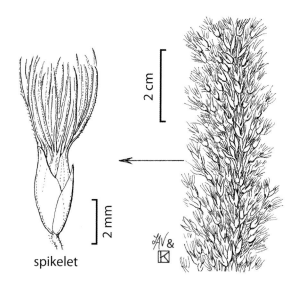

spikelet

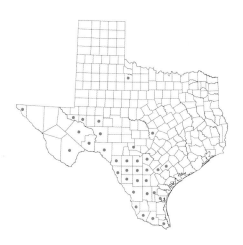

2. *Pappophorum vaginatum* Buckley (whiplash pappusgrass). Perennial with erect or sometimes geniculate culms to 1 m long. Panicles tightly contracted. Awns about twice as long as lemma body, spreading. Similar habitats to those of *P. bicolor*, and they are often found growing together. Fair forage value for cattle (Powell 1994).

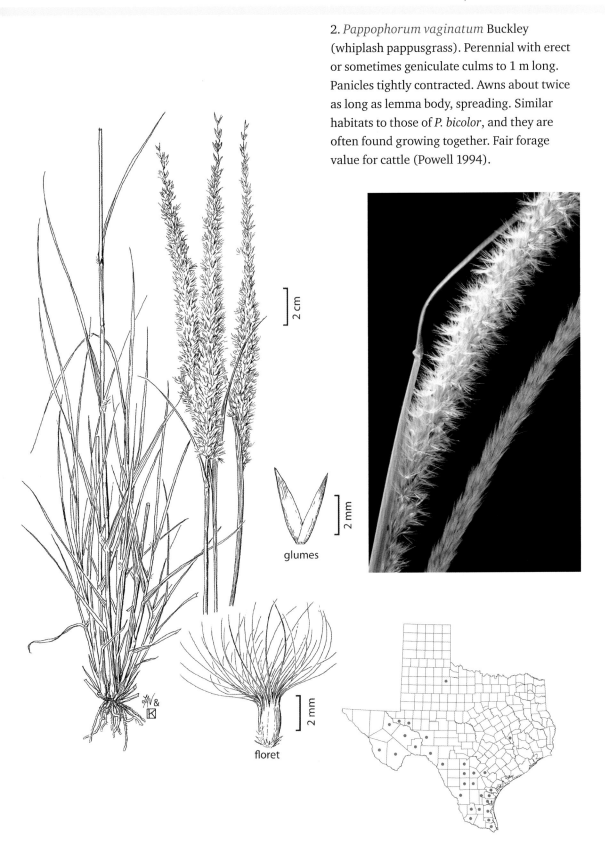

glumes

floret

Parapholis C. E. Hubb.
(Poöideae: Poeae)

Low, tufted, much-branched annuals, with slender, curved, cylindric spikes at each of numerous branch tips. Spikelets 1- or 2-flowered, solitary at the nodes of a thickened rachis, partially embedded in, and falling attached to, the rachis joint. Glumes large, subequal, coriaceous, 3- or 5-nerved, somewhat asymmetrical, tapering to a point, the 2 placed in front of the spikelet and appearing as halves of a single glume. Lemma thin, hyaline, 1-nerved, awnless, shorter than the glumes but exceeding the hyaline palea. Basic chromosome number, $x = 7$. Photosynthetic pathway, C_3. Represented in Texas by a single species.

(Worley 2007)

1. *Parapholis incurva* C. E. Hubb. (curved sicklegrass, sicklegrass). Tufted annual with spreading erect or decumbent culms. Sheaths of the upper leaves inflated, enclosing the lower spikelets. Spikes solitary, curved, and twisted. Grows at or above high-tide mark along coastal areas.

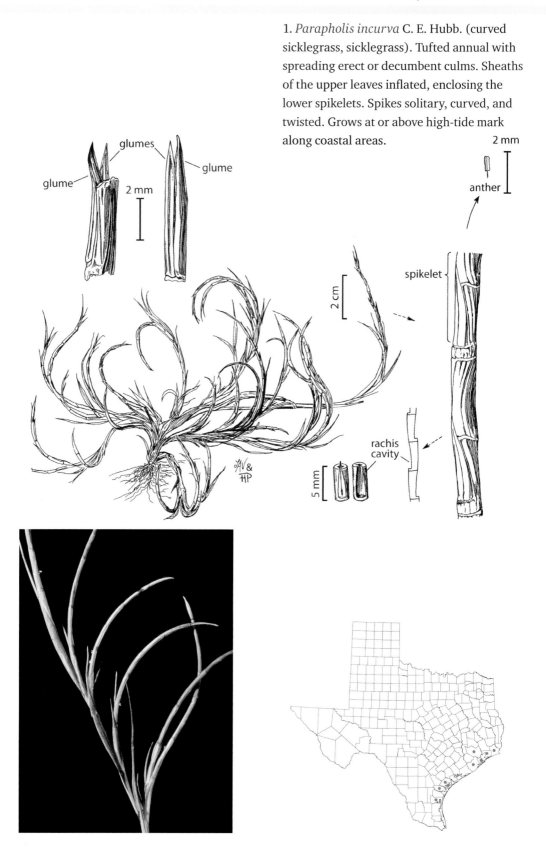

Pascopyrum Á. Löve
(Poöideae: Triticeae)

Perennial, rhizomatous. Inflorescence a bilateral spike, spikelets 1 at a node or occasionally paired at the lower nodes. Disarticulation is above the glumes and beneath each floret. Spikelets with 2–12 florets, ascending but not appressed. Glumes ½ to ⅔ the length of the spikelet, narrowly lanceolate, stiff, curved. Lemmas lanceolate, rounded on the back, acute, mucronate, or short-awned. Paleas slightly shorter than the lemmas. Basic chromosome number, $x = 7$. Photosynthetic pathway, C_3. Represented in Texas by the only species in the genus.

(Barkworth 2007g)

1. *Pascopyrum smithii* (Rydb.) Barkworth and D. R. Dewey (western wheatgrass). Rhizomatous perennial with numerous glaucous culms up to 1 m in height. Auricles up to 1 mm long, purplish. Inflorescence a spike with relatively few spikelets. Glumes typically narrowed below the middle and slightly curved. Lemmas unawned. Generally found in heavier soils but will occur in sandy soils as well. Most common on wet flats and floodplains and in depressions and swales. A segregate of *Agropyron*, occasionally listed as *A. smithii* Rydb., *Elymus smithii* (Rydb.) Gould, or *Elytrigia smithii* (Rydb.) Nevski. Listed as fair for wildlife and good for livestock by Hatch and Pluhar (1993), as *E. smithii* (Rydb.) Nevski.

Paspalidium Stapf
(Panicoideae: Paniceae)

Rhizomatous perennials. Culms to 1 m tall, thick and succulent, often decumbent and stoloniferous at the base. Ligule a ring of hairs. Inflorescence a panicle of subsessile spikelets on short, unbranched primary branches, these widely spaced on the main rachis. Spikelets borne singly in 2 rows on the flattened branch rachis, oriented with the back of the lemma of the fertile floret turned toward the rachis as in *Paspalum*. Disarticulation below the glumes. Glumes both present, the first short, acute to broadly obtuse or truncate at the apex, the second slightly shorter than the pointed lemma of the sterile floret, obtuse to acute at the apex. Lemma of the fertile floret firm or hard, moderately rugose, broadly pointed at the tip. Palea flat or slightly convex on the back. Basic chromosome number, $x = 9$. Photosynthetic pathway, C_3. Represented in Texas by a single species.

(Allen 2003c)

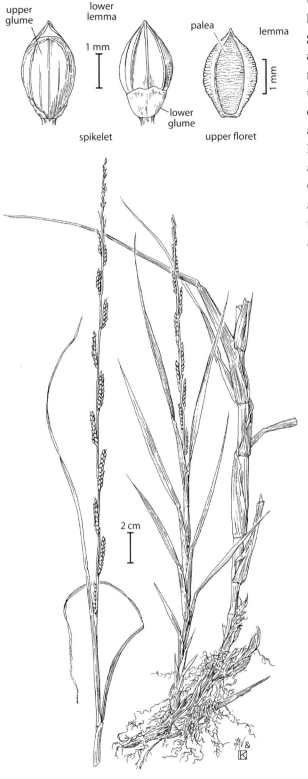

upper
glume

lower
lemma

palea

lemma

1 mm

1 mm

lower
glume

spikelet

upper floret

2 cm

1. *Paspalidium germinatum* (Forssk.) Stapf (Egyptian paspalidium, water paspalidium, Egyptian water-crown grass, Egyptian water grass). Rhizomatous perennial. Culms erect and to 1 m long. Ligules a short, ciliated membrane. Panicles composed of 5–20 short, unilateral spicate branches. Found in brackish or freshwater habitats along streams, lakes, and ditches, often in standing water. Poor livestock and wildlife values (Hatch, Schuster, and Drawe 1999). Gould (1975b); Hatch, Gandhi, and Brown (1990); and Jones, Wipff, and Montgomery (1997) listed 2 varieties, but the distinction between them is not consistent, and they do not warrant recognition.

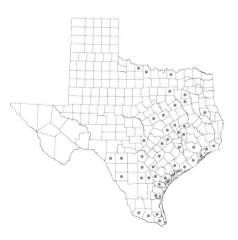

Paspalum L.
(Panicoideae: Paniceae)

Annuals and perennials, many with rhizomes or stolons. Blades usually flat, often thin and broad. Inflorescence with 1 to many unilateral spikelike branches, these scattered or, in a few species, paired at the culm apex. Spikelets subsessile or short-pediceled, solitary or in pairs on a flattened, occasionally broadly winged rachis, with the rounded back of the lemma of the fertile floret turned toward the rachis. Disarticulation at the base of the spikelet. Lower glume typically absent but irregularly present in a few species. Upper glume and lemma of the sterile floret usually about equal, broad and rounded at the apex, infrequently acute. Lemma of the fertile floret firm or indurate, rounded on the back, usually obtuse, with inrolled margins. Palea broad, flat or slightly convex, the margins entirely enfolded by the lemma. Basic chromosome number, $x = 10$. Photosynthetic pathway, C_4. Represented in Texas by 32 species and 10 varieties.

1. Spikelets solitary.
    2. Panicles with 1–70 branches, not composed of a terminal pair.
        3. Branches 7–70, axes extending beyond the distal spikelets............................ 25. *P. repens*
        3. Branches 1–6, terminating in a spikelet.
            4. Upper florets olive to dark brown..26. *P. scrobiculatum*
            4. Upper florets pale or stramineous.
                5. Axes of panicle branches 0.6–1.3 mm wide ....14. *P. laeve*
                5. Axes of panicle branches 1.8–3.3 mm wide.
                    6. Spikelets 1.7–2.1 mm long... 9. *P. dissectum*
                    6. Spikelets 3.2–4.0 mm long ....1. *P. acuminatum*
    2. Panicle with a terminal pair of branches, sometimes with 1–5 additional branches below.
        7. Upper glumes pubescent on the margins or back.
            8. Spikelets 1.3–1.9 mm long; upper glumes pilose on the margins ...5. *P. conjugatum*
            8. Spikelets 2.4–3.2 mm long; upper glumes short-pubescent on the back ....
            .... 10. *P. distichum*
        7. Upper glumes glabrous.
            9. Spikelets elliptic, apices acute to acuminate.
                10. Plants rhizomatous, culms usually solitary, in brackish to salt marsh habitat.....
                ....30. *P. vaginatum*
                10. Plants with short, weak rhizomes, appearing cespitose; usually in disturbed
                inland habitats ....2. *P. almum*
            9. Spikelets ovate, apices obtuse to broadly acute.
                11. Spikelet 2.5–4.0 mm long .... 21. .*P. notatum*
                11. Spikelets 1.9–2.3 mm long... ....18. *P. minus*
1. Spikelets paired, or if only 1 spikelet functional, a naked pedicel or rudimentary spikelet present.
    12. Margins of upper glumes and lower lemmas pilose.
        13. Panicle branches 2–7; spikelets 2.3–4.0 mm long ....8. *P. dilatatum*

13. Panicle branches (4–)10–30; spikelets 1.8–2.8 mm long29. *P. urvillei*

12. Margins of upper glumes and lower lemmas not pilose.

    14. Upper florets olive to dark brown.

        15. Lower glumes often present; aquatic with decumbent culms, rooting at lower nodes .... 19. *P. modestum*

        15. Lower glumes absent; terrestrial or aquatic with erect culms

            16. Plants annual.

                17. Spikelets 1.3–1.8 mm wide, glabrous; panicles with 1–10(–28) branches, the axes 0.7–2.3 mm wide............................4. *P. boscianum*

                17. Spikelets 1.7–2.4 mm long, pubescent; panicles with 1–5 branches, axes 0.8–1.3 mm wide .... 7. *P. convexum*

            16. Plants perennial.

                18. Plants cespitose, short rhizomes present or not; culms stout, 1–2 m tall.

                    19. Spikelets 2.5–3.0 mm long................................22. *P. plicatulum*

                    19. Spikelets <2.5 mm long.

                        20. Spikelets 1.8–2.4 mm long; axes of branches 1.0–1.7 mm wide ....31. *P. virgatum*

                        20. Spikelets 1.1–1.8 mm long; axes of branches 0.5–1.2 mm wide .... 6. *P. conspersum*

                18. Plants not cespitose, rhizomatous; culms to 1.5 m tall, varying thickness.

                    21. Rhizomes distinct, long ....*P. wrightii*

                    21. Rhizomes indistinct, short................................*P. plicatulum*

    14. Upper florets white, stramineous, or golden brown.

                22. Lower lemmas with well-developed ribs over the veins; upper glumes absent .... 17. *P. malacophyllum*

                22. Lower lemmas not ribbed; upper glumes present.

                    23. Panicles with 15–100 branches .... 13. *P. intermedium*

                    23. Panicles with 1–15 branches.

                        24. Spikelet pairs not imbricate .... 3. *P. bifidum*

                        24. Spikelet pairs imbricate.

                            25. Spikelets 1.3–2.5 mm long .... Key A

                            25. Spikelets 2.5–4.1 mm long .... Key B

Key A

1. Upper glumes, and usually lower lemmas, short-pubescent.
    2. Lower glumes present .... 15. *P. langei*
    2. Lower glumes absent...27. *P. setaceum*
1. Upper glumes and lower lemma glabrous.
        3. Panicles both terminal and axillary, axillary panicles partially or completely enclosed by subtending leaf sheaths .... 27. *P. setaceum*
        3. Panicles all terminal.
            4. Upper panicle branches erect 20. *P. monostachyum*
            4. Upper panicle branches ascending to spreading.
                5. Upper glumes and lower lemmas 3-veined .... 23. *P. praecox*
                5. Upper glumes and lower lemmas 5-veined ....16. *P. lividum*

Key B

1. Upper glumes, and usually lower lemmas, pubescent.
    2. Lower glumes present .... 15. *P. langei*
    2. Lower glumes absent.
        3. Upper glumes and lower lemmas densely pubescent, hairs >0.1 mm long; leaf blades 2–5 mm wide .... 12. *P. hartwegianum*
        3. Upper glumes and lower lemmas glabrous or sparsely pubescent, hairs <0.1 mm long; leaf blades 4–18 mm wide .... 24. *P. pubiflorum*
1. Upper glumes and lower lemmas glabrous.
        4. Upper florets golden brown.
            5. Plants rhizomatous; culms erect; spikelets 1.9–1.6 mm wide.
                6. Upper glumes 5-veined ....11. *P. floridanum*
                6. Upper glumes 3-veined .... 28. *P. unispicatum*
            5. Plants not rhizomatous; culms decumbent and root at lower nodes; spikelets 1.3–1.6 mm wide .... 19. *P. modestum*
        4. Upper florets stramineous to pale.
                7. Terminal panicle branches erect.
                    8. Upper glumes 1-veined; blades involute .... 20. *P. monostachyum*
                    8. Upper glumes 3-veined; blades flat............................. 28. *P. unispicatum*
                7. Terminal panicle branches spreading or ascending.
                        9. Spikelets 2.2–2.6 mm long.
                            10. Spikelets 1.2–1.5 mm wide, elliptical to obovate ....16. *P. lividum*
                            10. Spikelets 2.1–3.1 mm wide, orbicular to suborbicular ...23. *P. praecox*

9. Spikelets 2.6–4.1 mm long.

    11. Plants decumbent, root at lower nodes...........24. *P. pubiflorum*

    11. Plants rhizomatous.

        12. Upper glumes 3-veined; blades conduplicate.....................
....23. *P. praecox*

        12. Upper glumes 5-veined; blades flat..... 11. *P. floridanum*

(Allen and Hall 2003; Gould 1975b)

1. *Paspalum acuminatum* Raddi (brook paspalum, brook crowngrass, canoegrass). Rhizomatous perennial with sprawling culms. Erect portion of culms not more than 20 cm tall. Ligules 1.0–2.5 mm long, membranous. Panicles terminal, with 2–5 racemosely arranged branches. Found on the edges of lakes, ponds, rice fields, and wet ditches.

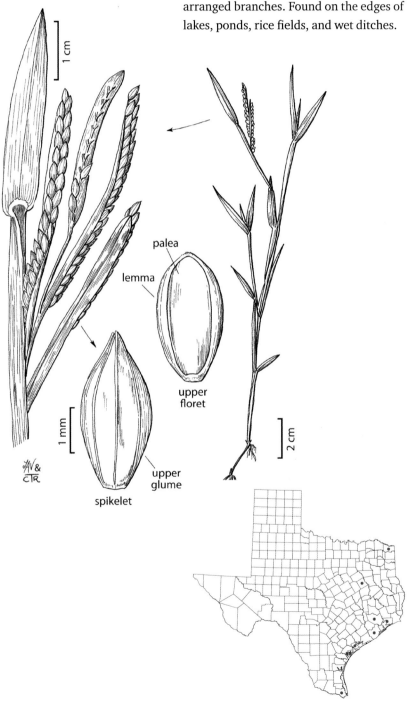

palea

lemma

upper floret

1 mm

spikelet

upper glume

1 cm

2 cm

2. *Paspalum almum* Chase (comb's paspalum, comb's crowngrass). Perennial with short rhizomes. Culms to 50 cm long, erect. Ligules 0.2–5.2 mm long, membranous. Panicles terminal, composed of a digitate pair of branches with 1–5 more below. Probably introduced from South America as a forage species; now escaped and established in a few localities.

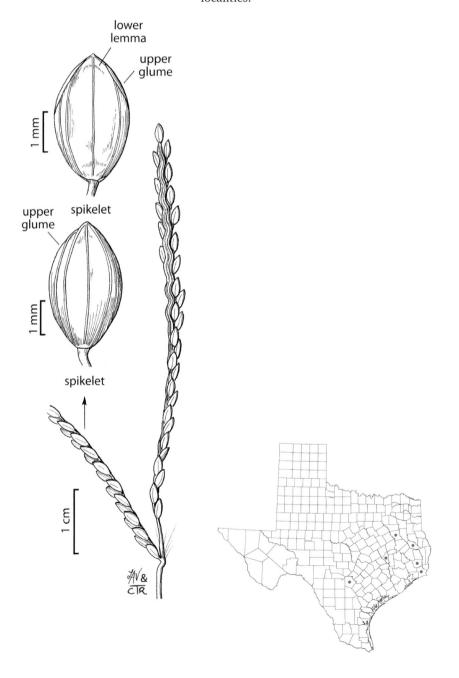

3. *Paspalum bifidum* (Bertol.) Nash
(pitchfork paspalum, pitchfork crowngrass).
Erect perennial with distinct rhizomes.
Culms to 1.4 m long. Ligules 1–6 mm
long, membranous. Panicles terminal with
15–45 racemosely arranged branches. A
fire-adapted species of the longleaf pine
ecosystem. Also found in sandy, open areas
in oak woodlands and savannahs.

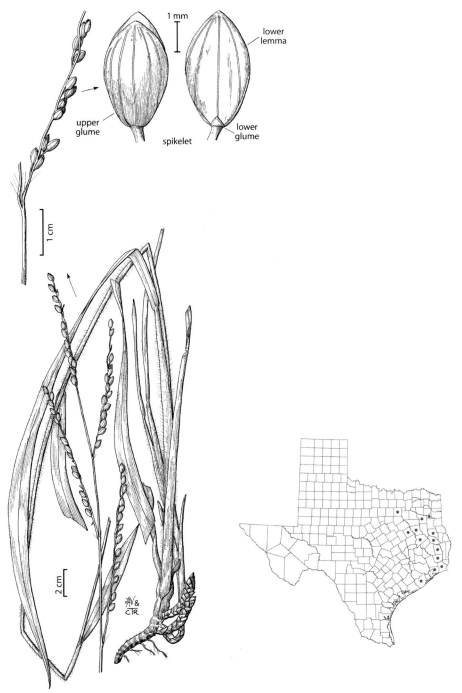

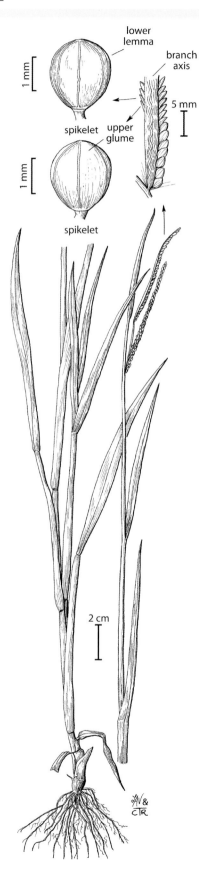

4. *Paspalum boscianum* Flüggé (bull paspalum). Annual with culms to 1 m long, erect or prostrate and rooting at the nodes. Ligules 1–3 mm long, membranous. Panicles terminal with 1–10(–28) racemosely arranged branches. Found in dry or moist disturbed sites, cutover forests, ditches, and field borders. Telfair (2006) listed it as important for wildlife.

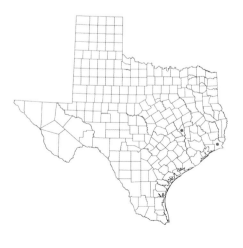

5. *Paspalum conjugatum* P. J. Bergius (sour paspalum, sour crowngrass). Perennial, stoloniferous. Culms to about 50 cm long. Ligules 0.5–0.8 mm long, membranous. Panicles terminal, branches 2, a third sometimes below. Found in disturbed areas, waste places, and edge of forests. Sometimes used as a turfgrass.

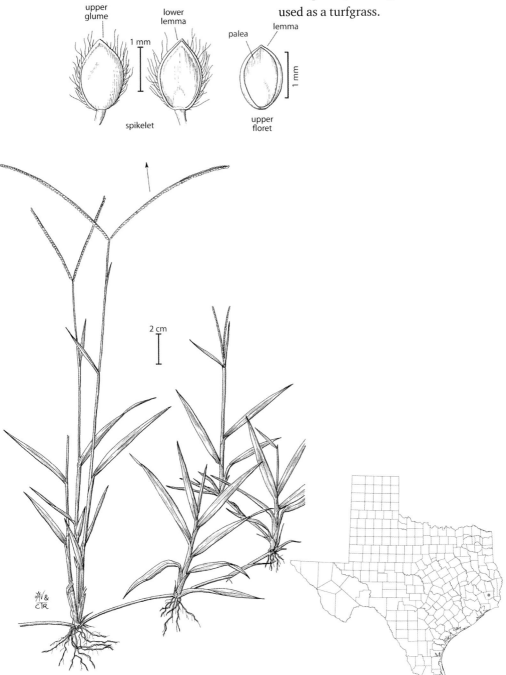

upper glume

lower lemma

palea

lemma

1 mm

1 mm

spikelet

upper floret

2 cm

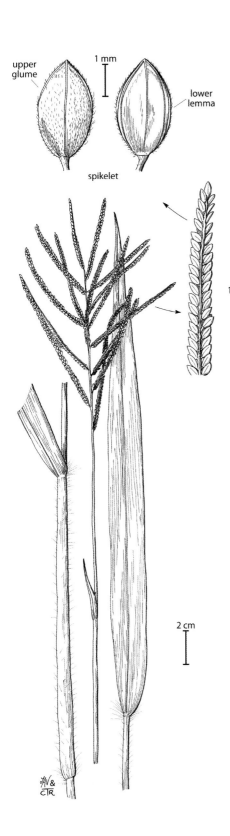

upper
glume

1 mm

lower
lemma

spikelet

1 cm

2 cm

6. *Paspalum conspersum* Schrad. (scattered paspalum). Cespitose perennial. Culms stout, erect, to 2 m long. Ligules 1–2 mm long, membranous. Panicles terminal with 4–13 racemosely arranged branches. A native of Mexico and Argentina introduced for forage. Established in disturbed areas and along roadsides. Not reported in Gould (1975b); Hatch, Gandhi, and Brown (1990); or Turner et al. (2003).

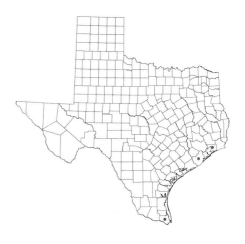

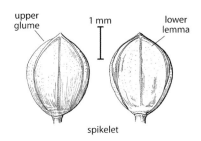

upper glume • 1 mm • lower lemma

spikelet

7. *Paspalum convexum* Humb. & Bonpl. *ex* Flüggé (Mexican paspalum, Latin American crowngrass). Erect annual to about 50 cm tall. Ligules 2–4 mm long, membranous. Panicles terminal with 1–5 racemosely arranged branches. An introduced weedy species found in disturbed sites. Not recently collected and perhaps not a constant member of the Texas flora.

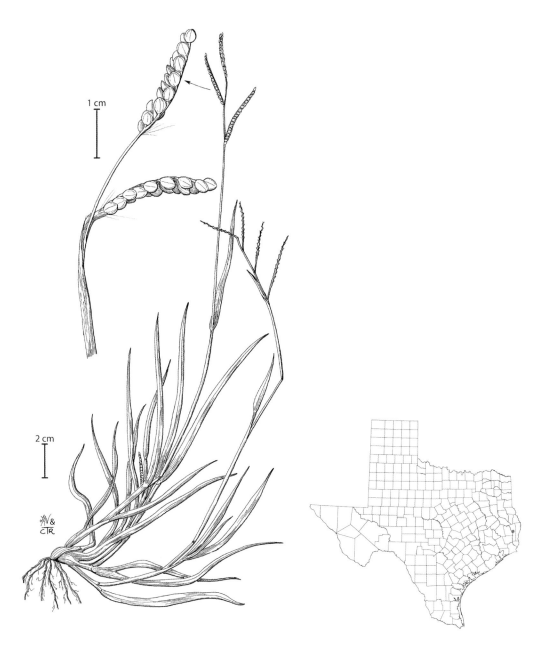

1 cm

2 cm

8. *Paspalum dilatatum* Poir. (dallisgrass, paspalum grass, water paspalum, golden crowngrass). Perennial with short rhizomes forming a knotty base. Culms to 1.75 m long, erect. Ligules 1.5–4.0 mm long, membranous. Panicles terminal with 2–7 racemosely arranged branches. A native of Brazil and Argentina. Widely seeded as a forage species, now a weed of disturbed areas, roadsides, lawns, and waste places. Pubescence along the margins of the lower lemmas and upper glumes is distinctive and is also found on *P. urvillei*. Ergot alkaloids are occasionally present, causing "paspalum staggers" in livestock (Burrows and Tyrl 2001). Listed as fair for wildlife and good for livestock (Hatch and Pluhar 1993; Powell 1994).

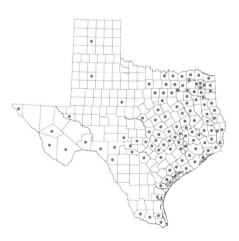

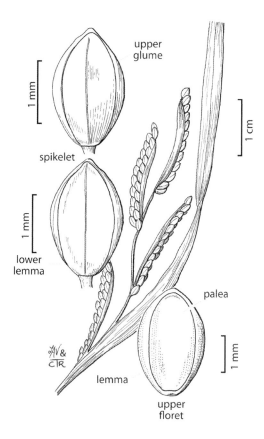

upper
glume

1 mm

spikelet

1 mm

lower
lemma

1 cm

palea

lemma

1 mm

upper
floret

9. *Paspalum dissectum* (L.) L. (mudbank
paspalum, mudbank crowngrass).
Rhizomatous perennial, forming mats.
Culms to 50 cm long, decumbent. Ligules
membranous, 2.0–2.5 mm long. Panicles with
2–6 racemosely arranged branches. Plants
can have cleistogamous spikelets with small
anthers (Diggs et al. 1999). Found along
lakeshores, rice fields, stream banks, and wet
roadside ditches.

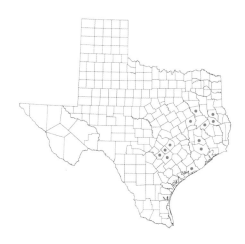

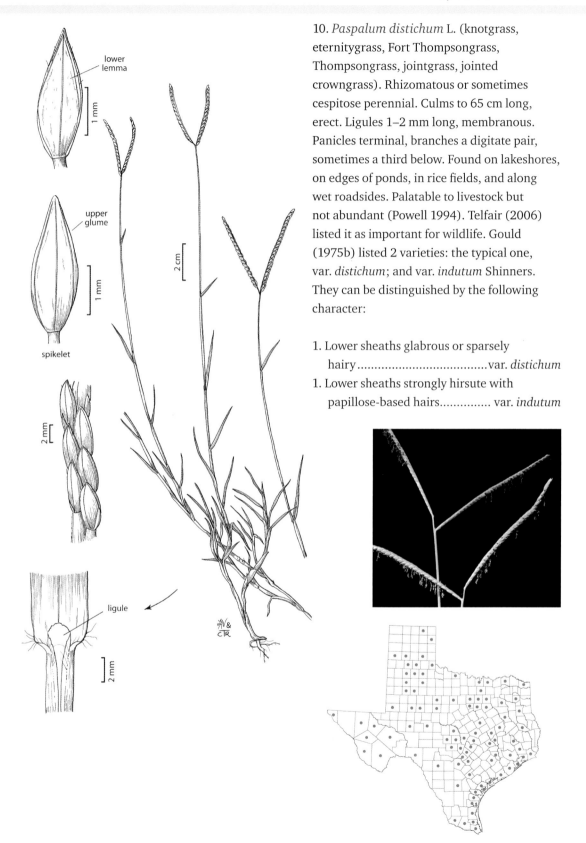

10. *Paspalum distichum* L. (knotgrass, eternitygrass, Fort Thompsongrass, Thompsongrass, jointgrass, jointed crowngrass). Rhizomatous or sometimes cespitose perennial. Culms to 65 cm long, erect. Ligules 1–2 mm long, membranous. Panicles terminal, branches a digitate pair, sometimes a third below. Found on lakeshores, on edges of ponds, in rice fields, and along wet roadsides. Palatable to livestock but not abundant (Powell 1994). Telfair (2006) listed it as important for wildlife. Gould (1975b) listed 2 varieties: the typical one, var. *distichum*; and var. *indutum* Shinners. They can be distinguished by the following character:

1. Lower sheaths glabrous or sparsely hairy ....var. *distichum*
1. Lower sheaths strongly hirsute with papillose-based hairs............... var. *indutum*

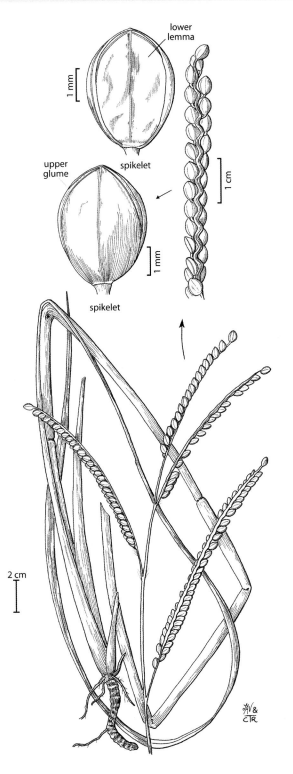

lower
lemma

1 mm

upper
glume

spikelet

1 cm

1 mm

spikelet

2 cm

11. *Paspalum floridanum* Michx. (Florida
paspalum, big paspalum, Florida crowngrass).
Rhizomatous perennial. Culms erect, to 2 m
long. Ligules 1–4 mm long. Panicles terminal
with 1–6 racemosely arranged branches. A
component of the longleaf pine ecosystem;
grows in oak savannahs, pine flatwoods, open
woods, and roadside ditches and swales. Fair
to good livestock and wildlife values (Hatch,
Schuster, and Drawe 1999). Telfair (2006)
listed the seeds as especially important for
wildlife. Varieties found in Gould (1975b);
Hatch, Gandhi, and Brown (1990); and
Jones, Wipff, and Montgomery (1997) are not
recognized here.

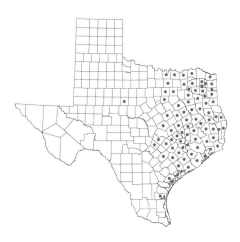

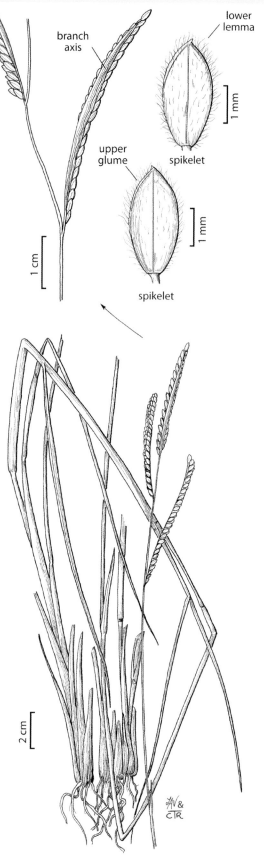

12. *Paspalum hartwegianum* E. Fourn. (Hartweg's paspalum, Hartweg's crowngrass). Perennial. Culms to 1.2 m long, erect. Ligules 2–5 mm long, membranous. Panicles terminal with 4–9 racemosely arranged branches. Found in wet prairies, swales, and ditches. Similar to *P. pubiflorum*. Fair livestock and wildlife values (Hatch, Schuster, and Drawe 1999).

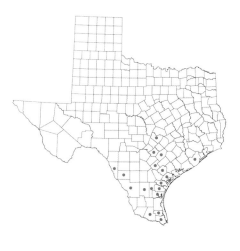

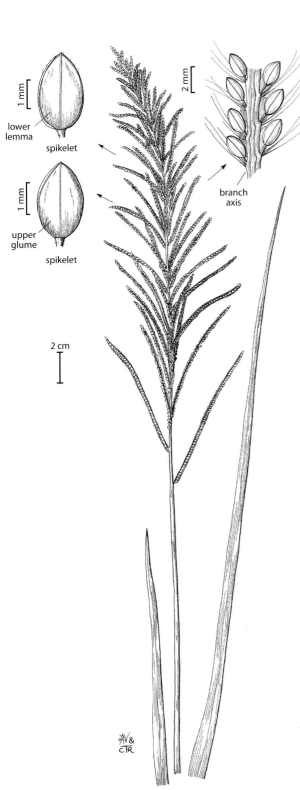

lower
lemma

spikelet

1 mm

upper
glume

spikelet

1 mm

2 mm

branch
axis

2 cm

13. *Paspalum intermedium* Munro *ex* Morong & Britton (intermediate paspalum). Perennial with short rhizomes. Culms to 2 m long, erect. Ligules 2–3 mm long. Panicles terminal with 60–100 racemosely arranged branches. An introduced roadside weed found in disturbed sites. Not reported in Gould (1975b); Hatch, Gandhi, and Brown (1990); Jones, Wipff, and Montgomery (1997); or Turner et al. (2003).

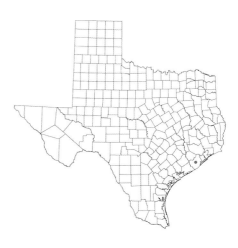

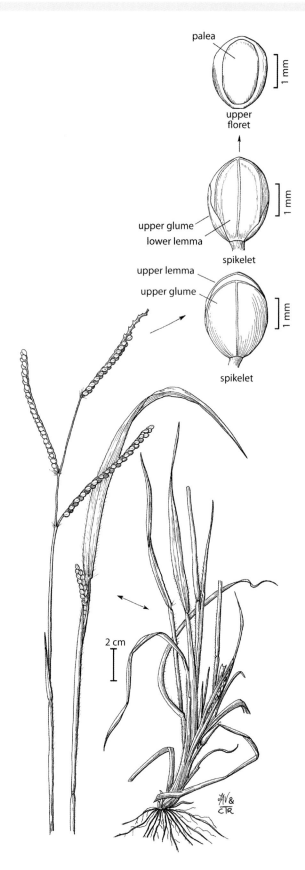

palea

upper
floret

1 mm

upper glume
lower lemma

spikelet

1 mm

upper lemma
upper glume

1 mm

spikelet

2 cm

AV &
CTR

14. *Paspalum laeve* Michx. (field paspalum, field crowngrass). Perennial with short rhizomes. Culms to 1.2 m long, erect. Ligules 1.5–4.0 mm long, membranous. Panicles terminal with 1–6 racemosely arranged branches. Found on the edges of woodlands, generally in sandy soils or disturbed areas. Gould (1975b) recognizes 3 varieties: var. *circulare* (Nash) Fern.; the typical one, var. *laeve*; and var. *pilosum* Scribn. They can be distinguished by the following characters:

1. Spikelets orbicular or suborbicular...............
.... var. *circulare*
1. Spikelets ovate, oval, or obovate.
   2. Lower leaf sheaths and often blades
      strongly pilose ....var. *pilosum*
   2. Lower leaf sheaths and blades glabrous
      or sparsely hirsute ....var. *laeve*

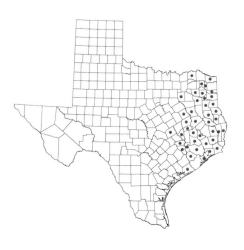

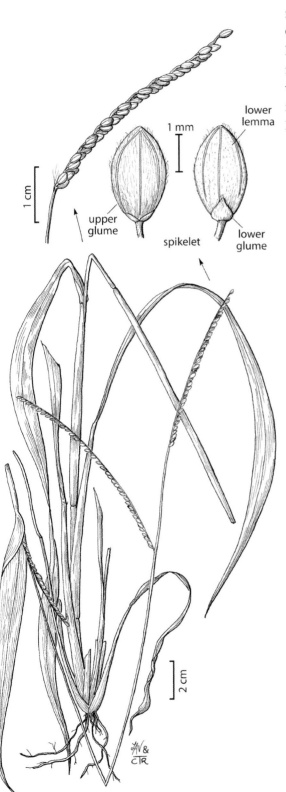

1 cm

1 mm

lower
lemma

upper
glume

spikelet

lower
glume

2 cm

15. *Paspalum langei* (E. Fourn.) Nash (rustyseed paspalum, Lange's paspalum, rustyseed crowngrass). Cespitose perennial. Culms erect, to 1.3 m long. Ligules 0.5–2.0 mm long, membranous. Panicles terminal with 1–4 racemosely arranged branches. Found in moist woods and shaded ditch banks. Good forage for livestock and wildlife (Hatch, Schuster, and Drawe 1999).

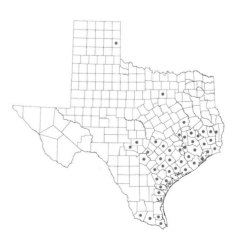

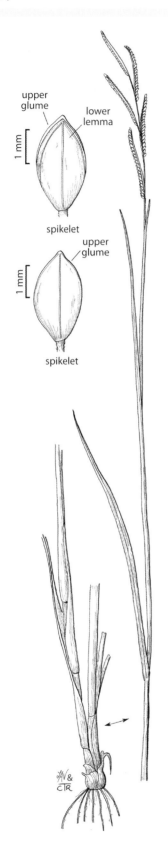

upper
glume

lower
lemma

1 mm

spikelet

upper
glume

1 mm

spikelet

16. *Paspalum lividum* Trin. *ex* Schltdl.
(longtom, pull-and-be-damned). Perennial.
Culms to 1 m long, erect. Ligules 2–5 mm
long, membranous. Panicles terminal with
3–11 racemosely arranged branches. Found
in brackish and freshwater marshes, ditches,
and wet disturbed sites. Mainly coastal.
This species has a fair forage value for both
livestock and wildlife (Hatch and Pluhar
1993).

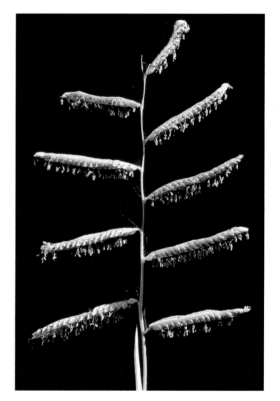

2 cm

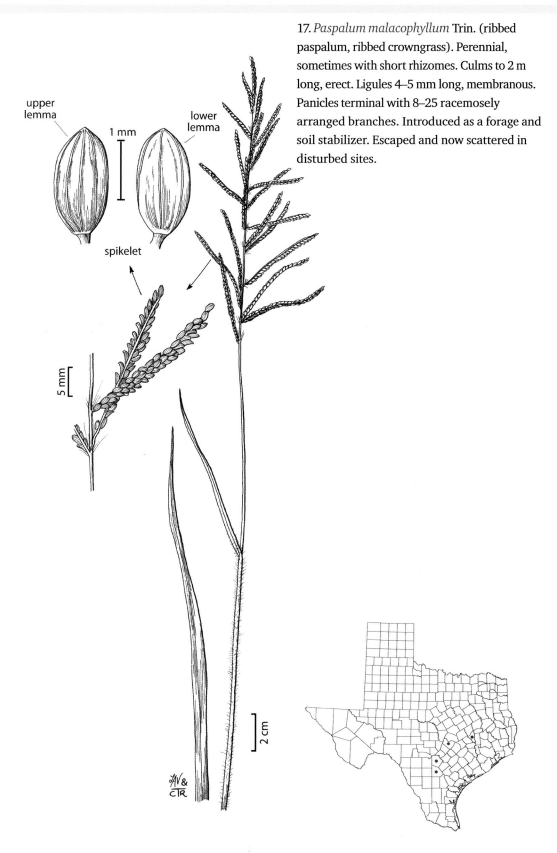

upper
lemma

1 mm

lower
lemma

spikelet

5 mm

2 cm

17. *Paspalum malacophyllum* Trin. (ribbed
paspalum, ribbed crowngrass). Perennial,
sometimes with short rhizomes. Culms to 2 m
long, erect. Ligules 4–5 mm long, membranous.
Panicles terminal with 8–25 racemosely
arranged branches. Introduced as a forage and
soil stabilizer. Escaped and now scattered in
disturbed sites.

18. *Paspalum minus* E. Fourn. (mat paspalum, matted paspalum, matted crowngrass). Perennial with short rhizomes. Culms rarely over 30 cm long, usually forming a dense mat. Ligules 0.2–0.7 mm long, membranous. Panicles terminal, usually composed of a digitate pair of branches, sometimes a third below. Found on forest edges and in disturbed sites.

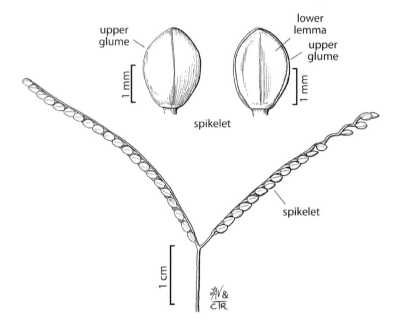

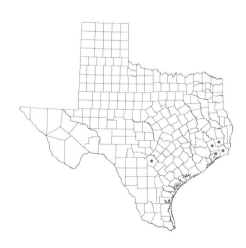

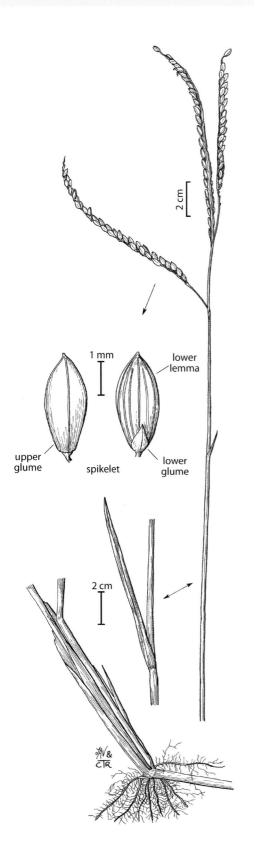

upper
glume

spikelet

lower
lemma

lower
glume

19. *Paspalum modestum* Mez (water paspalum). Perennial with sprawling culms that root at the nodes. Ligules 1.0–2.5 mm long, membranous. Panicles with 2–10 racemosely arranged branches. A recent South American introduction found in rice fields and roadside ditches. Not reported in Gould (1975b); Hatch, Gandhi, and Brown (1990); Jones, Wipff, and Montgomery (1997); or Turner et al. (2003).

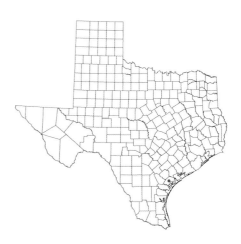

20. *Paspalum monostachyum* Vasey (gulfdune paspalum, single-spike paspalum, gulfdune grass). Rhizomatous perennial. Culms to 1.2 m long, erect. Ligules 0.3–5.0 mm long. Panicles with 1–3 racemosely arranged branches. Found on coastal dunes, wet prairies, and marshes. Highly desirable for livestock and fair for wildlife (Hatch, Schuster, and Drawe 1999).

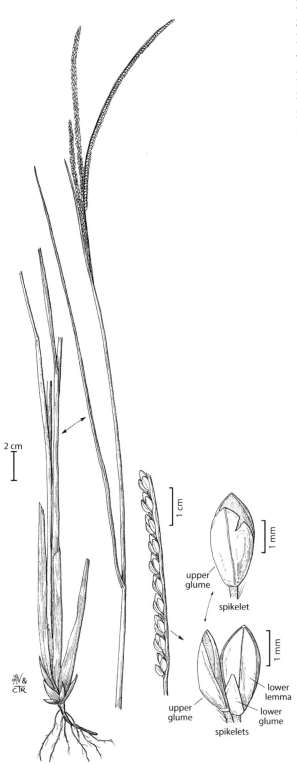

2 cm

1 cm

1 mm

upper
glume

spikelet

1 mm

upper
glume

lower
lemma

lower
glume

spikelets

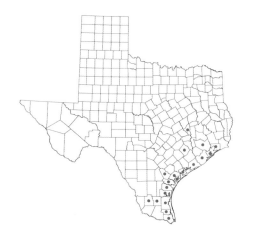

21. *Paspalum notatum* Flüggé (bahiagrass, common bahiagrass, Pensacola bahiagrass). Rhizomatous perennial. Culms to 1.5 m long, erect. Ligules 0.2–0.5 mm long. Panicles terminal, usually composed of a digitate pair with 1–3 additional branches below. An introduced species for forage, turf, and erosion control. Now widely distributed in the eastern half of the state in disturbed sites and along roadsides. Many cultivars available for "improved" pastures and turf. Numerous infraspecific taxa have been presented, but none are recognized here. Listed as poor for wildlife and fair for livestock (Hatch and Pluhar 1993).

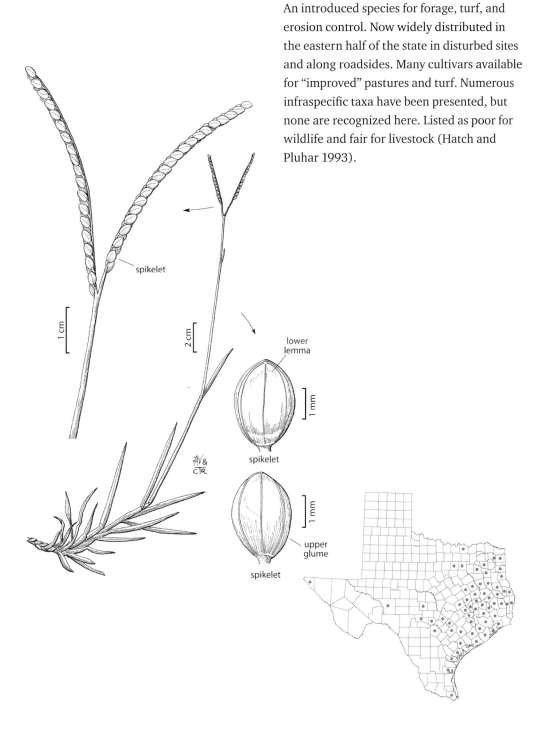

upper glume

1 mm

lower lemma

spikelet

1 mm

spikelet

1 cm

2 cm

22. *Paspalum plicatulum* Michx. (brownseed paspalum, plaited paspalum, brownseed crowngrass). Perennial, rhizomes indistinct. Culms stout, erect, to 1.1 m tall. Ligules 2–3 mm long, membranous. Panicles terminal with 2–7 racemosely arranged branches. Frequent on sandy to sandy loam soils along forest margins, disturbed sites, and prairies. Listed as fair for wildlife and livestock (Hatch and Pluhar 1993). Telfair (2006) listed the seeds as especially important for wildlife.

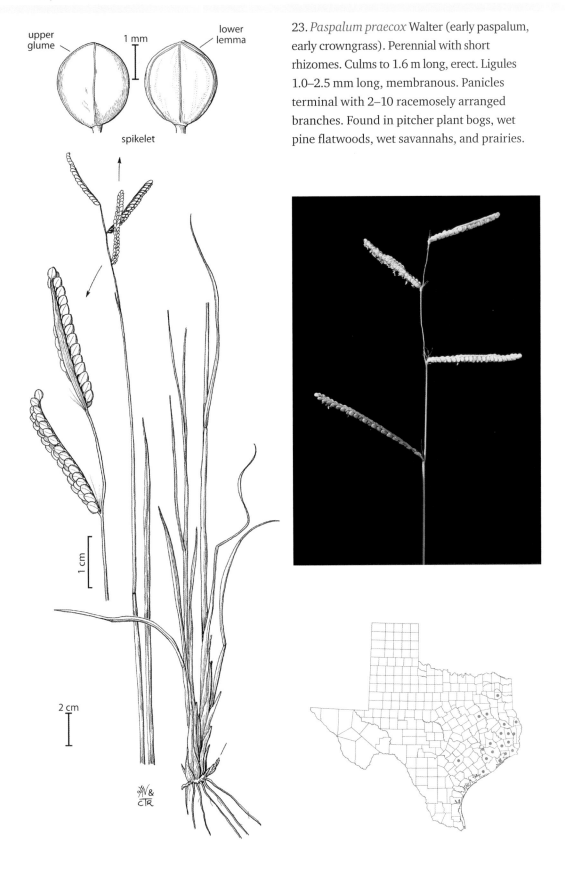

upper
glume

1 mm

lower
lemma

spikelet

1 cm

2 cm

23. *Paspalum praecox* Walter (early paspalum, early crowngrass). Perennial with short rhizomes. Culms to 1.6 m long, erect. Ligules 1.0–2.5 mm long, membranous. Panicles terminal with 2–10 racemosely arranged branches. Found in pitcher plant bogs, wet pine flatwoods, wet savannahs, and prairies.

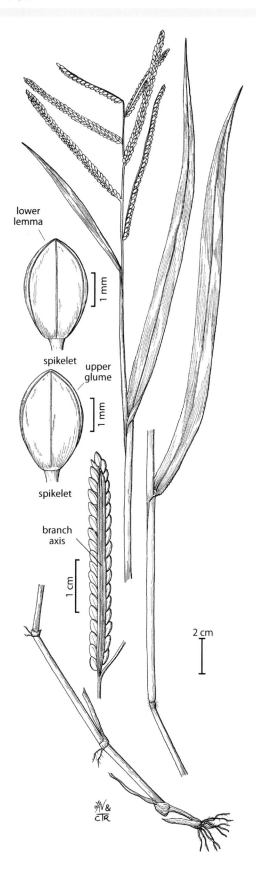

lower
lemma

1 mm

spikelet

upper
glume

1 mm

spikelet

branch
axis

1 cm

2 cm

24. *Paspalum pubiflorum* Rupr. *ex* E. Fourn.
(hairyseed paspalum, hairyseed crowngrass).
Perennial, decumbent, roots at lower nodes.
Culms to 1.3 m long. Ligules 1–3 mm
long, membranous. Panicles terminal with
2–7 racemosely arranged branches. Found
in shady, low moist areas, ditches, and
swales. Fair forage and fair seed producer
(Hatch, Schuster, and Drawe 1999). Gould
(1975b) listed 2 varieties based primarily on
pubescence patterns. The variation seems
almost continuous and does not warrant
varietal recognition.

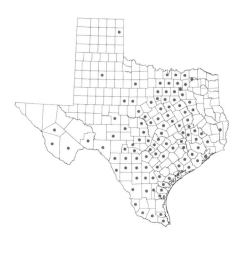

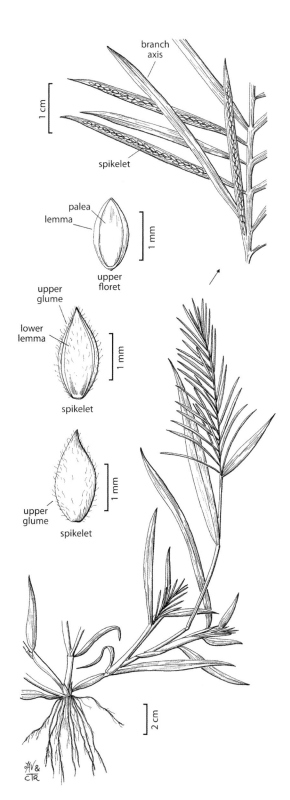

branch axis

spikelet

palea
lemma
upper floret

upper glume
lower lemma
spikelet

upper glume
spikelet

2 cm

1 cm

1 mm

1 mm

1 mm

25. *Paspalum repens* P. J. Bergius (water paspalum, horsetail crowngrass, horsetail paspalum). Annual, aquatic, floating or rhizomatous. Culms to 55 cm long, erect. Ligules 1–4 mm long, membranous. Panicles terminal with 20–70 racemosely arranged branches. Found along lakeshores, stream banks, and roadside ditches. Gould (1975b) and Hatch, Gandhi, and Brown (1990) listed this species as *P. fluitans* (Ell.) Kunth.

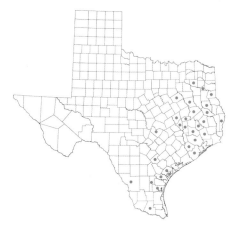

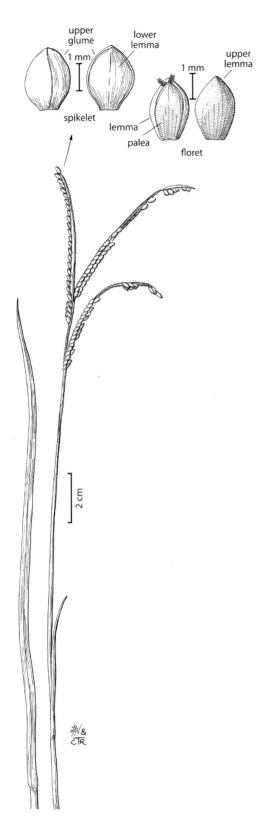

upper glume

lower lemma

1 mm

upper lemma

1 mm

spikelet

lemma

palea

floret

2 cm

JAV & CTR

26. *Paspalum scrobiculatum* L. (Indian paspalum, India paspalum, kodo-millet, ditchmillet). Annual. Culms erect, to 1.5 m long. Ligules 0.3–1.2 mm long, membranous. Panicles terminal with 1–5 branches. Found as a scattered weed. An introduction from India where it is grown as a cereal (*kodo*). Occasionally escapes. The species, like several other members of the genus, can be toxic under certain conditions.

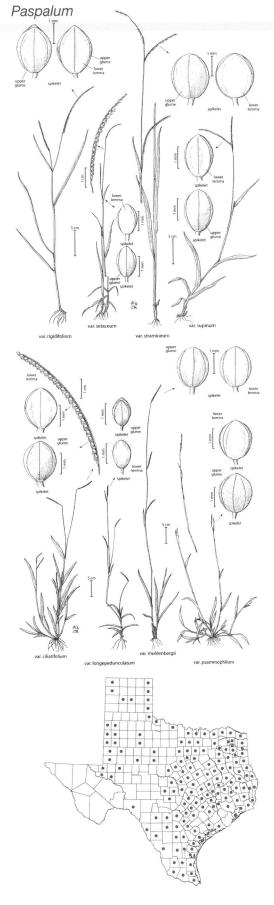

var. rigidifolium / var. setaceum / var. stramineum / var. supinum / var. ciliatifolium / var. longepedunculatum / var. muhlenbergii / var. psammophilum

27. *Paspalum setaceum* Michx. (thin paspalum, slender crowngrass, fringed-leaf paspalum, hurrah grass, yellow sand paspalum). Perennial, cespitose or with short rhizomes. Culms to 1.25 m long, erect to prostrate and spreading. Ligules 0.2–0.5 mm long, membranous. Panicles terminal and axillary with 1–6 racemosely arranged branches, axillary branches partially or wholly enclosed in subtending leaf sheaths. A variable species of disturbed sandy sites, prairies, open woodlands, pine forests, and roadsides. Listed as fair for wildlife and livestock (Hatch and Pluhar 1993). Telfair (2006) listed seeds as especially important for wildlife. Of the 9 varieties, 5 occur in Texas: var. *ciliatifolium* (Michx.) Vasey (fringeleaf paspalum); var. *muhlenbergii* (Nash) D. J. Banks (hurrahgrass); var. *setaceum*; var. *stramineum* (Nash) D. J. Banks (yellow sand paspalum); and var. *supinum* (Bosc *ex* Poir.) Trin. (supine thin paspalum). They can sometimes be distinguished by the following characters:

1. Leaf blades glabrous (or almost so) on the surfaces but ciliate on the margins .... var. *ciliatifolium*
1. Leaf blades hirsute on the surfaces and margins.
  2. Plants widely spreading to prostrate .... var. *supinum*
  2. Plants erect to spreading.
    3. Lower lemmas with evident midvein; spikelets usually glabrous...var. *muhlenbergii*
    3. Lower lemmas with indistinct midvein; spikelets usually pubescent.
      4. Leaves grayish-green, blades 1.5–7.0 mm wide..............var. *setaceum*
      4. Leaves yellowish-green to dark green, blades 3–14 mm wide .... var. *stramineum*

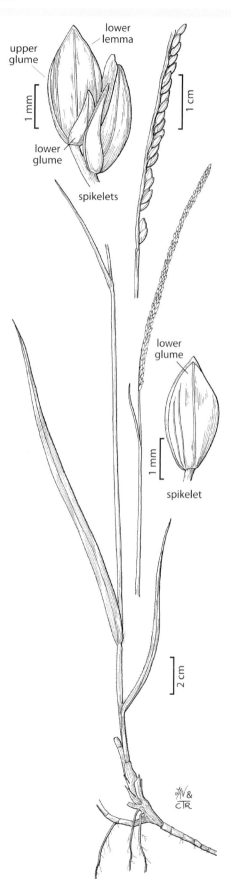

upper glume
lower lemma
1 mm
lower glume
spikelets
1 cm

lower glume
1 mm
spikelet

2 cm

28. *Paspalum unispicatum* (Scribn. and Merr.) Nash (one-spike paspalum). Rhizomatous perennial. Culms to 80 cm long, erect. Ligules 1–2 mm long, membranous. Inflorescence terminal, erect, a spicate raceme or a panicle with 1–2 subterminal spicate branches wholly or partially enclosed in subtending sheaths. Uncommon in sandy soils of the South Texas Plains.

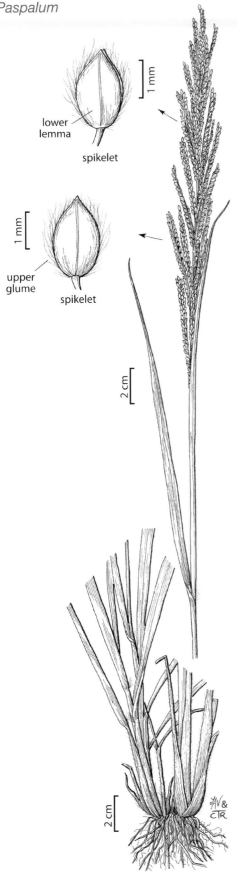

lower lemma

spikelet

1 mm

upper glume

spikelet

1 mm

2 cm

2 cm

JAV & CTR

29. *Paspalum urvillei* Steud. (vaseygrass, Urville's grass). Perennial with short rhizomes forming a knotty base. Culms to 2.2 m tall. Ligules 1–7 mm long, membranous. Panicles terminal with 10–30 racemosely arranged branches. A South American introduction now found in moist to wet disturbed sites, roadside ditches, and swales. Pubescence along the margins of the lower lemmas and upper glumes is distinctive and similar to that in *P. dilatatum*. Poor livestock and wildlife values (Hatch, Schuster, and Drawe 1999).

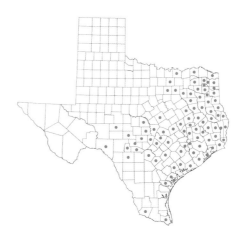

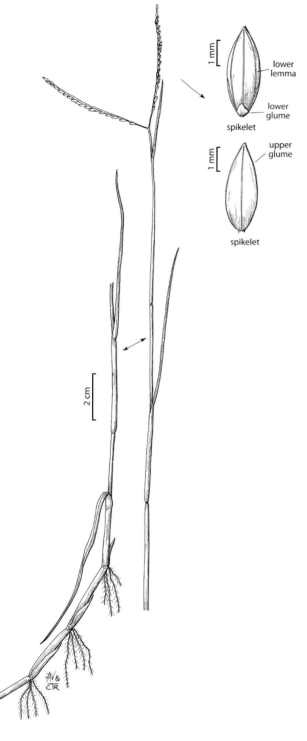

1 mm

lower
lemma

lower
glume

spikelet

1 mm

upper
glume

spikelet

30. *Paspalum vaginatum* Sw. (seashore paspalum, seashore crowngrass). Perennial with rhizomes and/or stolons. Erect portions of culms to 65 cm. Ligules 1–2 mm long, membranous. Panicles terminal, branches paired, a third sometimes below. Found in coastal regions in brackish and salt marshes. Closely related to *P. distichum*. Poor to fair for livestock and wildlife (Hatch, Schuster, and Drawe 1999).

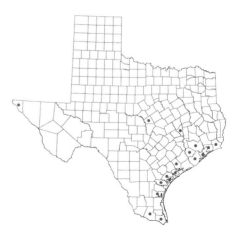

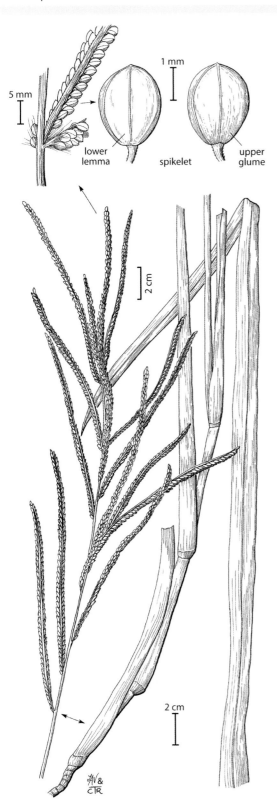

lower lemma

spikelet

upper glume

31. *Paspalum virgatum* L. (talquezal). Stout perennial with erect culms to 2 m long. Ligules about 2 mm long, membranous. Panicles terminal with 10–20 racemosely arranged branches. An introduction from Mexico and South America. Found in disturbed areas and cultivated fields.

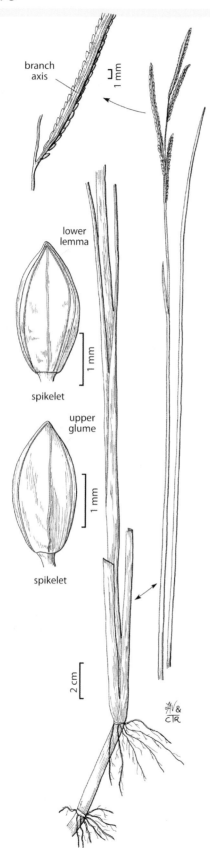

branch
axis

1 mm

lower
lemma

1 mm

spikelet

upper
glume

1 mm

spikelet

2 cm

32. *Paspalum wrightii* Hitchc. & Chase (Wright's paspalum). Aquatic to semiaquatic perennial, rhizomatous and/or stoloniferous. Culms to 1.5 m long, erect. Ligules 1–3 mm long. Panicles terminal with 5–8 racemosely arranged branches. An introduction from Mexico and South America. Grows along or in water. Not listed in Gould (1975b); Hatch, Gandhi, and Brown (1990); or Jones, Wipff, and Montgomery (1997).

Pennisetum L. C. Richard
(Panicoideae: Paniceae)

Annuals or perennials with generally tall, erect culms, usually tightly cespitose in Texas species. Leaf blades usually thin and flat, occasionally folded. Inflorescence a dense, bristly, tightly contracted, spicate panicle. Spikelets solitary to 2 to several in fascicles of numerous bristles. Bristles usually free at the base in Texas species and in 3 distinct series: outer, inner, and primary. Fascicle disarticulating from the rachis and containing the spikelets. Lower glume small or vestigial. Upper glume and lemma of the sterile floret about equal or the glume slightly shorter. Lemma and palea of the fertile floret thin, smooth, and shiny, the margins of the lemma thin and flat. Basic chromosome numbers, $x = 5$, 7, 8, and 9. Photosynthetic pathway, C_4. Represented in Texas by 13 species and 1 subspecies.

Genus is poorly collected because plants are mostly ornamentals or cultivated.

1. Bristles scabrous, occasionally primary bristles with a few inconspicuous long hairs.
    2. Primary and other bristles about the same length.
        3. Panicles erect; fascicles with a stipelike base 1.5–6.0 mm long .... 2. *P. alopecuroides*
        3. Panicles drooping; fascicles subsessile, base <1 mm long.
            4. Plants green; upper glumes 7–9-veined .... 7. *P. nervosum*
            4. Plants purple; upper glumes 1–3-veined.. 6. *P. macrostachys*
    2. Primary bristles noticeably longer than other bristles in the fascicles.
        5. Panicles dense; 20–40 fascicles per centimeter of rachis; paleas of lower florets present .... 10. *P. purpureum*
        5. Panicles less dense; paleas of lower florets absent .... 2. *P. alopecuroides*
1. Bristles, at least primary, long-ciliate.
        6. Spikelets 9–12 mm long .... 13. *P. villosum*
        6. Spikelets 2.5–7.0 mm long.
            7. Fascicles not disarticulating from the rachises; upper lemma margins pubescent .... 5. *P. glaucum*
            7. Fascicles disarticulating from the rachises at maturity; upper lemma margins glabrous.
                8. Upper lemmas smooth and shiny, conspicuously different from the lower..9. *P. polystachion*
                8. Lower and upper lemmas similar in texture.
                    9. Rachises, at least basally, glabrous or sometimes scabrous.
                        10. Inner bristles grooved and fused; spikelets sessile.
                            11. Many outer bristles longer than the spikelets; terminal bristles 10–25 mm long... 3. *P. ciliare*
                            11. Outer bristles not longer than the spikelets; terminal bristles 3–7 mm long .... 12. *P. setigerum*
                        10. Inner bristles not grooved or fused; spikelets pedicellate ....4. *P. flaccidum*
                    9. Rachises, at least basally, pubescent.

12. Plants 2–3 m tall; 30–40 fascicles per centimeter of rachis.. .... 10. *P. purpureum*

12. Plants 0.5–2.0 m tall; 5–17 fascicles per centimeter of rachis.

  13. Midculm leaves 2.0–3.5 mm wide, midvein thickened ....11. *P. setaceum*

  13. Midculm leaves 3–11 mm wide, midvein not noticeably thickened.

    14. Plants with short rhizomes; nodes pubescent ....8. *P. orientale*

    14. Plants without rhizomes; nodes glabrous.....1. *P. advena*

(Wipff 2003g)

1. *Pennisetum advena* Wipff and Veldcamp (purple fountaingrass). Cespitose perennial. Culms to 1.5 m long. Ligules <1 mm long. Panicles 20–35 cm long, drooping. A species of uncertain origin but sold as the ornamental *P. setaceum* "Rubrum." Not included in Gould (1975b); Hatch, Gandhi, and Brown (1990); or Turner et al. (2003).

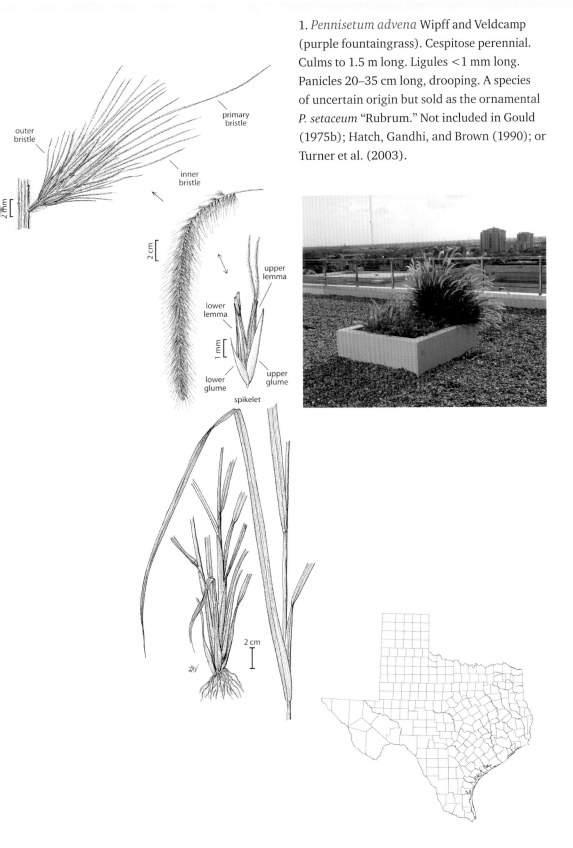

2. *Pennisetum alopecuroides* (L.) Spreng. (foxtail fountaingrass). Cespitose perennial. Culms to 1 m long. Ligules 0.2–0.5 mm long. Panicles 6–20 cm long, erect. Introduced from Southeast Asia as an ornamental. Not included in Gould (1975b); Hatch, Gandhi, and Brown (1990); or Turner et al. (2003).

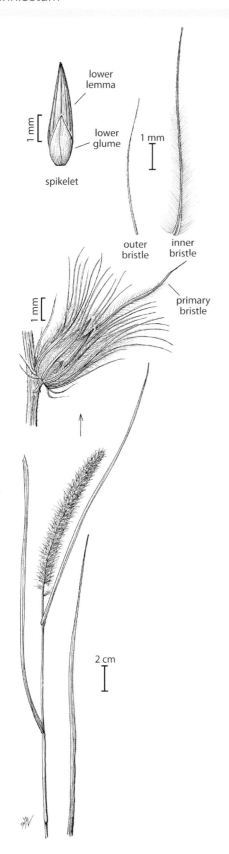

lower
lemma

1 mm

lower
glume

spikelet

1 mm

outer
bristle

inner
bristle

primary
bristle

1 mm

2 cm

3. *Pennisetum ciliare* (L.) Link (buffel grass). Perennial with or without rhizomes. Culms to 1.5 m long. Ligules 0.5–3.0 mm long. Panicles 2–20 cm long. Introduced from Africa and Asia. Grown as a forage species. It is extremely tolerant of drought and grazing. Gould (1975b) and Hatch, Gandhi, and Brown (1990) included it as *Cenchrus ciliaris* L. because of the slightly fused bristles. See Diggs et al. (2006) for a review of controversy surrounding the placement of this species in either *Cenchrus* or *Pennisetum*. Listed as poor for wildlife and good for livestock (Hatch and Pluhar 1993).

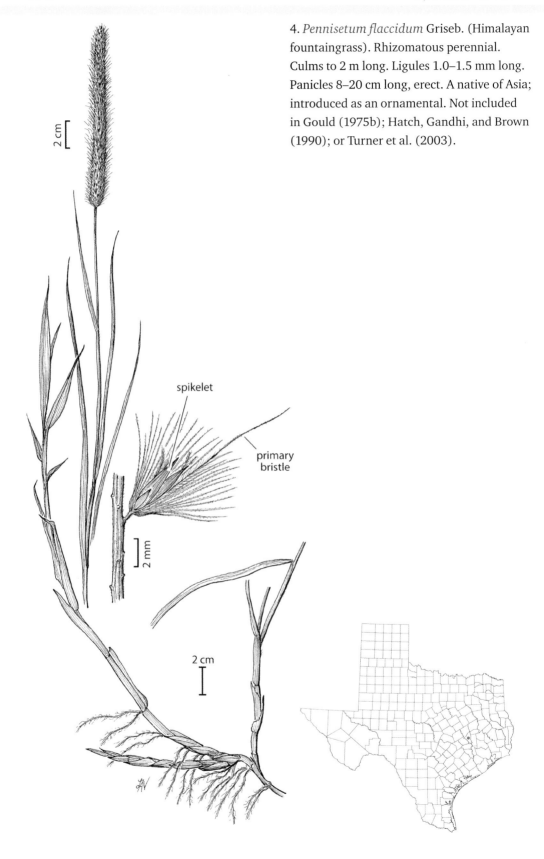

4. *Pennisetum flaccidum* Griseb. (Himalayan fountaingrass). Rhizomatous perennial. Culms to 2 m long. Ligules 1.0–1.5 mm long. Panicles 8–20 cm long, erect. A native of Asia; introduced as an ornamental. Not included in Gould (1975b); Hatch, Gandhi, and Brown (1990); or Turner et al. (2003).

2 cm

spikelet

primary bristle

2 mm

2 cm

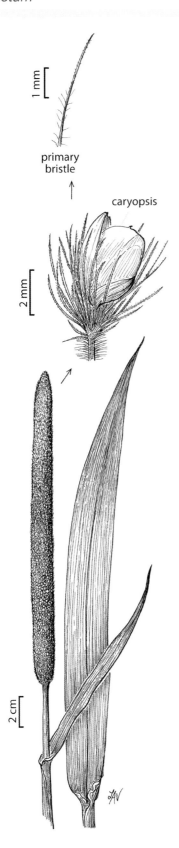

1 mm

primary
bristle

↑

caryopsis

2 mm

2 cm

5. *Pennisetum glaucum* (L.) R. Br. (pearl millet). Coarse annual. Culms 0.5–3.0 m tall. Ligules 2–5 mm long. Panicles 4–200 cm long, erect. Introduced from Asia and cultivated for grain and forage. Superficially looks cattail-like. Sometimes used in soil stabilization, perhaps because it seldom persists for more than a few years. Not included in Gould (1975b) or Hatch, Gandhi, and Brown (1990). Known to accumulate nitrates and poison livestock (Burrows and Tyrl 2001).

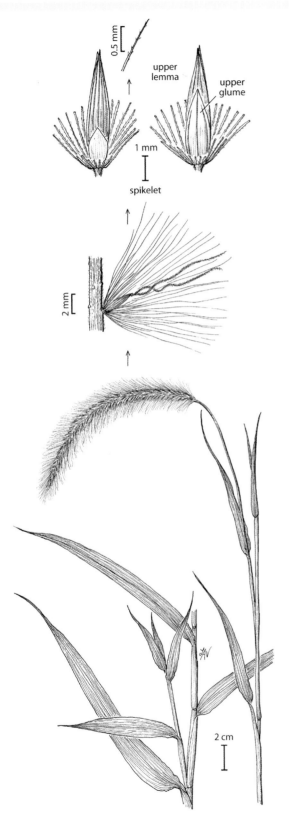

6. *Pennisetum macrostachys* (Brongn.) Trin. (Pacific fountaingrass). Perennial. Culms 1–3 m long. Ligules 0.1–0.3 mm long. Panicles 15–40 cm long, drooping. Introduced from the South Pacific, used as an ornamental, and commonly sold under the name "Burgundy Giant." Not included in Gould (1975b); Hatch, Gandhi, and Brown (1990); or Turner et al. (2003).

lower
lemma

upper
glume

1 mm

lower
glume

spikelet

primary
bristle

2 mm

2 cm

7. *Pennisetum nervosum* (Nees) Trin. (bentspike fountaingrass). Cespitose perennial. Culms 1–4 m long. Ligules 0.5–1.5 mm long. Panicles 15–25 cm long, drooping. A South American introduction.

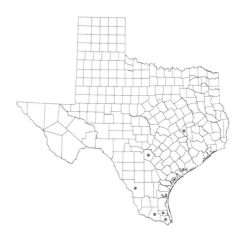

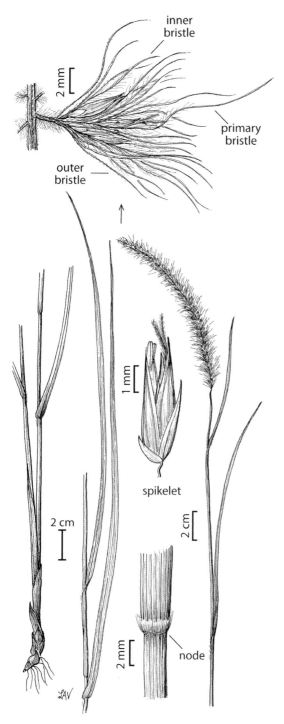

8. *Pennisetum orientale* Willd. *ex* Rich. (white fountaingrass, laurisa grass). Rhizomatous perennial. Culms to 2 m long. Ligules 1–2 mm long. Panicles 10–40 cm long, erect to slightly arching. A native found from Africa to India and introduced to Texas as an ornamental. Not included in Gould (1975b) or Hatch, Gandhi, and Brown (1990).

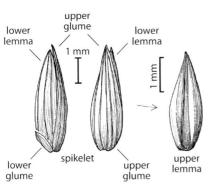

lower lemma | upper glume | lower lemma

1 mm

1 mm

lower glume | spikelet | upper glume | upper lemma

9. *Pennisetum polystachion* (L.) Schult. (mission grass). Annual or perennial. Culms to 2 m tall. Ligules 1.5–3.0 mm long. Panicles 10–25 cm long, erect. An introduced African species. The U.S. Department of Agriculture considers it a noxious weed. The Texas species belongs to the subsp. *setosum* (Sw.) Brunken. Not included in Gould (1975b) or Hatch, Gandhi, and Brown (1990). Turner et al. (2003) reported this as a separate species.

primary bristle

2 mm

2 cm

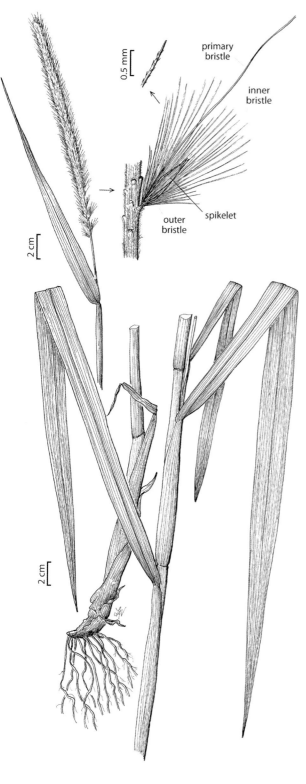

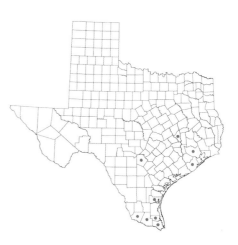

10. *Pennisetum purpureum* Schumach. (elephant grass, napier grass). Large perennial, sometimes with rhizomes and/or stolons. Culms 2–8 m tall, erect. Ligules 1.5–5.0 mm long. Panicles 8–35 cm long, erect. An African introduction used as an ornamental or infrequently as forage. Not included in Gould (1975b) or Hatch, Gandhi, and Brown (1990). High producing but unpalatable at maturity to grazing animals (Hatch, Schuster, and Drawe 1999).

11. *Pennisetum setaceum* (Forssk.) Chiov. (tender fountaingrass, purple fountaingrass, crimson fountaingrass). Cespitose perennial. Culms to 1.5 m tall. Ligules 0.5–1.0 mm long. Panicles 5–35 cm long, erect to arching. Introduced from the Mediterranean region as an ornamental. Becoming an invasive weed in many places. Not included in Gould (1975b); Hatch, Gandhi, and Brown (1990); or Turner et al. (2003).

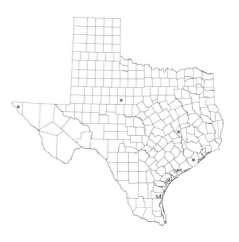

12. *Pennisetum setigerum* (Vahl) Wipff
(birdwood grass). Perennial. Culms to 1 m long.
Ligules 0.5–1.5 mm long. Panicles 2–15 cm long,
erect. A native of Africa, India, and Arabia.
It has been introduced as a forage species.
Sometimes reported as *Cenchrus setigerum*
Valh because of the fused bristles. Not included
in Gould (1975b); Hatch, Gandhi, and Brown
(1990); or Turner et al. (2003).

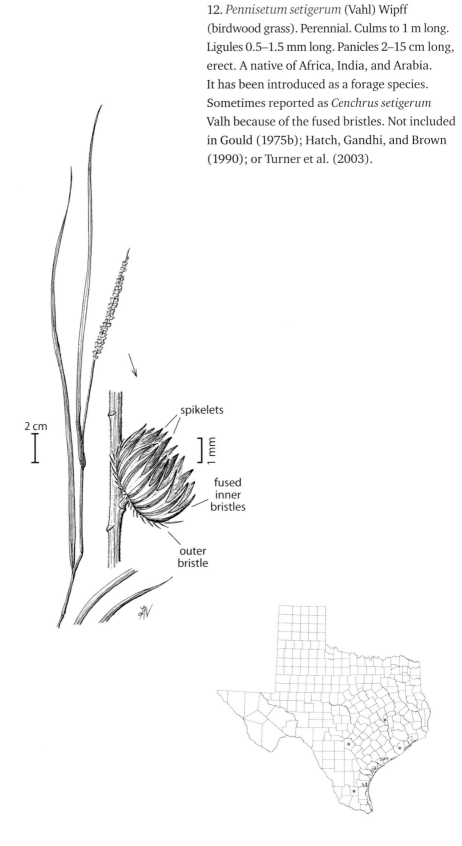

2 cm

spikelets

1 mm

fused
inner
bristles

outer
bristle

13. *Pennisetum villosum* R. Br. *ex* Fresen.
(feathertop, feathertop pennisetum).
Rhizomatous perennial. Culms to 75 cm
long. Ligules 1.0–1.3 mm long. Panicles
4–12 cm long. Introduced from Africa as an
ornamental.

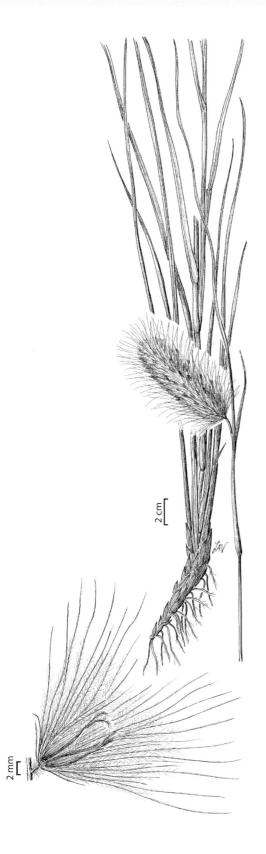

Phalaris L.
(Poöideae: Poeae)

Annuals and perennials. Leaves mostly glabrous, with membranous ligules and flat blades. Inflorescence a contracted, usually spikelike panicle. Spikelets with 1 terminal perfect floret and 1 or 2 reduced florets below, the latter reduced to scales. Disarticulation above the glumes. Glumes about equal, large, awnless, usually laterally flattened and dorsally keeled, the keel often with a thin, membranous wing. Lemma of the fertile floret awnless, coriaceous and glossy, shorter and firmer than the glumes, often more or less hairy, permanently enclosing the faintly 2-nerved palea and plump caryopsis. Basic chromosome numbers, x = 6 and 7. Photosynthetic pathway, C_3. Represented in Texas by 6 species.

1. Sterile florets usually 1, or if 2, then lower florets to 0.7 mm long and upper florets 1–3 mm long.
    2. Plants annual; sterile florets 1, glabrous or nearly so... 6. *P. minor*
    2. Plants perennial; sterile florets usually 1, sometimes 2, hairy......................... 2. *P. aquatica*
1. Sterile florets 2.
        3. Panicles cylindric, sometimes lobed ....1. *P. angusta*
        3. Panicles ovoid to oblong-ovoid, not lobed.
            4. Sterile florets 0.6–1.2 m long...3. *P. brachystachys*
            4. Sterile florets 1.5–4.5 mm long.
                5. Glumes 7–10 m long, 2.0–2.5 mm wide ....4. *P. canariensis*
                5. Glumes 4–6(–8) mm long; 0.8–1.5 mm wide .... 5. *P. caroliniana*

(Barkworth 2007h)

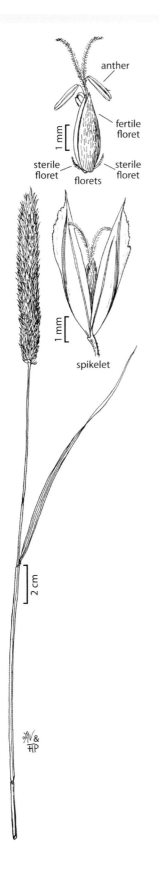

anther

fertile
floret

1 mm

sterile
floret

sterile
floret

florets

1 mm

spikelet

2 cm

1. *Phalaris angusta* Nees *ex* Trin. (Timothy canarygrass, narrow canarygrass). Annual with culms to 1.7 m long. Ligules 4–7 mm long, lacerated. Panicles 2–20 cm long, 0.6–1.5 mm wide. Spikelets borne singularly, not clustered. Found in open grasslands and prairies, along roadsides, in ditches, and in swales. Included in Gould (1975b) as *P. angustata*. Often planted in game food plots (Hatch, Schuster, and Drawe 1999).

2. *Phalaris aquatica* L. (bulbous canarygrass).
Cespitose perennial with short rhizomes.
Culms to 2 m long. Ligules 2–12 mm long,
membranous, lacerate. Panicles 2–15 cm long,
1.0–2.5 cm wide, spikelets borne singularly,
not clustered. Glume wings consistently
entire. Introduced as a forage species from the
Mediterranean region. An excellent forage, but
it does become weedy and is most frequently
found in disturbed areas that are periodically
flooded. Excluded by Gould (1975b) because it
was cultivated.

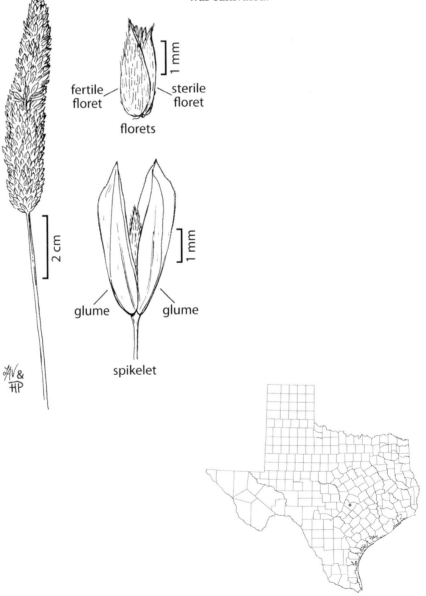

3. *Phalaris brachystachys* Link. (shortspike canarygrass). Annual. Culms to 1 m long. Ligules 4–6 mm long, membranous, lacerated. Panicles 1.5–5.0 cm long, 1–2 cm wide. Spikelets borne singularly, not clustered. Glume wings entire, abruptly pointed. A ruderal species from the Mediterranean region. Found in disturbed sites and waste areas.

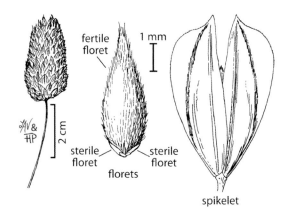

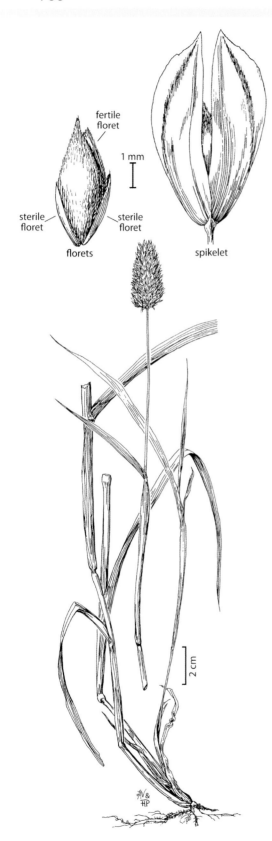

fertile floret

sterile floret

sterile floret

1 mm

florets

spikelet

4. *Phalaris canariensis* L. (canarygrass). Annual with culms to 1 m long. Ligules 3–6 mm long, lacerated. Panicles 1.5–5.0 cm long, 1–2 cm wide; spikelets borne singularly, not clustered. Glume wings widening distally; semicircular in outline. An introduction from southern Europe frequently used in birdseed and found around feeders.

2 cm

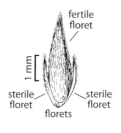

fertile
floret

sterile
floret

sterile
floret

florets

1 mm

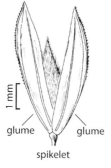

glume

glume

1 mm

spikelet

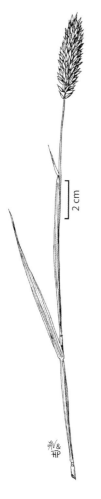

2 cm

5. *Phalaris caroliniana* Walt. (Carolina canarygrass, wild canarygrass). Annual with culms to 1.5 m long. Ligules 1.5–7.0 mm long, membranous. Panicles 0.5–8.0 cm long, 0.8–2.0 cm wide. Spikelets borne singularly, not clustered. Glume wings entire, acute. A native species that occurs in wet, marshy, and swampy ground. Found specifically along fence rows, ditches, swales, open woodlands, and roadsides. Fair spring feed for livestock and seed for wildlife (Hatch, Schuster, and Drawe 1999).

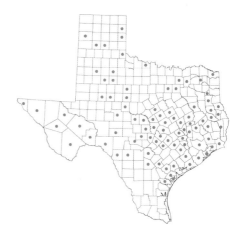

6. *Phalaris minor* Retz. (littleseed
canarygrass, lesser canarygrass). Annual
with culms to 1 m tall. Ligules 5–12 mm long,
membranous, lacerated. Panicles 1–8 cm long,
1–2 cm wide. Spikelets borne singularly, not
clustered. Glume wings variable, irregularly
dentate, crenate or entire. Gould (1975b)
referred to it as an introduction from the
Mediterranean region, and Barkworth (2007h)
called it a native. Regardless, it is a plant of
disturbed soils or waste areas.

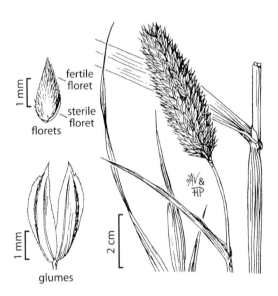

Phalaroides Wolf
(Poöideae: Poeae)

Perennials, rather coarse, densely cespitose or in small clusters from rhizomatous bases. Leaves mostly glabrous, with membranous ligules and flat blades. Inflorescence a contracted, usually spikelike panicle. Spikelets with 1 terminal perfect floret and 1 or 2 reduced florets below, the latter reduced to scales. Disarticulation above the glumes. Glumes about equal, large, awnless, usually laterally flattened and dorsally keeled, the keel often with a thin, membranous wing. Lemma of the fertile floret awnless, coriaceous and glossy, shorter and firmer than the glumes, often more or less hairy, permanently enclosing the faintly 2-nerved palea and plump caryopsis. Basic chromosome numbers, $x = 6$ and 7. Photosynthetic pathway, C_3. Represented in Texas by a single species. A segragate of *Phalaris*.

1. *Phalaroides arundinacea* (L.) Rauschert (reed canarygrass). Perennial with stout, scaly rhizomes. Plants can grow to over 2 m in height. Auricles absent; ligules up to 1 cm; blades with very serrate margins. Inflorescence a dense panicle. Sterile florets 2, about ½ the size of the perfect floret. A segregate of *Phalaris*. Found in or near water, along lakes, streams, and irrigation ditches. Often forms dense stands and can be a noxious weed in some situations. Not included in Gould (1975b); Turner et al. (2003); or Hatch, Gandhi, and Brown (1990). Usually an ornamental in Texas.

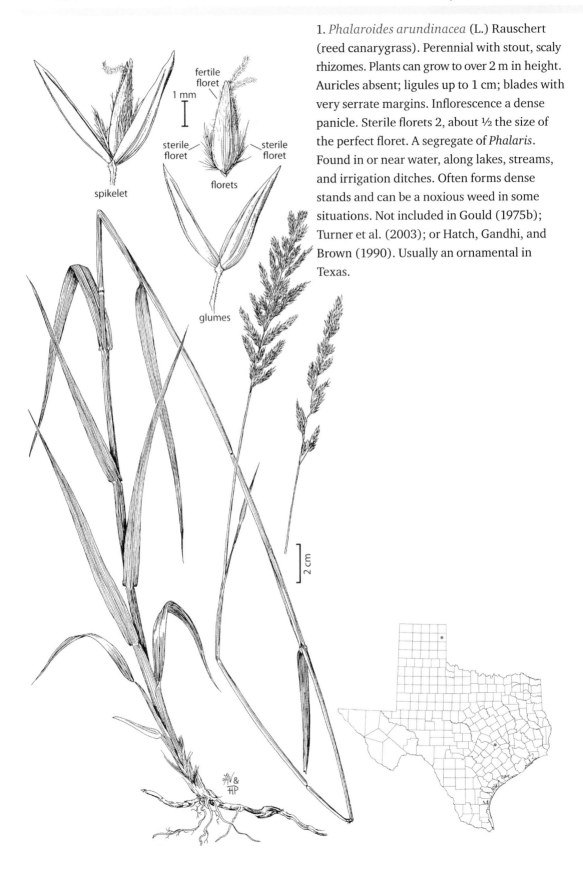

fertile floret

1 mm

sterile floret

sterile floret

florets

spikelet

glumes

2 cm

Phanopyrum Nash
(Panicoideae: Paniceae)

Stout perennial with erect culms from creeping stoloniferous bases, these rooting freely at the nodes. Ligule a membrane. Blades long, broad, and flat with auriculate or cordate bases. Panicles large with numerous stiffly erect or spreading primary branches, with small clusters of spikelets on short pedicels, many on short, appressed lateral branches. Both glumes present, the lowermost only slightly shorter than the second. These, along with the lemma of the lower floret, more or less dorsally compressed. Lower floret neuter with a palea shorter than the lemma. Lemma and palea of the upper floret narrowly oblong, much shorter than the glumes and lower lemma. Basic chromosome number $x = 10$. Photosynthetic pathway, C_3. Represented in Texas by the only member of the genus.

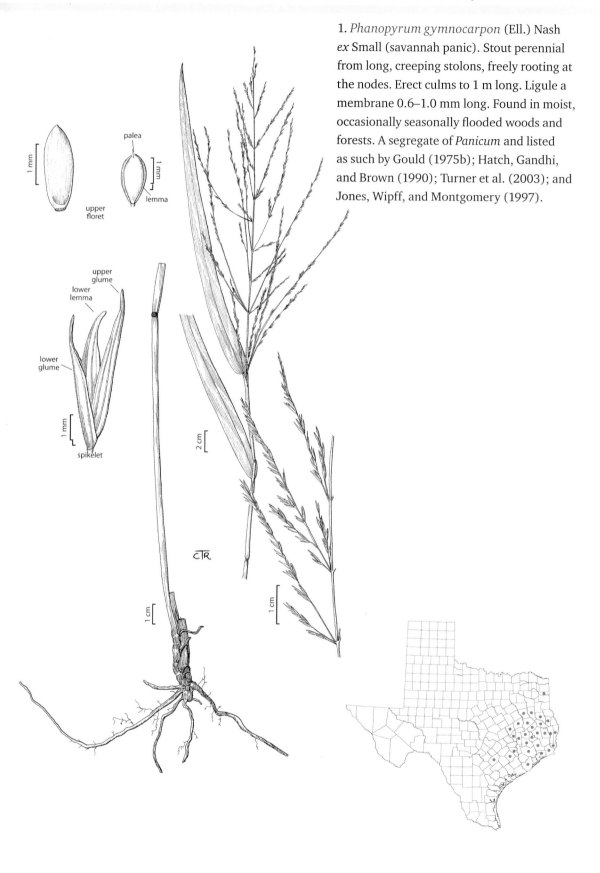

1. *Phanopyrum gymnocarpon* (Ell.) Nash *ex* Small (savannah panic). Stout perennial from long, creeping stolons, freely rooting at the nodes. Erect culms to 1 m long. Ligule a membrane 0.6–1.0 mm long. Found in moist, occasionally seasonally flooded woods and forests. A segregate of *Panicum* and listed as such by Gould (1975b); Hatch, Gandhi, and Brown (1990); Turner et al. (2003); and Jones, Wipff, and Montgomery (1997).

upper floret

palea

1 mm

1 mm

lemma

upper glume

lower lemma

lower glume

1 mm

spikelet

2 cm

CTR

1 cm

1 cm

Phleum L.
(Poöideae: Poeae)

Perennial, cespitose or tufted. Sheaths open, occasionally inflated and separating from the culm; ligules up to 6 mm long, membranous, often with small rounded auricles. Inflorescence a cylindric, tightly contracted panicle. Spikelets with 1 floret, perfect. Disarticulation above, or occasionally below, the glumes. Glumes 2, large, subequal, laterally flattened, mostly 3-nerved, broad, abruptly narrowed at the apex to a mucro or short, stout awn. Lemma membranous, 3–7-nerved, broad and blunt, awned, much shorter than the glumes. Palea membranous, narrow, about as long as the lemma. Basic chromosome number, $x = 7$. Photosynthetic pathway, C_3. Represented in Texas by a single species .

(Barkworth 2007i)

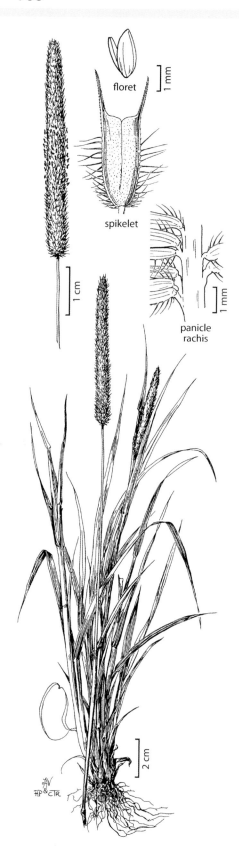

floret

1 mm

spikelet

panicle
rachis

1 cm

1 mm

2 cm

1. *Phleum pratense* L. (Timothy, herdsgrass).
Erect perennial up to 1.5 m tall, often with
swollen bulbous culm base. Auricles small;
ligules to 4 mm long. Panicles spikelike,
branches obscure and found only at or near the
base. Glumes awned and enclosing the perfect
florets. An early European introduction, it is
now cultivated as a hay or pasture grass in
most temperate areas of the world. Found
in pastures, hay meadows, roadside ditches,
and mesic disturbed sites. The Texas species
belongs to the typical subspecies, subsp.
pratense.

Phragmites Adans.
(Arundinoideae: Arundineae)

Tall, rhizomatous and stoloniferous perennial reed grasses with broad leaves and large, plumose panicles. Ligule a ring of short hairs. Spikelets several-flowered, disarticulating above the glumes and between the florets. Lower floret staminate or sterile, the terminal floret also reduced. Rachilla villous with long hairs. Glumes and lemmas thin, acute or acuminate, glabrous, the lower 1-nerved, the upper and the lemmas usually 3-nerved. Basic chromosome number, $x = 6$. Photosynthetic pathway, C_3. Represented in Texas by a single species.

(Allred 2003e)

1. *Phragmites australis* (Cav.) Trin *ex* Steud. (common reed, reed, Danube grass). Rhizomatous or sometimes a stoloniferous perennial with culms up to 4 m tall. Similar to *Arundo donax* but shorter, with narrower leaf blades and long villous hairs along the rachilla. Found most frequently in tidal or mudflats, along streams and rivers, and in marshes. Abundant and forms large colonies. Fair to good forage and nesting habitat (Hatch, Schuster, and Drawe 1999). Telfair (2006) listed it as important for wildlife. Several subspecies have been described, but none are recognized here.

floret

spikelet

ligule

Phyllostachys Siebold. and Zucc.
(Bambusoideae: Bambuseae)

Perennials, shrubby or treelike, usually in spreading clumps or thickets. Rhizomes leptomorphic. Culms 3–20 m tall, 3–15 cm thick, erect or slightly nodding. Nodes slightly swollen. Internodes strongly flattened their entire length, doubly sulcate above the branches. Branches 2–3 per node, unequal. Culm leaves coriaceous, early deciduous. Spikelets with 2 to several florets, upper reduced. Anthers 3, style branches 3. Basic chromosome number, $x = 12$. Photosynthetic pathway, C_3. Jones, Wipff, and Montgomery (1997) listed 26 species cultivated in Texas. Two of the most common are included here. *Phyllostachys nigra* (Lodd.) Munro has black culms with a single collection in Tarrant County (Turner et al. 2003).

1. Auricles and fimbriae absent from culm leaves .... 1. *P. aurea*
1. Auricles present on upper culm leaves; fimbriae present on all culm leaves .... 2. *P. bambusoides*

(Stapleton and Barkworth 2007)

1. *Phyllostachys aurea* Carrière *ex* Rivière and
C. Rivière (fishpole bamboo, golden bamboo,
yellow bamboo). Perennial. Culms erect, to
10 m tall, 1–4 cm thick. Internodes glabrous,
initially green turning golden or grayish.
Auricles and fimbriae lacking on foliage
leaves. A cultivar that forms large clumps from
extensive and aggressive rhizomes. This is
not the only species of the genus with golden
culms.

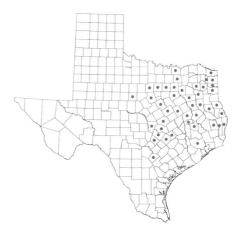

2. *Phyllostachys bambusoides* Siebold & Zucc. (giant timber bamboo, giant bamboo). Culms erect or leaning toward the light, to 22 m long and 15 cm thick. Internodes glabrous, green, golden, or striped in many cultivars. Auricles and fimbriae usually well developed on foliage leaves. Widely cultivated and the largest member of the genus in Texas.

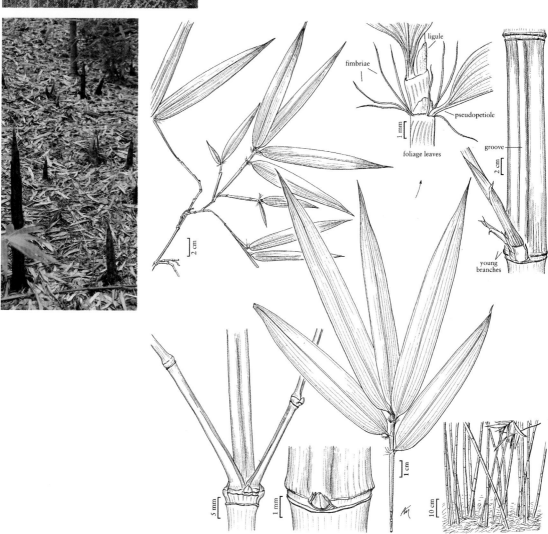

P. bambusoides

Piptatherum P. Beauv.
(Poöideae: Stipeae)

Perennials, cespitose or occasionally rhizomatous. Culms 10–150 cm long, erect. Sheaths open, glabrous, smooth; auricles absent; ligules 0.2–1.5 mm long, membranous, hyaline. Inflorescence a closed or occasionally open, terminal panicle of 1-flowered spikelets; pedicels often appressed to the branches. Disarticulation above the glumes and between the florets. Glumes about equal and as long as or longer than the floret, 1–9-veined. Florets dorsally compressed. Lemma firm and becoming indurate, with a straight, persistent awn, 3–7-veined, margins flat. Palea not covered by the lemma, as long as the lemma, with 2 veins. Basic chromosome number, $x = 11$. Photosynthetic pathway, C_3. Represented in Texas by a single species.

Barkworth (2007j)

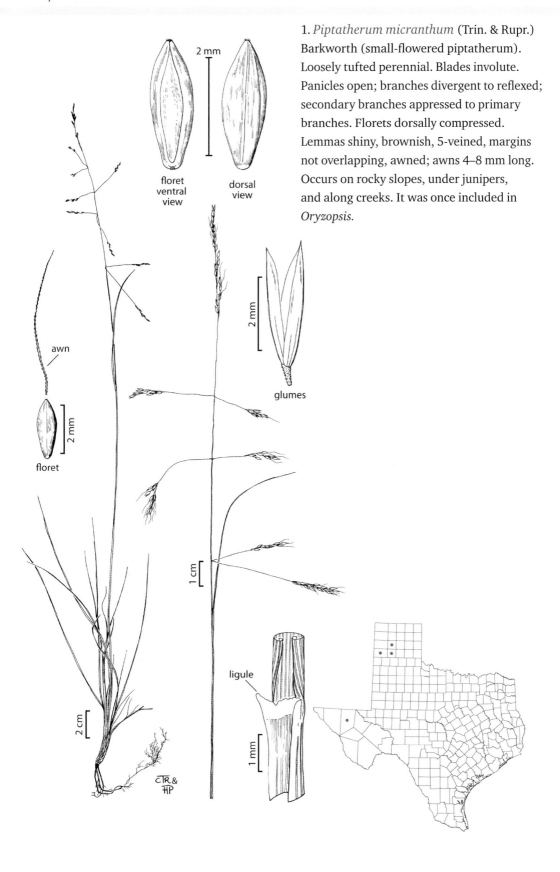

1. *Piptatherum micranthum* (Trin. & Rupr.) Barkworth (small-flowered piptatherum). Loosely tufted perennial. Blades involute. Panicles open; branches divergent to reflexed; secondary branches appressed to primary branches. Florets dorsally compressed. Lemmas shiny, brownish, 5-veined, margins not overlapping, awned; awns 4–8 mm long. Occurs on rocky slopes, under junipers, and along creeks. It was once included in *Oryzopsis*.

2 mm

floret
ventral
view

dorsal
view

awn

2 mm

floret

2 mm

glumes

1 cm

ligule

1 mm

Piptochaetium J. Presl
(Poöideae: Stipeae)

Perennials, cespitose. Culms 4–150 cm long, usually erect, occasionally decumbent, glabrous, not branched above the base. Sheaths open, margins glabrous; auricles absent; ligules membranous, decurrent, truncate to acute, sometimes highest at the sides. Blades convolute or flat, translucent between the veins. Inflorescence a terminal panicle, open or contracted, spikelets typically found on the upper ½ of branches. Spikelets with 1 perfect floret. Rachilla not prolonged beyond the base of the floret. Disarticulation above the glumes, beneath the floret. Glumes subequal, longer than the floret, lanceolate, 3–7-veined. Florets globose to fusiform, terete to slightly laterally compressed. Calluses well developed, sharp or blunt, glabrous, or with yellow to brown antrorsely strigose hairs. Lemmas coriaceous to indurate, glabrous or pubescent, often smooth distally and papillate or tuberculate basally, margins involute, fitting into the grooved paleas, apices fused into a crown, awned. Awns usually twice geniculate. Paleas longer than the lemmas, sulcate between the veins. Anthers 3. Styles 2. Caryopses terete, globose, or lenticular. Basic chromosome numbers, $x = 7$ and 11. Photosynthetic pathway, C_3. Represented in Texas by 3 species. A segregate of *Stipa*.

1. Plants of East and Central Texas .... 1. *P. avenaceum*
1. Plants of the Trans-Pecos region of West Texas.
    2. Calluses sharp-pointed; glumes about 5 mm long; blades rolled and threadlike, elongate,
        drooping .... 2. *P. fimbriatum*
    2. Calluses blunt; glumes about 10 mm long; blades flat or loosely rolled, firm and somewhat
        erect .... 3. *P. pringlei*

(Allred 2005; Barkworth 2007k)

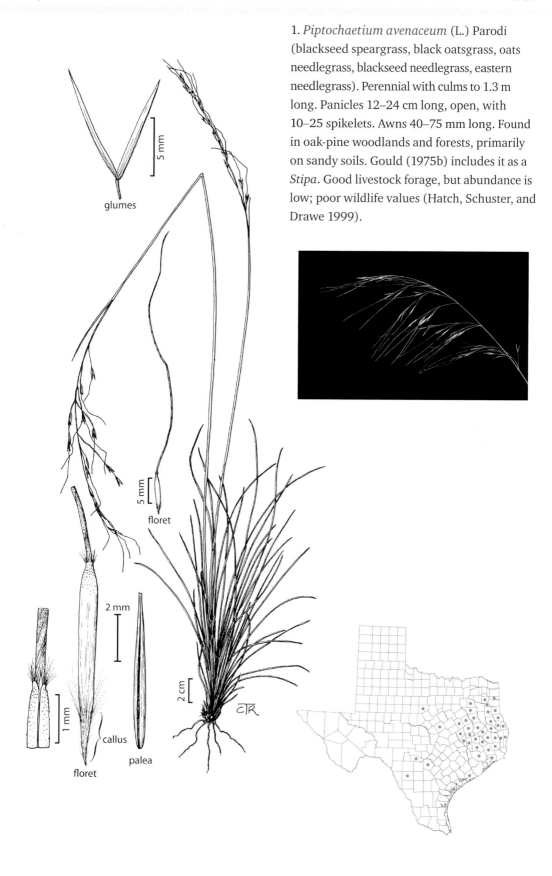

glumes

5 mm

floret

5 mm

callus

palea

floret

1 mm

2 mm

2 cm

CTR

1. *Piptochaetium avenaceum* (L.) Parodi (blackseed speargrass, black oatsgrass, oats needlegrass, blackseed needlegrass, eastern needlegrass). Perennial with culms to 1.3 m long. Panicles 12–24 cm long, open, with 10–25 spikelets. Awns 40–75 mm long. Found in oak-pine woodlands and forests, primarily on sandy soils. Gould (1975b) includes it as a *Stipa*. Good livestock forage, but abundance is low; poor wildlife values (Hatch, Schuster, and Drawe 1999).

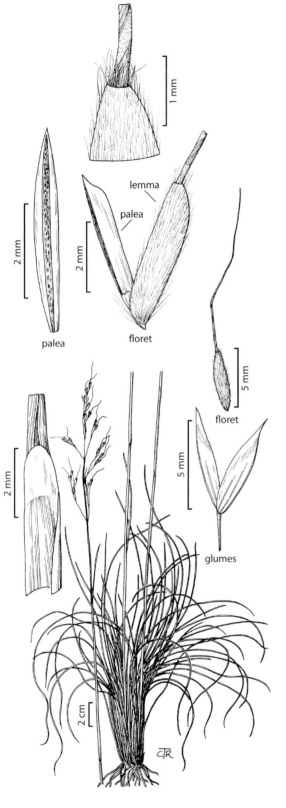

lemma

palea

floret

palea

floret

glumes

2. *Piptochaetium fimbriatum* (Kunth) Hitchc. (pinyon ricegrass). Perennial with culms to about 1 m long. Panicles 6–25 cm long, open, partially enclosed in subtending sheaths, 20–60 spikelets. Awns 10–20 mm long. Found in oak-pinyon woodlands. Powell (1994) reported this species as highly palatable to livestock but not abundant enough to be of much importance.

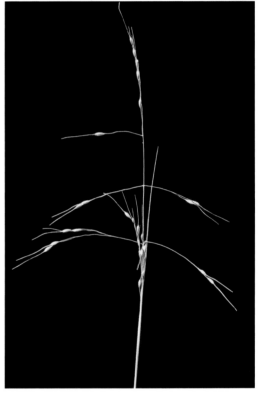

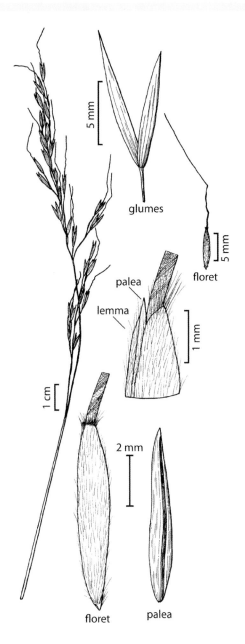

glumes

floret

palea

lemma

floret

palea

ligule

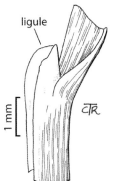

3. *Piptochaetium pringlei* (Beal) Parodi (Pringle's speargrass, Pringle's needlegrass). Perennial with mostly glabrous culms to 1.2 m tall. Panicles 6–20 cm long, open, with 10–25 spikelets. Awns 15–35 mm long. Found in oak woodlands on rocky soils. Gould (1975b) included it as a *Stipa*. Considered good forage for grazing animals where locally abundant (Powell 1994).

Pleuraphis Torr.
(Chloridoideae: Cynodonteae)

Perennials, mostly rhizomatous or occasionally stoloniferous. Sheaths open; ligule a short membrane; auricles absent. Inflorescence a 2-sided spike or spikelike panicle, the spikelets in clusters of 3 at each node of a zigzag rachis, the clusters deciduous as a whole. Disarticulation at the base of the branches, leaving the zigzag rachis. Spikelets of the cluster dissimilar. The 2 lateral spikelets of each branch short-pedicellate with 1–5 staminate or sterile florets. Glumes nearly as long as the florets, deeply cleft into 2 or more lobes, with 1 or more dorsal awns. The central spikelet sessile with 1 fertile floret. Glumes firm, flat, indurate, and usually asymmetrical, bearing an awn on one side from about the middle. Lemmas thin, 3-nerved, awned or awnless. Palea about as large as the lemma and similar in texture. Basic chromosome number, $x = 9$. Photosynthetic pathway, C_4. Represented in Texas by 2 species. A segregate of *Hilaria*.

1. Glumes fan-shaped, broadest at the apex .... 2. *P. mutica*
1. Glumes lanceolate, broadest at the middle .... 1. *P. jamesii*

(Allred 2005)

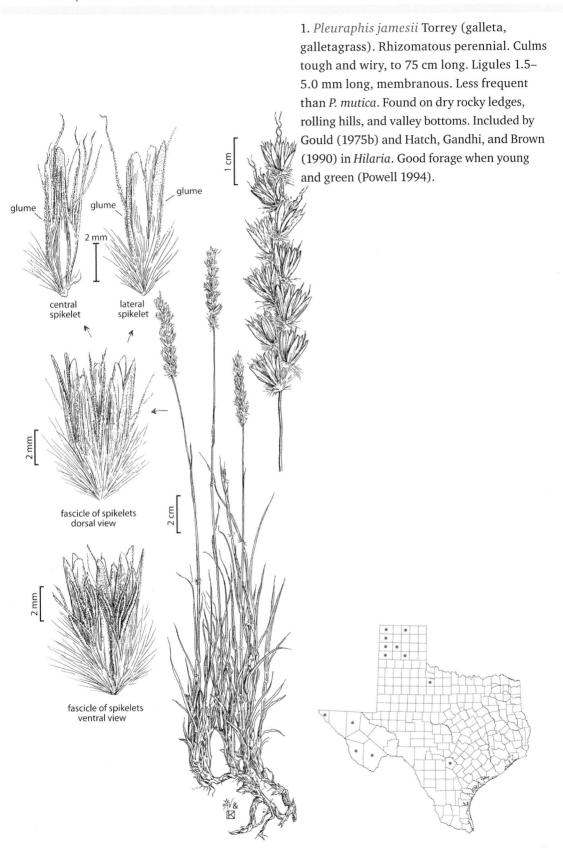

1. *Pleuraphis jamesii* Torrey (galleta, galletagrass). Rhizomatous perennial. Culms tough and wiry, to 75 cm long. Ligules 1.5–5.0 mm long, membranous. Less frequent than *P. mutica*. Found on dry rocky ledges, rolling hills, and valley bottoms. Included by Gould (1975b) and Hatch, Gandhi, and Brown (1990) in *Hilaria*. Good forage when young and green (Powell 1994).

glume

glume

glume

2 mm

central spikelet

lateral spikelet

1 cm

2 mm

fascicle of spikelets
dorsal view

2 cm

2 mm

fascicle of spikelets
ventral view

central spikelet

2 mm

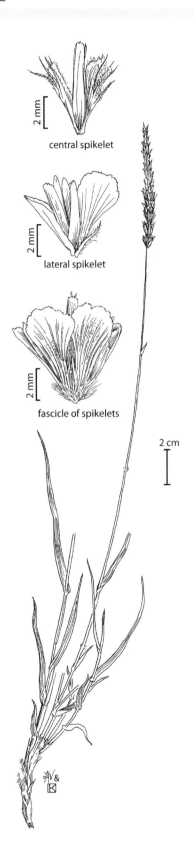

lateral spikelet

2 mm

fascicle of spikelets

2 mm

2 cm

2. *Pleuraphis mutica* (Buckley) Benth. (tobosa, tobosagrass). Perennial. Culms tough, erect, to 60 cm long. Ligules 0.5–2.0 mm long. More common than P. *jamesii*. Found in level upland areas and in valley bottoms that are occasionally flooded. Included by Gould (1975b) and Hatch, Gandhi, and Brown (1990) in *Hilaria*. Fair forage for cattle and horses when it is young and green (Powell 1994).

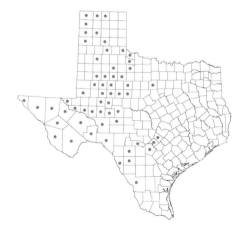

Poa L.
(Poöideae: Poeae)

Low to moderately tall annuals and perennials, many with rhizomes. Blades mostly flat or folded, with boat-shaped tips. Inflorescence an open or contracted panicle or occasionally reduced to a raceme. Spikelets mostly small, 2–7-flowered, awnless. Disarticulation above the glumes and between the florets. Glumes relatively broad, 1–3-nerved. Lemmas thin, broad, usually keeled and with a membranous border, obtuse or broadly acute at the apex, 5-nerved, the veins often puberulent below. Base of the lemma glabrous or with long, kinky, cottony hairs. Palea glabrous. Basic chromosome number, $x = 7$. Photosynthetic pathway, C_3. Represented in Texas by 16 species and 1 subspecies.

1. Culms with bulbous bases; florets modified and forming small, leafy plantlets.......... 6. *P. bulbosa*
1. Culms without bulbous bases; florets not modified into small, leafy plantlets.
    2. Anthers 1 mm long or less, nearly all well developed.
        3. Plants annual.
            4. Panicle branches erect; inflorescences narrow, contracted; sheaths scabrous...5. *P. bigelovii*
            4. Panicle branches, at least the lowermost, spreading; inflorescence not narrow; sheaths glabrous.
                5. Lemmas with long, kinky hairs at base and with 3 strong veins and 2 faint veins; rare..7. *P. chapmaniana*
                5. Lemmas without long, kinky hairs at base and with 5 strong veins; common ....1. *P. annua*
        3. Plants perennial.
            6. Sheaths densely scabrous with retrorsely pointed hairs, rarely glabrous; panicles 13–40 cm long, rachis internodes >4 cm long.......12. *P. occidentalis*
            6. Sheaths glabrous to sparsely with retrorsely pointed hairs; panicles mostly <12 cm long, rachis internodes <3.5 cm long.............................. 11. *P. leptocoma*
    2. Anthers >1 mm long, or vestigial and rudimentary.
        7. Culms and nodes strongly compressed; plants strongly rhizomatous.
            8. Lemmas 5–6 mm long; plants dioecious ....2. *P. arachnifera*
            8. Lemmas 2–3 mm long; plants hermaphroditic...............8. *P. compressa*
        7. Culms and nodes terete; plants cespitose or rhizomatous.
            9. Floret base with a tuft of cobwebby hairs ....Key A
            9. Floret base without a tuft of cobwebby hairs ....Key B

Key A

1. Plants dioecious; panicle branches densely flowered ....2. *P. arachnifera*
1. Plants hermaphroditic; panicle branches sparsely flowered.
    2. Plants with strong rhizomes; panicle branches glabrous to scabrous, terete...... 13. *P. pratensis*
    2. Plants without rhizomes; panicle branches distinctly scabrous, angled.
        3. Ligules 3–10 mm long; lower glume narrow, sickle-shaped .... 16. *P. trivialis*
        3. Ligules 0.5–2.0 mm long; lower glume broad, not sickle-shaped.
            4. Lower panicle branches widely spreading at maturity; plants of East and Central Texas.. 15. *P. sylvestris*
            4. Lower panicle branches erect or somewhat spreading at maturity; plants of West Texas .... 10. *P. interior*

Key B

1. Plants dioecious; uppermost culm blade very reduced .... 9. *P. fendleriana*
1. Plants hermaphroditic; upper culm blade not greatly reduced.
    2. Plants rhizomatous... 3. *P. arida*
    2. Plants cespitose.
        3. Palea keels softly puberulent to short-villous for much of their length; anthers 1.0–1.4 mm long .... 4. *P. autumnalis*
        3. Palea keels scabrous; anthers to 2.5 mm long
            4. Basal branching almost all intravaginal .... 14. *P. strictiramea*
            4. Basal branching almost all extravaginal... 10. *P. interior*

(Allred 2005; Gould 1975b; Soreng 2007)

1. *Poa annua* L. (annual bluegrass, dwarf meadowgrass, low speargrass). Annual, rarely extending for a second growing season, roots at the lower nodes. Culms to about 30 cm long. Panicles ascending, spreading or reflexed. A native of Eurasia. One of the first plants to flower in the fall (October and November) and continues until late spring (May). A common weed of lawns and lawn borders. Poor wildlife value and low forage producer (Hatch, Schuster, and Drawe 1999; Powell 1994).

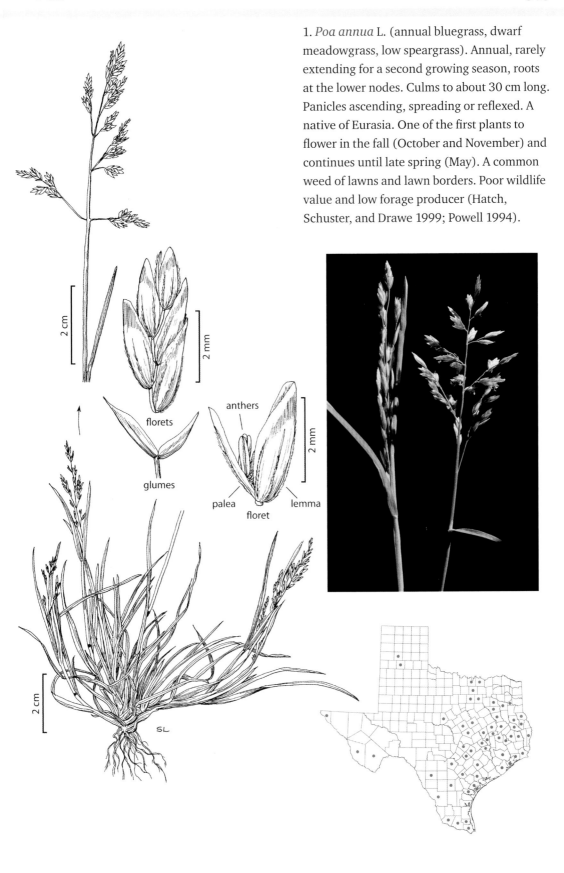

florets

glumes

anthers

palea   lemma
floret

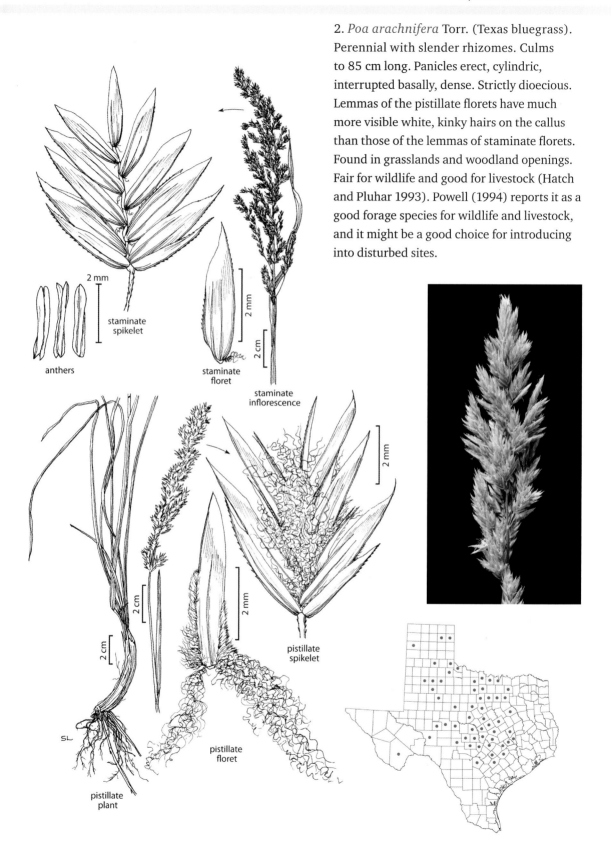

2. *Poa arachnifera* Torr. (Texas bluegrass).
Perennial with slender rhizomes. Culms
to 85 cm long. Panicles erect, cylindric,
interrupted basally, dense. Strictly dioecious.
Lemmas of the pistillate florets have much
more visible white, kinky hairs on the callus
than those of the lemmas of staminate florets.
Found in grasslands and woodland openings.
Fair for wildlife and good for livestock (Hatch
and Pluhar 1993). Powell (1994) reports it as a
good forage species for wildlife and livestock,
and it might be a good choice for introducing
into disturbed sites.

2 mm

staminate
spikelet

anthers

2 mm

2 cm

staminate
floret

staminate
inflorescence

2 mm

pistillate
spikelet

2 cm

2 cm

2 mm

SL

pistillate
floret

pistillate
plant

palea

lemma

spikelet

SL

3. *Poa arida* Vasey (plains bluegrass).
Rhizomatous perennial. Culms to 80 cm
long, erect or decumbent, sometimes weakly
compressed. Panicles about 12 cm long, erect,
narrow, contracted, interrupted basally.
Calluses usually glabrous, occasionally short-
webbed. Found primarily in riparian areas of
varying salinity and/or alkalinity.

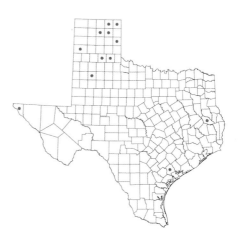

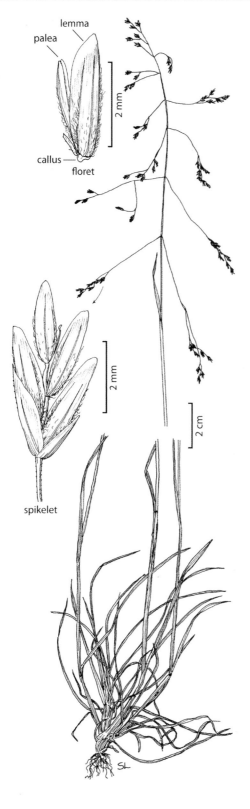

lemma

palea

2 mm

callus

floret

spikelet

2 mm

2 cm

4. *Poa autumnalis* Muhl. *ex* Elliott (autumn bluegrass, flexuous speargrass). Cespitose perennial. Culms to 90 cm long, often decumbent. Panicles broadly pyramidal at maturity, branches spreading to reflexed at maturity. Found in pine or mixed pine-hardwood forests. Contrary to its vernacular name, it flowers from March until May. Fair spring forage (Hatch, Schuster, and Drawe 1999).

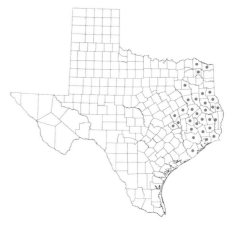

5. *Poa bigelovii* Vasey and Scribn. (Bigelow's bluegrass). Annual. Culms to 60 cm long, erect. Panicles erect, contracted, sometimes interrupted at the base. Found in arid upland areas on shady, rocky slopes. More frequent in the western portion of the state. One of the most widespread bluegrass species in the Trans-Pecos region, and it could be of some limited winter forage value (Powell 1994).

lemma

palea

floret

spikelet

SL

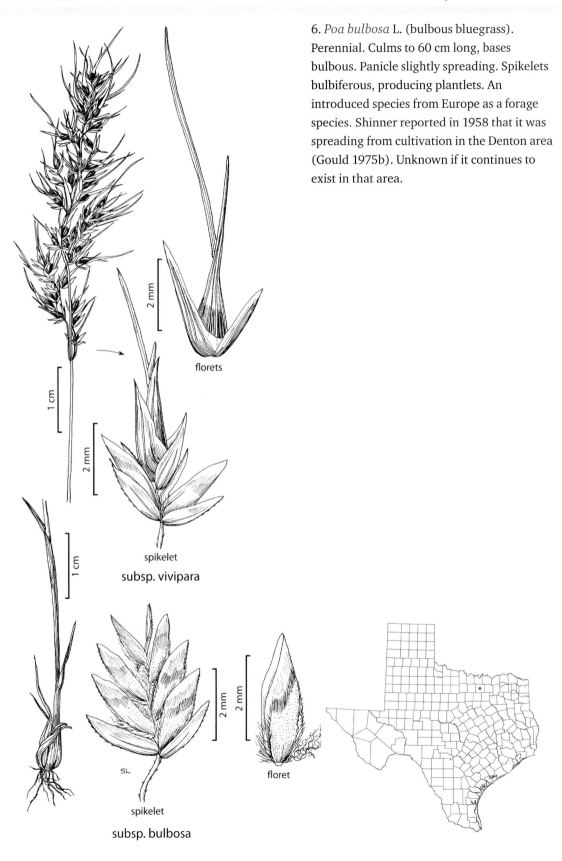

6. *Poa bulbosa* L. (bulbous bluegrass). Perennial. Culms to 60 cm long, bases bulbous. Panicle slightly spreading. Spikelets bulbiferous, producing plantlets. An introduced species from Europe as a forage species. Shinner reported in 1958 that it was spreading from cultivation in the Denton area (Gould 1975b). Unknown if it continues to exist in that area.

florets

2 mm

1 cm

2 mm

spikelet

subsp. vivipara

1 cm

SL

spikelet

subsp. bulbosa

2 mm

2 mm

floret

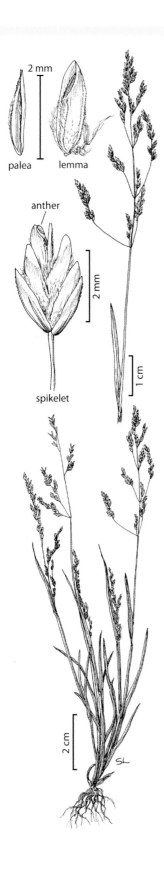

palea   lemma

anther

spikelet

7. *Poa chapmaniana* Scribn. (Chapman's bluegrass, Chapman's poa). Tufted annual. Culms to only about 40 cm long, erect or geniculate. Panicles open, moderately to densely congested. Similar to *P. annua* but with cobwebby hairs at base of lemmas. Diggs et al. (1999) suggested that florets do not open during flowering and that most seed production is cleistogamous. Found in woodlands and disturbed sites. Infrequent.

8. *Poa compressa* L. (Canada bluegrass).
Rhizomatous perennial. Culms to 60 cm long,
wiry. Culms, nodes, and sheaths strongly
compressed. Panicles usually <10 cm long,
erect, narrow. Branches 1–3 per node, erect to
sometimes spreading. Gould (1975b) reported
that this species was probably seeded as
forage and does not exist outside cultivation.
However, the scattered occurrence in the state
suggests that it probably is established.

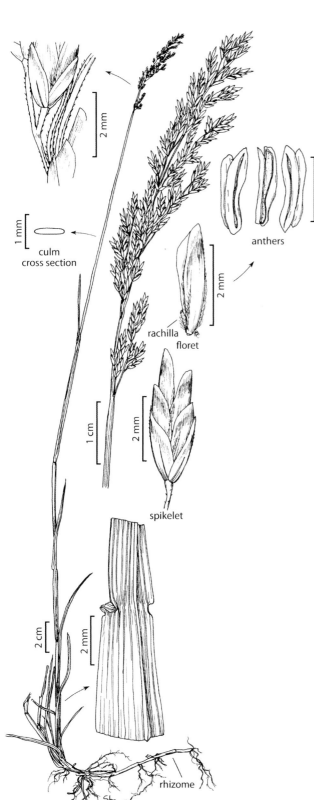

culm
cross section

anthers

rachilla
floret

spikelet

1 cm

2 mm

2 cm

2 mm

rhizome

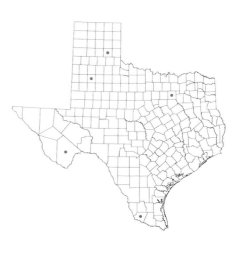

9. *Poa fendleriana* (Steud.) Vasey (Fendler's muttongrass, Vasey's muttongrass, muttongrass, mutton bluegrass). Perennial with inconspicuous rhizomes, dioecious. Culms to 70 cm long, erect or occasionally decumbent. Panicles erect, contracted, congested. Found on rocky mountain slopes and in canyons. Most frequent in the central mountains of the Trans-Pecos, where it provides forage for mostly wildlife (Powell 1994). The Texas species belongs to the typical subspecies, subsp. *fendleriana*.

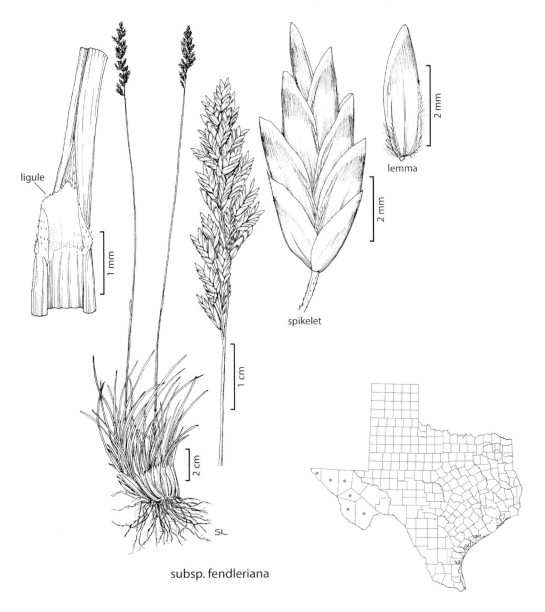

ligule

1 mm

spikelet

lemma

2 mm

2 mm

1 cm

2 cm

SL

subsp. fendleriana

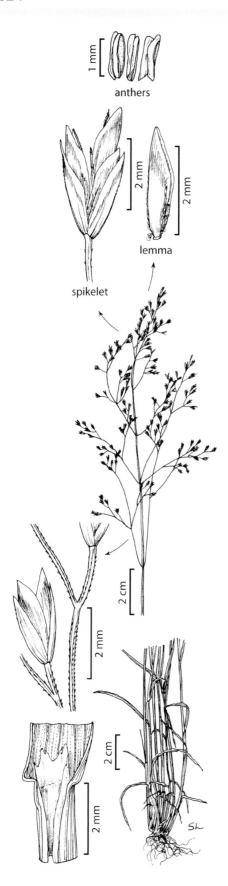

anthers

spikelet

lemma

10. *Poa interior* Rydb. (interior bluegrass, inland bluegrass). Cespitose perennial. Culms to 80 cm long, slender, erect to ascending. Panicles to about 15 cm long, narrow and dense. Branches usually erect or erect-spreading. Calluses usually webbed. Found in dry to moist habitats at higher altitudes. Gould (1975b) had not seen specimens, but the species was reported from the High Plains and Trans-Pecos regions of the state. Soreng (2007) does not show this species in Texas. Most likely at higher elevations of the Guadalupe and perhaps Chisos mountains.

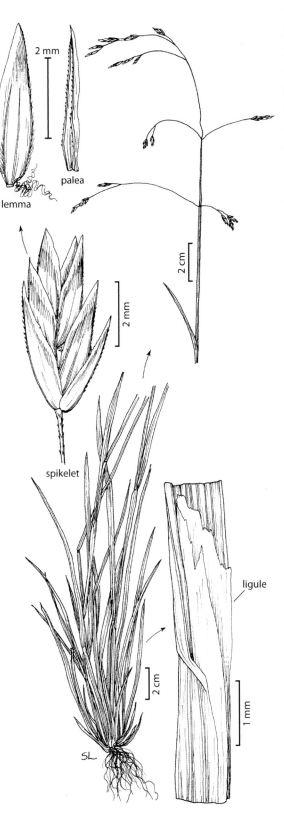

2 mm

palea

lemma

2 cm

2 mm

spikelet

ligule

2 cm

1 mm

SL

11. *Poa leptocoma* Trin. (western bog bluegrass). Perennial, rarely with slender rhizomes. Culms to 1 m long, erect to decumbent. Panicles 5–15 cm long, open, sparse. Branches spreading and sometimes reflexed. Calluses sparsely webbed. Found around lakes and ponds and along streams. According to Soreng, as reported by Jones, Wipff, and Montgomery (1997), this is the species most likely to occur within the state rather than the closely related species *P. reflexa* Vasey and Scribn.

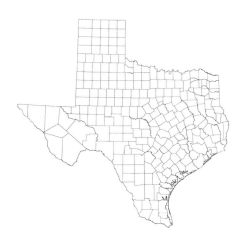

12. *Poa occidentalis* Vasey (New Mexico bluegrass). Cespitose perennial. Culms to 1.2 m long, compressed-keeled. Panicles open, nodding; branches spreading and drooping. Not included in Gould (1975b). As the vernacular name suggests, this species is to be expected along the border with New Mexico.

palea

lemma

florets

2 mm

lower glume

upper glume

spikelet

2 mm

2 cm

blade

2 mm

sheath

ligule

2 mm

keel

2 cm

SL

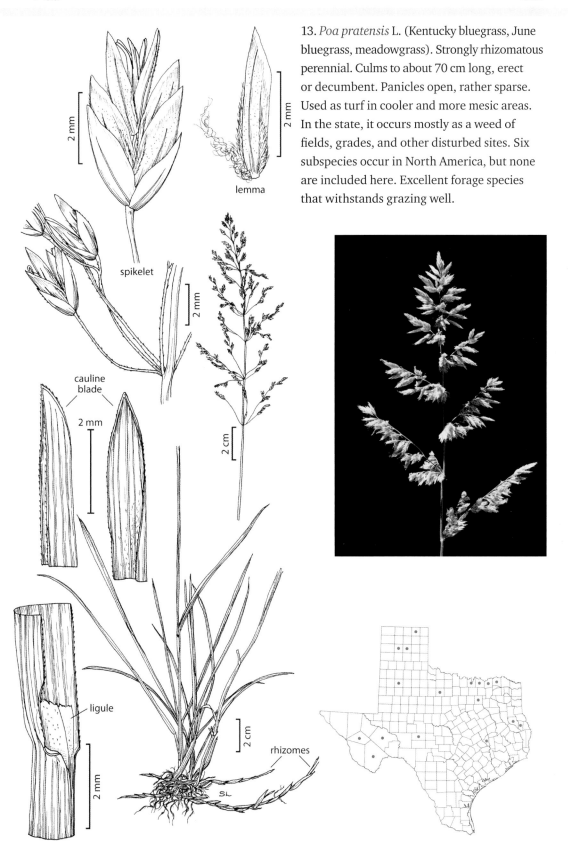

13. *Poa pratensis* L. (Kentucky bluegrass, June bluegrass, meadowgrass). Strongly rhizomatous perennial. Culms to about 70 cm long, erect or decumbent. Panicles open, rather sparse. Used as turf in cooler and more mesic areas. In the state, it occurs mostly as a weed of fields, grades, and other disturbed sites. Six subspecies occur in North America, but none are included here. Excellent forage species that withstands grazing well.

lemma

spikelet

cauline blade

2 mm

2 cm

ligule

rhizomes

2 mm

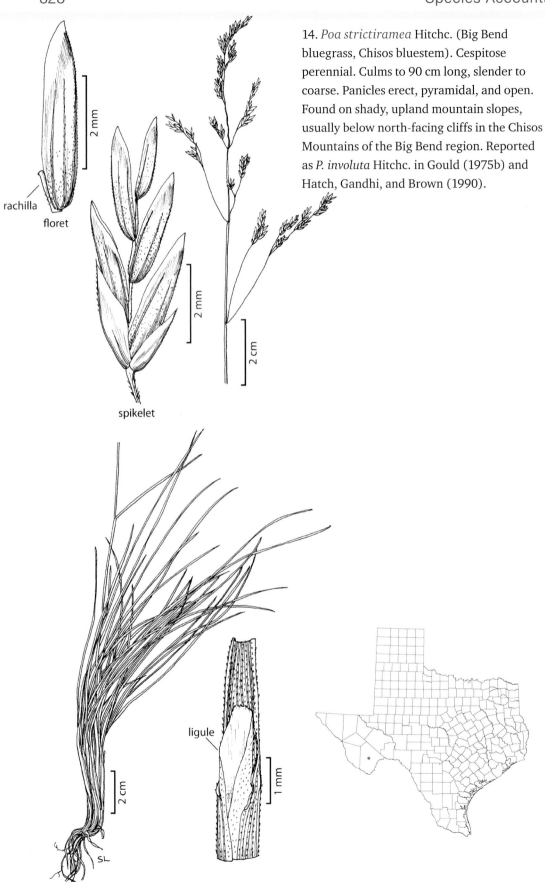

rachilla

floret

spikelet

ligule

14. *Poa strictiramea* Hitchc. (Big Bend bluegrass, Chisos bluestem). Cespitose perennial. Culms to 90 cm long, slender to coarse. Panicles erect, pyramidal, and open. Found on shady, upland mountain slopes, usually below north-facing cliffs in the Chisos Mountains of the Big Bend region. Reported as *P. involuta* Hitchc. in Gould (1975b) and Hatch, Gandhi, and Brown (1990).

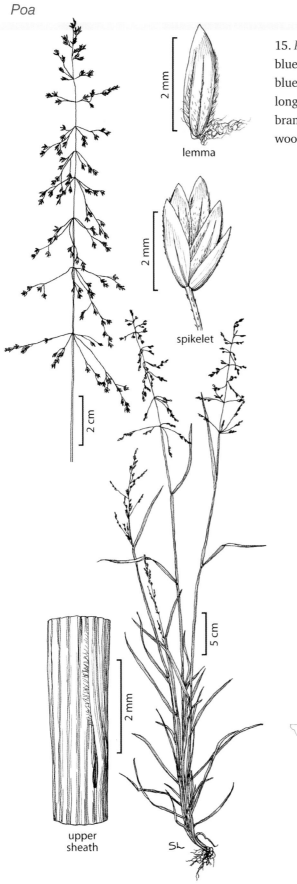

lemma

spikelet

2 mm

2 cm

5 cm

2 mm

upper
sheath

15. *Poa sylvestris* A. Gray (woodland bluegrass, sylvan's speargrass, sylvan bluegrass). Tufted perennial. Culms to 1.2 m long, often decumbent. Panicle open, lower branches reflexed at maturity. Found in moist woodland sites and shaded low areas.

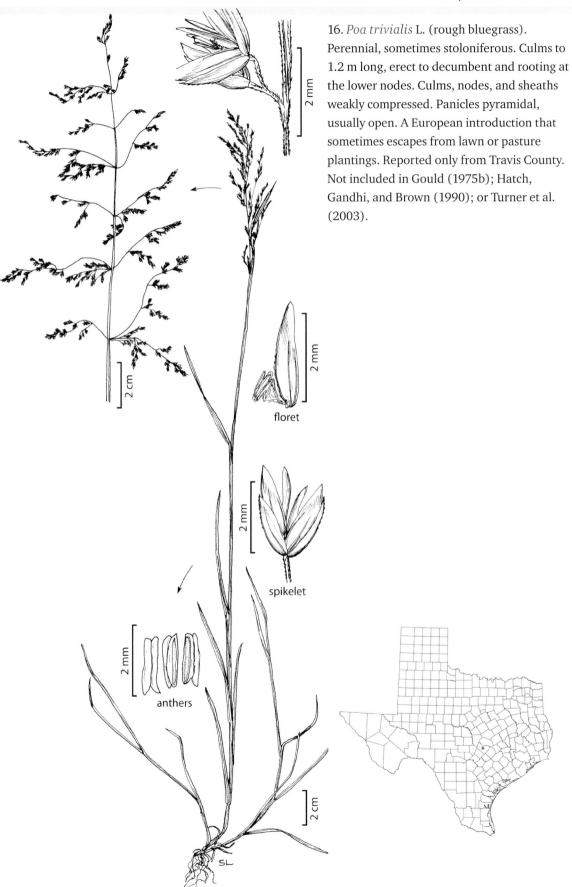

2 mm

2 mm

floret

2 mm

spikelet

2 mm

anthers

2 cm

2 cm

SL

16. *Poa trivialis* L. (rough bluegrass). Perennial, sometimes stoloniferous. Culms to 1.2 m long, erect to decumbent and rooting at the lower nodes. Culms, nodes, and sheaths weakly compressed. Panicles pyramidal, usually open. A European introduction that sometimes escapes from lawn or pasture plantings. Reported only from Travis County. Not included in Gould (1975b); Hatch, Gandhi, and Brown (1990); or Turner et al. (2003).

Polypogon Desf.
(Poöideae: Poeae)

Low to moderately tall annuals and perennials, with weak, decumbent-erect culms that often root at the lower nodes. Blades thin and flat. Inflorescence a dense, contracted panicle of small, 1-flowered spikelets that disarticulate below the glumes and fall entire. Glumes about equal, 1-nerved, abruptly awned from an entire or notched apex. Lemma broad, smooth, and shining, mostly 5-nerved, much shorter than the glumes, awnless or with a short, delicate awn from the broad, often minutely toothed apex. Palea slightly shorter than the lemma. Stamens 1–3. Basic chromosome number, $x = 7$. Photosynthetic pathway, C_3. Represented in Texas by 5 species.

1. Plants annual.
    2. Glumes deeply lobed; lemmas awnless or awn 1 mm or shorter .... 3. *P. maritimus*
    2. Glumes not deeply lobed; lemmas awned, awn to 0.5–4.5 mm long .... 4. *P. monospeliensis*
1. Plants perennial.
    3. Glumes unawned ....5. *P. viridis*
    3. Glumes awned, awns 0.2–3.2 mm long.
        4. Stipes 1.5–2.5 mm long; glumes tapering from about midlength to the acute unlobed apices ....1. *P. elongatus*
        4. Stipes 0.2–1.5 mm long; glumes not tapering to the apices, the apices usually rounded, sometimes acute, often lobed ....2. *P. interruptus*

(Barkworth 2007l)

1. *Polypogon elongatus* Kunth (southern beardgrass). Perennial but often flowering the first year. Culms to 1 m long, erect to decumbent. Ligules 4–8 mm long. Panicles erect to nodding, dense, interrupted. A Mexican species extending into the Big Bend area of the Trans-Pecos.

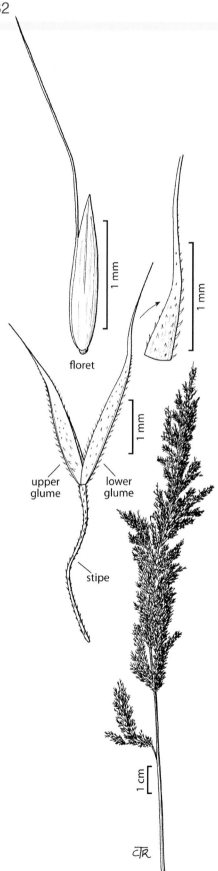

floret

1 mm

1 mm

upper glume

lower glume

1 mm

stipe

1 cm

CTR

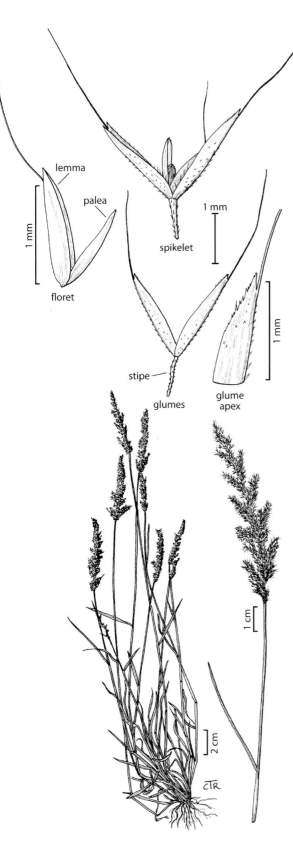

lemma

palea

1 mm

floret

spikelet

1 mm

stipe

glumes

glume apex

1 mm

2 cm

CTR

1 cm

2. *Polypogon interruptus* Kunth (ditch beardgrass). Perennial, often flowering the first year. Culms to 90 cm tall, decumbent. Ligules 2–6 mm long, membranous. Panicles interrupted, branches slightly spreading. Found in mesic sites where water periodically stands or accumulates. Uncommon. Allred (2005) reported that this is a male-sterile hybrid between *Agrostis stolonifera* and *P. monspeliensis*. Not listed in Gould (1975b); Hatch, Gandhi, and Brown (1990); Jones, Wipff, and Montgomery (1997); or Turner et al. (2003).

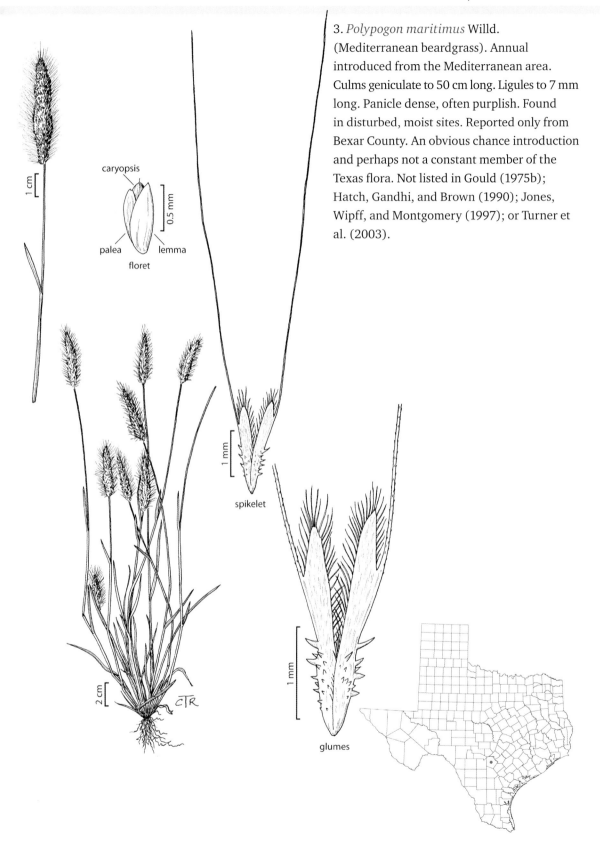

caryopsis

0.5 mm

palea   lemma

floret

1 cm

spikelet

1 mm

glumes

1 mm

2 cm   CTR

3. *Polypogon maritimus* Willd.
(Mediterranean beardgrass). Annual
introduced from the Mediterranean area.
Culms geniculate to 50 cm long. Ligules to 7 mm
long. Panicle dense, often purplish. Found
in disturbed, moist sites. Reported only from
Bexar County. An obvious chance introduction
and perhaps not a constant member of the
Texas flora. Not listed in Gould (1975b);
Hatch, Gandhi, and Brown (1990); Jones,
Wipff, and Montgomery (1997); or Turner et
al. (2003).

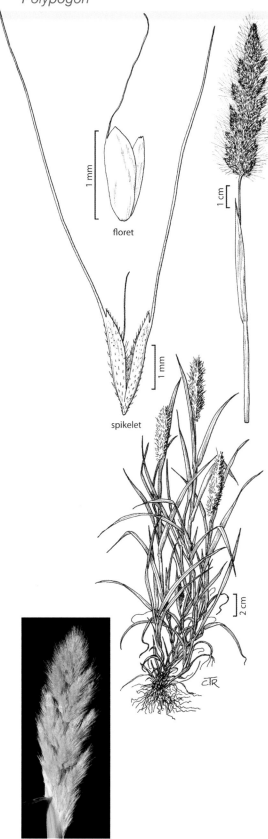

floret

spikelet

4. *Polypogon monspeliensis* (L.) Desf. (rabbitfoot grass, rabbitfoot polypogon, annual beardgrass, annual rabbitfoots grass, rabbit's-foot). Annual. Culms can grow to 1 m tall but are usually about half that height. Ligules 2.5–16 mm long. Panicles dense, greenish. An introduced species from southern Europe, now a widespread ruderal species. Likely to occur in any moist or wet disturbed sites. Particularly fond of alkaline soils. Poor livestock and wildlife values (Hatch, Schuster, and Drawe 1999). Powell (1994) reported it was probably palatable to livestock and wildlife but not abundant.

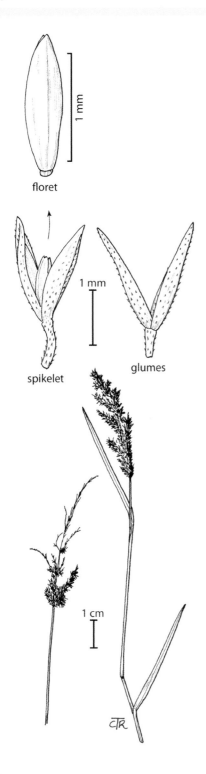

floret

spikelet

glumes

1 mm

1 mm

1 cm

CTR

5. *Polypogon viridis* (Gouan) Breistr. (water beardgrass, water bentgrass). Perennial, but usually flowering the first year. Culms to 90 cm tall, decumbent and rooting at the lower nodes. Ligules 2–5 mm long, membranous. Panicles dense, interrupted at base, greenish to purplish. A European introduction now found in mesic habitats associated with streams, rivers, and ditches. Included in Gould (1975b) and Hatch, Gandhi, and Brown (1990) as *Agrostis semiverticillata* (Forssk.) Christ.

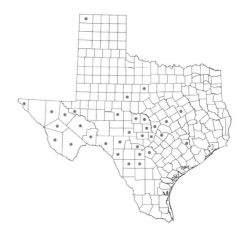

Psathyrostachys Nevski
(Poöideae: Triticeae)

Perennial, cespitose, or occasionally rhizomatous or stoloniferous. Basal sheaths closed, upper sheaths open. Inflorescence a bilateral spike. Disarticulation at the rachis nodes. Spikelets 2–3 per node, sessile, with 1–3 florets, often with reduced florets above. Glumes equal to subequal, side by side, subulate, free to the base, weakly 1-nerved. Lemmas narrowly elliptic, rounded, 5–7-nerved, veins usually prominent, apices acute to awned. Paleas equal or slightly longer than the lemma, membranous, bifid. Basic chromosome number, $x = 7$. Photosynthetic pathway, C_3. Represented in Texas by a single species.

(Baden 2007)

1. *Psathyrostachys juncea* (Fisch.) Nevski
(Russian wildrye). Densely cespitose
perennial. Culms erect and up to 1.2 m long.
Auricles 0.2–2.0 mm long; ligules 0.1–0.2 mm
long, membranous. Spikes to 20 cm long.
Rachises hirsute on margins, puberulent
elsewhere. Spikelets usually 3 per node, lateral
spikelets usually larger than central one.
Lemmas awnless or with awns up to 4 mm
long. Introduced from Asia as a forage species;
sometimes used for roadside restoration.
Escaped from cultivation and found along
roadsides and abandoned fields. Tolerant of
saline environments. Previously included in
Elymus. Not listed in Gould (1975b) or Hatch,
Gandhi, and Brown (1990).

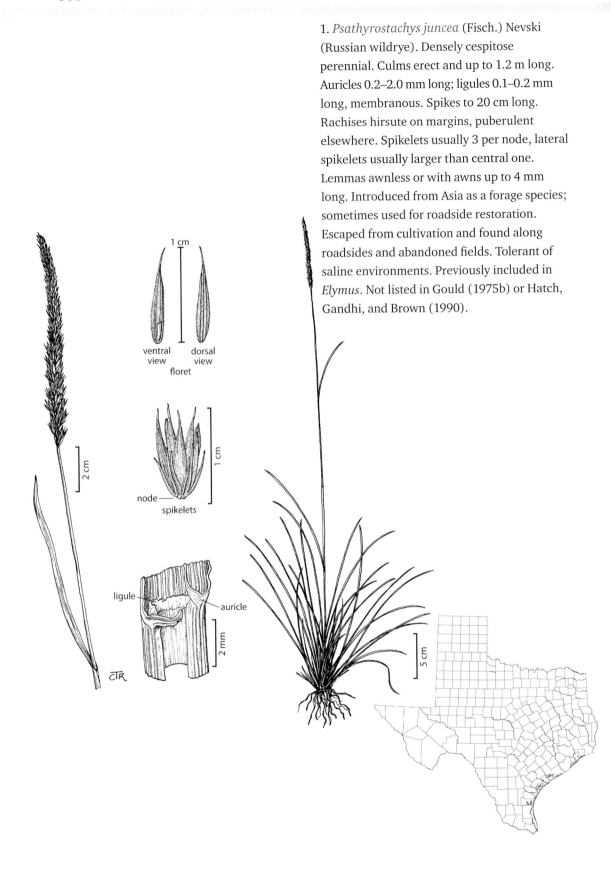

Pseudoroegneria (Nevski) Á. Löve
(Poöideae: Triticeae)

Cespitose perennial occasionally with short rhizomes. Sheaths open, auricle well developed, ligules membranous. Inflorescence a bilateral spike with widely spaced, solitary spikelets; rachis internodes concave adjacent to the spikelets. Spikelets with 3–9 florets, appressed; disarticulation above the glumes and beneath the florets. Glumes unequal, lanceolate or narrowly lanceolate, acuminate, glabrous, 3–7-nerved. Lemmas 5-nerved, usually glabrous, lanceolate, unawned or awned, awns strongly divergent at maturity; paleas equal to the lemmas. Basic chromosome number, $x = 7$. Photosynthetic pathway, C_3. Represented in Texas by a single species.

(Carlson 2007)

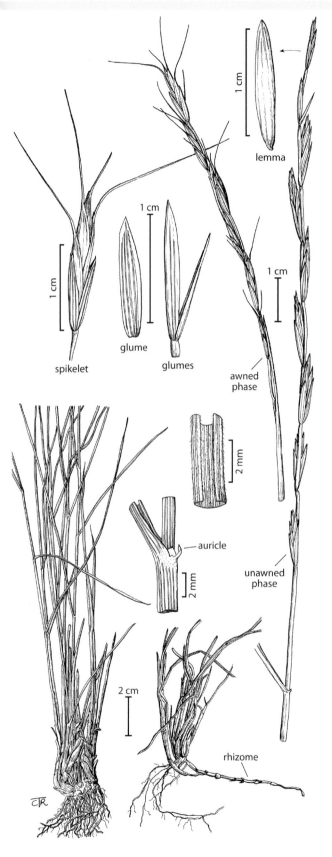

lemma

1 cm

glume

glumes

spikelet

awned phase

auricle

unawned phase

rhizome

1. *Pseudoroegneria spicata* (Pursh) Á. Löve (bluebunch wheatgrass). Cespitose perennial that is occasionally rhizomatous. Culms can grow to 1 m tall and are usually glaucous. Auricles prominent; ligules short, 0.1–0.4 mm long; blades involute when dry. Spike to 15 cm long. Spikelets with 4–9 florets. Glumes about ½ as long as spikelet. Lemmas awnless or with a strongly divergent awn to 3 cm long. This is primarily a western species that grows only in the far western counties of Texas. It prefers medium-textured soils and occurs in steppe, shrub-steppe, and open woodland areas. Rhizomatous plants are generally found in more mesic sites; cespitose individuals are in xeric sites. Previously included in *Agropyron*, *Elymus*, and *Leymus*. Gould listed it as *A. spicatum* Scribn. & Smith.

Redfieldia Vasey
(Chloridoideae: Cynodonteae)

Perennials with long creeping or vertical rhizomes. Sheaths open, longer than the internodes; ligules a fringe of hairs; auricles absent. Inflorescence terminal, an open panicle, $\frac{1}{3}$ to $\frac{1}{2}$ as long as the culms. Spikelets (1–)2–6-flowered, rachilla pronounced between florets, reduced floret (when present) above the fertile florets, on long flexuous pedicels, laterally compressed, grayish in color. Disarticulation above the glumes and between the florets. Glumes 2, unequal, lanceolate, 1-nerved, upper shorter than the lowermost lemma. Lemmas 3-nerved, veins forming 3 minute teeth, hairy, chartaceous, callus with a tuft of hairs. Palea glabrous, slightly shorter than the lemma. Basic chromosome number, $x = 10$. Photosynthetic pathway, C_4. Represented in Texas by the only species in the genus.

(Hatch 2003b)

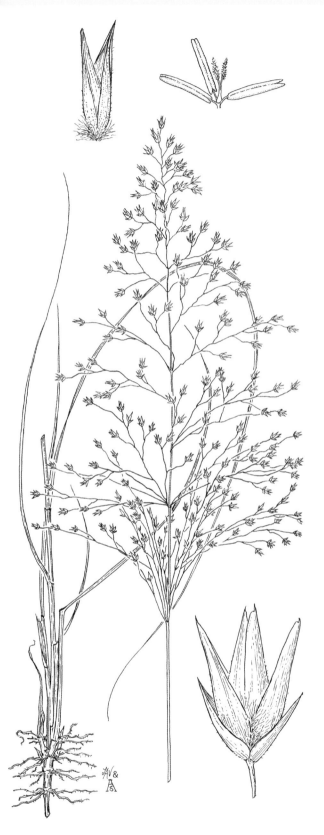

1. *Redfieldia flexuosa* (Thurb. *ex* A. Gray) Vasey (blowout grass). A strongly rhizomatous perennial that can grow to 1.3 m tall. Blades elongate and narrow. Panicles large and diffuse. Spikelets only slightly longer than wide. Lemmas tend to curve outward at the tip. As the vernacular name implies, the species is found on sand hills and dunes and helps bind the soils in such areas. Restricted to the upper Panhandle region of the state.

Rostraria Trin.
(Poöideae: Poeae)

Annual. Sheaths open, pubescent; auricles absent; ligules membranous; blades stiff, flat or involute. Inflorescence a panicle, spikelike, dense. Spikelets laterally compressed, 2–7 florets. Disarticulation above the glumes and between the florets. Lower glumes shorter than the first, 1-veined; upper glumes about equal to the lowest lemmas, 3-veined. Lemmas thin, membranous, apices bifid, awned from just below the teeth. Paleas about as long as lemmas, hyaline, veins sometimes awnlike. Basic chromosome number, $x = 7$. Photosynthetic pathway, C_3. Represented in Texas by a single species.

(Standley 2007c)

1. *Rostraria cristata* (L.) Trin. (annual koeleria, Mediterranean hairgrass, annual junegrass, cat-tail grass). Annual with culms up to 50 cm long. Panicles wide, cylindric, not interrupted at the base. Spikelets laterally flattened, 3–6 florets. Lower lemmas awned, awns 1–3 mm long; upper lemmas with shorter awn or awnless. An introduced European species. Included as *Koeleria gerardii* (Vill.) Skinner by Gould (1975b) and Hatch, Gandhi, and Brown (1990). Poor livestock and wildlife values (Hatch, Schuster, and Drawe 1999). No variety is recognized here.

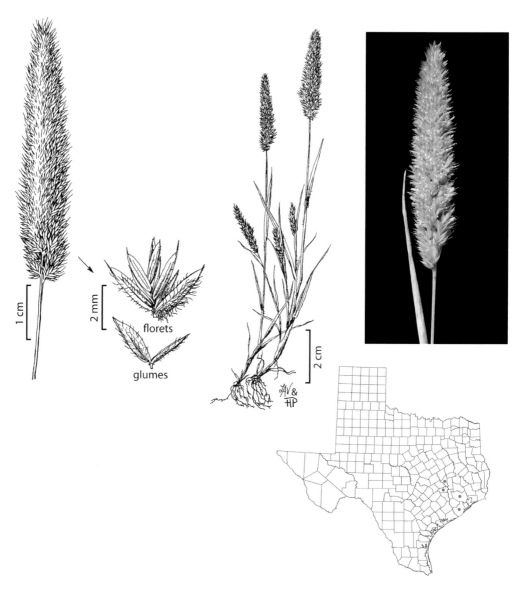

florets

glumes

Rottboellia L. F.
(Panicoideae: Andropogoneae)

Cespitose annuals. Culms 30–300 cm long, glabrous or sparsely pubescent below the nodes. Sheaths with or without papillose-based hairs. Auricles absent. Ligules membranous, ciliate. Blades flat. Inflorescences terminal and axillary, solitary rames with more than 1 spikelet unit; spikelets partially embedded in the rame axes. Disarticulation in the rame axes. Spikelets in pairs of 1 sessile and 1 pedicellate, unawned. Sessile spikelet with 2 florets; the upper florets perfect, lemma and palea hyaline; the lower floret staminate or sterile, anthers 3. Pedicels thick, fused to the rame axes. Pedicellate spikelet sterile or staminate. Caryopses with a hard endosperm. Basic chromosome numbers, $x = 9$ and 10. Photosynthetic pathway, C_4. Represented in Texas by a single species.

(Wipff 2003h)

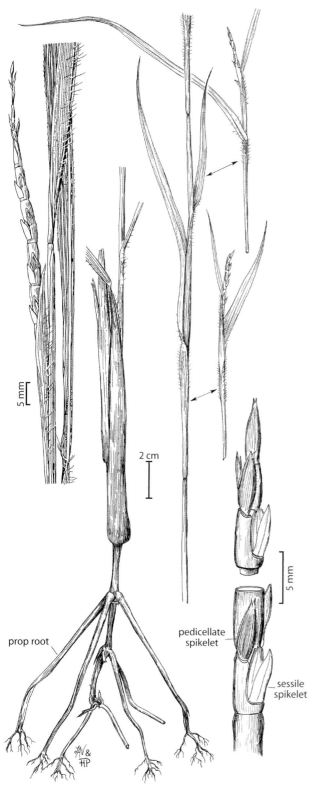

5 mm

2 cm

5 mm

prop root

pedicellate
spikelet

sessile
spikelet

1. *Rottboellia cochinchinensis* (Lour.) Clayton (itchgrass). Annual. Culms with prop roots from the lowermost nodes. Sheaths and blades with papillose-based hairs. An introduction from Southeast Asia that is considered one of the world's worst weeds. The U.S. Department of Agriculture classifies it as a noxious weed. The hairs from the leaves are a skin irritant. Not listed in Gould (1975b) or Hatch, Gandhi, and Brown (1990).

Saccharum L.
(Panicoideae: Andropogoneae)

Tall, coarse perennials, mostly with culms in dense clumps from short, stout rhizomes. Leaves large, the blades flat and broad. Inflorescence a large, densely flowered, plumose panicle as much as 60 cm in length. Spikelets all alike and perfect, in pairs of 1 sessile and 1 pediceled, with a tuft of long, silky hair at the base, disarticulating below the glumes or the sessile spikelet falling attached to the pedicel and a section of the rachis. Glumes large and firm. Lemma of the sterile floret and lemma and palea of the fertile floret hyaline, sometimes absent. Basic chromosome number, $x = 10$. Photosynthetic pathway, C_4. Represented in Texas by 8 species and 2 varieties. Species with awned lemmas are sometimes placed in a separate genus, *Erianthus*.

1. Spikelets unawned, or if awned, awns <5 mm long.
  2. Spikelets with visible awns 2–5 mm long .... 8. *S. ravennae*
  2. Spikelets unawned, or if awned, awns contained within glumes.
    3. Lower glumes of sessile spikelet pubescent .... 3. *S. bengalense*
    3. Lower glumes of sessile spikelet glabrous, or sometimes ciliate distally .... 7. *S. officinarum*
1. Spikelets awned, the awns 10–26 mm long.
      4. Awns spirally coiled at base.
        5. Callus hairs 3–7 mm long, shorter than the spikelet .... 4. *S. brevibarbe*
        5. Callus hairs 9–14 mm long, exceeding the spikelet .... 1. *S. alopecuroides*
      4. Awns straight to curved at base.
        6. Callus hairs longer than the spikelets; lower panicle nodes densely pilose .... 6. *S. giganteum*
        6. Callus hairs absent or not longer than the spikelets; lower panicle nodes glabrous or sparsely pilose.
        7. Calluses glabrous or with hairs to 2 mm long; panicles 1.0–2.5 cm wide .... 2. *S. baldwinii*
        7. Callus hairy, hairs 3–7 mm long; panicles 3–10 cm wide
          8. Awns flat basally; lower lemmas of sessile spikelets not or indistinctly veined .... 4. *S. brevibarbe*
          8. Awns terete basally; lower lemmas of sessile spikelets typically 3-veined .... 5. *S. coarctatum*

(Webster 2003)

1. *Saccharum alopecuroides* (L.) Nutt. (silver plumegrass). Rhizomatous perennial with culms to 2.5 m tall. Auricles absent. Ligules 1–3 mm long. Panicles 3–10 cm wide. Found in damp woods, open areas, and field margins. Included by Gould (1975b) and Hatch, Gandhi, and Brown (1997) in *Erianthus, E. alopecuroides* (L.) Ell. Specific epithet is sometimes erroneously spelled *alopecuroideum* (Turner et al. 2003).

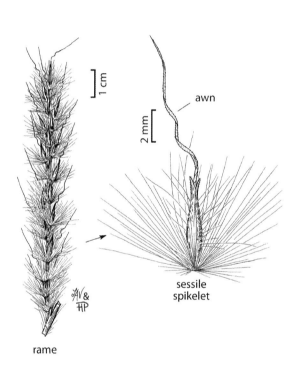

rame

awn

2 mm

1 cm

sessile spikelet

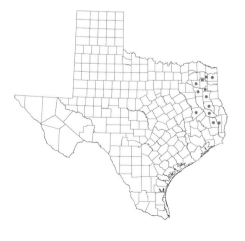

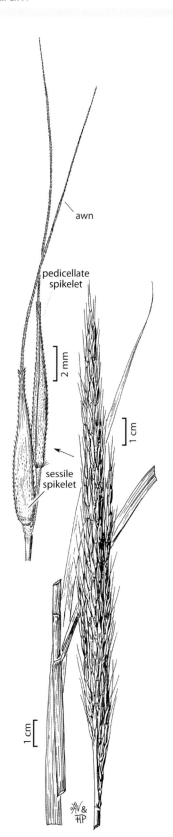

2. *Saccharum baldwinii* Spreng. (narrow plumegrass). Perennial. Plants rarely stoloniferous. Culms to 2 m long. Ligules 1–3 mm long with lateral lobes. Panicles 1.0–2.5 cm wide. Found along shaded river bottoms and the sandy shores of streams. Included in Gould (1975b) and Hatch, Gandhi, and Brown (1997) as *Erianthus strictus* Baldw.

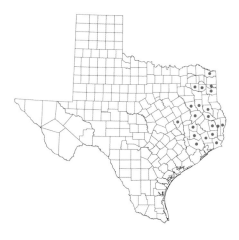

3. *Saccharum bengalense* Retz. (tall cane).
Nonrhizomatous perennial. Culms to 5 m long.
Auricles absent. Panicles to 90 cm long. An
introduced species used as an ornamental. Not
included in Gould (1975b); Hatch, Gandhi,
and Brown (1997); or Turner et al. (2003).

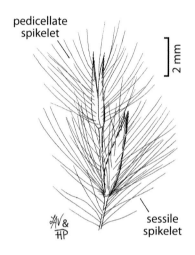

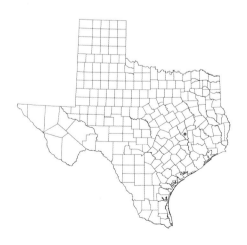

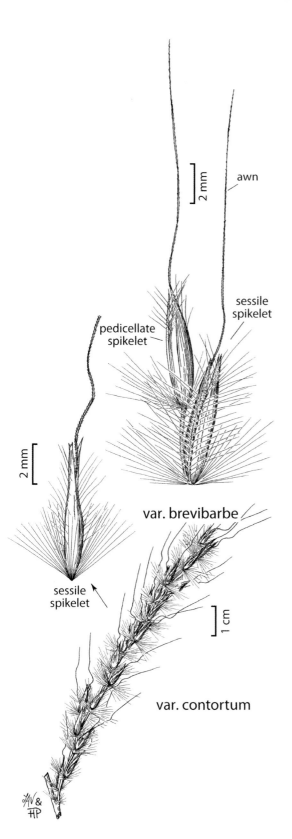

var. brevibarbe

var. contortum

4. *Saccharum brevibarbe* (Michx.) Pers. (shortbeard plumegrass, bent arm plumegrass). Perennial. Plants rhizomatous with culms to 2.5 m long. Auricles absent. Ligules 1–2 mm long. Panicles 3–10 cm wide. Found along lakeshores and stream banks and in swales and ditches. Palatable to cattle and horses, but poor wildlife values except for cover (Hatch, Schuster and Drawe 1999). Two varieties occur in the state: the typical one, var. *brevibarbe*; and var. *contortum* (Baldw.) R. D. Webster. The latter was included in Gould (1975b) and Hatch, Gandhi, and Brown (1997) as *Erianthus contortus* Baldw. The varieties can be distinguished by the following characters:

1. Awns 15–22 mm long, straight...................................var. *brevibarbe*
1. Awns 10–18, spirally coiled........var. *contortum*

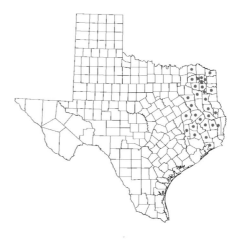

5. *Saccharum coarctatum* (Fernald) R. D. Webster (compressed plumegrass, bunched plumegrass). Perennial, with short rhizomes or without rhizomes. Culms to 2.5 m long. Auricles 0.3–3.0 mm long. Ligules 1–2 mm long. Panicles 3–7 cm wide. Found in wet areas, along stream banks and lakeshores, and in meadows of the Coastal Prairies. Not included in Gould (1975b) or Hatch, Gandhi, and Brown (1997).

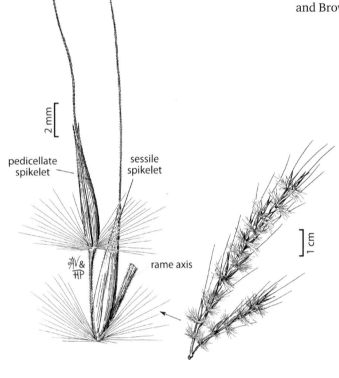

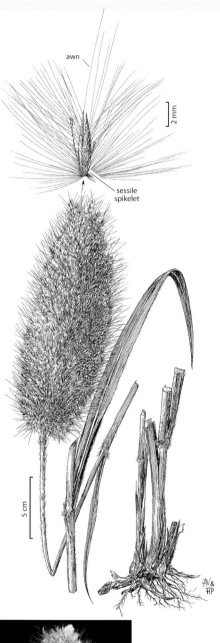

6. *Saccharum giganteum* (Walter) Pers. (sugarcane plumegrass, giant plumegrass). Rhizomatous perennial. Culms to 2.5 m long. Auricles absent. Ligules 2–6 mm long. Panicles to 15 cm wide. A native species of wet soils, bogs, swamps, and swales. Long callus hairs and straight awns are distinctive. Included by Gould (1975b) and Hatch, Gandhi, and Brown (1997) in *Erianthus, E. giganteus* (Walt.) Muhl.

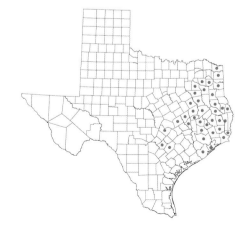

7. *Saccharum officinarum* L. (sugarcane). Perennial. Plants with short rhizomes. Culms to 6 m tall, 2–5 cm thick. Auricles present. Ligules 2–3 mm long. Panicles to 20 cm wide. A native of Asia and the Pacific, now cultivated for sugar in the Rio Grande Valley of South Texas. It is being used frequently as an ornamental. Species provides about half the world's sugar supply, and it also is used for molasses, syrup, rum, and ethanol.

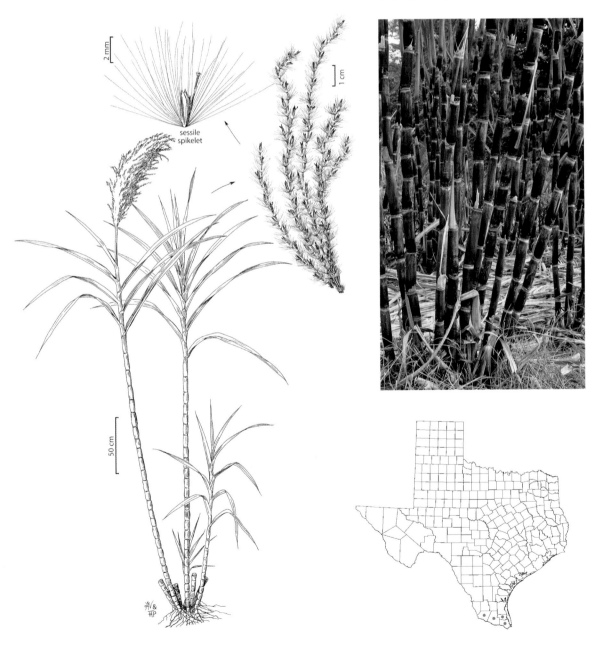

8. *Saccharum ravennae* (L.) L. (ravennagrass). Plants cespitose perennials, nonrhizomatous. Culms to 4 m long. Auricles absent. Ligules 0.5–1.1 mm long. Panicles to 70 cm long. A native of southern Europe that has been introduced as an ornamental. Not included in Gould (1975b); Hatch, Gandhi, and Brown (1997); or Turner et al. (2003).

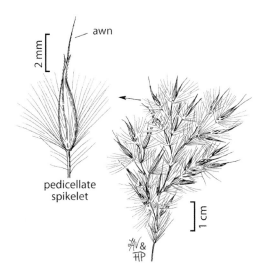

Sacciolepis Nash
(Panicoideae: Paniceae)

Medium-sized to tall annuals and perennials, with rather weak culms and thin, flat blades. Inflorescence a contracted and slender, usually long and spikelike, panicle. Spikelets slender, broad and gibbous at the base, disarticulating below the glumes. First glume short, acute, usually much narrower than the spikelet. Second glume broad, inflated-saccate at the base, strongly several-nerved, awnless. Lemma of the sterile floret narrow, usually 3- or 5-nerved, awnless, about as long as the second glume. Lemma and palea of the fertile floret smooth, indurate, rounded at the apex, much shorter than the second glume and lemma of the sterile floret. Basic chromosome number, $x = 9$. Photosynthetic pathway, C_3. Represented in Texas by 2 species.

1. Primary branches fused to the rachises for at least ¾ their length; upper glumes 9-veined .... 1. *S. indica*
1. Primary branches ascending but free from the rachises; upper glumes 11–12-veined .... 2. *S. striata*

(Wipff 2003i)

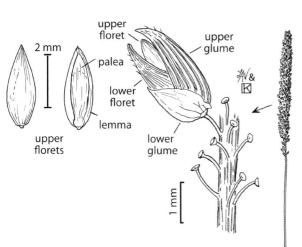

upper floret

upper glume

palea

lower floret

lower glume

lemma

upper florets

2 mm

1 mm

1. *Sacciolepis indica* (L.) Chase (India cupscale, glenwood grass, Chase's Glenwoodgrass). Annual. Culms to 1 m long, trailing, often rooting at the lower nodes. Ligules <1 mm long. Spikelets green to dark purple. Lower paleas ½ or less as long as lower lemmas. An introduced species now found in or around streams, ponds, lakes, and ditches. Gould (1975b) reported it only from Newton County, but it appears to be moving westward and is now found in Tyler and Liberty counties.

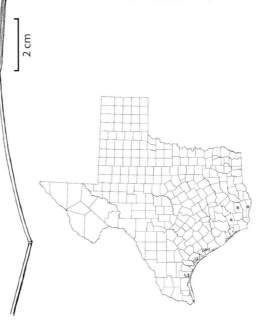

2 cm

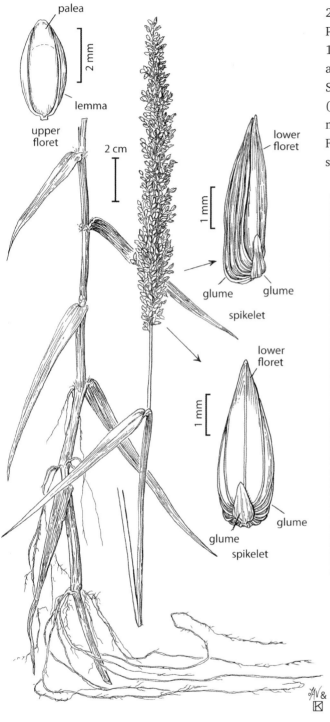

2. *Sacciolepis striata* (American cupscale). Perennial with or without rhizomes. Culms to 1.5 m long, erect to sprawling, often rooting at the lower nodes. Ligules <1 mm long. Spikelets green and often purple-tipped (sometimes mostly purple). Lower paleas ¾ to nearly equaling length of the lower lemmas. Found along and/or growing in ponds, lakes, streams, and ditches.

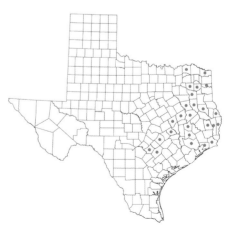

Schedonnardus Steud.
(Chloridoideae: Cynodonteae)

Low, tufted perennial with slender, wiry, erect, or decumbent culms. Leaves mostly in a basal clump, the blades short, flat, usually spirally twisted and wavy-margined, about 1 mm broad. Ligule membranous, 2–3 mm long, decurrent. Panicles ½ to ¾ the entire length of the culm, with a stiff, curved, wiry central axis and slender, spreading, undivided primary branches. At maturity the panicle breaks off at the base and becomes a tumbleweed. Spikelets slender, 1-flowered, sessile, widely spaced, and appressed on the branches and at the tip of the main axis. Disarticulation above the glumes. Glumes narrow, lanceolate or acuminate, 1-nerved, the upper about as long as the lemma, the lower shorter. Lemma narrow, rigid, 3-nerved, awnless or with a minute awn tip. Palea similar to the lemma and about as long. Basic chromosome number, $x = 10$. Photosynthetic pathway, C_4. Represented in Texas by the only species in the genus.

(Snow 2003c)

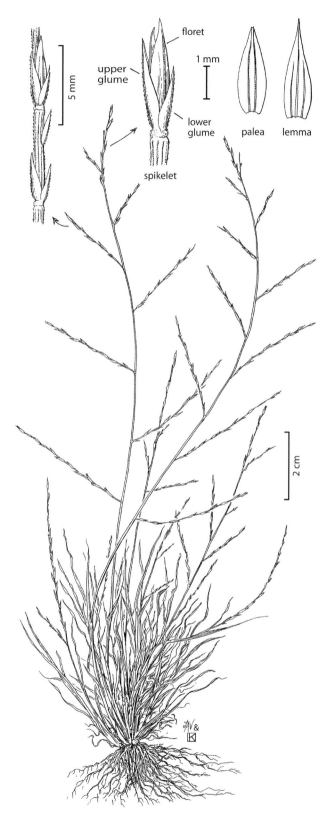

floret

upper
glume

1 mm

lower
glume

palea   lemma

5 mm

2 cm

spikelet

1. *Schedonnardus paniculatus* (Nutt.) Trel.
(tumblegrass, Texas crabgrass). Short, rather
spindly perennial. Panicles with spikelets
appressed along the branches and main axis.
Rachises become curved at maturity. A plant
of disturbed and waste areas. As the first
vernacular name suggests, the panicle breaks
at the base and serves as a tumbleweed,
dispersing seeds as it bounds along. Listed
as poor for wildlife and livestock (Hatch and
Pluhar 1993; Powell 1994).

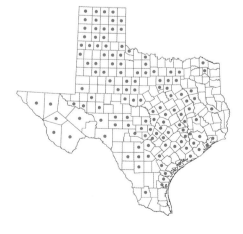

Schedonorus P. Beauv.
(Poöideae: Poeae)

Perennials with short rhizomes, often tufted and can grow to 2 m tall. Sheaths open, round, smooth or scabrous. Ligules short, membranous. Auricles present, prominent, ciliate or not. Inflorescence panicles, terminal, large but simple; branches erect, glabrous, smooth or scabrous. Disarticulation above the glumes and below the florets. Spikelets large, 2–22 florets, terete or laterally compressed. Glumes subequal, shorter than the lowermost floret, unawned. Lemmas firm, with 5–7 faint veins, awnless or with a short awn that rarely exceeds 2 mm in length. Paleas slightly shorter and narrower than the lemma, occasionally with cilia along the upper margins. Basic chromosome number, $x = 7$. Photosynthetic pathway, C_3. Represented in Texas by 2 species.

Gould (1975b) included these in *Festuca*. *Schedonorus* is more closely related to *Lolium* than it is to *Festuca*.

1. Auricles glabrous; lemmas unawned or with a mucro to 0.2 mm long................ 2. *S. pratensis*
1. Auricles ciliate on some sheaths, with at least 1 or 2 hairs along the margins; lemmas unawned or with an awn to 4 mm long.. 1. *S. arundinaceus*

(Darbyshire 2007; Gould 1975b)

1. *Schedonorus arundinaceus* (Schreb.)
Dumort. (tall fescue). Perennial, sometimes with
rhizomes. Culms to 2 m tall. Ligules 1–2 mm
long. Panicle branches at the lowest nodes
usually 2. A Eurasian introduction grown for
forage, hay, reclamation, and turf. It escapes
to roadsides, mesic swales, and other moist
sites. Frequently infected with a fungi that
produces ergotlike alkaloids that are toxic to
livestock. Listed by Gould (1975b); Hatch,
Gandhi, and Brown (1997); and Diggs et al.
(2006) as *Festuca arundinacea*. Listed as fair
for wildlife and good for livestock (Hatch and
Pluhar 1993, as *F. arundinacea* Schreb.).

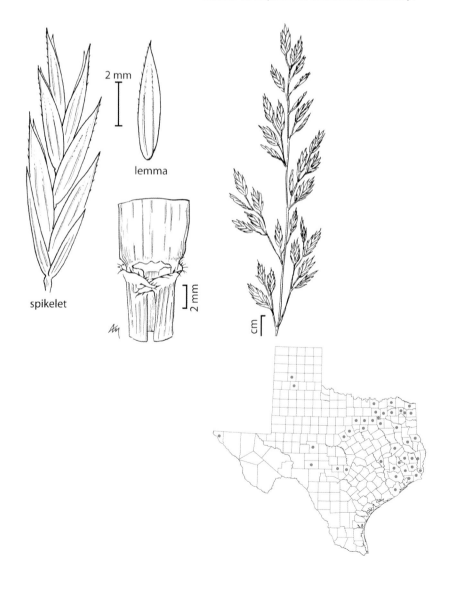

spikelet

2 mm

lemma

2 mm

cm

2. *Schedonorus pratensis* (Huds.) P. Beauv. (meadow fescue). Perennial with culms to 1.3 m long. Ligules to 0.5 mm long. Panicle branches 1 or 2 at lowest nodes. A Eurasian introduction that is widely established in the United States. Gould (1975b) reported that it did not occur in Texas (as *Festuca pratensis*), but specimens have been collected recently.

2 mm

lemma

auricle

2 mm

spikelet

1 cm

Schismus P. Beauv.
(Danthonioideae: Danthonieae)

Low, tufted annuals with narrow, short blades and small, usually contracted panicles. Spikelets several-flowered, the uppermost floret usually reduced. Rachilla disarticulating above the glumes and between the florets. Glumes subequal, large, lanceolate, several-nerved, slightly shorter than to about equaling the terminal floret. Lemmas several-nerved, broadly rounded and notched at the apex, the midnerve extended into a short mucro. Palea broad, 2-keeled, as long as the lemma or slightly shorter. Basic chromosome number, $x = 6$. Photosynthetic pathway, C_4. Represented in Texas by 2 species.

1. Lower glumes as long as or longer than the lowest floret; paleas shorter than the lemmas ....
   ....1. *S. arabicus*
1. Lower glumes shorter than the lowest floret; paleas as long as or longer than the lemmas ....
   .... 2. *S. barbatus*

(Kellogg 2003)

1. *Schismus arabicus* Nees (Arabian schismus). Annual with culms to 16 cm long. Lemmas with dense spreading hairs, lobes longer than wide, acute to acuminate. A native of Asia that is now established as a weed of open, disturbed sites. Not included in Gould (1975b) or Hatch, Gandhi, and Brown (1997).

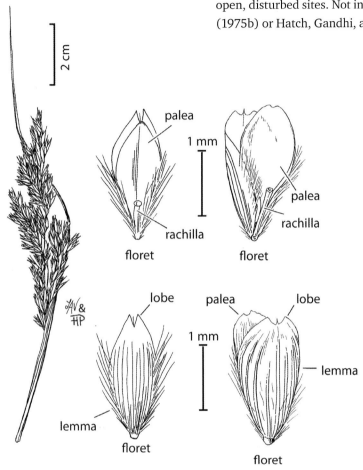

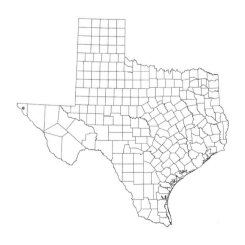

2. *Schismus barbatus* (Loefl. *ex* L.) Thell.
(common Mediterranean grass). Annual with
culms to 30 cm long. Lemmas with appressed
pubescence, or glabrous with spreading hairs
along margin, lobes wider than long, acute to
obtuse. An introduction from Asia that is now a
weed of sandy areas, disturbed sites, fields, or
dry riverbeds. Powell (1994) reported limited
occurrence and of minor forage value.

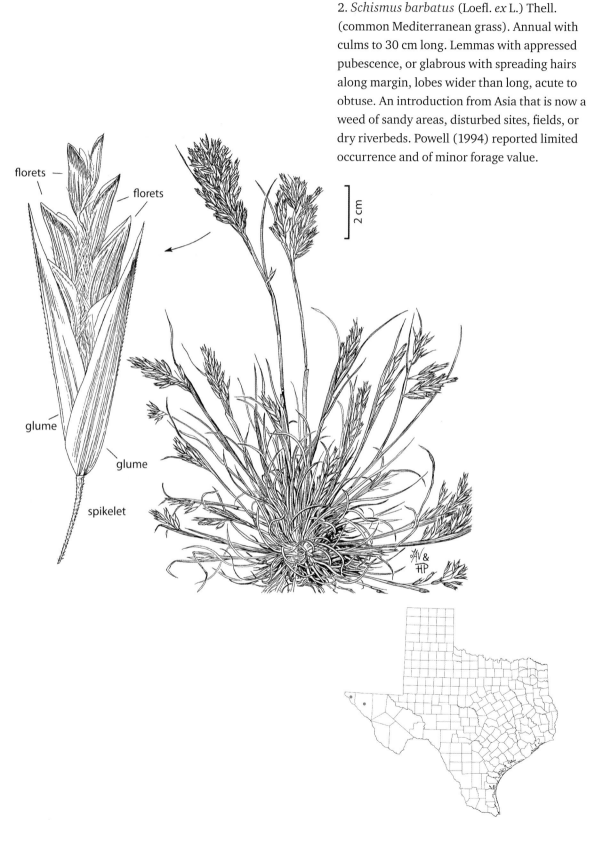

florets

florets

glume

glume

spikelet

2 cm

Schizachyrium Nees
(Panicoideae: Andropogoneae)

Cespitose perennials, some with short, stout rhizomes. Leaves with rounded or compressed-keeled sheaths, and flat or folded, infrequently terete, blades. Flowering culms much-branched, each leafy branch terminating in a single pedunculate raceme. Spikelets appressed to the rachis or somewhat divergent at maturity. Disarticulation in the rachis, the sessile spikelet falling attached to a pedicel and a section of the rachis. Sessile spikelet fertile, awned, the pediceled spikelet reduced but usually present. Glumes of the sessile spikelet large and firm. Lemmas of the sterile and fertile florets thin and membranous, the latter usually with a geniculate arm. Basic chromosome number, $x = 10$. Photosynthetic pathway, C_4. Represented in Texas by 6 species and 3 varieties.

1. Peduncles with 2 rames ....5. *S. spadiceum*
1. Peduncles with a single rame.
    2. Leaf blades 0.5–2.0 mm wide, with stripe of white, spongy tissue on their adaxial surfaces ....6. *S. tenerum*
    2. Leaf blades 1.5–9.0 mm wide, without a stripe of white, spongy tissue on their adaxial surfaces.
        3. Plants root and branch at the lower nodes..2. *S. littorale*
        3. Plants do not root or branch at the lower nodes.
            4. Pedicellate spikelets usually staminate, unawned .... 1. *S. cirratum*
            4. Pedicellate spikelets usually sterile, unawned or the awn to 6 mm long.
                5. Upper lemmas cleft for about ½ their length; lower glumes glabrous......4. *S. scoparium*
                5. Upper lemmas cleft from ⅔ to ¾ their length; lower glumes pubescent or glabrous .... 3. *S. sanguineum*

(Wipff 2003j)

1. *Schizachyrium cirratum* (Hack.) Wooton and Standl. (Texas beardgrass, Texas bluestem). Cespitose or short-rhizomatous perennial. Culms to 0.75 m long. Ligules 1.0–2.5 mm long. Awns 13–24 mm long. Pedicellate spikelets usually staminate, unawned. Found on rocky slopes at elevations above 5,000 ft (1,550 m) in western Texas. It should be good forage for livestock and wildlife (Powell 1994).

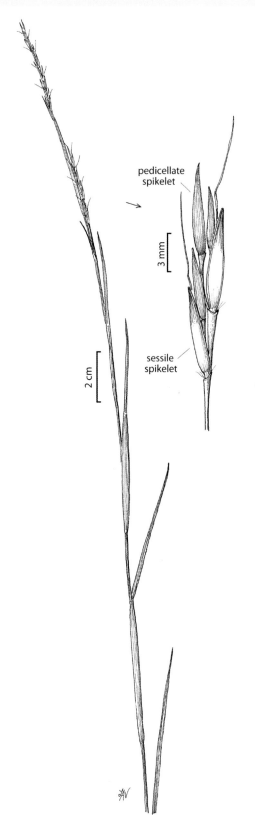

pedicellate spikelet

3 mm

sessile spikelet

2 cm

2. *Schizachyrium littorale* (Nash) E. P. Bicknell (seashore bluestem). Perennial, sometimes rhizomatous. Culms to 1.6 m long. Auricles yellow, flexible. Ligules 1.5–2.0 mm long. Pedicellate spikelets often staminate, unawned or awned, awns to 3.5 mm long. Found on shifting coastal sand dunes. Appears rhizomatous because the lower nodes are covered by blowing sand. A variety of *S. scoparium* in Gould (1975b); Hatch, Gandhi, and Brown (1997); Jones, Wipff, and Montgomery (1997); and Turner et al. (2003).

2 cm

pedicellate spikelets

sessile spikelet

2 mm

3. *Schizachyrium sanguineum* (Retz.) Alston (crimson bluestem). Cespitose perennial with culms to 1.2 m long. Ligules 0.7–2.0 mm long. Awns 15–25 mm long. Pedicellate spikelets sterile or staminate, awned, awns to 6 mm long. Found on rocky slopes and well-drained soils in West and South Texas. The Texas species belongs to var. *hirtiflorum* (Nees) S. L. Hatch. Listed as a distinct species in Gould (1975b) and Turner et al. (2003).

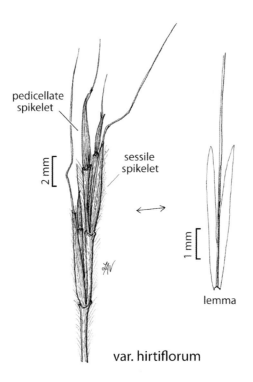

pedicellate spikelet

sessile spikelet

2 mm

1 mm

lemma

var. hirtiflorum

4. *Schizachyrium scoparium* (Michx.) Nash (little bluestem). Perennial, cespitose or rhizomatous. Culms to 2 m long. Sheaths rounded or keeled. Ligules 0.5–2.0 mm long. Awns 2.5–17 mm long. Pedicellate spikelets staminate or sterile, unawned or awned, awn to 4 mm long. Widespread and common species of open woodlands, grasslands, and prairies. One of the more abundant and recognizable species along roadsides throughout the state. Listed as poor for wildlife and good for livestock (Hatch and Pluhar 1993). Considered a good forage, especially early in the season (Powell 1994). Telfair (2006) listed it as important for wildlife. Two varieties are included here: var. *divergens* (Hack.) Gould; and the typical one, var. *scoparium*. The former is sometimes referred to as pinehill bluestem. The varieties can be distinguished by the following characters:

1. Pedicellate spikelets 5–10 mm long, usually staminate; sheaths and blades usually tomentose ....var. *divergens*
1. Pedicellate spikelets 1–6 mm long, usually sterile; sheaths and blades usually glabrous ....var. *scoparium*

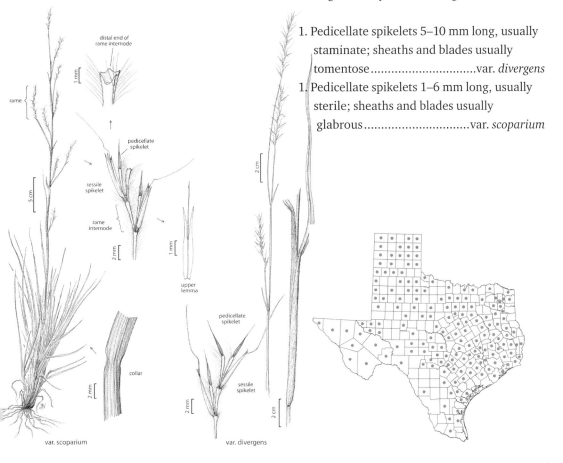

var. scoparium   var. divergens

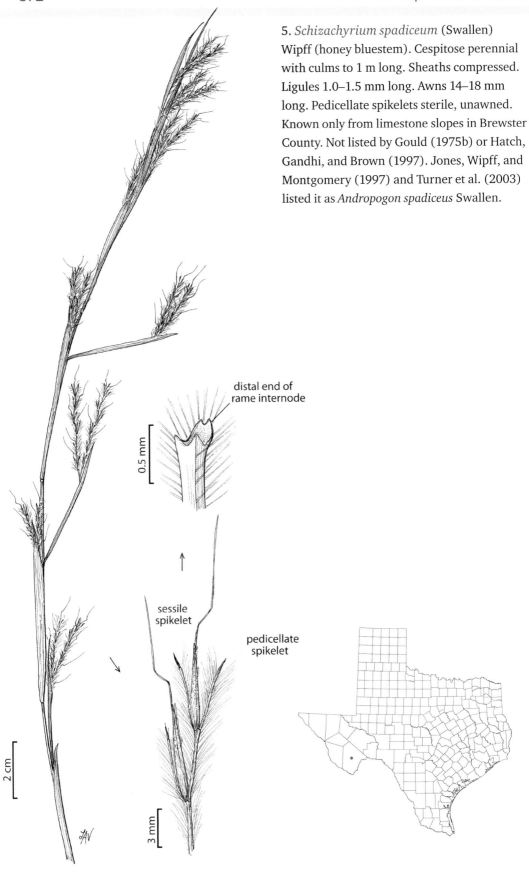

5. *Schizachyrium spadiceum* (Swallen) Wipff (honey bluestem). Cespitose perennial with culms to 1 m long. Sheaths compressed. Ligules 1.0–1.5 mm long. Awns 14–18 mm long. Pedicellate spikelets sterile, unawned. Known only from limestone slopes in Brewster County. Not listed by Gould (1975b) or Hatch, Gandhi, and Brown (1997). Jones, Wipff, and Montgomery (1997) and Turner et al. (2003) listed it as *Andropogon spadiceus* Swallen.

distal end of
rame internode

0.5 mm

sessile
spikelet

pedicellate
spikelet

2 cm

3 mm

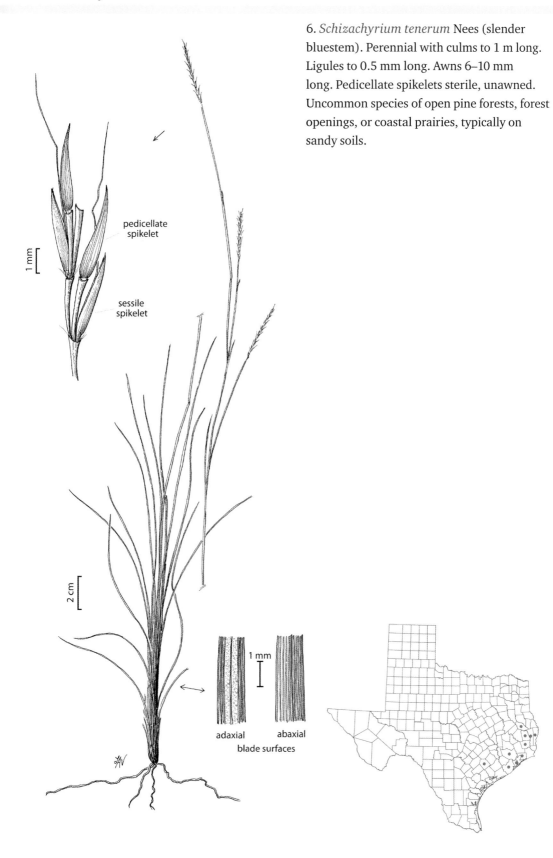

pedicellate
spikelet

1 mm

sessile
spikelet

2 cm

1 mm

adaxial   abaxial
blade surfaces

6. *Schizachyrium tenerum* Nees (slender
bluestem). Perennial with culms to 1 m long.
Ligules to 0.5 mm long. Awns 6–10 mm
long. Pedicellate spikelets sterile, unawned.
Uncommon species of open pine forests, forest
openings, or coastal prairies, typically on
sandy soils.

Sclerochloa P. Beauv.
(Poöideae: Poeae)

Low, tufted annual with culms mostly <10 cm long and inflorescences partially included in greatly enlarged upper leaf sheaths. Leaves scarcely differentiated into sheath and blade, without a ligule or thickened area at the junction. Inflorescence a spicate raceme or contracted panicle, 1–2 cm long. Spikelets awnless, 3-flowered, subsessile, and crowded on the main inflorescence axis and on short branches. Disarticulation below the glumes, the spikelet falling entire. Glumes broad, rather firm, the lower short, 3-nerved, the upper about ½ as long as the spikelet, 7-nerved. Lemmas broad, rounded on the back and obtuse at the apex, membranous on the margin above, 5-nerved. Palea hyaline. Basic chromosome number, $x = 7$. Photosynthetic pathway, C_3. Represented in Texas by a single species.

(Brandenburg 2007c)

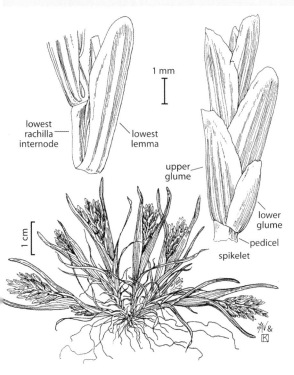

lowest
rachilla
internode

lowest
lemma

1 mm

upper
glume

lower
glume

pedicel

spikelet

1 cm

1. *Sclerochloa dura* (L.) P. Beauv. (hardgrass). Annual. Very low-growing, matted or prostrate plant. Sheaths overlapping, open or closed for about ½ their length; blades usually exceeding the inflorescences, apices bow-shaped as in *Poa*. Inflorescences partially contained in upper leaf sheaths. A European introduction that has become a common weed of yards, roadsides, and other disturbed sites. Handles heavy traffic well.

Scleropogon Phil.
(Chloridoideae: Cynodonteae)

Low, mat-forming perennial, stoloniferous. Leaves tufted with flat, firm blades. Sheaths open; ligules a fringe of hairs; auricles absent. Plants staminate or pistillate (dioecious) or less frequently both sexes present in different spikelets but on the same plant (monoecious). Spikelets large, in spicate racemes or reduced panicles, the pistillate ones bristly with long awns. Staminate spikelets mostly with 5–10(–20) persistent florets. Glumes 2, 1-nerved or rarely 3-nerved, thin, pale, awnless, the lower and upper separated by a short internode. Lemmas of the staminate spikelets similar to the glumes. Palea obtuse, shorter than the lemma. Pistillate spikelets with 3–5 florets and 1 to several rudiments (reduced to awns) above. Rachilla disarticulating above the glumes, the florets falling together. Glumes of the pistillate spikelets 2, unequal, usually 3-nerved. Lemmas of the pistillate spikelet firm, rounded on the back, 3-nerved, the veins extending to slender, spreading awns. Paleas narrow, the 2 veins extending into short awns. Basic chromosome number, $x = 10$. Photosynthetic pathway, C_4. Represented in Texas by the only species in the genus.

(Reeder 2003d)

1. *Scleropogon brevifolius* Phil. (Burrograss). Perennial. Stoloniferous plants monoecious, dioecious, and/or synoecious. Spikelets either unisexual (staminate and pistillate) or perfect with the staminate florets below the pistillate. Staminate lemmas with 3 veins, awnless or each vein extending into a short awn (3 mm). Pistillate lemmas strongly 3-veined with each vein extending in an awn up to 15 cm long. A plant of dry, often disturbed areas or overgrazed rangeland. Listed as poor for wildlife and livestock (Hatch and Pluhar 1993) and has low palatability (Powell 1994).

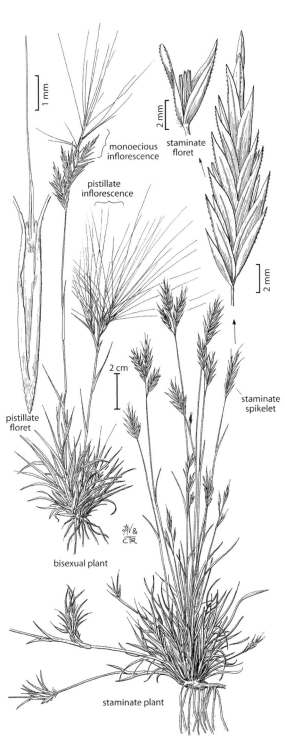

1 mm

2 mm

monoecious inflorescence

staminate floret

pistillate inflorescence

2 mm

2 cm

pistillate floret

staminate spikelet

bisexual plant

staminate plant

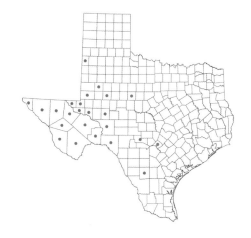

Secale L.
(Poöideae: Triticeae)

Texas species are annual. Inflorescence a dense bilateral spike with spikelets solitary at the nodes. Disarticulation above the glumes and between the florets. Spikelets usually 2-flowered, the rachilla extending behind the upper floret as a minute bristle. Glumes narrow, rigid, and subulate. Lemmas broad, 5-nerved, sharply keeled, ciliate on the keel, tapering to a stout awn. Basic chromosome number, $x = 7$. Photosynthetic pathway, C_3. Represented in Texas by a single species.

(Barkworth 2007m)

1. *Secale cereale* L. (rye). Cultivated annual with culms to 3 m tall. Spikelets distinctly laterally compressed. Glumes with awns 1–3 mm long. Lemmas with awns 7–50 mm long. An important cereal grain sometimes used to make whiskey. Sometimes cultivated across the state on sandy soils. It also is used in roadside stabilization for a rapid, initial ground cover. Very susceptible to infection by ergot fungi and has historically caused deaths in livestock and humans. However, if not infected, it has good livestock and wildlife values (Hatch, Schuster, and Drawe 1999).

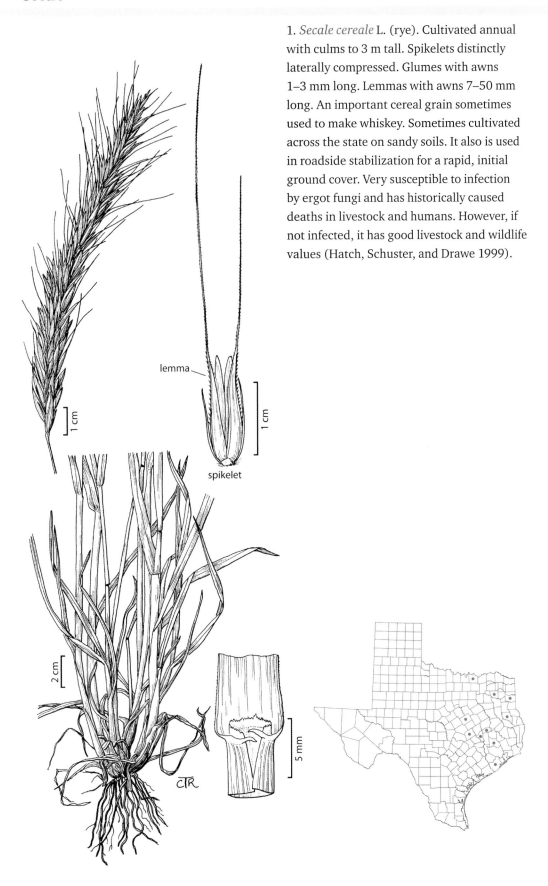

lemma

1 cm

spikelet

1 cm

2 cm

5 mm

CTR

Setaria P. Beauv.
(Panicoideae: Paniceae)

Cespitose annuals and perennials, with erect or geniculate culms, these often branching at the base. Leaf blades typically flat and thin. Inflorescence a slender, usually contracted and densely flowered, bristly panicle, the spikelets subsessile on the main axis and on short branches. Some or all of the spikelets subtended by 1 to several persistent bristles (reduced branches), the spikelets disarticulating above the bristles. Glumes and lemma of sterile floret typically glabrous, prominently nerved, acute or obtuse, the lower glume short, the upper glume and lemma of the sterile floret equal, or more frequently the upper glume ½ to ⅔ as long. Lemma and palea of the fertile floret indurate, rounded at the apex, usually finely or coarsely transverse-rugose. Basic chromosome number, $x = 9$. Photosynthetic pathway, C_4. Represented in Texas by 16 species, 4 subspecies, and 2 varieties. Jones, Wipff, and Montgomery (1997) listed 2 additional perennial cultivars for the state: *S. paniculifera* (Steud.) Fourn. and *S. poiretiana* (Schult.) Kunth, which are included in the checklist but not described here.

1. Terminal spikelet of each panicle branch subtended by a single bristle, single bristle sometimes present below other spikelets.
    2. Blades plicate, >10 mm wide ....8. *S. palmifolia*
    2. Blades not plicate, <10 mm wide ....11. *S. reverchonii*
1. All spikelets subtended by 1 to several bristles.
        3. Bristles 4–12 below each spikelet.
            4. Plants perennial ....9. *S. parviflora*
            4. Plants annual ....10. *S. pumila*
        3. Bristles 1–3(–6) below each spikelet.
            5. Bristles retrorsely scabrous.
                6. Sheath margins glabrous.. 1. *S. adhaerans*
                6. Sheath margins ciliate, at least above .... 14. *S. verticillata*
            5. Bristles antrorsely scabrous.
                7. Plants perennial ....Key A
                7. Plants annual ....Key B

Key A

1. Spikelets 2.8–3.2 mm long... 15. *S. villosissima*
1. Spikelets 1.9–2.8(–3.0) mm long.
    2. Panicles 2–6 cm long; spikelets 1.9–2.1 mm long; culms branching at upper nodes............
    ....13. *S. texana*
    2. Panicles 5–30 cm long; spikelets 2.0–2.8(–3.0) mm long; culms seldom branching at upper nodes.
        3. Lower paleas narrow, ½ to ¾ as long as lemmas.
            4. Blades <5 mm wide; panicles 6–15 cm long; bristles ascending ....5. *S. leucopila*

4. Blades >5 mm wide; panicles 15–25 cm long; bristles diverging .... 12. *S. scheelei*

3. Lower paleas broad, subequal to lemmas .... 6. *S. macrostachya*

Key B

1. Upper lemmas shiny, smooth.

    2. Spikelets about 2 mm long; lower paleas about equal to lower lemmas .... 7. *S. magna*

    2. Spikelets about 3 mm long; lower paleas absent or to ½ as long as lower lemmas ....4. *S. italica*

1. Upper lemmas dull, distinctly transversely rugose.

    3. Upper lemmas coarsely rugose..2. *S. corrugata*

    3. Upper lemmas finely rugose.

        4. Panicles loosely spicate; rachises visible, hispid .... 3. *S. grisebachii*

        4. Panicles densely spicate; rachises not visible, villous .... 16. *S. viridis*

(Rominger 2003; Gould 1975b)

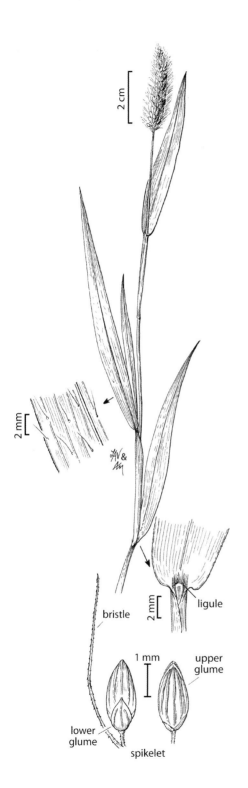

2 cm

2 mm

ligule

2 mm

bristle

1 mm

upper
glume

lower
glume

spikelet

1. *Setaria adhaerans* (Forssk.) Chiov. (bur bristlegrass, tropical barred bristlegrass). Annual with weak, geniculate or decumbent culms, to 75 cm long. Ligules 1–2 mm long, hairs white. Panicles 2–6 cm long, verticillate, green to purple. Lower paleas less than ½ as long as lower spikelets. A ruderal species often found in the shade of trees or tall weeds.

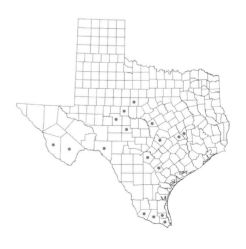

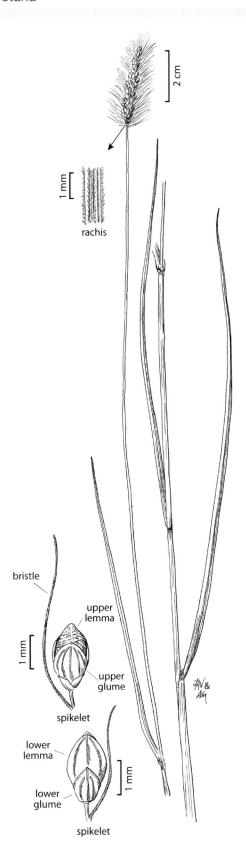

2 cm

1 mm

rachis

bristle

upper
lemma

1 mm

upper
glume

spikelet

lower
lemma

lower
glume

1 mm

spikelet

2. *Setaria corrugata* (Elliott) Schult. (coastal bristlegrass, coastal foxtail). Annual with culms to 1 m long. Ligules about 1 mm long. Panicles 3–15 cm long, densely spicate. Lower paleas about ¾ as long as the upper paleas. Infrequent; found on sandy, disturbed soils and pinelands.

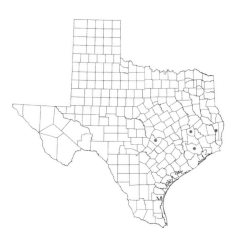

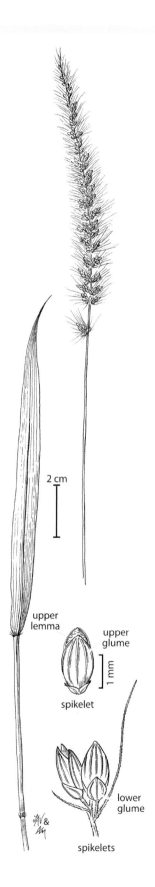

2 cm

upper
lemma

upper
glume

1 mm

spikelet

lower
glume

spikelets

3. *Setaria grisebachii* Fourn. (Grisebach's bristlegrass). Annual. Culms to 1 m long. Ligules 1 mm long, ciliate. Panicles 3–18 cm long, loosely spicate, interrupted, often purple. Lower paleas about ⅓ as long as upper paleas. Found on open ground primarily in the Big Bend region; scattered elsewhere.

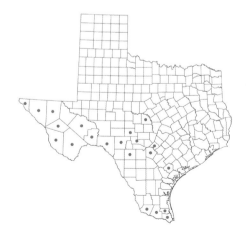

4. *Setaria italica* (L.) P. Beauv. (foxtail millet, Italian millet, German millet, Hungarian millet). Annual with culms to 1 m tall. Ligules 1–2 mm long. Panicles 8–30 cm long, dense, spikelike, interrupted at the base, nodding. Lower paleas absent or to ½ as long as lower lemmas. Sometimes cultivated or used to initially stabilize roadsides; occasionally escapes but rarely persists. Found in fields, along roadsides, and in waste places. Gould (1975b) included this with *S. viridis*, but *S. italica* has a shiny and smooth upper lemma and palea, and the entire inflorescence nods. Often planted as part of mixtures for game plots (Hatch, Schuster, and Drawe 1999).

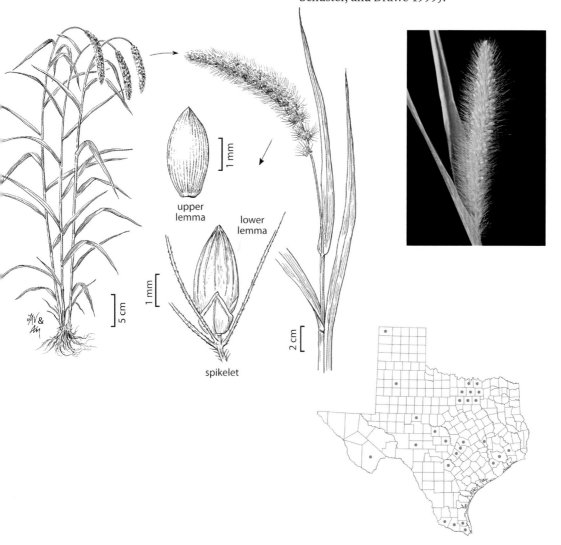

2 cm

1 mm

lower
palea

↑

1 mm

lower
glume

upper
glume

spikelet

1 mm

5. *Setaria leucopila* (Scribn. and Merr.) K. Schum. (plains bristlegrass, streambed bristlegrass). Cespitose perennial. Culms to 1 m long. Ligules 1.0–2.5 mm long. Panicles 6–15 cm long, spikelike, interrupted distally as well as basally. Lower paleas ½ to ¾ as long as upper paleas. Found throughout the drier portions of the state, usually associated with areas that occasionally have abundant moisture. Most economically important of all the perennial species of *Setaria*. Hatch and Pluhar (1993) listed the economic value of this species as fair for wildlife and good for livestock. Powell (1994) reported it an important forage species because it is palatable and widespread.

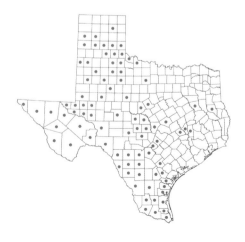

6. *Setaria macrostachya* Kunth (plains bristlegrass). Cespitose perennial. Culms to 1.2 m long. Ligules 2–4 mm long. Panicles 10–30 cm long, dense, rarely lobed. Lower paleas nearly equaling the upper paleas. Found in arid grasslands of Central and South Texas. Fair livestock and wildlife values (Hatch, Schuster, and Drawe 1999).

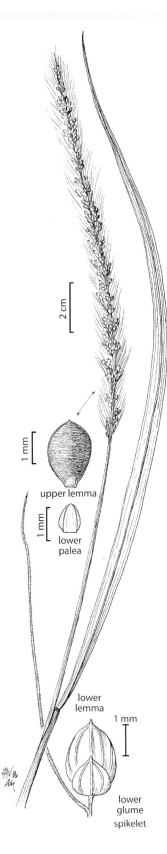

2 cm

1 mm

upper lemma

1 mm

lower palea

lower lemma

1 mm

lower glume

spikelet

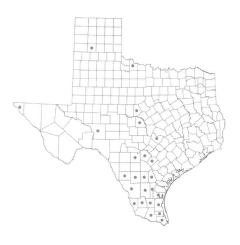

2 cm

2 cm

upper
lemma

1 mm

lower
glume

1 mm

spikelet

7. *Setaria magna* Griseb. (giant bristlegrass, giant foxtail, saltmarsh foxtail). Annual. Culms to 6 m tall, 2–3 cm thick at the base, with prop roots. Ligules 1–2 mm long. Panicles to 50 cm long, 5 cm wide, densely spikelike. Lower paleas equaling the lower lemmas. Found in saline or brackish marshes and one collection in the Trinity River bottom, Tarrant County. Largest of the native *Setaria* species (Diggs et al. 1999). Large seed producer and good for birds (Hatch, Schuster, and Drawe 1999).

8. *Setaria palmifolia* (J. König) Stapf (palmgrass). Perennial. Culms to 2 m long. Ligules about 1 mm long. Blades to 50 cm long and 8 cm wide, plicated. Panicles to 40 cm long, open. Introduced from Asia, where it is sometimes used as a source of grain and the shoots are used as a vegetable. Sometimes grown as an ornamental; occasionally escapes.

5 cm

upper
lemma

lower
lemma

upper
glume

1 mm

lower
glume

spikelet

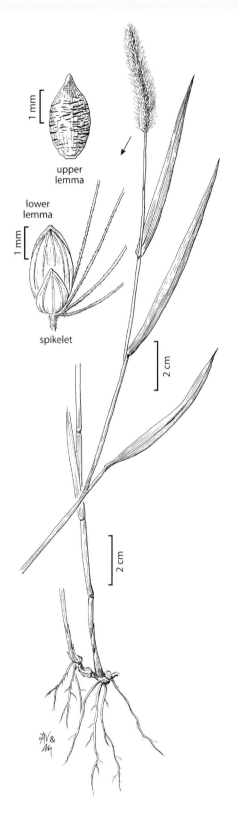

upper
lemma

lower
lemma

spikelet

2 cm

2 cm

JAV &
AM

9. *Setaria parviflora* (Poir.) Kerguélen (knotroot bristlegrass, knotroot foxtail). Rhizomatous perennial with short, knotty rhizomes. Culms to 1.2 m tall. Ligules <1 mm long. Panicles 3–8 cm long, densely spicate. Lower paleas equaling the lower lemmas. Found in moist areas along the shoreline of lakes, in ditches, along streams, or in swales. Gould (1975b) and Hatch, Gandhi, and Brown (1990) listed this species as *S. geniculata* (Lam.) P. Beauv. Reported by Hatch and Pluhar (1993, as *S. geniculata*) and Powell (1994) to be of fair value to wildlife and livestock.

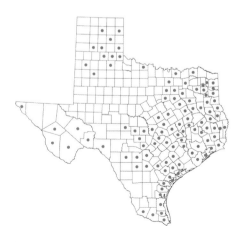

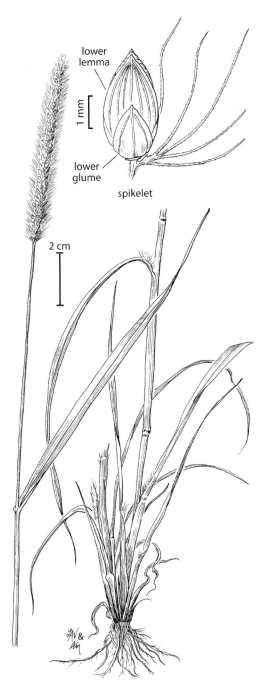

lower
lemma

1 mm

lower
glume

spikelet

2 cm

10. *Setaria pumila* (Poir.) Roem. & Schult.
(yellow foxtail, pigeon grass). Annual. Culms
to 1.3 mm long. Ligules <1 mm long. Panicles
3–15 cm long, erect, densely spicate. Lower
paleas equaling the lower lemmas. A weed
of fields, yards, roadsides, and waste places.
Gould (1975b) and Hatch, Gandhi, and Brown
(1990) referred to this species as *S. glauca*
(L.) P. Beauv. The Texas species belongs
to the typical subspecies, subsp. *pumila*.
Bristles known to cause mechanical damage
to the mouths of grazing animals. Good seed
producer (Hatch, Schuster, and Drawe 1999)
but of limited forage value (Powell 1994).

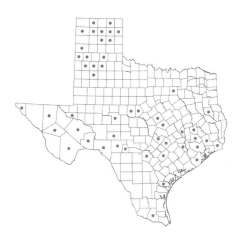

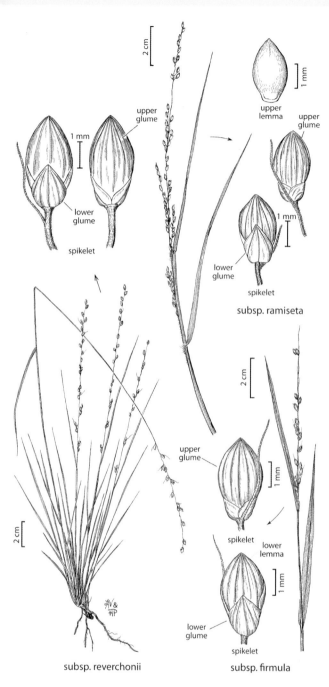

subsp. reverchonii

subsp. ramiseta

subsp. firmula

11. *Setaria reverchonii* (Vasey) Pilger (Reverchon bristlegrass). Rhizomatous perennial, rhizomes short, knotty. Culms to 90 cm long, erect. Ligules composed of stiff hairs 1–2 mm long. Panicles 5–20 cm long, erect, interrupted. Lower paleas absent. Found on sandy plains and prairies, limestone hills and outcrops, and gravelly soils. Fair forage but good seed producer (Hatch, Schuster, and Drawe 1999; Powell 1994). Three subspecies occur in Texas: subsp. *firmula* (Hitchc. and Chase) W. E. Fox (Knotgrass); subsp. *ramiseta* (Scribn.) W. E. Fox (Rio Grande bristlegrass); and the typical one, subsp. *reverchonii*. Gould (1975b); Hatch, Gandhi, and Brown (1990); and Jones, Wipff, and Montgomery (1997) recognized the latter 2 subspecies as species. The subspecies can be distinguished by the following characters:

1. Spikelets 3.5–4.5 mm long ....subsp. *reverchonii*
1. Spikelets 2.1–3.2 mm long.
   2. Spikelets about 2.5 mm long; blades 2–4 mm wide .... subsp. *ramiseta*
   2. Spikelets about 3.0–3.2 mm long; blades 4–7 mm wide................... subsp. *firmula*

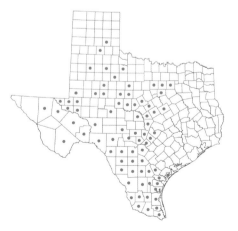

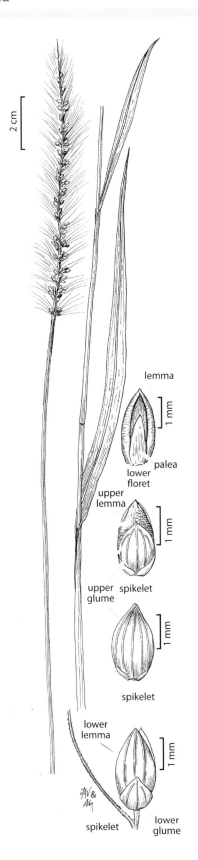

lemma

palea

lower floret

upper lemma

1 mm

upper glume

spikelet

1 mm

spikelet

1 mm

lower lemma

1 mm

spikelet

lower glume

12. *Setaria scheelei* (Steud.) Hitchc. (southwestern bristlegrass, Scheele's bristlegrass). Stout perennial. Culms geniculate-spreading, to 1.2 m long. Ligules 1–2 mm long. Panicles 15–25 cm long, lower branches spreading. Lower paleas about ½ as long as upper paleas. A shade-tolerant species that is common in canyons and river bottoms, along fencerows, and in open woods. Good livestock forage and large seeds for wildlife (Hatch, Schuster, and Drawe 1999).

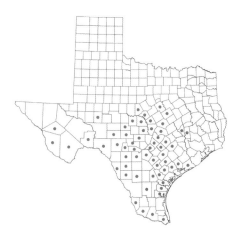

2 cm

bristle

1 mm

spikelet

13. *Setaria texana* Emery (Texas bristlegrass). Perennial. Culms to 70 cm long, wiry, branched distally. Ligules to 1 mm long. Panicles 2–6 cm long, spikelike, not interrupted. Lower paleas about ½ the length of the upper paleas. Found on sandy loam soils in partial shade beneath trees and shrubs. Fair livestock and wildlife values (Hatch, Schuster, and Drawe 1999). Powell (1994) thought it of good forage value where abundant.

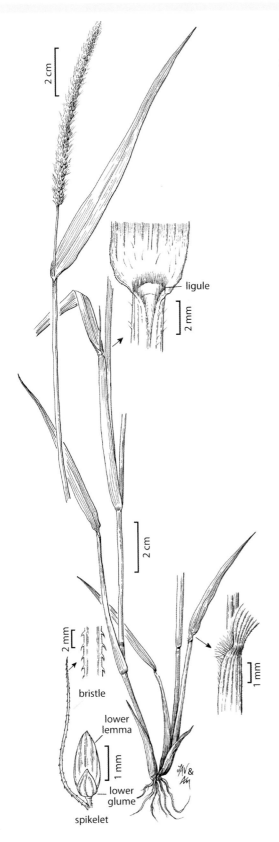

14. *Setaria verticillata* (L.) P. Beauv. (hooked bristlegrass, bur bristlegrass, foxtail grass, rough foxtail grass, bristly foxtail grass). Annual with culms to 1 m long. Ligules to 1 mm long. Panicles 5–15 mm long. Lower paleas about ½ as long as the spikelet. An introduction from Europe that is a common weed. Scattered throughout the state in disturbed habitats. Jones, Wipff, and Montgomery (1997) listed 2 varieties for this species, but none are recognized here.

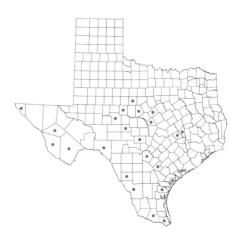

15. *Setaria villosissima* (Scribn. & Merr.) K. Schum. (hairyleaf bristle-grass). Cespitose perennial. Culms to 1 m long. Ligules about 1 mm long, hairs white. Sheaths and blades villous. Panicles 10–20 cm long, loosely spicate. Lower paleas about ⅕ as long as upper paleas. A rare species typically found on granitic soils.

1 mm

spikelet

2 cm

16. *Setaria viridis* (L.) P. Beauv. (green bristlegrass, green foxtail grass). Annual. Culms to 2.5 m long. Ligules 1–2 mm long. Panicles 3–20 cm long, densely spicate, nodding. Lower paleas about $\frac{1}{3}$ as long as the lower lemmas. A native of Eurasia that is an aggressive weed. Scattered throughout the state in waste places, disturbed sites, roadsides, and fields. Gould (1975b) included S. *italica* here, but S. *viridis* has a dull and transversely rugose upper lemma and palea, and drupes only near the inflorescence tip. Two varieties occur within the state: var. *major* (Gaudin) Peterm.; and the typical one, var. *viridis*. They can be distinguished by the following character:

1. Culms 1.0–2.5 m long .... var. *major*
1. Culms 0.2–1.0 m long .... var. *viridis*

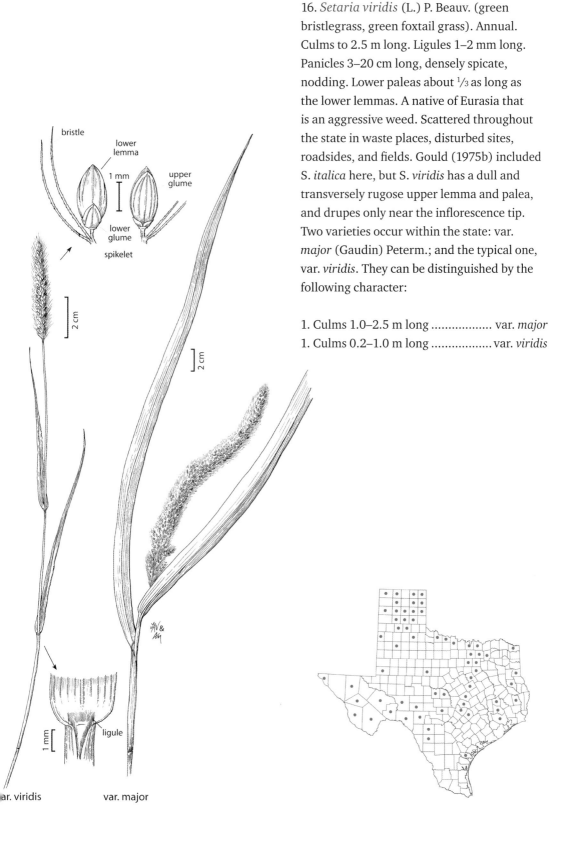

bristle
lower lemma
1 mm
upper glume
lower glume
spikelet
2 cm
2 cm
ligule
1 mm
var. viridis
var. major

Sorghastrum Nash
(Panicoideae: Andropogoneae)

Slender, rather tall perennials with flat blades and narrow, contracted panicles. Spikelets in pairs of 1 sessile and perfect and 1 pediceled and rudimentary, but the pediceled spikelets completely reduced and represented only by the slender, hairy pedicel. Disarticulation beneath the sessile spikelets. Glumes firm, subequal, awnless. Lemma of the lower (sterile) floret absent or membranous and rudimentary. Lemma of the upper (fertile) floret membranous, with a stout, geniculate, and twisted awn. Basic chromosome number, $x = 10$. Photosynthetic pathway, C_4. Represented in Texas by 2 species.

1. Plants rhizomatous; awns once-geniculate, 10–20(–30) mm long .... 2. *S. nutans*
1. Plants cespitose; awns twice-geniculate, 21–40 mm long .... 1. *S. elliottii*

(Dávila Aranda and Hatch 2003)

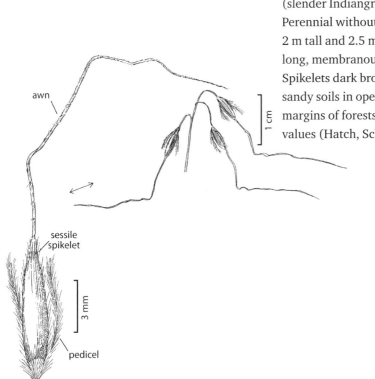

awn

sessile
spikelet

1 cm

3 mm

pedicel

1. *Sorghastrum elliottii* (C. Mohr) Nash
(slender Indiangrass, long-bristle Indiangrass).
Perennial without rhizomes. Culms to nearly
2 m tall and 2.5 mm thick. Ligules 2–4 mm
long, membranous. Sheath auricles present.
Spikelets dark brown at maturity. Found in
sandy soils in open woodlands and on the
margins of forests. Good livestock and wildlife
values (Hatch, Schuster, and Drawe 1999).

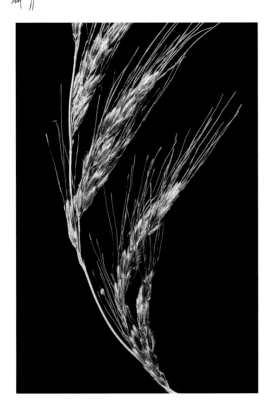

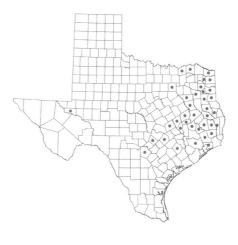

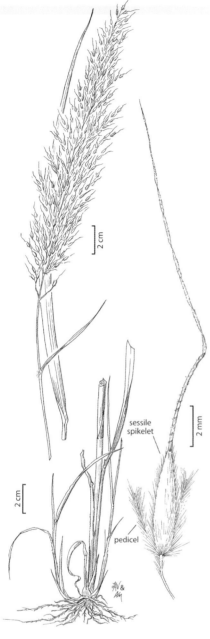

sessile
spikelet

pedicel

2. *Sorghastrum nutans* (L.) Nash (yellow Indiangrass). Perennial with short, stout, scaly rhizomes. Culms to 2.5 m tall and 4.5 mm thick. Ligules 2–6 mm long, membranous. Sheath auricles prominent. Spikelets golden brown or straw-colored. One of the 4 dominant grasses of the "true or tallgrass prairie." Also found in open woodlands, savannahs, and shrublands. It is used in erosion control, roadside stabilization, and land restoration. Several cultivars are available as ornamentals. Good livestock and fair wildlife values (Hatch and Pluhar 1993; Powell 1994). Young plants are a source of cyanide poisoning (Burrows and Tyrl 2001). State grass of Oklahoma.

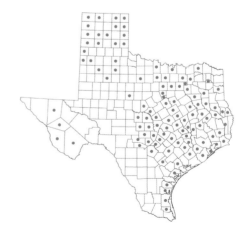

Sorghum Moench
(Panicoideae: Andropogoneae)

Annuals and perennials, many with tall, stout culms. Blades long, flat, narrow, or broad. Inflorescence a large, open or contracted panicle, the spikelets clustered on short racemose branchlets. Spikelets in threes at the branchlet tips, 1 sessile and fertile and 2 pediceled and staminate or neuter, and below the tips in pairs of 1 sessile and perfect and 1 pediceled and reduced. Disarticulation below the sessile spikelet, the rachis section and the pedicel or pedicels falling attached to the sessile spikelet. Glumes coriaceous, awnless, about equal in length. Lemma of the sterile floret and lemma and palea of the fertile floret membranous, the lemma of the fertile floret usually with a geniculate and twisted awn, this readily deciduous in S. *halepense*. Basic chromosome number, $x = 5$. Photosynthetic pathway, C_4. Represented in Texas by 2 species and 2 subspecies.

Gould (1975b) and Hatch, Gandhi, and Brown (1990) included S. *almum* as a distinct species. Here it is considered to be an annual hybrid (*S.* x *almum* Parodi). It has wider sessile spikelets (2.0–2.8 mm) and more veins in the lower glumes (13–15) than S. *halepense*. This hybrid has been collected in the Lower Rio Grande Valley and Sutton County.

1. Plants annual or short-lived perennial; caryopses exposed at maturity .... 1. *S. bicolor*
1. Plants rhizomatous perennial; caryopses not exposed at maturity .... 2. *S. halepense*

(Barkworth 2003m; Allred 2005)

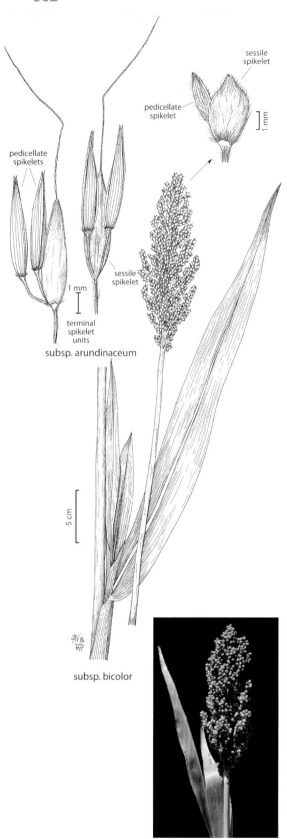

pedicellate spikelets

sessile spikelet

pedicellate spikelet

sessile spikelet

1 mm

1 mm

terminal spikelet units

subsp. arundinaceum

5 cm

subsp. bicolor

1. *Sorghum bicolor* (L.) Moench (sorghum, milo, sweet sorghum, broomcorns, giant sorghum). Typically annual. Culms can grow over 5 m tall and 5 cm thick. Often with prop roots. Cultivated and found along field margins, abandoned fields, and roadsides. Two subspecies are recognized here: subsp. *bicolor*; and subsp. *arundinaceum* (Desv.) de Wet & J. R. Harlan. The former is the common sorghum cultivar. "Grain sorghums" have short panicles and branches, "broomcorns" have elongated panicles and branches (used for making whiskbrooms), and "sweet sorghums" have an abundance of sugar in the culms. The caryopses are also used to make beer. A new cultivar, "giant sorghum," is being developed for use as a biomass plant for cellulosic ethanol production. Good forage and grain species (Hatch, Schuster, and Drawe 1999). Plants can poison livestock by accumulation of cyanide, nitrates, and tannins (Burrows and Tyrl 2001). The subspecies can be distinguished by the following characters:

1. Inflorescence branches remaining intact and caryopses exposed at maturity .... subsp. *bicolor*
1. Inflorescence branches disarticulating and caryopses not exposed at maturity ....subsp. *arundinaceum*

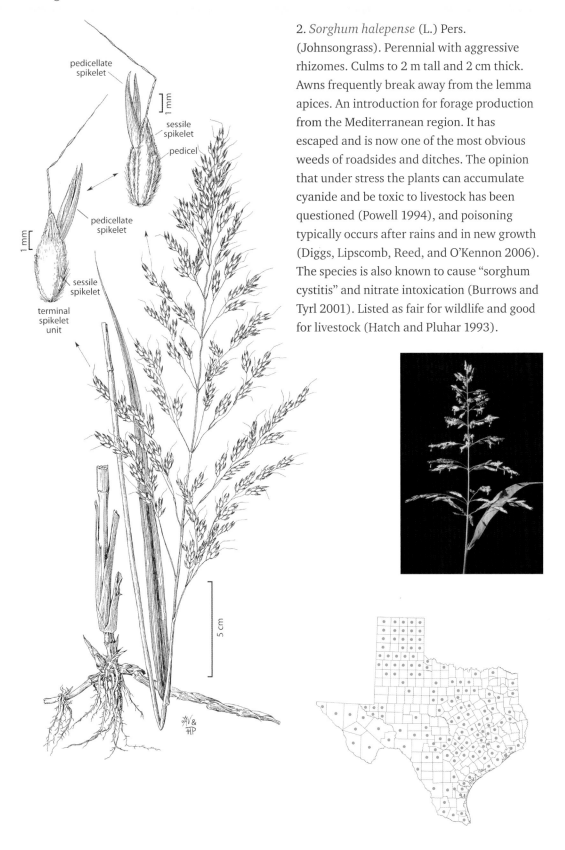

pedicellate spikelet

sessile spikelet

pedicel

1 mm

pedicellate spikelet

1 mm

sessile spikelet

terminal spikelet unit

5 cm

2. *Sorghum halepense* (L.) Pers. (Johnsongrass). Perennial with aggressive rhizomes. Culms to 2 m tall and 2 cm thick. Awns frequently break away from the lemma apices. An introduction for forage production from the Mediterranean region. It has escaped and is now one of the most obvious weeds of roadsides and ditches. The opinion that under stress the plants can accumulate cyanide and be toxic to livestock has been questioned (Powell 1994), and poisoning typically occurs after rains and in new growth (Diggs, Lipscomb, Reed, and O'Kennon 2006). The species is also known to cause "sorghum cystitis" and nitrate intoxication (Burrows and Tyrl 2001). Listed as fair for wildlife and good for livestock (Hatch and Pluhar 1993).

Spartina Schreb.
(Chloridoideae: Cynodonteae)

Perennials, frequently rhizomatous. Leaves tough and firm, the blades long, flat or involute. Ligule a ring of long or short hairs. Inflorescence of few to numerous, racemosely arranged, short, usually appressed branches, bearing closely spaced, sessile spikelets. Disarticulation below the glumes. Spikelets 1-flowered, laterally flattened. Glumes unequal, keeled, usually 1-nerved or the upper with 3 closely placed veins, acute or the second short-awned. Lemma firm, keeled, strongly 1- or 3-nerved and often with additional indistinct lateral veins, tapering to a narrow but usually rounded, awnless tip. Palea as long as, or longer than, the lemma, with broad, membranous margins on either side of the closely placed veins. Basic chromosome number, $x = 10$. Photosynthetic pathway, C_4. Represented in Texas by 7 species.

1. Leaf blades with smooth or slightly scabrous margins .... 1. *S. alternifolia*
1. Leaf blades with strongly scabrous margins.
   2. Plants without rhizomes, or plants clumped from short rhizomes.
      3. Panicles with 15–75 appressed branches...7. *S. spartinae*
      3. Plants with 2–16 branches.
         4. Upper glumes 1-veined...4. *S. densiflora*
         4. Upper glumes 3–4-veined... 2. *S. bakeri*
   2. Plants with long, creeping rhizomes.
      5. Rhizomes whitish ....5. *S. patens*
      5. Rhizomes purplish or brown.
         6. Upper glumes awned, awns 3–8 mm long....................................6. *S. pectinata*
         6. Upper glumes unawned or if awned, the awn up to 2 mm long .... 3. *S. cynosuroides*

(Barkworth 2003n; Gould 1975b)

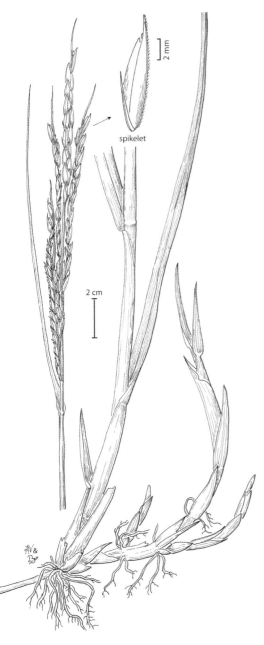

spikelet

2 mm

2 cm

1. *Spartina alterniflora* Loisel. (smooth cordgrass, saltmarsh grass). Rhizomatous perennial. Rhizomes elongated, white, scales inflated and only slightly imbricate. Culms to 2 m tall, succulent with an unpleasant odor when fresh. Ligules 1–2 mm long. Panicles 10–40 cm long. Only member of the genus with smooth and glabrous glume keels. Found on muddy soils of tidal flats and bayou margins. Usually found in saturated soils or standing water. Good livestock and wildlife values (Hatch, Schuster, and Drawe 1999).

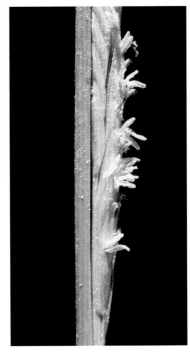

2. *Spartina bakeri* Merr. (sand cordgrass).
Cespitose perennial, without rhizomes. Culms
to 2 m tall in large clumps. Ligules 0.5–2.0 mm
long. Panicles 8–25 cm long. Glumes with
hispid keels. Lower glumes about ½ as long
as spikelet. Infrequent; reported only from
marshes and bayous in Orange and Jefferson
counties. Fair forage values (Hatch, Schuster,
and Drawe 1999).

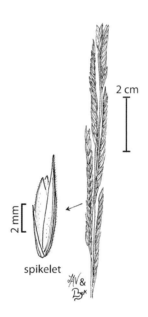

2 cm

2 mm

spikelet

3. *Spartina cynosuroides* (L.) Roth (big cordgrass, salt reedgrass). Rhizomatous perennial, rhizomes elongated, brown-purplish to tan, scales imbricate. Culms to 3.5 m tall. Ligules 1–3 mm long. Panicles 15–40 cm long. Glumes with hispid keels. Lower glume from ½ to ⅔ as long as spikelet. Grows mainly in shallow water along bayous and tidal flats. Can be mistaken for *Arundo* and *Phragmites* in vegetative state. Fair livestock and wildlife values (Hatch, Schuster, and Drawe 1999).

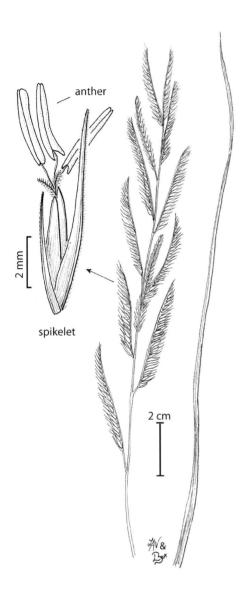

anther

2 mm

spikelet

2 cm

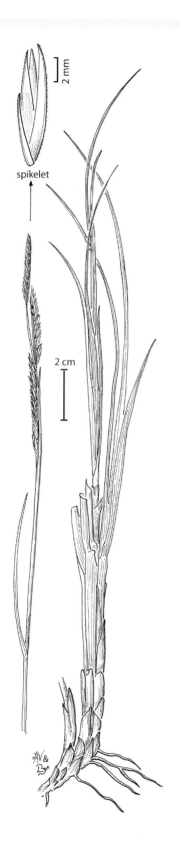

2 mm

spikelet

2 cm

4. *Spartina densiflora* Brongn. (denseflower cordgrass). Cespitose perennial, rarely rhizomatous, rhizomes, when present, short. Culms to 1.5 m tall, forming large clumps. Ligules 1–2 mm long. Panicles 10–30 cm long. Glume keels hispid. Lower glumes about ½ as long as adjacent lemmas. An introduced species from South America. Found in coastal marshes and inland sites. Not included in Gould (1975b); Hatch, Gandhi, and Brown (1990); or Jones, Wipff, and Montgomery (1997).

2 mm

spikelet

2 cm

5. *Spartina patens* (Aiton) Muhl. (saltmeadow cordgrass, marshhay cordgrass, rush saltgrass). Strongly rhizomatous perennial, rhizomes elongated, whitish, scales not imbricate. Culms to 1.5 m tall, usually solitary. Ligules about 0.5 mm long. Panicles 3–15 cm long. Glume keels scabrous to hispid. Lower glumes about ½ as long as spikelet. Found on beaches, sandy flats, and dunes and in muddy bayous and brackish marshes. Listed as poor for wildlife and good for livestock (Hatch and Pluhar 1993), but good wildlife cover (Hatch, Schuster, and Drawe 1999).

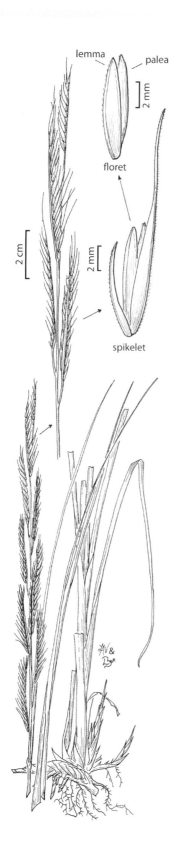

lemma   palea

2 mm

floret

2 cm

2 mm

spikelet

6. *Spartina pectinata* Link. (prairie cordgrass, freshwater cordgrass, sloughgrass, tall marshgrass). Strongly rhizomatous perennial, rhizomes elongate, purplish-brown to light brown (drying white), scales closely imbricate. Culms to 2.5 m tall. Ligules 1–3 mm long. Panicles 10–50 cm long. Glumes with hispid keels. Upper glumes awned, awns 4–10 mm long. Found in fresh or brackish marshy, low wet meadows, ditches, and swales.

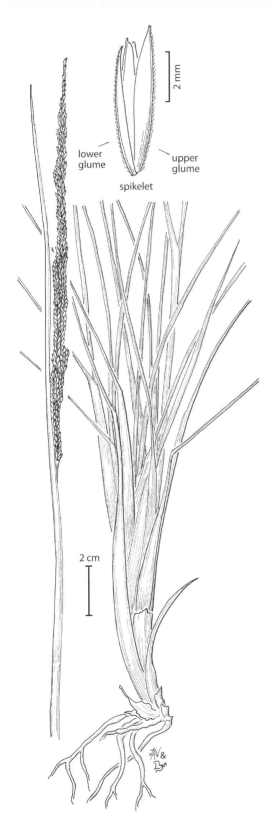

lower glume

upper glume

spikelet

2 mm

2 cm

7. *Spartina spartinae* (Trin.) Merr. *ex* Hitchc. (gulf cordgrass, sacahuista, sacahuista grass, coastal sacahuista). Plants grow in large clumps, rhizomes absent. Culms to 2 m long. Ligules 1–2 mm long. Panicles 6–70 cm long, cylindric. Lower glume nearly as long as spikelet, hispid on keels. Grows in marshes and ditches and along roadsides, sandy beaches, and coastal salt flats. Can be dominant over large areas. Typically prefers soils that are periodically inundated but usually above sea level. Listed as poor for wildlife and fair for livestock (Hatch and Pluhar 1993).

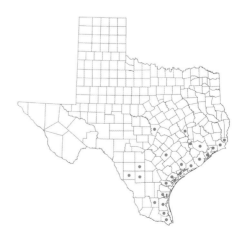

Sphenopholis Scribn.
(Poöideae: Poeae)

Low, tufted annuals or short-lived perennials with soft, flat blades and usually contracted panicles of 2- (rarely 1-) to 3-flowered spikelets, these disarticulating below the glumes. Lower glume narrow, acute, 1-nerved (rarely 3-nerved), upper glume broad, obovate, 3–5-nerved, obtuse or broadly acute at the apex, usually slightly shorter than the lowermost lemma. Lemmas firm, faintly 5-nerved or the veins not visible, rounded on the smooth or rugose back, awnless or less frequently with an awn from just below the apex. Palea thin, membranous, colorless. Basic chromosome number, $x = 7$. Photosynthetic pathway, C_3. Represented in Texas by 5 species.

1. Upper lemmas scabrous on the sides; anthers 0.5–2.0 mm long.
    2. Lower glumes less than ⅓ as wide as the upper glumes; blade <2 mm wide, involute or filiform .... 1. *S. filiformis*
    2. Lower glumes more than ⅓ as wide as the upper glumes; blades 1–7 mm wide, flat or slightly involute ....4. *S. nitida*
1. Upper lemmas smooth on the sides, sometimes slightly scabrous on the lowermost part.
    3. Upper glumes cucullate; panicles erect, spikelike; spikelets densely arranged ....
    ....5. *S. obtusata*
    3. Upper glumes not cucullate; panicles nodding, not spikelike; spikelets often loosely arranged.
        4. Spikelets 2–4 mm long; lowest lemmas 2–3 mm long .... 2. *S. intermedia*
        4. Spikelets 4.0–5.2 mm long; lowest lemmas 3.1–4.2 mm long.............. 3. *S. longiflora*

(Daniel 2007; Gould 1975b)

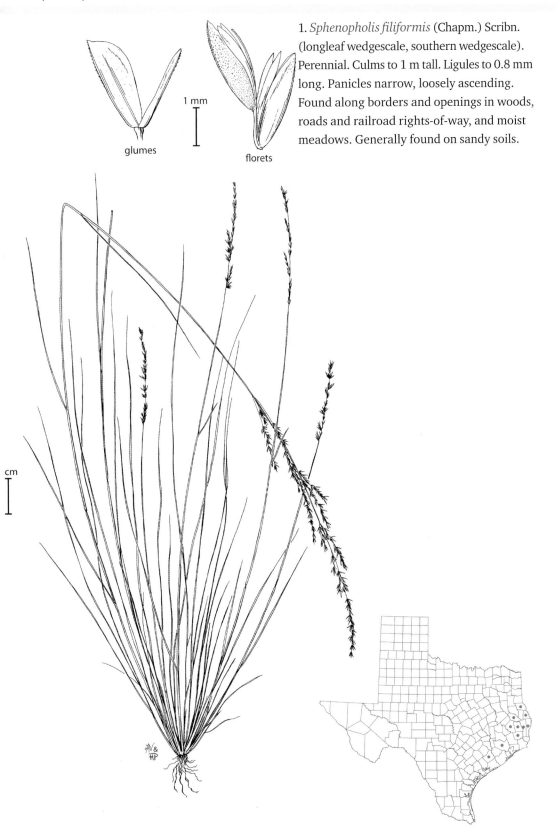

1 mm

glumes

florets

1. *Sphenopholis filiformis* (Chapm.) Scribn. (longleaf wedgescale, southern wedgescale). Perennial. Culms to 1 m tall. Ligules to 0.8 mm long. Panicles narrow, loosely ascending. Found along borders and openings in woods, roads and railroad rights-of-way, and moist meadows. Generally found on sandy soils.

cm

2. *Sphenopholis intermedia* (Rydb.) Rydb.
(slender wedgescale). Tufted perennial. Culms
to 1.2 m long. Ligules 1.5–2.5 mm long. Panicles
usually nodding, not spikelike, spikelets
loosely arranged. Found in forests, meadows,
and waste places. Listed by Gould (1975b);
Hatch, Gandhi, and Brown (1990); Jones,
Wipff, and Montgomery (1997); and Turner
et al. (2003) as a synonym of S. *obtusata* var.
major (Torr.) Erdmann.

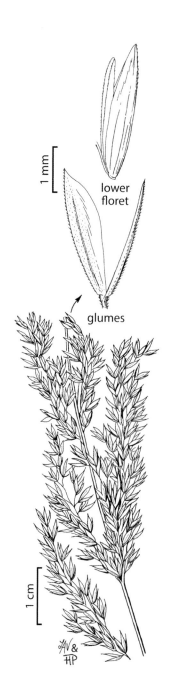

lower
floret

glumes

1 mm

1 cm

3. *Sphenopholis longiflora* (Vasey *ex* L. H. Dewey) Hitchc. (bayou wedgegrass). Cespitose perennial. Culms to 70 cm long. Ligules 0.7–2.6 mm long. Panicles nodding, not spikelike, loosely arranged. Found in forest bottoms and along streams and bayous. Listed by Gould (1975b); Hatch, Gandhi, and Brown (1990); and Jones, Wipff, and Montgomery (1997) as a synonym of *S. obtusata* var. *major* (Torr.) Erdmann.

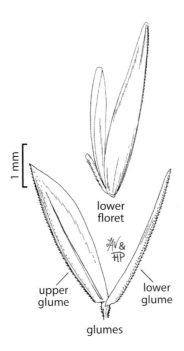

1 mm

lower
floret

upper
glume

lower
glume

glumes

4. *Sphenopholis nitida* (Biehler) Scribn. (shiny wedgescale, shining wedgescale). Perennial. Culms to 80 cm long. Ligules 0.2–1.0 mm long. Panicles erect to nodding, spikelets loosely to densely arranged. Found in hardwood forests, in pine flatwoods, and along stream banks.

1 mm

florets

upper glume

lower glume

2 cm

distal floret

lower floret

glumes

1 mm

2 cm

5. *Sphenopholis obtusata* (Michx.) Scribn. (prairie wedgescale, wedgegrass). Tufted perennial, often flowering the first season and appearing as an annual. Culms to 1.3 m long. Ligules 1.0–2.5 mm long. Panicles narrow, erect, often spikelike, spikelets usually dense. Found in prairies, marshes, dunes, forests, and waste places. Poor livestock and wildlife values (Hatch, Schuster, and Drawe 1999). Powell (1994) reported it as good forage.

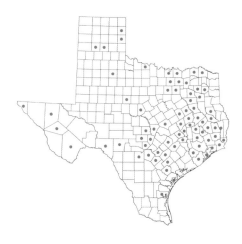

Sporobolus R. Br.
(Chloridoideae: Cynodonteae)

Annuals and perennials, mostly cespitose but a few with creeping rhizomes. Sheaths open; ligule a fringe of hairs or a ciliate membrane; auricles absent. Inflorescence an open or, less frequently, contracted panicle. Spikelets with 1 fertile floret, small, awnless. Disarticulation above the glumes. Glumes 2, unequal, 1-nerved, and usually shorter than the lemma. Lemma 1-nerved, thin, awnless. Palea well developed, as long as to longer than the lemma. Basic chromosome numbers, x = 6 and 9. Photosynthetic pathway, C_4. Represented in Texas by 23 species and 5 varieties.

1. Plants annual.
    2. Lower panicle nodes with 7–15 branches; glumes very unequal.
        3. Pedicels 0.1–1.0 mm long, appressed ....18. *S. pyramidatus*
        3. Pedicels 2–8 mm long, widespreading .... 4. *S. coahuilensis*
    2. Lower panicle nodes with 1–3 branches; glumes about equal.
        4. Lemmas stigose ....21. *S. vaginiflorus*
        4. Lemmas glabrous .... 16. *S. neglectus*
1. Plants perennial.
        5. Pedicels 5–25 mm long.
            6. Spikelets 2–3 mm long; second glume 1-veined ....20. *S. texanus*
            6. Spikelets 5–6 mm long; second glume 3-veined ....19. *S. silveanus*
        5. Pedicels up to 4 mm long.
            7. Plants with long, creeping, scaly rhizomes.
                8. Spikelets <3.2 mm long; seashore plants ....22. *S. virginicus*
                8. Spikelets 4–5 mm long; woodland or prairie plants ....5. *S. compositus*
            7. Plants without rhizomes.
                9. Spikelets <2.8 mm long .... Key A
                9. Spikelets, at least some, >3 mm long .... Key B

Key A

1. Lower sheaths strongly compressed-keeled .... 2. *S. buckleyi*
1. Lower sheaths round.
    2. Glumes about equal, ½ to ⅔ as long as floret.
        3. Spikelets 2.0–2.7 mm long; upper glumes usually ½ to ⅔ as long as the florets, acute to obtuse, entire.. 13. *S. indicus*
        3. Spikelets 1.3–1.8 mm long; upper glume usually less than ½ as long as the florets, truncate, erose to denticulate...8. *S. diandrus*
    2. Glumes unequal, the lower short, the upper about as long as florets.
        4. Panicles spikelike.
            5. Culms 1–2 m tall, 4–10 mm thick near base; mature panicles 1–4 cm wide ....
            ....11. *S. giganteus*

5. Culms about 1 m tall or less, 2–4 mm thick near base; mature panicles <1 cm wide ....6. *S. contractus*

4. Panicles open, at least above, the lower portion often enclosed in subtending sheaths.

    6. Panicles 10–30 cm wide, open.

        7. Sheaths with copious long hairs at corners; hairs erect....... 7. *S. cryptandrus*

        7. Sheaths glabrous or with only a few hairs at corners.

            8. Secondary panicle branches spikelet-bearing to the base; pedicels mostly appressed, 0.2–0.5 mm long.. 23. *S. wrightii*

            8. Secondary panicle branches without spikelets on lower ¼ to ½; pedicels mostly spreading, 0.5–2.0 mm long.....................................1. *S. airoides*

    6. Panicles mostly <10 cm wide.

        9. Primary panicle branches without spikelets on the lower ⅛ to ½ of their length.

            10. Plants with hard, knotty bases; blades spreading at right angles to the culms, 1.0–1.5 mm wide.....................................15. *S. nealleyi*

            10. Plant bases not hard, knotty; blades erect or ascending, 2–6 mm wide.

                11. Pedicels appressed to the secondary branches; pulvini glabrous; rachises straight, erect .... 7. *S. cryptandrus*

                11. Pedicels spreading to the secondary branches; pulvini pubescent; rachises drooping or nodding...........10. *S. flexuosus*

        9. Primary panicle branches spikelet-bearing to the base.

            12. Lower glumes usually 1-veined; lemmas and upper glumes 2.0–3.2 mm long...6. *S. contractus*

            12. Lower glumes usually veinless; lemmas and upper glumes 1.1–2.0 mm long ....9. *S. domingensis*

Key B

1. Panicle branches in distinct whorls, lower nodes with 3 or more branches.

    2. Mature panicles pyramidal, 2–6 cm wide; blades 0.8–2.0 mm wide .... 14. *S. junceus*

    2. Mature panicles cylindric, 0.4–1.6 cm wide; blades 2–5 mm wide ....17. *S. purpurascens*

1. Panicle branches not in whorls, lower nodes with 1–2, rarely 3, branches.

    3. Panicles spikelike, contracted, the branches appressed along rachises.

        4. Spikelets 4–6(–10) mm long; panicles terminal and axillary; sheaths without conspicuous hairs at corners.

            5. Lemmas pubescent ....3. *S. clandestinus*

            5. Lemmas glabrous...5. *S. compositus*

        4. Spikelets 1.7–3.5 mm long; panicles terminal only; sheaths with conspicuous hairs at corners.

            6. Culms 1–2 m tall, 4–10 mm thick near base; mature panicles 1–4 cm wide... 11. *S. giganteus*

6. Culms about 1 m tall or less, 2–4 mm thick near base; mature panicles <1 cm wide.. 6. *S. contractus*
3. Panicles pyramidal to ovate, open, the branches ascending or spreading along rachises.
7. Spikelets 2.3–3.0 mm long; panicles diffuse, about as wide as long.. 20. *S. texanus*
7. Spikelets 3–7 mm long; panicles not diffuse, longer than wide.
8. Mature spikelets plumbeous (lead-colored); sheath bases dull and fibrous.. 12. *S. heterolepis*
8. Mature spikelets purplish-brown to purplish; sheath bases shiny, indurate ....19. *S. silveanus*

(Peterson, Hatch, and Weakley 2003)

1. *Sporobolus airoides* (Torr.) Torr. (alkali sacaton, fine-top saltgrass). Coarse perennial with numerous culms (to 1.5 m long) arising from a hard base. Clumps can grow to over 1 m in width. Panicles open, diffuse, pyramidal. Spikelets 1.3–2.5 mm long, greenish to purple. Found on dry, rocky slopes, playas, and floodplains and around saline and alkaline flats. *Sporobolus tharpii* Hitchc. reported in Gould (1975b) and Jones, Wipff, and Montgomery (1997) is included here. Listed as poor for wildlife and fair for livestock grazing (Hatch and Pluhar 1993) but regarded as a good forage grass in alkaline areas.

1 mm

spikelet

2 cm

1 mm

spikelet

2. *Sporobolus buckleyi* Vasey (Buckley's dropseed). Cespitose perennial. Culms to 1 m long. Sheaths keeled and flattened basally. Panicles diffuse, ovate. Spikelets 1–2 mm long, purplish or brownish. Found in shaded habitats of extreme South Texas.

2 cm

spikelet

3. *Sporobolus clandestinus* (Biehler) Hitchc. (hidden dropseed, purple-flowered dropseed, rough dropseed). Perennial, occasionally with rhizomes. Culms to 1.5 m long, often glaucous. Panicles narrow, spikelet, included in the upper leaf sheaths. Spikelets 4–9 mm long. Found in sandy soils, along roadsides and fencerows, and in mesic longleaf pine communities and oak-pine forests. Jones, Wipff, and Montgomery (1997) listed this as *S. compositus* var. *clandestinus* (Biehler) J. Wipff & S. D. Jones.

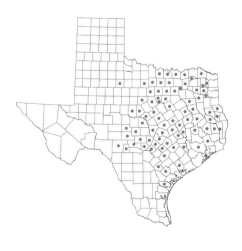

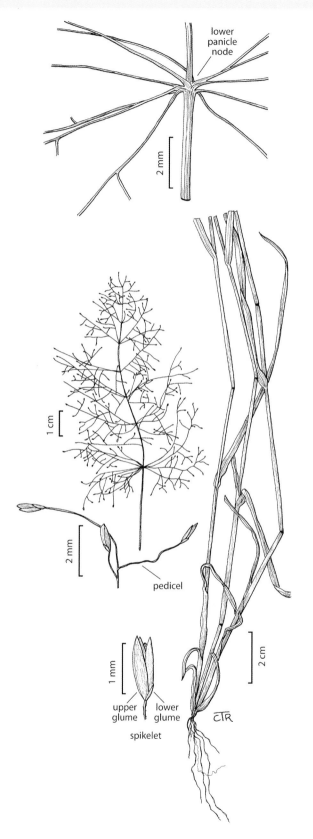

4. *Sporobolus coahuilensis* Valdés-Reyna (Coahuila dropseed). Perennial, occasionally with rhizomes. Culms to 60 cm long, erect to ascending. Panicles usually open, lowest branches whorled. Spikelets 1.1–1.5 mm long. Only recently collected in Brewster and Hudspeth counties. Not reported in Gould (1975b); Hatch, Gandhi, and Brown (1990); Jones, Wipff, and Montgomery (1997); or Peterson, Hatch, and Weakley (2003).

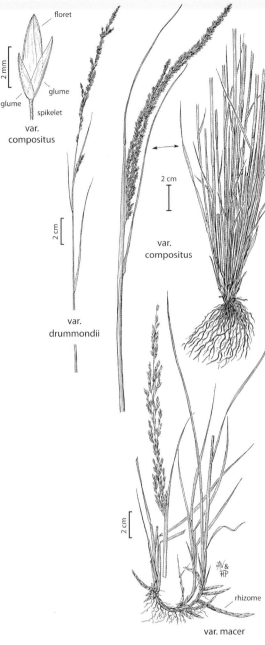

floret

2 mm

glume

glume

spikelet

var. compositus

2 cm

var. drummondii

var. compositus

2 cm

var. drummondii

2 cm

2 cm

rhizome

var. macer

5. *Sporobolus compositus* (Poir.) Merr. (head-like dropseed, rough dropseed, flaggrass). Perennial, sometimes with rhizomes. Culms to 1.5 m tall. Panicles to 30 cm long, spikelike, contracted, partially included in upper sheaths. Lower nodes with 1–2(–3) branches. Spikelets 4–6(–10) mm long. A common species of roadsides, prairies, open woodlands, grasslands, beaches, and live oak–pine forests. Poor for wildlife and fair for livestock (Hatch and Pluhar 1993). Gould (1975b) and Hatch, Gandhi, and Brown (1990) listed this species as *S. asper* (Michx.) Kunth. Three varieties are recognized: var. *compositus*; var. *drummondii* (Trin.) Kartesz & Gandhi (Mississippi dropseed); & var. *macer* (Trin.) Kartesz & Gandhi (tall dropseed). Jones, Wipff, and Montgomery (1997) listed var. *drummondii* as a synonym of var. *compositus*. The varieties can be distinguished by the following characters:

1. Rhizomes present .... var. *macer*
1. Rhizomes absent.
    2. Culms stout, 2–5 mm thick; upper sheaths 2.6–6.0 mm wide ....var. *compositus*
    2. Culms slender, 1–2 mm thick; upper sheaths <2.5 mm wide .... var. *drummondii*

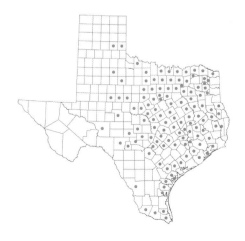

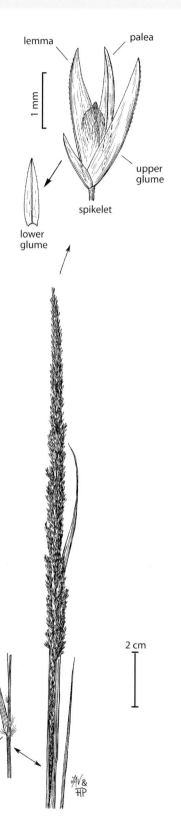

6. *Sporobolus contractus* Hitchc. (spike dropseed). Cespitose perennial. Culms to 1.2 m long. Sheaths with tufts of white, erect, hairs 2.3 mm long on each side of the collar. Panicles contracted, spikelike, dense, usually included. Spikelets 1.7–3.2 mm long, whitish to gray. Found in sandy soils in desert shrublands and grasslands and in juniper woodlands. Not nearly as common as *S. cryptandrus*, which it resembles. Powell (1994) suspected that this would be a good forage species wherever it was abundant.

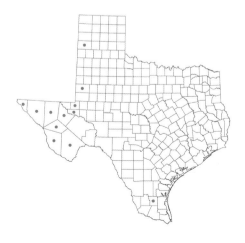

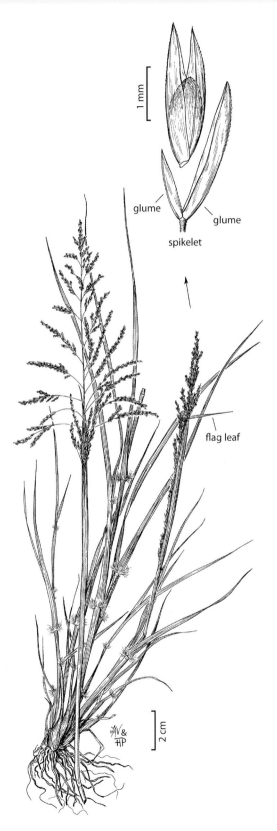

1 mm

glume

glume

spikelet

flag leaf

2 cm

7. *Sporobolus cryptandrus* (Torr.) A. Gray (sand dropseed, covered spike dropseed). Cespitose perennial. Culms to 1.2 m long, erect. Sheaths with distinctive tufts of white, erect, hairs 2–4 mm long on either side of the collar. Panicle open or closed depending upon how much is enclosed by the elongated upper sheaths. Initially contracted and spikelike and becoming open and pyramidal. Spikelets 1.5–2.5 mm long, gray, brown, or purplish. One of the most common and widely distributed grasses in the state. Found along roadsides, in pastures and grasslands, and in sandy areas in deciduous and evergreen forests. Listed as poor for wildlife and fair for livestock (Hatch and Pluhar 1993). Powell (1994) ranked this as an important forage species because of its abundance.

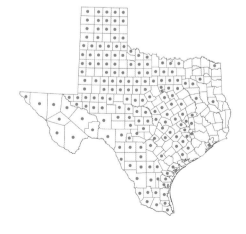

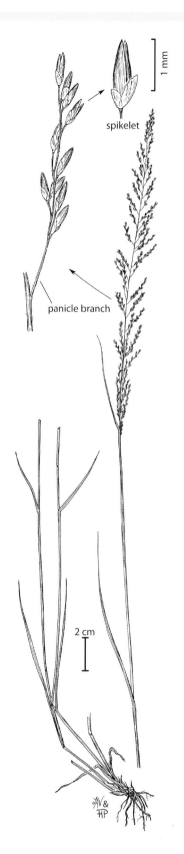

spikelet

1 mm

panicle branch

2 cm

8. *Sporobolus diandrus* (Retz.) P. Beauv. (tussock dropseed). Cespitose perennial. Culms to 80 cm long. Panicles contracted to open. Spikelets 1.3–1.8 mm long. Both glumes less than ½ as long as the spikelets, truncate. A recent introduction from India. Not reported in Gould (1975b); Hatch, Gandhi, and Brown (1990); or Jones, Wipff, and Montgomery (1997).

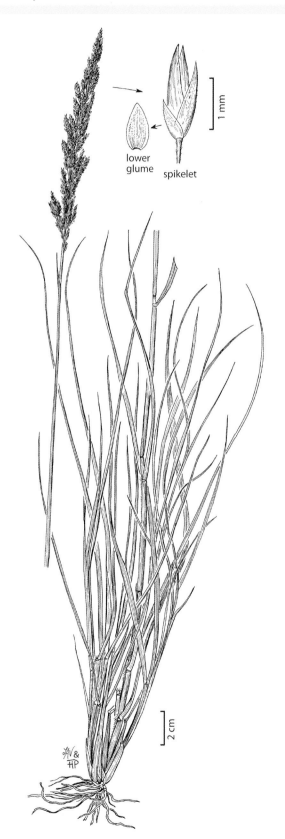

lower
glume

spikelet

1 mm

2 cm

9. *Sporobolus domingensis* (Trin.) Kunth (coral dropseed). Cespitose perennial. Culms to 1 m long. Panicles contracted, sometimes spikelike, interrupted below. Spikelets 1.6–2.0 mm long. Found in sandy or rocky disturbed sites adjacent to the coast. Not reported in Gould (1975b); Hatch, Gandhi, and Brown (1990); or Jones, Wipff, and Montgomery (1997).

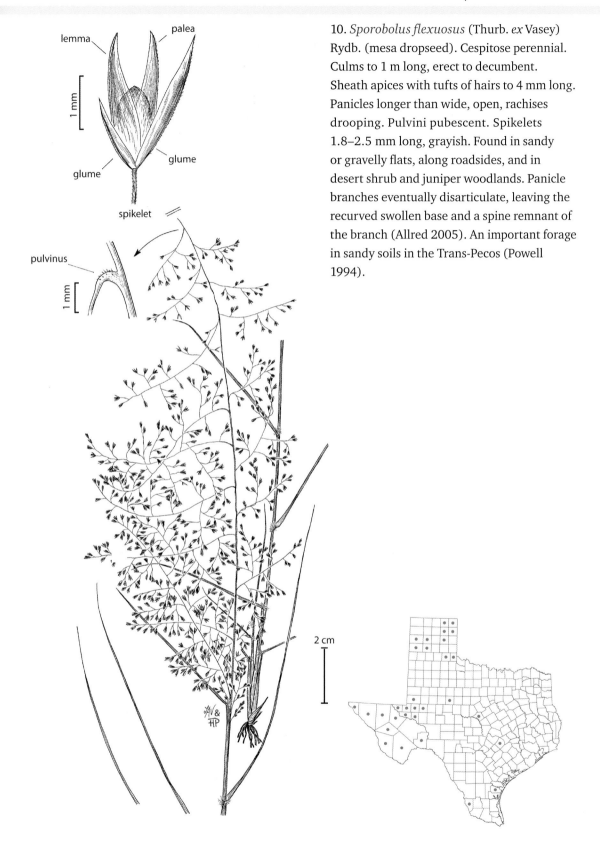

10. *Sporobolus flexuosus* (Thurb. *ex* Vasey) Rydb. (mesa dropseed). Cespitose perennial. Culms to 1 m long, erect to decumbent. Sheath apices with tufts of hairs to 4 mm long. Panicles longer than wide, open, rachises drooping. Pulvini pubescent. Spikelets 1.8–2.5 mm long, grayish. Found in sandy or gravelly flats, along roadsides, and in desert shrub and juniper woodlands. Panicle branches eventually disarticulate, leaving the recurved swollen base and a spine remnant of the branch (Allred 2005). An important forage in sandy soils in the Trans-Pecos (Powell 1994).

spikelet

11. *Sporobolus giganteus* Nash (giant dropseed). Cespitose perennial, robust with culms to 2 m long. Panicles spikelike, dense, usually included in uppermost sheaths. Spikelets 2.5–3.5 mm long, whitish to grayish. Found on sand dunes and along sandy riverbanks and roadsides. Excellent as both a forage species and soil binder (Powell 1994).

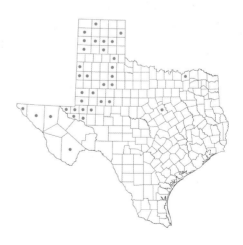

spikelet

2 cm

12. *Sporobolus heterolepis* (A. Gray) A. Gray
(prairie dropseed). Cespitose perennial.
Culms stiffly erect, to 90 cm long. Sheaths dull
and fibrous basally. Panicles open or slightly
contracted, narrowly pyramidal. Spikelets
3–6 mm long, grayish. Palea splitting between
veins as fruit matures. Found in prairies
and along woodland borders and roadsides.
Sometimes used as an ornamental. Diggs et al.
(2006) reported that all collections identified
as S. *heterolepis* were really *S. silveanus* and
that the former does not occur in the state.
Both are included here until the situation is
resolved.

spikelet

1 mm

2 cm

13. *Sporobolus indicus* (L.) R. Br. (smutgrass). Perennial. Culms to 1 m long, usually much shorter. Panicles narrow, tightly contracted, sometimes included. Spikelets 2.0–2.6 mm long. Found in disturbed sites in both clayey and sandy soils. Spikelets often infected with a fungus (*Bipolaris* spp.) that leaves a black "smut" on hands and clothing. Listed as poor for wildlife and livestock (Hatch and Pluhar 1993).

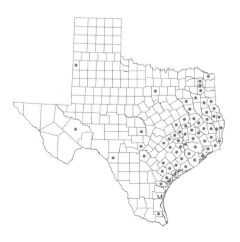

14. *Sporobolus junceus* (P. Beauv.) Kunth (pineywoods dropseed, purple dropseed, wiregrass). Cespitose perennial with long basal leaves. Culms to 1 m long. Panicles open, narrow, pyramidal. Branches whorled or verticellate. Spikelets 2.6–3.8 mm long, purplish-red. Found in mixed forests, coastal prairies, and pine barrens. Usually in sandy to sandy loam soils. Fair livestock and wildlife values (Hatch, Schuster, and Drawe 1999).

spikelet

inflorescence branches

1 mm

1 cm

2 cm

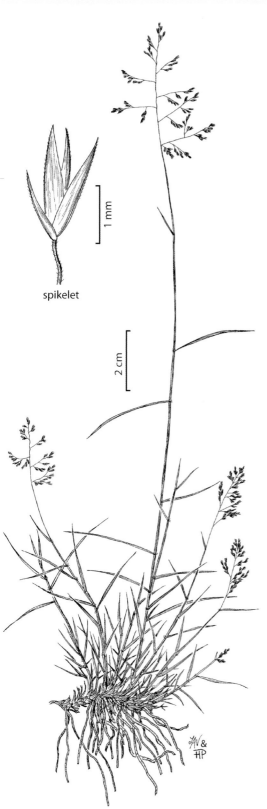

spikelet

1 mm

2 cm

15. *Sporobolus nealleyi* Vasey (gypgrass, Nealley's dropseed). Cespitose perennial from a hard, knotty base. Culms slender, to 50 cm long. Sheaths pubescent on margins and collar with long (to 4 mm), kinky hairs. Panicles longer than wide, open, interrupted basally, lower portions sometimes included. Spikelets 1.3–2.1 mm long, purplish. Grows in rocky and sandy soils derived from gypsum, or near alkaline habitats. According to Allred (2005) the middle of clumps tend to die, forming "bird nests." Probably a good forage in its limited habitats (Powell 1994).

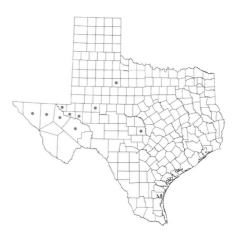

16. *Sporobolus neglectus* Nash (puffsheath dropseed, small dropseed, povertygrass). Delicate annual. Culms to 45 cm long, erect or decumbent, wiry. Sheaths inflated. Panicles contracted, cylindric, included. Spikelets 1.6–3.0 mm long. Found on disturbed sites and sandy fields. Jones, Wipff, and Montgomery (1997) listed this as a variety of *S. vaginiflorus*.

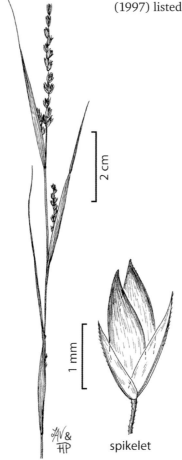

spikelet

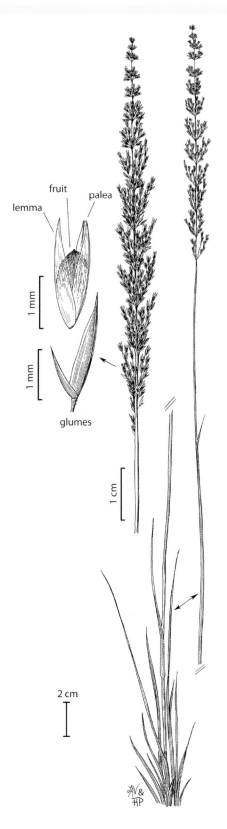

17. *Sporobolus purpurascens* (Sw.) Ham. (purple dropseed). Cespitose perennial. Culms stiffly erect, to 90 cm long. Sheath apices hairy, hairs to 5 mm long. Panicles narrow, contracted, lower nodes with 3–5 branches. Spikelets 2.8–3.8 mm long, purplish. Found in sandy soils, mostly in oak mottes and coastal prairies. Fair livestock and wildlife values (Hatch, Schuster, and Drawe 1999).

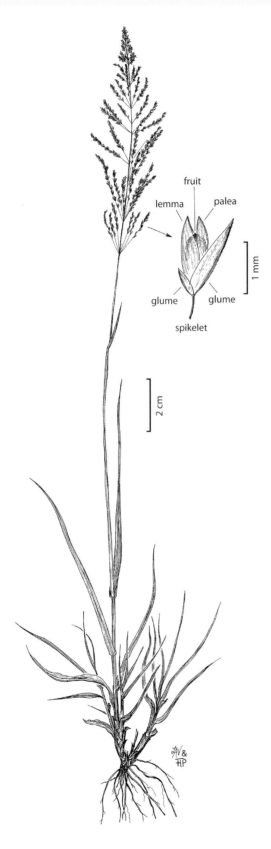

fruit

lemma   palea

1 mm

glume   glume

spikelet

2 cm

18. *Sporobolus pyramidatus* (Lam.) Hitchc.
(whorled dropseed, target dropseed). Annual
or very short-lived perennial. Culms usually
about 35 cm long, erect or decumbent.
Panicles contracted when immature,
pyramidal when mature, open. Branches
whorled at lower nodes. Found in open
disturbed sites, often in coastal sands, on
sandy or saline clayey soils, or on alkaline
inland soils. Gould (1975b) listed the annual
form as *S. pulvinatus* Swallen.

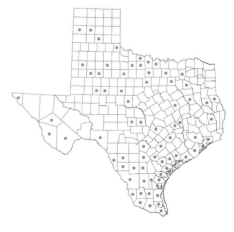

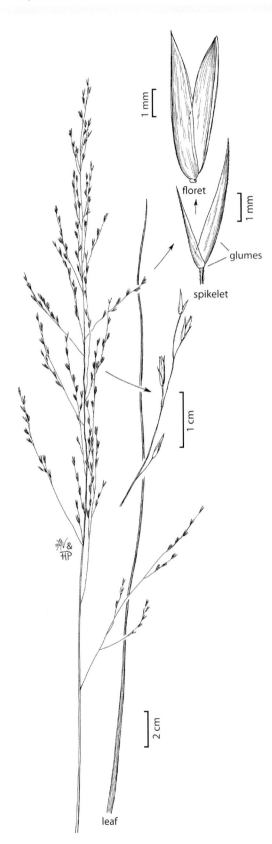

1 mm

floret

1 mm

glumes

spikelet

1 cm

leaf

2 cm

19. *Sporobolus silveanus* Swallen (Silveus' dropseed). Cespitose perennial with slender erect culms to 1.2 m long. Panicles loosely open to slightly contracted, longer than wide, pyramidal, few-flowered. Spikelets 4.5–7.0 mm long, purple. Found in pine woodlands and forest openings and in the Blackland Prairies.

glume

glume

spikelet

1 mm

2 cm

20. *Sporobolus texanus* Vasey (Texas dropseed). Perennial, often appearing annual. Culms to 70 cm long, decumbent at base. Panicles open diffuse, as wide as long, partially included by uppermost sheaths. Pedicels 6–25 mm long. Spikelets 2.3–3.0 mm long. Lower glumes often without midrib. Found around rivers, ponds, tanks, and wet alkaline areas.

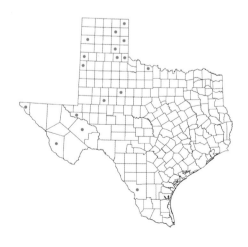

21. *Sporobolus vaginiflorus* (Torr. *ex* A. Gray). Alph. Wood (poverty dropseed, southern povertygrass). Annual with wiry culms to 60 cm. Culms usually erect or decumbent. Sheaths usually inflated. Panicles cylindric, contracted, and contained in leaf sheaths. Lower nodes with 1–2(–3) branches. Spikelets 2.3–6.0 mm long. Found in disturbed sites on limestone outcrops and calcareous soils. Gould (1975b) spelled the specific epithet *vaginaeflorus*. Two varieties are recognized: var. *ozarkanus* (Fernald) Shinners (Ozark dropseed); and the typical one, var. *vaginiflorus*. Gould (1975b) and Hatch, Gandhi, and Brown (1990) listed the former as a separate species; and Jones, Wipff and Mongomery (1997) listed it as a synonym of var. *neglectus* (Nash) Scribn. The varieties can be distinguished by the following characters:

1. Lemmas faintly 3-veined; sheath bases sparsely hairy; glumes usually longer than florets ....var. *ozarkanus*
1. Lemmas 1-veined; sheath bases glabrous; glumes usually shorter than the florets ....var. *vaginiflorus*

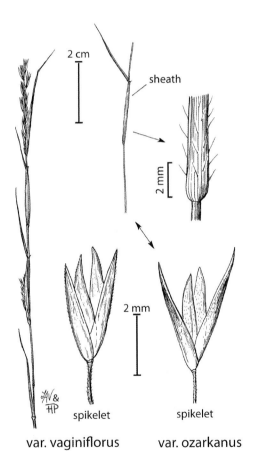

2 cm

sheath

2 mm

2 mm

AV & HP

spikelet

spikelet

var. vaginiflorus

var. ozarkanus

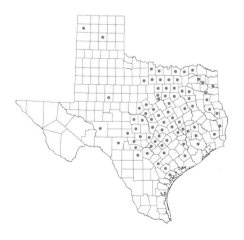

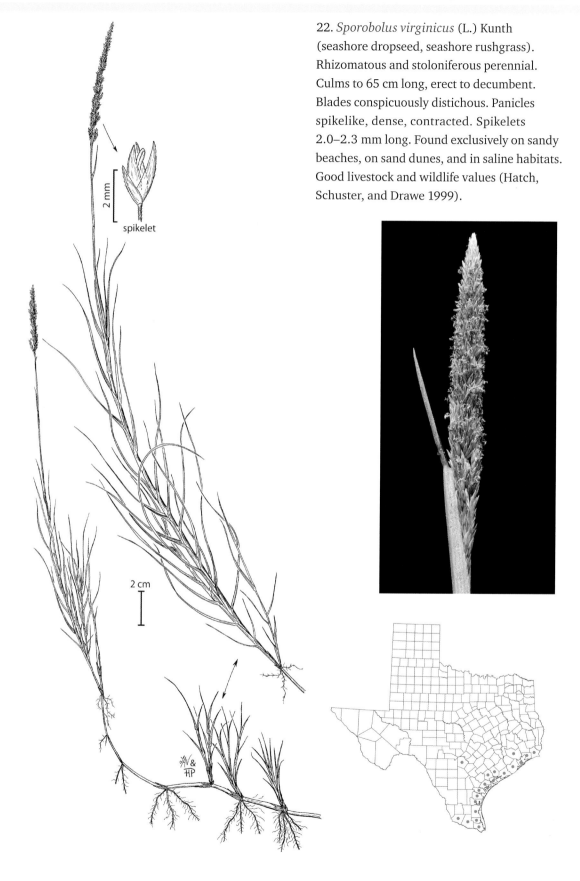

22. *Sporobolus virginicus* (L.) Kunth
(seashore dropseed, seashore rushgrass).
Rhizomatous and stoloniferous perennial.
Culms to 65 cm long, erect to decumbent.
Blades conspicuously distichous. Panicles
spikelike, dense, contracted. Spikelets
2.0–2.3 mm long. Found exclusively on sandy
beaches, on sand dunes, and in saline habitats.
Good livestock and wildlife values (Hatch,
Schuster, and Drawe 1999).

2 mm

spikelet

2 cm

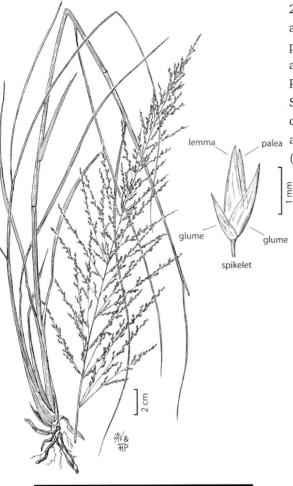

lemma   palea

glume   glume

spikelet

1 mm

2 cm

23. *Sporobolus wrightii* Munro *ex* Scribn. (big alkali sacaton, Wright's sacaton). Robust, erect perennial. Numerous culms, up to 2.5 m long, arising from a hard, densely cespitose base. Panicles open, broadly lanceolate, exserted. Spikelets 1.5–2.5 mm long. Found on moist clay flats and on borders of alkaline and saline areas, playas, and swales. Fair forage species (Powell 1994).

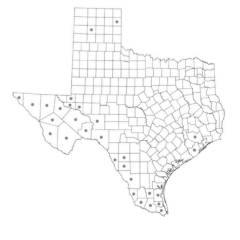

Steinchisma Raf.
(Panicoideae: Paniceae)

Perennials with slender, tufted culms that are often freely branching. Ligule a fringed scale. Panicles small, few-flowered, variable in branching and general aspect, with spikelets clustered along erect-spreading primary and short secondary branches. Spikelets glabrous and at first broadly oblong but later widely gaping at the apex between the florets. Glumes both present, the lowermost reduced. Lower floret neuter, the palea becoming inflated, obovate, apiculate, and much broader than the lemma. Lemma of the upper floret smooth but not shining. Basic chromosome number, $x = 10$. Photosynthetic pathway, intermediate between C_3 and C_4. Represented in Texas by a single species.

(Freckmann and Lelong 2003c; Gould and Shaw 1983)

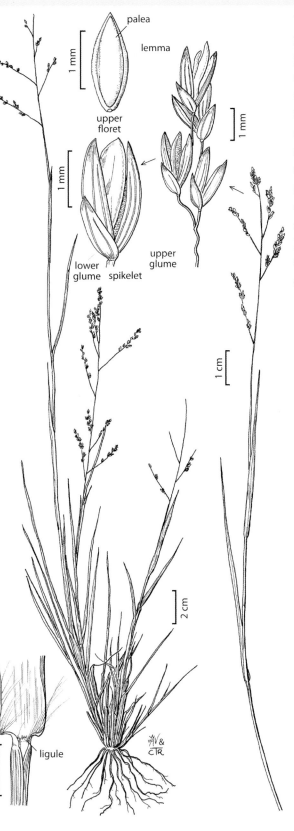

palea
lemma
upper floret
1 mm
1 mm
1 mm
lower glume
spikelet
upper glume
1 cm
2 cm
ligule
JAV & CTR

1. *Steinchisma hians* (Elliot) Nash (gaping panicgrass, gaping grass). Rhizomatous perennial. Culms to 75 cm long. Sheaths compressed. Ligules 0.2–0.5 mm long, an erose-ciliate membrane. Paleas of the lower floret enlarged, indurate, and giving the spikelet the appearance of "gaping." Found in swales in grasslands, roadside ditches, moist areas in open woodlands, and pinelands. A segregate of *Panicum*. One of the few examples of intermediate photosynthetic pathways, suggesting that it is evolving from a C_3 to a C_4. Included in *Panicum* by Gould 1975b; Hatch, Gandhi, and Brown 1990; and Jones, Wipff, and Montgomery (1997).

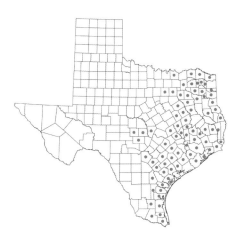

Stenotaphrum Trin.
(Panicoideae: Paniceae)

Low, mat-forming, stoloniferous perennials with thick, succulent, flat blades and (in Texas species) spikelets partially embedded in the flattened and thickened rachis of a short, unilateral, spicate inflorescence. Disarticulation at the nodes of the rachis, the spikelets falling attached to the sections. (Note: The spikelets are actually borne on rudimentary branch axes that are extended as a point beyond the insertion of the 1–3 spikelets present at each node of the main rachis.) First glume short, irregularly rounded. Second glume and lemma of the sterile floret about equal, glabrous, pointed at the apex, faintly nerved. Reduced (lower) floret often staminate. Lemma of the fertile floret chartaceous. Basic chromosome number, $x = 9$. Photosynthetic pathway, C_4. Represented in Texas by a single species.

(Allred 2003f)

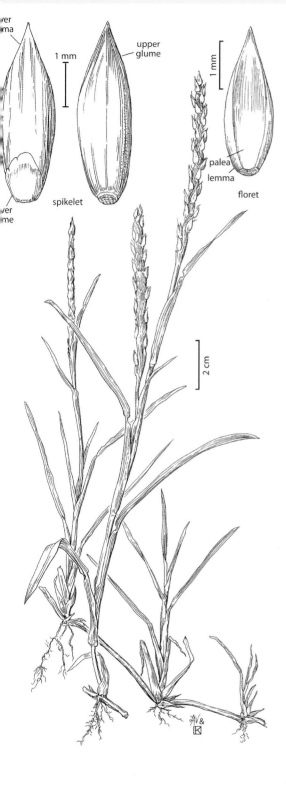

lver
ma

1 mm

upper
glume

1 mm

spikelet

palea
lemma

floret

lver
me

2 cm

1. *Stenotaphrum secundatum* (Walt.)
Kuntze (St. Augustine grass, carpetgrass).
Aggressively stoloniferous perennial. Culms
decumbent, rooting at the lower nodes,
producing a dense matted tuft. Prophylls
prominent. Spikelets embedded in the
flattened, succulent rachis. A major turfgrass
species that is infrequently collected. Also
used as an ornamental in hanging baskets.
Considered by some to be introduced but was
collected on the Atlantic coast prior to 1800
and now considered native. Poor livestock and
wildlife values (Hatch, Schuster, and Drawe
1999).

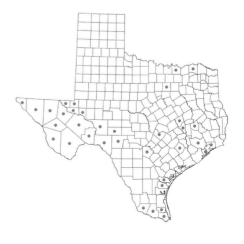

Themeda Forssk.
(Panicoideae: Andropogoneae)

Usually cespitose annuals or perennials. Culms from 30–310 cm long and erect. Ligules membranous, sometimes ciliate; auricles absent; sheaths open. Inflorescences numerous, both terminal and axillary, false panicles. Peduncles shorter than the subtending leaf sheaths, with 1–8 rames. Rames with homogamous sessile-pedicellate spikelet pairs at the base and 1–4 smaller, heterogamous sessile-pedicellate spikelet pairs above. Disarticulation in the rames below the sessile spikelets of the heterogamous units. Homogamous spikelet pairs forming an involucre around the rame bases, spikelets staminate or sterile, unawned. Heterogamous spikelet pairs, sessile spikelet perfect and awned; pedicellate spikelets staminate, sterile, and unawned. Basic chromosome number, $x = 10$. Photosynthetic pathway, C_4. Represented in Texas by a single species.

(Barkworth 2003o)

1. *Themeda triandra* Forssk. (kangaroograss, rooigrass). Cespitose perennial with culms up to 3 m tall. False panicles up to 50 cm long. Rames 15–20 mm long, with 1 heterogamous spikelet pair. An introduced species from Asia that is commonly used as an ornamental; perhaps has escaped and become established in Texas. A single record from Travis County has been reported. Not listed by Gould (1975b) or Hatch, Gandhi, and Brown (1990).

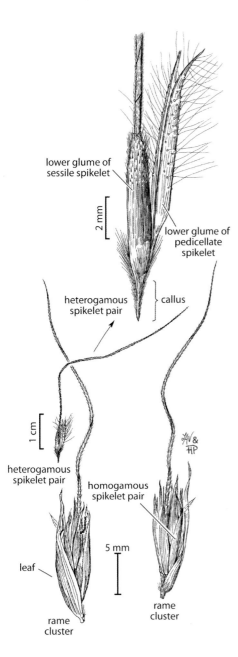

lower glume of
sessile spikelet

2 mm

lower glume of
pedicellate
spikelet

heterogamous
spikelet pair

callus

1 cm

heterogamous
spikelet pair

homogamous
spikelet pair

5 mm

leaf

rame
cluster

rame
cluster

Thinopyrum Á. Löve
(Poöideae: Triticeae)

Cespitose or rhizomatous perennials. Inflorescence a terminal spike that persists at maturity. Sheaths open; auricles present or absent; ligules membranous. Disarticulation typically above the glumes and beneath the florets, sometimes in the rachis. Spikelets 1–3 times the length of the middle internodes, solitary, often arching outward at maturity, 3–10 florets. Glumes stiff, coriaceous to indurate, apices truncate to acute, sometimes mucronate. Lemma 5-nerved, coriaceous, truncate, obtuse or acute, sometimes mucronate or awned (to 3 cm). Basic chromosome number x = 7. Photosynthetic pathway, C_3. Represented in Texas by 2 species and 2 subspecies. Members of this genus have been included in *Agropyron*, *Elymus*, and *Elytrigia*.

1. Plants rhizomatous ....1. *T. intermedium*
1. Plants not rhizomatous ....2. *T. ponticum*

(Barkworth 2007n)

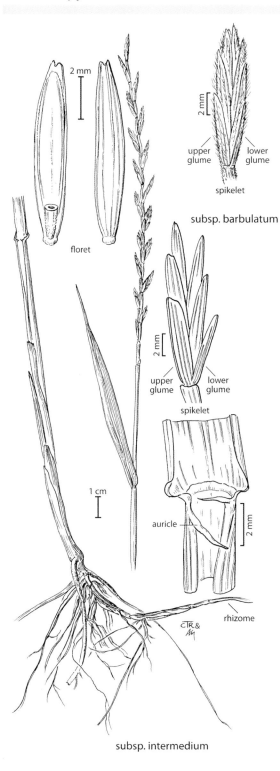

floret

subsp. barbulatum

spikelet

upper glume

lower glume

spikelet

auricle

rhizome

subsp. intermedium

1. *Thinopyrum intermedium* (Host) Barkworth and D. R. Dewey (intermediate wheatgrass). Rhizomatous perennial. Culms to 1.2 m tall. Auricles to 2 mm long. Glume midveins more prominent and longer than lateral veins. An introduction from Europe that is used extensively for erosion control, revegetation, and forage. Two subspecies occur in Texas: subsp. *barbulatum* (Schur) Barkworth & D. R. Dewey; and subsp. *intermedium*. They can be distinguished by the following character:

1. Lemmas and glumes glabrous ....subsp. *intermedium*
1. Lemmas with hairs, sometimes only on margin; glumes usually hairy .... subsp. *barbulatum*

2. *Thinopyrum ponticum* Barkworth and D. R.
Dewey (tall wheatgrass, rush wheatgrass).
Cespitose perennial. Culms to 2 m tall. Auricles
to 1.5 mm long. Midveins about as prominent
and long as lateral veins. A Eurasian
introduction used along roadsides for erosion
control.

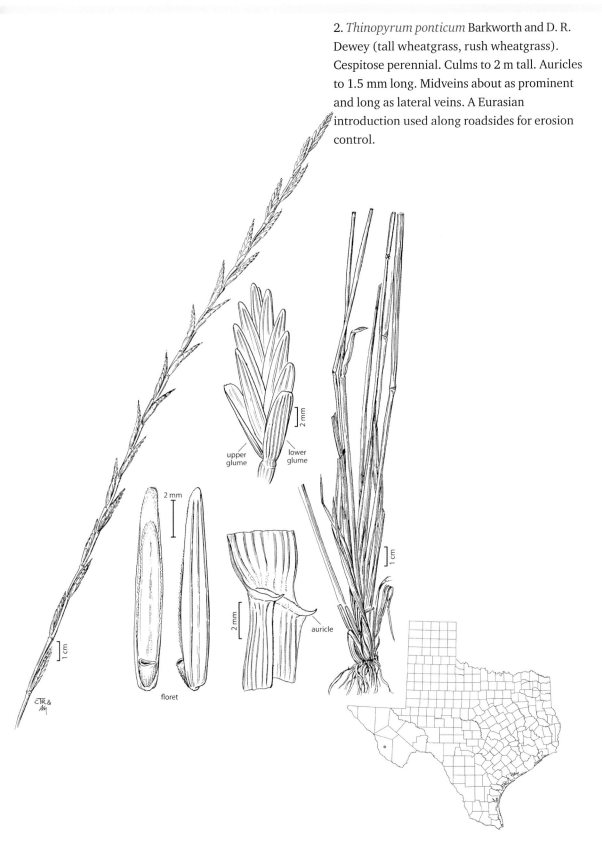

Trachynia (L.) Link
(Poöideae: Brachypodieae)

The Texas species an annual with short-pediceled or subsessile spikelets borne singly at the nodes of a stiffly erect, spicate raceme. Inflorescence sometimes reduced to a single, terminal spikelet. Spikelets large, several- to many-flowered. Disarticulation above the glumes and between the florets. Glumes unequal, stiffly pointed, 5–7-nerved. Lemmas firm, rounded or flattened on back, 7-nerved, the midnerve extending into a short or long awn from an entire apex. Palea about as long as the lemma, 2-keeled, often with stiff, pectinate-ciliate hairs on the veins. Basic chromosome number, $x = 7$. Photosynthetic pathway, C_3. Represented in Texas by a single species.

(Piep 2007)

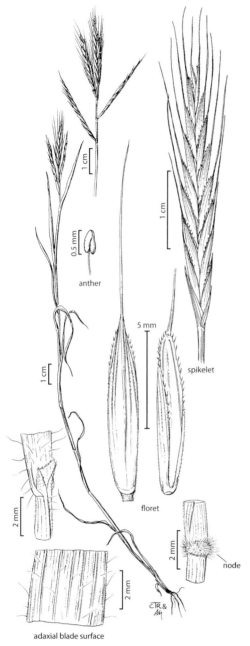

anther

spikelet

floret

node

adaxial blade surface

1. *Trachynia distachya* (L.) Link (purple falsebrome). Annual. Culms erect or occasionally geniculate, up to 1 m tall. Spikelets up to 40 mm long, strongly laterally compressed, and with 7–15 florets. Florets cleistogamous with much-reduced anthers. Lemmas awned, awns up to 17 mm long, straight or curved. A southern European introduction and now a weed of disturbed sites. This species is sometimes treated as the only annual member of the genus *Brachypodium* (Gould 1975b; Hatch, Gandhi, and Brown 1990; and Jones, Wipff, and Montgomery 1997).

Trachypogon Nees
(Panicoideae: Andropogoneae)

Moderately tall, cespitose perennials with culms terminating in a spikelike raceme or, less frequently, 2 to a few spicate branches. Spikelets in pairs on a continuous rachis, 1 subsessile and 1 with a slightly longer pedicel. Subsessile spikelet staminate, as large as the perfect spikelet but awnless and persistent on the rachis. Longer-pediceled spikelet perfect, with a firm, rounded, several-nerved first glume; a firm, few-nerved second glume; and a thin, narrow lemma (of the fertile floret) bearing a stout, twisted, geniculate or flexuous awn. Palea of fertile floret usually absent. Basic chromosome number, $x = 10$. Photosynthetic pathway, C_4. Represented in Texas by a single species.

(Allred 2003g)

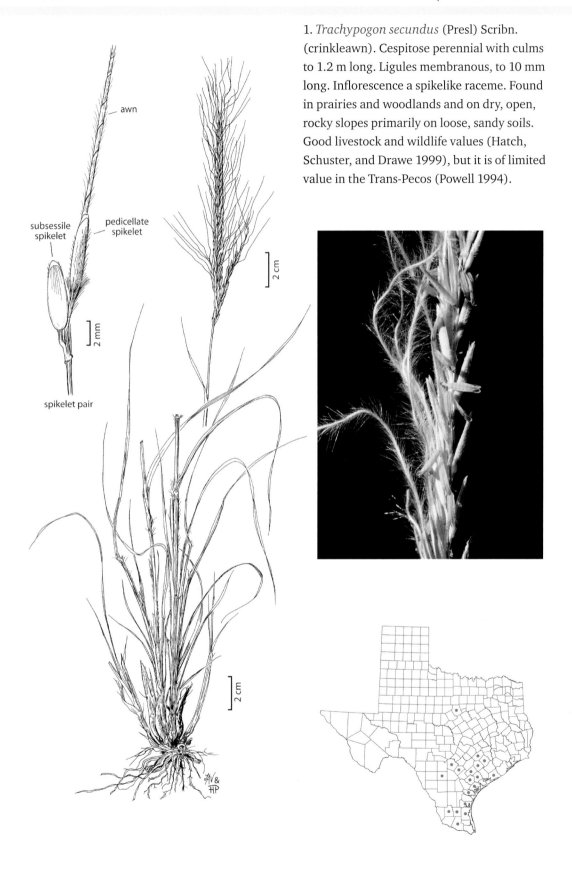

awn

subsessile
spikelet

pedicellate
spikelet

2 mm

spikelet pair

2 cm

2 cm

1. *Trachypogon secundus* (Presl) Scribn. (crinkleawn). Cespitose perennial with culms to 1.2 m long. Ligules membranous, to 10 mm long. Inflorescence a spikelike raceme. Found in prairies and woodlands and on dry, open, rocky slopes primarily on loose, sandy soils. Good livestock and wildlife values (Hatch, Schuster, and Drawe 1999), but it is of limited value in the Trans-Pecos (Powell 1994).

Tragus Hall
(Chloridoideae: Cynodonteae)

Low annuals with weak stems; soft, flat blades; and slender, spikelike inflorescences of bristly, burlike clusters of 2–5 spikelets. Ligule a ring of short, woolly hairs. Disarticulation at the base of each spikelet cluster, the inflorescence axis persistent. Spikelets 1-flowered. First glume small, thin, much-reduced or absent. Second glume of the lower 2 spikelets of a cluster large and firm, bearing 3 rows of stout, hooked spines. Lemmas of the lower spikelets thin and flat. Upper 1–3 spikelets sterile, the uppermost usually rudimentary. Basic chromosome number, $x = 10$. Photosynthetic pathway, C_4. Represented in Texas by a single species. Gould (1975b) and Hatch, Gandhi, and Brown (1990) reported *T. racemosus* (L.) All. in Texas, but this appears to reflect confusion with *T. berternianus*.

(Wipff 2003k).

1. *Tragus berteronianus* Schult. (spike burgrass, goatgrass, pricklegrass, spur burgrass). Annual with geniculate culms. Ligules a hyaline membrane fringed with fine, soft hairs. Spikelets in a burlike cluster that disarticulates from the plant. Generally found in loose, sandy soils in waste places and disturbed sites. It does not offer significant forage to be of significance (Powell 1994).

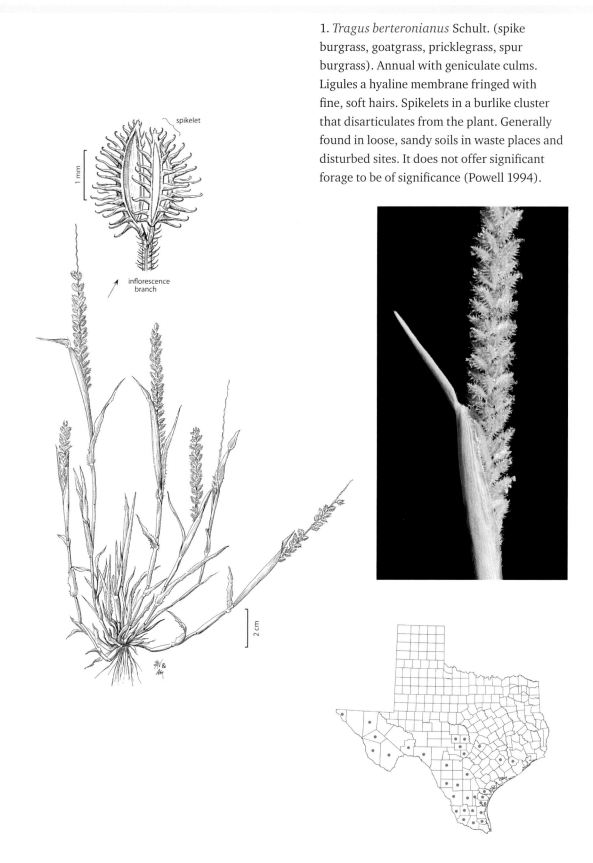

Trichloris E. Fourn. *ex* Benth.
(Chloridoideae: Cynodonteae)

Cespitose perennials, occasionally stoloniferous. Culms up to 1.5 m tall, herbaceous. Sheaths open, auricles absent. Ligules membranous, ciliate, cilia longer than membranous base. Distinctive tufts of hairs at the edge of the throat on each side of the ligules. Blades linear and flat or folded. Inflorescences panicles of nondisarticulating branches. Branches in 1 or more whorls, spikelets in 2 rows on the abaxial side of the branches. Spikelets laterally compressed, 2–5 florets, lowest 1–2 perfect, upper 1–3 florets progressively reduced and sterile. Glumes much shorter than the spikelets; upper glume awned. Lower lemmas 3-veined, veins prolonged into 3 awns. Central awns 8–12 mm long, lateral awns 0.5–12 mm long. Basic chromosome number, $x = 10$. Photosynthetic pathway, C_4. Closely related to *Chloris* and sometimes included there (Gould 1975b; Hatch, Gandhi, and Brown 1990). It differs, however, in the 3-awned lemmas. Represented in Texas by 2 species.

1. Lowest lemma awns about equal ....1. *T. crinita*
1. Lowest lemma awns distinctly different in length, central awn greatly exceeds lateral
   awns.. 2. *T. pluriflora*

(Barkworth 2003p)

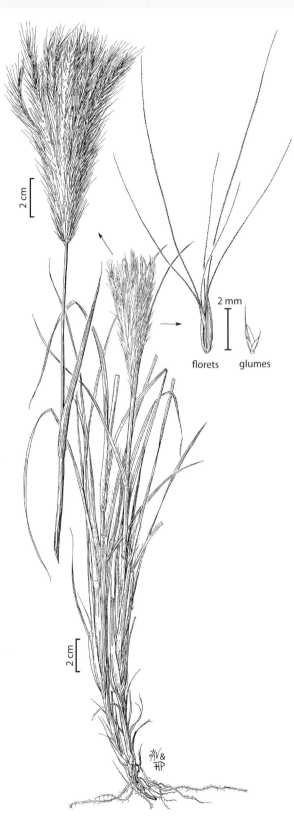

1. *Trichloris crinita* (Lag.) Parodi (false rhodesgrass, multiflowered chloris). A cespitose perennial that occasionally has stolons. Culms up to 1 m tall. Panicles with numerous branches in whorls that appear as a single terminal cluster. Found mostly in clayey, alluvial soils near streams. Included in *Chloris* by Gould (1975b) and Hatch, Gandhi, and Brown (1990).

florets   glumes

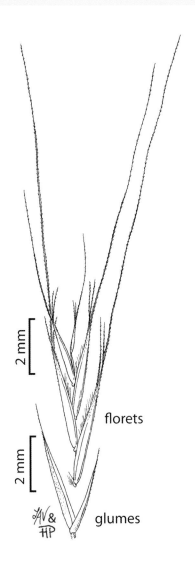

florets

2 mm

2 mm

glumes

AV & HP

2. *Trichloris pluriflora* E. Fourn. Clayton (multiflower false rhodesgrass). A cespitose perennial, occasionally stoloniferous. Culms up to 1.5 m tall. Panicles with 7–20 branches in distinctive whorls. Found in silty and clayey, low, brushy areas. Good livestock and wildlife values (Hatch, Schuster, and Drawe 1999). Included in *Chloris* by Gould (1975b) and Hatch, Gandhi, and Brown (1990).

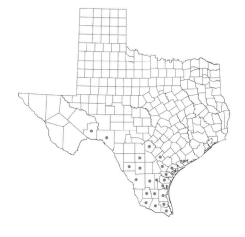

Trichoneura Anderss.
(Chloridoideae: Cynodonteae)

Cespitose annuals and perennials with spikelets borne on few to several spicate primary branches, these widely scattered on the upper portion of the culm or clustered near its apex. Spikelets 3–9-flowered, disarticulating above the glumes and between the florets. Glumes about equal, longer than the lemmas, thin, 1-nerved, acuminate or short-awned. Lemmas 3-nerved, the lateral nerves conspicuously long-ciliate, the midnerve glabrous or short-pubescent. Apex of lemma narrow, obtuse or notched, often short-awned. Base of lemma with a short, hairy callus. Palea broad, well developed. Basic chromosome number, $x = 10$. Photosynthetic pathway, C_4. Represented in Texas by a single species.

(Wipff 2003l)

1. *Trichoneura elegans* Swallen (Silveus' grass, hairy-nerve grass). A robust annual. Culms to 1.2 cm long, usually rooting at the lower nodes. Ligules 1–3 mm long, membranous. Glumes with 1 vein. Infrequent on dry, deep, loose, sandy soils; along fields and roadways; and in prairies.

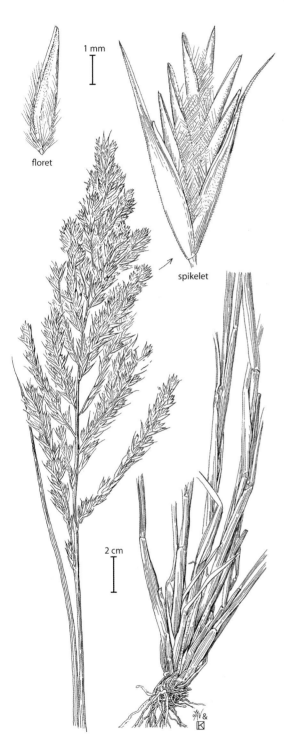

floret

spikelet

2 cm

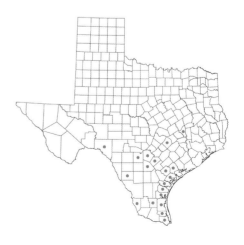

Tridens Roem. and Schult.
(Chloridoideae: Cynodonteae)

Perennials, cespitose or infrequently rhizomatous. Sheaths open; ligules a ciliate fringe of hairs; auricles absent or occasionally with short, membranous lateral auricles. Inflorescence an open or closed panicle. Disarticulation above the glumes and between the florets. Spikelets with 3–9 florets, florets strongly imbricate. Glumes 2, subequal, lower 1-nerved, the upper 1–3-nerved. Lemma broad, thin, 3-nerved, short-hairy on the veins below, rounded at the back, mostly bidentate at the apex, the midnerve and often the lateral veins extended as minute mucros. Palea slightly shorter than the lemma. Basic chromosome number, $x = 10$. Photosynthetic pathway, C_4. Represented in Texas by 9 species and 4 varieties.

1. Panicles open, not spikelike.
  2. Pedicels all <1 mm long..2. *T. ambiguus*
  2. Pedicels, at least some, >1 mm long.
    3. Lemma lateral veins rarely excurrent.
      4. Lemmas 4–6 mm long; ligules 0.4–1.0 mm long ....3. *T. buckleyanus*
      4. Lemmas 2–3 mm long; ligules 1–3 mm long .... 5. *T. eragrostoides*
    3. Lemma lateral veins excurrent as short mucros.
      5. Blades mostly 1–3 mm wide; panicles mostly 5–16 cm long ....9. *T. texanus*
      5. Blades mostly 3–10 mm wide; panicles mostly 15–35 cm long............. 6. *T. flavus*
1. Panicles closed, spikelike.
      6. Lemma veins glabrous or hairy only at base................................1. *T. albescens*
      6. Lemma veins hairy to well above the base.
        7. Glumes much exceeding lower lemmas and usually as long as spikelet ....
        .... 8. *T. strictus*
        7. Glumes shorter than to only slightly as long as lower lemmas, much shorter than entire spikelet.
          8. Lemmas excurrent at apex as a short mucro, pubescent from ½ their length or less.. 4. *T. congestus*
          8. Lemmas not excurrent, pubescent well above ½ their length.................
          ....7. *T. muticus*

(Gould 1975b; Valdes-Reyna 2003e)

1. *Tridens albescens* (Vasey) Wooton & Standl. (white tridens). Perennial with short, knotty, hard rhizomatous bases. Culms to 1 m tall. Panicles dense. Branches appressed. Spikelets purple-tinged. Found in low prairies, ditches, swales, and open woods on clayey soils that are periodically inundated. Commonly found in partial shade of shrubs. Listed as fair for wildlife and livestock (Hatch and Pluhar 1993). Reported as good forage for livestock (Powell 1994).

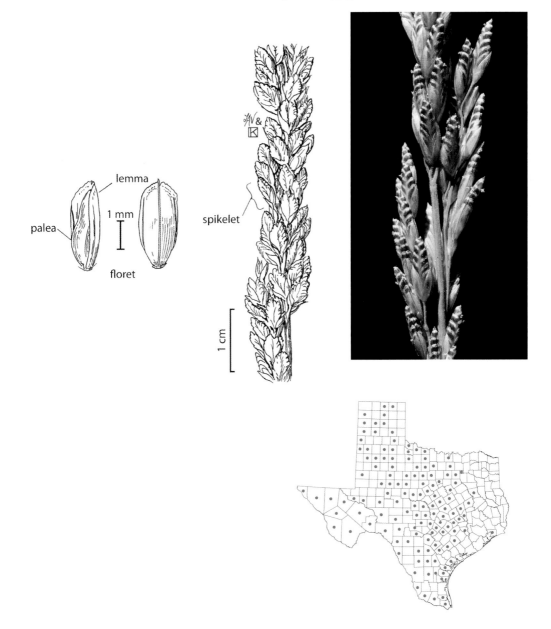

2. *Tridens ambiguus* (Elliott) Schult. (pine-barren tridens, pine-barren fluffgrass). Perennial with short rhizomes. Culms to 1.3 m long. Panicles not dense. Branches not appressed. Found in pine flatwoods, pine-oak savannahs, and roadways next to forested areas. Infrequent.

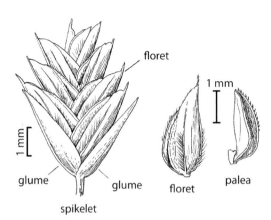

floret

1 mm

1 mm

glume   glume   floret   palea

spikelet

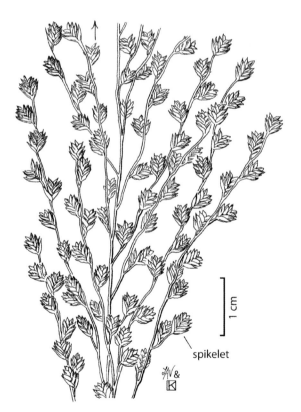

1 cm

spikelet

3. *Tridens buckleyanus* (L. H. Dewey) Nash
(Buckley's tridens, Buckley's fluffgrass).
Cespitose perennial. Culms to 80 cm tall.
Panicles not dense. Branches widely spaced,
ascending to slightly spreading. Endemic to the
eastern edge of the Edwards Plateau. Found
along shaded stream banks and woodland
borders on rocky slopes. Texas Organization
of Endangered Species places this species on
its Watch List (Category V) (Jones, Wipff, and
Montgomery 1997). Not listed in *Rare Plants of
Texas* (Poole et al. 2007).

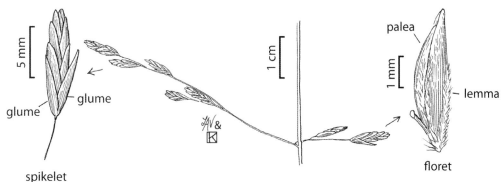

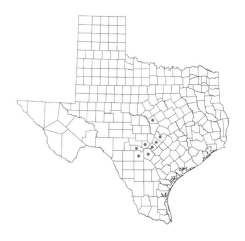

4. *Tridens congestus* (L. H. Dewey) Nash (pink tridens). Tufted perennial with short rhizomes. Culms to 75 cm long. Panicles dense. Branches erect or ascending. Spikelets more or less evenly pinkish. Found in moist depressions of otherwise dry, rocky hills. Listed as a Texas endemic (Diggs et al. 2006), but it should be expected in the grasslands and woodlands of south-central Oklahoma. Not listed as rare by Jones, Wipff, and Montgomery (1997) or Poole et al. (2007).

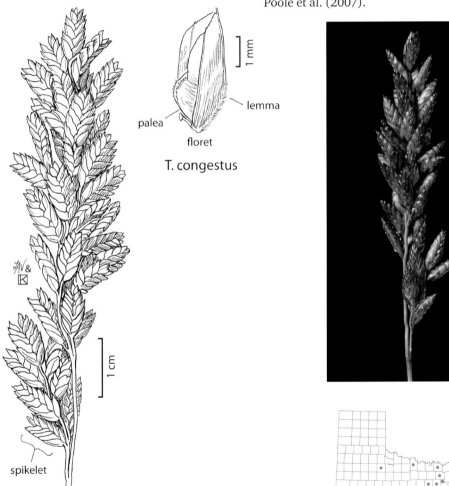

palea lemma

floret

T. congestus

spikelet

5. *Tridens eragrostoides* (Vasey & Scribn.) Nash (lovegrass tridens). Densely tufted perennial with short rhizomes. Culms to 1 m tall. Panicle not dense. Branches ascending, spreading or reflexed. Found typically in brushy grasslands under shrubs in partial shade. Good livestock and wildlife values (Hatch, Schuster, and Drawe 1999).

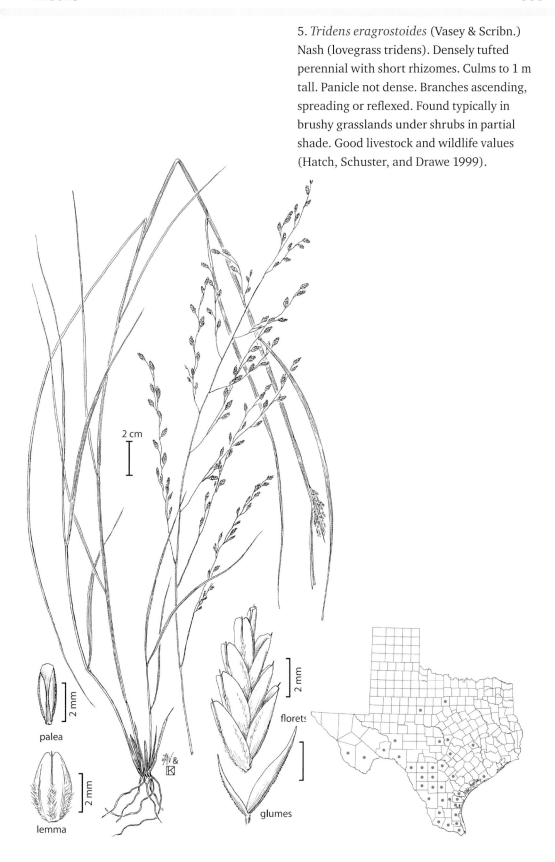

2 cm

palea

2 mm

lemma

2 mm

florets

2 mm

glumes

6. *Tridens flavus* (L.) Hitchc. (purpletop, purpletop tridens, tall redtop, greasegrass, Chapman's tridens). Perennial with short rhizomes. Culms to 1.8 m tall. Panicles not dense, nodding. Branches strongly divergent, drooping. Grows in open woods, old fields, and pine and oak woodlands. Common along roadsides in the "lost pines" area in Bastrop County. Diggs et al. (2006) explained that the vernacular name "greasegrass" comes from the "greasy feel" one gets from passing the hand over the inflorescence. Apparently a greasy substance is secreted at the base of the panicle branches. Listed as fair for wildlife and livestock (Hatch and Pluhar 1993). Telfair (2006) listed it as important for wildlife. Two varieties occur in Texas: var. *chapmanii* (Small) Shinners; and the typical variety, var. *flavus*. They can be distinguished by the following characters:

1. Panicles nodding; pulvini, if hairy, hairs restricted to the adaxial surface of branches ....var. *flavus*
1. Panicles erect; pulvini always hairy, hairs around the base of branches .... var. *chapmannii*

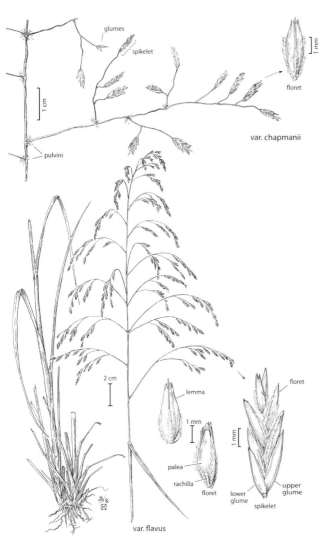

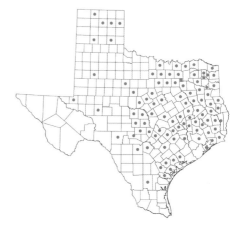

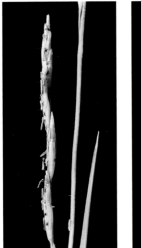

7. *Tridens muticus* (Torr.) Nash (slim tridens, awnless fluffgrass). Perennial with short, knotty, rhizomatous bases. Culms to 80 cm long. Panicles not dense. Branches appressed. Found on dry, open, sandy or clayey sites in grasslands, roadsides, and gravelly slopes. Fair livestock and wildlife values (Hatch, Schuster, and Drawe 1999). Two varieties occur in Texas: var. *elongatus* (Buckley) Shinners (sometimes called rough tridens); and the typical one, var. *muticus*. They can be distinguished by the following characters:

1. Upper glumes 1-veined, usually but not always 5 mm or shorter.............var. *muticus*
1. Upper glumes 3–7-veined, usually 6–8 mm long .... var. *elongatus*

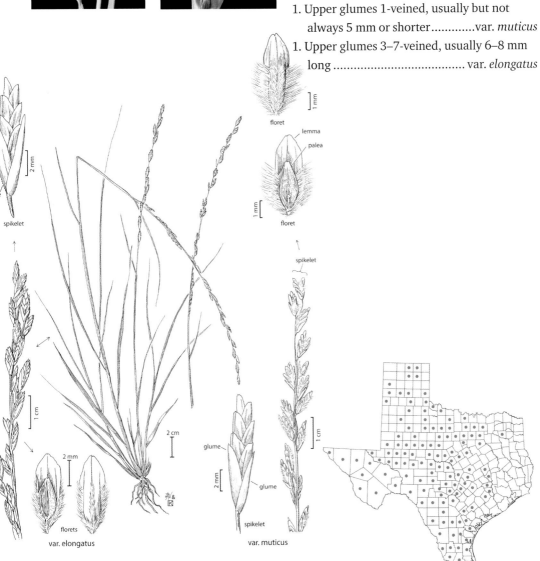

floret

lemma
palea

floret

spikelet

spikelet

2 mm

1 cm

2 cm

2 mm

glume

2 mm

glume

spikelet

florets

var. elongatus

var. muticus

1 cm

8. *Tridens strictus* (Nutt.) Nash (longspike tridens). Perennial with short, knotty, hard rhizomatous bases. Culms to 1.7 m long. Panicles dense. Branches appressed. Found in coastal grasslands, open woods, old fields, and roadsides, mostly on sandy soils, but occurs on clayey soils as well.

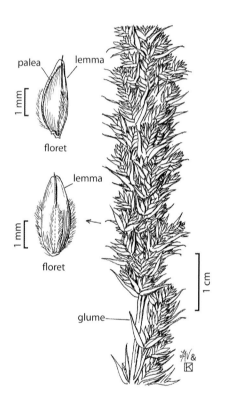

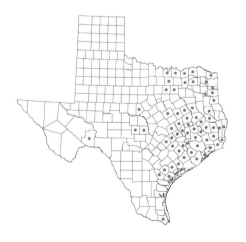

9. *Tridens texanus* (S. Watson) Nash (Texas tridens, long-spike fluffgrass). Tufted perennial with short rhizomes. Culms to 75 cm tall. Panicles not dense. Branches lax, strongly divergent to drooping. Infrequent under shrubs and along fencerows and roadsides.

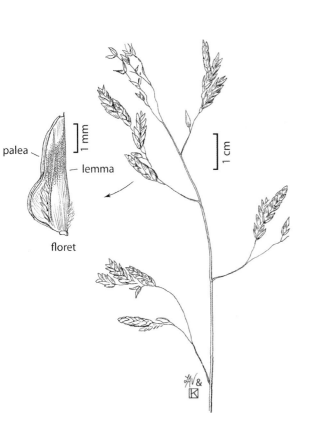

palea

lemma

floret

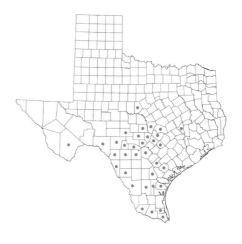

Triplasis P. Beauv.
(Chloridoideae: Cynodonteae)

Annuals or perennials, cespitose or rarely rhizomatous. Culms with many nodes and short internodes, eventually breaking up at the nodes. Sheaths open; ligules a fringe of hairs or a ciliated membrane; auricles absent. Inflorescence a panicle, open, exserted or partially included in the upper sheath. Clusters of cleistogamous spikelets regularly borne in the axils of the upper leaf sheaths. Spikelets of terminal inflorescence with 2–4 florets, those of the sheath axils usually with a single floret. Glumes 2, unequal, 1-nerved. Lemmas narrow, 3-nerved, the veins ciliate or densely pubescent, the midnerve extended as a mucro or a short awn from a notched apex. Palea strongly 2-nerved and 2-keeled, silky-villous on the keels. Basic chromosome number $x = 10$. Photosynthetic pathway, C_4. Represented in Texas by a single species.

(Hatch 2003c)

1. *Triplasis purpurea* (Walter) Chapman (purple sandgrass). Cespitose annual or rhizomatous perennial. Culms to 1 m tall. Ligule a ring of hairs to 1 mm long. Sheaths inflated. Blades shorter than the sheaths. Glumes with erose apices. The Texas species belongs to the typical variety, var. *purpurea*. Found in sandy areas throughout much of the state. Small reduced inflorescences are contained within the inflated sheaths. When the old culms break apart, these are blown about and serve as a dispersal mechanism (Allred 2005). Of poor forage value, but it does stabilize sandy soils (Powell 1994).

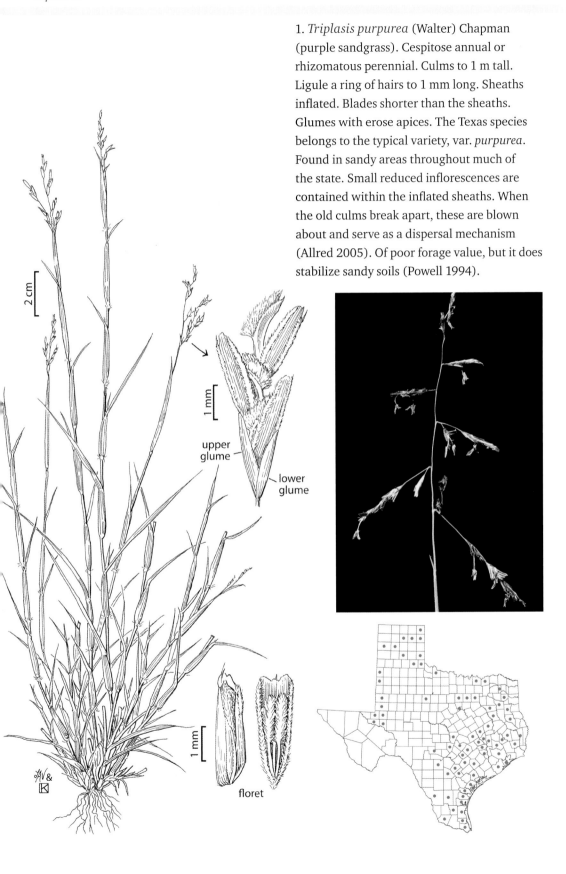

upper glume

lower glume

floret

Tripogon Roem. & Schult.
(Chloridoideae: Cynodonteae)

Low, tufted perennials (the American species) with filiform leaves, these mostly in a basal clump. Inflorescence a slender spike, the spikelets sessile or subsessile and solitary at the nodes of a straight or somewhat flexuous rachis. Spikelets several-flowered, disarticulating above the glumes and between the florets. Glumes unequal, acute or acuminate, 1-nerved, shorter than the lemmas. Lemmas 3-nerved, with a tuft of hair at the base, short-awned from between the lobes of a minutely notched apex. Basic chromosome number, $x = 10$. Photosynthetic pathway, C_4. Represented in Texas by a single species.

(Wipff 2003m)

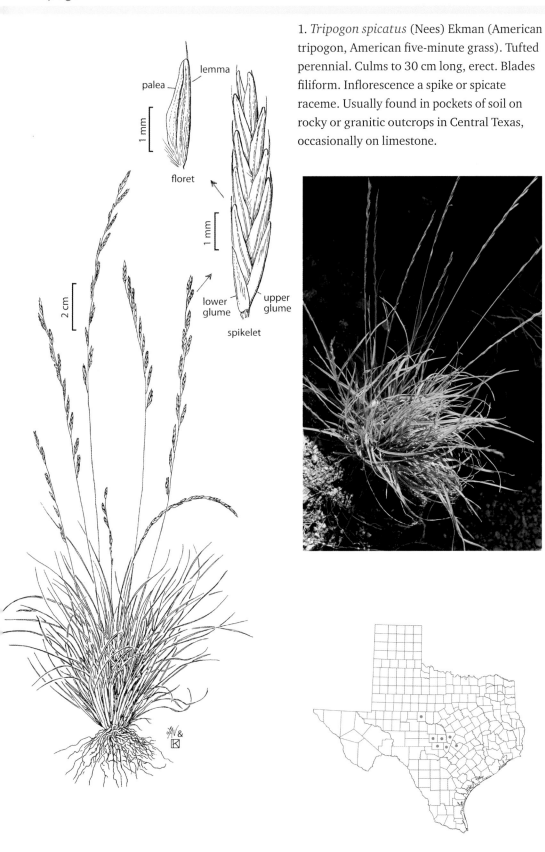

palea
lemma
1 mm
floret

1 mm

lower glume
upper glume
spikelet

2 cm

1. *Tripogon spicatus* (Nees) Ekman (American tripogon, American five-minute grass). Tufted perennial. Culms to 30 cm long, erect. Blades filiform. Inflorescence a spike or spicate raceme. Usually found in pockets of soil on rocky or granitic outcrops in Central Texas, occasionally on limestone.

Tripsacum L.
(Panicoideae: Andropogoneae)

Large cespitose perennials with stout, thick-based culms and usually broad, flat blades. Inflorescence a spikelike raceme or series of 2 to few spikelike racemose branches bearing staminate spikelets above and pistillate spikelets below. Staminate spikelets 2-flowered, in pairs on one side of a continuous rachis. Pistillate spikelets below the staminate and on the same rachis, single, sessile, and partially embedded on the rachis. Glumes of the staminate spikelet flat, several-nerved, relatively thin. Glumes of the pistillate spikelet hard and bony, fused with the rachis and tightly enclosing the rest of the spikelet. Lemmas of the sterile and fertile florets thin and membranous, awnless, often reduced. Staminate portion of the rachis deciduous as a whole, the pistillate portion breaking up at the nodes into beadlike units. Basic chromosome number, $x = 9$. Photosynthetic pathway, C_4. Represented in Texas by a single species.

(Barkworth 2003q)

1. *Tripsacum dactyloides* (L.) L. (eastern gamagrass, eastern mockgrass). Stout, monoecious perennial in clumps from thick, knotty rhizomes. Culms to over 3 m tall. Blades to 1.2 m long and 5 cm wide. Pistillate spikelets below the staminate ones (androgynous spicate racemes). Found on moist to periodically flooded areas in grasslands, swales, and ditches. The Texas species belongs to the typical variety, var. *dactyloides*. A valued native grass for hay production and grazing. Hatch and Pluhar (1993) rated this species good for livestock and fair for wildlife.

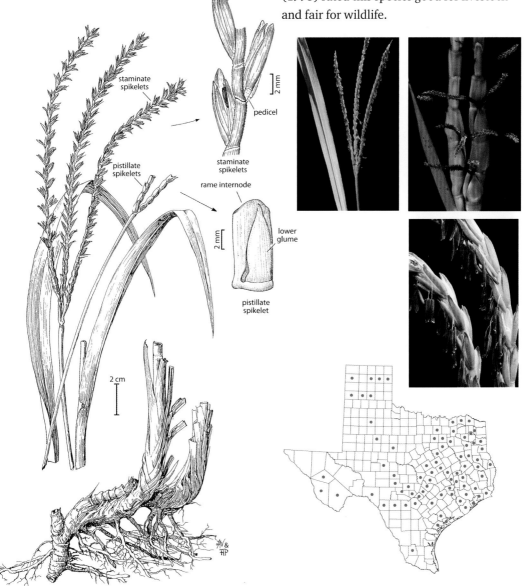

Triraphis R. Br.
(Chloridoideae: Cynodonteae)

Cespitose annuals and perennials with culms up to 1.4 m tall. Leaves cauline; auricles absent; ligules of hairs or membranous and long-ciliate. Inflorescences are terminal and open or contracted panicles. Spikelets are laterally compressed, with 3–9 perfect florets, reduced florets above if present. Disarticulation is above glumes and between the florets. Glumes about equal, shorter than the lemmas, 1-veined. Lemmas 3-veined, each extending into an awn. Basic chromosome number, $x = 10$. Photosynthetic pathway, C_4. Represented in Texas by a single species.

(Wipff 2003n)

1. *Triraphis mollis* R. Br. (purple needlegrass).
Annual or short-lived perennial. Culms generally
>1 m long, and panicles up to 30 cm long.
Panicle branches usually appressed. Glumes
mucronate, lemmas 3-awned. An introduction
from Australia that is reported only from
Dimmitt County. Likely to spread. Not reported
in Gould (1975b); Hatch, Gandhi, and Brown
(1990); or Jones, Wipff, and Montgomery
(1997).

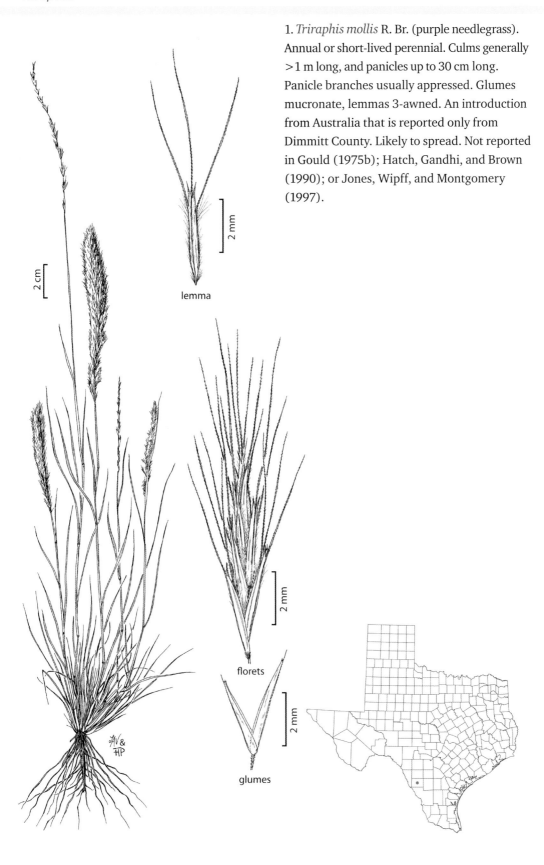

2 mm

lemma

2 cm

2 mm

florets

2 mm

glumes

Trisetum Pers.
(Poöideae: Poeae)

Tufted perennials and a few annuals, with slender culms, flat blades, and usually narrow panicles of 2-flowered (rarely 3–4-flowered) spikelets. Rachilla usually villous, prolonged above the uppermost floret, disarticulating above the glumes and between the florets or, in a few species, below the glumes. Glumes thin, nearly equal to very unequal, 1–3-nerved, acute, awnless, one or both usually equaling or exceeding the florets in length. Lemmas 5-nerved, bifid at the membranous apex, mostly with a straight or bent awn from the base of the notch but occasionally awnless, usually puberulent at the base. Basic chromosome number, $x = 7$. Photosynthetic pathway, C_3. Represented in Texas by 2 species.

1. Plants annual .... 1. *T. interruptum*
1. Plants perennial .... 2. *T. spicatum*

(Rumely 2007)

1. *Trisetum interruptum* Buckley (prairie trisetum, prairie false oat). Annual. Culms to about 50 cm long. Ligules 1.0–2.5 mm long, membranous. Lemma awns 4–8 mm long. Widespread in a number of different habitats. Reported from deserts, plains, arid shrublands, and riparian woodlands. Sometimes placed in *Sphenopholis*. Palatable to grazing animals but not abundant (Powell 1994).

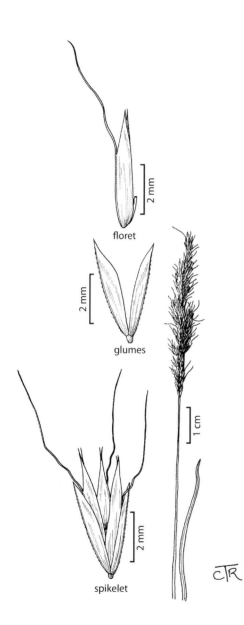

floret

glumes

spikelet

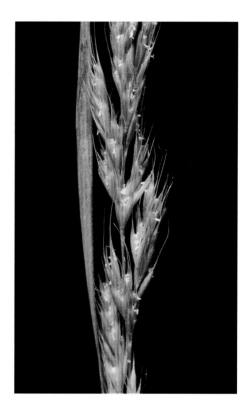

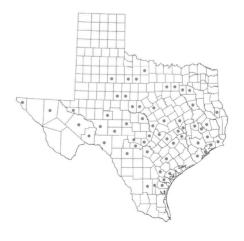

2. *Trisetum spicatum* (L.) K. Richt. (spike
trisetum). Cespitose perennial. Culms to 1.2 m
tall, erect. Ligules 1–4 mm long, membranous.
Lemma awns 3–8 mm long. Found at higher
elevations in the Guadalupe Mountains. Not
included in Gould (1975b) or Hatch, Gandhi,
and Brown (1990).

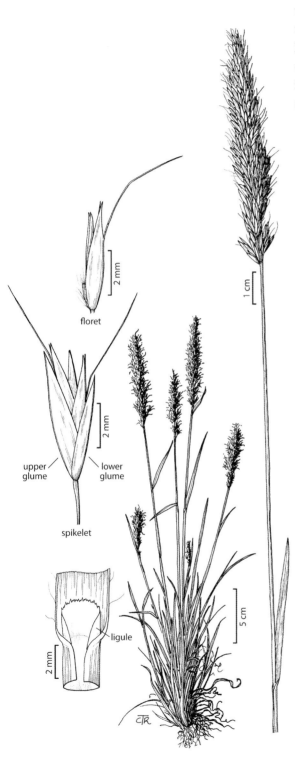

floret

2 mm

upper
glume

lower
glume

2 mm

spikelet

ligule

2 mm

5 cm

1 cm

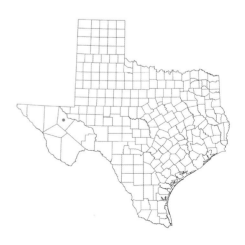

Triticum L.
(Poöideae: Triticeae)

Annuals with broad, flat blades. Inflorescence terminal, a thick bilateral spike. Spikelets solitary at the nodes on a continuous rachis; does not disarticulate in Texas species or only under pressure. Spikelets with 2–5 florets, laterally flattened and oriented with flat surface facing the rachis. Glumes thick and firm, 3- to several-nerved, toothed, mucronate or with a short awn at the apex. Lemmas similar to the glumes in texture, keeled, asymmetric, many-nerved, awnless or with a single awn. Basic chromosome number, $x = 7$. Photosynthetic pathway, C_3. Represented in Texas by 2 species.

1. Glumes loosely appressed to the lower florets; rachises not disarticulating with pressure .... 1. T. *aestivum*
1. Glumes tightly appressed to the lower florets; rachises disarticulating with little pressure .... 2. T. *spelta*

(Morrison 2007)

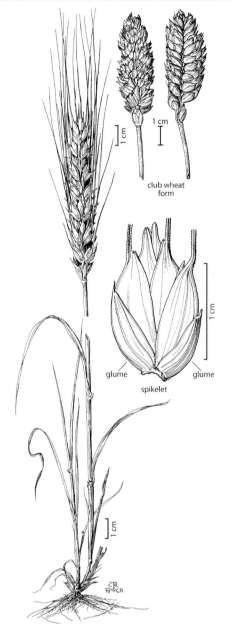

club wheat
form

glume   glume
spikelet

1. *Triticum aestivum* L. (wheat, common wheat, bread wheat, soft wheat). Cultivated annual. Culms to 1.5 m tall. Lemma awns to 12 cm long. Rachises not disarticulating. The most common cultivated wheat. Both an awned and unawned form are cultivated. Sometimes used for a rapid cover for erosion control. *Triticum asestivum* hybridizes with *Secale cereale* (rye) to produce *Triticale*. This hybrid is sometimes cultivated and occurs frequently around old fields and roadsides. It is most often confused with wheat because they both have broad glumes, but *Triticale* is usually glaucous and has lemma veins that converge toward the apex (Allred 2005). The distribution map is fairly meaningless except to demonstrate that cultivated plants are poorly collected. Good livestock and wildlife values (Hatch, Schuster, and Drawe 1999).

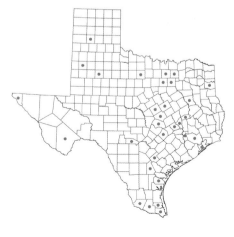

2. *Triticum spelta* L. (spelt, dinkel). Cultivated annual. Culms to 1.2 m tall. Lemma awns to 2 cm long. Rachises readily disarticulate into little "barrel-shaped" segments reminding one of *Aegilops cylindrica*. Not listed in Gould (1975b) or Hatch, Gandhi, and Brown (1990).

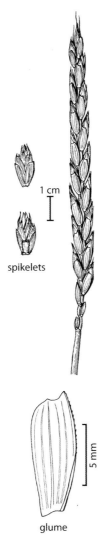

spikelets

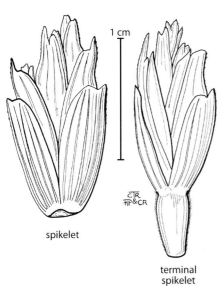

glume

spikelet

terminal spikelet

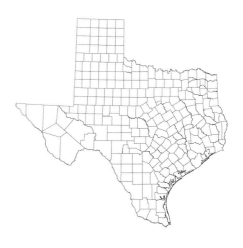

Uniola L.
(Chloridoideae: Cynodonteae)

Tall, coarse perennials, one rhizomatous, the other stoloniferous. Leaves long, the blades flat but involute on drying. Ligule a fringe of hairs. Inflorescence an open or a contracted panicle of 5- to many-flowered spikelets. Lower 2–6 florets of the spikelet sterile. Spikelets laterally compressed, disarticulating below the glumes, falling entire. Glumes subequal, acute, awnless or slightly mucronate, 3-nerved, keeled, the keel serrulate. Lemmas acute to narrowly obtuse, awnless or mucronate, 3–9-nerved, keeled, the keels serrulate. Palea shorter to longer than the lemma, 2-keeled, the keels winged and serrate to ciliate. Flowers perfect, with 3 stamens and 2 lodicules, these fleshy, cuneate. Ovary glabrous, with a single style and 2 plumose stigmas. Caryopsis linear, with an embryo less than ½ the length of the grain. Basic chromosome number, $x = 10$. Photosynthetic pathway, C_4. Represented in Texas by a single species.

(Yates 2003)

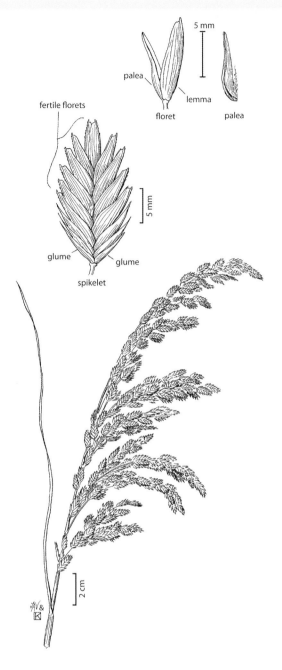

1. *Uniola paniculata* L. (seaoats). Stout perennial from long, thick rhizomes. Culms to 2 m tall. Ligules a dense ring of hairs 1–3 mm long. Narrow blades to 80 cm long and 1 cm wide. Panicles contracted. Spikelets strongly flattened. Found on beaches, sand dunes, and sand flats. Frequently used in floral arrangements and often overcollected for that purpose. Many areas restrict collection of this species. Poor forage but some wildlife use (Hatch, Schuster, and Drawe 1999).

Urochloa P. Beauv.
(Panicoideae: Paniceae)

Annuals or perennials, occasionally stoloniferous or geniculate and rooting at the lower nodes. Culms up to 5 m long, herbaceous. Sheaths open; auricles rarely present; ligules a ring of hairs; blades ovate to ovate-lanceolate, flat. Inflorescences terminal or axillary panicles. Primary branches usually spikelike. Disarticulation beneath the spikelets. Spikelets dorsally compressed, with 2 florets, the lower sterile or staminate; the upper perfect. Upper lemmas indurate, transversely rugose and verrucose. Upper paleas rugose and shiny. Basic chromosome numbers, $x = 7, 8, 9$, and 10. Photosynthetic pathway, C_4. Represented in Texas by 13 species and 1 variety. All species were variously placed in *Bracharia* and/or *Panicum* (Gould 1975b; Hatch, Gandhi, and Brown 1990).

1. Lowest node of panicle with 2 or more branches .... see *Megathyrsus maximus*
1. Lowest node of panicle with solitary branch.
  2. Spikelets paired at middle of primary panicle branches.
    3. Primary panicle branches flat; lower glumes 0–3-veined.
      4. Plants annual; spikelets 1.8–2.2 mm long .... 11. *U. reptans*
      4. Plants perennial; spikelets 2.5–5.0 mm long.
        5. Upper lemmas awned .... 5. U. *mosambicensis*
        5. Upper lemmas unawned .... 6. *U. mutica*
    3. Primary panicle branches triquetrous; lower glumes 3–7-veined.
      6. Spikelets 4.8–6.0 mm long .... 13. *U. texana*
      6. Spikelets 2.0–4.2 mm long.
        7. Primary panicle branches densely hairy, hairs papillose-based ....
        .... 1. *U. arizonica*
        7. Primary panicle branches occasionally densely hairy, few if any hairs papillose-based.
          8. Lower lemma 7-veined .... 4. *U. fusca*
          8. Lower lemma 5-veined .... 10. *U. ramosa*
  2. Spikelets solitary at middle of primary panicle branches.
          9. Panicle branches triquetrous, 0.2–0.4 mm wide ......3. *U. ciliatissima*
          9. Panicle branches flat or crescent-shaped, 0.5–2.5 mm wide.
            10. Upper lemmas awned, awns 3.0–1.2 mm long.
              11. Plants perennial; lower floret staminate .... 5. *U. mosambicensis*
              11. Plants annual; lower floret sterile .... 7. *U. panicoides*
            10. Upper lemmas awnless, with or without mucronate tip.
              12. Panicle branches crescent-shaped in cross section; spikelets in a single row along the branches .... 2. *U. brizantha*
              12. Panicle branches flat in cross section; spikelets in 2 rows along the branches.
                13. Internodes between the glumes <0.3 mm long ....
                .... 9. *U. platyphylla*

13. Internodes between the glumes >0.3 mm long, conspicuously separated.
    14. Spikelets 3.3–3.8 mm long; base of leaf blades not clasping culm .... 12. *U. subquadripara*
    14. Spikelets 4–6 mm long; base of blade clasping culm .... 8. *U. plantaginea*

(Wipff and Thompson 2003)

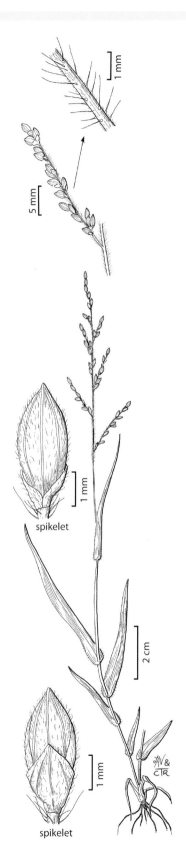

spikelet

spikelet

1. *Urochloa arizonica* (Scribn. & Merr.) Morrone & Zuloaga (annual signalgrass, Arizona signalgrass). Annual with freely branching culms. Culms somewhat geniculate at base. Nodes glabrous or hispid. Primary panicle branches 7–15, spikelike. Lower florets staminate or sterile. Found on rocky and sandy open sites. Gould (1975b) reported this species as *Panicum arizonicum* Scribn. & Merr.; Hatch, Gandhi, and Brown (1990) listed it as *Brachiaria arizonica* (Scribn. & Merr.) S. T. Blake.

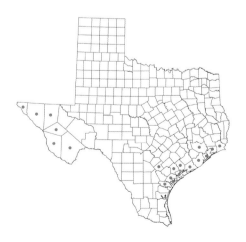

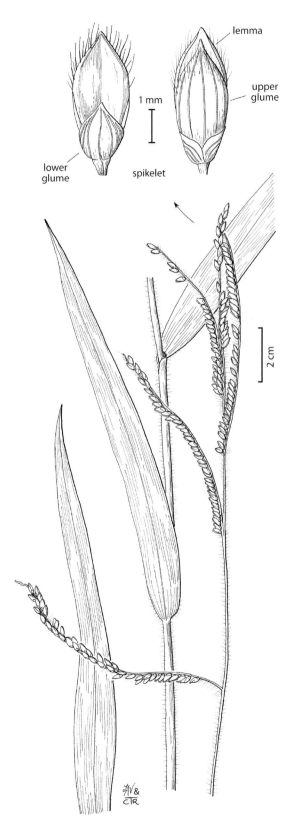

2. *Urochloa brizantha* (Hochst. *ex* A. Rich.) R. D. Webster (palisada signalgrass). Perennial with short rhizomes. Culms to 2 m tall. Nodes glabrous. Primary panicle branches 1–7(–16), spikelike. Lower florets staminate. An African introduction that is scattered in Texas. Not reported by Gould (1975b) or Hatch, Gandhi, and Brown (1990).

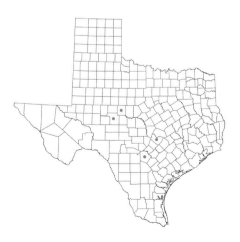

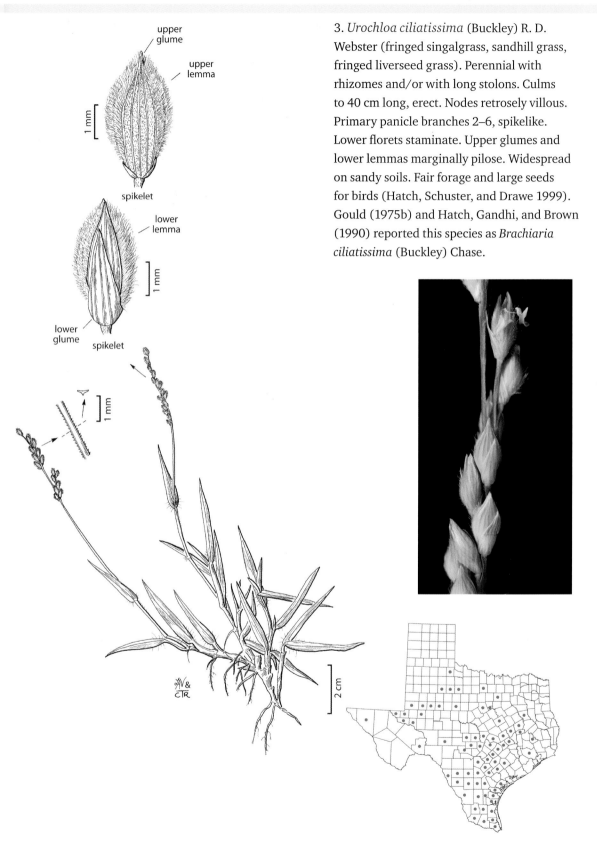

upper glume

upper lemma

1 mm

spikelet

lower lemma

1 mm

lower glume

spikelet

1 mm

2 cm

AV & CTR

3. *Urochloa ciliatissima* (Buckley) R. D. Webster (fringed singalgrass, sandhill grass, fringed liverseed grass). Perennial with rhizomes and/or with long stolons. Culms to 40 cm long, erect. Nodes retrosely villous. Primary panicle branches 2–6, spikelike. Lower florets staminate. Upper glumes and lower lemmas marginally pilose. Widespread on sandy soils. Fair forage and large seeds for birds (Hatch, Schuster, and Drawe 1999). Gould (1975b) and Hatch, Gandhi, and Brown (1990) reported this species as *Brachiaria ciliatissima* (Buckley) Chase.

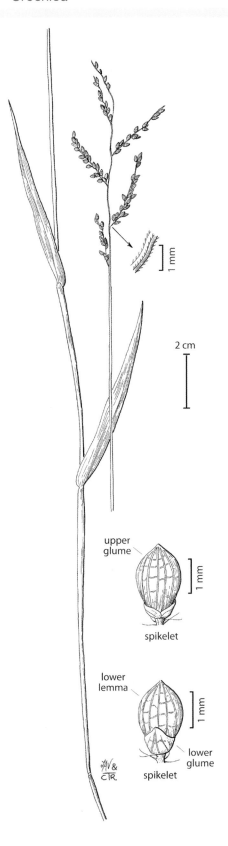

2 cm

1 mm

upper glume

spikelet

1 mm

lower lemma

spikelet

lower glume

1 mm

4. *Urochloa fusca* (Swartz) B. F. Hansen & Wunderlin (browntop signal-grass, browntop liverseed grass, hurrah grass). Annual, tufted, with geniculate culms to 1.5 m tall. Nodes glabrous or occasionally short-pilose. Primary panicle branches 5–30, spikelike. Lower florets staminate or sometimes sterile. A weedy species of moist, disturbed sites such as ditches and field borders. Gould (1975b) reported this species as *Panicum fasciculatum* Swartz; Hatch, Gandhi, and Brown (1990) listed it as *Brachiaria fasciculata* (Swartz) S. T. Blake; Turner et al. (2003) listed it as *U. fasciculata* (Swartz) R. D. Webster.

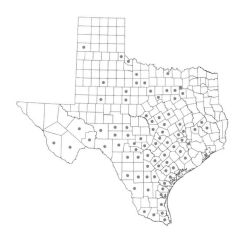

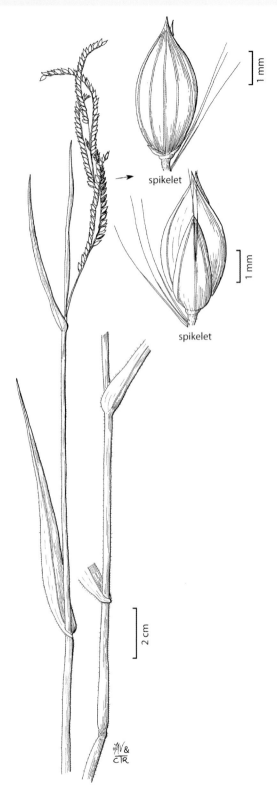

spikelet

spikelet

1 mm

1 mm

2 cm

5. *Urochloa mosambicensis* (Hack.) Dandy (sabi grass). Perennial with or without stolons. Culms to 1.5 m long. Nodes pubescent. Primary panicle branches 2–6(–15), spikelike. Lower florets staminate. An African species that is grown for forage and hay in its native area. A few reported collections in South Texas, and it is expected to spread. Not reported by Gould (1975b) or Hatch, Gandhi, and Brown (1990).

upper glume

1 mm

lower lemma

lower glume

spikelet

6. *Urochloa mutica* (Forssk.) T. Q. Nguyen (papagrass, para liverseed grass). Stoloniferous perennial, culms to 5 m long. Nodes villous. Panicle branches 10–30, spikelike. Lower florets staminate. An African species introduced as a forage crop. Now escaped and a weed in moist, disturbed sites. Found along watercourses and in coastal marshes. Gould (1975b) and Hatch, Gandhi, and Brown (1990) listed this as *Panicum purpurascens* Raddi. Good livestock and wildlife values (Hatch, Schuster, and Drawe 1999).

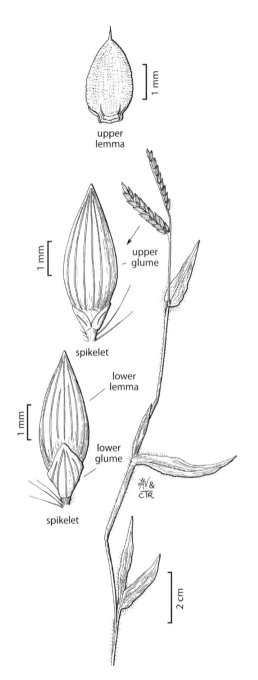

upper
lemma

upper
glume

spikelet

lower
lemma

lower
glume

spikelet

1 mm

1 mm

1 mm

2 cm

AV &
CTR

7. *Urochloa panicoides* P. Beauv. (liverseed grass). Annual. Culms to 1 m long, decumbent and usually rooting at the lower nodes. Nodes glabrous. Primary panicle branches 2–10, spikelike. Lower florets sterile. An introduced species from Africa that is now on the federal noxious weed list. A weed of sandy areas. The Texas species belongs to the typical variety, var. *panicoides*. Not listed by Gould (1975b); Hatch, Gandhi, and Brown (1990); or Turner et al. (2003).

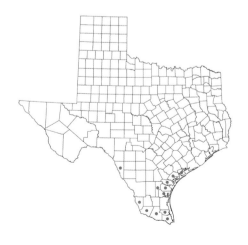

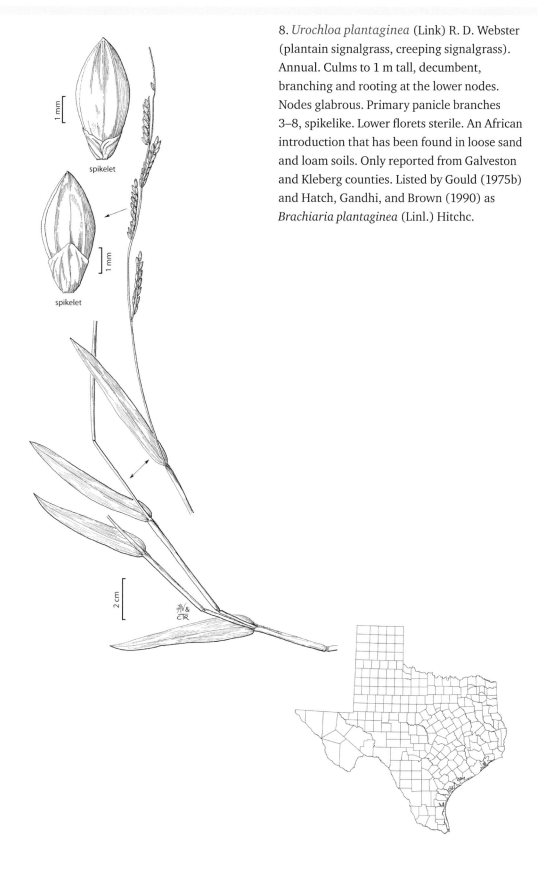

spikelet

spikelet

8. *Urochloa plantaginea* (Link) R. D. Webster (plantain signalgrass, creeping signalgrass). Annual. Culms to 1 m tall, decumbent, branching and rooting at the lower nodes. Nodes glabrous. Primary panicle branches 3–8, spikelike. Lower florets sterile. An African introduction that has been found in loose sand and loam soils. Only reported from Galveston and Kleberg counties. Listed by Gould (1975b) and Hatch, Gandhi, and Brown (1990) as *Brachiaria plantaginea* (Linl.) Hitchc.

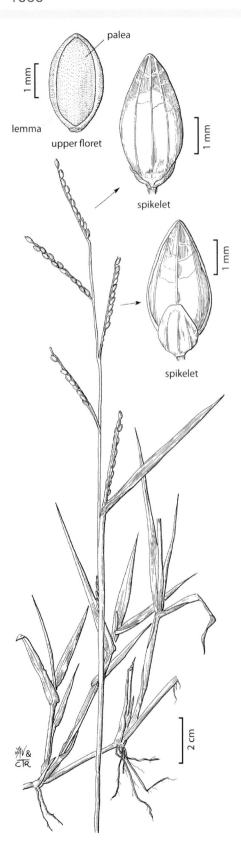

palea

lemma

upper floret

1 mm

spikelet

1 mm

spikelet

1 mm

2 cm

JAV &
CTR

9. *Urochloa platyphylla* (Munro *ex*
C. Wright) R. D. Webster (broadleaf
signalgrass). Annual. Culms to 1 m long,
rooting at the lower nodes. Nodes glabrous.
Primary panicle branches 2–8, spikelike.
Lower florets sterile. A weedy species in
woodland openings, in ditches, along fields,
and in other sandy disturbed areas. Included in
Gould (1975b) and Hatch, Gandhi, and Brown
(1990) as *Brachiaria platyphylla* (Munro *ex*
C. Wright) Nash. Poor livestock and wildlife
values (Hatch, Schuster, and Drawe 1999).

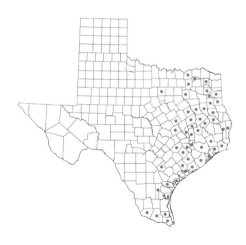

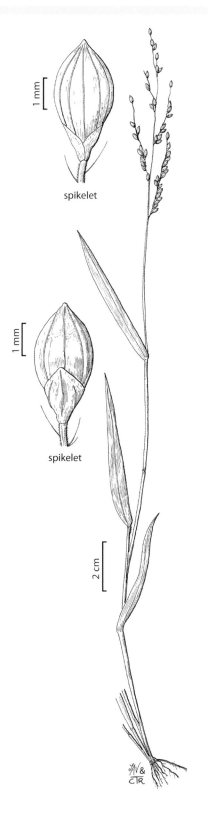

spikelet

spikelet

10. *Urochloa ramosa* (L.) T. Q. Nguyen (browntop millet, Dixie liverseed grass). Tufted annual. Culms decumbent and often rooting at the lower nodes. Nodes pubescent. Primary panicle branches 3–15, spikelike. Lower florets sterile. An introduction from Asia, where it is grown as forage and for grain. A weed of moist, disturbed sites in Texas. Not listed in Gould (1975b) or Hatch, Gandhi, and Brown (1990).

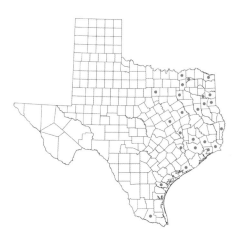

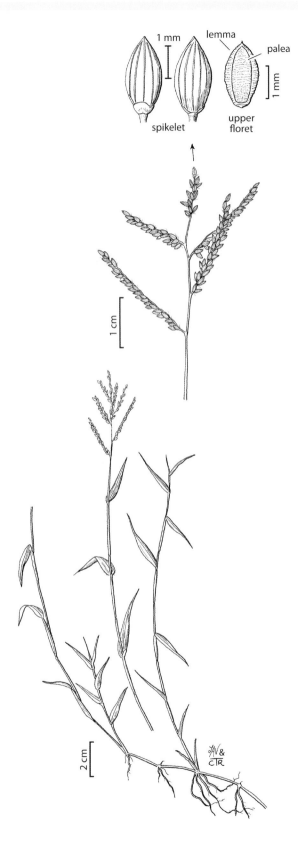

1 mm

lemma

palea

1 mm

spikelet

upper floret

1 cm

2 cm

11. *Urochloa reptans* (L.) Stapf. (sprawling signalgrass). Annual mat-forming species. Culms decumbent, sprawling, rooting at the nodes. Nodes glabrous or puberulent. Panicle branches 4–16, spikelike. Lower florets sterile or staminate. An introduced weedy species found primarily in moist, disturbed sites. Reported by Gould (1975b) and Hatch, Gandhi, and Brown (1990) as *Bracharia reptans* (L.) Gard. &. C. E. Hubb.

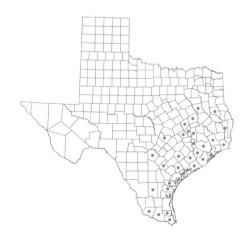

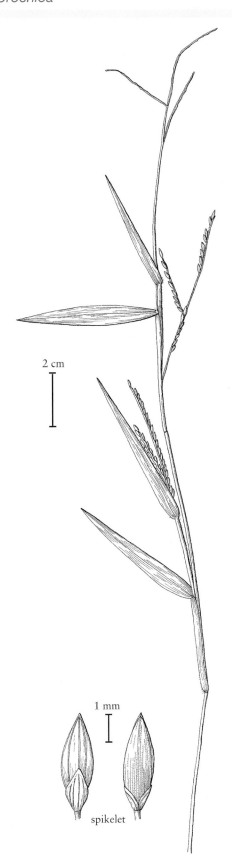

2 cm

1 mm

spikelet

12. *Urochloa subquadripara* (Trin.) R. D. Webster (armgrass millet). Annual or short-lived perennial, decumbent with culms rooting at lower nodes. Nodes glabrous. Panicle branches 3–9, spikelike, secondary branches absent. Spikelet appressed to branches in 2 rows. Lower florets sterile, lower palea present. A drought-tolerant forage species from Asia. Known from Florida, but recently collected and tentatively identified as a new record for Texas. Growing as a weed and probably an accidental introduction. Collected at Texas Agrilife Research and Extension Center in Weslaco. Its growth habitat and drought tolerance indicate that it has the potential to become a troublesome weed.

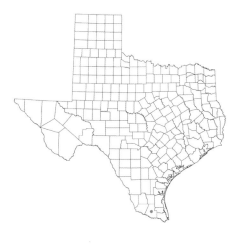

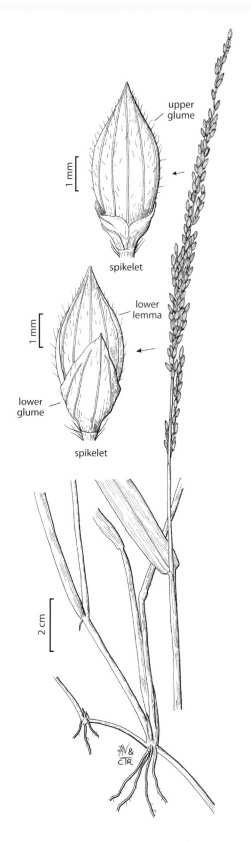

upper glume

1 mm

spikelet

lower lemma

1 mm

lower glume

spikelet

2 cm

13. *Urochloa texana* (Buckley) R. D. Webster (Texas signalgrass, Texas millet, Texas panicum, Colorado grass, Texas liverseed grass). Annual with culms to 2 m tall. Nodes puberulent. Panicle branches spikelike. Lower florets staminate or rarely sterile. A weedy species of ditches, abandoned fields, and sandy, moist waste areas. Fair livestock and good wildlife values (Hatch, Schuster, and Drawe 1999). Listed as *Panicum texanum* Buckley by Gould (1975b) and *Brachiaria texana* (Buckley) S. T. Blake by Hatch, Gandhi, and Brown (1990).

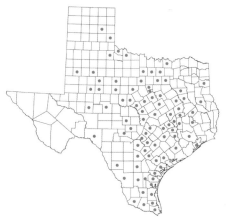

Vaseyochloa Hitchc.
(Chloridoideae: Cynodonteae)

Perennial with moderately tall, erect culms in small clumps. Slender, creeping rhizomes frequently developed. Sheaths pilose externally at the apex. Ligule a short, lacerate, densely pilose membrane. Blades long, narrow, flat, or folded. Inflorescence a panicle, with several-flowered spikelets on slender, erect or spreading branches. Disarticulation above the plumes and between the florets. Glumes firm, acute, shorter than the lemmas, the first narrow, 3- or 5-nerved, the second broader, 7- or 9-nerved. Lemmas broad, firm, rounded, and hairy on the back, 7- or 9-nerved, tapering to a narrow, obtuse apex. Palea shorter than the lemma, broad, splitting down the middle at maturity. Caryopsis dark brown or black, oval, concave-convex, with 2 persistent hornlike style bases at the rounded apex. Basic chromosome number not known; counts of $2n = 60$ and 68 reported. Photosynthetic pathway, C_4. Represented in Texas by the only member of the genus.

(Lonard 2003)

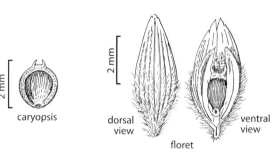

caryopsis

dorsal view

floret

ventral view

1. *Vaseyochloa multinervosa* (Vasey) Hitchc. (Texasgrass). Perennial with or without rhizomes. Ligules a short-fringed membrane. Caryopses dark brown to black with 2 persistent hornlike style bases. Found on coastal barrier islands, in live oak mottes, on sand dunes, and throughout the coastal sand plain. Collections from Bexar and Uvalde counties are uncharacteristic. A Texas endemic. Good livestock and wildlife values (Hatch, Schuster, and Drawe 1999).

spikelet

Vulpia C. C. Gmel.
(Poöideae: Poeae)

Tufted annuals with narrow blades, usually contracted, spikelike panicles, with 3 to many florets. Disarticulation above the glumes and between the florets. Glumes narrow, lanceolate or acuminate, 1–3-nerved, the lower often very short. Lemmas rounded on the back, inconspicuously 5-nerved, tapering to a fine awn or merely acuminate. Caryopsis cylindric and elongate. Anthers usually 1, infrequently 3, per flower. Basic chromosome number, $x = 7$. Photosynthetic pathway, C_3. Represented in Texas by 4 species and 3 varieties.

1. Lower glumes less than ½ length of upper glumes...2. *V. myuros*
1. Lower glumes ½ or more the length of upper glumes.
    2. Spikelets with 4–17 florets; awn of lowermost florets 0.3–6.0 mm long.......... 3. *V. octoflora*
    2. Spikelets with 1–7 florets; awn of lowermost florets 3–12 mm long.
        3. Lemmas of lowermost floret 2.5–3.5 mm long; caryopses 1.5–2.0 mm long ....4. *V. sciurea*
        3. Lemmas of lowermost floret 3.5–7.5 mm long; caryopses 3.5–5.5 mm long ....
        .... 1. *V. bromoides*

(Gould 1975b; Lonard 2007)

1. *Vulpia bromoides* (L.) Gray (brome fescue,
brome sixweeksgrass). Annual to 50 cm tall.
Culms solitary or tufted, usually erect or slightly
decumbent. Ligules <1 mm long, membranous.
Awn of lowermost lemmas 2–13 mm long. A
European introduction found in disturbed
habitats. Infrequent.

2 mm

spikelet

2 cm

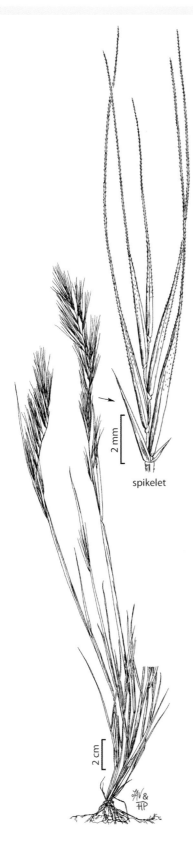

spikelet

2 mm

2 cm

2. *Vulpia myuros* (L.) C. C. Gmel. (foxtail fescue, rattail fescue, rattail sixweeksgrass). Annual to almost 1 m tall, but usually much shorter. Culms solitary or loosely tufted. Ligules 0.3–0.5 mm long, membranous. Awn of lowermost lemmas 5–22 mm long. A Mediterranean introduction that occurs in dry, sandy, disturbed sites. Varieties listed by Gould (1975b) have been reduced to forma and are not included here.

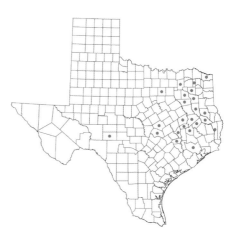

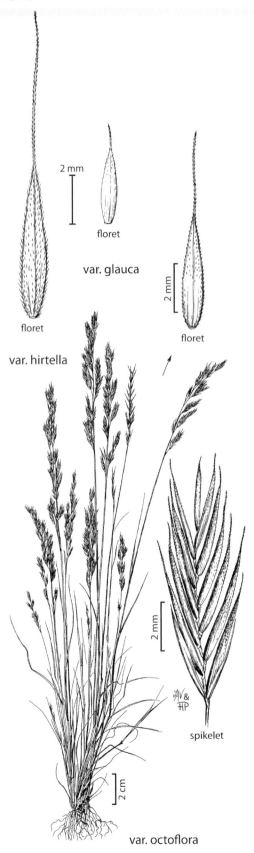

2 mm

floret

var. glauca

2 mm

floret

floret

var. hirtella

2 mm

spikelet

2 cm

var. octoflora

3. *Vulpia octoflora* (Walt.) Rydb. (common sixweeksgrass, sixweeks fescue). Annual to 60 cm tall. Culms solitary or tufted, erect. Ligules 0.3–1.0 mm long, membranous. Awn of lowermost lemmas 0.3–9.0 mm long. Most common member of the genus in Texas. Can be found in almost any disturbed site. Three varieties have been included: var. *glauca* (Nutt.) Fernald; var. *hirtella* (Piper) Henrard; and var. *octoflora*. They can sometimes be distinguished by the following characters:

1. Spikelets 4.0–6.5 mm long; awn of
   lowermost lemmas 0.3–3.0 mm long...........
   ....var. *glauca*
1. Spikelets 5.5–13 mm long; awn of the
   lowermost lemmas 2.5–9.0 mm long.
     2. Lemmas scabrous to pubescent...............
     .... var. *hirtella*
     2. Lemmas usually smooth, sometimes
       scabrous on upper part .... var. *octoflora*

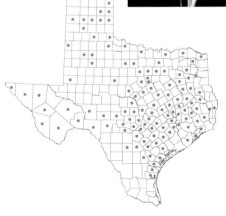

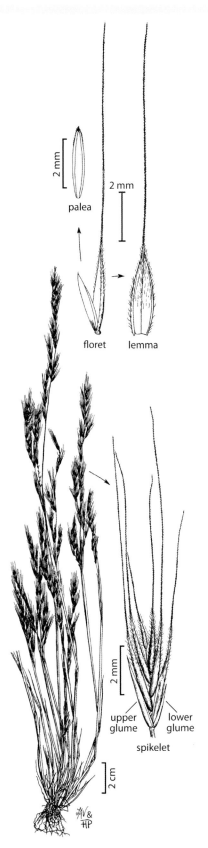

palea

floret   lemma

upper   lower
glume   glume

spikelet

4. *Vulpia sciurea* (Nutt.) Henrard (squirreltail fescue, squirrel sixweeksgrass). Annual to 60 cm tall. Culms solitary or tufted, erect to drooping. Ligules 0.5–1.0 mm long, membranous. Awn of lowermost lemmas 4–10 mm long. Native species found in deep, sandy soils of open woodlands, old fields, roadsides, and sand hills.

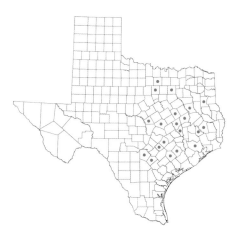

Willkommia Hack.
(Chloridoideae: Cynodonteae)

Low annuals or perennials, with flat or somewhat involute blades. Inflorescence with few to several short, closely flowered branches. Spikelets appressed and closely imbricated in 2 rows from the base to the apex of the branch rachis. Spikelets 1-flowered, disarticulating above the glumes. Glumes unequal, thin, the first short, rounded at the apex, narrow, and nerveless, and the second about as long as the lemma, 1-nerved, acute at the apex. Lemma 3-nerved, awnless, rounded dorsally, pubescent between the nerves and on the margins. Palea well developed, densely pubescent on the 2 nerves. Basic chromosome number, $x = 10$. Photosynthetic pathway, C_4. Represented in Texas by a single species and variety.

(Wipff 2003o)

lower glume

spikelet

upper glume

1 mm

ventral view

floret

dorsal view

1 cm

2 cm

var. texana

1. *Willkommia texana* Hitchc. (willkommia). A low, tufted, short-lived perennial. Culms to 40 cm long. Ligules a minute fringed membrane. Inflorescence a spikelike panicle. It is occasional in hard, clayey soils bordering swales, ponds, and small lakes in southern and southeastern counties. Diggs et al. (2006) reported this as an endemic species to Texas; however, it has been collected in Oklahoma and Argentina (although a different variety) (Wipff 2003o). The Texas species belongs to the typical variety, var. *texana*. Not listed as rare by Jones, Wipff, and Montgomery (1997) or Poole et al. (2007). Poor livestock and wildlife values (Hatch, Schuster, and Drawe 1999).

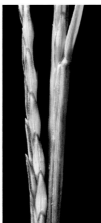

Zea L.
(Panicoideae: Andropogoneae)

Monoecious plants with tall, thick, usually succulent culms and broad, flat blades. Staminate spikelets in unequally pediceled pairs on spikelike branches, these forming large panicles at the culm apex. Glumes of the staminate spikelet broad, thin, several-nerved. Lemma and palea hyaline. Pistillate spikelets sessile in pairs on a thickened woody or corky axis (cob). Glumes of the pistillate spikelets broad, thin, rounded at the apex, much shorter than the mature caryopsis. Lower floret sterile or occasionally fertile. Lemma of lower floret and lemma and palea of upper (fertile) floret membranous and hyaline. Basic chromosome number, $x = 10$. Photosynthetic pathway, C_4. Represented in Texas by 2 species and 1 subspecies.

1. Plants annual .... 1. *Z. mays*
1. Plants perennial, with rhizomes .... 2. *Z. perennis*

(Iltis 2003)

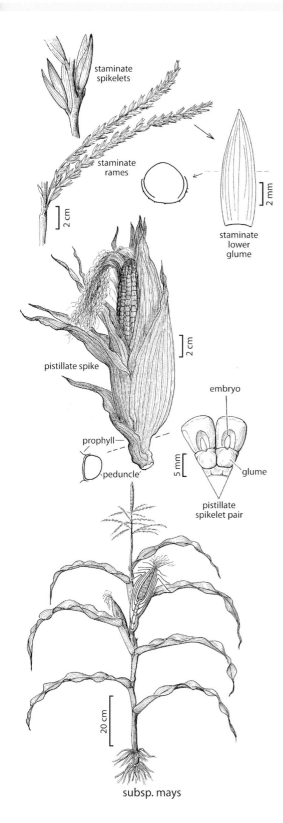

staminate
spikelets

staminate
rames

2 cm

staminate
lower
glume

2 mm

pistillate spike

2 cm

prophyll

peduncle

embryo

glume

5 mm

pistillate
spikelet pair

subsp. mays

1. *Zea mays* L. (corn, Indian corn, maize).
Annual, often with prop roots at the lowermost
nodes. Culms to 6 m long and 5 cm thick.
Blades to 90 cm long and 12 cm wide. Pistillate
inflorescences axillary, spikes tightly and
permanently enclosed in subtending leaf
sheaths and an enlarged prophyll (the ear).
Rachises thickened and tough (the cob).
Style branches long (the silk). Staminate
inflorescences terminal (the tassel). This is
the cultivated corn, which is the subsp. *mays,*
that has so many uses, such as an alternative
energy source for petroleum products. Again,
the distribution map is meaningless in the case
of cultivated plants. Corn is probably grown in
every county in the state. Good livestock and
wildlife values (Hatch, Schuster, and Drawe
1999).

2. *Zea perennis* (Hitchc.) Reeves & Mangelsd. (perennial teosinte). Perennial with long rhizomes. Culms to 2.5 m long, 2 cm thick. Blades to 80 cm long and 4.5 cm wide. Pistillate spikelets in axillary inflorescences. Staminate panicles terminal. Reported from Brazoria County. Not included in Gould (1975b); Hatch, Gandhi, and Brown (1990); Jones, Wipff, and Montgomery (1997); or Turner et al. (2003).

fruit case

styles

5 cm

1 cm

2 mm

staminate
lower glume

Zizania L.
(Ehrhartoideae: Oryzeae)

Tall, perennial, reedlike marsh grasses with broad leaves and large panicles of 1-flowered, unisexual spikelets. Ligule membranous, often large. Staminate spikelets pendulous on the lower inflorescence branches. Pistillate spikelets erect on the short, stiffly erect upper panicle branches. Disarticulation below the spikelet. Glumes absent. Pistillate spikelets slender, about 2 cm long, with a firm, 3-nerved, long-awned lemma tightly clasping a narrow, 2-nerved palea. Staminate spikelets with a thin, 5-nerved, acuminate or short awn-tipped lemma, 3-nerved paleas, and 6 stamens. Basic chromosome number, $x = 15$. Photosynthetic pathway, C_3. Represented in Texas by a single species.

(Gould 1975b; Terrell 2007c)

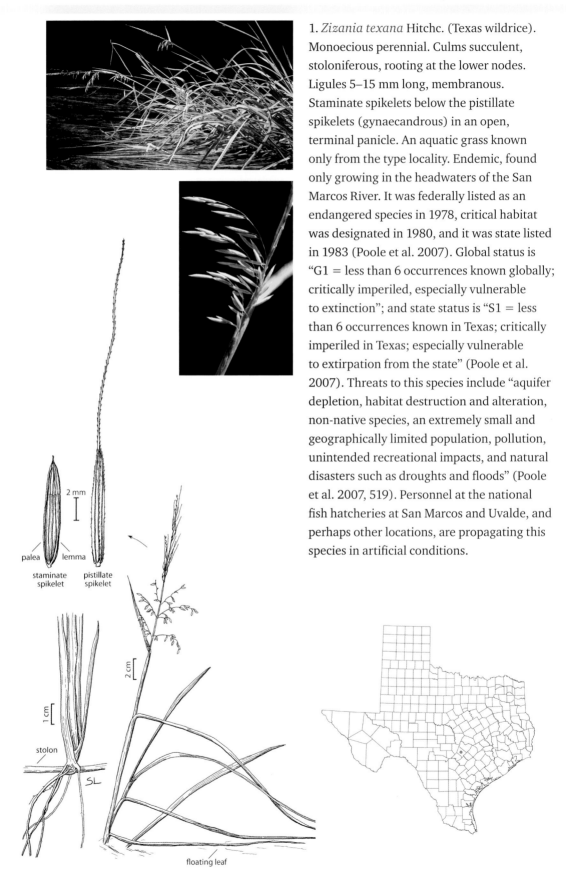

1. *Zizania texana* Hitchc. (Texas wildrice). Monoecious perennial. Culms succulent, stoloniferous, rooting at the lower nodes. Ligules 5–15 mm long, membranous. Staminate spikelets below the pistillate spikelets (gynaecandrous) in an open, terminal panicle. An aquatic grass known only from the type locality. Endemic, found only growing in the headwaters of the San Marcos River. It was federally listed as an endangered species in 1978, critical habitat was designated in 1980, and it was state listed in 1983 (Poole et al. 2007). Global status is "G1 = less than 6 occurrences known globally; critically imperiled, especially vulnerable to extinction"; and state status is "S1 = less than 6 occurrences known in Texas; critically imperiled in Texas; especially vulnerable to extirpation from the state" (Poole et al. 2007). Threats to this species include "aquifer depletion, habitat destruction and alteration, non-native species, an extremely small and geographically limited population, pollution, unintended recreational impacts, and natural disasters such as droughts and floods" (Poole et al. 2007, 519). Personnel at the national fish hatcheries at San Marcos and Uvalde, and perhaps other locations, are propagating this species in artificial conditions.

2 mm

palea   lemma

staminate   pistillate
spikelet   spikelet

2 cm

1 cm

stolon

SL

floating leaf

Zizaniopsis Döll and Asch.
(Ehrhartoideae: Oryzeae)

Large, robust, succulent perennials with stout rhizomes; broad, flat leaves; and large, open panicles. Ligule thin, membranous, to over 1 cm long. Spikelets large, 1-flowered, unisexual, the staminate and pistillate on the same branches, the pistillate above and the staminate below. Disarticulation below the spikelet. Glumes absent. Pistillate spikelet with a thin, 7-nerved, short-awned lemma and a large, 3-nerved palea. Caryopsis obovate, asymmetrical, beaked with a persistent style. Staminate spikelet with a thin, awnless, 5-nerved lemma, and a 3-nerved paleas of similar texture. Stamens 6. Basic chromosome number, $x = 12$. Photosynthetic pathway, C_3. Represented in Texas by a single species.

(Gould 1975b; Terrell 2007d)

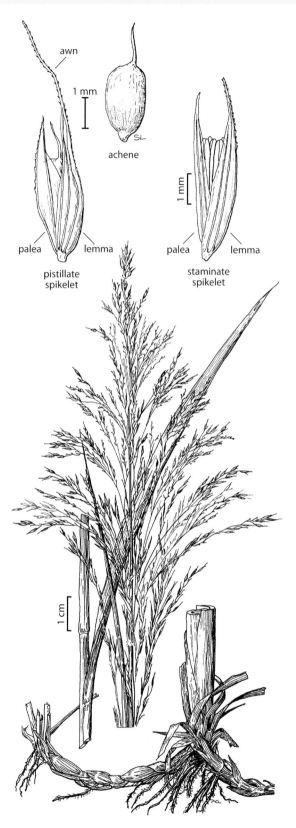

awn

1 mm

achene

palea   lemma   palea   lemma

pistillate   staminate
spikelet   spikelet

1. *Zizaniopsis miliacea* (Michx.) Döll and Asch. (giant cutgrass, water millet, southern wildrice). Coarse, monoecious, rhizomatous perennial. Culms thick, glabrous, to 3 m long. Ligules membranous with numerous veins, 6–20 mm long. Staminate and pistillate spikelets on same inflorescence branches, staminate below. Found growing in shallow water along streams, lakes, and marshes. Good livestock and wildlife values (Hatch, Schuster, and Drawe 1999). Telfair (2006) listed it as important for wildlife.

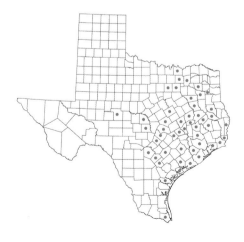

Zoysia Willd.
(Chloridoideae: Cynodonteae)

Low, sod-forming perennials, mostly with rhizomes and slender stolons. Inflorescence a slender, few-flowered spike, the spikelets solitary at the nodes of a slender, zigzag rachis. Disarticulation at the base of the 1-flowered, laterally compressed spikelets. First glume absent. Second glume firm, acute, mucronate or short-awned. Lemma thin, membranous, awnless, narrow at the apex, shorter than the second glume and enclosed by it. Palea present or absent. Basic chromosome number, $x = 10$. Photosynthetic pathway, C_4. Represented in Texas by 3 species. Gould (1975b) and Hatch, Gandhi, and Brown (1990) did not include this genus.

1. Blades to 0.5 mm wide; racemes with 3–12 spikelets..3. *Z. pacifica*
1. Blades 0.5–5.0 mm wide; racemes with 10–50 spikelets.
    2. Pedicels 1.6–3.5 mm long; spikelets 1.0–1.4 mm wide ....1. *Z. japonica*
    2. Pedicels 0.6–1.6 mm long; spikelets 0.6–1.0 mm wide ....2. *Z. matrella*

(Anderson 2003)

1. *Zoysia japonica* Steud. (Japanese lawngrass, Korean lawngrass). Rhizomatous perennial. Ligules minute, to 0.25 mm long. Peduncles exserted. Spikelet with short awn, awn to 1.1 mm long. A turfgrass found in scattered counties accross Texas.

spikelet

2. *Zoysia matrella* (L.) Merr. (manilagrass). Rhizomatous perennial. Ligules minute, to 0.25 mm long. Peduncles exserted. Spikelet awnless or awned, awn to 1 mm long. A turfgrass reported from Brewster County.

1 mm

spikelet

1 mm

caryopsis

lemma

2 cm

3. *Zoysia pacifica* (Goudswaard) M. Hota &
Kuroki (Korean velvetgrass). Rhizomatous
perennial. Ligules minute, to 0.25 mm long.
Peduncles usually included or shortly exserted.
Spikelets awned or awnless, awn to 0.5 mm
long. Turfgrass but not common. Reported
by Jones, Wipff, and Montgomery (1997) as
occurring in the state. No county distribution
information available.

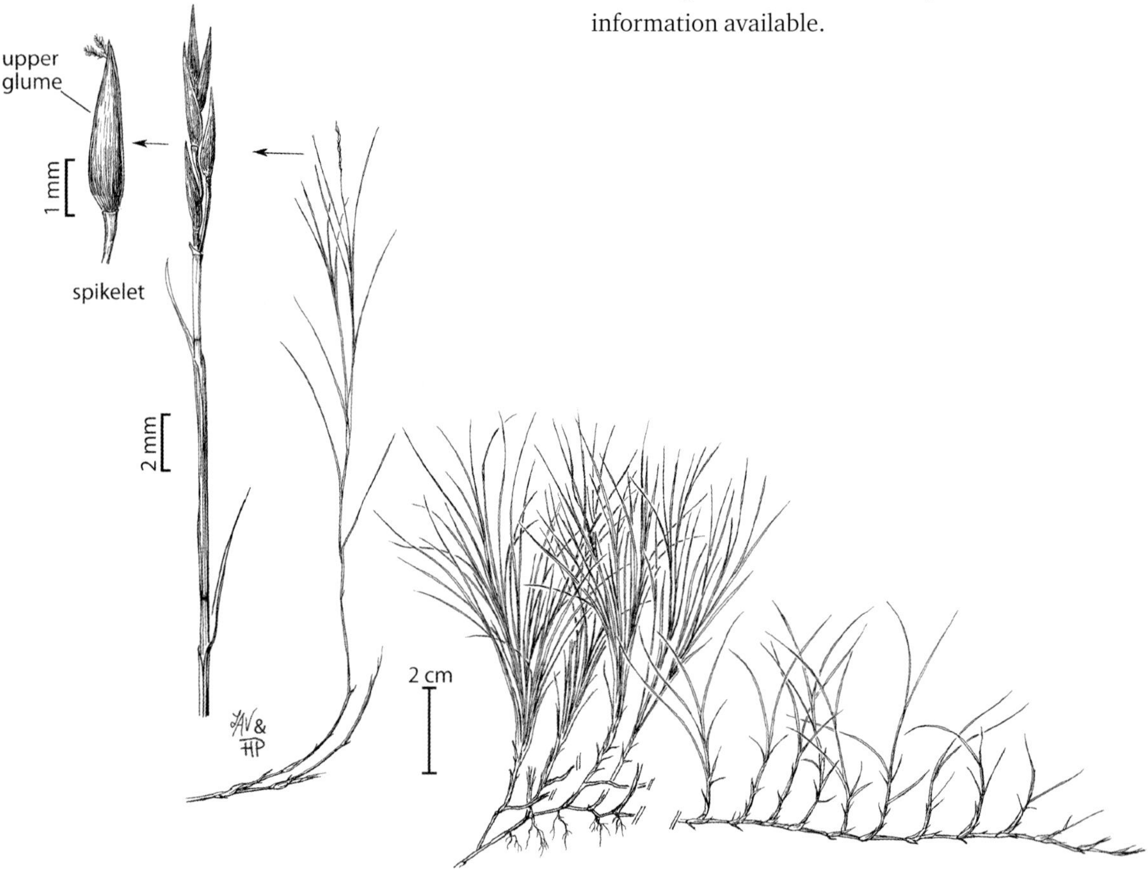

upper
glume

1 mm

spikelet

2 mm

2 cm

Zuloagaea Bess
(Panicoideae: Paniceae)

Rhizomatous perennials with hard, cormlike culm bases. Sheaths shorter than internodes, keeled; ligules membranous and ciliated. Panicles open, pyramidal, lowest node with a single branch. Spikelets 2.5–5.5 mm long, greenish to purple, glabrous. Lower glumes about $^2/_3$ as long as upper glumes, 3–5-veined; upper glumes about as long as lower lemma; 5–7-veined. Lower florets sterile or staminate; upper florets perfect, finely transverse rugose. Basic chromosome number, $x = 9$. Photosynthetic pathway, C_4. Represented in Texas by the only member of the genus.

(Bess et al. 2006)

1. *Zuloagaea bulbosa* (Kunth) Bess (bulbous panicgrass). Perennial with short, thin rhizomes. Culms can be up to 2 m tall and usually have a hard, cormlike base. Sheaths are characteristically shorter than the internodes and are keeled; blades decilinear. Ligules 0.5–2.0 mm long, membranous. Cormlike culm base is distinctive; however, plants growing on very dry sites might lack the corms. A recent segregate of *Panicum*. Gould (1975b); Hatch, Gandhi, and Brown (1990); Jones, Wipff, and Montgomery (1997); and Turner et al. (2003) listed this species as *P. bulbosum* Kunth. *Panicum plenum* Hitchc. is included here. Listed as fair for wildlife and good for livestock (Hatch and Pluhar 1993).

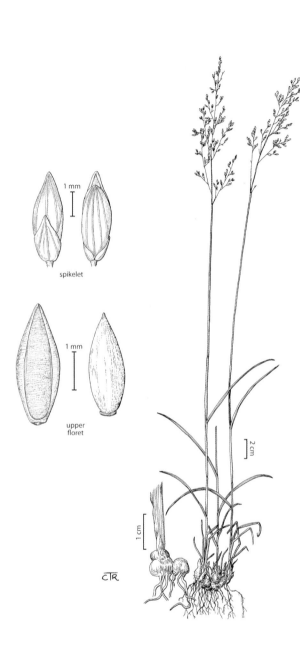

spikelet

upper floret

CTR

Late Additions

During final manuscript preparation it was discovered that 1 genus (*Hakonechloa*) and 6 species (*H. macra, Phyllostachys aureosulcata, Poa reflexa, P. secunda, Secale strictum,* and *Setaria megaphyllum*) had been overlooked or recently found to occur within Texas. These 6 species are added here to make the guide as complete as possible in hopes that any grass, other than some ornamentals, can be found in it. These species are not keyed, but they are briefly described and illustrated. They are in the checklist and are among the 723 species listed for the state.

Hakonechloa Makino *ex* Honda
(Arundinoideae: Arundineae)

Cespitose perennial, with extensive scaly rhizomes and sometimes stolons. Culms to 1 m tall. Sheaths open. Auricles absent. Abaxial ligule a line of hairs, membranous. Blades are unique in that they are resupinate, that is, the blade is twisted, the abaxial surface is facing upward, and the adaxial surface is downward. Inflorescence a rather loose panicle. Spikelets long-pedicellate, laterally compressed, with 5–10 florets. Rachilla segments pilose. Disarticulation below the glumes. Glumes shorter than the lowermost spikelet; lower slightly shorter than the upper. Lemmas 3-veined, margins with papillose-based hairs. Basic chromosome number, $x = 10$. Photosynthetic pathway, C_3. Represented in Texas by a single introduced species. Genus not included by Gould (1975b) or Hatch, Gandhi, and Brown (1990), but it is reported as occurring in the state by Jones, Wipff, Montgomery (1997).

(Thieret 2003e)

Hakonechloa macra (Munro) Makino
(Japanese forest grass, hakone grass).
Rhizomatous and stoloniferous perennial.
Leaf blade twisted so that the adaxial surface
faces downward (resupinate); adaxial surface
turns orange in the fall. An ornamental species
reported to occur within the state, but no
location cited (Jones, Wipff, and Montgomery
1997). No locations reported.

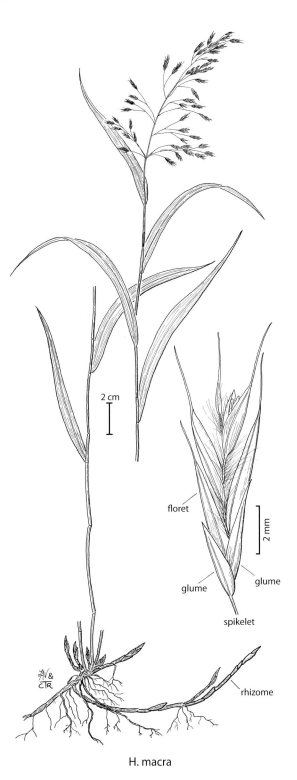

2 cm

floret

2 mm

glume

glume

spikelet

rhizome

H. macra

(Bambusoideae: Bambuseae)
Phyllostachys aureosulcata McClure (yellow grove bamboo, golden crookstem). A rhizomatous perennial species typically forming dense stands. Culms to 10 m tall, with a distinct yellowish-colored groove or sulcus; about 10% of the culms exhibit a distinctive zigzag in the lower part. Culms characteristically erect, and the plant makes an excellent hedge or screen. Listed as occurring in the state by Jones, Wipff, and Montgomery (1997) and by an unverified report indicating that the species has been collected in Harris County (www. herbarium .usu.edu/webmanual). The form "Spectabilis" has yellow culms and green grooves; the form "Aureocaulis" has lemon yellow culms that are often bright magenta in the spring for a short time. The hardy and well-adapted species is probably much more common than reported.

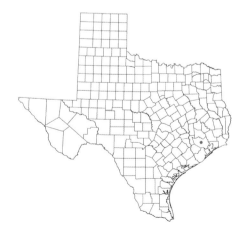

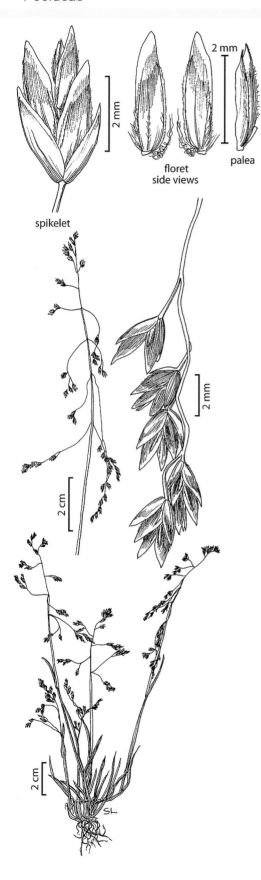

spikelet

floret
side views

palea

(*Poöideae: Poeae*)

Poa reflexa Vasey & Scribn. (nodding bluegrass). A short-lived, tufted perennial typically found on drier and often disturbed sites. As the specific epithet suggests, at least the lower panicle branches are reflexed at maturity. Jones, Wipff, and Montgomery (1997) reported the species as present in the state based on a specimen housed at the herbarium at Texas Tech University, although they had not seen or verified the collection. Soreng (2007) suggested that the specimens are most likely the closely related *P. leptocoma* Trin., which prefers wetter sites and has more scabrous panicle branches, shorter anthers, and glabrous or pectinately ciliate palea keels. Also, the southernmost distribution of *P. reflexa* is northern New Mexico, whereas *P. leptocoma* is found in southern New Mexico in counties immediately across the border with Texas.

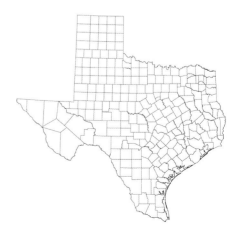

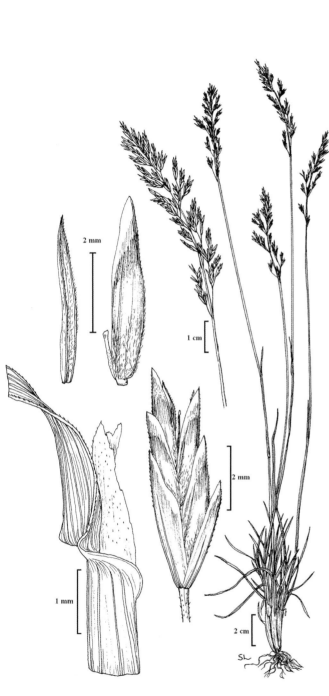

(Poöideae: Poeae)

Poa secunda J. Presl (secund bluegrass). A densely tufted perennial that is widespread in temperate western North America and furnishes valuable spring forage. Not reported for Texas by Jones, Wipff, and Montgomery (1997) or Soreng (2007), but it has been reported from 2 counties (Brazos and Edwards) by Turner et al. (2003). Further study is needed to determine if the species does indeed occur within the state boundaries or represents a misidentification.

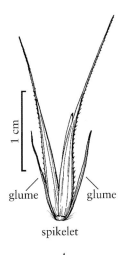

glume / glume

spikelet

S. strictum

(Poöideae: Triticeae)

Secale strictum (C. Presl) C. Presl (perennial rye). Cespitose perennial with culms to 1 m or more in length. Rachis disarticulates easily, and the lemmas are 8–16 mm long; the annual species (S. *cereale*) has more persistent rachises and lemmas that are 14–18 mm long. Barkworth (2007m) stated it is not established in North America, but it was used in crossing trials. Probably not a member of the Texas flora except in cultivation.

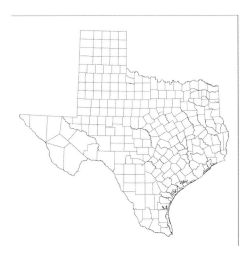

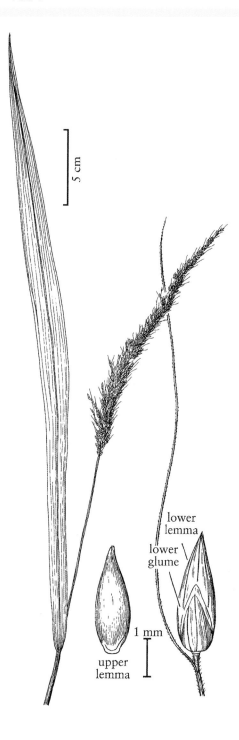

(Panicoideae: Paniceae)

Setaria megaphylla (Steud.) T. Durand &
Schinz (bigleaf bristlegrass). A large perennial
with culms up to 2 m long, nodes villous.
Ligules of hairs about 2 mm long. Blades to
60 cm long and 8 cm wide and strongly plicate.
Panicles narrow, branches erect. Bristles
solitary, typically only present below the
terminal spikelet on each branch. Differs from
S. palmifolia, which it resembles because of the
large and plicated leaves, by having a narrow,
closed panicle and reduced or scalelike palea
of the lower floret. A tropical species that has
become established in Florida. Jones, Wipff,
and Montgomery (1997) reported it cultivated
in Texas as *S. poiretiana* (Schult.) Kunth, and
it should be expected as an ornamental. No
specific locations given.

As noted in the ecoregion descriptions, many counties lack information on grass diversity. The following description of taxonomic practices is included to familiarize those who may not know the techniques for collecting and documenting plant species. Hopefully, this will encourage more collecting, particularly of grasses, throughout the state.

Herbarium specimens are flattened, dried, permanently preserved, adequately documented, and properly stored plant materials. They are housed in a herbarium (plural, herbaria). Herbarium specimens are typically stored in cabinets. Historically, cabinets were made of wood, but they are mostly of metal construction today. The cabinets are partitioned into spaces designed to hold standard herbarium sheets. In some herbaria, specimens are stored in specially designed plastic containers with secure lids. These are particularly useful in tropical and humid areas where insects damage specimens. Sheets are grouped by species, placed in heavy paper folders, alphabetized, and stored accordingly.

These specimens are the historical foundation of plant taxonomy; and as such, they are extremely valuable and, for all practical purposes, irreplaceable. Plant collections also are vital to such fields as plant anatomy, plant morphology, plant genetics, biosystematics, agronomy, biogeography, plant ecology, forestry, and range science. Figure A.1 shows the distribution of active herbaria across the state, and Table A.1 lists information about each herbarium. Following are some important uses of herbarium specimens.

1. Voucher specimens: This is perhaps one of the most important uses of herbarium specimens and relates to the plant nomenclatural process of typification. Beginning with Linnaeus's *Species Plantarium* in 1753, plant names have been permanently fixed with specific herbarium specimens. When a taxonomist names a new taxon (i.e., genus, species, variety, etc.), often a single specimen is being described or the author may designate a specimen from all those examined that exemplifies the distinguishing features of the new taxon. The designated specimen is the holotype, also known as the nomenclatural type. As long as this specimen exists, the plant name is permanently fixed to it. Other forms of type specimens (isotype, lectotype, paratype, topotype, etc.) depend upon the known existence of the holotype and/or circumstances surrounding the description of the taxon (i.e., the paratype is a specimen cited in the original description but is not the holotype) (Radford et al. 1974). Because of the importance of types, they are often

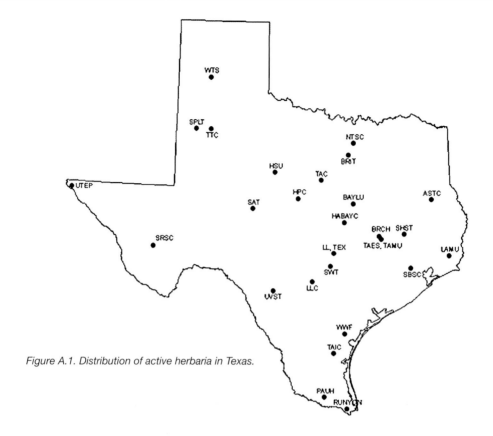

Figure A.1. Distribution of active herbaria in Texas.

separated from the rest of the collections in herbaria to reduce unnecessary handling, and the types are usually more securely stored to guarantee their safety.

Specimens may also validate records of plant structure, form, and characteristics used for taxonomic identification. They are especially important as voucher records in studies of chromosome number, polyploidy series, micromorphological characteristics, genetic relationships, and host plants. They enable other investigators to examine the exact specimens previously studied to ensure accuracy and validate results, an essential step in the scientific process. Voucher specimens are also used to document the occurrence of taxa from particular locations (e.g., military installations, conservation areas, wildlife preserves).

2. Records of distribution: Plant distributions found in floras, manuals, and checklists are almost always based on herbaria collections. Records of occurrence by county, plant community, ecosystem, ecoregion, state, and so forth are also based on permanently mounted specimens stored in herbaria. Collections provide a valuable historical record of where plants have occurred in both space and time. Specimens can document when and where foreign weeds were introduced (Diggs, Lipscomb, and O'Kennon 1999). Distributional data are extremely important to plant biogeographers and plant ecologists.

Poole et al. (2007) concluded that the flora of Texas remains poorly understood compared to the flora of other states. Texas is a large state, and over 90% of

Table A.1. Herbaria in Texas.

Herbarium code	Address	Number	Specialty
ASTC	Biology Department Stephen F. Austin University P.O. Box 13003 SFA Station Nacogdoches, TX 75962-3003	78,000	Eastern Texas
BAYLU	Biology Department Baylor University Baylor Sciences Building A254 Waco, TX 76798-7388	61,000	Texas
BRCH	Botanical Research Center P.O. Box 6717 Bryan, TX 77805-6717	9,000	Texas; southern U.S.; northeastern Mexico
BRIT	Botanical Research Institute 509 Pecan Street Fort Worth, TX 76102-4060	1,000,000	Worldwide, especially Texas; southeastern U.S.
HABAYC	Biology Department University of Mary Hardin-Baylor Belton, TX 76513-2599	2,249	Texas
HPC	Biology Department Howard Payne University Brownwood, TX 76801	40,000	Central Texas, especially Edwards Plateau; Sonora and Baja, Mexico
HSU	Biology Department Hardin-Simmons University Box N, HSU Station Abilene, TX 79698	5,500	Mostly Texas, especially Taylor County
LAMU	Biology Department Lamar University P.O. Box 10037 Beaumont, TX 77710	5,000	Southeastern Texas, especially Big Thicket; National Preserve
LL	Plant Resources Center University of Texas at Austin 1 University Station F0404 Austin, TX 78712-0471	65,000	Texas; Central America; Mexico
LLC	Biology Department Our Lady of the Lake University 4111 Southwest 24th Street San Antonio, TX 78207-4689	10,000	Primarily Texas
NTSC	Biological Sciences Department University of North Texas Box 305220 Denton, TX 76203-5220	16,000	Local; southwestern U.S.

Herbarium code	Address	Number	Specialty
PAUH	Biology Department University of Texas–Pan American Edinburg, TX 78539-2999	8,000	Lower Rio Grande Valley, Texas
RUNYON	Biological Sciences Department University of Texas, Southmost College 80 Fort Brown Brownsville, TX 78520	5,000	Lower Rio Grande Valley, Texas; adjacent Mexico
SAT	Biology Department Angelo State University P.O. Box 10890, ASU Station San Angelo, TX 76909	45,000	Western Texas
SBSC	Robert A. Vines Environmental Center 8856 Westview Drive Houston, TX 77055	40,000	U.S., especially southeastern Texas; Mexico
SHST	Department of Biological Sciences Sam Houston State University Huntsville, TX 77341-2116	16,000	Eastern Texas
SPLT	Science Department South Plains College Box 62 Levelland, TX 79336	20,000	Texas
SRSC	Department of Biology Sul Ross State University Alpine, TX 79832	100,000	Trans-Pecos Texas; northern Chihuahuan Desert
SWT	Department of Biology Texas State University–San Marcos San Marcos, TX 78666-4616	30,000	Texas
TAC	Biological Sciences Department Tarleton State University P.O. Box T-0100 Stephenville, TX 76402	1,287	Texas; Oklahoma
TAES	Department of Ecosystem Science and Management Texas A&M University 2138 TAMU College Station, TX 77843-2138	220,000	Primarily Texas and Mexico
TAIC	Department of Biology Texas A&M University–Kingsville Kingsville, TX 78363-8202	8,000	Primarily southern Texas
TAMU	Biology Department	35,000	Domesticated and

Herbarium code	Address	Number	Specialty
	Texas A&M University 3258 TAMU College Station, TX 77845-3258		cultivated plants
TEX	Plant Resources Center University of Texas at Austin 1 University Station F0404 Austin, TX 78712-0471	1,006,000	Texas; Latin America
TTC	Biological Sciences Department Texas Tech University Lubbock, TX 79409-3131	22,625	Arid lands
UTEP	Laboratory for Environmental Biology Centennial Museum University of Texas 500 West University El Paso, TX 79968-0519	65,000	Desert mountain ranges of southern New Mexico and western Texas; Chihuahua and Durango, Mexico
UVST	Biology Department Southwest Texas Junior College 2401 Garner Field Road Uvalde, TX 78801-6297	14,372	Texas
WTS	Department of Life, Earth, and Sciences West Texas A&M University Canyon, TX 79016-0001	75,000	Southwestern U.S.
WWF	Rob & Bessie Welder Wildlife Foundation P.O. Box 1400 Sinton, TX 78387-1400	5,600	Texas

the land is privately owned. Thus, the few botanists in the state have not had access to much of the land area. Although new county records are reason for celebration in some states, in Texas dozens of new county records can be found in a single day of collecting.

Some who find this subject interesting will look at the distribution map for a well-known species or read the list of species from a particular county or ecoregion. They will be inclined to say, "I can walk right out back and find _____ growing in this county!" Fill in the blank with any species you are absolutely certain occurs in your county but is not included on the maps. Unfortunately, you are the now the only one who knows for certain that the species does indeed occur in the area. To inform the rest of us, you must properly collect, press, identify, label, and submit a specimen to an active herbarium where the new data are accessible. New and expanded distributional data, based on voucher species, should be published as notes or short articles in local, regional, and even statewide scientific publications.

Expanding the information about grass distribution affords excellent educational opportunities for all ages. Numerous 4-H activities can be designed around locating, identifying, and collecting new county records. Also, collecting the grass flora from a particular county could be an excellent class project, "special problems" topic, or even the foundation of a master's thesis. Undoubtedly, numerous new state records and even new species are yet undiscovered within the state. Expanded distributional data are valuable and needed to further

the knowledge base concerning the grass flora of Texas. The number of documented species by county is presented in the ecoregion descriptions. These numbers illustrate that almost all areas of Texas are poorly collected.

3. Identification and verification of plant materials: Herbarium specimens are used to identify plants and verify collections. Although species descriptions, illustrations, and photographs found in taxonomic literature are excellent, nothing can replace comparing an unknown specimen with a herbarium mount to verify identification or to distinguish between two closely related taxa.

4. Education: Synoptic collections, especially dedicated to teaching or identification, are often separated from the general collection. Teaching collections are an essential tool in courses covering plant taxonomy, plant ecology, crop sciences, and range science. Agrostology is especially dependent upon mass collections of specimens for demonstration of plant characters, practice keying, and representation of species, genera, and other taxa.

5. Records of host plants: Herbarium specimens are valuable as host plant records in studies of rusts, smuts, fungi, lichens, and algae, as well as animal organisms such as aphids, beetles, and scale insects (Harrington 1977).

Collecting and Pressing Grass Specimens

One of the first considerations in collecting grasses is to secure the appropriate permission to collect on private or public lands (Diggs, Lipscomb, and O'Kennon 1999). Gaining permission from private landowners is important in maintaining a positive working relationship and allowing botanists to continue to collect on private lands. Also, public lands (parks, refuges, conservation areas, etc.) may have collecting regulations. There may be strict legal sanctions for not following collecting guidelines. Federally listed threatened and/or endangered species must never be collected without the proper permits from the U.S. Fish and Wildlife Service.

Figure A.2. Equipment necessary for field collection of grasses: a digging implement, plant press, and collection book.

Equipment necessary for field collection of grasses is fairly simple, consisting of a digging implement, plant press, and collection book (fig. A.2). Almost any digging implement may be used to remove the grass specimen from the ground, ensuring that roots and the lower portion of the plant are intact. Implements vary from the toe of one's boot, a common table knife, pocket or hunting knife (a good way to break the blade is to use it to pry!), dandelion digger, Japanese hori-hori knife, trowel, geologic pick, or small shovel. In my experience, the best all-around tool to collect grasses is a small, garden-size pick mattock (fig. A.2). Beware of inexpensive "army surplus" models; they are generally heavier and bulkier, and the head easily breaks when prying. Better-quality pick mattocks can be found at garden-supply shops and nurseries. The pick or pointed end is excellent for rocky and hardened caliche or clayey soils, whereas the broad, chiseled, hoelike end is perfect for looser soils. Either end is suitable for prying root systems from the ground. Grass roots have little diagnostic value; however, the base of the shoot system is immensely important in determining the life cycle (annual or perennial) of the grass, the development or presence of rhizomes and/or stolons, the pattern of basal branching, and the character of basal leaves. Upper portions of the grass plant with the inflorescences and spikelets might be adequate for identification, but without the basal portion of the grass, the collection is wholly inadequate for a proper herbarium specimen.

If only one collection is being made, be certain to select an "average" representative of the species in the area. Several specimens can be collected to show the range of variation (Harrington 1977). Sometimes it is important to collect "information-rich" specimens; that is, specimens that have been grazed, have insect damage, harbor insect galls, or have other organisms associated with them (Diggs, Lipscomb, and O'Kennon 1999). Keys for grass identification are almost always based on inflorescence type and mature spikelets, so that is the minimum necessary for a specimen. The best specimens always have fully developed inflorescences, flowers, and fruits, as well as the base of the plant with roots and culms with mature leaves.

Once collected, specimens should be flattened under pressure and dried as soon as possible. This is not as critical for grasses as for more delicate groups. However, to preserve the collection in the most realistic form, do not unduly delay pressing. Specimens may be placed in plastic "lawn-size" trash bags (clear or white is preferable to black) and stored in a cool, shaded, dry location until processed.

Flattening and drying of specimens are accomplished in a single step known as pressing. A plant press is utilized during this process to flatten the specimens. Place the collections in a folded piece of newspaper that is next placed between absorbent blotters inserted between 2 pieces of corrugated cardboard (fig. A.3). Blotters and newspaper facilitate the movement, or wicking, of moisture from the specimen, and the corrugated cardboard helps remove the moisture from the press. Daily changing of blotters can help retain specimen color and prevent mold. Group together the "sandwiches" of cardboard-blotter-newspaper-specimen-newspaper-blotter-cardboard, place them between heavy wooden frames, and secure with straps, ropes, or belts. Retighten the straps each day. Presses can be placed in specially designed drying ovens or left to dry unaided. Adequate drying of grass specimens without the aid of an oven takes about 3–7 days. Drying time, obviously, is dependent upon the amount of moisture in specimens at time of pressing. Plant presses (frames, straps, corrugates, and blotters) can be purchased from biological-supply companies.

A standard herbarium mounting sheet is 11.5 in × 16.5 in (26 cm × 42 cm). Grass specimens should be folded to a size of no more than 14 in (36 cm) in length (fig. A.4). Specimens can be bent or folded to form a V, N, or even M shape (Diggs, Lipscomb, and O'Kennon 1999). To keep the grasses from extending outside the press and maintain the folded form, push the points through a slight tear in the newspaper or place a slotted piece of heavy paper or cardboard over the points. Manufactured presses, cardboards, and blotters are a standard 12 in × 18 in (30 cm × 46 cm). Try to use half sheets of newspaper (full newspaper page torn longitudinally); when folded, they come closest to matching the size of the press. Some grasses, particularly bamboos and

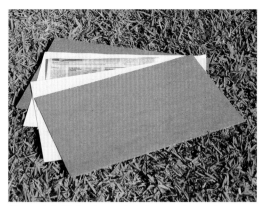

Figure A.3. Flattening and drying specimens in a folded piece of newspaper placed between absorbent blotters inserted between 2 pieces of corrugated cardboard.

canes, are too large and rigid to bend and fit on a single sheet. In this case, cut the specimen into segments and press each segment separately. The collection number should be the same, but letter each part sequentially (*2545*a, *2545*b, *2545*c, etc.).

Figure A.4. Grass specimen folded for herbarium specimen.

All data needed for an adequate herbarium label should be written in a field notebook at the time of collecting. Plant collectors typically number specimens consecutively throughout their lifetime, and generally only the collection number is written on the newspaper containing the specimen. Duplicate specimens should have the same number. Occasionally, a tag or small piece of paper on a string can be attached to the specimen with the collection number written on the tag. After the species name, the collection location is probably the most critical information on the label. The location data should be detailed enough to enable a stranger unfamiliar with the area to return to the collection point. Essential data include state, county, landmarks, accurate distances, nearby towns or cities, highway and county road numbers, collector(s) name, number, and date (Diggs, Lipscomb, and O'Kennon 1999). In this day of advanced technology, the use of a GPS (geographic positioning system) is becoming more and more common. Accurate latitude and longitude or UTMs from a GPS unit are the best location data possible. Other important data to include on labels are elevation, habitat, soil type, associated species, relative abundance, insect visitors, and any information not obvious from the specimen (height, color of stamens, stickiness of foliage, odor, annual/perennial, etc.). Synonyms and vernacular names may also be included if space permits. With the increased use of digital handheld recorders (data loggers), all data necessary for a herbarium label can be entered directly in the field and then downloaded to a computer and printed; thus, transcription errors are eliminated. Such devices can be programmed to "prompt" the user if he or she forgets to enter some information or enters data incorrectly. If a field log is used, it should be a pocket-sized, permanently bound, hard-covered notebook with waterproof pages if possible. Writing in pencil is recommended to prevent loss of data due to moisture damage.

Labels (fig. A.5a–c) should be printed on acid-free, 100% rag, bonded paper. They should be approximately 3 in × 5 in (8 cm × 13 cm) in size. Handwritten labels should be in permanent ink, but labels are best typed or printed if possible. Many herbaria have professionally produced labels with the name, location, and other information preprinted. Labels are usually placed in the lower right-hand corner of the herbarium sheet. Labels should be affixed with only herbarium-grade glue. Adhesive tape should never be used on herbarium sheets because it yellows, cracks, and loses its adhesiveness.

A standard herbarium mounting sheet is made of heavy, high-grade paper designed to last for hundreds of years without yellowing or damaging the specimen. It is acid-free and usually 100% rag, bonded paper. Specimens can be attached to the sheet by sewing, using gummed cloth strips, or gluing with specially designed, herbarium-grade adhesive. Sewing is labor intensive and not used much today. Mounting with gummed cloth strips is rapid, clean, and easy; however, it is difficult to prevent movement of the specimens on the sheet. It is not as popular as in the past. Gluing is the most common method of attaching specimens to sheets today. It is the best method for securing the specimen to the sheet; however, it is labor and space intensive and messy. All products necessary for mounting specimens can be acquired from herbarium-supply companies.

On label of dupl. in M.B.G. is entered:
Hab. Camanche Spring on fertile soil
in large bunches

TRACY HERBARIUM, TEXAS A&M UNIVERSITY

<u>Eriochloa sericea</u> (Scheele) Munro ex
Vasey

Det. by: R. B. Shaw

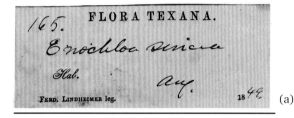

165.

FLORA TEXANA.

Eriochloa sericea

Hab.   *Aug.*

FERD. LINDHEIMER leg.   18 *49*   (a)

20

TRACY HERBARIUM, A. & M. COLLEGE OF TEXAS

PLANTS OF TEXAS

<u>Bouteloua</u> <u>curtipendula</u> (Michx.) Torr.

On rocky roadcut, black clayey soil;
4 miles north of Valley Mills . Chromosome number, 2n=52.

Bosque County

Frank W. Gould 9753b   12 October 1961   (b)

20

S. M. TRACY HERBARIUM (TAES)
TEXAS A&M UNIVERSITY
PLANTS OF TEXAS
BELL COUNTY

POACEAE

Eriochloa sericea (Scheele) Munro *ex* Vasey

GPS Coordinates: N 31° 13' 02.00", W 97° 30' 36.86". Elevation: 620 ft. Site
description: Owl Creek Park, Lake Belton. Soil description: sandy clay loam,
somewhat rocky in areas. Area near boat ramps and campsites manicured, edges of
trees and further than about 100 yards from shore not mowed or grazed. Associated
Species: *Panicum virgatum, Setaria parviflora, Echinochloa colona, Chloris
cucullata,* and *Eragrostis curtipedicellata.*

Louis Lee 030   12 July 2003   (c)

Figure A.5 a–c. Specimen labels

GLOSSARY

A

Abaxial. Located on the side away from the axis.

Achene. A small, dry, hard, indehiscent, 1-locular, 1-seeded fruit. See *caryopsis*.

Acicular. Needlelike.

Acuminate. Gradually tapering to a point.

Acute. Sharply pointed, not abruptly or long-tapering, but making an angle of less than 90°; less tapered than acuminate.

Adaxial. Located on the side toward the axis.

Adnate. Referring to fusion or attachment of unlike structures, such as a palea fused with the lemma. See *connate*.

Adventitious root. A root that arises from any organ other than the primary root or its branches.

Adventive. Introduced by chance; referring to accidental seedlings that are imperfectly naturalized.

Agrostology. The branch of systematic botany that deals with grasses.

Allopatric. Not growing together; occupying different geographic areas; opposite of sympatric.

Alpine. Referring to the area above the timberline.

Alveolate. Honeycombed, with angular depressions separated by thin partitions.

Androecium. The stamens of the flower referred to collectively.

Androgynous. Having staminate and pistillate spikelets in the same inflorescence, staminate above the pistillate.

Annual. Of one season's or year's duration, from seed to maturity and death.

Anther. The pollen-bearing part of a stamen.

Anthesis. The period during which the flower is open and functional.

Antrorse. Directed forward or toward the apex; opposite of retrorse.

Apical meristem. Terminal growing point.

Apiculate. Terminating abruptly in a small, short point.

Appressed. Lying flat or close against something.

Arborescent. Treelike in size and habit.

Arcuate. Arched or bowed; like an arc.

Aristate. Awned; with a stiff bristle at the tip or terminating nerves at the back or margins of organs such as the glumes, lemmas, and paleas.

Articulation. A joint or node. See *disarticulation*.

Ascending. Growing obliquely upward, like a culm.

Asperous. Rough or harsh to the touch.

Asperity. Short macrohair; also termed *pricklehair*.

Asymmetrical. Without symmetry.

Attenuate. Tapering gradually to a tip or base.

Auricle. An ear-shaped appendage; a structure that occurs in pairs laterally at the base of the leaf blade in some grasses and laterally at the sheath apex in others.

Awn. A bristle or stiff, hairlike projection; in the grass spikelet, usually the prolongation of the midnerve or lateral nerves of the glumes, lemmas, or paleas.

Axil. The upper angle formed between two structures, such as the leaf or spikelet and stem axis.

Axillary. In an axil.

Axis. The central stem or branch upon which the parts or organs are arranged, as in the culm or inflorescence.

B

Basic chromosome number. The lowest, actual or theoretical, haploid chromosome number, designated by x, as $x = 7$, 9, or 10; usually applied to a group of species.

Beaked. Ending in a firm, prolonged, slender tip.

Bearded. Bearing stiff, usually long hairs.

Bidentate. Having 2 teeth.

Bifid. Deeply 2-cleft.

Bilobed. Having 2 lobes.

Blade. The expanded portion of a flattened structure such as a leaf or flower petal. The blade of the grass leaf is the usually flattened, expanded portion above the sheath.

Bract. A modified leaf subtending a flower or belonging to an inflorescence; sometimes used in reference to the scales of a vegetative bud or other shoot structure.

Bracteal. Of or pertaining to bracts.

Bracteate. Having bracts.

Bracteole. Small bract.

Branch. A division or subdivision of an axis.

Bristle. A stiff hair or hairlike projection.

Bulb. An underground or partially underground bud with swollen, fleshy scales, such as an onion.

Bur. A structure containing seeds and covered with spines or prickles (as in *Cenchrus*).

C

Callus. The hard, usually pointed base of the spikelet (as in *Heteropogon*, *Andropogon*, and related genera) or of the floret (as in *Aristida* and *Stipa*), just above the point of disarticulation. In the former case, the callus is a portion of the rachis; in the latter, it is a portion of the rachilla. In *Eriochloa* the callus is the thickened node and remnant lower glume; in *Chrysopogon* it is part of the pedicel.

Canescent. With very short white or gray hairs, giving the surface a gray or white appearance.

Capillary. As applied to hairs, very slender and fine.

Capitate. Head-shaped; collected into a head or dense cluster.

Capsule. A dry, more or less dehiscent fruit composed of more than one carpel.

Carinate. Keeled with one or more longitudinal ridges.

Carpel. A unit of the pistil; a simple pistil is formed from a single carpel and a compound one from 2 or more carpels.

Cartilaginous. Firm and tough but flexible; like cartilage.

Caryopsis. A dry, hard, indehiscent, 1-seeded fruit with the thin pericarp adnate to the seed coat; the characteristic grass fruit. This differs from the achene only in the fusion of the pericarp and seed coat.

Cespitose (or caespitose). In tufts or dense clumps.

Chartaceous. With the texture of stiff writing paper, such as parchment used to make charts.

Chasmogamy. Pollination and fertilization of open flowers or florets; opposite of *cleistogamy*.

Ciliate. With marginal fringe of hairs.

Clavate. Club-shaped; thickened or enlarged at the apex from a slender base.

Cleistogamy. Pollination and fertilization within closed flowers or florets; opposite of *chasmogamy*.

Coleoptile. The sheath of the shoot of the embryo that protects the developing shoot.

Coleorhiza. The sheath of the primary root of the embryo that protects the developing radicle.

Collar. The outer side of a grass leaf at the junction of the blade and sheath.

Column. The modified base of an awn (as in *Aristida*, *Andropogon*, *Bothriochloa*).

Compressed. Flattened, usually laterally.

Compressed-keeled. Flattened laterally with a ridge or keel.

Conduplicate. Folded lengthwise down the middle.

Connate. Referring to fusion or attachment of like structures, such as the margins of the leaf sheath. See adnate.

Continuous. Referring to an axis that does not disarticulate at the nodes at maturity, usually applied to a rachis or inflorescence branch.

Contracted. Referring to an inflorescence that is narrow and dense with typically short, appressed branches or pedicels.

Convolute. Rolled up longitudinally; used primarily to describe leaf blades; same as involute.

Cordate. Heart-shaped, with a broad, notched base and a pointed tip.

Coriaceous. Leathery in texture.

Corm. The enlarged, fleshy base of a culm.

Cormous. With a corm.

Cotyledon. The leaf in a grass embryo.

Crown. The persistent base of a perennial grass.

Cucullate. Hooked, or having the shape of a hood.

Culm. The stem of a grass.

Cuneate. Wedge-shaped; narrowly triangular.

Cyme. A usually broad and flattish, determinate inflorescence, with the central or uppermost flower maturing first; grass inflorescences are cymose in that they are determinate, with the terminal spikelet maturing first.

D

Deciduous. Falling naturally, like the awns in *Sorghum halepense*.

Decilinear. With 10 or numerous parallel lines or veins.

Decumbent. Lying upon the surface, but ascending at the tip.

Decurrent. Extending downward from the point of insertion.

Dense. Referring to inflorescences in which the spikelets are crowded.

Depauperate. Impoverished; stunted.

Dichotomous. Divided in 2 equal parts, as the inflorescence branches of *Achnatherum hymenoides*.

Diffuse. Widely spreading.

Digitate. Like the fingers of a hand, usually referring to an inflorescence in which the branches all originate from a common point (as in *Chloris*, *Digitaria*).

Dioecious. Unisexual, with staminate and pistillate flowers on separate plants.

Diploid. With respect to polyploid chromosome series, having twice the basic (*x*) number of chromosomes.

Disarticulation. Separation at the joints or nodes at maturity.

Distal. Referring to the end opposite the point of attachment.

Distichous. Distinctly 2-ranked; in 2 vertical rows.

Distinct. Separate, as parts that are not united.

Diverging. Widely spreading; distinct.

Dorsal. The back side or surface; the surface turned away from the central stalk or axis; abaxial.

Dorsally compressed. Flattened from the front and back, said of spikelets, florets, glumes, lemmas, and/or panicles. Dorsal compression is common in the Paniceae.

Drooping. Inclined downward.

E

Elliptic. In the form of a flattened circle; more than twice as long as broad; widest in the middle and the 2 ends equal.

Elongate. Narrow; the length many times the width.

Embryo. The undeveloped plant within a seed.

Endemic. Indigenous or native in a restricted locality.

Endosperm. Nutritive tissue arising in the embryo sac of most angiosperms following the fertilization of the 2 fused polar nuclei (primary endosperm nuclei) by 1 of the 2 male gametes. In most organisms the cells of the endosperm have a *3n* chromosome number.

Entire. Undivided; the margin continuous, without teeth or lobes.

Erose. Irregular and uneven, as if gnawed or worn away.

Excurrent. Extending out or beyond.

Exotic. Not native; introduced.

Extravaginal. Referring to a type of branching where the tip breaks through the enclosing sheath; opposite of *intravaginal*.

F

Falcate. Flat and curved to one side, tapering toward the apex.

Fascicle. A cluster or close bunch, usually used in reference to culms, leaves, or branches of the inflorescence.

Fertile floret. A floret capable of producing a seed; may be perfect or pistillate.

Fibrous. Composed of fibers; fiberlike.

Filiform. Long, slender, and terete.

Flabellate. Fan-shaped; broadly wedge-shaped.

Flag leaf. The final leaf blade to emerge; the leaf immediately below the inflorescence.

Flexuous. Bent alternately in opposite directions.

Floret. As applied to grasses, the lemma and palea with the enclosed flower; may be perfect, pistillate, staminate, or sterile.

Flower. A plant reproductive structure composed of an ovary, stamens, and enveloping bracts (lemmas and paleas). The flower can be perfect, pistillate, staminate, or sterile.

Foliaceous. Leaflike, or with leaves.

Fruit. As applied to grasses, the ripened ovary. The typical grass fruit is a caryopsis.

G

Geniculate. Abruptly bent, as at the elbow or knee joint; frequently used to describe an awn or culm.

Gibbous. Swollen on one side; like a pouchlike swelling.

Glabrous. Without hairs; about the same as smooth.

Gland. A protuberance, depression, or appendage that secretes or appears to secrete a fluid.

Glandular. Glandlike or bearing glands.

Glaucous. Covered with a whitened or bluish waxy bloom, as a cabbage leaf or a plum.

Globose. Spherical or rounded; globelike.

Glomerate. In densely contracted, headlike clusters.

Glume. The pair of bracts usually present at the base of the grass spikelet.

Grain. In respect to grass, the threshed or unhusked fruit, usually a caryopsis; used to refer to the mature ovary alone or the ovary enclosed in persistent bracts (palea, lemma, or glumes).

Gross morphology. The external structure and form.

Gynaecandrous. With pistillate and staminate spikelets on the same inflorescence, pistillate above.

Gynoecium. The female portion of a flower as a whole; the pistil of the grass flower.

H

Haploid. With respect to polyploid chromosome series, the basic (x) number of chromosomes. With respect to sporophyte-gametophyte generations, the gametic ($1n$) chromosome number; half the number of chromosomes of the somatic (diploid) cells.

Hastate. Triangular or arrowhead-shaped, with 2 spreading lobes at base.

Herb. A plant without persistent, woody stems, at least above the ground.

Herbaceous. Without persistent woody stems; dying to the ground each year.

Herbarium. A collection of dried and pressed plant specimens, usually mounted for permanent preservation; the room, building, or institution in which such a collection is kept or to which it belongs.

Hermaphroditic. Referring to a spikelet with both male and female reproductive parts; same as perfect and bisexual.

Hilum. Scar where the seed was attached to the ovary wall.

Hirsute. With rather coarse and stiff hairs, these long, straight, and erect or ascending.

Hirsutulous. Somewhat hirsute.

Hispid. With erect, rigid, stiff bristles or bristle-like hairs.

Hooked. Curved at the tip.

Hyaline. Thin in texture and transparent or translucent.

Hydrophytic. Water-loving, wet; in respect to moisture, habitats are classified as *hydrophytic* (wet), *mesic* (moist), or *xeric* (dry).

Hygroscopic. Altering form or position due to changes in humidity; like the awn in *Achnatherum* and *Hesperostipa*.

I

Imbricate. Partly overlapping, as the shingles of a roof.

Imperfect. Unisexual flowers or florets; with either male or female reproductive structures but not both.

Included. Not protruding, as the awn in some *Agrostis*.

Indehiscent. Closed, not opening; like a caryopsis.

Indigenous. Native.

Indurate. Hard.

Inflated. Bladderlike; blown up like a balloon.

Inflorescence. The flowering portion of a shoot; in grasses, the spikelets and the axis or branch system that supports them, the inflorescence being delimited at the base by the uppermost leafy node of the shoot.

Intercalary meristem. An actively growing primary tissue region somewhat removed from the apical meristem; present at the base of the internodes in young grass shoots.

Intercostal. Referring to the area between the nerves.

Internode. Portion of the stem or other structure between 2 nodes.

Interrupted. With the order broken; used to describe inflorescences with major gaps in spikelets or branches.

Intravaginal. Referring to growth of a shoot when the tip does not break through the enclosing sheath but emerges at the top; opposite of extravaginal, and the more common condition in grasses.

Introduced. Referring to a plant intentionally brought in for a specific purpose. The plant may escape and persist.

Involucre. A circle or cluster of bracts or reduced branchlets that surround a flower or floret, or a group of flowers or florets.

Involute. Rolled inward from the edges.

J

Joint. The node of a grass culm; an articulation.

K

Keel. A prominent dorsal ridge, like the keel of a boat. Glumes and lemmas of laterally compressed spikelets are often sharply keeled; the palea of some florets is 2-keeled.

L

Lacerate. Irregularly cleft or torn.

Lamina. The blade or expanded part of a leaf.

Lanate. With long, tangled, woolly hairs.

Lanceolate. Lance-shaped; relatively narrow, tapering to both ends from a point below the middle.

Lateral. Borne on the sides of a structure.

Laterally compressed. Flattened from the sides, said of spikelets, florets, glumes, lemmas, and/or paleas; most common type of spikelet compression in the grasses.

Lax. Loose or drooping.

Lemma. The lowermost of the 2 bracts enclosing the flower in the grass floret; most important structure for grass identification.

Lignified. Woody; converted into lignin.

Ligule. A membranous or hairy appendage on the adaxial surface of the grass leaf at the junction of sheath and blade; often one of the few vegetative characters used in identification.

Linear. Long and narrow with parallel margins, like the majority of grass leaf blades.

Lobed. With lobes or segments of an organ cut less than halfway to the middle; apex of segments rounded.

Lodicule. Small, scalelike processes, usually 2 or 3 in number, at the base of the stamens in grass flowers. Lodicules are generally interpreted as rudimentary perianth segments.

Loose. Open, like a panicle with spreading branches.

M

Membranous. With the character of a membrane; thin, soft, and pliable.

Meristem. Embryonic or undifferentiated tissue, the cells of which are capable of active division.

Mesic. Referring to a moist habitat; in respect to moisture, habitats are classified as *hydrophytic* (wet), *mesic* (moist), or *xeric* (dry).

Midnerve. The central nerve of an oddly nerved structure; same as midrib.

Midrib. The central rib of a structure.

Monoecious. Flowers unisexual, with male and female flowers borne on the same plant; like corn, the ears being female and the tassels being male.

Monotypic. Having a single type or representative, as a genus with only one species.

Morphology. The science that deals with the form and structure of plants (or animals). Broadly interpreted, morphology includes anatomy, histology, and the nonphysiological aspects of cytology and embryology. In plant taxonomy, morphology is concerned mainly with external form.

Mucro. A short, small, abrupt tip of an organ, as the projection of a nerve of the leaf.

Mucronate. With a mucro.

Muriculate. Finely covered with short, sharp points.

N

Naked. Lacking some structure.

Nerve. A simple vein or slender rib of a leaf or bract (glumes, lemma, and/or palea).

Neuter. Without functional stamens or pistils; same as sterile.

Node. The joint of a stem; the region of attachment of the leaves.

Nodding. Inclined from an upright position; used primarily to describe panicles, as in several species of *Bromopsis*.

O

Oblong. From 2 to 3 times longer than broad and with nearly parallel sides.

Obovate. Inverted ovate.

Obtuse. Blunt or rounded at the apex.

Open. Loose; like a panicle with spreading branches.

Opposite. Referring to two structures at a node and across from each other.

Ovary. Part of the pistil that contains the ovules.

Ovate. Egg-shaped, with the broadest end toward the base.

Ovule. The structure that develops into the caryopsis.

P

Palea. The uppermost of the 2 bracts enclosing the grass flower in the floret; usually 2-nerved and 2-keeled and often enclosed by the lemma.

Panicle. As applied to grasses, any inflorescence where the spikelets are not sessile or individually pediceled on the main axis.

Paniculate. Borne in a panicle.

Papillate, papillose. Bearing minute nipple-shaped projections (papillae).

Pappus. Tuft of hairs or bristles at the basal tip of a structure.

Pectinate. With narrow, closely set, and divergent segments or units, like the teeth of a comb.

Pedicel. The stalk of a single flower; in grasses, applied to the stalk of a single spikelet.

Peduncle. A primary flower stalk, usually applied to the stalk of a flower cluster or the stalk of an inflorescence.

Pedunculate. With a peduncle.

Pendulous. Suspended or hanging.

Perennial. Lasting year after year.

Perfect. A flower (or spikelet) with both functional male and female reproductive structures.

Perianth. The floral envelope, undifferentiated or consisting of calyx and corolla.

Pericarp. The fruit wall developed from the ovary wall; the ripened outer wall of the mature fruit.

Persistent. Remaining attached.

Petiole. A leaf stalk.

Phenotype. The aggregate of visible characteristics of an organism.

Phytomere. The basic structural unit of the grass shoot; an internode together with the leaf sheath, ligule, blade and portion of the node at the upper end, and a bud and portion of the node at the lower end.

Pilose. With soft, straight hairs; about the same as villous.

Pistil. The female (seed-bearing) structures of the flower, ordinarily consisting of the ovary, stigma, and style.

Pistillate. Having a pistil but not stamens.

Planoconvex. Flat on one side and rounded on the other.

Plicate. Folded into pleats, usually lengthwise, as in a fan.

Plumose. Feathery; having fine, elongate hairs on either side.

Plumule. Embryonic culm- and leaf-producing structure.

Pollen grain. Microgametophyte at the stage when it is shed from the anther; usually 2- or 3-nucleate.

Polyploid. Referring to a plant with 3 or more basic sets of chromosomes; any ploidy level above the diploid.

Prickle-hair. Short macrohair; also termed *asperity*.

Primary shoot. The shoot developing from the plumule of the embryo.

Prophyll (prophyllum). The first leaf of a lateral shoot or vegetative culm branch. The prophyll is a sheath, usually with 2 strong, lateral nerves and numerous fine, intermediate nerves; a blade is never developed.

Prostrate. Lying flat on the ground.

Proximal. Referring to the end of a structure by which it is attached.

Pseudopetiolate. Referring to constriction and sometimes twisting of the basal portion of the leaf blade (rarely the sheath apex), giving the appearance of a petiole; most common in bamboos.

Pseudospikelet. The highly reduced inflorescence of some bamboos that consists of a shortened axis bearing a single spikelet at the apex and few to several bracts below.

Puberulent. Minutely pubescent.

Pubescent. Hairy; hairy with short, soft hairs.

Pulvinus. A swelling at the base of a leaf or of a branch of the inflorescence.

Pungent. Terminating in a rigid, sharp point.

Pustulate. With irregular, blisterlike swellings or pustules.

Pyramidal. Like a pyramid; referring to a cone-shaped inflorescence.

R

Raceme. As applied to grasses, an inflorescence in which all the spikelets are borne on pedicels inserted directly on the main (undivided) inflorescence axis, or some are sessile and some pediceled on the main axis.

Racemose. Bearing or like a raceme.

Rachilla. The axis of a grass spikelet.

Rachis. Main axis of an inflorescence.

Radicle. The primary root of the grass embryo.

Rame. Disarticulating inflorescence branch; the basic unit of the typically Andropogoneae inflorescence; an inflorescence branch composed of spikelet units (usually a fertile, sessile spikelet; a reduced pedicellate spikelet; and the internode that extends from the sessile spikelet to the next distal sessile spikelet).

Reduced floret. A staminate or neuter floret; if highly reduced, then termed *rudimentary floret*.

Reflexed. Curved outward or backward

Reniform. Kidney-shaped.

Retrorse. Directed backward or downward.

Revolute. Rolled backward from the margin.

Rhizomatous. With rhizomes.

Rhizome. An underground stem, usually with scale leaves and adventitious roots borne at regularly spaced nodes.

Root. The descending axis of the grass, without nodes and internodes, and absorbing water and nutrients from the soil.

Rosette. A dense basal cluster of leaves arranged in a circle; like the winter rosette of *Dichanthelium*.

Ruderal. A plant growing in highly disturbed areas such as road rights-of-way and old fields.

Rudiment. An imperfectly developed organ or part, as the rudimentary floret or florets of some spikelets.

Rugose. Wrinkled.

S

Saccate. Sac- or pouch-shaped.

Scaberulent. Slightly scabrous.

Scaberulous. Slightly scabrous.

Scabrous. Rough to the touch, usually because of the presence of minute prickle-hairs in the epidermis.

Scarious. Thin, dry, and membranous, not green.

Scutellum. A band of tissue between the embryonic shoot and the endosperm.

Secund. Borne on one side of the axis.

Seed. A ripened ovule.

Seminal root. Seed root or the primary root and all other roots that arise from embryonic tissue below the scutellar node.

Sericeous. Covered with long, straight, soft, appressed hairs.

Serrate. Saw-toothed, with sharp teeth pointing forward.

Serrulate. Finely serrate.

Sessile. Attached directly at the base; without a stalk.

Seta. A bristle or a rigid, sharp-pointed, bristle-like organ.

Setaceous. Bristly or bristlelike.

Sheath. The tubular basal portion of a leaf that encloses the stem, as in grasses and sedges.

Sinuous (sinuate). With a strong, wavy margin.

Spathe. A large bract enclosing an inflorescence as in some genera of the Andropogoneae.

Spicate. Spikelike.

Spicule. Short, stout, pointed projection of the leaf epidermis; often grades into a prickle-hair.

Spike. An inflorescence with flowers or spikelets sessile on an elongated, unbranched rachis.

Spikelet. The basic unit of the grass inflorescence, usually consisting of a short axis, the rachilla, bearing 2 "empty" bracts, the glumes, at the basal nodes and 1 or more florets above. Each floret consists usually of 2 bracts, the lemma (lower) and the palea (upper), which enclose a flower. The flower usually includes 2 lodicules (vestigial perianth segments), 3 stamens, and a pistil.

Spikelike. Similar to a spike inflorescence.

Sprawling. Spread out in an irregular fashion; viney.

Stamen. The male organ of the flower, consisting of a pollen-bearing anther on a filiform filament. Collectively, the stamens of a flower are referred to as the *androecium*.

Staminate. Having stamens but not pistils.

Stigma. The part of the ovary or style that receives the pollen for effective fertilization.

Stipe. A stalk.

Stipitate. Having a stalk or stipe, as an elevated gland.

Stolon. A modified horizontal stem that loops or runs along the surface of the ground and serves to spread the plant by rooting at the nodes.

Stoloniferous. Bearing stolons.

Strigillose. Similar to strigose, but hairs shorter.

Strigose. With appressed, stiff, short hairs.

Style. The contracted portion of the pistil between the ovary and the stigma.

Sub-. Latin prefix meaning "almost," "somewhat," "of inferior rank," "beneath."

Subtending. Below or beneath something, often enclosing it.

Subulate. Awl-shaped.

Succulent. Fleshy or juicy.

Sucker. A vegetative shoot of subterranean origin.

Sulcate. With long, narrow grooves or channels.

Sympatric. Growing together; occupying the same geographic area; opposite of allopatric.

Syninflorescence. Generally in bamboos, the aggregation of spikelets or pseudospikelets terminal to the culm or lateral branches.

Synoecious. Having male and female flowers in the same inflorescence.

T

Taxon (plural, taxa). Any taxonomic unit, as species, genus, or tribe.

Teeth. Pointed divisions of a structure.

Terete. Cylindric; round in cross section.

Tiller. A subterranean or ground-level lateral shoot, usually erect, as contrasted with horizontally spreading shoots referred to as either stolons or rhizomes.

Translucent. Allowing the passage of light rays, but not transparent.

Transverse. Extending across a structure, such as the transverse rugose lemma of some panicoid genera (e.g., as in some *Eriochloa*, *Setaria*).

Trifid. Divided into 3 parts.

Triquetrous. With 3 salient angles, each side concave or channeled.

Truncate. Terminating abruptly as if cut off transversely.

Tuberculate. Having small, pimplelike structures or covered with tubercules.

Turgid. Swollen; tightly drawn by pressure from within.

U

Uncinate. Hooked at the tip.

Unilateral. One-sided, developed or hanging on one side.

Unisexual. With either male or female sex structures but not both.

Utricle. A small, bladdery, 1-seeded fruit.

V

Ventral. Inner side of an organ; upper surface of a leaf; adaxial.

Verticil. A whorl; with 3 or more members or parts attached at the same node of the supporting axis.

Vestigial. Rudimentary and almost completely reduced; with only a vestige remaining.

Villous. Bearing long and soft, not matted hairs; shaggy; same as pilose.

Viscid. Sticky; glutinous.

W

Weed. Any plant growing where it is not wanted.

Wing. Any membranous expansion bordering a structure.

Winter annual. A plant that germinates in the fall and completes its life cycle the following spring or summer.

Woolly. With long, soft, interwoven hairs.

X

Xeric. Characterized by, or pertaining to, conditions of scanty moisture supply; referring to a very arid region or environment; in respect to moisture, habitats are classified as *hydrophytic* (wet), *mesic* (moist), or *xeric* (dry).

Allen, C. M. 2003a. Coelorachis. In Barkworth et al., Flora of North America, vol. 25.

Allen, C. M. 2003b. Hemarthria. In Barkworth et al., Flora of North America, vol. 25.

Allen, C. M. 2003c. Paspalidium. In Barkworth et al., Flora of North America, vol. 25.

Allen, C. M., and D. W. Hall. 2003. Paspalum. In Barkworth et al., Flora of North America, vol. 25.

Allred, K. W. 1982. Describing the grass inflorescence. J. Range Manage. 35:672–675.

Allred, K. W. 2003a. Aristida. In Barkworth et al., Flora of North America, vol. 25.

Allred, K. W. 2003b. Arundo. In Barkworth et al., Flora of North America, vol. 25.

Allred, K. W. 2003c. Bothriochloa. In Barkworth et al., Flora of North America, vol. 25.

Allred, K. W. 2003d. Cortaderia. In Barkworth et al., Flora of North America, vol. 25.

Allred, K. W. 2003e. Phragmites. In Barkworth et al., Flora of North America, vol. 25.

Allred, K. W. 2003f. Stenotaphrum. In Barkworth et al., Flora of North America, vol. 25.

Allred, K. W. 2003g. Trachypogon. In Barkworth et al., Flora of North America, vol. 25.

Allred, K. W. 2005. A Field Guide to the Grasses of New Mexico. 3rd ed. Las Cruces: Agriculture Experiment Station, New Mexico State University.

Allred, K. W. 2007a. Apera. In Barkworth et al., Flora of North America, vol. 24.

Allred, K. W. 2007b. Dactylis. In Barkworth et al., Flora of North America, vol. 24.

Allred, K. W., and M. E. Barkworth. 2007. Anthoxanthum. In Barkworth et al., Flora of North America, vol. 24.

Anderson, S. J. 2003. Zoysia. In Barkworth et al., Flora of North America, vol. 25.

Arriaga, M. O. 2007. Amelichloa. In Barkworth et al., Flora of North America, vol. 24.

Aulbach, C. 2003. Eustachys. In Barkworth et al., Flora of North America, vol. 25.

Baden, C. 2007. Psathyrostachys. In Barkworth et al., Flora of North America, vol. 24.

Bailey, R. G. 1995. Description of the ecoregions of the United States. 2nd ed. USDA Forest Service Misc. Publ. No. 1391. Washington, D.C.: Government Printing Office.

Barkworth, M. E. 2003a. Axonopus. In Barkworth et al., Flora of North America, vol. 25.

Barkworth, M. E. 2003b. Chloris. In Barkworth et al., Flora of North America, vol. 25.

Barkworth, M. E. 2003c. Ctenium. In Barkworth et al., Flora of North America, vol. 25.

Barkworth, M. E. 2003d. Cynodon. In Barkworth et al., Flora of North America, vol. 25.

Barkworth, M. E. 2003e. Dichanthium. In Barkworth et al., Flora of North America, vol. 25.

Barkworth, M. E. 2003f. Distichlis. In Barkworth et al., Flora of North America, vol. 25.

Barkworth, M. E. 2003g. Elionurus. In Barkworth et al., Flora of North America, vol. 25.

Barkworth, M. E. 2003h. Enteropogon. In Barkworth et al., Flora of North America, vol. 25.

Barkworth, M. E. 2003i. Heteropogon. In Barkworth et al., Flora of North America, vol. 25.

Barkworth, M. E. 2003j. Hilaria. In Barkworth et al., Flora of North America, vol. 25.

Barkworth, M. E. 2003k. Hyparrhenia. In Barkworth et al., Flora of North America, vol. 25.

Barkworth, M. E. 2003l. Miscanthus. In Barkworth et al., Flora of North America, vol. 25.

Barkworth, M. E. 2003m. Sorghum. In Barkworth et al., Flora of North America, vol. 25.

Barkworth, M. E. 2003n. Spartina. In Barkworth et al., Flora of North America, vol. 25.

Barkworth, M. E. 2003o. Themeda. In Barkworth et al., Flora of North America, vol. 25.

Barkworth, M. E. 2003p. Trichloris. In Barkworth et al., Flora of North America, vol. 25.

Barkworth, M. E. 2003q. Tripsacum. In Barkworth et al., Flora of North America, vol. 25.

Barkworth, M. E. 2007a. Achnatherum. In Barkworth et al., Flora of North America, vol. 24.

Barkworth, M. E. 2007b. Ehrharta. In Barkworth et al., Flora of North America, vol. 24.

Barkworth, M. E. 2007c. Hesperostipa. In Barkworth et al., Flora of North America, vol. 24.

Barkworth, M. E. 2007d. Leymus. In Barkworth et al., Flora of North America, vol. 24.

Barkworth, M. E. 2007e. Melica. In Barkworth et al., Flora of North America, vol. 24.

Barkworth, M. E. 2007f. Nassella. In Barkworth et al., Flora of North America, vol. 24.

Barkworth, M. E. 2007g. Pascopyrum. In Barkworth et al., Flora of North America, vol. 24.

Barkworth, M. E. 2007h. Phalaris. In Barkworth et al., Flora of North America, vol. 24.

Barkworth, M. E. 2007i. Phleum. In Barkworth et al., Flora of North America, vol. 24.

Barkworth, M. E. 2007j. Piptatherum. In Barkworth et al., Flora of North America, vol. 24.

Barkworth, M. E. 2007k. Piptochaetium. In Barkworth et al., Flora of North America, vol. 24.

Barkworth, M. E. 2007l. Polypogon. In Barkworth et al., Flora of North America, vol. 24.

Barkworth, M. E. 2007m. Secale. In Barkworth et al., Flora of North America, vol. 24.

Barkworth, M. E. 2007n. Thinopyrum. In Barkworth et al., Flora of North America, vol. 24.

Barkworth, M. E., and L. K. Anderton. 2007. Glyceria. In Barkworth et al., Flora of North America, vol. 24.

Barkworth, M. E., J. J. N. Campbell, and B. Salomon. 2007. Elymus. In Barkworth et al., Flora of North America, vol. 24.

Barkworth, M. E., K. C. Capels, S. Long, L. K. Anderton, and M. B. Piep, eds. 2007. Flora of North America. Vol. 24, Magnoliophyta: Commelinidae (in part): Poaceae, Part 1. New York: Oxford University Press.

Barkworth, M. E., K. M. Capels, S. Long, and M. B. Piep, eds. 2003. Flora of North America. Vol. 25, Magnoliophyta: Commelinidae (in part): Poaceae, Part 2. New York: Oxford University Press.

Barkworth, M. E., and E. E. Terrell. 2007. Oryza. In Barkworth et al., Flora of North America, vol. 24.

Baum, B. R. 2007. Avena. In Barkworth et al., Flora of North America, vol. 24.

Bess, E. C., A. N. Doust, G. Davidse, and E. A. Kellogg. 2006. Zuloagaea, a new genus of neotropical grass within the "bristle clade." Systematic Botany 31:656–70.

Bhat, R. V., V. Nagarajan, and P. G. Tulpule. 1978. Health hazards of mycotoxins in India. New Delhi: Indian Council of Medical Research.

Bothmer, R. von, C. Baden, and N. H. Jacobsen. 2007. Hordeum. In Barkworth et al., Flora of North America, vol. 24.

Brandenburg, D. M. 2007a. Cinna. In Barkworth et al., Flora of North America, vol. 24.

Brandenburg, D. M. 2007b. Diarrhena. In Barkworth et al., Flora of North America, vol. 24.

Brandenburg, D. M. 2007c. Sclerochloa. In Barkworth et al., Flora of North America, vol. 24.

Burrows, G. E., and R. J. Tyrl. 2001. Toxic plants of North America. Ames: Iowa State University Press.

Campbell, C. S. 2003. Andropogon. In Barkworth et al., Flora of North America, vol. 25.

Carlson, J. R. 2007. Pseudoroegneria. In Barkworth et al., Flora of North America, vol. 24.

Chen, S.-L., et al. 2006. Poaceae (Gramineae). In Z.-Y Wu, P. H. Raven, and D.-Y. Hong, eds., Flora of China, vol. 22 (Poaceae). Beijing: Science Press; St. Louis: Missouri Botanical Garden Press.

Clark, L. G. 2007. Lamarckia. In Barkworth et al., Flora of North America, vol. 24.

Clark, L. G., and E. A. Kellogg. 2007. Poaceae. Gramineae. Grass family. In Barkworth et al., Flora of North America, vol. 24.

Clark, L. G., and J. K. Triplett. 2007. Arundinaria. In Barkworth et al., Flora of North America, vol. 24.

Correll, D. S., and M. C. Johnston. 1970. Manual of the vascular plants of Texas. Renner: Texas Research Foundation.

Cory, V. L., and H. B. Parks. 1937. Catalogue of the flora of the state of Texas. Bull. No. 550. College Station: Texas Agricultural Experiment Station.

Crins, W. J. 2007. Alopecurus. In Barkworth et al., Flora of North America, vol. 24.

Dallas Morning News. 2008. Texas Almanac 2008–2009. Dallas: The Dallas Morning News.

Daniel, T. F. 2007. Sphenopholis. In Barkworth et al., Flora of North America, vol. 24.

Darbyshire, S. J. 2003. Danthonia. In Barkworth et al., Flora of North America, vol. 25.

Darbyshire, S. J. 2007. Schedonorus. In Barkworth et al., Flora of North America, vol. 24.

Darbyshire, S. J., and L. E. Pavlick. 2007. Festuca. In Barkworth et al., Flora of North America, vol. 24.

Dávila Aranda, P. D., and S. L. Hatch. 2003. Sorghastrum. In Barkworth et al., Flora of North America, vol. 25.

Diggs, G. M., B. L. Lipscomb, and R. J. O'Kennon. 1999. Shinners and Mahler's illustrated flora of North Central Texas. Sida, Botanical Miscellany 16. Fort Worth: Botanical Research Institute of Texas.

Diggs, G. M., B. L. Lipscomb, M. D. Reed, and R. J. O'Kennon. 2006. Illustrated flora of East Texas. Sida, Botanical Miscellany 26. Fort Worth: Botanical Research Institute of Texas.

Epstein, W., K. Gerber, and R. Karler. 1964. The hypnotic constituent of Stipa vaseyi, sleepy grass. Experientia (Basel) 20:390.

Everitt, J. H., D. L. Drawe, C. R. Little, and R.I. Lonard. 2011. Grasses of South Texas: A guide to identification and value. Texas Tech University Press. Lubbock.

Freckmann, R. W., and M. G. Lelong. 2003a. Dichanthelium. In Barkworth et al., Flora of North America, vol. 25.

Freckmann, R. W., and M. G. Lelong. 2003b. Panicum. In Barkworth et al., Flora of North America, vol. 25.

Freckmann, R. W., and W. G. Lelong. 2003c. Steinchisma. In Barkworth et al., Flora of North America, vol. 25.

Frye, R. G., K. L. Brown, and C. A. McMahan. 1984. Vegetation types of Texas. Map prepared by GIS Lab. Austin: Texas Parks and Wildlife. 222.tpwd.state.tx.us/admin/veg/.

Gabel, M. L. 2003. Imperata. In Barkworth et al., Flora of North America, vol. 25.

Godfrey, C. L., G. S. McKee, and H. Oakes. 1973. General soils map of Texas. College Station: Texas Agricultural Experiment Station.

Good, R. 1953. The geography of flowering plants. 2nd ed. London: Longmans, Green.

Gould, F. W. 1962. Texas plants—a checklist and ecological summary. MP-585. College Station: Texas Agricultural Experiment Station.

Gould, F. W. 1969. Texas plants—a checklist and ecological summary. MP-585 rev. College Station: Texas Agricultural Experiment Station.

Gould, F. W. 1975a. The grasses of Texas. College Station: Texas A&M University Press.

Gould, F. W. 1975b. Texas plants—a checklist and ecological summary. MP-585 rev. College Station: Texas Agricultural Experiment Station.

Gould, F. W., and T. W. Box. 1965. Grasses of the Texas Coastal Bend. College Station: Texas A&M University Press.

Gould, F. W., G. O. Hoffman, and C. A. Rechenthin. 1960. Vegetational areas of Texas. Leaflet 492. College Station: Texas Agricultural Experiment Station.

Gould, F. W., and R. B. Shaw. 1983. Grass systematics. 2nd ed. College Station: Texas A&M University Press.

Grass Phylogeny Working Group. 2001. Phylogeny and subfamilial classification of the grasses (Poaceae). Annals of the Missouri Botanical Garden 88:373–457.

Griffith, G. E., S. A. Bryce, J. M. Omernik, J. A. Comstock, A. C. Rodgers, B. Harrison, S. L. Hatch, and D. Bezanson. 2004. Ecoregions of Texas (color poster with map, descriptive text, and photographs). Reston, Va.: U.S. Geological Survey. (map scale 1:2,500,000)

Griffiths, J. F., and R. Orton. 1968. Agroclimatic atlas of Texas—Part 1. MP-888. College Station: Texas Agricultural Experiment Station.

Hall, D. W., and J. W. Thieret. 2003. Chrysopogon. In Barkworth et al., Flora of North America, vol. 25.

Harrington, H. D. 1977. How to identify grasses and grasslike plants. Athens: Swallow Press, Ohio University Press.

Hartley, J. R. 1954. The agrostological index. Australian Journal of Botany 2:1–21.

Harvey, M. J. 2007a. Agrostis. In Barkworth et al., Flora of North America, vol. 24.

Harvey, M. J. 2007b. Lachnagrostis. In Barkworth et al., Flora of North America, vol. 24.

Hatch, S. L. 2003a. Dactyloctenium. In Barkworth et al., Flora of North America, vol. 25.

Hatch, S. L. 2003b. Redfieldia. In Barkworth et al., Flora of North America, vol. 25.

Hatch, S. L. 2003c. Triplasis. In Barkworth et al., Flora of North America, vol. 25.

Hatch, S. L., K. N. Gandhi, and L. E. Brown. 1990. Checklist of the vascular plants of Texas. MP-1655. College Station: Texas Agricultural Experiment Station.

Hatch, S. L., and J. Pluhar. 1993. Texas range plants. College Station: Texas A&M University Press.

Hatch, S. L., J. L. Schuster, and D. L. Drawe. 1999. Grasses of the Texas Gulf Prairies and Marshes. College Station: Texas A&M University Press.

Highight, K. W., J. K. Wipff, and S. L. Hatch. 1988. Grasses (Poaceae) of the Texas Cross Timbers and Prairies. MP-1657. College Station: Texas Agricultural Experiment Station.

Hilu, K. W. 2003. Eleusine. In Barkworth et al., Flora of North America, vol. 25.

Hiser, K. M. 2003. Ixophorus. In Barkworth et al., Flora of North America, vol. 25.

Hitchcock, A. S. 1935. Manual of the grasses of the United States. USDA Misc. Publ. 200. Washington, D.C.: Government Printing Office.

Hitchcock, A. S. 1950. Manual of the grasses of the United States. 2nd ed., rev. by A. Chase. USDA Misc. Publ. 200. Washington, D.C.: Government Printing Office.

Iltis, H. H. 2003. Zea. In Barkworth et al., Flora of North America, vol. 25.

Johnson School of Public Affairs. 1978. Preserving Texas' natural heritage. Research Project 31. Austin: University of Texas.

Jones, S. D., J. K. Wipff, and P. M. Montgomery. 1997. Vascular plants of Texas: A comprehensive checklist including synonymy, bibliography, and index. Austin: University of Texas Press.

Kellogg, E. A. 2003. Schismus. In Barkworth et al., Flora of North America, vol. 25.

Kral, R. 2004. An evaluation of Anthenantia (Poaceae). Sida 21:293–310.

Krishnamachari, K. A., and R. Y. Bhat. 1976. Poisoning by ergot bajra (pearl millet) in man. Indian Journal of Medical Research 64:1624–28.

Loflin, B., and S. Loflin. 2006. Grasses of the Texas Hill Country: A field guide. College Station: Texas A&M University Press.

Lonard, R. I. 1993. Guide to grasses of the Lower Rio Grande Valley, Texas. Edinburg: University of Texas–Pan American Press.

Lonard, R. I. 2003. Vaseyochloa. In Barkworth et al., Flora of North America, vol. 25.

Lonard, R. I. 2007. Vulpia. In Barkworth et al., Flora of North America, vol. 24.

Long, S. 2007. Cynosurus. In Barkworth et al., Flora of North America, vol. 24.

Lorenz, K. 1979. Ergot of cereal grains. CRC. Critical Review of Food Science and Nutrition 11:311–54.

Lundell, C. L. 1961. Flora of Texas. Vol. 3. Renner: Texas Research Foundation.

Lundell, C. L. 1969. Flora of Texas. Vol. 2. Renner: Texas Research Foundation.

Michael, P. W. 2003. Echinochloa. In Barkworth et al., Flora of North America, vol. 25.

Morrison, L. A. 2007. Triticum. In Barkworth et al., Flora of North America, vol. 24.

National Research Council. 1993. Vetiver grass: A thin green line against erosion. Washington, D.C.: Nation Academy Press.

Pavlick, L. E., and L. K. Anderton. 2007. Bromus. In Barkworth et al., Flora of North America, vol. 24.

Peterson, P. M. 2003a. Eragrostis. In Barkworth et al., Flora of North America, vol. 25.

Peterson, P. M. 2003b. Muhlenbergia. In Barkworth et al., Flora of North America, vol. 25.

Peterson, P. M., and C. R. Annable. 2003. Blepharoneuron. In Barkworth et al., Flora of North America, vol. 25.

Peterson, P. M., S. L. Hatch, and A. S. Weakley. 2003. Sporobolus. In Barkworth et al., Flora of North America, vol. 25.

Piep, M. B. 2007. Brachypodium. In Barkworth et al., Flora of North America, vol. 24.

Poole, J. M, W. R. Carr, D. M. Price, and J. R. Singhurst. 2007. Rare plants of Texas. College Station: Texas A&M University Press.

Powell, A. M. 1994. Grasses of the Trans-Pecos and adjacent areas. Austin: University of Texas Press.

Pyrah, G. L. 2007. Leersia. In Barkworth et al., Flora of North America, vol. 24.

Radford, A. E., W. C. Dickison, J. R. Massey, and C. E. Bell. 1974. Vascular plant systematics. New York: Harper and Row.

Reeder, C. G. 2003. Lycurus. In Barkworth et al., Flora of North America, vol. 25.

Reeder, J. R. 2003a. Cottea. In Barkworth et al., Flora of North America, vol. 25.

Reeder, J. R. 2003b. Enneapogon. In Barkworth et al., Flora of North America, vol. 25.

Reeder, J. R. 2003c. Pappophorum. In Barkworth et al., Flora of North America, vol. 25.

Reeder, J. R. 2003d. Scleropogon. In Barkworth et al., Flora of North America, vol. 25.

Rominger, J. M. 2003. Setaria. In Barkworth et al., Flora of North America, vol. 25.

Rumely, J. H. 2007. Trisetum. In Barkworth et al., Flora of North America, vol. 24.

Sánchez, E. 2003. Microchloa. In Barkworth et al., Flora of North America, vol. 25.

Sánchez-Ken, J. G., and L. G. Clark. 2003. Chasmanthium. In Barkworth et al., Flora of North America, vol. 25.

Schantz, H. L. 1954. The place of grasslands in the earth's cover of vegetation. Ecology 35:143–45.

Scott, H. D., and J. M. McKimmey. 2000. Major land resource areas. In Water and Chemical Transport in Soils of the Southeastern USA. Southern Cooperative Series Bulletin #395.

Schouten, Y., and J. F. Veldcamp. 1985. A revision of *Anthoxanthum* including *Hierochlo* (Gramineae) in Malesia and Thailand. Blumea 30:319–51.

Schuster, J. L., and S. L. Hatch. 1990. Texas plants—an ecological summary. In Hatch, Gandhi, and Brown, Checklist of the vascular plants of Texas.

Shaw, R. B. 2008. Grasses of Colorado. Boulder: University Press of Colorado.

Shaw, R. B., B. S. Rector, and A. M. Dube. 2011. Distribution of Texas Grasses. Brit. Press 33. Sida, Botanical Miscellany.

Shaw, R. B., R. D. Webster, and C. M. Bern. 2003. Eriochloa. In Barkworth et al., Flora of North America, vol. 25.

Simon, B. K. and S. W. L. Jacobs. 2003. *Megathyrsus,* a new generic name for *Panicum* subgenus *Megathyrsus.* Austrobaileya 6:571–574.

Smith, J. P., Jr. 2003. Gymnopogon. In Barkworth et al., Flora of North America, vol. 25.

Smith, J. P., Jr. 2007. Hainardia. In Barkworth et al., Flora of North America, vol. 24.

Snow, N. 2007a. Briza. In Barkworth et al., Flora of North America, vol. 24.

Snow, N. 2007b. Limnodea. In Barkworth et al., Flora of North America, vol. 25.

Snow, N. 2003a. Buchloë. In Barkworth et al., Flora of North America, vol. 25.

Snow, N. 2003b. Leptochloa. In Barkworth et al., Flora of North America, vol. 25.

Snow, N. 2003c. Schedonnardus. In Barkworth et al., Flora of North America, vol. 25.

Soreng, R. J. 2007. Poa. In Barkworth et al., Flora of North America, vol. 24.

Spearing, D. 1991. Roadside geology of Texas. Missoula, Mont.: Mountain Press Publishing.

Standley, L. A. 2007a. Holcus. In Barkworth et al., Flora of North America, vol. 24.

Standley, L. A. 2007b. Koeleria. In Barkworth et al., Flora of North America, vol. 24.

Standley, L. A. 2007c. Rostraria. In Barkworth et al., Flora of North America, vol. 24.

Stapleton, C. M .A. 2007. Bambuseae. In Barkworth et al., Flora of North America, vol. 24.

Stapleton, C. M. A., and M. E. Barkworth. 2007. Phyllostachys. In Barkworth et al., Flora of North America, vol. 24.

Stephens, A. R., and W. H. Holmes. 1989. Historical atlas of Texas. Norman: University of Oklahoma Press.

Stephenson, S. N., and J. M. Saarela. 2007. Brachyelytrum. In Barkworth et al., Flora of North America, vol. 24.

Stieber, M. T., and J. K. Wipff. 2003. Cenchrus. In Barkworth et al., Flora of North America, vol. 25.

Stubbendieck, J. S., S. L. Hatch, and L. M. Landholt. 2003. North American wildland plants. Lincoln: University of Nebraska Press.

Telfair, R. C., II. 2006. Native plants important to wildlife in East Texas. In Diggs et al., Illustrated flora of East Texas.

Terrell, E. E. 2007a. Lolium. In Barkworth et al., Flora of North America, vol. 24.

Terrell, E. E. 2007b. Luziola. In Barkworth et al., Flora of North America, vol. 24.

Terrell, E. E. 2007c. Zizania. In Barkworth et al., Flora of North America, vol. 24.

Terrell, E. E. 2007d. Zizaniopsis. In Barkworth et al., Flora of North America, vol. 24.

Thieret, J. W. 2003a. Arthraxon. In Barkworth et al., Flora of North America, vol. 25.

Thieret, J. W. 2003b. Calamovilfa. In Barkworth et al., Flora of North America, vol. 25.

Thieret, J. W. 2003c. Coix. In Barkworth et al., Flora of North America, vol. 25.

Thieret, J. W. 2003d. Eremochloa. In Barkworth et al., Flora of North America, vol. 25.

Thieret, J. W. 2003e. Hackelochloa. In Barkworth et al., Flora of North America, vol. 25.

Thieret, J. W. 2003f. Hackonechloa. In Barkworth et al., Flora of North America, vol. 25.

Thieret, J. W. 2003g. Microstegium. In Barkworth et al., Flora of North America, vol. 25.

Thieret, J. W. 2003h. Monanthochloë. In Barkworth et al., Flora of North America, vol. 25.

Trewartha, G. T. 1968. An introduction to weather and climate. New York: McGraw-Hill.

Tucker, G. C. 2007. Desmazeria. In Barkworth et al., Flora of North America, vol. 24.

Turner, B. L., H. Nichols, G. Denny, and O. Doron. 2003. Atlas of the vascular plants of Texas. Vol. 2. Sida, Botanical Miscellany 24. Fort Worth: Botanical Research Institute of Texas.

United States Department of Agriculture. 2006. National Agriculture Statistics Service. www .nass.usda.gov/statistics_by_state/Texas.

United States Department of Agriculture. 2007. National Agriculture Statistics Service. www .nass.usda.gov/statistics_by_state/Texas.

Valdes-Reyna, J. 2003a. Blepharidachne. In Barkworth et al., Flora of North America, vol. 25.

Valdes-Reyna, J. 2003b. Dasyochloa. In Barkworth et al., Flora of North America, vol. 25.

Valdes-Reyna, J. 2003c. Erioneuron. In Barkworth et al., Flora of North America, vol. 25.

Valdes-Reyna, J. 2003d. Munroa. In Barkworth et al., Flora of North America, vol. 25.

Valdes-Reyna, J. 2003e. Tridens. In Barkworth et al., Flora of North America, vol. 25.

Veldkamp, J. F. 2004. Miscellaneous notes on mainly southeast Asian Gramineae. Reinwardtia 12:135–40.

Vicari, M., and D. R. Bazely. 1993. Do grasses fight back? The case for antiherbivore defenses. Trends in Ecology and Evolution 8:137–41.

Watson, L., and M. J. Dallwitz. 1992. The grass genera of the world. Wallingford, UK: CAB International.

Webster, R. D. 2003. Saccharum. In Barkworth et al., Flora of North America, vol. 25.

Webster, R. D., and S. L. Hatch. 1990. Variation in the morphology of the lower lemma in the Digitaria section Aequiglumae (Poaceae: Paniceae). Sida 14:145–67.

Wipff, J. K. 2003a. Allolepis. In Barkworth et al., Flora of North America, vol. 25.

Wipff, J. K. 2003b. Anthenantia. In Barkworth et al., Flora of North America, vol. 25.

Wipff, J. K. 2003c. Bouteloua. In Barkworth et al., Flora of North America, vol. 25.

Wipff, J. K. 2003d. Digitaria. In Barkworth et al., Flora of North America, vol. 25.

Wipff, J. K. 2003e. Melinis. In Barkworth et al., Flora of North America, vol. 25.

Wipff, J. K. 2003f. Oplismenus. In Barkworth et al., Flora of North America, vol. 25.

Wipff, J. K. 2003g. Pennisetum. In Barkworth et al., Flora of North America, vol. 25.

Wipff, J. K. 2003h. Rottboellia. In Barkworth et al., Flora of North America, vol. 25.

Wipff, J. K. 2003i. Sacciolepis. In Barkworth et al., Flora of North America, vol. 25.

Wipff, J. K. 2003j. Schizachyrium. In Barkworth et al., Flora of North America, vol. 25.

Wipff, J. K. 2003k. Tragus. In Barkworth et al., Flora of North America, vol. 25.

Wipff, J. K. 2003l. Trichoneura. In Barkworth et al., Flora of North America, vol. 25.

Wipff, J. K. 2003m. Tripogon. In Barkworth et al., Flora of North America, vol. 25.

Wipff, J. K. 2003n. Triraphis. In Barkworth et al., Flora of North America, vol. 25.

Wipff, J. K. 2003o. Willkommia. In Barkworth et al., Flora of North America, vol. 25.

Wipff, J. K. 2007a. Aira. In Barkworth et al., Flora of North America, vol. 24.

Wipff, J. K. 2007b. Gastridium. In Barkworth et al., Flora of North America, vol. 24.

Wipff, J. K., and R. A. Thompson. 2003. Urochloa. In Barkworth et al., Flora of North America, vol. 25.

Worley, T. 2007. Parapholis. In Barkworth et al., Flora of North America, vol. 24.

Yates, H. O. 2003. Uniola. In Barkworth et al., Flora of North America, vol. 25.

Zuloaga, F. O., L. M. Giussani, and O. Morrone. 2007. *Hopia,* a new monotypic genus segregated from *Panicum* (Poaceae). Taxon 56:145–156.

Achnatherum aridum, 161, 162
Achnatherum curvifolium, 161, 163
Achnatherum eminens, 161, 164
Achnatherum hymenoides, 161, 165
Achnatherum lobatum, 161, 166
Achnatherum nelsonii, 161, 167
Achnatherum perplexum, 161, 166, 168
Achnatherum robustum, 161, 169
Achnatherum scribneri, 161, 166, 170
Aegilops cylindrica, 172, 987
African Bermudagrass, 371
African dogstooth grass, 373
African windmillgrass, 329
Agropyron cristatum, 174
Agropyron spicatum, 840
Agrostis elliottiana, 175, 176
Agrostis exarata, 175, 177
Agrostis gigantea, 175, 178, 182
Agrostis hyemalis, 175, 176, 179
Agrostis perennans, 175, 180
Agrostis scabra, 175, 176, 179, 180, 181
Agrostis stolonifera, 175, 178, 182, 833
Aira caryophyllea, 184
Aira elegans, 184
alkali grass, 441
alkali muhly, 643
alkali sacaton, 921
Allolepis texana, 186
Alopecurus carolinianus, 187, 188
Alopecurus myosuroides, 187, 189
Amazon sprangletop, 596
Amelichloa clandestina, 191
American barnyard grass, 447
American beakgrain, 388
American cupscale, 858
American five-minute grass, 977
American mannagrass, 542
American tripogon, 977
Andropogon elliottii, 192, 195

Andropogon gerardii, 192, 193, 196
Andropogon glomeratus, 192, 194
Andropogon gyrans, 192, 195
Andropogon hallii, 192, 196
Andropogon spadiceus, 872
Andropogon ternarius, 192, 197
Andropogon virginicus, 192, 198
Andropogoneae: *Andropogon,* 192–98; *Arthraxon,* 234–35; *Bothriochloa,* 255–67; *Chrysopogon,* 345–47; *Coelorachis,* 350–52; *Coix,* 353–54; *Dichanthium,* 419–22; *Elionurus,* 455–57; *Eremochloa,* 514–15; *Hackelochloa,* 549–50; *Hemarthria,* 553–54; *Heteropogon,* 558–60; *Hyparrhenia,* 570–72; *Imperata,* 573–75; *Microstegium,* 627–28; *Miscanthus,* 629–30; *Rottboellia,* 845–46; *Saccharum,* 847–55; *Schizachyrium,* 867–73; *Sorghastrum,* 898–900; *Sorghum,* 901–3; *Themeda,* 948–49; *Trachypogon,* 955–56; *Tripsacum,* 978–79; *Zea,* 1014–16
Angleton bluestem, 421
Anisantha diandrus, 199, 200
Anisantha rubens, 199, 201
Anisantha sterilis, 199, 202
Anisantha tectorum, 199, 203
annual beardgrass, 835
annual bluegrass, 815
annual hairgrass, 184
annual junegrass, 844
annual koeleria, 844
annual muhly, 662
annual rabbitfoots grass, 835
annual ryegrass, 605
annual signalgrass, 992
annual ticklegrass, 176
Anthenantia rufa, 204, 205
Anthenantia texana, 206
Anthenantia villosa, 204, 206
Anthoxanthum aristatum, 207, 208, 209

Anthoxanthum odoratum, 209
aparejograss, 686
Apera spica-venti, 211
Arabian schismus, 865
Argentine fingergrass, 531
Argentine sprangletop, 591
Aristida adscensionis, 212, 215
Aristida arizonica, 214, 216
Aristida barbata, 222
Aristida basiramea, 212, 217
Aristida brownii, 230
Aristida desmantha, 212, 218
Aristida dichotoma, 212, 217, 219
Aristida divaricata, 213, 220, 222
Aristida gypsophila, 212, 221, 227
Aristida havardii, 213, 220, 222
Aristida lanosa, 213, 223, 226
Aristida longespica, 212, 224
Aristida oligantha, 212, 225
Aristida palustris, 214, 226
Aristida pansa, 213, 221, 227
Aristida purpurascens, 214, 228
Aristida purpurea, 213, 214, 230
Aristida ramosissima, 212, 231
Aristida schiedeana, 212, 232
Aristida ternipes, 212, 213, 233
Aristideae, *Aristida,* 212–33
Aristidoideae, *Aristida,* 212–33
Arizona brome, 312
Arizona cottontop, 427
Arizona fescue, 533
Arizona signalgrass, 992
Arizona threeawn, 216
Arizona wheatgrass, 460
armgrass millet, 1003
arrowfeather threeawn arrowgrass, 228
Arthraxon hispidus, 235
Arundinaria gigantea, 236, 237, 238
Arundinaria tecta, 236, 237, 238
Arundineae: *Arundo,* 239–40; *Hakonechloa,* 1028–29; *Phragmites,* 799–800
Arundinoideae: *Arundo,* 239–40; *Hakonechloa,* 1028–29; *Phragmites,* 799–800
Arundo donax, 240
Asian crabgrass, 426
Asian rice, 698
Australian bluestem, 258

Australian brome, 292
autumn bent, 180
autumn bluegrass, 818
Avena fatua, 241, 242
Avena sativa, 241, 243
awned dichanthium, 421
awnless barnyard grass, 443
awnless bluestem, 260
awnless fluffgrass, 971
awnless wildrye, 462
Axonopus affinis, 246
Axonopus compressus, 244, 245
Axonopus fissifolius, 244, 246
Axonopus furcatus, 244, 247

bahiagrass, 759
Bambusa bambos, 248, 249
Bambusa multiplex, 248, 250
Bambuseae: *Arundinaria,* 236–38; *Bambusa,* 248–50; *Phyllostachys,* 801–3, 1030
Bambusoideae: *Arundinaria,* 236–38; *Bambusa,* 248–50; *Phyllostachys,* 801–3, 1030
barley, 569
barnyard grass, 444
barren brome, 202
basketgrass, 696
bayou wedgegrass, 915
beaked panicgrass, 703
beaked panicum, 703
bearded shorthusk, 277
bearded skeletongrass, 547
bearded sprangletop, 593
beardgrass, 547
beardless wildrye, 601
beggarstick threeawn, 232
bent arm plumegrass, 851
bent spike fountaingrass, 779
Berg's witchgrass, 705
Bermudagrass, 370
big alkali sacaton, 943
Big Bend bluegrass, 828
big bluestem, 193
big carpetgrass, 247
big cordgrass, 907
big paspalum, 749
big quackgrass, 279
big sandbur, 309

big sandreed, 302
Bigelow's bluegrass, 819
Bigelow's desertgrass, 252
bigleaf bristlegrass, 1034
bigtop lovegrass, 492
birdwood grass, 784
bitter beachgrass, 702
bitter panicum, 702
black grama, 338
black oatsgrass, 807
blackgrass, 189
blackseed needlegrass, 807
blackseed speargrass, 807
bladygrass, 575
Blepharidachne bigelovii, 252
Blepharoneuron tricholepis, 254
blood rosettegrass, 398
blowout grass, 842
blue grama, 339
blue panicgrass, 704
blue panicum, 704
bluebunch wheatgrass, 840
blue-joint panicgrass, 722
bog muhly, 685
Bosc's panicgrass, 395
Bothriochloa alta, 255, 256, 257
Bothriochloa barbinodis, 255, 256, 257, 266
Bothriochloa bladhii, 255, 258
Bothriochloa caucasica, 258
Bothriochloa edwardsiana, 255, 259, 261, 266
Bothriochloa exaristata, 255, 260
Bothriochloa hybrida, 255, 259, 261
Bothriochloa ischaemum, 255, 262
Bothriochloa laguroides, 255, 263
Bothriochloa longipaniculata, 255, 260, 264
Bothriochloa pertusa, 255, 265
Bothriochloa saccharoides, 263, 264
Bothriochloa springfieldii, 255, 257, 266
Bothriochloa wrightii, 255, 257, 267
Bouteloua aristidoides, 268, 269
Bouteloua breviseta, 342
Bouteloua chondrosioides, 268, 270
Bouteloua curtipendula, 268, 271
Bouteloua pectinata, 340
Bouteloua repens, 268, 272
Bouteloua rigidiseta, 268, 273
Bouteloua uniflora, 268, 274

Bouteloua warnockii, 268, 275
Brachyelytreae, *Brachyelytrum,* 276–77
Brachyelytrum erectum, 277
Brachypodieae, *Trachynia,* 953–54
bread wheat, 986
bristle basketgrass, 696
bristle-leafed muhly, 675
bristly dogtail, 376
bristly foxtail grass, 895
bristly wolfstail, 614
Briza maxima, 278, 279
Briza media, 278, 280
Briza minor, 278, 281
broadleaf carpetgrass, 245
broad-leaf chasmanthium, 317
broadleaf panicgrass, 402
broadleaf rosettegrass, 402
broadleaf signalgrass, 1000
brome fescue, 1008
brome sixweeksgrass, 1008
Bromeae: *Anisantha,* 199–203; *Bromopsis,* 282–90; *Bromus,* 291–98; *Ceratochloa,* 311–15
Bromopsis anomalus, 282, 283, 287
Bromopsis ciliatus, 282, 284
Bromopsis inermis, 282, 285
Bromopsis lanatipes, 282, 286
Bromopsis nottowayanus, 288
Bromopsis porteri, 282, 283, 287
Bromopsis pubescens, 282, 288
Bromopsis richardsonii, 282, 289
Bromopsis texensis, 282, 290
Bromus alopercurus, 296
Bromus arenarius, 291, 292, 295
Bromus arizonicus, 312
Bromus caroli-henrici, 292
Bromus catharticus, 314
Bromus commutatus, 291, 293
Bromus hordeaceus, 291, 294
Bromus japonicus, 291, 292, 295
Bromus lanceolatus, 291, 296
Bromus macrostachys, 296
Bromus mollis, 294
Bromus nottowayanus, 288
Bromus racemosus, 291, 297, 298
Bromus secalinus, 291, 293, 295, 297, 298
Bromus unioloides, 314

brook crowngrass, 739
brook paspalum, 739
broomcorn millet, 718
broomcorns, 902
broomsedge bluestem, 198
broomsedge, 198
brownseed crowngrass, 760
brownseed paspalum, 760
browntop liverseed grass, 995
browntop millet, 1001
browntop signalgrass, 995
Buchloë dactyloides, 300
Buckley's dropseed, 922
Buckley's fluffgrass, 967
Buckley's tridens, 967
buffalograss, 300
buffel grass, 775
bulbous bluegrass, 820
bulbous canarygrass, 788
bulbous panicgrass, 1026
bull paspalum, 742
bullgrass, 653
bunch cutgrass, 587
bunched plumegrass, 852
bur bristlegrass, 882, 895
Burrograss, 877
buryseed umbrellagrass, 477
bush muhly, 666
bush-grass, 666
bushy beardgrass, 194
bushy bluestem, 194
bushybeard bluestem, 194

Calamovilfa gigantea, 302
California brome, 313
California cottontop, 427
Canada bluegrass, 822
Canada brome, 288
Canada wildrye, 461
canarygrass, 790
cane bluestem, 257
canoegrass, 739
canyongrass, 587
Caribbean cupgrass, 519
Caribbean fingergrass, 528
Carolina canarygrass, 791
Carolina crabgrass, 429

Carolina jointgrass, 351
carpetgrass, 947
Catapodium rigidum, 386
catchfly grass, 586
Cathestecum erectum, 304
cat-tail grass, 844
cedar panicgrass, 409
Cenchrus brownii, 305, 306
Cenchrus ciliaris, 775
Cenchrus echinatus, 305, 307
Cenchrus longispinus, 305, 308, 310
Cenchrus myosuroides, 305, 309
Cenchrus spinifex, 305, 308, 310
centipedegrass, 515
Centotheceae, *Chasmanthium,* 316–19
Centothecoideae, *Chasmanthium,* 316–19
Ceratochloa arizonica, 311, 312
Ceratochloa carinata, 311, 313, 315
Ceratochloa cathartica, 311, 314
Ceratochloa polyantha, 311, 315
Chapman's bluegrass, 821
Chapman's poa, 821
Chapman's tridens, 970
Chase's Glenwoodgrass, 857
Chasmanthium latifolium, 316, 317
Chasmanthium laxum, 316, 318
Chasmanthium sessiliflorum, 316, 319
cheatgrass, 203
chickenfoot grass, 528
Chihuahua lovegrass, 491
Chinese silvergrass, 630
chino grama, 342
Chisos bluestem, 828
Chlorideae, *Gymnopogon,* 546–48
Chloridoideae: *Allolepis,* 185–86; *Blephari-
    dachne,* 251–52; *Blepharoneuron,* 253–54;
    Bouteloua, 268–75; *Buchloë,* 299–300;
    Calamovilfa, 301–2; *Cathestecum,* 303–4;
    Chloris, 320–33; *Chondrosum,* 334–44;
    Cottea, 358–59; *Ctenium,* 366–67; *Cyn-
    odon,* 368–73; *Dactyloctenium,* 378–79;
    Dasyochloa, 383–84; *Distichlis,* 440–41;
    Eleusine, 452–54; *Enneapogon,* 474–75;
    Enteropogon, 476–77; *Eragrostis,* 478–513;
    Erioneuron, 523–26; *Eustachys,* 527–31;
    Gymnopogon, 546–48; *Hilaria,* 561–63;
    Leptochloa, 590–98; *Lycurus,* 612–14; *Mi-*

crochloa, 625–26; *Monanthochloë,* 631–32; *Muhlenbergia,* 635–88; *Munroa,* 689–90; *Pappophorum,* 726–28; *Pleuraphis,* 810–12; *Redfieldia,* 841–42; *Schedonnardus,* 859–60; *Scleropogon,* 876–77; *Spartina,* 904–11; *Sporobolus,* 918–43; *Tragus,* 957–58; *Trichloris,* 959–61; *Trichoneura,* 962–63; *Tridens,* 964–73; *Triplasis,* 974–75; *Tripogon,* 976–77; *Triraphis,* 980–81; *Uniola,* 988–89; *Vaseyochloa,* 1005–6; *Willkommia,* 1012–13; *Zoysia,* 1021–24
Chloris andropogonoides, 321, 322
Chloris barbata, 320, 323
Chloris canterae, 320, 324
Chloris ciliata, 320, 325
Chloris cucullata, 321, 326
Chloris divaricata, 320, 327
Chloris gayana, 320, 321, 328
Chloris inflata, 323
Chloris pilosa, 320, 321, 329
Chloris submutica, 321, 330
Chloris texensis, 321, 331
Chloris verticillata, 321, 332
Chloris virgata, 320, 321, 333
Chondrosum barbatum, 334, 336
Chondrosum brevisetum, 334, 337
Chondrosum eriopodum, 334, 338
Chondrosum gracile, 334, 339
Chondrosum hirsutum, 334, 340
Chondrosum kayi, 334, 341, 344
Chondrosum parryi, 335
Chondrosum ramosa, 334, 342
Chondrosum simplex, 334, 343
Chondrosum trifidum, 334, 341, 344
Chrysopogon pauciflorus, 345, 346
Chrysopogon zizanioides, 345, 347
churchmouse threeawn, 219
Cinna arundinacea, 349
cliff muhly, 665
clubhead cutgrass, 585
cluster fescue, 535
Coahuila dropseed, 924
coast barnyardgrass, 449
coastal bristlegrass, 883
coastal foxtail, 883
coastal lovegrass, 503
coastal sacahuista, 911
coastal saltgrass, 441
coastal sandbur, 310
cockspur, 447
Coelorachis cylindrica, 350, 351
Coelorachis rugosa, 350, 352
cogongrass, 575
Coix lacryma-jobi, 354
Colorado brome, 315
Colorado grass, 1004
comb's crowngrass, 740
comb's paspalum, 740
common bahiagrass, 759
common carpetgrass, 246
common crabgrass, 437
common Mediterranean grass, 866
common millet, 718
common reed, 800
common sandbur, 310
common sixweeksgrass, 1010
common wheat, 986
common windgrass, 211
common witchgrass, 707
common wolfstail, 613
compressed plumegrass, 852
coral dropseed, 929
corm-based panicgrass, 409
corn, 1015
Cortaderia jubata, 355, 356
Cortaderia selloana, 355, 357
cotta grass, 359
Cottea pappophoroides, 359
cottongrass, 575
couchgrass, 468
covered spike dropseed, 927
crane grass, 577
creek-oats, 317
creeping bent, 182
creeping lovegrass, 504
creeping muhly, 669
creeping river grass, 448
creeping signalgrass, 999
creeping softgrass, 565
creeping wildrye, 601
crested wheatgrass, 174
crimson bluestem, 870
crimson fountaingrass, 783
crinkleawn, 956

Critesion brachyantherum, 360, 361
Critesion jubatum, 360, 362
Critesion marinum, 360, 363
Critesion murinum, 360, 364
Critesion pusillum, 360, 365
crowfoot, 326, 379
Ctenium aromaticum, 367
cultivated oats, 243
cultivated rice, 698
curly mesquite, 562
curly oatgrass, 382
curly threeawn, 218
curlyleaf muhly, 675
curlyleaf needlegrass, 163
curved sicklegrass, 730
cushion-tufted panicgrass, 416
cutover muhly, 654
Cynodon aethiopicus, 368, 369
Cynodon dactylon, 368, 370
Cynodon nlemfuënsis, 368, 371
Cynodon plectostachyus, 368, 372
Cynodon transvaalensis, 368, 373
Cynodonteae: *Allolepis,* 185–86; *Blephari-dachne,* 251–52; *Blepharoneuron,* 253–54; *Bouteloua,* 268–75; *Buchloë,* 299–300; *Calamovilfa,* 301–2; *Cathestecum,* 303–4; *Chloris,* 320–33; *Chondrosum,* 334–44; *Ctenium,* 366–67; *Cynodon,* 368–73; *Dactyloctenium,* 378–79; *Dasyochloa,* 383–84; *Distichlis,* 440–41; *Eleusine,* 452–54; *Enteropogon,* 476–77; *Eragrostis,* 478–513; *Erioneuron,* 523–26; *Eustachys,* 527–31; *Gymnopogon* 546–48; *Hilaria,* 561–63; *Leptochloa,* 590–98; *Lycurus,* 612–14; *Microchloa,* 625–26; *Monanthochloë,* 631–32; *Muhlenbergia,* 635–88; *Munroa,* 689–90; *Pleuraphis,* 810–12; *Redfieldia,* 841–42; *Schedonnardus,* 859–60; *Scleropogon,* 876–77; *Spartina,* 904–11; *Sporobolus,* 918–43; *Tragus,* 957–58; *Trichloris,* 959–61; *Trichoneura,* 962–63; *Tridens,* 964–73; *Triplasis,* 974–75; *Tripogon,* 976–77; *Triraphis,* 980–81; *Uniola,* 988–89; *Vaseyochloa,* 1005–6; *Willkommia,* 1012–13; *Zoysia,* 1021–24
Cynosurus echinatus, 375
cypress rosettegrass, 400

Dactylis glomerata, 377
Dactyloctenium aegyptium, 379
dallisgrass, 746
Danthonia sericea, 380, 381
Danthonia spicata, 380, 382
Danthonieae: *Cortaderia,* 355–57; *Danthonia,* 380–82; *Schismus,* 864–66
Danthonioideae: *Cortaderia,* 355–57; *Danthonia,* 380–82; *Schismus,* 864–66
Danube grass, 800
darnel lovegrass, 482
darnel ryegrass, 608
darnel, 608
Dasyochloa pulchella, 252, 384
deer muhly, 671
deergrass, 671
deer-tongue grass, 396
deer-tongue panic, 396
delicate hairgrass, 184
delicate muhly, 656
densely flowered cordgrass, 908
desert muhly, 659
desert saltgrass, 441
Desmazeria rigida, 386
Diarrhena americana, 387, 388
Diarrhena obovata, 387, 388, 389
Diarreheneae, *Diarrhena,* 387–89
Dichanthelium aciculare, 391, 393
Dichanthelium acuminatum, 392, 394
Dichanthelium angustifolium, 398
Dichanthelium boscii, 391, 395, 402, 412
Dichanthelium clandestinum, 391, 396
Dichanthelium commutatum, 391, 397
Dichanthelium consanguineum, 391, 398
Dichanthelium depauperatum, 390, 399
Dichanthelium dichotomum, 392, 400
Dichanthelium ensifolium, 392, 401, 417
Dichanthelium latifolium, 391, 395, 402
Dichanthelium laxiflorum, 390, 403
Dichanthelium linearifolium, 390, 404
Dichanthelium malacophyllum, 392, 405
Dichanthelium nodatum, 390, 406, 409
Dichanthelium ovale, 392, 408
Dichanthelium pedicellatum, 390, 406, 409
Dichanthelium polyanthes, 391, 410
Dichanthelium portoricense, 391, 411
Dichanthelium ravenelii, 392, 412

Dichanthelium scabriusculum, 391, 396, 413
Dichanthelium scoparium, 391, 414
Dichanthelium sphaerocarpon, 391, 415
Dichanthelium strigosum, 390, 416
Dichanthelium tenue, 392, 401, 417
Dichanthelium wrightianum, 392, 418
Dichanthium annulatum, 419, 420
Dichanthium aristatum, 419, 421
Dichanthium sericeum, 419, 422
Digitaria arenicola, 423, 425
Digitaria bicornis, 424, 426
Digitaria californica, 423, 427
Digitaria ciliaris, 424, 428
Digitaria cognata, 423, 425, 429, 436
Digitaria filiformis, 423, 424, 430
Digitaria hitchcockii, 423, 431
Digitaria insularis, 423, 432
Digitaria ischaemum, 424, 433, 439
Digitaria milanjiana, 424, 434
Digitaria patens, 423, 435
Digitaria pubiflora, 423, 436
Digitaria sanguinalis, 424, 437
Digitaria texana, 424, 438
Digitaria violascens, 424, 439
dinkel, 987
Distichlis spicata, 186, 441
ditch beardgrass, 833
ditch millet, 764
Dixie liverseed grass, 1001
Dore's needlegrass, 167
downy chess, 203
downy oatgrass, 381
downy wildrye, 472
dune muhly, 674
Durban crowfoot, 379
dwarf meadowgrass, 815

ear muhly, 641
early crowngrass, 761
early paspalum, 761
early wildrye, 466
eastern gamagrass, 979
eastern mannagrass, 544
eastern mockgrass, 979
eastern needlegrass, 807
eastern riverbank wildrye, 469
Echinochloa colona, 442, 443

Echinochloa crus-galli, 442, 444
Echinochloa crus-pavonis, 442, 445
Echinochloa esculenta, 442, 446
Echinochloa muricata, 442, 447
Echinochloa polystachya, 442, 448
Echinochloa walteri, 442, 449
egg-leaved rosettegrass, 408
Egyptian crowfoot, 379
Egyptian paspalidium, 734
Egyptian water grass, 734
Egyptian water-crown grass, 734
Ehrharta calycina, 451
Ehrharteae, *Ehrharta*, 450–51
Ehrhartoideae: *Ehrharta*, 450–51; *Leersia*, 584–89; *Luziola*, 609–11; *Oryza*, 697–98; *Zizania*, 1017–18; *Zizaniopsis*, 1019–20
elephant grass, 782
Eleusine indica, 452, 453
Eleusine tristachya, 452, 454
Elionurus barbiculmis, 455, 456
Elionurus tripsacoides, 455, 457
Elliott's beardgrass, 195
Elliott's bent, 176
Elliott's bluestem, 195
Elliott's lovegrass, 490
Elymus arizonicus, 458, 460
Elymus canadensis, 458, 461
Elymus curvatus, 459, 462
Elymus elymoides, 458, 463
Elymus glabriflorus, 459, 464
Elymus interruptus, 458, 465
Elymus macgregorii, 459, 466
Elymus pringlei, 458, 467
Elymus repens, 458, 468
Elymus riparius, 459, 469
Elymus texensis, 458, 470
Elymus trachycaulus, 458, 471
Elymus villosus, 458, 472
Elymus virginicus, 459, 466, 473
English wildrye, 606
Enneapogon desvauxii, 475
Enteropogon chlorideus, 477
Eragrostis airoides, 480, 482
Eragrostis amabilis, 478, 483
Eragrostis barrelieri, 479, 484
Eragrostis capillaris, 478, 485
Eragrostis cilianensis, 478, 479, 486

Eragrostis ciliaris, 478, 487
Eragrostis curtipendicellata, 479, 488
Eragrostis curvula, 480, 481, 489
Eragrostis elliottii, 480, 490
Eragrostis erosa, 480, 491
Eragrostis hirsuta, 480, 481, 492
Eragrostis hypnoides, 478, 493
Eragrostis intermedia, 480, 494, 497
Eragrostis japonica, 478, 495
Eragrostis lehmanniana, 480, 481, 496
Eragrostis lugens, 480, 481, 494, 497
Eragrostis mexicana, 478, 498
Eragrostis minor, 479, 499
Eragrostis palmeri, 480, 490
Eragrostis pectinacea, 478, 501
Eragrostis pilosa, 479, 502
Eragrostis refracta, 480, 503
Eragrostis reptans, 478, 504
Eragrostis secundiflora, 480, 505
Eragrostis sessilispica, 480, 506
Eragrostis silveana, 479, 507
Eragrostis spectabilis, 479, 508
Eragrostis spicata, 479, 509
Eragrostis superba, 479, 510
Eragrostis swallenii, 480, 511
Eragrostis tef, 478, 512
Eragrostis trichodes, 480, 513
Eremochloa ophiuroides, 515
Eriochloa acuminata, 516, 517
Eriochloa contracta, 516, 518
Eriochloa polystachya, 516, 519
Eriochloa pseudoacrotricha, 516, 520
Eriochloa punctata, 516, 521
Eriochloa sericea, 516, 522
Erioneuron avenaceum, 523, 524
Erioneuron nealleyi, 523, 525
Erioneuron pilosum, 523, 526
eternitygrass, 748
Ethiopian dogstooth grass, 369
eulalia, 628, 630
European dunegrass, 600
Eustachys caribaea, 527, 528
Eustachys neglecta, 527, 529
Eustachys petraea, 527, 530
Eustachys retusa, 527, 531
everlasting grass, 521
eyelash grass, 339

fall panicum, 710
fall witchgrass, 429
false buffalograss, 690
false grama, 304
false rhodesgrass, 960
false saltgrass, 186
feather beardgrass, 197
feather fingergrass, 333
feather windmillgrass, 333
feathertop pennisetum, 785
feathertop, 785
Fendler's muttongrass, 823
fern grass, 386
Festuca arizonica, 532, 533
Festuca arundinacea, 862
Festuca ligulata, 532, 534
Festuca paradoxa, 532, 535
Festuca rubra, 532, 536
Festuca subverticillata, 532, 537
Festuca versuta, 532, 538
few-flowered panicgrass, 407
few-flowered threeawn, 225
field crowngrass, 752
field paspalum, 752
fineleaved nassell, 693
fineleaved needlegrass, 693
fine-top saltgrass, 921
fishpole bamboo, 802
flaggrass, 925
flat crabgrass, 247
flexible sasa grass, 628
flexuous speargrass, 818
floating mannagrass, 544
Florida crowngrass, 749
Florida paspalum, 749
Florida rhaphis, 346
fluffgrass, 384
fly-away grass, 179
forest muhly, 679
forked threeawn, 217
forker panicgrass, 400
Fort Thompsongrass, 748
four-spike fingergrass, 529
fowl mannagrass, 545
foxtail barley, 362
foxtail brome, 201
foxtail chess, 201

foxtail fescue, 1009
foxtail fountaingrass, 774
foxtail grass, 895
foxtail millet, 885
foxtail muhly, 640
foxtail, 188
freshwater cordgrass, 910
fringed brome, 284
fringed chloris, 325
fringed liverseed grass, 994
fringed singalgrass, 994
fringed windmillgrass, 325
fringed-leaf paspalum, 765

galleta, 811
galletagrass, 811
gaping grass, 945
gaping panicgrass, 945
Gastridium phleoides, 540
German millet, 885
Ghiesbreght's panicum, 713
Ghiesbreght's witchgrass, 713
giant bamboo, 803
giant bristlegrass, 888
giant cane, 237, 240
giant cutgrass, 1020
giant dropseed, 931
giant foxtail, 888
giant plumegrass, 853
giant sandreed, 302
giant sorghum, 902
giant thorny bamboo, 249
giant timber bamboo, 803
giant witchgrass, 716
giantreed, 240
glenwoodgrass, 857
Glyceria grandis, 541, 542
Glyceria notata, 541, 543
Glyceria septentrionalis, 541, 544
Glyceria striata, 541, 545
goatgrass, 958
golden bamboo, 802
golden crookstem, 1030
golden crowngrass, 746
goldentop, 583
goosegrass, 453
gophertail lovegrass, 487

gravelbar muhly, 652
greasegrass, 970
Great Basin brome, 315
great brome, 200
Great Plains wildrye, 461
green bristlegrass, 897
green foxtail grass, 897
green muhly, 668
green nassella, 694
green needlegrass, 694
green sandbur, 306
green silkyscale, 206
green sprangletop, 592
Grisebach's bristlegrass, 884
Guadalupe fescue, 534
Guinea grass, 616
Gulf Coast barnyardgrass, 445
Gulf cockspur grass, 445
gulf cordgrass, 911
gulfdune grass, 758
gulfdune paspalum, 758
gummy lovegrass, 488
Gymnopogon ambiguus, 546, 547
Gymnopogon brevifolius, 546, 548
gypgrass, 935
gypsum grama, 337
gypsum threeawn, 221

Hackelochloa granularis, 550
Hainardia cylindrica, 552
hairy beakgrain, 389
hairy chess, 293
hairy crabgrass, 437
hairy dropseed, 254
hairy grama, 340
hairy muhly, 687
hairy panicgrass, 394
hairy panicum, 716
hairy tridens, 526
hairy woodland brome, 288
hairy woollygrass, 526
hairyawn muhly, 646
hairyleaf bristle-grass, 896
hairy-nerve grass, 963
hairyseed crowngrass, 762
hairyseed paspalum, 762
hakone grass, 1029

Hakonechloa macra, 1029
Hall's bluestem, 196
Hall's panicum, 714
Hall's witchgrass, 714
hardgrass, 552, 875
hare barley, 364
Hartweg's crowngrass, 750
Hartweg's paspalum, 750
Harvard's grama, 270
Havard's threeawn, 222
head-like dropseed, 925
hedge bamboo, 250
hedgehog grass, 307
Heller's rosettegrass, 407
Hemarthria altissima, 554
hemlock witchgrass, 411
herdsgrass, 798
Hesperostipa comata, 555, 556
Hesperostipa neomexicana, 555, 557
Heteropogon contortus, 558, 559
Heteropogon melanocarpus, 558, 560
hidden dropseed, 923
Hilaria belangeri, 300, 561, 562
Hilaria swallenii, 561, 563
Himalayan fountaingrass, 776
hog millet, 718
Holcus lanatus, 565
honey bluestem, 872
hooded fingergrass, 326
hooded windmillgrass, 326
hook threeawn, 233
hooked bristlegrass, 895
Hopia obtusa, 567
Hordeum vulgare, 569
horsetail crowngrass, 763
horsetail paspalum, 763
hurrah grass, 765, 995
hybrid bluestem, 261
Hyparrhenia hirta, 570, 571
Hyparrhenia rufa, 570, 572

Imperata brevifolia, 573, 574
Imperata cylindrica, 573, 575
India cupscale, 857
India lovegrass, 502
India paspalum, 764
Indian corn, 1015

Indian goosegrass, 453
Indian paspalum, 764
Indian ricegrass, 165
Indian wood-oats, 317
inland bluegrass, 824
inland muhly, 658
inland saltgrass, 441
inland sea-oats, 317
innocent-weed, 308
interior bluegrass, 824
intermediate paspalum, 751
intermediate wheatgrass, 951
Italian millet, 885
Italian ryegrass, 605
itchgrass, 846
Ixophorus unisetus, 577

Japanese brome, 295
Japanese forest grass, 1029
Japanese grass, 628
Japanese lawngrass, 1022
Japanese lovegrass, 483, 495
Japanese millet, 446
Japanese stilt grass, 628
jaragua grass, 572
Job's tears, 354
Johnsongrass, 903
jointed crowngrass, 748
jointed goatgrass, 172
jointgrass, 748
jointhead, 235
June bluegrass, 827
jungle rice, 443

kangaroograss, 949
Kay's grama, 341
Kentucky bluegrass, 827
King Ranch bluestem, 262
Kleberg bluestem, 420
kleingrass, 709
knotgrass, 748
knotroot bristlegrass, 890
knotroot foxtail, 890
kodo-millet, 764
Koeleria gerardii, 844
Koeleria macrantha, 579
Korean lawngrass, 1022

Korean velvetgrass, 1024
KR bluestem, 262
Kunth's panicgrass, 398
Kunth's smallgrass, 626

lacegrass, 485
Lachnagrostis filiformis, 581
Lamarckia aurea, 583
lanceolate brome, 266
Lange's paspalum, 753
large-flowered tridens, 524
large-mesquite grama, 272
Latin American crowngrass, 745
laurisa grass, 780
leafy rosettegrass, 410
least muhly, 662
Leersia hexandra, 584, 585
Leersia lenticularis, 584, 586
Leersia monandra, 584, 587
Leersia oryzoides, 584, 588
Leersia virginica, 584, 589, 628
Lehmann's lovegrass, 496
Leptochloa chloridiformis, 590, 591
Leptochloa dubia, 590, 592
Leptochloa filiformis, 595
Leptochloa fusca, 590, 593
Leptochloa nealleyi, 590, 594
Leptochloa panicea, 590, 595
Leptochloa panicoides, 590, 596
Leptochloa virgata, 590, 597
Leptochloa viscida, 590, 598
lesser canarygrass, 792
Leymus arenarius, 599, 600
Leymus triticoides, 599, 601
Limnodea arkansana, 603
limpograss, 554
Lindheimer's muhly, 660
linear-leaved panicgrass, 404
little barley, 365
little bluestem, 871
little lovegrass, 499
little quakinggrass, 281
littleseed canarygrass, 792
liverseed grass, 998
lobed needlegrass, 166
Lolium multiflorum, 604, 605
Lolium perenne, 604, 605, 606

Lolium rigidum, 604, 607
Lolium temulentum, 604, 608
long-awn cockspur, 449
longawn muhly, 677
longawned hairgrass, 646
long-beaked witchgrass, 708
long-bristle Indiangrass, 899
long-leaf basketgrass, 696
long-leaf chasmanthium, 319
longleaf threeawn, 226
longleaf wedgescale, 913
long-leaf wood-oats, 319
long-spike fluffgrass, 973
longspike silver bluestem, 264
longspike tridens, 972
long-spine sandbur, 308
longtom, 754
loose silkybent, 211
lopgrass, 294
Louisiana cupgrass, 521
lovegrass tridens, 969
lovegrass, 510
low speargrass, 815
Luziola fluitans, 609, 610
Luziola peruviana, 609, 611
Lycurus phleoides, 612, 613
Lycurus setosus, 612, 614

maidencane, 715
maize, 1015
manilagrass, 1023
many-flowered panicgrass, 410
many-flowered rosettegrass, 410
many-stemmed muhly, 665
marked glyceria, 543
marsh muhly, 668
marshhay cordgrass, 909
mat grama, 343
mat paspalum, 756
mat sandbur, 308
matted crowngrass, 756
matted paspalum, 756
meadow barley, 361
meadow brome, 293
meadow fescue, 863
meadowgrass, 827
Mediterranean barley, 363

Mediterranean beardgrass, 834
Mediterranean hairgrass, 844
Mediterranean lovegrass, 484
Megathyrsus maximus, 616
Melica bulbosa, 617, 618
Melica montezumae, 617, 619
Melica mutica, 617, 620
Melica nitens, 617, 621
Melica porteri, 617, 622
Meliceae: *Glyceria,* 541–45; *Melica,* 617–22
Melinis repens, 624
Merrill's bluestem, 259
mesa dropseed, 930
mesquitegrass, 273, 666
Mexicali muhly, 647
Mexican brome, 283
Mexican lovegrass, 498
Mexican muhly, 661
Mexican needlegrass, 191
Mexican paspalum, 745
Mexican sprangletop, 593
Mexican wildrye, 467
Mexican windmillgrass, 330
Microchloa kunthii, 626
Microstegium vimineum, 628
milo, 902
Miscanthus sinensis, 630
mission grass, 781
Mississippi sprangletop, 595
mixedglume muhly, 650
Monanthochloë littoralis, 632
Montezuma melic, 619
Moorochloa eruciformis, 634
Mormon needlegrass, 162
mountain muhly, 663
mourning lovegrass, 497
mouse barley, 364, 365
mouse foxtail, 189
mudbank crowngrass, 747
mudbank paspalum, 747
Muhlenbergia andina, 637, 640
Muhlenbergia arenacea, 636, 641
Muhlenbergia arenicola, 638, 642, 684
Muhlenbergia asperifolia, 636, 643
Muhlenbergia brevis, 635, 644, 649
Muhlenbergia bushii, 637, 645
Muhlenbergia capillaris, 638, 646, 670

Muhlenbergia crispiseta, 635, 647
Muhlenbergia cuspidata, 638, 648
Muhlenbergia depauperata, 635, 644, 649
Muhlenbergia diversiglumis, 636, 650
Muhlenbergia dubia, 639, 651
Muhlenbergia eludens, 636, 652
Muhlenbergia emersleyi, 638, 653
Muhlenbergia expansa, 638, 654
Muhlenbergia filiformis, 635, 638, 655
Muhlenbergia filipes, 674
Muhlenbergia fragilis, 635, 656
Muhlenbergia frondosa, 637, 657
Muhlenbergia glabrifloris, 637, 658
Muhlenbergia glauca, 636, 637, 659
Muhlenbergia lindheimeri, 639, 660
Muhlenbergia mexicana, 637, 661
Muhlenbergia minutissima, 635, 662
Muhlenbergia montana, 638, 663
Muhlenbergia pauciflora, 636, 639, 664
Muhlenbergia polycaulis, 636, 639, 665
Muhlenbergia porteri, 638, 666
Muhlenbergia pungens, 636, 667
Muhlenbergia racemosa, 637, 668
Muhlenbergia repens, 637, 669
Muhlenbergia reverchonii, 638, 670
Muhlenbergia rigens, 638, 671
Muhlenbergia rigida, 638, 672
Muhlenbergia schreberi, 636, 639, 673
Muhlenbergia sericea, 638, 674
Muhlenbergia setifolia, 638, 675
Muhlenbergia sobolifera, 637, 676
Muhlenbergia spiciformis, 639, 677
Muhlenbergia straminea, 637, 678
Muhlenbergia sylvatica, 637, 679
Muhlenbergia tenuiflora, 637, 680
Muhlenbergia tenuifolia, 636, 639, 681
Muhlenbergia texana, 635, 636, 682
Muhlenbergia thurberi, 637, 683
Muhlenbergia torreyi, 638, 642, 684
Muhlenbergia uniflora, 635, 685
Muhlenbergia utilis, 637, 686
Muhlenbergia villiflora, 637, 687
Muhlenbergia wrightii, 638, 688
multiflower false rhodes-grass, 961
multiflowered chloris, 960
Munro's grass, 690
Munroa squarrosa, 690

mutton bluegrass, 823
muttongrass, 823

napier grass, 782
narrow canarygrass, 787
narrow melic, 620
narrow plumegrass, 849
narrowleaf chasmanthium, 319
narrow-leaved panicgrass, 393
Nassella leucotricha, 691, 692
Nassella tenuissima, 691, 693
Nassella viridula, 691, 694
Nealley's dropseed, 935
Nealley's erioneuron, 525
Nealley's grama, 274
Nealley's sprangletop, 594
Nealley's woollygrass, 525
needle grama, 269
needle-and-thread, 556
needleleaf rosettegrass, 393
Nelson's needlegrass, 167
Nepalese browntop, 628
nerved mannagrass, 545
New Mexico bluegrass, 826
New Mexico muhly, 664
New Mexico needlegrass, 557
nimblewill, 673
nineawn pappusgrass, 475
nitgrass, 540
nodding bluegrass, 1031
nodding brome, 283, 287
nodding fescue, 537
nodding muhly, 645
nodding wildrye, 461
northern mannagrass, 544

oatmeal grass, 586
oats needlegrass, 807
obovate beakgrain, 389
oldfield threeawn, 225
one-flowered grama, 274
one-spike paspalum, 766
oniongrass, 618
open-flower rosettegrass, 403
Oplismenus hirtellus, 696
orchardgrass, 377
Oryza sativa, 698

Oryzeae: *Leersia,* 584–89; *Luziola,* 609–11; *Oryza,* 697–98; *Zizania,* 1017–18; *Zizaniopsis,* 1019–20
Ozarkgrass, 603

Pacific bent, 581
Pacific fountaingrass, 778
palisada signalgrass, 993
palmgrass, 889
pampasgrass, 357
Pan-American balsamscale, 457
panic millet, 718
Paniceae: *Anthenantia,* 204–6; *Axonopus,* 244–47; *Cenchrus,* 305–10; *Dichanthelium,* 390–418; *Digitaria,* 423–39; *Echinochloa,* 442–49; *Eriochloa,* 516–22; *Hopia,* 566–67; *Ixophorus,* 576–77; *Megathyrsus,* 615–16; *Melinis,* 623–24; *Moorochloa,* 633–34; *Oplismenus,* 695–96; *Panicum,* 699–725; *Paspalidium,* 733–34; *Paspalum,* 735–70; *Pennisetum,* 771–85; *Phanopyrum,* 795–96; *Sacciolepis,* 856–58; *Setaria,* 880–97, 1034; *Steinchisma,* 944–45; *Stenotaphrum,* 946–47; *Urochloa,* 990–1004; *Zuloagaea,* 1025–26
Panicoideae: *Andropogon,* 192–98; *Anthenantia,* 204–6; *Arthraxon,* 234–35; *Axonopus,* 244–47; *Bothriochloa,* 255–67; *Cenchrus,* 305–10; *Chrysopogon,* 345–47; *Coelorachis,* 350–52; *Coix,* 353–54; *Dichanthelium,* 390–418; *Dichanthium,* 419–22; *Digitaria,* 423–39; *Echinochloa,* 442–49; *Elionurus,* 455–57; *Eremochloa,* 514–15; *Eriochloa,* 516–22; *Hackelochloa,* 549–50; *Hemarthria,* 553–54; *Heteropogon,* 558–60; *Hopia,* 566–67; *Hyparrhenia,* 570–72; *Imperata,* 573–75; *Ixophorus,* 576–77; *Megathyrsus,* 615–16; *Melinis,* 623–24; *Microstegium,* 627–28; *Miscanthus,* 629–30; *Moorochloa,* 633–34; *Oplismenus,* 695–96; *Panicum,* 699–725; *Paspalidium,* 733–34; *Paspalum,* 735–70; *Pennisetum,* 771–85; *Phanopyrum,* 795–96; *Rottboellia,* 845–46; *Saccharum,* 847–55; *Sacciolepis,* 856–58; *Schizachyrium,* 867–73; *Setaria,* 880–97, 1034; *Sorghastrum,* 898–900; *Sorghum,* 901–3; *Steinchisma,* 944–45; *Stenotaphrum,*

Panicoideae (*cont.*)
    946–47; *Themeda*, 948–49; *Trachypogon*,
    955–56; *Tripsacum*, 978–79; *Urochloa*, 990–
    1004; *Zea*, 1014–16; *Zuloagaea*, 1025–26
Panicum amarum, 700, 702
Panicum anceps, 699, 703
Panicum antidotale, 699, 704
Panicum arizonicum, 992
Panicum bergii, 700, 705
Panicum brachyanthum, 699, 706, 724
Panicum bulbosum, 1026
Panicum capillare, 701, 707
Panicum capillarioides, 700, 708
Panicum coloratum, 699, 709
Panicum dichotomiflorum, 700, 710
Panicum diffusum, 700, 711
Panicum flexile, 700, 712
Panicum ghiesbreghtii, 700, 713
Panicum hallii, 700, 714
Panicum hemitomon, 699, 715
Panicum hirsutum, 700, 716
Panicum hirticaule, 700, 717
Panicum miliaceum, 700, 718
Panicum philadelphicum, 701, 719
Panicum plenum, 1026
Panicum repens, 699, 720
Panicum rigidulum, 699, 721
Panicum tenerum, 699, 722
Panicum trichoides, 700, 723
Panicum verrucosum, 699, 706, 724
Panicum virgatum, 700, 725
papagrass, 997
Pappophoreae: *Cottea*, 358–59; *Enneapogon*,
    474–75; *Pappophorum*, 726–28
Pappophorum bicolor, 726, 727, 728
Pappophorum vaginatum, 726, 728
para liverseed grass, 997
Paraguayan windmillgrass, 324
Parapholis incurva, 730
Pascopyrum smithii, 732
Paspalidium germinatum, 734
Paspalum acuminatum, 735, 739
Paspalum almum, 735, 740
Paspalum bifidum, 737, 741
Paspalum boscianum, 736, 742
Paspalum conjugatum, 735, 743
Paspalum conspersum, 736, 744

Paspalum convexum, 736, 745
Paspalum dilatatum, 735, 746, 767
Paspalum dissectum, 735, 747, 768
Paspalum distichum, 735, 748
Paspalum floridanum, 737, 738, 749
paspalum grass, 746
Paspalum hartwegianum, 737, 750
Paspalum intermedium, 736, 751
Paspalum laeve, 735, 752
Paspalum langei, 737, 753
Paspalum lividum, 737, 738, 754
Paspalum malacophyllum, 736, 755
Paspalum minus, 735, 756
Paspalum modestum, 736, 757
Paspalum monostachyum, 737, 758
Paspalum notatum, 735, 759
Paspalum plicatulum, 736, 760
Paspalum praecox, 737, 738, 761
Paspalum pubiflorum, 737, 750, 762
Paspalum repens, 735, 763
Paspalum scrobiculatum, 735, 764
Paspalum setaceum, 737, 765
Paspalum unispicatum, 737, 766
Paspalum urvillei, 736, 746, 767
Paspalum vaginatum, 735, 768
Paspalum virgatum, 736, 769
Paspalum wrightii, 736, 770
pearl millet, 777
Pennisetum advena, 772, 773
Pennisetum alopecuroides, 771, 774
Pennisetum ciliare, 771, 775
Pennisetum flaccidum, 771, 776
Pennisetum glaucum, 771, 777
Pennisetum macrostachys, 771, 778
Pennisetum nervosum, 771, 779
Pennisetum orientale, 772, 780
Pennisetum polystachion, 771, 781
Pennisetum purpureum, 771, 772, 782
Pennisetum setaceum, 772, 783
Pennisetum setigerum, 771, 784
Pennisetum villosum, 771, 785
Pensacola bahiagrass, 759
perennial crabgrass, 434
perennial quackgrass, 280
perennial rye, 1033
perennial ryegrass, 606
perennial teosinte, 1016

perennial veldtgrass, 451
perennial wildrye, 606
perplexing needlegrass, 168
Peruvian watergrass, 611
Phalaris angusta, 786, 787
Phalaris aquatica, 786, 788
Phalaris brachystachys, 786, 789
Phalaris canariensis, 786, 790
Phalaris caroliniana, 786, 791
Phalaris minor, 786, 792
Phalaroides arundinacea, 794
Phanopyrum gymnocarpon, 796
Philadelphia witchgrass, 719
Phleum pratense, 798
Phragmites australis, 800
Phyllostachys aurea, 801, 802
Phyllostachys aureosulcata, 1030
Phyllostachys bambusoides, 801, 803
pigbutt threeawn, 219
pigeon grass, 891
pimple panicum, 706
pine dropseed, 254
pine muhly, 651
pine-barren fluffgrass, 966
pine-barren tridens, 966
pineland muhly, 651
pinewoods fingergrass, 530
pineywoods dropseed, 934
pinhole bluestem, 257
pink muhly, 646
pink pappusgrass, 727
pink tridens, 968
pinyon ricegrass, 808
Piptatherum micranthum, 805
Piptochaetium avenaceum, 806, 807
Piptochaetium fimbriatum, 806, 808
Piptochaetium pringlei, 806, 809
pitchfork crowngrass, 741
pitchfork paspalum, 741
pitscale grass, 550
pitted bluestem, 265
plains bluegrass, 817
plains bristlegrass, 886, 887
plains lovegrass, 494
plains muhly, 648
plaited paspalum, 760
plantain signalgrass, 999

Pleuraphis jamesii, 810, 811, 812
Pleuraphis mutica, 810, 811, 812
plumegrass, 630
Poa annua, 813, 815, 821
Poa arachnifera, 813, 816
Poa arida, 814, 817
Poa autumnalis, 814, 818
Poa bigelovii, 813, 819
Poa bulbosa, 813, 820
Poa chapmaniana, 813, 821
Poa compressa, 813, 822
Poa fendleriana, 814, 823
Poa interior, 814, 824
Poa involuta, 828
Poa leptocoma, 813, 825, 1031
Poa occidentalis, 813, 826
Poa pratensis, 813, 827
Poa reflexa, 1031
Poa secunda, 1032
Poa strictiramea, 814, 828
Poa sylvestris, 814, 829
Poa trivialis, 814, 830
Poeae: *Agrostis,* 175–82; *Aira,* 183–84; *Alopecurus,* 187–89; *Anthoxanthum,* 207–9; *Apera,* 210–11; *Avena,* 241–43; *Briza,* 278–81; *Cinna,* 348–49; *Cynosurus,* 374–75; *Dactylis,* 376–77; *Desmazeria,* 385–86; *Festuca,* 532–38; *Gastridium,* 539–40; *Hainardia,* 551–52; *Holcus,* 564–65; *Koeleria,* 578–79; *Lachnagrostis,* 580–81; *Lamarckia,* 582–83; *Limnodea,* 602–3; *Lolium,* 604–8; *Parapholis,* 729–30; *Phalaris,* 786–92; *Phalaroides,* 793–94; *Phleum,* 797–98; *Poa,* 813–30, 1031–32; *Polypogon,* 831–36; *Rostraria,* 843–44; *Schedonorus,* 861–63; *Sclerochloa,* 874–75; *Sphenopholis,* 912–17; *Trisetum,* 982–84; *Vulpia,* 1007–11
poison darnel, 608
Polypogon elongatus, 831, 832
Polypogon interruptus, 831, 833
Polypogon maritimus, 831, 834
Polypogon monspeliensis, 831, 833, 835
Polypogon viridis, 831, 836
pond lovegrass, 496
Poöideae: *Achnatherum,* 161–70; *Aegilops,* 171–72; *Agropyron,* 173–74; *Agrostis,* 175–82; *Aira,* 183–84; *Alopecurus,* 187–89;

Poöideae (*cont.*)

Amelichloa, 190–91; Anisantha, 199–203; Anthoxanthum, 207–9; Apera, 210–11; Avena, 241–43; Brachyelytrum, 276–77; Briza, 278–81; Bromopsis, 282–90; Bromus, 291–98; Ceratochloa, 311–15; Cinna, 348–49; Critesion, 360–65; Cynosurus, 374–75; Dactylis, 376–77; Desmazeria, 385–86; Diarrhena, 387–89; Elymus, 458–73; Festuca, 532–38; Gastridium, 539–40; Glyceria, 541–45; Hainardia, 551–52; Hesperostipa, 555–57; Holcus, 564–65; Hordeum, 568–69; Koeleria, 578–79; Lachnagrostis, 580–81; Lamarckia, 582–83; Leymus, 599–601; Limnodea, 602–3; Lolium, 604–8; Melica, 617–22; Nassella, 691–94; Parapholis, 729–30; Pascopyrum, 731–32; Phalaris, 786–92; Phalaroides, 793–94; Phleum, 797–98; Piptatherum, 804–5; Piptochaetium, 806–9; Poa, 813–30, 1031–32; Polypogon, 831–36; Psathyrostachys, 837–38; Pseudoroegneria, 839–40; Rostraria, 843–44; Schedonorus, 861–63; Sclerochloa, 874–75; Secale, 878–79, 1033; Sphenopholis, 912–17; Thinopyrum, 950–52; Trachynia, 953–54; Trisetum, 982–84; Triticum, 985–87; Vulpia, 1007–11

Porter's melic, 622
poverty dropseed, 941
poverty grass, 219, 220
poverty oatgrass, 382
poverty threeawn, 220
povertygrass, 936
prairie cordgrass, 910
prairie cupgrass, 518
prairie dropseed, 932
prairie false oat, 983
prairie junegrass, 579
prairie panicgrass, 706
prairie threeawn, 225
prairie trisetum, 983
prairie wedgescale, 917
pricklegrass, 958
Pringle's needlegrass, 809
Pringle's speargrass, 809
proso millet, 718
Psathyrostachys juncea, 838
Pseudoroegneria spicata, 840
puffsheath dropseed, 936

pull-and-be-damned, 754
pull-up muhly, 655
pungent muhly, 667
purple dropseed, 934, 937
purple falsebrome, 954
purple fountaingrass, 773, 783
purple lovegrass, 508
purple muhly, 672, 674
purple needlegrass, 230, 981
purple pampasgrass, 356
purple sandgrass, 975
purple silkyscale, 205
purple threeawn, 230
purple-flowered dropseed, 923
purpletop tridens, 970
purpletop, 970

quackgrass, 468
Queensland bluegrass, 422

rabbit's-foot, 835
rabbitfoot grass, 835
rabbitfoot polypogon, 835
rattail fescue, 1009
rattail sixweeksgrass, 1009
Ravenel's panicgrass, 412
ravennagrass, 845
red brome, 201
red fescue, 536
red grama, 344
red lovegrass, 505
red muhly, 669
red natal grass, 624
red rice, 698
red sprangletop, 595
Redfieldia flexuosa, 842
redtop bent, 178
redtop panicum, 721
reed canarygrass, 794
reed, 800
rescue brome, 314
rescue grass, 314
Reverchon bristlegrass, 892
Reverchon's muhly, 670
Rhodesgrass, 328
ribbed crowngrass, 755
ribbed paspalum, 755

rice cutgrass, 588
rice, 698
Richardson's brome, 289
ridged mannagrass, 545
ring muhly, 684
ringed dichanthium, 420
ringgrass, 684
Rio Grande lovegrass, 500
ripgut brome, 200
river cane, 237
rock dropseed, 676
rock muhly, 676
rooigrass, 949
Rostraria cristata, 844
Rothrock's grama, 336
Rottboellia cochinchinensis, 846
rough barnyardgrass, 447
rough bent, 181
rough bluegrass, 830
rough dogtail, 376
rough dropseed, 923, 925
rough foxtail grass, 895
rough-leafed muhly, 643
roughstalked witchgrass, 717
round-fruited panicgrass, 415
round-seeded rosettegrass, 415
rush saltgrass, 909
rush wheatgrass, 952
Russian millet, 718
Russian wildrye, 838
rustyseed crowngrass, 753
rustyseed paspalum, 753
rye brome, 298
rye, 879

sabi grass, 996
sacahuista grass, 911
sacahuista, 911
Saccharum alopecuroides, 847, 848
Saccharum baldwinii, 847, 849
Saccharum bengalense, 847, 850
Saccharum brevibarbe, 847, 851
Saccharum coarctatum, 847, 852
Saccharum giganteum, 847, 853
Saccharum officinarum, 847, 854
Saccharum ravennae, 847, 855
Sacciolepis indica, 856, 857

Sacciolepis striata, 856, 858
salt reedgrass, 907
saltgrass, 441
saltmarsh foxtail, 888
saltmarsh grass, 905
saltmeadow cordgrass, 909
sand bluestem, 196
sand cordgrass, 906
sand dropseed, 927
sand lovegrass, 513
sand muhly, 641, 642
sand witchgrass, 425
sandhill grass, 994
sandhill muhly, 667
Sarita panicgrass, 406
Sarita rosettegrass, 406
satingrass, 668, 673
satintail, 574, 575
savannah hairgrass, 654
savannah panic, 796
sawtooth lovegrass, 510
scattered paspalum, 744
Schedonnardus paniculatus, 860
Schedonorus arundinaceus, 861, 862
Schedonorus pratensis, 861, 863
Scheele's bristlegrass, 893
Schismus arabicus, 864, 865
Schismus barbatus, 864, 866
Schizachyrium cirratum, 867, 868
Schizachyrium littorale, 867, 869
Schizachyrium sanguineum, 867, 870
Schizachyrium scoparium, 867, 869, 871
Schizachyrium spadiceum, 867, 872
Schizachyrium tenerum, 867, 873
Schraders-grass, 314
Schreber's muhly, 673
Sclerochloa dura, 875
Scleropoa rigida, 386
Scleropogon brevifolius, 877
scratchgrass, 643
screwleaf muhly, 678
Scribner's needlegrass, 170
Scribner's panicgrass, 407
S-curve threeawn, 231
sea barley, 363
seaoats, 989
seashore bluestem, 869

seashore crowngrass, 768
seashore dropseed, 942
seashore paspalum, 768
seashore rushgrass, 942
Secale cereale, 879, 986
Secale strictum, 1033
secund bluegrass, 1032
seep muhly, 670
Setaria adhaerans, 880, 882
Setaria corrugata, 881, 883
Setaria glauca, 891
Setaria grisebachii, 881, 884
Setaria italica, 881, 885, 897
Setaria leucopila, 880, 886
Setaria macrostachya, 881, 887
Setaria magna, 881, 888
Setaria megaphylla, 1034
Setaria palmifolia, 880, 889, 1034
Setaria parviflora, 880, 890
Setaria poiretiana, 1034
Setaria pumila, 880, 891
Setaria reverchonii, 880, 892
Setaria scheelei, 881, 893
Setaria texana, 880, 894
Setaria verticillata, 880, 895
Setaria villosissima, 880, 896
Setaria viridis, 881, 897
shama millet, 443
shining wedgescale, 916
shiny wedgescale, 916
shoregrass, 632
short muhly, 644
shortbeard plumegrass, 851
shortleaf cottontop, 431
shortleaf skeletongrass, 548
shortleaf woollygrass, 524
shortspike canarygrass, 789
short-stalked lovegrass, 488
showy chloris, 333
sicklegrass, 730
sideoats grama, 271
silky bluestem, 422
silky wildoatgrass, 381
silver beardgrass, 263
silver bluestem, 263
silver hairgrass, 184
silver plumegrass, 848

silvergrass, 630
silverleaf grass, 610
silvery beardgrass, 197
Silveus' dropseed, 939
Silveus' grass, 963
Silveus' lovegrass, 507
Simpson's grass, 715
single threeawn, 232
singleawn aristida, 232
single-spike paspalum, 758
sixweeks fescue, 1010
sixweeks grama, 269, 336
sixweeks muhly, 649, 662
sixweeks threeawn, 215
skunktail grass, 362
sleepygrass, 169
slender bluestem, 873
slender chasmanthium, 318
slender crabgrass, 430
slender crowngrass, 765
slender fingergrass, 430
slender grama, 272
slender Indiangrass, 899
slender meadow foxtail, 189
slender muhly, 646, 681
slender panicgrass, 417, 708
slender wedgescale, 914
slender wheatgrass, 471
slender woodoats, 318
slim tridens, 971
slimbristle sandbur, 306
slimflowered muhly, 680
slim-leaf rosettegrass, 404
slimspike foxtail, 189
slimspike threeawn, 224
slimspike windmill grass, 322
sloughgrass, 910
small carpgrass, 235
small dropseed, 936
small-flowered panicgrass, 723
small-flowered piptatherum, 805
smooth barley, 364
smooth brome, 285, 297
smooth cordgrass, 905
smooth crabgrass, 433
smooth creeping lovegrass, 493
smutgrass, 933

soft wheat, 986
soft-leaved panicgrass, 405
soft-leaved rosettegrass, 405
soft-tufted panicgrass, 403
Sonoran sprangletop, 598
Sorghastrum elliotti, 898, 899
Sorghastrum nutans, 898, 900
Sorghum bicolor, 901, 902
Sorghum halepense, 901, 903
sorghum, 902
sour cottontop, 432
sour crowngrass, 743
sour paspalum, 743
sourgrass, 432
southeastern panic, 722
southeastern wildrye, 464
southern beardgrass, 832
southern cane, 237
southern crabgrass, 428
southern cutgrass, 585
southern povertygrass, 941
southern sandbur, 307
southern shorthusk, 277
southern watergrass, 610
southern wedgescale, 913
southern wildrice, 1020
southern witchgrass, 708
southwestern bristlegrass, 893
southwestern cupgrass, 517
southwestern needlegrass, 164
southwestern wildrye, 465
Spartina alterniflora, 904, 905
Spartina bakeri, 904, 906
Spartina cynosuroides, 904, 907
Spartina densiflora, 904, 908
Spartina patens, 904, 909
Spartina pectinata, 904, 910
Spartina spartinae, 904, 911
speargrass, 692
spelt, 987
Sphenopholis filiformis, 912, 913
Sphenopholis intermedia, 912, 914
Sphenopholis longiflora, 912, 915
Sphenopholis nitida, 912, 916
Sphenopholis obtusata, 912, 915, 917
spidergrass, 233
spike bent, 177

spike burgrass, 958
spike dropseed, 925
spike grass, 441
spike lovegrass, 509
spike muhly, 688
spike trisetum, 984
spiny dogtail grass, 376
split bluestem, 197
splitbeard beard-grass, 197
Sporobolus airoides, 919, 921
Sporobolus buckleyi, 918, 922
Sporobolus clandestinus, 919, 923
Sporobolus coahuilensis, 918, 924
Sporobolus compositus, 918, 919, 923, 925
Sporobolus contractus, 919, 920, 926
Sporobolus cryptandrus, 919, 926, 927
Sporobolus diandrus, 918, 928
Sporobolus domingensis, 919, 929
Sporobolus flexuosus, 919, 930
Sporobolus giganteus, 918, 919, 931
Sporobolus heterolepis, 920, 932
Sporobolus indicus, 918, 933
Sporobolus junceus, 919, 934
Sporobolus nealleyi, 919, 935
Sporobolus neglectus, 918, 936
Sporobolus purpurascens, 919, 937
Sporobolus pyramidatus, 918, 938
Sporobolus silveanus, 918, 920, 932, 939
Sporobolus texanus, 918, 920, 940
Sporobolus vaginiflorus, 918, 936, 941
Sporobolus virginicus, 918, 942
Sporobolus wrightii, 919, 943
sprawling signalgrass, 1002
spreading windmill grass, 327
spreading witchgrass, 710
Springfield bluestem, 266
sprucetop grama, 270
spur burgrass, 958
squirrel sixweeksgrass, 1011
squirreltail barley, 362
squirreltail fescue, 1011
squirreltail grass, 362
squirreltail, 463
St. Augustine grass, 947
stargrass, 372
starved panicgrass, 399
starved rosettegrass, 399

Steinchisma hians, 945
Stenotaphrum secundatum, 947
stiff muhly, 672
stiff ryegrass, 607
stiff-leaf fingergrass, 530
stiff-leaved panicgrass, 408
stinkgrass, 486
Stipa eminens, 164
Stipa tenuissima, 693
Stipeae: *Achnatherum,* 161–70; *Amelichloa,*
    190–91; *Hesperostipa,* 555–57; *Nassella,*
    691–94; *Piptatherum,* 804–5; *Piptochae-*
    tium, 806–9
stout lovegrass, 492
stout woodgrass, 349
streambed bristlegrass, 886
sugarcane plumegrass, 853
sugarcane, 854
Swallen's curly mesquite, 563
Swallen's lovegrass, 511
sweet signalgrass, 634
sweet sorghum, 902
sweet tanglehead, 560
sweet vernalgrass, 209
switch cane, 237, 238
switchgrass, 725
swollen windmillgrass, 323
sword-leaf panicgrass, 401
sword-leaf rosettegrass, 401
sylvan bluegrass, 829
sylvan's speargrass, 829

tall bluestem, 256
tall cane, 850
tall fescue, 862
tall grama, 340
tall marshgrass, 910
tall melic, 621
tall redtop, 970
tall wheatgrass, 952
tall-swamp panicgrass, 413
talquezal, 769
tanglehead, 559
target dropseed, 938
teal lovegrass, 493
teel lovegrass, 493
teff, 512

tender fountaingrass, 783
Texas beardgrass, 868
Texas bluegrass, 816
Texas bluestem, 868
Texas bristlegrass, 894
Texas brome, 290
Texas cottontop, 435
Texas crabgrass, 438, 860
Texas crowfoot, 592
Texas cupgrass, 522
Texas dropseed, 940
Texas fescue, 538
Texas fingergrass, 438
Texas grama, 273
Texas liverseed grass, 1004
Texas millet, 1004
Texas muhly, 682
Texas nassella, 692
Texas needlegrass, 692
Texas panicum, 1004
Texas signalgrass, 1004
Texas tridens, 973
Texas tussockgrass, 692
Texas wildrice, 1018
Texas wildrye, 470
Texas windmillgrass, 331
Texas wintergrass, 692
Texasgrass, 1006
thatching grass, 571
Themeda triandra, 949
thin paspalum, 765
Thinopyrum intermedium, 950, 951
Thinopyrum ponticum, 950, 952
thintail, 552
Thompsongrass, 748
three-flower melic, 621
threespike goosegrass, 454
Thurber's muhly, 683
ticklegrass, 179, 181
Timothy canarygrass, 787
Timothy, 798
tiny lovegrass, 485
toothache grass, 367
torpedo grass, 720
Trachynia distachya, 954
Trachypogon secundus, 956
Tragus berteronianus, 958

Trichloris crinita, 959, 960
Trichloris pluriflora, 959, 961
Trichoneura elegans, 963
Tridens albescens, 964, 965
Tridens ambiguus, 964, 966
Tridens buckleyanus, 964, 967
Tridens congestus, 964, 968
Tridens eragrostoides, 964, 969
Tridens flavus, 964, 970
Tridens muticus, 964, 971
Tridens strictus, 964, 972
Tridens texanus, 964, 973
Triplasis purpurea, 975
Tripogon spicatus, 977
Tripsacum dactyloides, 979
Triraphis mollis, 981
Trisetum interruptum, 982, 983
Trisetum spicatum, 982, 984
Triticeae: *Aegilops,* 171–72; *Agropyron,* 173–74; *Critesion,* 360–65; *Elymus,* 458–73; *Hordeum,* 568–69; *Leymus,* 599–601; *Pascopyrum,* 731–32; *Psathyrostachys,* 837–38; *Pseudoroegneria,* 839–40; *Secale,* 878–79, 1033; *Thinopyrum,* 950–52; *Triticum,* 985–87
Triticum aestivum, 985, 986
Triticum spelta, 985, 987
tropical barred bristlegrass, 882
tropical panicgrass, 723
tropical sprangletop, 597
tufted foxtail, 188
tufted lovegrass, 501
tumble lovegrass, 506
tumble windmillgrass, 332
tumblegrass, 860
turkey grass, 577
tussock dropseed, 928
two-flower melic, 620

Uniola paniculata, 989
upland bent, 180
Urochloa arizonica, 990, 992
Urochloa brizantha, 990, 993
Urochloa ciliatissima, 990, 994
Urochloa fusca, 990, 995
Urochloa mosambicensis, 990, 996
Urochloa mutica, 990, 997

Urochloa panicoides, 990, 998
Urochloa plantaginea, 991, 999
Urochloa platyphylla, 990, 1000
Urochloa ramosa, 990, 1001
Urochloa reptans, 990, 1002
Urochloa subquadripara, 991, 1003
Urochloa texana, 990, 1004
Urville's grass, 767

variable panicgrass, 397
variable rosettegrass, 397
Vasey's muttongrass, 823
vaseygrass, 767
Vaseyochloa multinervosa, 1006
velvetgrass, 565
velvety panicgrass, 414
vernal cupgrass, 520
vernalgrass, 208
vetiver, 347
vine mesquite, 567
violet crabgrass, 439
Virginia bluestem, 198
Virginia wildrye, 473
Vulpia bromoides, 1007, 1008
Vulpia myuros, 1007, 1009
Vulpia octoflora, 1007, 1010
Vulpia sciurea, 1007, 1011

Warnock's grama, 275
warty panicgrass, 724
water beardgrass, 836
water bentgrass, 836
water millet, 1020
water paspalidium, 734
water paspalum, 746, 757, 763
wedgegrass, 917
weeping lovegrass, 489
western bog bluegrass, 825
western threeawn, 218
western tripleawn grass, 218
western wheatgrass, 732
western witchgrass, 436
wheat, 986
whiplash pappusgrass, 728
white cutgrass, 589
white fountaingrass, 780
white tridens, 965

white-grass, 587
whitegrass, 589
whorled dropseed, 938
wild canarygrass, 791
wild oat, 242
Willkommia texana, 1013
willkommia, 1013
winter bent, 179
wiregrass, 934
wirestem muhly, 657, 661
wiry witchgrass, 712
witchgrass, 707
wolftail, 613, 614
woodland bluegrass, 829
woodland muhly, 679
woolly brome, 286
woolly rosettegrass, 413
woolly spiked grama, 270
woolly threeawn, 223
woolly tripleawn, 223
woollyfoot grama, 338
woollysheath threeawn, 223
woolspike balsamscale, 456

Wooton's threeawn, 227
Wright's bluestem, 267
Wright's panicgrass, 418
Wright's paspalum, 770
Wright's sacaton, 943
wrinkled jointgrass, 352

yellow bamboo, 802
yellow foxtail, 891
yellow grove bamboo, 1030
yellow Indiangrass, 900
yellow sand paspalum, 765
yellowsedge bluestem, 198
Yorkshire fog, 565

Zea mays, 1014, 1015
Zea perennis, 1014, 1016
Zizania texana, 1018
Zizaniopsis miliacea, 1020
Zoysia japonica, 1021, 1022
Zoysia matrella, 1021, 1023
Zoysia pacifica, 1021, 1024
Zuloagaea bulbosa, 1025, 1026